···Introductory···
Plant Science
second edition

Investigating the Green World

Cynthia McKenney • Ursula Schuch • Amanda Chau
Texas Tech University · University of Arizona · Blinn College

Kendall Hunt
publishing company

Book Team

Chairman and Chief Executive Officer   Mark C. Falb
President and Chief Operating Officer   Chad M. Chandlee
Vice President, Higher Education   David L. Tart
Director of Publishing Partnerships   Paul B. Carty
Project/Development Supervisor   Lynnette M. Rogers
Vice President, Operations   Timothy J. Beitzel
Project Coordinator   Torrie Johnson
Permissions Editor   Jade Sprecher
Cover Designer   Faith Walker

Cover image © Shutterstock, Inc.

Kendall Hunt
publishing company

www.kendallhunt.com
Send all inquiries to:
4050 Westmark Drive
Dubuque, IA 52004-1840

Printed in the United States of America
10 9 8 7 6 5 4 3 2 1

Dedication

In memory of my beloved parents, David and Carolyn, who enabled my love of plants.

—Cynthia McKenney

With gratitude in honor of my mother Maria and in memory of my father Heinrich.

—Ursula Schuch

To my mom, Joyce, and my husband, Andrew.

—Amanda Chau

Brief Contents

Preface   xxiii

About the Book   xxv

Acknowledgments   xxvii

About the Authors   xxix

Chapter 1   The Science of Plants   001

Section I   Structure of the Green World   023

Chapter 2   Essential Molecules of Life   025

Chapter 3   Cells and Tissues   045

Chapter 4   Roots   077

Chapter 5   Stems   095

Chapter 6   Leaves   119

Section II   Reproduction in the Green World   141

Chapter 7   Flowers and Sexual Reproduction   143

Chapter 8   Fruits, Seeds, Dissemination, and Germination   163

Chapter 9   Asexual Reproduction and Plant Propagation   181

Section III   Physiological Processes of the Green World   199

Chapter 10   Photosynthesis   201

Chapter 11   Respiration   229

Chapter 12   Plant Responses to Hormonal and Environmental Stimuli   243

Chapter 13   Soils, Plant Nutrition, and Transport in Plants   271

Section IV   Continuity of the Green World   299

Chapter 14   Cell Cycle and Plant Life Cycle   301

Chapter 15   Patterns of Inheritance   315

Chapter 16   Molecular Basis of Inheritance   335

Chapter 17   Biotechnology and Genetically Modified Plants   347

Chapter 18   Evolution   381

Section V	Diversity of the Green World	403
Chapter 19	Phylogeny and Taxonomy	405
Chapter 20	Cyanobacteria and Algae	423
Chapter 21	The Plant Kingdom	449
Chapter 22	Bryophytes	463
Chapter 23	Lycophytes and Ferns	479
Chapter 24	Gymnosperms	499
Chapter 25	Angiosperms	519
Chapter 26	Fungi: Friends or Foes of the Green World	547
Section VI	The Green World in the Web of Life	567
Chapter 27	Ecosystems and Biomes	569
Chapter 28	Dynamics of Plant Communities and Populations	589
Chapter 29	Plants as Food, Commercial Products, and Pharmaceuticals	615
Glossary		647
Index		671

Contents

Preface   xxiii

About the Book   xxv

Acknowledgments   xxvii

About the Authors   xxix

Chapter 1   The Science of Plants   1
Learning Objectives   1
The Importance of Plants   2
Plant Domestication and Global Agriculture   5
Characteristics of Life   7
        Living Organisms Have Cell(s) and Are Organized   7
        Living Organisms Acquire Energy and Materials   9
        Living Organisms Grow and Develop   9
        Living Organisms Reproduce   10
        Living Organisms Respond to Stimuli   11
        Living Organisms Adapt to Their Environment   11
Diversity of Life   12
        Classification of Living Organisms   13
        Unique Characteristics of Plants   15
The Science of Plants   15
        Disciplines of Plant Science   16
        Scientific Processes   16
Key Terms   19
Summary   19
Reflect   20
References   21

Section I   Structure of the Green World   23
Chapter 2   Essential Molecules of Life   25
Learning Objectives   25
Water: The Main Component of Life   26
        Water Is a Polar Molecule   26

Water Molecules Stick Together and to Other Polar Molecules   27
Water Is an Important Solvent   27
Water Provides Evaporative Cooling   28
Four Classes of Organic Molecules: The Foundations and Structures of Life   28
Carbohydrates Provide Fuel and Structure for Life   29
Nucleic Acids Make up the Blueprint of Life   32
Proteins Are the Substance of Life   33
Lipids Provide Fuel and Barrier for Life   38
Key Terms   42
Summary   43
Reflect   44
References   44

Chapter 3   Cells and Tissues   45
Learning Objectives   45
History of Cell Discovery   47
Scientists Who Discovered Cells   47
Development of the Cell Theory   48
Cells Are Tiny Units of Life   48
Smaller the Better   49
Observing Cells with Microscopes   49
Basic Cell Types   50
The Two Basic Types: Prokaryotic and Eukaryotic Cells   50
Common Cell Structures and Functions   51
The Plasma Membrane: Selective Barrier   53
The Cytoplasm-Metabolic Center and Internal Transport   55
Deoxyribonucleic Acid (DNA): Blueprint of Life   55
The Nucleus: Control Center   56
The Endoplasmic Reticulum–Biomolecule Factories   58
Dictyosomes or Golgi Apparatus: Packaging and Shipping Center   58
Vacuoles: Storage Facilities   58
Mitochondria: Energy Powerhouses   59
Ribosomes: Protein Factories   60
Plant Cell Structures and Functions   60
Chloroplasts: The Green Plastids That Make Food   63
Chromoplasts: Colorful Plastids That Add a Touch of Color to the Green World   65
Leucoplasts: Colorless Plastids That Store Stuffs   65
The Cell Wall: Support and Protection   65
Differences and Similarities between Plant and Animal Cells   66
Plant Tissue Systems and Functions   66
Ground Tissue System: The Storage and Support System   68
Vascular Tissue System: The Transport System   69

Dermal Tissue System: The Protective Covering   71
Key Terms   74
Summary   74
Reflect   76
References   76

Chapter 4   Roots   77
Learning Objectives   77
Root Function   78
Types of Root Systems   79
Specialized Roots   80
   Roots for Stabilization   80
   Roots for Extra Storage   81
   Roots for Extra Anchorage   82
Root Structures   83
   The Region of Cell Division   84
   The Region of Cell Elongation   85
   The Region of Maturation   85
Monocot and Eudicot Roots   90
Associations and Symbiotic Relationships   90
   Association with Fungi   92
   Association with Nitrogen-Fixing Bacteria   92
Key Terms   93
Summary   93
Reflect   94
References   94

Chapter 5   Stems   95
Learning Objectives   95
What Is a Stem?   96
Stem Functions   96
Structural Differences in Stem Development   98
   Primary Growth   98
   Organization of Vascular Tissues   98
   Secondary Growth of Eudicots and Gymnosperms   99
   Monocots Lack Secondary Growth   100
Stem Classification   100
Branch and Stem Components   102
Leaf Attachment   103
Specialized Stems   108
   Bulbs   109
Key Terms   116

Summary   116
Reflect   117
References   118

Chapter 6   Leaves   **119**
Learning Objectives   119
Leaf Morphology   120
    External Morphology   120
    Internal Morphology   121
Leaf Functions   123
    Processes   123
    Water Retention   124
    Plant Protection   126
    Support   126
    Pollination   127
    Trapping Food   128
Identification and Classification   129
    Leaf Complexity   129
    Leaf Venation   130
    Leaf Shape and Color   131
Response of Leaves to the Environment   136
    Heat and Drought   136
    Changing Seasons   136
    Air Pollution   137
Key Terms   138
Summary   138
Reflect   139
References   140

Section II   Reproduction in the Green World   **141**
Chapter 7   Flowers and Sexual Reproduction   **143**
Learning Objectives   143
Sexual versus Asexual Reproduction   144
The Complete Flower   144
    Sepals and Tepals   144
    Petals   145
    Stamens   146
    Pistils   147
Flowers and Their Pollinators   147
Identification and Classification of Flowers   149
    Symmetry   149
    Corolla Types   150

Inflorescence Morphology   152
Constructing a Floral Formula   153
Key Terms   159
Summary   160
Reflect   161
References   161

Chapter 8   Fruits, Seeds, Dissemination, and Germination   163

Learning Objectives   163
Fruit Development and Structure   164
Fruit Development   164
Structural Components   164
Fruits and Their Seed Dispersal Mechanisms   165
Identification and Classification of Fruiting Structures   165
Propagation by Seed   167
Germination   168
Dormancy   176
Quiescence   176
Physiological Dormancy   176
Physical Dormancy   176
Double Dormancy   177
Seed Production and Storage   177
Seed Certification   177
Seed Preservation   177
Key Terms   178
Summary   178
Reflect   180
References   180

Chapter 9   Asexual Reproduction and Plant Propagation   181

Learning Objectives   181
Advantages and Disadvantages of Asexual Reproduction   182
Advantages of Asexual Propagation   183
Disadvantages of Asexual Propagation   184
Types of Asexual Propagation   184
Vegetative Cuttings   184
Grafting   190
Budding   191
Layering   192
Underground Plant Parts   193
Micropropagation   194
Apomixis   194

Key Terms   195
Summary   195
Reflect   197
References   197

Section III   Physiological Processes of the Green World   199

Chapter 10   Photosynthesis   201
Learning Objectives   201
The Role of Photosynthesis for Life on Earth   203
The Nature and Function of Light in Photosynthesis   207
Pigments and Their Role in Photosynthesis   208
    Chlorophyll   208
    Accessory Pigments   208
Chloroplasts and Photosystems   210
    Photosystems   210
The Light Reactions and the Calvin–Benson Cycle   212
    The Light Reactions   212
    The Calvin–Benson Cycle   216
    C_4 Photosynthesis   218
Comparing the C_3 and C_4 Pathways   220
Crassulacean Acid Metabolism (CAM)   221
Environmental Factors Affecting Photosynthesis   225
    Light Intensity   225
    Temperature   226
    Carbon Dioxide   226
Key Terms   226
Summary   226
Reflect   227
References   228

Chapter 11   Respiration   229
Learning Objectives   229
Aerobic Respiration   230
    Glycolysis   231
    Pyruvate Conversion to Acetyl Coenzyme A   232
    The TCA or Krebs Cycle   233
    The Electron Transport Chain and Oxidative Phosphorylation   234
    Chemiosmosis and Oxidative Phosphorylation   236
    Control of Respiration Through Feedback Mechanisms   237
Anaerobic Respiration   237
Respiration and Fresh Produce   239
Key Terms   240

Summary   240
Reflect   241
References   241

Chapter 12   Plant Responses to Hormonal and Environmental Stimuli   243
Learning Objectives   243
Signal Reception and Transduction   244
Plant Hormones   247
    Auxin   248
    Cytokinins   252
    Gibberellins (GA)   253
    Abscisic Acid   255
    Ethylene   255
    Brassinosteroids and Other Plant Hormones   258
Responses to Light   259
    Blue Light   259
    Red and Far-Red Light   259
Responses to Gravity   262
Responses to Mechanical Stimuli   263
Responses to Environmental Signals   265
Responses to Pathogens, Insects, and Herbivores   265
    Hypersensitive Response   266
    System-Acquired Resistance   267
    Defenses against Herbivores   268
Key Terms   268
Summary   269
Reflect   269
References   270

Chapter 13   Soils, Plant Nutrition, and Transport in Plants   271
Learning Objectives   271
Soils   272
    Soil Horizons   272
    Soil Texture and Structure   273
Essential Elements for Plant Growth   275
    Nutrient Deficiency   278
    Nutrient Availability and Cation Exchange   280
Nutrient and Water Uptake   281
Movement of Molecules across Membranes   282
    Passive Transport: Diffusion, Channels, and Carriers   282
    Active Transport: Proton Pumps   283
    Passive Ion Exclusion   283

Active Ion Exclusion   283

Nutrient Uptake through Leaves   284

Symbiotic Microorganisms Providing Nutrients   284

Water and Solute Transport in the Plant   284

Water Potential and Water Transport   285

Root Pressure   288

Capillary Action   289

The Cohesion-Tension Theory   290

Plant Water Uptake in Dry or Saline Soils   291

Translocation   291

The Pressure-Flow Hypothesis   292

Phloem Loading and Unloading   293

Key Terms   295

Summary   295

Reflect   297

References   297

Section IV   Continuity of the Green World   **299**

Chapter 14   Cell Cycle and Plant Life Cycle   **301**

Learning Objectives   301

Prokaryotic Cell Cycle   302

Eukaryotic Cell Cycle   303

Interphase: Preparation between Cell Division   303

Mitosis: Cell Division That Produces Growth   303

Cytokinesis: Physical Separation of Cells   306

Meiosis: Essential Cell Division for Sexual Reproduction   306

Meiosis and Sexual Reproduction Generates Genetic Variation   309

Key Role of Cell Division in The Unique Life Cycle of Plants   310

Key Terms   310

Summary   311

Reflect   313

Reference   313

Chapter 15   Patterns of Inheritance   **315**

Learning Objectives   315

Patterns of Inheritance   316

Mendelian Inheritance   316

One-Character Inheritance or Single-Factor Crosses   318

Two-Character Inheritance or Two-Factor Crosses   321

Mendel's Principles of Inheritance   322

Complex Pattern of Inheritance   325

Importance and Impact of Plant Breeding   328

Traditional Breeding and Hybridization   329
Somatic Fusion and Tissue Culture   330
Genetic Modification or Engineering   330
Key Terms   331
Summary   331
Reflect   332
References   333

Chapter 16   Molecular Basis of Inheritance   335
Learning Objectives   335
Molecular Basis of Inheritance   336
DNA Structure and Organization   336
DNA Replication   338
Gene Expression   339
Mutations   343
Key Terms   344
Summary   344
Reflect   345
References   345

Chapter 17   Biotechnology and Genetically Modified Plants   347
Learning Objectives   347
Combining DNA from Different Sources with Recombinant DNA Technology   348
Recombinant DNA Technology   348
Restriction Enzymes Cut DNA into Fragments   349
DNA Ligases Glue DNA Fragments Together   351
Ways to Deliver Foreign DNA Molecules into Cells   352
Use of Reporter Genes to Identify Transformed Cells   354
Techniques for Genetic Modification of Plants   355
Using *Agrobacterium tumefaciens* to Transform Plant Cells   356
Using *Agrobacterium*-mediated Transformation Without Tissue Culture   359
Using Biolistics to Transform Plant Cells   359
Using Biolistics to Transform Plant Plastids   359
Genetically Modified Plants for Human Benefit   360
Genetically Modified (GM) Crops in the United States   360
GM Traits that Confer Tolerance to Herbicides   360
Stacking GM Traits   363
GM Traits that Confer Resistance to Insects   363
GM Traits that Confer Resistance to Viruses   365
GM Traits that Modified Product Quality   365
GM Traits that Confer Tolerance to Abiotic Stress   368
GM Traits that Produce Therapeutic Products   369

Benefits and Risks of Genetically Modified Plants 370

    Increased Economic Benefits to Growers 370

    Reduced Environmental Impacts from Pesticides 371

    Enhanced Product Quality and Improved Post-Harvest Processing 371

    Potential Health Risks 372

    Potential Impact on Nontarget Organisms 372

    Potential Contamination of Foreign Genes into Wild Species 373

Key Terms 374

Summary 374

Reflect 376

References 376

Chapter 18   Evolution 381

Learning Objectives 381

History of Evolution 382

    Evolution by Natural Selection According to Darwin and Wallace 383

Evidence in Support of Evolution 385

    Basis of Natural Selection 386

Genetic Composition and Evolution 387

The Hardy-Weinberg Law 388

Processes of Evolution 389

    Mutation 390

    Gene Flow 390

    Genetic Drift 391

    Nonrandom Mating 391

    Natural Selection 391

    Microevolution and Macroevolution 392

    The Pace of Evolution 393

    Adaptive Radiation 394

    Convergent Evolution 395

    Coevolution 396

Speciation 396

    Reproductive Isolation 397

    Allopatry 398

    Sympatry 399

    Rejoining of Isolated Populations 399

Key Terms 400

Summary 400

Reflect 401

References 402

Section V   Diversity of the Green World   403

Chapter 19   Phylogeny and Taxonomy   405

Learning Objectives   405
Classification Systems   406
Hierarchical Classification and Taxonomy   406
Binomial Nomenclature   406
   Rules for Writing Scientific Names   409
   How to Handle Hybrids   410
Systematics and Cladistics   410
   Systematics   410
   Cladistics   411
Domains   411
Key Terms   421
Summary   421
Reflect   422

Chapter 20   Cyanobacteria and Algae   423

Learning Objectives   423
Characteristics of the Inhabitants of the Green World   424
Taking a Closer Look at Cyanobacteria   424
   Differences in Body Forms and Structures   426
   Differences in Cell Division   430
Different Types of Cyanobacteria   430
   Cyanobacteria that Fix Nitrogen   430
   Cyanobacteria Acting as Chloroplasts   431
   The "Other" Cyanobacteria: Prochlorophytes   432
Taking a Closer Look at Algae   432
   What Are Protists?   432
Different Types of Algae   433
   Green Algae Are the Closest Relatives of Plants   434
   Red Algae Are Close Relatives of Green Algae and Plants   436
   Brown Algae Are the Giants of the Algal World   437
   Diatoms Are Algae with Glass Shells   439
   Dinoflagellates Are Troublemakers of the Algal World   441
   Euglenoids Have Characteristics of Plants and Animals   442
Importance of Cyanobacteria and Algae to Humans   443
   Important Phytoplankton   443
   Toxic Blooms and Dead Zones   443
   Biofuel Producers   444
   Food and Industrial Products   444

Key Terms   445
Summary   445
Reflect   447
References   447

Chapter 21   The Plant Kingdom   **449**
Learning Objectives   449
Evolution of Land Plants (Embryophytes)   450
   Major Plant Groups   450
   Ancestors of Plants   451
   Emergence and Diversification of Land Plants   453
Characteristics of Land Plants (Kingdom Plantae)   455
   Complex Polymers   456
   Multicellular Structures   457
   Unique Life Cycle Alternation of Generations   458
   Associations with Mycorrhizal Fungi   459
Key Terms   459
Summary   459
Reflect   460
References   460

Chapter 22   Bryophytes   **463**
Learning Objectives   463
Evolutionary Relationships among Bryophytes   464
Characteristics of Bryophytes   466
   Taking a Closer Look at Liverworts   466
   The Gametophyte Generation   467
   The Sporophyte Generation   468
   Asexual Reproduction   470
Taking a Closer Look at Hornworts   470
   The Gametophyte Generation   470
   The Sporophyte Generation   471
   Asexual Reproduction   471
Taking a Closer Look at Mosses   471
   The Gametophyte Generation   472
   The Sporophyte Generation   472
   Asexual Reproduction   474
Importance of Bryophytes to Humans   474
   Horticultural Uses   474
   Household and Industrial Uses   475
   Fuel Production   475
   Medical Uses   476

Ecological Importance and Uses   476

Key Terms   476

Summary   477

Reflect   477

References   478

Chapter 23   Lycophytes and Ferns   479

Learning Objectives   479

Evolution of Vascular Plants (Eutracheophytes)   480

   The Sporophyte Became Dominant   481

   Emergence of Vascular Tissues, Stems, and Roots   481

   Evolution of Leaves   481

Characteristics of Seedless Vascular Plants   482

   Taking a Closer Look at Lycophytes   482

   Taking a Closer Look at Ferns   485

   Taking a Closer Look at Horsetails   491

   Taking a Closer Look at Whisk Ferns   492

Importance of Seedless Vascular Plants to Humans   493

   Horticultural and Agricultural Uses   493

   Household and Industrial Uses   494

   Food   494

   Medical Uses   494

   Coal Formation   494

Key Terms   495

Summary   495

Reflect   496

References   497

Chapter 24   Gymnosperms   499

Learning Objectives   499

Evolution of Seed Vascular Plants (Spermatophytes)   500

   Sporangia Became Indehiscent   500

   Gametophytes Became Dependent   500

   Pollen Development   501

   Seed Development   501

   Sporophytes Became Woody   502

Characteristics of Gymnosperms   502

   Evolutionary Relationships among Gymnosperms   504

   Taking a Closer Look at Cycads and Their Uses   504

   Taking a Closer Look at Ginkgo and Its Uses   506

   Taking a Closer Look at Conifers and Their Uses   507

   Taking a Closer Look at Gnetophytes and Their Uses   514

Key Terms   515
Summary   515
Reflect   516
References   517

Chapter 25   Angiosperms   519
Learning Objectives   519
Characteristics of Angiosperms   520
    Flower Development   521
    Evolutionary Trends among Flowers   522
    Seed Development   524
    Fruit Development   524
Evolutionary Relationships among Angiosperms   525
    Basal Angiosperms   525
    Core Angiosperms   528
Taking a Closer Look at Magnoliids and Their Importance   531
    The Magnolia Family (Magnoliaceae)   531
Taking a Closer Look at Monocots and Their Importance   531
    The Lily Family (Liliaceae)   532
    The Orchid Family (Orchidaceae)   533
    The Grass Family (Poaceae)   534
Taking a Closer Look at Eudicots and Their Importance   536
    The Bean or Legume Family (Fabaceae)   537
    The Rose Family (Rosaceae)   538
    The Pumpkin Family (Cucurbitaceae)   539
    The Mustard Family (Brassicaceae)   540
    The Nightshade Family (Solanaceae)   541
    The Sunflower Family (Asteraceae)   542
Key Terms   543
Summary   544
Reflect   544
References   545

Chapter 26   Fungi: Friends or Foes of the Green World   547
Learning Objectives   547
Characteristics of Fungi   548
    Evolutionary Relationships   548
    Unique Cell Structure and Body Form   549
    Unique Reproduction   550
Different Types of Fungi   552
    Taking a Closer Look at Chytridiomycota   553
    Taking a Closer Look at Glomeromycota   553

Taking a Closer Look at Ascomycota 554
Taking a Closer Look at Basidiomycota 555
Importance of Fungi to the Green World 556
Mycorrhizae: Partnerships between Fungi and Plant Roots 557
Endophytes: Partnerships between Fungi and Plants 558
Pathogens: Parasites of Plants 559
Lichens: Partnerships between Fungi and Algae or Cyanobacteria 561
Important Decomposers and Biogeochemical Transformers:
Recycling Nutrients for Plants 563
Key Terms 563
Summary 563
Reflect 564
References 565

Section VI   The Green World in the Web of Life 567
Chapter 27   Ecosystems and Biomes 569
Learning Objectives 569
Global Climate Patterns 570
Biomes of the World 573
Arctic Tundra 575
Boreal Forest 577
Temperate Coniferous Forest 578
Temperate Broadleaf and Mixed Forests 579
Temperate Grasslands, Savannas, and Shrublands 580
Deserts and Xeric Shrublands 583
Tropical and Subtropical Moist Broadleaf Forests 584
Key Terms 586
Summary 586
Reflect 587
References 587

Chapter 28   Dynamics of Plant Communities and Populations 589
Learning Objectives 589
Levels of Ecological Studies 590
Interactions between Organisms 591
Mutualism 591
Commensalism 592
Competition 593
Parasitism or Predation 593
Population Ecology 594
Ecosystem Dynamics and Human Activity 597
Energy Flow in Ecosystems 600

The Food Chain and Food Web   600
Energy Transfer in Ecosystems   603
Biogeochemical Cycles   604
The Carbon Cycle   605
The Nitrogen Cycle   605
Ecological Succession   608
Succession on Mount St. Helens Following a Volcanic Eruption   611
Key Terms   612
Summary   613
Reflect   613
References   614

Chapter 29   Plants as Food, Commercial Products, and Pharmaceuticals   **615**
Learning Objectives   615
History of Plant Domestication for Food and Other Uses   616
Food Plants Essential to Humans   618
Grasses: Maize, Rice, Wheat, and Other Grasses   618
Legumes   624
Potatoes, Cassava, Sweet Potatoes, and Other Starches   626
Fruits, Nuts, and Vegetables   628
Plant Oils and Sugar   634
Commercial Products   636
Flavored and Fermented Beverages   636
Herbs and Spices   638
Paper, Cloth, and Wood   638
Medicinal Plants   643
Key Terms   644
Summary   644
Reflect   645
References   645

Glossary   647
Index   671

Preface

When you take a moment to observe the web of life surrounding you, it becomes apparent how precious plants truly are to our existence. As concerns increase globally with regard to our environment, the study of plants has evolved from a course focusing on the traditional review of plant classification, evolution, and function to a study also including the impact of the environment on plant response, methods of propagation, and unique uses for plants. To answer many of the problems plaguing our daily lives, it is essential to also have a clear understanding of the physiology of plant responses. Carbon sequestration, remediation of contaminants, food for a rapidly growing population, reduction of temperatures, and the development of pharmaceuticals and neutraceuticals for health are only a few of the solutions provided by plants.

The dramatic changes in our world have created the need for a plant science book with this evolving focus on plants. We have made every attempt to provide the most up-to-date information available. It has been an honor and privilege to coauthor this book. The diversity in our plant science backgrounds has provided a unique focus for this exploration of the green world.

Cynthia McKenney
Ursula Schuch
Amanda Chau

About the Book

THEMATIC PRESENTATION

This book is organized thematically to highlight the diversity of the different plant systems while focusing on the plant functions that ultimately determine their unique positions in the web of life. The first unit, "Structure of the Green World", concentrates on the building blocks such as cells and tissues that provide the foundation for the introduction to the organs and systems of a plant. Once the students are familiar with these components, the diversity of specific plant functions is more easily addressed. "Reproduction in the Green World" is the second unit and it introduces common methods of sexual and asexual reproduction. A natural progression of complexity following the identification of plant structures and reproduction is a review of the physiological processes and pathways in the third unit, "Physiological Processes of the Green World". This unit includes photosynthesis and respiration along with plant responses to environmental conditions.

Once the study of the physiology and morphology of plants is completed, the focus of the book turns to evolution, genetics, and classification. The fourth unit is "Continuity of the Green World". Here evolution and genetics are explored as the driving forces behind the diversity of plant life. Concomitantly, a review of phylogeny and taxonomy provide the platform for the review of different classifications used for plants outlined in the fifth unit, "Diversity of the Green World". The concluding unit VI, "The Green World in the Web of Life", provides a rich exploration of biomes, ecosystems, and plant populations along with the diverse uses found for plants.

COURSE FEATURES

As quickly as technology advances and changes our understanding of plants, technology has an equivalent impact on pedagogy. No longer is "chalk talk" the standard for presentation of information, nor should it be. Online resources, enhanced imagery, online interactive activities, increased access to a constant stream of new information, and enhanced communication all provide a rich resource for course development and delivery. With 80 percent of the university campuses in the United States offering online instruction, course materials to facilitate distributed learning are essential. To address these concerns, *Introductory Plant Science: Investigating the Green World* includes a plethora of resources for both the novice and experienced educator. These resources include an instructor's manual with chapter development content, PowerPoint presentations for each of the chapters, a test bank of questions with appropriate answers, sample syllabi for either the semester or quarter system, clearly identified goals and learning objectives for each chapter, directions to aid students in chapter reading, outlines to guide the students through the highlights of each chapter, assignments, boldface text for new terms being introduced, questions for in-class polling systems, and a glossary to aid in managing the technical jargon of the discipline. In addition, focus boxes provide application information regarding topics included in the chapters to enhance long-term retention of key topics. The heavy inclusion of imagery to the book and website provides course enhancement to assist visual learners and alt tags have been included in the online materials for visually impaired students.

Online resources include an adaptive learning platform which allows each student to have a personal experience as they move through the material. In addition, there are alternative exercises . . . interactive exercises to engage students in the review of new concepts, links to additional articles and other resources related to the course content, relevant current issues and recent publications, self-review questions for students, and an online glossary. The well-designed layout of the course website enhances the navigation to different activities for online learners and instructors.

Acknowledgments

We have many people to thank for their help with this book. We are grateful to all the editorial and production staff at Kendall/Hunt. Special thanks to our project coordinator Torri Johnson, Lynne Rogers, our Production Editor, Traci Vaske, and to Paul Carty, Director of Publishing Partnerships, for helping us discover a new adaptive course platform. We also thank Michael Clayton for his photographs and micrographs.

We are also extremely grateful to our students, colleagues, families, and friends who have inspired and supported us during the long writing and revision process. Thank you.

About the Authors

Cynthia McKenney is the former Rockwell Professor of Horticulture in the Department of Plant and Soil Science at Texas Tech University. During her 35 years on faculty at Texas Tech, she taught numerous courses in horticulture, including principles of horticulture, arboriculture, plant propagation, greenhouse crop production, and both herbaceous and woody plants. Dr. McKenney has been very involved in distance education since the late 1990's developing numerous undergraduate courses that included laboratory exercises, as well as, multiple graduate courses. She has also participated in a USDA grant to create a graduate course on global horticulture and human nutrition to enhance community resilience and food security. Ultimately, Dr. McKenney taught over 10 courses via interactive video conferencing, online instruction, and blended formats. This resulted in her having student in several different countries in Europe and Asia. Dr. McKenney's experience in distance education has provided her skills to aid in the development of many of the on-line features for this textbook. In addition to teaching, she has conducted research on water-conserving landscapes, native plants, and alternative food production systems. Her environmental viewpoint has added an additional dimension to this introductory plant science textbook.

Ursula Schuch is a professor in the School of Plant Sciences at the University of Arizona. Her educational background is in forestry and horticulture. She has taught horticulture classes including herbaceous plant materials and several plant production classes for temperate and tropical climates. Dr. Schuch has conducted research on the physiology of plants in many different climates. She is interested in understanding the tolerance of plants to environmental stresses such as drought, high and low temperatures, and salinity. Whether teaching plant physiology principles or production practices, she incorporates relevant examples students can relate to and improve their understanding of the subject matter.

Amanda Chau is a professor in the Division of Agricultural and Natural Sciences at Blinn College in Texas. She received her M.Sc. in forestry from University of Toronto and Ph.D. in biological sciences from Simon Fraser University. At Texas A&M University, she conducted her postdoctoral studies on manipulating fertilization as a management tactic against pest insects on greenhouse crops. She taught courses in entomology, horticultural and floricultural entomology, and principles of biological control. In 2007, Dr. Chau joined the faculty at Blinn College and has been teaching freshmen courses in biology and botany. Dr. Chau believes that we learn best as a community and strives to cultivate a learner-centered community with her students. She looks forward to exploring the fascinating world of plants with both her students and those who read this textbook.

Chapter 1

The Science of Plants

© Smit/Shutterstock.com

Learning Objectives

- Discuss how and why most life depend on green organisms
- Discuss how plants and other organisms differ from non-living things
- Summarize the key features of plants
- Distinguish among the three domains and six major groups and give representative organisms from each group
- Discuss how the six groups fit into the new classification of supergroups
- Define the term *plant science* and briefly describe at least five disciplines within plant science
- Explain the scientific process and outline the steps of the scientific method
- Distinguish between inductive and deductive reasoning

Congratulations! You are about to embark on a fantastic journey to explore and study the most important organisms on planet Earth, plants. We are totally dependent on plants for survival. Next time when you go to a grocery store or supermarket, take a good look around and you will be amazed at how important plants are to us. All the vegetables and fruits we enjoy are from plants (Figure 1.1). Plants provide food for livestock, poultry, and fish, which in turn provide us with essential sources of protein. Look inside your refrigerator and you will notice without plants we will not have eggs or dairy products such as cheese, butter, yogurt, ice cream, and milk. All the spices and herbs used for cooking are from plant parts and seeds. The beverages you enjoy such as coffee, tea, juice, Coca-Cola, and many others are from plants (Figure 1.2). The chocolate candies and hot cocoa you crave are from plants. Plants also provide us with shelter and clothing. Your home and furniture are mostly made of wood (Figure 1.3). Look inside your closets and you will realize most of your clothing is made of plant fibers. Even your textbooks are made of plant pulp.

Plants are not only the most important group of organisms on planet Earth but also the most diverse and fascinating. Why are plants so important? What are plants? How do they differ from other organisms? What is plant science? In this chapter, we begin our journey by taking a closer look at plants and the science of plants.

THE IMPORTANCE OF PLANTS

Plants are the most important green organisms on Earth! They provide food not only for humans but for all other organisms. Green organisms like plants are the only organisms capable of using solar energy to power life. Green organisms carry out a complex **metabolic process** called **photosynthesis** to capture solar energy and store it in organic molecules we call food. Green organisms are referred to as producers or **autotrophs** because they make their own food and do not rely on other organisms to survive. Consumers

metabolic process
Series of chemical reactions carried out by living organisms to build or break down organic molecules and to store or release energy to power life.

photosynthesis
Complex metabolic process that green organisms use to capture solar energy and convert it into chemical energy of organic molecules such as glucose or food.

autotrophs
Organisms that are able to make their own food using photosynthesis.

(a)

(b)

(c)

© Supri Suharjoto/Shutterstock.com

© Subbotina Anna/Shutterstock.com

Figure 1.1 Plants provide us with (a) fruits, (b) vegetables, (c) spices and herbs.

(a) (b) (c)

Figure 1.2 Plants provide beverages such as (a) tea, (b) coffee, (c) juice, and more.

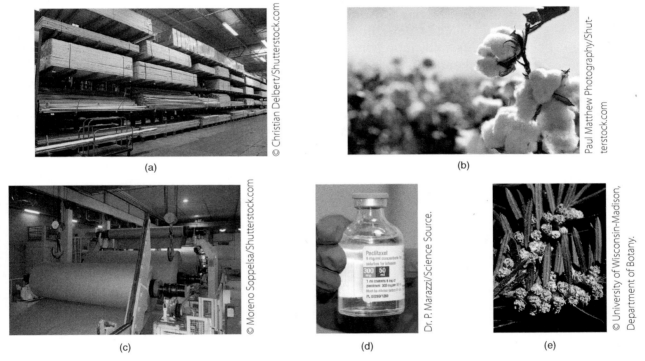

(a) (b)

(c) (d) (e)

Figure 1.3 Plants provide (a) building materials, (b) cotton, (c) paper, and pulp, and (d & e) medicines such as Taxol.

or **heterotrophs** like us, on the other hand, need to acquire organic molecules and energy from other organisms to survive. To release energy stored in food, all organisms carry out **cellular respiration**. Photosynthesis and cellular respiration are the two essential processes that power life and we will take a closer look at them in Chapters 10–11.

Plants not only produce food but also generate oxygen, which is important to aerobic or oxygen-requiring organisms like us. Plants also prevent carbon dioxide from accumulating in the air by using it to make organic molecules. Excess carbon dioxide in the lower atmosphere traps thermal radiation from the earth's surface. This "greenhouse effect" is responsible for global warming and significant climate changes (see Box 10.1). Excess

heterotrophs
Organisms that cannot produce their own food but have to depend on other organisms for food.

cellular respiration
Complex metabolic process that all organisms use to break down organic molecules and release the stored energy to sustain life.

BOX 1.1   Plant Domestication and Its Impacts

Plant domestication altered the course of human history by shifting the once hunter-gatherer societies into agricultural ones. Hunter-gatherer societies relied exclusively on gathering wild plants and hunting wild animals for food. Human societies in different parts of the world began plant domestication and food production between 13,000 and 5,000 years ago (Ross-Ibarra, Morrell, & Gaut, 2007). This transition in human history is often referred to as the "Neolithic Revolution" or "Origins of Agriculture." Plant domestication led to food production economies that stimulated the rise of cities and modern civilization. Surprisingly, humans rely on a very small number of crop plants for food. As much as 70% of the calories consumed by humans are from only 15 crops. Five crops (rice, wheat, maize, sugarcane, and barley), in particular, contribute more than 50% of the calories consumed.

Over time, humans began to select particular plant characteristics for cultivation, which eventually led to morphological and physiological changes in these crop plants. As a result, domesticated species became very different from their wild ancestor and relatives. This process of change or selection is often referred to as **domestication**. For example, domestication of cereals such as wheat, rice, maize, and barley led to more and larger fruits or grains (Figure Box 1.1), thicker stalks, and seeds easily separated from the chaff. Instead of dispersing their seeds, domesticated species retain their mature seeds, allowing them to be harvested for food and replanting. As the domestication process continues, crop plants become more dependent on the humans who cultivate them, just as humans become more dependent on these crop plants.

domestication
Selection of particular plant characteristics for cultivation that eventually leads to morphological and physiological changes in a crop plant.

© University of Wisconsin-Madison, Department of Botany.

Figure Box 1.1   Domestication of maize or corn led to more and larger fruits or kernels.

carbon dioxide also negatively affects the ocean and marine organisms. Carbon dioxide is readily absorbed by the ocean, which acts as a reservoir for carbon. But excess carbon dioxide may alter the ocean's natural chemistry and ability to absorb and exchange the gas (Caldeira & Wickett, 2005). Green organisms are extremely important because they remove excess carbon dioxide and replenish oxygen in the atmosphere.

Plants are important components of all ecosystems. They provide food and shelter for many organisms. Plants in coastal wetlands help absorb excess water and control floods. Plant roots help stabilize soils and prevent erosion. Most plant roots form partnerships with nitrogen-fixing bacteria, which convert atmospheric nitrogen into usable forms such as nitrates. This unique partnership between plants and nitrogen-fixing bacteria makes nitrogen available for many other organisms. Plants are not only the dominant primary producers on land but also important contributors to the global carbon and nitrogen cycles.

Plants are an important source of energy for human society. Fossil fuels (natural gas, oil, and coal) that supply over 85% of our energy today were mostly produced from dead plants. As an alternative to fossil fuels, biofuels such as ethanol are produced by fermenting carbohydrates from corn and switchgrass. People living in many developing countries still use plant materials as firewood to cook and heat their homes.

Plants provide important pharmaceutical products for managing human health (Levetin & McMahon, 2006). Many active compounds in medicines today were first discovered in plants. Aspirin, for example, is a popular medicine for reducing pain, fever, and inflammation. The active compound salicylic acid was first isolated from the bark of willow (*Salix*) trees and later replaced by a similar compound called *acetylsalicylic acid*. The letter *a* in the word *aspirin* is from acetylsalicylic acid, and *spirin* from *Spirea*, the plant from which salicylic acid was first isolated. Digoxin and Digitoxin from the purple foxglove (*Digitalis purpurea*) are important for treatment and management of congestive heart failure. The burn plant *Aloe vera* contains active compounds such as aloin, which are effective in treating burns, minor cuts, skin conditions, and constipation. *Aloe* sap also has a moisturizing effect and is now a common ingredient in lotions, shampoos, and cosmetics. Taxol is an effective anticancer drug that was first extracted from the bark of the Pacific yew (*Taxus brevifolia*) (see Figure 1.3). Taxol slows tumor growth and is an effective treatment for ovarian cancer and breast cancer. Many additional plant compounds are useful for treatment and management of human diseases. We explore how plants are used in foods, commercial products, and pharmaceuticals in Chapter 29.

PLANT DOMESTICATION AND GLOBAL AGRICULTURE

When did plant domestication begin? Where did cultivated plants originate? In 1882, Alphonse de Candolle, a Swiss botanist, published his work "Origin of Cultivated Plants," which marked the beginning of crop geography. By 1940, Nikolai I. Vavilov, a Russian plant geographer, had identified a total of seven primary centers of origin based on areas of greatest diversity (Vavilov, 1992). Since the 1950s, molecular and archaeological data suggest that some crops do not have centers of origin. Jack R. Harlan revised

Vavilov's centers of origin to six major regions of plant domestication: Near Eastern, Africa, North America, Central and South America, South Asia and Pacific Islands, and Chinese regions. Currently, 11 different centers of plant domestication or agriculture have been identified throughout these six regions (Smith, 2006) (Figure 1.4).

The oldest center of plant domestication is the Near East region, which includes present-day Iran, Iraq, Turkey, Syria, Lebanon, Jordan, and Israel. Plant domestication began in this region about 13,000 years ago. Wild barley, emmer wheat, and einkorn wheat were the first domesticated plants and the ancestors of our modern barley and bread wheat. The most recent center of plant domestication is in North America, which is now the United States. Sunflower, pepo squash, marsh elder, and chenopod were domesticated between 5,000 and 4,000 years ago in eastern North America (Smith, 2006). Only sunflower and squash are still being grown here. Although many crops were first cultivated in certain areas of the world, they eventually spread to other areas. For the past 500 years, many crops like wheat, rice, and maize have been cultivated throughout the world in areas where they grow best. For example, rice is now grown in more than 111 countries in the world and also six U.S. states: Arkansas, California, Louisiana, Mississippi, Missouri, and Texas. Arkansas is the top rice producer in the United States and accounts for 40% of rice production in this country (De Datta, 1981). In addition to sunflower and squash, over 100 food crops are now grown in the United States and the market value of these crops sold in 2007 was over $143 billion (U.S. Department of Agriculture, 2009).

The global human population has been increasing at an alarming rate. About 10,000 years ago, the global population was only 5 million; however,

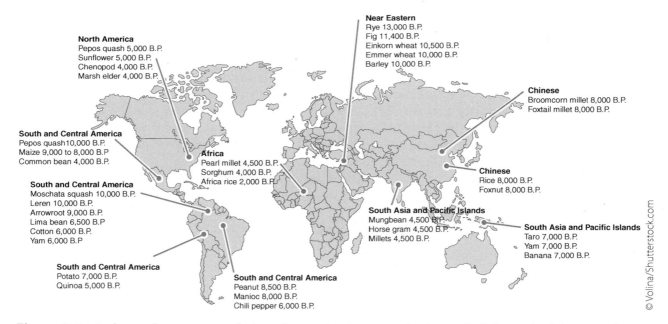

Figure 1.4 Independent centers of plant domestication or agriculture, their first principle crop plants, and estimated times of domestication. B.P., before present, a time scale used in archeology and geology to specify when events occurred in the past. Standard practice uses in 1950 as the origin of the age scale. Adapted from Smith (2006).

it reached 2 billion by 1930. The global population doubled again to 4 billion in 1975. At the start of the 21st century, the global population was about 6 billion (Buhr & Sinclair, 1998). In 2011, it reached 7 billion (Population Reference Bureau, 2011). The need to increase plant production to feed such a rapidly growing population is paramount. Major efforts are being made to improve cultivation practices as well as quality and yield of existing crops. The pressure to increase plant production results in tremendous pressure on the environment. Major environmental concerns associated with increased plant production include loss of biodiversity due to deforestation, loss of crop genetic diversity due to selection, and degradation of land, soil, and fresh water (Buhr & Sinclair, 1998). Efforts are in progress to preserve genetic diversity of crops and their wild relatives by creating seed and clonal banks. With the advances in molecular technology and genetic engineering, crop plants can be genetically modified to enhance their yield and nutritional quality, resistance to insect pests and pathogens, and tolerance to herbicides. We explore plant genetics and its impact on plant breeding in Chapters 15 & 17.

CHARACTERISTICS OF LIFE

Plants are very different from animals like us. Plants do not move, run, or swim from one place to another. Most plants do not eat other plants or animals to get their nourishment. Surprisingly, plants and animals are very much alike. They are living organisms and have a number of unique characteristics that distinguish living organisms from nonliving things.

Living Organisms Have Cell(s) and Are Organized

All living organisms are made up of one or more **cells**. Although cells are microscopic and cannot be seen with the naked eye, they are the basic units of life. All life-sustaining processes take place in a cell and it is also where **deoxyribonucleic acid (DNA)**, the genetic blueprint, resides. Some organisms such as bacteria and some algae are single-celled (**unicellular**) whereas organisms such as plants and humans are made up of many cells (**multicellular**). Cells making up a plant body come in a mind-boggling array of sizes, shapes, colors, and functions. We take a closer look at cells and their functions in the next chapter.

Living organisms are highly organized. All cells are made up of smaller components such as cell membrane, ribosomes, DNA, and cytosol, and many have organelles, which are membrane-bound structures. We investigate cellular components that make up plant cells in the following chapter. These cellular components are made up of **macromolecules** or large organic molecules such as proteins, lipids, carbohydrates, and nucleic acids. These macromolecules, in turn, are made up of smaller units such as amino acids, fatty acids, glycerol, monosaccharides, and nucleotides (see Appendix). These smaller units are made up of **atoms**, which are the smallest functional units of elements. Atoms cannot be broken down further into other substances by ordinary chemical or physical means. In the biological world, atoms form the simplest level of organization (Figure 1.5).

Cells are dynamic environments where molecules are constantly being built or broken down and where energy is being stored or released. These life-sustaining processes help unicellular organisms grow and reproduce.

cell
Basic unit of life; makes up all living organisms; where all life-sustaining processes take place and where the genetic blueprint, DNA, resides.

deoxyribonucleic acid (DNA)
"Blueprint" of life that is responsible for the storage, expression, and transmission of genetic information.

unicellular
(*uni-* in Latin means "one") Organisms that are made up of just one cell.

multicellular
Organisms that are made up of many cells (*multi* in Latin means "many").

macromolecules
Large organic molecules that are made by living organisms; the four major classes are proteins, lipids, carbohydrates, and nucleic acids.

atoms
The smallest functional units of elements and the simplest level of biological organization; cannot be broken down further into other substances by ordinary chemical or physical means.

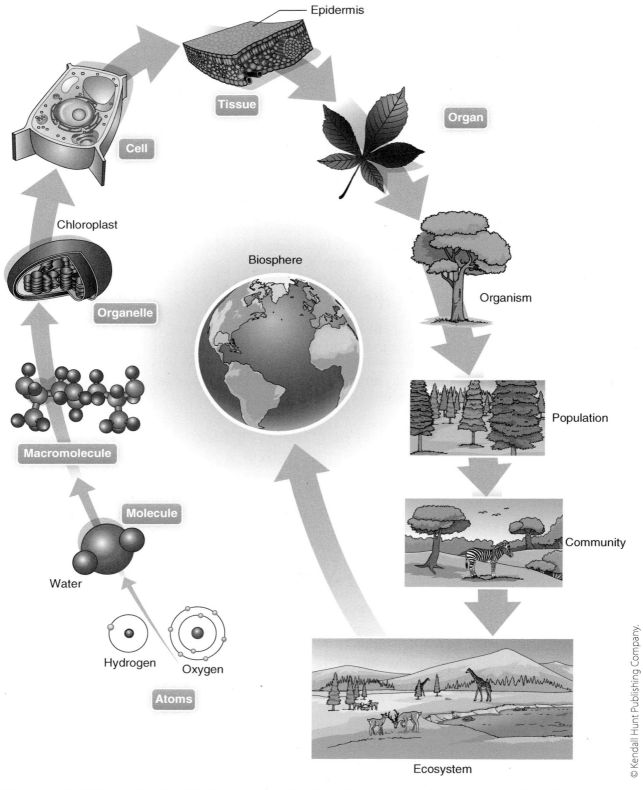

Epidermis

Tissue

Cell

Organ

Chloroplast

Organelle

Biosphere

Organism

Macromolecule

Population

Molecule

Community

Water

Hydrogen   Oxygen

Atoms

Ecosystem

Figure 1.5   Different levels of biological organization—from atoms, the simplest and smallest level, to the biosphere.

For multicellular organisms like plants, cells come together to form **tissues**, performing tasks individual cells cannot. Epidermis, xylem, and phloem are some examples of plant tissues. We investigate different tissues making up the plant body in the next chapter. When different tissues come together, they form **organs** that carry out complex tasks individual tissues cannot. Roots and leaves are examples of plant organs. We take a closer look at the four major plant organs in Chapters 4–7. When different organs come together, they make up a multicellular **organism** such as a plant (Figure 1.5).

When the same kind of organisms come together and live in the same area, they form a **population** such as a patch of bluebonnets in a field (Figure 1.6a). When populations of different organisms live and interact in the same area, they form a **community** (Figure 1.6b). For example, a forest community consists of not only plants such as trees, shrubs, and vines, but also pollinating insects, the herbivores that feed on them, and the predators that feed on herbivores. When the community is considered together with its physical environment, they form an **ecosystem** such as a forest, grassland, or lake. **Biomes** are defined as major communities classified according to the predominant vegetation and characterized by adaptations of organisms to that particular environment. For example, tundra, desert, grasslands, and rainforest are some of Earth's biomes. When all the ecosystems on planet Earth are considered together, they form the **biosphere**, which is the highest and most complex level of organization in the biological world. We explore biomes and ecosystems further in Chapter 27.

Living Organisms Acquire Energy and Materials

All living organisms require energy and materials to sustain life and they are broadly divided into two main types: autotrophs and heterotrophs. Autotrophs make their own food and do not rely on other organisms to survive. They are further divided into two main types depending on the energy source used for food production. **Photoautotrophs** such as plants and green organisms use solar energy whereas **chemoautotrophs** use chemical energy. Heterotrophs need to feed on other organisms to survive. Some heterotrophs such as animals acquire materials and energy by ingestion while others such as fungi and bacteria do so by absorption. In Chapters 10–11, we take a closer look at photosynthesis, a process carried out by photoautotrophs to make food, and cellular respiration, a process carried out by all organisms to power life.

Living Organisms Grow and Develop

All living organisms grow, allowing them to increase their size and mass. In unicellular organisms, cell multiplication or division (see the next chapter) increases the number of individuals in a population. In multicellular organisms, cell division and cell enlargement enable the organism to grow in size. Multicellular organisms such as animals and plants do not have the same type of growth. Animals have **determinate growth** and they stop growing after they reach a certain size. Many plants like trees, on the other hand, have **indeterminate growth**, allowing them to grow throughout their lives. Some plant parts such as fruits, flowers, and leaves, however, have determinate growth.

---

tissues
Group(s) of cells coming together to perform specific functions.

organs
Group(s) of tissues that come together to perform specific functions.

organism
Different organs come together to make up a distinct living entity.

population
All individuals of the same species that live and interact in the same area.

community
Populations of all organisms that live and interact in the same area.

ecosystem
Community and its physical environment in a particular area.

biome
Community of animals, plants, and other organisms living in an environment classified by the type of vegetation and adaptations of organisms to the environment.

biosphere
All the ecosystems on Earth; the highest and most complex level of organization in the biological world.

photoautotrophs
Autotrophic organisms that use solar energy for food production; compare with chemoautotrophs.

chemoautotrophs
Autotrophic organisms that use chemical energy for food production; compare with photoautotrophs.

determinate growth
After an organism reaches a certain size, growth stops.

indeterminate growth
Growth continues throughout an organism's life.

(a)   (b)

Figure 1.6   A population of bluebonnets and a community of wildflowers and trees.

differentiation
Process when cells take on different shape and form, allowing them to perform different functions.

development
Includes both growth and differentiation.

sexual reproduction
Union of the gametes produced by two individuals produces offspring that are genetically different from their parents.

Multicellular organisms are made up of many cells; cells in different parts of the body are responsible for different tasks. Some cells specialize in storage and others in transportation of water and food. Growth only produces more cells of the same type. **Differentiation** refers to the process when cells take on different shape and form, allowing them to perform different functions. Growth and differentiation are collectively referred to as **development**. In plants, development begins as a fertilized egg, which develops into a multicellular embryo. The embryo (seed) germinates and develops into a seedling with roots, shoots, and leaves (Figure 1.7). The seedling further develops into a mature adult plant.

Living Organisms Reproduce

Reproduction is the most distinctive characteristic of living organisms and enables the genetic blueprint, DNA, to pass from one generation to the next. DNA governs the organization, development, characteristics, and functions of living organisms and are examined further in Chapter 14. Some organisms reproduce sexually, which involves gametes or sex cells from two individuals. **Sexual reproduction**, the union of the gametes, produces

(a)   (b)

Figure 1.7   Development: unique characteristics of multicellular organisms.

offspring that are genetically different from their parents. Some organisms reproduce asexually, which does not involve gametes from two individuals. **Asexual reproduction** produces offspring that are genetically identical to their parents. Some organisms such as humans can only reproduce sexually whereas other organisms such as plants can reproduce both sexually and asexually (Figure 1.8). We explore reproduction, a crucial life-continuing process, further in Chapters 7 and 9.

Living Organisms Respond to Stimuli

Living organisms respond to changes in the environment by either moving toward or away from the changes or stimuli. When you touch something hot, your hands quickly move away from the heat. Like animals, plants also respond to stimuli, but their responses may not be as fast or dramatic. For example, when you place a potted plant in front of a window, the leaves will bend toward the light in a few days. Plants often respond to stimuli by changing the direction of growth and growth takes time. However, not all plant responses are slow. Leaves of the sensitive plant, *Mimosa pudica*, fold quickly when touched and unfold themselves when they are left alone. Leaves of the Venus flytrap, *Dionaea muscipula*, resemble tiny foothold or bear traps. When an insect lands on a leaf and brushes against its trigger hairs, the leaf quickly slams shut and traps the insect inside (Figure 1.9). The leaf will then secrete enzymes and digest the insect. Sensitive plants and Venus flytraps manipulate water pressure in their cells to bring about rapid leaf movement.

Living Organisms Adapt to Their Environment

Living organisms change or adapt to their environment in order to survive. They may change in structure, form, and function in response to predation, herbivory (the act of eating a plant or plant-like organism), or availability of oxygen, light, and water. Prickly pear cacti in the genus *Opuntia* are examples of plants adapted to survive in desert environments. Their thick stems

asexual reproduction
Does not involve gametes from two individuals and produces offspring that are genetically identical to their parent.

(a)

(b)

Figure 1.8a–b Asexual and sexual reproduction: (a) houseleeks, *Sempervivum*, producing smaller plants by asexual reproduction or vegetation growth; (b) dandelion, *Taraxacum*, producing seeds by sexual reproduction but also by apomixis, a type of asexual reproduction.

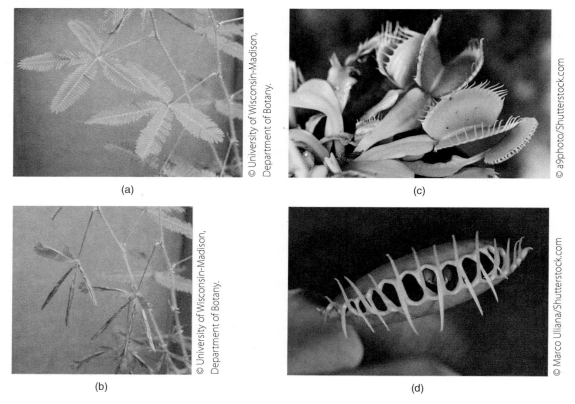

Figure 1.9a–d The sensitive plant, *Mimosa pudica,* responding to touch (a and b), and the Venus flytrap, *Dionaea muscipula,* capturing a fly (c and d), are examples of living organisms responding to stimuli.

pneumatophores
Roots growing above the water level to allow for oxygen uptake.

are modified for water storage and their leaves are modified into spines to reduce water loss and deter herbivory (Figure 1.10a). Mangrove trees produce specialized roots called **pneumatophores**, which grow out of the soil in order for the plants to breathe (Figure 1.10b). Pneumatophores allow mangrove trees to thrive in swampy environments where their roots are submerged. Pitcher plants in the genus *Sarracenia* produce pitcher-shaped leaves with fluid at the bottom to trap and digest insects (Figure 1.10c). Insect-trapping leaves allow plants such as pitcher plants, Venus flytraps, and sundews to acquire nitrogen from insects, which enables them to grow in nitrogen-deficient bogs and swampy areas. Changes in the characteristics of organisms from one generation to the next will allow organisms to be more successful at survival and reproduction. This process of change is often referred to as evolution. We examine the theory of evolution closely in Chapter 18.

DIVERSITY OF LIFE

All living organisms display the six characteristics discussed earlier, but they vary greatly in their shape, size, form, and energy acquisition. Some are unicellular whereas others are multicellular. Some are autotrophs whereas others are heterotrophs. Scientists have been trying to develop a system to organize living organisms and elucidate their evolutionary relationships.

(a)

(b)

(c)

Figure 1.10a–c   Various plant adaptations: (a) modified stem and leaves of the prickly pear cactus, (b) pneumatophores of mangroves, and (c) modified leaves of pitcher plants for trapping insects.

Classification of Living Organisms

For a long time, living organisms were organized or classified into groups based on similarities in their morphological characteristics. With recent advances in molecular technology, we are now capable of comparing and analyzing DNA and even RNA (ribonucleic acid) sequences among organisms. In addition to morphological data, molecular data are being used to classify organisms and examine their evolutionary relationships. **Systematics** refers to the study of biological diversity and evolutionary relationships among organisms. **Taxonomy** is an aspect of systematics that focuses on the description, naming, and grouping of organisms. There are at least 10 schools of taxonomy and each school advocates its own classification concept and scheme (for review, see Christoffersen, 1995). Classification schemes are also influenced by the characteristics (morphological, molecular, or both) used to group organisms.

systematics
Study of biological diversity and evolutionary relationships among organisms.

taxonomy
Study of describing, naming, and grouping of organisms.

Ever-Changing Classification

Traditional or Linnaean classification organizes organisms into progressively smaller taxa or ranks. The system begins with the broadest and most inclusive taxon of kingdom and narrows to phylum, class, order, family, genus, and specific epithet or species. Species is the narrowest and most exclusive group, representing only a single group of organism. The taxon of domain was created in the late 1970s to include various kingdoms. Domain is now the broadest and most inclusive taxon. Some systematists (scientists who study systematics) even advocate replacing the Linnaean system with phylogenetic taxonomy, which classifies organisms into clades. We study systematics and taxonomy in greater depth in Chapter 14.

The Three Domains

The three domains of living organisms are Bacteria, Archaea, and Eukarya. Members in the domain Bacteria and Archaea are all unicellular **prokaryotes** (organisms without a nucleus in their cells). Members in the domain Eukarya, on the other hand, are all **eukaryotes** (organisms with a nucleus in their cells) and many of them are multicellular. In the past, the domain Eukarya consisted of four kingdoms: Protista, Plantae, Animalia,

prokaryote
Organism that is made up of a prokaryotic cell (cell without a nucleus).

eukaryote
Organism that is made up of one or more eukaryotic cells (cells with a nucleus).

and Fungi. Recent phylogenetic analyses reveal that protists do not share a common ancestor (or form a monophyletic group). Kingdom Protista has now been demolished and its members are now classified within several eukaryotic supergroups. The taxon of supergroup lies between domain and kingdom. Kingdom Plantae is now classified within the supergroup called *Archaeplastida* whereas kingdoms Animalia and Fungi are within the supergroup *Opisthokonta* (Figure 1.11).

Six Major Groups

Living organisms can be broadly divided into six major groups: bacteria, archaea, protists, plants, animals, and fungi. The six groups differ in their cell type, cell number, cell wall component, and energy acquisition (**Table 1.1**). For the rest of the book, we focus our attention on plants, the most important green organisms on planet Earth.

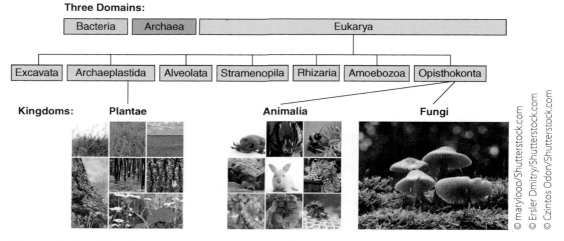

Figure 1.11 The domains, supergroups, and kingdoms of life. Adapted and modified from Brooker, Widmaier, Graham, & Stiling (2011).

Table 1.1 Key Characteristics of the Six Major Groups of Living Organisms

Group	Domain	Cell Type	Cell Number	Cell Wall Component	Energy Acquisition
Bacteria	Bacteria			Peptidoglycan	Mostly heterotrophic, some are autotrophic
		Prokaryotic	Unicellular		
Archaea	Archaea			No peptidoglycan	Heterotrophic
Protists	Eukarya	Eukaryotic	Mostly unicellular, some are simple multicellular	Cellulose, silica; some have no cell wall	Autotrophic, heterotrophic
Plants	Eukarya	Eukaryotic	Multicellular	Cellulose	Autotrophic
Animals	Eukarya	Eukaryotic	Multicellular	No cell wall	Heterotrophic
Fungi	Eukarya	Eukaryotic	Mostly multicellular	Chitin	Heterotrophic

Unique Characteristics of Plants

Not all green organisms are classified as plants in kingdom Plantae. Plant taxonomists today only recognize liverworts, hornworts, mosses, club mosses, ferns and their relatives, gymnosperms, and angiosperms as plants. Plants are also referred to as **embryophytes** because their embryos depend on the female plants for resources and protection. These green organisms share a number of characteristics distinguishing them from other green organisms such as cyanobacteria and green algae. To be classified as plants or embryophytes, the organisms must:

- Be multicellular eukaryotes
- Use chloroplasts to carry out photosynthesis
- Store energy or food as starch
- Have cell walls made up of cellulose
- Have multicellular embryos that develop within female sex organs
- Have a life cycle alternating between haploid and diploid generations

We examine these unique characteristics of plants throughout this book. We also take a look at the two other inhabitants of the green world, cyanobacteria and green algae, in Chapter 20.

Major Plant Groups

Kingdom Plantae consists of more than 300,000 species of living plants, which can be divided into informal groupings: **nonvascular plants** (bryophytes), **vascular seedless plants** (club mosses, ferns, and their relatives), and **vascular seed plants**. Vascular seed plants are subdivided into two major groups: **gymnosperms** (cone-bearing plants) and **angiosperms** (flowering plants). We explore the unique characteristics of these plant groups in Chapters 21–25. In chapters 4 to 7, we will be comparing morphology and anatomy of three major plant groups: monocotyledons (monocots), eudicotyledons (eudicots), and gymnosperms. Monocots and eudicots are both angiosperms. Throughout the book, we will use the term "eudicots" instead of "dicots". The term "dicots" or "Dicotyledoneae" traditionally referred to all non-monocot angiosperms and should no longer be used as a formal taxonomic unit (APG III 2009, Simpson 2010). We will explore the phylogenetic relationships among major groups of angiosperms in more details in chapter 25.

THE SCIENCE OF PLANTS

Simply put, plant science or plant biology is the scientific study of plants. The scope of plant science includes the origin, diversity, structure (cellular, molecular, and organismal levels), physiology, and genetics of plants as well as their interactions with other organisms and their physical environment. Our interest in plants, at first, was mostly practical and focused mainly on enhancing production of food and forage. Eventually, we became curious about what plants were made of and how they developed and reproduced.

embryophytes or land plants
Autotrophic organisms the embryos of which depend on maternal protection and resources for their development.

nonvascular plants
Group of plants that has no vascular tissues, true roots, true stems, or true leaves; has gametophytes that are dominant and sporophytes that are small and dependent on gametophytes for survival; includes liverworts, hornworts, and mosses.

vascular seedless plants
Group of plants that have vascular tissues, true roots and stems; use spores for dispersal; have sporophytes that are dominant and gametophytes that are small but independent from sporophytes; includes club mosses, ferns, and their relatives.

vascular seed plants
Group of plants that have vascular tissues, true roots, true stems, and true leaves; use seeds for dispersal; have sporophytes that are dominant and gametophytes that are small and dependent on sporophytes for survival; includes gymnosperms and angiosperms.

gymnosperms
Group of vascular seed plants that produce seeds and their seeds are not enclosed within a fruit (*gymnos* in Greek means "naked" and *sperma* means "seed"); includes cycads, ginkgo, gnetophytes, and conifers.

angiosperms
Group of vascular seed plants that produce seeds and their seeds are enclosed within a fruit, which is a mature or ripened ovary. *Angeion* in Greek means "vessel" and *sperma* means "seed."

Disciplines of Plant Science

Plant Science is divided into various disciplines or specialties:

- **Plant molecular biology**—the study of macromolecules (their structures and functions) and macromolecular mechanisms (gene replication, mutation, and expression) found in plants.
- **Plant biochemistry**—the study of chemical interactions within plants.
- **Plant cell biology**—the study of plant cell structure and function.
- **Plant anatomy**—the study of the internal structure of plants.
- **Plant morphology**—the study of the form and structure of the plant body.
- **Palynology**—the study of plant pollen and spores.
- **Plant physiology**—the study of plant function.
- **Plant genetics**—the study of plant heredity and variation that includes plant breeding and genetic engineering.
- **Plant systematics**—the study of evolutionary relationships among plant groups.
- **Plant taxonomy**—the study of describing, naming, and grouping of plants, viewed as one aspect of plant systematics.
- **Plant ecology**—the study of interactions among plants and between plants and their environment.
- **Paleobotany**—the study of fossil plants.
- **Ethnobotany**—the study of the traditional knowledge and customs of a people concerning plants and their medical, religious, and other uses.
- **Economic botany**—the study of relationships between plants and people.
- **Forensic botany**—the use of plant materials to help solve crimes or other legal problems.

In addition to these specialties, many botanists study or specialize in particular types of plants and their production systems.

- **Agronomy**—the science of soil management and crop production.
- **Horticulture**—the science of growing fruits, vegetables, flowers, or ornamental plants.
- **Forestry**—the science of cultivating, managing, and developing forests.
- **Bryology**—the study of bryophytes (mosses, liverworts, and hornworts).
- **Pteridology**—the study of ferns.

Scientific Processes

Scientific study is more than just a collection of facts about plants or the natural world. It is a dynamic process that involves observation, identification, investigation, analysis and interpretation of data, and reevaluation. The goal of scientific study, simply put, is to discover general principles that govern the operation of the natural world. These general principles will then help solve problems or provide new insights. Scientific study can be either discovery based or hypothesis testing. **Discovery-based science** is done without any preconceived hypothesis or expectation. It focuses on collecting and analyzing information that eventually lead to hypothesis testing. **Hypothesis testing** is often done in a series of steps leading to the rejection or acceptance of a hypothesis. The steps of hypothesis testing are often referred to as the scientific method.

The Scientific Method

In general, there are six steps in the scientific method and the goal is to develop and test a hypothesis.

1. **Ask a question**. Scientific study often begins with an observation that either stimulates our desire to know more about what we observed or to find a solution to the observed problem. We come up with a number of questions but eventually identify a single principal question.

2. **Formulate a hypothesis**. We review relevant scientific literature to find possible explanations to the observed phenomenon or solutions to the observed problem. Based on information we gathered from the literature, we formulate a number of hypotheses. Each **hypothesis** is basically an educated guess or a plausible solution to explain the problem. A specific prediction made by the hypothesis is then tested and possibly invalidated.

3. **Test with an experiment**. An experiment is designed and carried out to test the validity of the hypothesis. The goal of a carefully designed experiment is to gather information or data and evaluate if the results support or refute the predicted outcome. A good experiment must have the following components:

   a. **Independent variable** and **dependent variable**. An **independent variable** is the one variable, condition, or factor being manipulated and tested. All other variables, conditions, or factors are kept the same. The response to the independent variable being measured is called a **dependent variable**.

   b. **Control group** and **experimental group**. The two groups are identical, with one exception. The **experimental group** is exposed to the independent variable or the variable tested while the **control group** is not. When the experimental group responds differently from tahe control, it is most likely due to the independent variable.

4. **Analyze and interpret the results**. When the experiment is complete, the results are evaluated and a conclusion is made. The results either support or refute the predicted outcome from the hypothesis.

5. **Accept or reject the hypothesis**. If the results refute the hypothesis, it has to be modified and retested or rejected completely. If the results support the hypothesis, it simply means we have not found any contradicting evidence to reject the hypothesis. We can never prove whether a hypothesis is true or not because we do not know everything and have not tested all the variables. If new evidence appears later, the hypothesis has to be modified or abandoned.

6. **Communicate the finding(s)**. Scientists publish their findings in scientific journals and books. They also present their findings in conferences and meetings. Sharing new knowledge with the scientific

---

discovery-based science
Collection and analysis of data without any preconceived hypothesis or expectation; generally leads to hypothesis testing.

hypothesis testing
See *scientific methods*.

independent variable
One variable, condition, or factor being manipulated and tested in an experiment.

dependent variable
Response to the independent variable that is being measured.

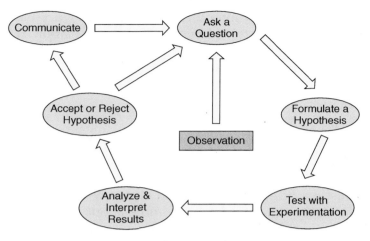

Figure 1.12 The scientific method cycle.

community enriches the scientific literature and also permits other scientists to repeat the experiment or design new experiments to confirm the validity of the hypothesis.

The scientific method is often described or presented as discrete linear steps. They actually form a cycle (Figure 1.12). At the end of the experiment, you often end up with more questions than before the start of the experiment, which lead to more hypotheses and experiments. Sometimes, the results can be confusing and seemingly contradictory. Scientific study is rarely a neat or straightforward process. However, a general principle or "big picture" emerges when results from different experiments are consistent.

Two Types of Reasoning

inductive reasoning Generates a unifying explanation or general principle after carefully evaluating specific studies; begins with specific studies and ends with a general principle.

Scientists use two types of reasoning or argument to interpret the data: inductive reasoning and deductive reasoning. **Inductive reasoning** is about generating a unifying explanation or general principle after carefully evaluating specific studies. Around 400 B.C., Hippocrates of Cos, a famous Greek physician and the father of modern medicine, used inductive reasoning to develop general theories about diseases (Killeffer, 1973).

An example of inductive reasoning:

Observation #1: My roses are pink.

Observation #2: My neighbor's roses are pink.

Observation #3: My cousin's roses are pink.

Conclusion: All roses are pink.

If you find a white rose, the argument "all roses are pink" needs to be revised. Inductive reasoning generates new knowledge but is prone to error. When contradictory results come to light, the general principle needs to be revised. However, many important theories are generated by inductive reasoning. For example, the cell theory, which we learn in the next chapter, was generated after cells were repeatedly observed in plants, animals, and microbes.

deductive reasoning Makes a specific prediction from a general principle and then tests it; begins with a general principle and ends with specific studies.

Deductive reasoning, on the contrary, makes a specific prediction from a general principle and then tests it.

An example of deductive reasoning:

General principle: All plants have waxy cuticles on their surfaces.
Observation: Cotton is a plant.
Conclusion: Cotton has a waxy cuticle on its surfaces.

Deductive reasoning begins with a general principle and ends with specific studies whereas inductive reasoning begins with specific studies and ends with a general principle. Deductive reasoning does not generate new knowledge but provides additional data to validate a general principle.

Key Terms

metabolic process
photosynthesis
autotrophs
heterotrophs
cellular respiration
domestication
cells
deoxyribonucleic acid
    (DNA)
unicellular
multicellular
macromolecules
atoms
tissues
organs
organism

population
community
ecosystem
biomes
biosphere
photoautotrophs
chemoautotrophs
determinate growth
indeterminate growth
differentiation
development
sexual reproduction
asexual reproduction
pneumatophores
systematics
taxonomy

prokaryotes
eukaryotes
embryophytes
nonvascular plants
vascular seedless plants
vascular seed plants
gymnosperms
angiosperms
discovery-based science
hypothesis testing
independent variable
dependent variable
control group
experimental group
inductive reasoning
deductive reasoning

Summary

- Plants are important to all organisms because they provide food, oxygen, and shelter. Plants also remove excess carbon dioxide from the air. In addition, plants are an important source of energy and medicine for humans.
- Organisms are divided into two main groups: autotrophs and heterotrophs.
- Photosynthesis and cellular respiration are two essential processes that power life.
- All living organisms display six characteristics distinguishing them from nonliving things. All living organisms:
  - Have cell(s) and distinct levels of organization
  - Acquire energy and materials
  - Grow and develop
  - Reproduce
  - Respond to stimuli
  - Adapt to their environments
- Systematics is the study of biological diversity and evolutionary relationships among organisms while taxonomy is the study of describing, naming, and grouping of organisms.

- Organisms are often organized into progressively smaller taxa or ranks. The system begins with the broadest and most inclusive taxon of domain and narrows to kingdom, phylum, class, order, family, genus, and species.

- Organisms are grouped into three domains: Bacteria, Archaea, and Eukarya.

- Based on differences in their cell type, cell number, cell wall component, and energy acquisition, organisms can be classified into six groups: bacteria, archaea, protists, plants, animals, and fungi.

- Plants or embryophytes are distinct from other organisms. To be classified as plants, the organisms must:
  - Be multicellular eukaryotes
  - Use chloroplasts to carry out photosynthesis
  - Store energy or food as starch
  - Have cell walls made up of cellulose
  - Have multicellular embryos developing within female sex organs
  - Have a life cycle alternating between haploid and diploid generations

- Plants are divided into informal groupings: nonvascular plants, vascular seedless plants, and vascular seed plants. Vascular seed plants are subdivided into two major groups: gymnosperms and angiosperms.

- Plant science or plant biology is the scientific study of plants. The scope of plant science includes the origin, diversity, structure (cellular, molecular, and organismal levels), physiology, and genetics of plants as well as their interactions with other organisms and their physical environment. These aspects are covered in different disciplines or specialties.

- Scientific study can be either discovery based or hypothesis testing and involves observation, identification, investigation, analysis and interpretation of data, and reevaluation. The goal of scientific study is to discover general principles that govern the operation of the natural world.

- The cycle of scientific method begins with:
  - Ask a question
  - Formulate a hypothesis
  - Test with an experiment
  - Analyze and interpret the results
  - Accept or reject the hypothesis
  - Communicate the finding(s)

- Inductive reasoning and deductive reasoning are used by scientists to interpret data.

Reflect

1. *Imagine one day when the sun no longer shines and Earth becomes dark.* What do you think will happen to all living organisms on earth?

2. *After watching the movie* **i, Robot,** *you and your friend debate whether robots will someday evolve into living entities.* Discuss the characteristics advanced robots must have to be classified as living organisms.

3. *You encounter a weird-looking organism on your way to class. You ask, "What type of organism is this?"* Recall the characteristics you use to determine whether an organism is a bacterium, protist, plant, animal, or fungus.

4. *What are plants?* Describe the characteristics all plants have.

5. *Convince me with numbers!* You are the production manager of a major greenhouse corporation and believe you can produce good-quality bell peppers using less fertilizer. You decide to run some experiments and show corporate headquarters your numbers. Using the scientific method, design an experiment that will show the effect of fertilizer on bell pepper production.

References

Angiosperm Phylogeny Group (APG III). (2009). An update of the Angiosperm Phylogeny Group classification for the orders and families of flowering plants: APG III. *Botanical Journal of the Linnean Society, 161,* 105–121.

Buhr, K. L., & Sinclair, T. R. (1998). Human population, plant production and evniornmental issues. In T. R. Sinclair & F. P. Gardner (Eds.), *Principles of ecology in plant production* (pp. 1–18). New York: CAB International.

Caldeira, K., & Wickett, M. E. (2005). Ocean model predictions of chemistry changes from carbon dioxide emissions to the atmosphere and ocean. *Journal of Geophysical Research, 110,* C09S04.

Christoffersen, M. L. (1995). Cladistic taxonomy, phylogenetic systematics, and evolutionary ranking. *Systematic Biology, 44,* 440–454.

De Datta, S. K. (1981). *Principles and practices of rice production.* New York: Wiley.

Killeffer, D. H. (1973). *How did you think of that?: An introduction to the scientific method.* Washington, DC: American Chemical Society.

Levetin, E., & McMahon, K. (2006). *Plants and society* (4th ed.). New York: McGraw-Hill.

Population Reference Bureau. (2011). *2011 World Population Data Sheet.* Retrieved July 12, 2012, from www.prb.org/Publications/Datasheets/2011/world-population-data-sheet/data-sheet.aspx.

Ross-Ibarra, J., Morrell, P. L., & Gaut, B. S. (2007). Plant domestication, a unique opportunity to identify the genetic basis of adaptation. *Proceedings of the National Academy of Sciences of the United States of America,104,* 8641–8648.

Simpson, M.G. (2010). *Plant Systematics.* 2nd ed. Burlington, MA: Elsevier Inc.

Smith, B. D. (2006). Eastern North America as an independent center of plant domestication. *Proceedings of the National Academy of Sciences of the United States of America, 103,* 12223–12228.

U.S. Department of Agriculture. (2009). *2007 Census of Agriculture, Volume 1, U.S. Summary and State Reports.* Retrieved December 30, 2018, from https://www.nass.usda.gov/Publications/AgCensus/2007/Full_Report/Volume_1,_Chapter_1_US/usv1.pdf

Vavilov, N. I. (1992). *The origin and geography of cultivated plants.* Cambridge, UK: Cambridge University Press.

STRUCTURE OF THE GREEN WORLD

Chapter 2   Essential Molecules of Life

Chapter 3   Cells and Tissues

Chapter 4   Roots

Chapter 5   Stems

Chapter 6   Leaves

Essential Molecules of Life

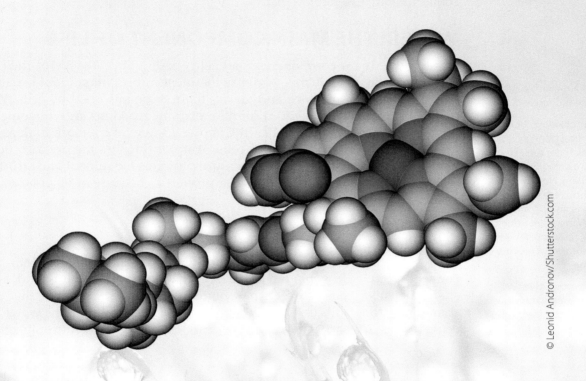

© Leonid Andronov/Shutterstock.com

Learning Objectives

- Discuss the properties of water and explain its importance to life.
- Distinguish the four classes of organic molecules that are essential to life.
- Compare and contrast major functions of these organic molecules and give examples.
- Discuss the importance of plant-based molecules to humans.

All living organisms are composed of one or more cells, the basic units of life. Cells come in an amazing array of sizes, shapes, colors, and functions. Have you ever wondered what made up these cells and their components? Did you realize all cells are not only made mostly of water but also surrounded by water? Life cannot exist without water. NASA scientists have been looking for water on Mars and the Moon as a prerequisite for life there. As you realized, water is not the only component of cells. Cells are composed of organic molecules such as carbohydrates, nucleic acids, proteins, and lipids. In this appendix, we investigate the unique properties and functions of these molecules that are essential to plants.

WATER: THE MAIN COMPONENT OF LIFE

solvent
Liquid in which a substance is dissolved.

polar covalent bond
Chemical bond in which atoms share their electrons instead of exchanging them. However, the electrons are not shared equally among atoms due to different electronegativity, resulting in uneven distribution of electric charge.

electronegative
The attraction of a given atom for the electrons of a covalent bond.

Water is the main component of all cells, making up 70–90% of their volume. For living organisms, water is the most important **solvent** because it dissolves many molecules and allows them to be absorbed by cells. Water provides an aqueous medium for all metabolic reactions and movement of materials into and out of cells. In Chapter 13, we take a close look at how molecules move into and out of cells. Water influences the organization of cell molecules. In plants, water pressure provides movement and structural support. Water has all these amazing properties because of its polarity.

Water Is a Polar Molecule

H_2O is the formula for a water molecule. Each water molecule contains two hydrogen atoms and one oxygen atom (Figure 2.1). A **polar covalent bond** connects each hydrogen atom to the oxygen atom. In a covalent bond, atoms share their electrons. Oxygen is a highly **electronegative** atom that pulls shared electrons. This uneven distribution of electrons within the water molecule results in a partial negative charge (δ^-) on the oxygen atom and a partial

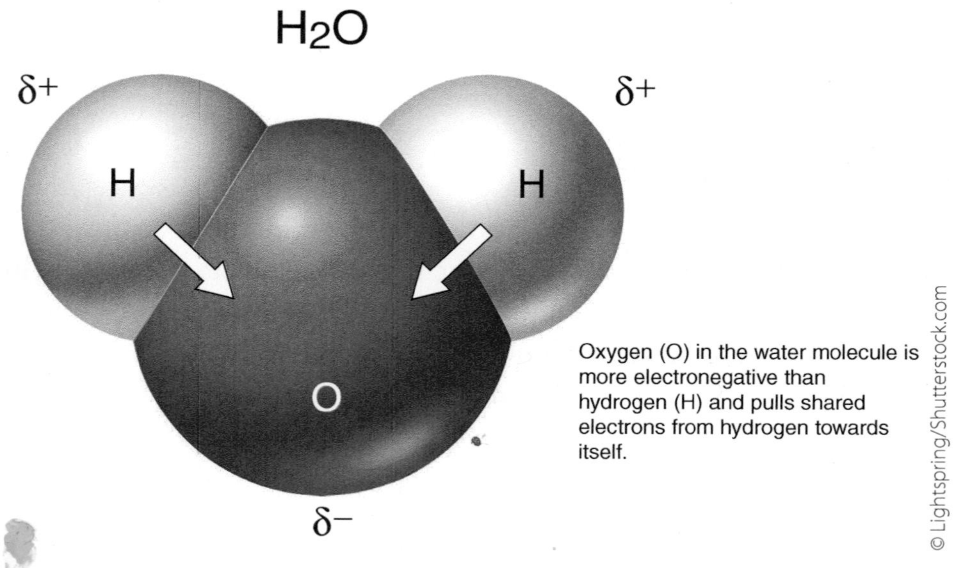

Oxygen (O) in the water molecule is more electronegative than hydrogen (H) and pulls shared electrons from hydrogen towards itself.

© Lightspring/Shutterstock.com

Figure 2.1   A water molecule is dipolar and has a partial negative charge (δ^-) on its oxygen and a partial positive charge (δ^+) on its hydrogen because of the uneven distribution of the electrons.

positive charge (δ^+) on each hydrogen atom (Figure 2.1). Water molecules are **polar** because of the uneven distribution of the electric charge. The polarity of water allows unique interactions among water molecules and between water molecules and other polar molecules. Water's polarity gives water many unique properties.

Water Molecules Stick Together and to Other Polar Molecules

The cohesiveness and adhesiveness of water are important to its transport in plants. **Cohesion** refers to the ability of water molecules to adhere or stick together. A hydrogen bond is formed when the negatively charged oxygen of a water molecule attracts the positively charged hydrogen of another water molecule. These hydrogen bonds help water molecules stick and move together. When transpiration occurs in the leaves, it creates a tension pulling water molecules up within the xylem (Figure 2.2). **Adhesion** refers to the ability of water molecules to stick to other polar molecules or surfaces. The adhesiveness of water helps its move along the xylem surface. In Chapter 13, we take a closer look at how plants transport water and minerals.

Water Is an Important Solvent

Water is often called the universal solvent because many substances (**solutes**) can easily dissolve in it to form **solutions**. Water molecules attract and surround dissolved ions (Figure 2.3). The adhesiveness of water allows water molecules to break up or dissolve polar molecules and ionic compounds such as sodium chloride (salt). Dissolved molecules and ions can react with one

dipolar
Molecule having an uneven distribution of electric charges in two different regions of the molecule.

cohesion
Ability of similar molecules to adhere or stick to each other.

adhesion
Ability of different molecules to stick to each other.

solutes
A substance that is dissolved in a solution.

solution
A liquid with one or more substances dissolved in it.

Claus Lunau/Science Source

Figure 2.2   Water molecules form hydrogen bonds between them allowing them to stick together. The cohesiveness of water is important to water transport within the xylem.

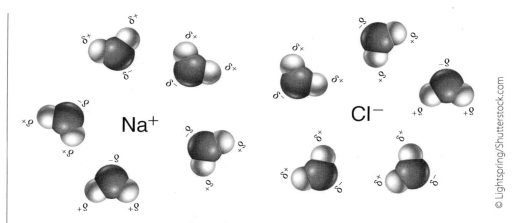

© Lightspring/Shutterstock.com

Figure 2.3 The adhesiveness of water allows water molecules to interact with other polar molecules and ions making water an excellent solvent.

another to form new molecules. Dissolved nutrients and waste products can easily be absorbed or excreted by cells. Plant cells sequester water-soluble waste products in their central vacuole instead of excreting them out of the cell.

A cell can manipulate its solute concentration to influence water movement. Guard cells in plants (see Chapter 6) manipulate their potassium ion concentration to open or close stomatal pores. To open stomatal pores, guard cells actively take in potassium ions to create a concentration gradient. The rise in potassium concentration causes water to move into the cells via osmosis. As a result, the guard cells become rigid and the pores open. To close stomatal pores, guard cells move potassium ions out of the cells. The drop in potassium concentration causes water to move out of the guard cells, which become flaccid and close the pores.

Water Provides Evaporative Cooling

Water exists in three states or forms: solid (ice), liquid (water), and gas (water vapor). When water changes form, it requires an input or a release of energy. For example, when you boil water, heat energy is used to change water from liquid to gas. This transformation from liquid to gas is called **vaporization** or **evaporation**. The **heat of vaporization** is the heat required to change 1 gram of a substance from liquid to gas. Water has a high heat of vaporization because a lot of energy is required to break the large number of hydrogen bonds among water molecules. As water evaporates from leaves, a process often referred to as *transpiration*, heat is removed from their tissues. Transpiration not only provides **evaporative cooling** to prevent plants from overheating but also creates a pulling force for water transport. For details regarding transpiration and water transport, see Chapter 13.

FOUR CLASSES OF ORGANIC MOLECULES: THE FOUNDATIONS AND STRUCTURES OF LIFE

In addition to water, four classes of organic molecules are essential to life: carbohydrates, nucleic acids, proteins, and lipids. **Organic molecules** contain carbon and almost all organic molecules found in living organisms also contain hydrogen. Unlike water, organic molecules are much more complex

vaporization
When a substance is transformed from its liquid form to gaseous form.

evaporation
See *vaporization*.

heat of vaporization
Heat required to change 1 gram of a substance from its liquid form to gaseous form.

evaporative cooling
Heat is removed as water evaporates, which helps to cool down surfaces or tissues.

organic molecules
Complex molecules found in living organisms and containing carbon and hydrogen.

and are made up of many different **elements** such as carbon, hydrogen, oxygen, nitrogen, phosphorus, and potassium. Elements that are essential to plants are discussed in detail in Chapter 13. Many of these organic molecules participate in chemical reactions that sustain life. Some provide energy, structure, and insulation; others make up the blueprint of life.

Carbohydrates Provide Fuel and Structure for Life

Many simple carbohydrates are important fuel molecules for living organisms; complex carbohydrates provide structure and energy storage. As you recall from Chapter 10, plants are autotrophs and can make their own carbohydrates using light energy via photosynthesis. Based on their size, carbohydrates are divided into three major types: **monosaccharides** (*mono* means "one" or "single" and *saccharides*, derived from a Greek word, mean "sugar"), **disaccharides** (mean "two sugars"), and **polysaccharides** (mean "many sugars").

Monosaccharides provide energy for living organisms. Monosaccharides consist of carbon, hydrogen, and oxygen in a 1:2:1 ratio and are the simplest carbohydrates. The molecular formula of monosaccharids is $(CH_2O)_n$ where n can be as small as 3, as in $C_3H_6O_3$ (Figure 2.4), or as large as 7, as in $C_7H_{14}O_7$. The most common monosaccharides are pentoses (five-carbon sugars) and

elements
Substances that cannot be broken down further by ordinary means.

monosaccharides
Simple carbohydrates or single sugars are building blocks or monomers of other carbohydrates.

disaccharides
Carbohydrates that make up of two monosaccharides.

polysaccharides
Complex carbohydrates made up of many monosaccharides.

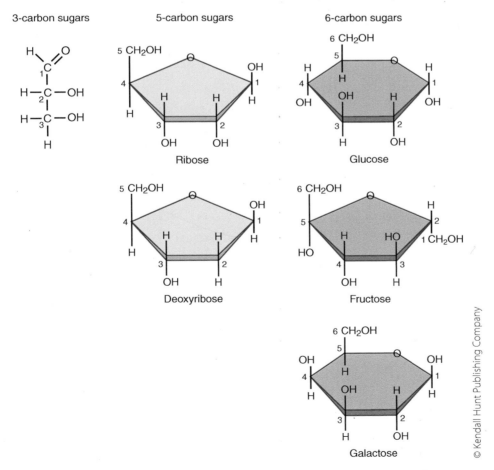

© Kendall Hunt Publishing Company

Figure 2.4   Three types of monosaccharides: 3-carbon sugars, 5-carbon sugars, and 6-carbon sugars.

hexoses (six-carbon sugars) (Figure 2.4). They tend to form a ring instead of a chain when dissolved in water. Glucose, a six-carbon sugar, is the primary source of energy for plants as well as other organisms. As we discuss in Chapter 10, plants use photosynthesis to make glucose and aerobic respiration to break it down for energy.

Monosaccharides function as building blocks for other carbohydrates. Monosaccharides such as glucose and fructose are building blocks for disaccharides and polysaccharides. In chemistry, carbohydrates, nucleic acids, and proteins are referred to as **polymers**. A polymer consists of similar or identical building blocks or units called **monomers**. For example, disaccharides such as lactose and sucrose consist of two monosaccharides (Figure 2.5). Lactose is the disaccharide found in milk and sucrose is the disaccharide found in plants.

Making and breaking apart polymers. When two or more monomers are joined together to form a polymer, water is also produced (Figure 2.6). This type of chemical reaction is known as **dehydration synthesis**. For example,

polymers
Complex and big molecules made up of many subunits called monomers.

monomers
Building blocks that make up a polymer.

dehydration synthesis
Chemical reaction combining two or more monomers together to form a polymer; water is also being produced.

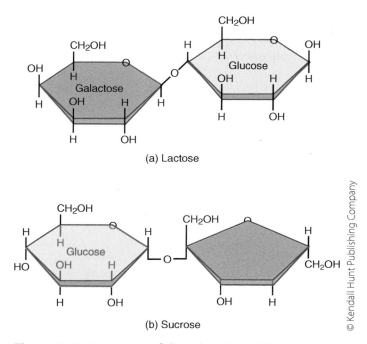

© Kendall Hunt Publishing Company

Figure 2.5 Examples of disaccharides: (a) lactose, a disaccharide made up of galactose and glucose; (b) sucrose, a disaccharide made up of glucose and fructose.

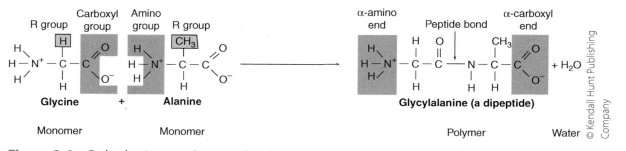

© Kendall Hunt Publishing Company

Figure 2.6 Dehydration synthesis is the chemical reaction that joins two or more monomers into a polymer. Water is produced as a by-product in this reaction.

glucose and fructose combine to form sucrose and water. When water is added to break apart a polymer, this reverse reaction is known as **hydrolysis** (*hydro* means "water" and *lysis* means "breaking apart").

Disaccharide sucrose transports glucose for plants. Unlike animals, plants do not transport glucose from one part to another. Instead, plants synthesize sucrose in leaves and use it to transport glucose from the leaves to the rest of the plant. Table sugar (sucrose) is harvested from sugar beets and sugarcane (see Chapter 29).

Polysaccharides provide structure and energy storage for plants. Polysaccharides are large and complex carbohydrates made from thousands of monosaccharides. Starch and cellulose are the two most important polysaccharides in plants (Figure 2.7). These polysaccharides consist of glucose chains that are insoluble in water. **Starch** is the primary storage molecule in plants and comes in two forms: **amylose** and **amylopectin**. Glucose chains are unbranched in amylose but highly branched in amylopectin. Many organisms have enzymes such as amylase and glycosidases that break down starch into glucose molecules for energy. In addition to starch, **fructans** are also storage molecules for grasses and cereals such as wheat, rye, and barley (Pollock & Cairns, 1991). Fructans are water-soluble polymers of fructose mainly found in leaves and stems.

Cellulose is the most abundant organic molecule on Earth and the most important structural molecule in plant cell walls. Although two other complex polysaccharides, hemicellulose and pectin, also contribute to the

hydrolysis
Chemical reaction that splits a polymer into individual monomers by adding water.

starch
Very large polysaccharide made up of glucose molecules and is the primary storage molecules in plants.

amylose
Form of starch made up of long and unbranched chains of glucose molecules.

amylopectin
Form of starch made up of long and branched chains of glucose molecules.

fructans
A polymer made up of fructose molecules, primary storage molecule in leaves and stems of wheat, rye, and barley.

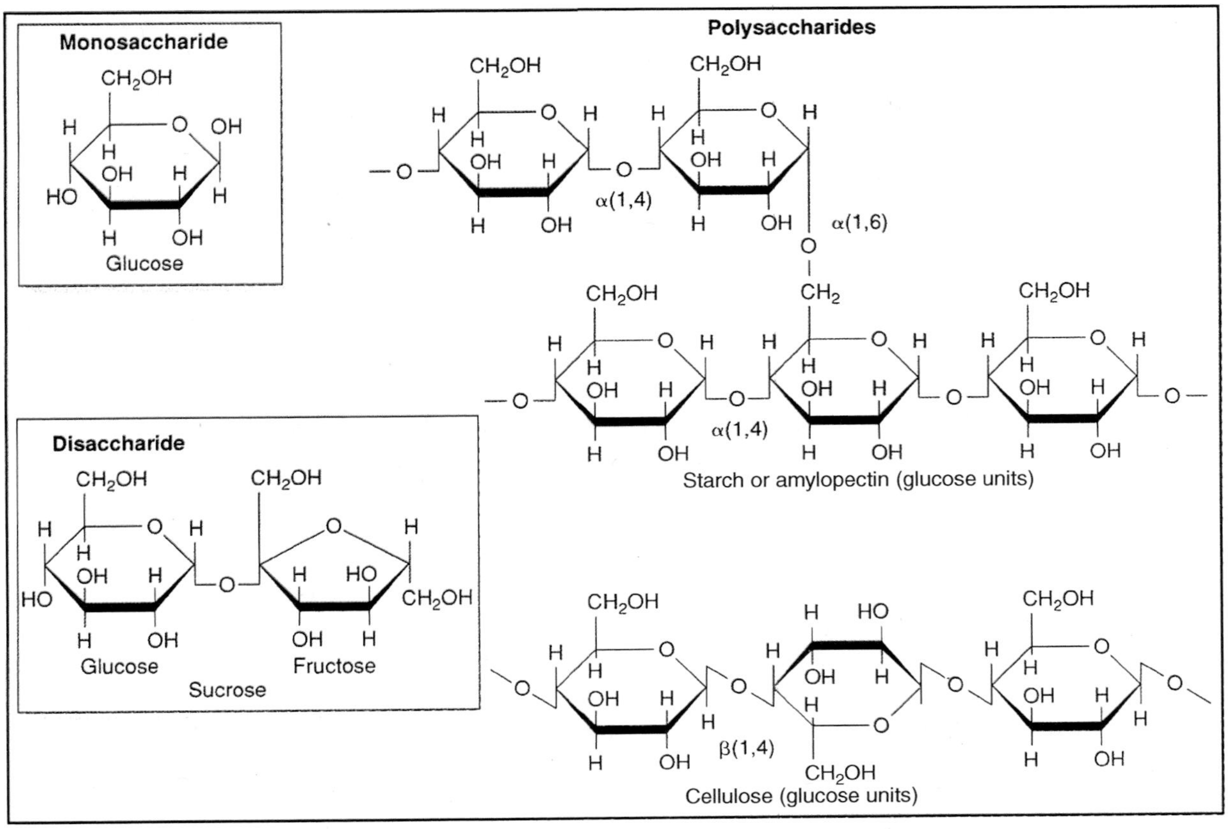

© Kendall Hunt Publishing Company

Figure 2.7   The two major polysaccharides of plants, starch and cellulose, are made up of many glucose molecules.

cellulose
A very large polysaccharide made up of glucose molecules which provides structure support for plants.

stability of cell walls (see Chapter 3), nearly 100 percent of cotton fibers and 50 percent of wood consist of cellulose.

Like amylose, cellulose is a large and water-insoluble polysaccharide made up of unbranched glucose chains (Figure 2.7). However, starch and cellulose differ in the orientation and bonding patterns of their glucose molecules (Figure 2.7). As a result, starch and cellulose have very different structure and physical properties. Unlike starch, most organisms lack enzymes to break down cellulose for energy. Termites have bacteria in their digestive tract that produce the enzyme cellulase to break down cellulose in wood. Wood-rot fungi and bacteria also produce cellulase to digest wood.

Nucleic Acids Make up the Blueprint of Life

As we will discuss in Chapters 3 and 15, there are two types of nucleic acids: deoxyribonucleic acid (DNA) and ribonucleic acid (RNA). DNA molecules are responsible for the storage, expression, and transmission of genetic information. RNA molecules, on the other hand, have diverse functions. Some RNAs are responsible for making proteins while others are responsible for regulating gene expression and cell development.

nucleotides
Basic units of building blocks that make up nucleic acids.

Nucleic acids are made up of many nucleotides. Like carbohydrates, nucleic acids are polymers made up of smaller and simpler units or monomers called **nucleotides**. Each nucleotide has three basic parts: a phosphate group, a pentose (five-carbon sugar), and a nitrogenous (nitrogen-containing) base (Figure 2.8). The pentose may be either ribose or deoxyribose (which contains one fewer oxygen atom than ribose) (Figure 2.8). A DNA or RNA molecule is made up of four types of nucleotides, which differ from one another only by their nitrogenous bases. Based on their structure, nitrogenous bases are divided into two major groups: purine and pyrimidine. Purine bases such as adenine and guanine have a double-ring structure (Figure 2.8) whereas pyrimidine bases such as cytosine, thymine, and uracil have a single-ring structure.

deoxyribonucleotides
Four nucleotides that make up DNA and have deoxyribose as their pentose.

DNA molecules provide instructions for life. Based on their respective nitrogenous bases, the four nucleotides of DNA are referred to as adenine (A), guanine (G), cytosine (C), and thymine (T) (Figure 2.8). These nucleotides are also known as **deoxyribonucleotides** because their pentose is deoxyribose. Each DNA molecule consists of two long strains of deoxyribonucleotides (= double) (Figure 2.9). The two strands of deoxyribonucleotides are held together by hydrogen bonds and coiled into a helical shape (= helix). Therefore, a DNA molecule is referred to as a "double helix." Genetic information is stored in genes, which are small sections of DNA that encode specific proteins or RNAs. Gene expression is the process of using information on the DNA to make proteins (see Chapter 15).

ribonucleotides
Four nucleotides that make up RNA and have ribose as their pentose.

messenger RNA (mRNA)
Delivers the genetic information or protein recipe to the ribosome, the protein factory of the cell.

RNA molecules help make proteins. The four nucleotides of RNA are adenine (A), guanine (G), cytosine (C), and uracil (U) (Figure 2.8). These nucleotides are also known as **ribonucleotides** because of their ribose. Unlike DNA molecules, RNA molecules consist of a single strand of ribonucleotides, which can be very short or long. The three main types of RNA involved in gene expression are **messenger RNA (mRNA)**, **transfer RNA (tRNA)**, and **ribosomal RNA (rRNA)**. In Chapter 15, we discuss the function and role of these three RNAs in gene expression.

transfer RNA (tRNA)
Transfers specific amino acids to the ribosome during protein synthesis.

ribosomal RNA (rRNA)
Ribonucleic acids that make up ribosomes.

ATP is the energy currency of life. Adenosine triphosphate (ATP) is the principal energy carrier for all living organisms. ATP shuttles energy from

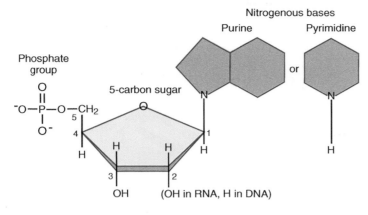

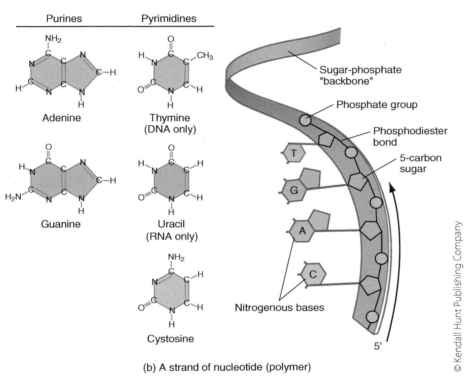

(a) A nucleotide (monomer)

(b) A strand of nucleotide (polymer)

Figure 2.8   Nucleic acid is made up of many nucleotides (a). In this example, nucleotides come together to form a strand of nucleotides (b).

one part of the cell to another. Like a nucleotide, ATP has three basic parts. However, ATP has three phosphate groups instead of one and an adenine base (Figure 2.10). In Chapter 10, we discuss how plants produce ATP during the light reactions of photosynthesis. ATP powers glucose production in the Calvin cycle. In Chapter 10, we also discuss how plants use aerobic respiration to convert glucose into ATP for cellular activities.

Proteins Are the Substance of Life

Proteins are the most abundant organic molecules in most living organisms and account for 50 percent or more of their dry mass. A single bacterial cell such as *Escherichia coli* contains more than 1,000 different types of proteins

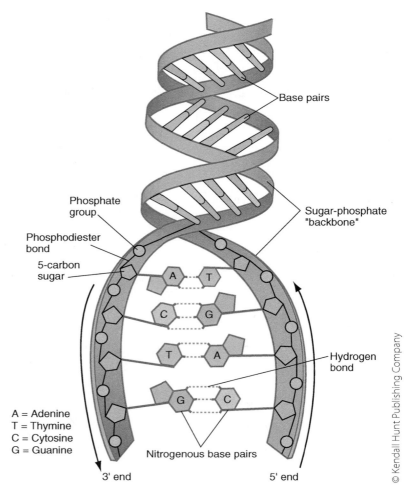

Base pairs

Phosphate group

Phosphodiester bond

5-carbon sugar

Sugar-phosphate "backbone"

A = Adenine
T = Thymine
C = Cytosine
G = Guanine

Hydrogen bond

Nitrogenous base pairs

3' end

5' end

© Kendall Hunt Publishing Company

Figure 2.9   Each DNA molecule is made up of two strands of nucleotides which are held together by hydrogen bonds and coiled into a helical shape thus the name "double helix".

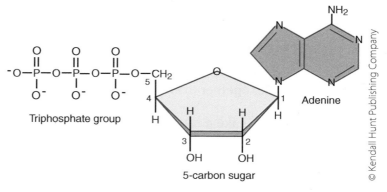

Triphosphate group

5-carbon sugar

Adenine

© Kendall Hunt Publishing Company

Figure 2.10   Adenosine triphosphate (ATP), the energy currency of life.

proteome
Sum of all the proteins found in a cell or an organism.

proteomics
Systematic study of proteomes encoded by genomes.

(Ishihama et al., 2008); a complex organism such as rice, *Oryza sativa*, has over 13,000 proteins (Komatsu & Tanaka, 2004). **Proteome** refers to the sum of all the proteins found in a cell or an organism. **Proteomics** is the systematic study of proteomes encoded by genomes. Next to cellulose, proteins are the second most abundant organic molecules of plant cells. Protein molecules are extremely diverse in size, shape, and function. Proteins are responsible for storing amino acids for growth, regulating cell development, structural support, and speeding up chemical reactions. Proteins are essential to life because of their many functions.

Protein molecules are made up of many amino acids. Like carbohydrates and nucleic acids, proteins are polymers made up of monomers called **amino acids**. Twenty types of amino acids are used by living organisms to make proteins (Figure 2.11). Although these 20 amino acids differ in their chemical properties, they have a very similar structure. All amino acids have an amino group ($-NH_2$), a carboxyl group ($-COOH$), a hydrogen atom, and an "R" group attached to a central carbon atom (Figure 2.11). The only difference among these amino acids is the "R" group, which can be an atom or a group of atoms. The "R" group determines the chemical property and identity of each amino acid.

amino acids
Small units that make up a polypeptide or protein.

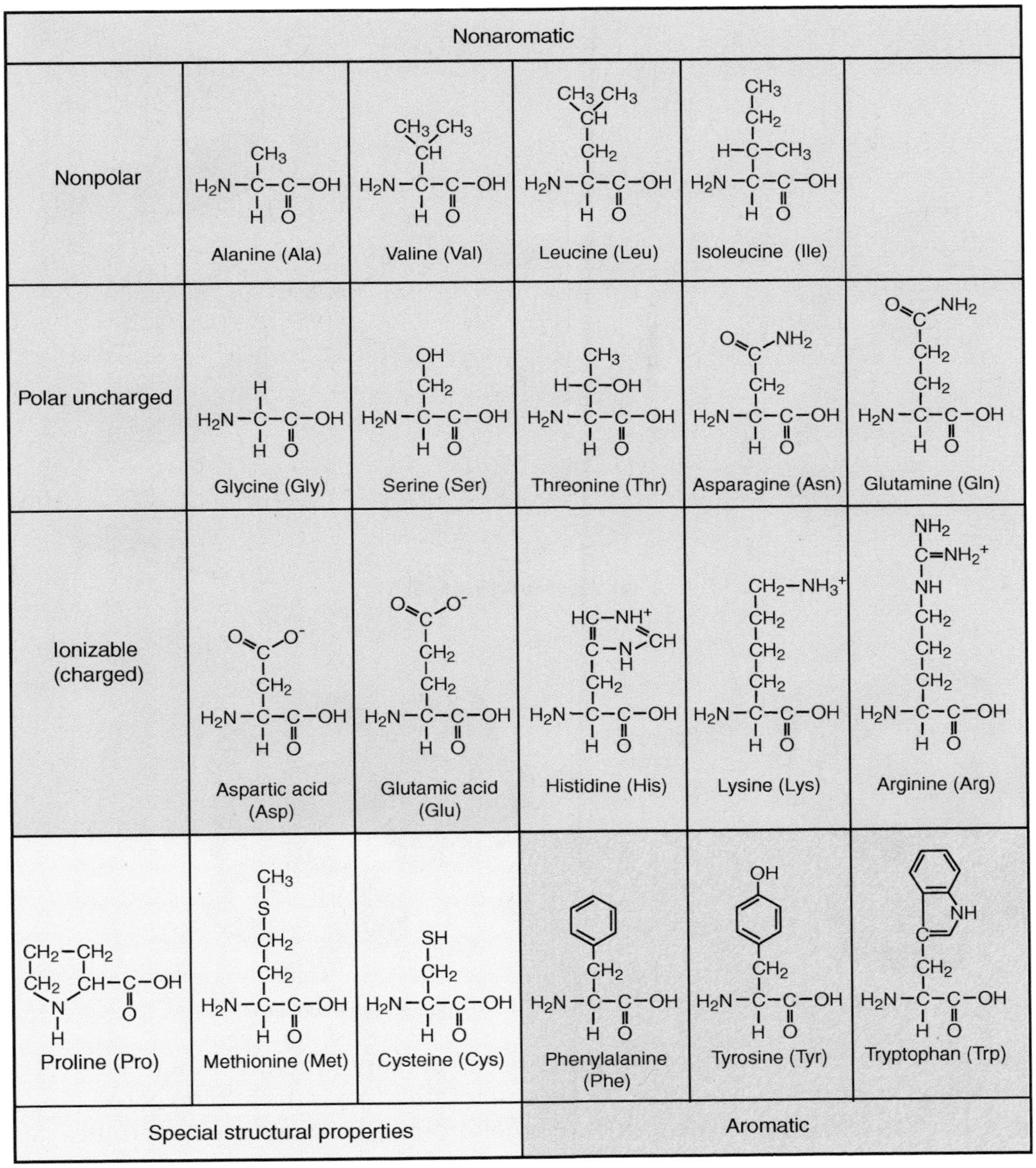

Figure 2.11 The twenty amino acids that are found in all living organisms.

© Kendall Hunt Publishing Company

polypeptide
A single long chain of amino acids

primary structure
Linear sequence of amino acids made up of a polypeptide.

Protein molecules are complex and have unique organizations. Proteins consist of one or more **polypeptides,** which are long chains of amino acids. The **primary structure** of a protein consists of a linear sequence of amino acids (Figure 2.12a). When the amino acids interact with each other, the chain folds into either a helical shape (α-helix) or a pleated sheet (β-pleated sheet) (Figure 2.12b). Hydrogen bonds maintain the shapes of α-helices and

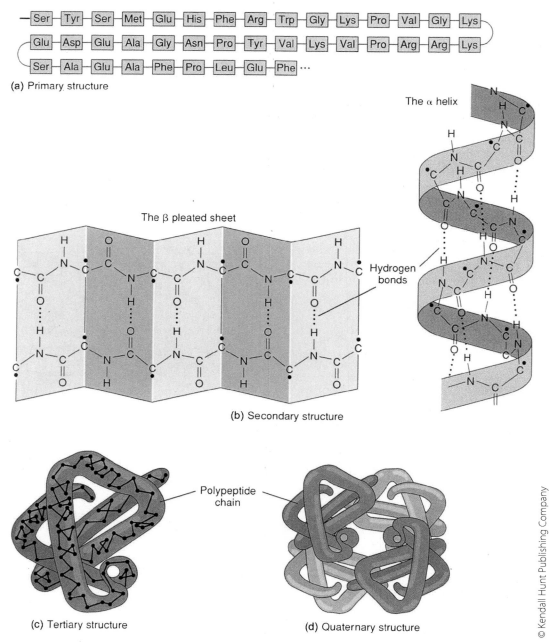

(a) Primary structure

The α helix

The β pleated sheet

Hydrogen bonds

(b) Secondary structure

Polypeptide chain

(c) Tertiary structure

(d) Quaternary structure

Figure 2.12a–d The four levels of protein structure: (a) primary structure is the linear sequence of amino acids or a polypeptide, (b) secondary structure is the folding of the amino acid chain into either a β pleated sheet of α helix form, (c) tertiary structure is the folding of the secondary structure, and (d) quaternary structure is formed when two or more polypeptides come together to form a functioning unit.

β-pleated sheets. These folded polypeptides form the **secondary structure** of proteins. **Fibrous proteins** that provide structural support and shape for organisms have mostly α-helical and β-pleated sheet folding. The **tertiary structure** of proteins is formed when α-helices or β-pleated sheets fold into a complex shape (Figure 2.12c). **Globular proteins** that are used as hormones or enzymes have these complex folds. Interactions among the amino acids maintain the tertiary shape of proteins. The most important chemical bond in these tertiary structures is the **disulfide bridge** formed between sulfur-containing amino acids. When two or more polypeptides come together to form a functional unit, they form a **quaternary structure** (Figure 2.12d). Most chemical bonds responsible for protein shape are relatively weak and are easily disrupted by changes in temperature and pH. **Denaturation** refers to the unfolding of proteins due to the disruption of the chemical bonds responsible for their shapes. When proteins unfold or lose their shape, they become unstable and nonfunctional. In some cases denaturation is reversible.

Storage proteins that supply amino acids for germination make some people ill. Some grasses such as wheat, rye, barley, spelt, kamut, and triticale have storage proteins in their seeds or grains to supply amino acids for embryonic growth during germination. In particular, a group of these storage proteins, commonly called "gluten," is important to bread and pasta making. Consuming gluten can make some people ill. Celiac disease, sometimes referred to as celiac sprue, is an autoimmune disorder that is triggered by the consumption of gluten, particularly gliadin (Braly & Hoggan, 2002) (Figure 2.13).

Enzymes speed up chemical reactions. **Enzymes** are large and complex globular proteins that act as **catalysts** to speed up chemical reactions. Catalysts are substances that accelerate the rate of chemical reactions by lowering their activation energy. They are not part of the chemical reaction and remain

secondary structure
When the amino acids in a polypeptide form hydrogen bonds between each other, the chain begins to fold into either a helical shape (α-helix) or a pleated sheet (β-pleated sheet).

fibrous proteins
Provide structure support and shape for organisms; have mostly α-helical and β-pleated sheet folding.

tertiary structure
When the secondary structure folds to form a complex shape, which is maintained mainly by disulfide bridges.

globular proteins
Proteins that have either a tertiary or quaternary structure.

disulfide bridge
Type of covalent bond formed between sulfur-containing amino acids.

quaternary structure
Consists of two or more polypeptides that come together to form a functional unit.

Figure 2.13   The molecular structure and formula of gliadin which is a component of gluten found in the grain of wheat and other cereals except oats. Gluten is important to the bread-making industry but can cause human diseases like celiac disease.

denaturation
Disruption of chemical bonds within a protein, causing it to unfold and lose its shape.

enzymes
Special class of proteins that speeds up chemical reactions.

catalysts
Substances that speed up a chemical reaction by lowering its activation energy but do not take part in the reaction.

substrate
Reactants that bind to the active site of an enzyme.

unchanged in the process. Therefore, enzymes can be used repeatedly. A **substrate** is the reactant on which an enzyme works. Enzymes are often named by adding *-ase* to the end of the root of their substrate names. For example, α-amylase is one of the enzymes that break down amylose (starch) into glucose (Figure 2.14) and sucrase is the enzyme that breaks down sucrose into glucose and fructose.

Lipids Provide Fuel and Barrier for Life

Lipids are fats and fatlike substances that are generally insoluble in water or hydrophobic (means "water-fearing"). Lipids contain large numbers of carbon and hydrogen atoms with a few oxygen atoms (Figure 2.15). Most lipid molecules consist of a glycerol, a three-carbon molecule, and two or three fatty acid chains. Fatty acid molecules are often referred to as hydrocarbon chains because of their numerous carbon and hydrogen atoms (Figure 2.15). Lipids are not polymers and are considerably smaller and less complex. Like the other organic molecules, lipids have many functions and are essential to life. Some lipids are storage molecules for energy, some are light-gathering pigments for photosynthesis, and others form barriers to reduce water loss. Two special classes of lipids, phospholipids and sterols, are essential to the structure and stability of cellular membranes. Some sterols also function as hormones.

Triglycerides such as fats and oils store energy. Fats and oils are both triglycerides and have similar structures. Each consists of three fatty acid molecules attached to a glycerol (Figure 2.15). Both the length and type

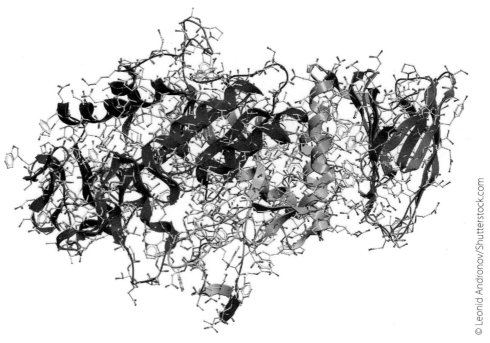

© Leonid Andronov/Shutterstock.com

Figure 2.14   α-Amylase, an enzyme that breaks down polysaccharides like starch and glycogen into glucose and maltose.

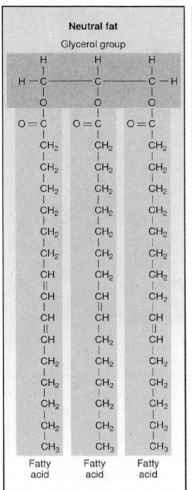

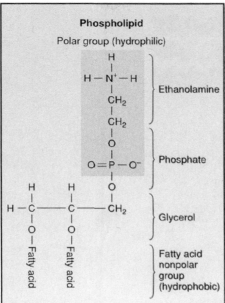

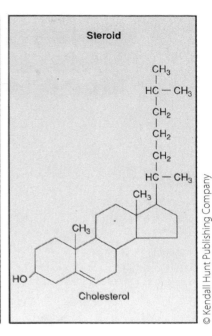

Figure 2.15 · Examples of lipid molecules: fat (triglycerides), phospholipid, and sterol.

of the fatty acids determine the physical nature of a triglyceride. Fatty acids can be **saturated** or **unsaturated**. A saturated fatty acid has no double bonds between carbon atoms and contains the maximum number of hydrogen atoms. In contrast, an unsaturated fatty acid has one or more double bonds among its carbon atoms. When a double bond forms between two carbon atoms, it reduces the number of hydrogen atoms attached to each carbon. The presence of double bonds in unsaturated fatty acids causes the fatty acids to kink and prevents tight packing of the fat molecules. When fat molecules are not tightly packed, they tend to have a lower melting point. Oils are unsaturated fats and liquids at room temperature (Figure 2.16). Plant triglycerides or oils such as corn oil, peanut oil, olive oil, soybean oil, and safflower oil are all extracted from oil-rich seeds. On the other hand, animal triglycerides or fats such as butter and lard are highly saturated fats and solids at room temperature (Figure 2.16). In comparison with storage carbohydrates such as starch, fats and oils contain a higher number of carbon–hydrogen bonds and twice the amount of energy.

saturated
When a fatty acid has no double bonds between carbon atoms and contains the maximum number of hydrogen atoms.

unsaturated
When a fatty acid has one or more double bonds among its carbon atoms and contains less hydrogen atoms.

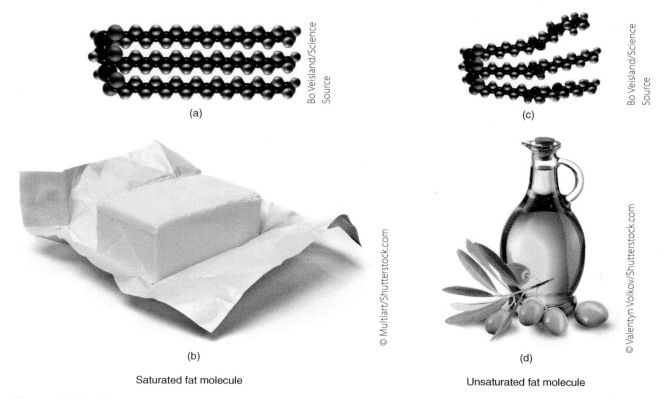

Bo Veisland/Science Source

(a)

(c)

Bo Veisland/Science Source

© Multiart/Shutterstock.com

(b)

Saturated fat molecule

© Valentyn Volkov/Shutterstock.com

(d)

Unsaturated fat molecule

Figure 2.16 At room temperature, saturated fat is solid whereas unsaturated fat is liquid.

cutin
Lipid used by plants to prevent desiccation found in plant cell walls and cuticle.

suberin
Major component of the walls of cork cells in woody plants and root endodermal cells; helps to restrict water and mineral movement in the roots.

waxes
Lipids used by plants to prevent desiccation found in plant cell walls and cuticles.

cuticle
Found on the surface of stems and leaves; made up of cutin and wax, which make plant surfaces waterproof and reflective.

Lipids form barriers to reduce water loss or restrict water movement. Recall from Chapter 21 that plants adapted to land-based living by producing lipids such as **cutin**, **suberin**, and **waxes** to prevent desiccation. Cutin and waxes secreted by epidermal cells form a protective and waterproof layer called the **cuticle**. This protective cuticle minimizes water loss from leaf and stem surfaces that are exposed to air (see Chapter 3).

Suberin is a major component of the walls of cork cells, the cells that form the outermost layer of bark in woody plants, and root endodermal cells (see Chapter 4). Within the endodermis, suberin forms the waterproof Casparian strip, which restricts water and mineral movement into the vascular cylinder.

Phospholipids and sterols provide structure and stability to cellular membranes. Most triglycerides are made up of three fatty acids attached to a glycerol backbone. Unlike triglyerides, **phospholipids** have two fatty acid chains and a modified phosphate group attached to the glycerol molecule (Figure 2.15). This modified phosphate has a negative charge allowing part of the phospholipid molecule to interact with water. The "head" of a phospholipid is hydrophilic and the "tails" are hydrophobic (Figure 2.17). When surrounded by water, phospholipid molecules form a two-layer or bilayer structure (Figure 2.17). The hydrophilic heads face outward where they interact with water and the hydrophobic tails face one another, forming a

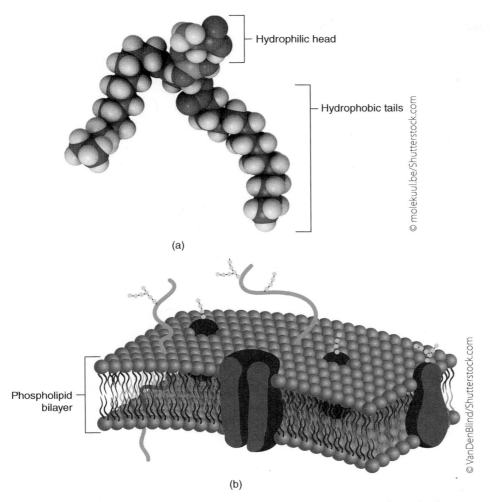

— Hydrophilic head

— Hydrophobic tails

© molekuul.be/Shutterstock.com

(a)

Phospholipid bilayer

© VanDenBlind/Shutterstock.com

(b)

Figure 2.17   A phospholipid molecule is shown on the left with the hydrophilic "head" pointing up and the two hydrophobic "tails" pointing down. Phospholipid molecules form a bilayer when surrounded by water. This bilayer or two-layer arrangement forms the structural basis of all cellular membranes.

hydrophobic layer in the middle. As we discuss in Chapter 3, this bilayer configuration is important to both the structure and functions of cellular membranes.

Sterols have unique chemical structures that distinguish them from other classes of lipids. All sterols have four interconnected hydrocarbon rings and sometimes a hydrocarbon chain is attached to one of the rings (Figure 2.15). Sitosterol is the most abundant sterol found in the membranes of green algae and plants, where it stabilizes the hydrophobic tails in the bilayer and regulates membrane fluidity. Cholesterol is the most common sterol found in animal cells but is present in only trace amounts in plant cells. Like sitosterol, cholesterol helps to stabilize and regulate the fluidity of cell membranes.

phospholipids
Structural basis of all membranes; contains two fatty acid chains and a modified phosphate group attached to the glycerol.

sterols
Made up of four interconnected hydrocarbon rings that may have a hydrocarbon chain attached to one of the rings.

Table 2.1 Distinguishing Features among the Four Classes of Organic Molecules

Class of Organic Molecules	Building Blocks	Selected Examples	Functions
Carbohydrates	Monosaccharides	Glucose, fructose	Provide energy for the cell and building blocks for other carbohydrates
	Disaccharides	Sucrose	Transport glucose to other parts of the plant
	Polysaccharides	Cellulose, starch	Starch stores glucose for plant cells
			Cellulose provides structure for plant cells
Nucleic acids	Deoxyribonucleotides	DNA	Store and transmits genetic information
	Ribonucleotides	RNA	Make proteins, regulate gene expression and cell development
Proteins	Amino acids	Gluten, enzymes	Storage proteins such as gluten supply amino acids for embryonic growth during germination
			Enzymes speed up chemical reactions in the cells
Lipids	Fatty acids and glycerol	Oils	Storage molecule for energy
		Cutin, suberin, waxes	Make plant surfaces waterproof to reduce water loss or regulate water movement inside the root
		Phospholipids, steroids	Provide cell membranes their structure and functions

Key Terms

solvent	vaporization	polysaccharides
polar covalent bond	evaporation	polymers
electronegative	heat of vaporization	monomers
polar	evaporative cooling	dehydration synthesis
cohesion	organic molecules	hydrolysis
adhesion	elements	starch
solutes	monosaccharides	amylose
solutions	disaccharides	amylopectin

fructans
cellulose
nucleotides
deoxyribonucleotides
ribonucleotides
messenger RNA (mRNA)
transfer RNA (tRNA)
ribosomal RNA (rRNA)
proteome
proteomics
amino acids

polypeptides
primary structure
secondary structure
fibrous proteins
tertiary structure
globular proteins
disulfide bridge
quaternary structure
denaturation
enzymes
catalysts

substrate
saturated
unsaturated
cutin
suberin
waxes
cuticle
phospholipids
sterols

Summary

- Life cannot exist without water.
  - Water is the main component of all cells, making up 70 to 90% of their volume.
  - Water molecules are polar making them stick to each other and to other polar molecules. The cohesiveness and adhesiveness of water molecules make other properties of water possible.
  - Water is an important solvent.
  - Water provides evaporative cooling.
- In addition to water, cells are composed of organic molecules such as carbohydrates, nucleic acids, proteins, and lipids.
- Carbohydrates
  - Carbohydrates are made up of many monosaccharides.
  - Monosaccharides such as glucose and fructose are simple carbohydrates that provide energy for living organisms and function as building blocks for other carbohydrates.
  - Disaccharide, such as sucrose, transports glucose for plants.
  - Polysaccharides, such as starch, provide structure and energy storage for plants.
  - Other polysaccharides, such as cellulose, provide structure and support for plants.
- Nucleic acids
  - Nucleic acids are made up of many nucleotides.
  - The two types of nucleic acids are deoxyribonucleic acid (DNA) and ribonucleic acid (RNA).
  - DNA molecules are responsible for the storage, expression, and transmission of genetic information.
  - RNA molecules on the other hand have diverse functions. Some RNAs are responsible for making proteins while others are responsible for regulating gene expression and cell development.
  - ATP, a modified nucleotide, is the energy currency of life.
- Proteins
  - Protein molecules are made up of many amino acids.
  - Protein molecules are extremely diverse in size, shape, and function.
  - Proteins are responsible for storing amino acids for growth, regulating cell development, structural support, and speeding up chemical reactions. Proteins are essential to life because of their many functions.

- Although these twenty amino acids differ in their chemical properties, they have a very similar structure.
- Four levels of protein structures are: primary, secondary, tertiary, and quaternary structure.
- Lipids
  - Lipids are not polymers and are considerably smaller and less complex.
  - Lipid molecules are made up of glycerol and fatty acids.
  - Lipids have many functions and are essential to life. Lipids such as fats and oils are storage molecules for energy, some lipids are light-gathering pigments for photosynthesis, while others form barriers to reduce water loss.
  - Two special classes of lipids, phospholipids and sterols, are essential to the structure and stability of cellular membranes. Some sterols also function as hormones.

Reflect

1. NASA scientists have been looking for water on Mars and the Moon as a prerequisite for life there. Describe the unique properties of water and explain why water is so important to living organisms.

2. How do you distinguish the four classes of organic molecules that are important to life? Do living organisms need all four classes of organic molecules to function? Explain your answer.

3. You encounter a weird organism that makes up of only two classes of organic molecules, nucleic acids and proteins. Can you figure out which type of organisms you are dealing with? Do an internet search and see if you can figure this out.

References

Braly, J., & Hoggan, R. (2002). *Dangerous grains: why gluten cereal grains may be hazardous to your health*. New York: Avery Books.

Ishihama, Y., Schmidt, T., Rappsilber, J., Mann, M., Hartl, F. U., Kerner, M. J., et al. (2008). Protein abundance profiling of the Escherichia coli cytosol. *BMC Genomics, 9*, 102.

Komatsu, S., & Tanaka, N. (2004). Rice proteome analysis: A step toward functional analysis of the rice genome. *Proteomics, 4*, 938–949.

Pollock, C. J., & Cairns, A. J. (1991). Fructan metabolism in grasses and cereals. *Annual Review of Plant Physiology and Plant Molecular Biology, 42*, 77–101.

Chapter 3

Cells and Tissues

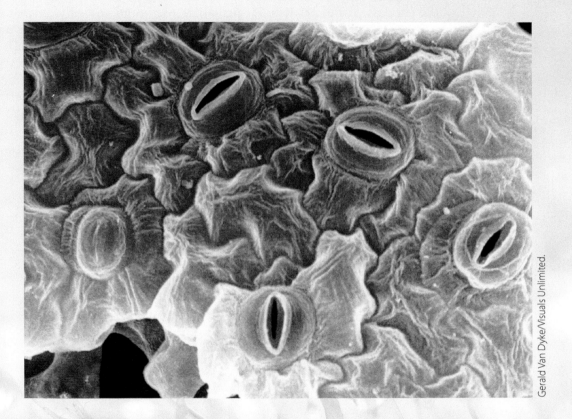

Gerald Van Dyke/Visuals Unlimited.

Learning Objectives

- Explain the principles and development of the cell theory
- Compare and contrast prokaryotic and eukaryotic cells
- Describe the functions of cell wall, plasma membrane, cytoplasm, DNA, ribosomes, nucleus, endoplasmic reticulum, dictyosomes, central vacuole, mitochondria, chloroplasts, chromoplasts, and leucoplasts of a plant cell
- Visually identify various cell structures
- Compare and contrast plant and animal cells
- Describe the functions and components of each tissue system

Plants are amazing organisms not only in their appearance and complexity but also in their importance to other living organisms. Like other living organisms, including us, plants are composed of cells, the basic units of life. Plants are multicellular, which means they are made up of many cells. Even though all cells in a plant body have the same genetic blueprint, these cells are not the same. Cells that make up a plant body come in a mind-boggling array of sizes, shapes, colors, and functions (Figure 3.1). Such diversity in cell shapes and functions allow specialization to occur. Most plant cells are able to produce food and generate oxygen using photosynthesis, which our cells cannot. Plant cells must be specially equipped to carry out this remarkable process. Some plant cells specialize in storage and others specialize in structure support. Some cells specialize in transportation of water and others specialize in transportation of food made by photosynthesis. This division of labor allows the entire organism to become more efficient.

Tissues are structural and functional units formed from specialized cells. They carry out tasks that individual cells cannot. The epidermis is a

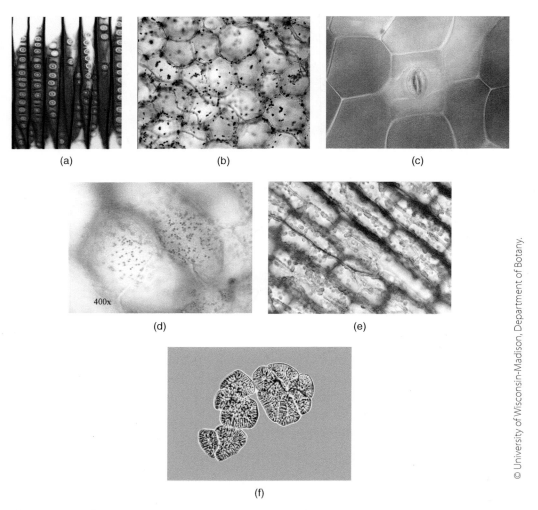

(a)   (b)   (c)

(d)   (e)

(f)

© University of Wisconsin-Madison, Department of Botany.

Figure 3.1a–f   A diversity of plant cells: (a) tracheids, (b) parenchyma cells with starch, (c) stoma, (d) red pepper cells with chromoplasts, (e) *Elodea* cells with chloroplasts, and (f) stone cells in pears.

complex tissue that protects the entire plant. Xylem is a complex tissue that transports water over great distances. When different tissues come together, they form organs that carry out complex tasks beyond the capability of individual tissues. Roots are organs found below ground. Their main functions are to absorb water and minerals, provide anchorage, and store food and water for the plant. Most higher plants have three groups of organs: roots, stems, and leaves. Some plants even have an additional group of organs called *flowers*. These organs are discussed in greater detail in Chapters 4–7.

HISTORY OF CELL DISCOVERY

The realm of the unseen world, the world of cells, was revealed in the mid-1600s following the invention of the microscope some 70 years earlier. Drawings made or later photos taken with the aid of the microscope are called *micrographs*. These micrographs reveal the intricate structures of cells and greatly increase our understanding of cell structure and function.

Scientists Who Discovered Cells

Cells are so small that they cannot be seen with the naked eye. The existence of these tiny structures was discovered and written about in 1665 by **Robert Hooke**, who used a primitive compound light microscope (compound refers to the use of two lenses, the eyepiece and the objective) to examine a thin slice of cork (Figure 3.2). He wrote:

Robert Hooke
discovered tiny pores in cork and coined the word "cells"

> *I took a good clear piece of Cork and with a Pen-knife sharpen'd as keen as a Razor, I cut a piece of it off, and thereby left the surface of it exceeding smooth, then examining it very diligently with a Microscope . . . . I could perceive it to appear a little porous . . . these pores, or cells . . . which were indeed the first microscopical pores I ever saw, and perhaps, that were ever seen, for I had not met with any Writer or Person, that had made any mention of them before this . . . .*

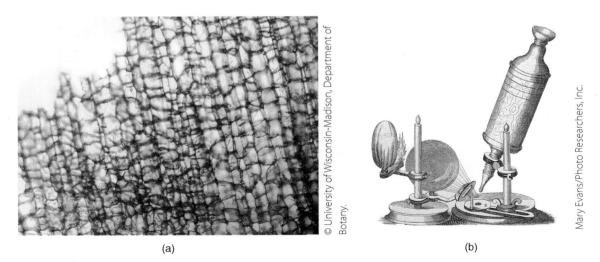

© University of Wisconsin-Madison, Department of Botany.

Mary Evans/Photo Researchers, Inc.

(a)

(b)

Figure 3.2a–b   (a) Hook's drawing of a thin slice of cork appeared in his book *Micrographia*, published in 1665, and (b) the microscope he used.

Hooke named these tiny pores "cells." During the 60 years following Hooke's discovery, cells were seen in a variety of plant and animal tissues by scientists such as Anton van Leeuwenhoek, Marcello Malpighi, and Nehemiah Grew. van Leeuwenhoek even discovered single-celled bacteria, which he called "animalcules" or "little animals." As microscopes continued to improve during the 19th century, even smaller structures were found within the cell. In 1802, Franz Bauer was the first to draw attention to a relatively large body inside cells of plant stigma. This large body was later referred to as the *nucleus* by Robert Brown in 1831. Following Bauer's and Brown's discovery, Rudolph Wagner observed a smaller body within the nucleus, which was later referred to as the "nucleolus" by Gabriel Gustav Valentin in 1839. (For more on history of cells, see Harris, 1999).

Development of the Cell Theory

The idea that all organisms are made up of cells originated in the mid-1800s when cells were discovered in different kinds of plant and animal tissues. Although botanist Matthias Schleiden and physiologist Theodor Schwann were not the first to understand the significance of cells, they were the first to explain this idea clearly. This idea is now called the *cell theory*. This theory basically states that all living organisms are made up of cells and cells are the basic units of life. In the late 1800s, Rudolf Virchow, a pathologist, extended the cell theory by stating that all cells come from preexisting cells (Figure 3.3).

CELLS ARE TINY UNITS OF LIFE

Most plant and animal cells are very small, ranging from 10 to 100 μm (one μm = 1/1000th of a centimeter), and cannot be seen with the unaided eye. Since there are 25,400 μm (2.54 cm) in an inch, it would take about 500 cells to extend across an inch of space. Some cells such as bacteria are even smaller, about 10 times smaller than plant and animal cells. The use of microscopes is vital to the field of **cell biology** or **cytology**, the study of cell structure and function.

cell biology
Study of cell structure and function.

cytology
See *cell biology*.

(a) Matthias Schleiden   (b) Rudolf Virchow   (c) Theodor Schwann

SPL/Science Source.   LOC/Science Source.   SPL/Science Source.

Figure 3.3a–c   (a) Matthias Schleiden, (b) Rudolf Virchow, and (c) Theodor Schwann and the Cell Theory.

Smaller the Better

Why are cells so small? Are there limits to how big a cell can get? Are there advantages for cells to stay small? Cells are very small because their surface area to volume ratio increases with reduction in size. A large surface area to volume ratio means there is more surface area relative to volume. Having a large surface area enables cells to more efficiently exchange nutrients and wastes with their environment through their plasma membranes. Having a small volume enables the faster distribution of oxygen and essential nutrients from the plasma membrane to the rest of the cell. When a cell becomes bigger, its volume increases more rapidly than its surface area. As a result, a very large cell would have too small a surface area to keep up with its metabolic needs and too large a volume for essential molecules to quickly reach its different parts (Figure 3.4).

Observing Cells with Microscopes

Hooke's discovery of cells and all the subsequent studies of cells and their components were made possible by the invention of the microscope in the late 1500s and its refinement during the 1600s. Today, microscopes are still important and powerful tools for studying cells, tissues, and organisms. There are basically two types of microscopes: light microscopes and electron microscopes (Figure 3.5).

A light microscope uses light as the source of illumination while an electron microscope uses electrons. Light microscopes magnify an object up to 1,000 times and can resolve structures that are as close as 0.2 μm (micrometer or micron). Electron microscopes are a lot more powerful than light microscopes and can magnify an object 200,000 times or more. They also have a much higher resolution and can resolve structures that are as close as 2 nm (nanometers), which is 100 times better than light microscopes. Light

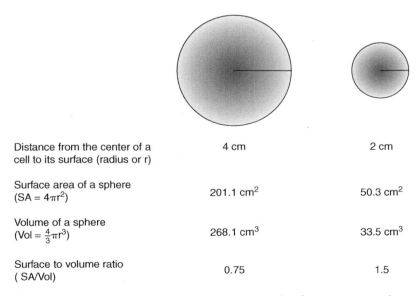

	4 cm	2 cm
Distance from the center of a cell to its surface (radius or r)	4 cm	2 cm
Surface area of a sphere ($SA = 4\pi r^2$)	201.1 cm^2	50.3 cm^2
Volume of a sphere ($Vol = \frac{4}{3}\pi r^3$)	268.1 cm^3	33.5 cm^3
Surface to volume ratio (SA/Vol)	0.75	1.5

Figure 3.4 Relationship between size and suface area to volume ratio for two spheres with different radii and calculations of their surface areas, volumes, and SA/Vol ratio.

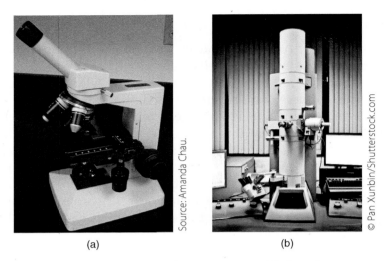

Source: Amanda Chau.

© Pan Xunbin/Shutterstock.com

(a)   (b)

Figure 3.5a–b (a) A modern compound light microscope and (b) an electron microscope.

microscopes can be used to observe living as well as preserved specimens. In comparison, electron microscopes can only be used to observe preserved specimens.

The invention of the electron microscope in the 1930s led to rapid advancements in the study of cells (cell biology). Although light microscopes have their limitations, they are useful for observing living cells. Some of the cell structures that are easily observed with a light microscope include the cell wall, nucleus, nucleolus, cytoplasm, chloroplasts, chromoplasts, leucoplasts, central vacuole, and even condensed chromosomes (Figure 3.6).

BASIC CELL TYPES

Cells are the basic units of life and all living organisms are made up of one or more cells. Organisms such as bacteria, amoeba, and paramecium are made up of just one cell and are referred to as **unicellular** (*uni-* in Latin means "one"). Other organisms such as plants and animals are made up of many cells and are referred to as **multicellular** (*multi-* in Latin means "many").

unicellular
(uni- in Latin means "one") Organisms that are made up of just one cell.

multicellular
Organisms that are made up of many cells (multi in Latin means "many").

The Two Basic Types: Prokaryotic and Eukaryotic Cells

Cells are divided into two basic types: prokaryotic and eukaryotic (Figure 3.7). The Greek word *kary* is found in both terms and means a "nut" or "nucleus." The Greek words *pro* and *eu* precede the word *kary* and mean "before" and "well," respectively. **Prokaryotic cells** are cells without (before) a nucleus whereas **eukaryotic cells** are cells with a well or true nucleus. Prokaryotic cells are found in domains Bacteria and Archaea and they are smaller and simpler than eukaryotic cells. Eukaryotic cells are found in domain Eukarya; notice the similarity between the two words, which includes plants, animals, fungi, and protists. Eukaryotic cells have not only a nucleus in their cells but also other membrane-bounded structures (organelles), which carry out diverse and complex functions (**Table 3.1**).

prokaryotic cells
Cells without (before) a nucleus (pro- in Greek means "before," kary means "nut" or "nucleus").

eukaryotic cells
Cells with a well or true nucleus (eu- in Greek means "well," kary means "nut" or "nucleus").

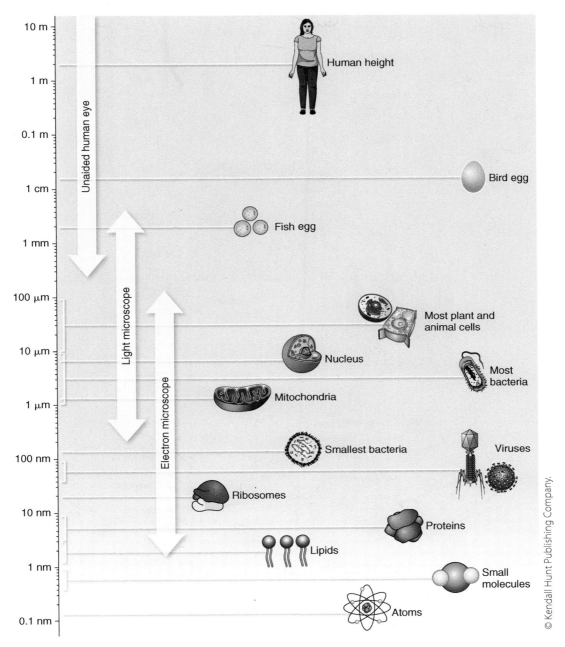

Figure 3.6 Relative sizes of chemical and biological molecules and the resolving powers of the naked eye, light microscope, and electron microscope.

COMMON CELL STRUCTURES AND FUNCTIONS

Although prokaryotic and eukaryotic cells are different in size and complexity, they both have four cell structures: the plasma membrane, the cytoplasm, DNA, and ribosomes. These four structures are crucial to the survival of a cell. Plant and animal cells are eukaryotic cells and they have organelles or membrane-bounded structures such as the nucleus, the endoplasmic reticulum, dictyosomes, vacuoles, and mitochondria. Plastids and the cell wall are structures that are unique to plant cells.

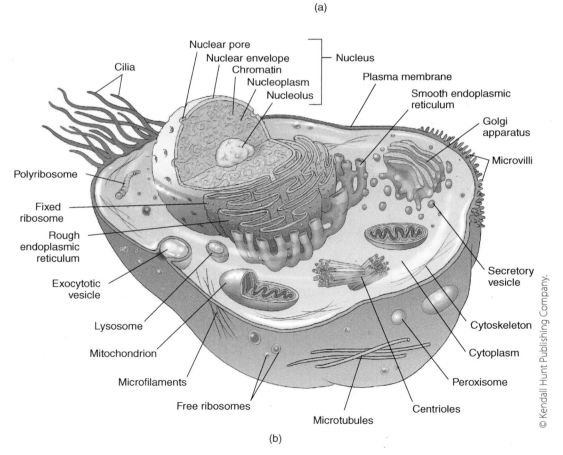

© Aila Sao Mai/Shutterstock.com

(a)

© Kendall Hunt Publishing Company.

(b)

Figure 3.7a–b   A typical prokaryotic (a) and a typical eukaryotic (b) cell.

Table 3.1 A Comparison of Cell Structures among Bacterial, Animal, and Plant Cells

Cell Structures	Bacteria (Prokaryotic)	Animal Cells (Eukaryotic)	Plant Cells (Eukaryotic)
Cell wall	✓ (Peptidoglycan)	✗	✓ (Cellulose)
Plasma membrane	✓	✓	✓
Cytoplasm	✓	✓	✓
Ribosomes	✓	✓	✓
DNA	✓	✓	✓
DNA associated with histones	✗	✓	✓
Nucleus	✗	✓	✓
Endoplasmic reticulum	✗	✓	✓
Dictyosomes or Golgi	✗	✓	✓
Mitochondria	✗	✓	✓
Chloroplasts	✗	✗	✓
Chromoplasts	✗	✗	✓
Leucoplasts	✗	✗	✓
Central vacuole	✗	✗	✓

The Plasma Membrane: Selective Barrier

In order for a cell to exist, it must be able to separate itself from the environment. The plasma membrane provides a physical boundary or barrier that keeps the cell's content isolated and together. However, a cell cannot survive if it is totally isolated from the environment. A cell needs to bring in water, oxygen, and nutrients from the outside and expel metabolic waste products. The plasma membrane is more than a passive barrier because it plays an important role in regulating the movement of molecules into and out of the cell.

The plasma membrane and other cell membranes are made up of two major components: **phospholipids** and **proteins**. Phospholipid molecules form two distinct layers that are often referred to as a bilayer. This phospholipid bilayer prevents the entry or exit of polar or charged molecules such as water and ions such as Na^+ or Cl^- from the cell. The protein molecules embedded in the bilayer are the ones that selectively allow the passage of polar molecules such as water and glucose into and out of the cell. These membrane proteins also permit cells to communicate with each other and detect changes in their environment. In 1972, S. Jonathan Singer and Garth Nicolson proposed the **fluid-mosaic model** to explain how phospholipids and proteins come together to form membranes (Figure 3.8). Phospholipids and proteins can move relative to each other within the membrane and are said to be fluid. Mosaic refers to how protein molecules intersperse throughout the bilayer. Some proteins extend right through the bilayer whereas others are loosely bound to the membrane surface, phospholipids, or other proteins.

phospholipid
Type of lipid that has a modified phosphate and two fatty acids attached to a glycerol backbone.

protein
One of four organic molecules that are essential to life.

fluid-mosaic model
Model proposed by Singer and Nicolson in 1972 to explain how phospholipids and proteins come together to form membranes.

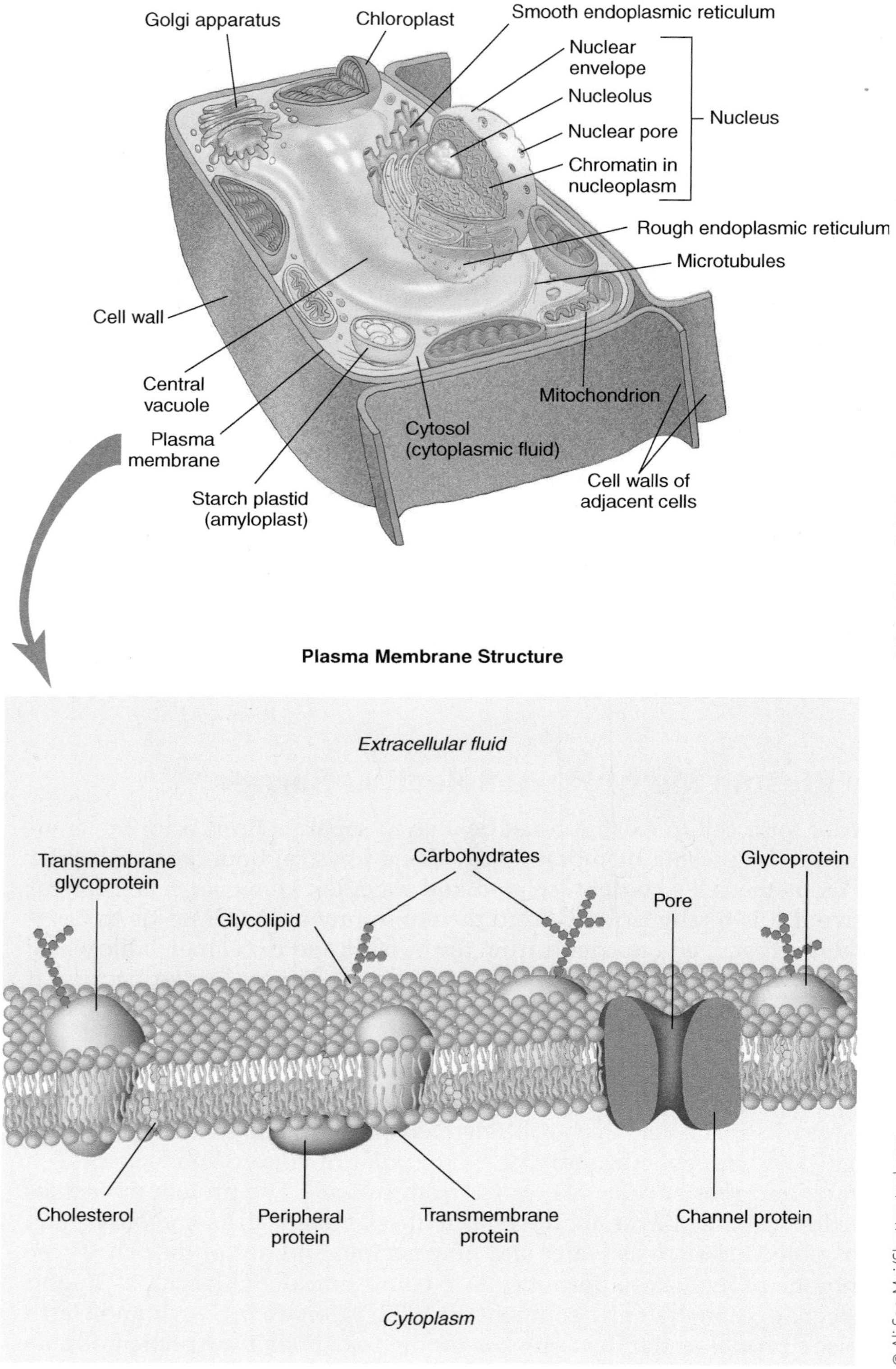

Plasma Membrane Structure

Figure 3.8 Fluid-mosaic model of membrane structure.

The Cytoplasm-Metabolic Center and Internal Transport

The region of the cell inside the plasma membrane is known as the cytoplasm. The aqueous portion of the cytoplasm is called the **cytosol** and it typically takes up 20–50% of the total cell volume. The rest of the cytoplasm is made up of a network of protein fibers forming the cytoskeleton and all the cellular structures within the plasma membrane.

The cytosol is an important region of the cell where many chemical reactions take place and life-sustaining materials are made. All the chemical reactions that take place inside a cell are collectively referred to as **metabolism**. **Catabolic** reactions are chemical reactions that break down complex molecules. **Anabolic** reactions are chemical reactions that build complex molecules. **Enzymes**, a special class of proteins that speeds up chemical reactions, are found here in the cytosol as well as building blocks for various metabolic activities.

The **cytoskeleton** is an intricate network of protein fibers that extends throughout the cytoplasm and provides structural support and internal transport for the cell. In eukaryotic cells, the two protein fibers that make up the cytoskeleton are **microtubules** and **actin filaments**. The cytoskeleton is also responsible for **cytoplasmic streaming**, sometimes referred to as cyclosis, the movement of cytoplasm within the cell. Cytoplasmic streaming facilitates movement of materials and organelles within the cell. In leaf cells, cytoplasmic streaming circulates chloroplasts around the cell for optimal exposure to sunlight. The cytoskeleton in plant cells also plays a major role in a number of cellular processes such as addition of cellulose to the cell wall, enlargement or growth of the cell, and movement of chromosomes during cell division.

Deoxyribonucleic Acid (DNA): Blueprint of Life

DNA molecules are called the "blueprints" of life and are responsible for the storage, expression, and transmission of genetic information. What is genetic information? It basically contains instructions or recipes for making all the proteins that are needed by a cell. These instructions or recipes are found in genes and each gene is a small section of DNA containing a specific sequence that makes a particular protein. Genes can be turned on to make proteins or turned off to stop protein production in order to direct cell activities.

DNA molecules are nucleic acids made up of smaller and simpler units called **nucleotides** (see Appendix). Each nucleotide has three basic parts: a phosphate group, a five-carbon sugar, and a nitrogen base. The four types of nucleotides that make up a DNA molecule are adenine (A), guanine (G), cytosine (C), and thymine (T). The DNA molecule is commonly referred to as a "double helix" because each DNA molecule consists of two long strains of nucleotides (= double). The two strains of nucleotides are held together by hydrogen bonds and coiled into a helical shape (= helix) (Figure 3.9). Although both prokaryotic and eukaryotic cells have DNA, the most noticeable difference between the two cell types is that only eukaryotic DNA are associated with a special group of protein called *histones*.

cytosol
Aqueous portion of the cytoplasm where many chemical reactions take place.

metabolism
All the chemical reactions that take place inside a cell.

catabolic
Reactions that break down complex molecules and release energy; opposite of anabolic.

anabolic
Process resulting in the biosynthesis or construction of larger molecules; opposite of catabolic.

enzymes
Special class of proteins that speeds up chemical reactions.

cytoskeleton
Intricate network of protein fibers that extends throughout the cytoplasm and provides structural support and internal transport for the cell.

microtubules
Thick protein fibers that make up the cytoskeleton.

actin filaments
Thin protein fibers that make up of the cytoskeleton.

cytoplasmic streaming
The movement of cytoplasm within the cell. Also called cyclosis.

nucleotides
Basic units of building blocks that make up nucleic acids.

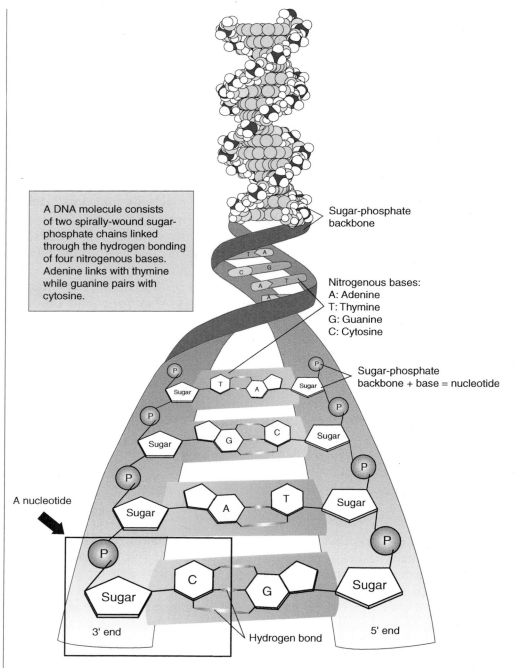

A DNA molecule consists of two spirally-wound sugar-phosphate chains linked through the hydrogen bonding of four nitrogenous bases. Adenine links with thymine while guanine pairs with cytosine.

Sugar-phosphate backbone

Nitrogenous bases:
A: Adenine
T: Thymine
G: Guanine
C: Cytosine

Sugar-phosphate backbone + base = nucleotide

A nucleotide

3' end

5' end

Hydrogen bond

Figure 3.9 A double-stranded structure of DNA made up of repeating subunits called *nucleotides*.

The Nucleus: Control Center

The nucleus is the most important organelle in eukaryotic cells because it contains and protects the genetic blueprints (DNA) of the cell. That is why the nucleus is often referred to as the brain or control center of the cell. The nucleus separates its content from the rest of the cell with two layers of membranes, called the **nuclear envelope**. Structurally complex

nuclear envelope
Separates the content of a nucleus from the rest of the cell with two layers of membranes.

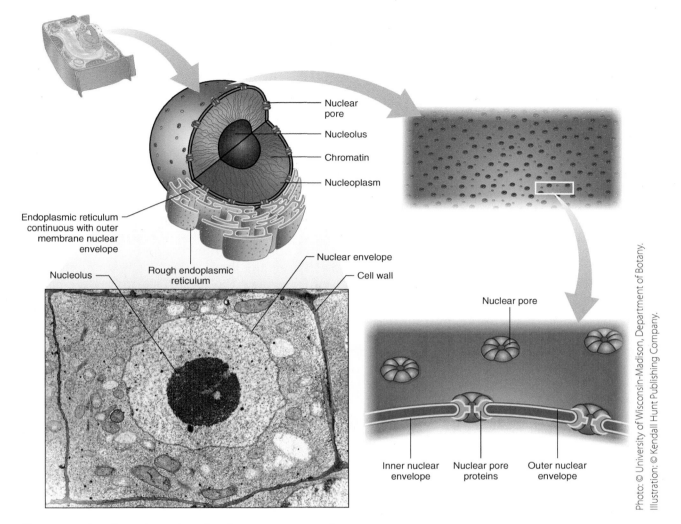

Labels in figure:
- Nuclear pore
- Nucleolus
- Chromatin
- Nucleoplasm
- Endoplasmic reticulum continuous with outer membrane nuclear envelope
- Rough endoplasmic reticulum
- Nuclear envelope
- Cell wall
- Nucleolus
- Nuclear pore
- Inner nuclear envelope
- Nuclear pore proteins
- Outer nuclear envelope

Photo: © University of Wisconsin-Madison, Department of Botany.
Illustration: © Kendall Hunt Publishing Company.

Figure 3.10   A nucleus showing its structures and components.

openings or **nuclear pores** are found all over the nuclear envelope. Nuclear pores selectively allow molecules to enter or leave the nucleus. Inside the nucleus are thin threads of chromatin, a network of protein fibers forming the **nucleoplasm**, and a prominent structure called *nucleolus* (plural: nucleoli) (Figure 3.10).

In all eukaryotic cells including plant cells, DNA molecules are always associated with proteins called *histones*; the DNA/histones complex is referred to as **chromatin**. Histone proteins help DNA to compact and condense during cell division. The condensed and visible form of DNA/histone complex is called **chromosome**. The number of chromosomes found in a nucleus varies tremendously among different organisms. For example, there are 46 chromosomes in a human nucleus but only 18 chromosomes in a radish nucleus. A tropical adder's tongue fern has 1,260 chromosomes in each nucleus while a jack jumper ant has only two. The number of chromosomes present in the nucleus seems to have no relation to the size and complexity of the organism.

nuclear pores
Structurally complex openings found all over the nuclear envelope and are responsible for selectively allowing molecules to enter or leave the nucleus.

nucleoplasm
Network of protein fibers found within the nucleus.

chromatin
Complex of DNA molecules and histones.

chromosome
Condensed and visible form of DNA/histone complex.

Nucleolus (plural: nucleoli) is the most noticeable round structure inside the nucleus of nondividing cells. DNA in the vicinity of the nucleolus is responsible for making ribosomal RNA (rRNA). These rRNA molecules and ribosomal proteins, which are imported from the cytoplasm, are being assembled into ribosomal subunits. The subunits then exit the nucleus via the nuclear pores to the cytoplasm where they are needed for protein synthesis.

The Endoplasmic Reticulum–Biomolecule Factories

The **endoplasmic reticulum (ER)** is a network of membranes that connects to the outer membrane of the nuclear envelope and extends outward into the cytoplasm. In some cells, the ER membrane makes up more than half of the total membrane in the cell. The main functions of the ER are to manufacture life-sustaining molecules and membrane components for the cell. The ER is also responsible for channeling these molecules to different parts of the cell. ER that has ribosomes attached to its surface, giving it a rough and bumpy look, is called **rough ER**. The main functions of rough ER are to produce and sort proteins for different destinations within or outside of the cell. The rough ER is also responsible for adding carbohydrates to proteins and lipids. ER that is not associated with any ribosomes is called **smooth ER**. The main function of smooth ER is to produce lipids in the form of oil droplets (Figure 3.11).

Dictyosomes or Golgi Apparatus: Packaging and Shipping Center

Dictyosomes are flattened disc-shaped sacs found stacked together in the cytoplasm. They are often referred to as the **Golgi apparatus**, which was named after physician Camillo Golgi, who discovered these structures in 1898. Dictyosomes collect, process, and deliver proteins for use outside of the cell. One might compare the functions of dictyosomes to carrier companies such as FedEx and UPS. Proteins produced by the rough ER are first transported to dictyosomes using vesicles (small membrane-bounded bodies). Modified proteins are then packed into vesicles and shipped from the dictyosomes to the plasma membrane. In plant cells, dictyosomes also produce and deliver cellulose to the cell wall. They even collect materials that are stored inside large membrane-bounded bodies called *vacuoles*.

Vacuoles: Storage Facilities

Vacuoles are large membrane-bounded bodies that store water, salts, water-soluble pigments such as anthocyanins, and even waste products. As much as 90% of the volume of a mature plant cell may be taken up by a single large vacuole, often referred to as the **central vacuole**. The central vacuole helps plant cells to maintain their shape by taking up water and making the cells turgid (*turg* in Latin means "swollen"). When the central vacuole swells, it exerts pressure against the cell wall, called *turgor pressure*. **Turgor pressure** provides the strength to keep nonwoody plants upright. If a plant loses water, it will collapse or wilt from the loss of turgor pressure.

---

endoplasmic reticulum (ER)
Network of membranes that connects to the outer membrane of the nuclear envelope; responsible for manufacturing life-sustaining molecules and membrane components of the cell.

rough ER
Endoplasmic reticulum that has ribosomes attached to its surface; responsible for producing and sorting proteins for different destinations within or outside of the cell.

smooth ER
Endoplasmic reticulum that is not associated with any ribosomes; responsible for producing lipids.

dictyosomes
Flattened disc-shaped sacs found stacked together in the cytoplasm; responsible for collecting, processing, and delivering proteins for use outside of the cell.

Golgi apparatus
See *dictyosomes.*

central vacuole
Single large vacuole that takes up as much as 90 percent of the volume of mature plant cells; a unique cell structure found only in plant cells.

turgor pressure
When water enters the central vacuole and causes it to swell and exert pressure against the cell wall; responsible for providing strength to keep nonwoody plants upright.

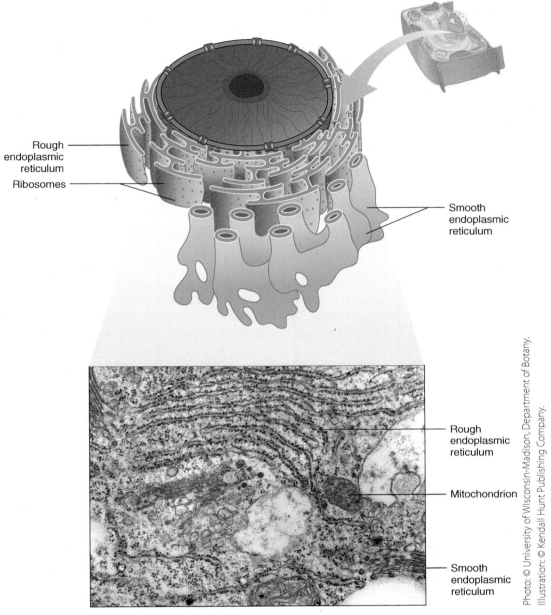

Rough endoplasmic reticulum

Ribosomes

Smooth endoplasmic reticulum

Rough endoplasmic reticulum

Mitochondrion

Smooth endoplasmic reticulum

Photo: © University of Wisconsin-Madison, Department of Botany.
Illustration: © Kendall Hunt Publishing Company.

Figure 3.11 A network of endoplasmic reticulum (ER) showing both rough and smooth ER.

Mitochondria: Energy Powerhouses

Mitochondria (singular: mitochondrion) are tiny rod-shaped membrane-bounded structures that carry out aerobic respiration (Figure 3.12). Many cells use aerobic respiration, a series of chemical reactions, to break down fuel molecules into energy in the form of ATP (adenosine triphosphate) that the cells can use (see Chapter 11 for more detail). It is not surprising that mitochondria are often referred to as the powerhouses of the cell. Each mitochondrion has two layers of membrane (inner and outer), similar to the nucleus. The inner membrane has a much larger surface area than the

mitochondria
(singular *mitochondrion*)
Tiny rod-shaped membrane-bounded structures that carry out aerobic respiration to produce energy for the cell.

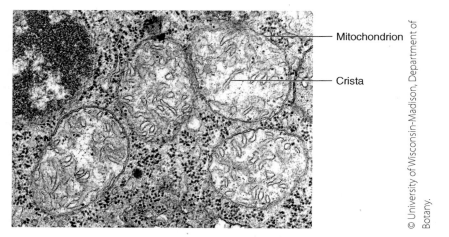

© University of Wisconsin-Madison, Department of Botany.

Figure 3.12   This micrograph shows four mitochondria and their cristae inside.

cristae
Fingerlike projections formed by folding of the inner mitochondrial membrane; responsible for creating a larger surface area for aerobic respiration to take place.

matrix
Fluid portion inside the mitochondrion that contains enzymes, proteins, mitochondrial DNA, and ribosomes.

ribosomal RNA (rRNA)
Ribonucleic acids that make up ribosomes.

plastids
Large organelles that are found in plant cells; the three main types are chloroplasts, chromoplasts, and leucoplasts.

outer membrane and folds inward to form many finger-like projections called **cristae**. These cristae provide room for large numbers of enzyme to carry out aerobic respiration. The interior of the mitochondrion is filled with a fluid called the **matrix**, which contains enzymes, proteins, mitochondrial DNA, and ribosomes. Mitochondria are highly unique because they have their own DNA and ribosomes, which are different from the cell. Mitochondrial DNA is a circular DNA molecule that does not have histones. Ribosomes found inside the mitochondrion are much smaller in size than those in the cell. Mitochondrial DNA and ribosomes resemble the ones found in prokaryotic cells. (See **Box 3.1**.)

Ribosomes: Protein Factories

Ribosomes are tiny particles found in all cells and are responsible for making all kinds of proteins. Proteins are made up of smaller and simpler units called *amino acids* (see Appendix). These tiny protein factories of the cell use instructions from DNA to make proteins by linking specific amino acids together. Some of these proteins are used by the cell and others are exported for use outside the cell. Each ribosome consists of two subunits made up of proteins and **ribosomal RNA (rRNA)**. Although there are many different types of RNA (ribonucleic acid), all RNA belong to the class "nucleic acids," the same class of organic molecule that DNA molecules belong to. DNA and RNA work closely together to make proteins for the cell in a process called *gene expression*. While some of the ribosomes are bound to the endoplasmic reticulum (only in eukaryotic cells), the majority of them are found free in the cytoplasm. Ribosomes lack membranes; therefore, they are not organelles.

PLANT CELL STRUCTURES AND FUNCTIONS

Plants are multicellular eukaryotes because they are made up of many eukaryotic cells. In addition to the common cell structures, a typical plant cell also has plastids and a cell wall (Figure 3.13). **Plastids** are large organelles that are found in plant cells and the three main types are chloroplasts,

BOX 3.1   Endosymbiosis and the Origins of Mitochondria and Chloroplasts

Mitochondria and chloroplasts are the only organelles with their own DNA and ribosomes, which are different from the nuclear DNA and cytoplasmic ribosomes. Their DNA and ribosomes are very similar to the ones prokaryotic cells have (Schnepf & Brown, 1971). These organelles are said to be semi-autonomous because they can grow and divide to reproduce on their own. When these organelles divide, they go through binary fission, a type of cell division used by prokaryotic cells, instead of mitosis. Mitochondria and chloroplasts are not completely autonomous because they require other parts of the cells for some essential components. For example, most of the proteins found in mitochondria are made by the cell and imported from the cytosol.

As early as the late 1800s, microscopists (J. Sachs, R. Altmann, and A. Schimper) already recognized the semiautonomous nature and bacteria-like staining properties of mitochondria and chloroplasts. But in 1905, it was Konstantin Mereschkowsky that synthesized these observations and proposed that chloroplasts are derived from cyanobacteria. In 1971, Lynn Margulis proposed the **theory of endosymbiosis** to explain the origins of mitochondria and chloroplasts (for review, see Gould, Waller, & McFadden, 2008) (Figure Box 3.1). *Endo* in Greek means "within" and *Symbio* means "living together." An endosymbiotic relationship is an intimate association between two different kinds of organisms, one living inside the other one. In the case of primary endosymbiosis, molecular and genetic evidence suggests that mitochondria and chloroplasts were once free-living prokaryotic cells that formed endosymbiotic relationships with larger prokaryotic (host) cells. Mitochondria are thought to have evolved from aerobic (oxygen-requiring) prokaryotes such as alpha-proteobacterium. Chloroplasts are thought to have evolved from cyanobacterium, a type of photosynthetic bacteria. Both mitochondria and chloroplasts are now fully integrated endosymbionts because they have lost the ability to survive on their own (for review, see Keeling, 2010).

All eukaryotes have either mitochondria or hydrogenosomes (modified, anaerobic [without oxygen] forms of mitochondria), which suggest that these organelles were integral to not only the origin of eukaryotic cells but also the diversity of heterotrophic eukaryotes. **Heterotrophs** are organisms that cannot produce their own food but have to depend on other organisms for food. The acquisition of chloroplasts by eukaryotic cells came later and led to the evolution of the autotrophic eukaryotes. **Autotrophs** are organisms that are able to make their own food using photosynthesis. Unlike mitochondria, chloroplast evolution was more complex. Secondary symbiosis and even tertiary endosymbiosis may explain the origin, diversification, and fate of chloroplasts in different autotrophic protist hosts (see Chapter 20 for more discussion). Secondary endosymbiosis is when an autotrophic eukaryotic cell forms an endosymbiotic relationship with another eukaryotic (host) cell. Secondary endosymbiosis and tertiary endosymbiosis offer explanations as to why some photosynthetic protists have chloroplasts with three to four membranes instead of two. Although the theory of endosymbiosis offers an explanation of how eukaryotic cells acquired organelles such as mitochondria and chloroplasts, it does not explain the origin of the nucleus found in all eukaryotic cells.

theory of endosymbiosis
Proposed by Margulis in 1971 to explain the origins of mitochondria and chloroplasts.

Heterotrophs
Organisms that cannot produce their own food but have to depend on other organisms for food.

Autotrophs
Organisms that are able to make their own food using photosynthesis.

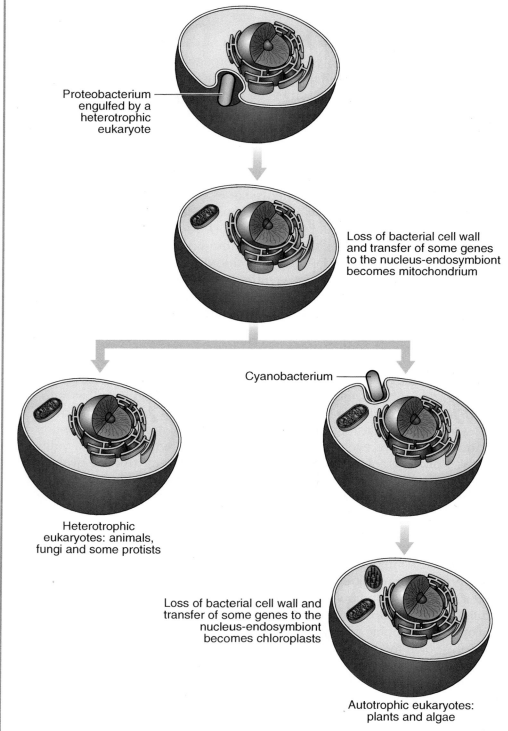

Proteobacterium engulfed by a heterotrophic eukaryote

Loss of bacterial cell wall and transfer of some genes to the nucleus-endosymbiont becomes mitochondrium

Cyanobacterium

Heterotrophic eukaryotes: animals, fungi and some protists

Loss of bacterial cell wall and transfer of some genes to the nucleus-endosymbiont becomes chloroplasts

Autotrophic eukaryotes: plants and algae

Figure Box 3.1   The theory of endosymbiosis offers explanations to the origins of eukaryotes and mitochondria and chloroplasts.

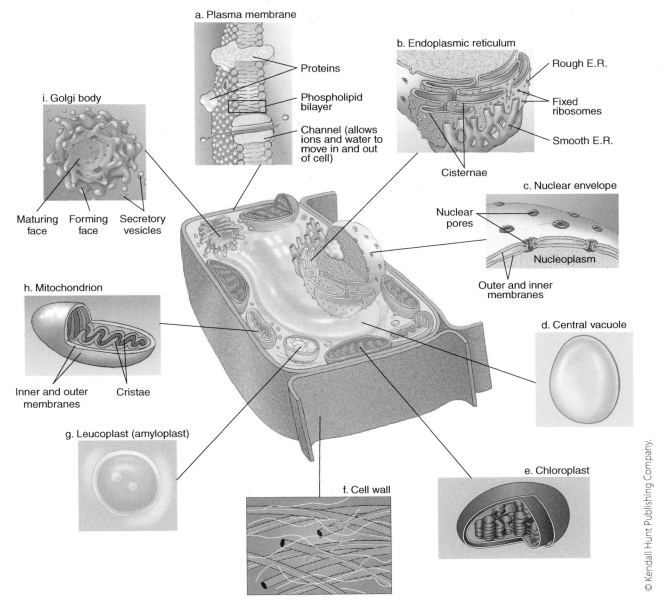

Figure 3.13a–i   A typical plant cell showing (a) plasma membrane, (b) endoplasmic reticulum, (c) nuclear envelope, (d) central vacuole, (e) chloroplast, (f) cell wall, (g) leucoplast, (h) mitochondrion, and (i) Golgi body.

chromoplasts, and leucoplasts. All plastids develop from **proplastids**, which are small, colorless or green, simple organelles found in the meristemic (stem) cells of roots and stems. Plastids are remarkably flexible and can change from one type to another with ease. All three types of plastids have two layers of membrane and each plastid performs a very specific function.

Chloroplasts: The Green Plastids That Make Food

Chloroplasts, the most common plastids, carry out photosynthesis and are found mostly in leaf and stem cells. Photosynthesis provides food not only for plants but also for other organisms. (This important process is

proplastids
Small, colorless or pale green, simple organelles found in the meristemic; responsible for producing various plastids.

chloroplasts
Most common plastids found mostly in leaf and stem cells; responsible for carrying out photosynthesis to produce food and oxygen.

grana
(singular *granum*)
Disc-shaped structures found within chloroplasts; have numerous interconnected discs that look like stacks of coins.

thylakoids
Hollow coin-like structures inside the chloroplasts.

stroma
Liquid portion of the chloroplast that contains enzymes essential for photosynthesis, chloroplast DNA, and ribosomes.

discussed in greater detail in Chapter 10.) Chloroplasts contain photosynthetic pigments such as chlorophylls and carotenoids that are used to capture light energy for photosynthesis. Light energy is then used to make organic or fuel molecules such as glucose (food). The green color of these plastids comes from the chlorophyll pigments. Chloroplasts are generally disc-shaped and have numerous interconnected **grana** (singular: granum), which look like stacks of coins. The coin-like structures are called **thylakoids**, which are hollow discs. The liquid portion of the chloroplast is called the **stroma**, which contains enzymes essential for photosynthesis, chloroplast DNA, and ribosomes. Similar to mitochondria, chloroplasts have their own DNA and ribosomes that are different from the cell and resemble the ones found in prokaryotic cells (Figure 3.14).

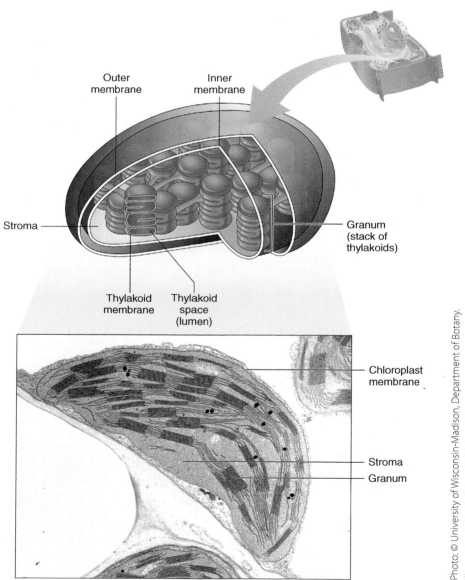

Figure 3.14   A chloroplast and its internal structures.

Photo: © University of Wisconsin-Madison, Department of Botany.
Illustration: © Kendall Hunt Publishing Company.

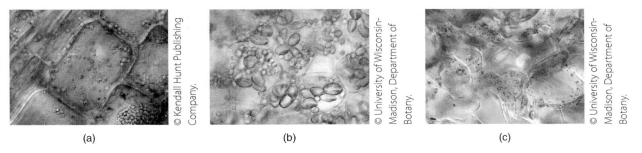

(a)   (b)   (c)

Figure 3.15a–c   Plastids in various plant cells: (a) *Elodea* cells with chloroplasts, (b) potato cells with amyloplasts, and (c) pepper cells with chromoplasts.

Chromoplasts: Colorful Plastids That Add a Touch of Color to the Green World

Chromoplasts (*chroma* in Greek means "color") are pigmented plastids (Figure 3.15). They have no chlorophylls but are able to synthesize and accumulate carotenoid pigments. Carotenoids are responsible for the yellow, orange, or red colors of many flowers, fruits, aging leaves, and some roots such as carrots. Chromoplasts often form when the chlorophylls and internal membrane structures of chloroplasts break down and disappear. This dramatic transformation can be seen when green tomatoes ripen and turn red. The main function of chromoplasts is to attract organisms to pollinate or disperse seeds for the plants.

Leucoplasts: Colorless Plastids That Store Stuffs

Leucoplasts are the third type of plastids found commonly in seeds, roots, and stems (Figure 3.15). These plastids are colorless and mainly responsible for producing and storing starch, oils, or proteins. Leucoplasts that synthesize and store starch are called **amyloplasts** and those that synthesize and store oils are called **elaioplasts**. When leucoplasts are exposed to light, they can synthesize chlorophylls and change into chloroplasts to carry out photosynthesis. This dramatic transformation can be seen when potato tubers turn green and sprout after exposure to light.

The Cell Wall: Support and Protection

Outside the plasma membrane of each plant cell, there is a rigid coating called a *cell wall*. The **cell wall** is made up of layers of long carbohydrate (**cellulose**); (see appendix) fibers. These fibers are held together by other complex carbohydrates including **pectin**, the substance that stiffens fruit jellies. The cell wall protects and supports the plant cell while allowing unrestricted movement of water and dissolved nutrients into the cell or between cells. Although bacterial cells and fungal cells also have cell walls, prokaryotic cell walls are made up of a modified carbohydrate called *peptidoglycan*, whereas fungal cell walls are made up of a complex carbohydrate called *chitin*.

The thickness of a plant cell's walls depends on the function of the cell as well as its age. The **primary cell wall** is the first layer of cell wall laid

chromoplasts
Plastids that synthesize and accumulate color pigments such as carotenoids (*chroma* in Greek means "color").

leucoplasts
Plastids found commonly in seeds, roots, and stems and are colorless; mainly responsible for producing and storing starch, oils, or proteins.

amyloplasts
Leucoplasts that synthesize and store starch.

elaioplasts
Leucoplasts that synthesize and store oils.

cell wall
Rigid coating made up of cellulose fibers that is found outside the plasma membrane of plant cells; responsible for protecting and supporting the plant cell.

cellulose
Complex carbohydrate or polysaccharide for structure support.

pectin
Gelatin-like substance found in the cell wall; responsible for holding cellulose fibers together.

primary cell wall
First layer of cell wall laid down by a growing cell.

middle lamella
Layer of pectin that acts as a glue to hold adjacent plant cells together.

secondary cell wall
Second layer of cell wall that has ligin and more cellulose than the primary cell wall.

lignin
Hard substance found in the secondary cell wall in which the cellulose fibers are embedded; found in tracheids, vessels, and sclereids; provides water-proofing and structure support for plants.

plasmodesmata
(singular *plasmodesma*)
Tiny channels through the cell walls that connect the plasma membrane and cytoplasm of adjacent plant cells; responsible for cell communication

down by a growing cell. It is very thin and flexible and can expand with the growth of the cell. The **middle lamella** is a layer of pectin that acts as a glue to hold adjacent cells together. In addition to the primary cell wall, some cells produce multiple layers of secondary cell walls. These additional layers are found between the primary cell wall and the plasma membrane. In sclerenchyma cells, the walls become so thick that there is no room for cytoplasm or nucleus and the cells are dead when they mature. The **secondary cell wall** has more cellulose than the primary cell wall and also contains **lignin**, a hard substance in which the cellulose fibers are embedded.

Plants are made up of many cells and these cells have to communicate with one another. **Plasmodesmata** (singular: plasmodesma) are tiny channels through the cell walls that connect the plasma membrane and cytoplasm of adjacent cells (Figure 3.16). These cytoplasmic connections allow signal molecules, nutrients, and ions to pass between the cells.

Differences and Similarities between Plant and Animal Cells

The cell wall, plastids (chloroplasts, chromoplasts, and leucoplasts), and a conspicuous central vacuole are distinctive structures that allow you to distinguish plant cells from animal cells. Unlike plant cells, animal cells have centrioles, which play a role in cell division. In addition, animal cells have lysosomes which are vesicles filled with enzymes for digestion (Table 3.1). Despite these differences, plant and animal cells are fundamentally very similar. They are both eukaryotic cells, which mean they both have a nucleus and organelles such as mitochondria, endoplasmic reticulum, and dictyosomes. Like all living cells including prokaryotic cells, plant and animal cells both have DNA, cytoplasm, ribosomes, and a plasma membrane (Figure 3.17).

PLANT TISSUE SYSTEMS AND FUNCTIONS

Plants are made up of many cells and these cells are not the same. Cells in different parts of plants become adapted for specific tasks and take on different shapes and forms. Some cells come together and specialize in transportation of water and others specialized in translocation of food made

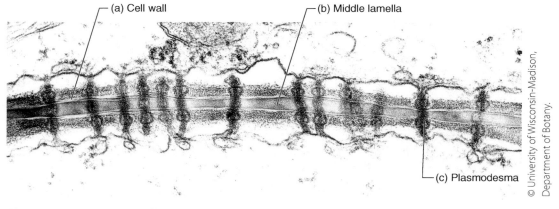

(a) Cell wall   (b) Middle lamella

(c) Plasmodesma

© University of Wisconsin-Madison, Department of Botany.

Figure 3.16a–c   (a) Cell walls, (b) middle lamella, and (c) plasmodesmata between adjacent cells.

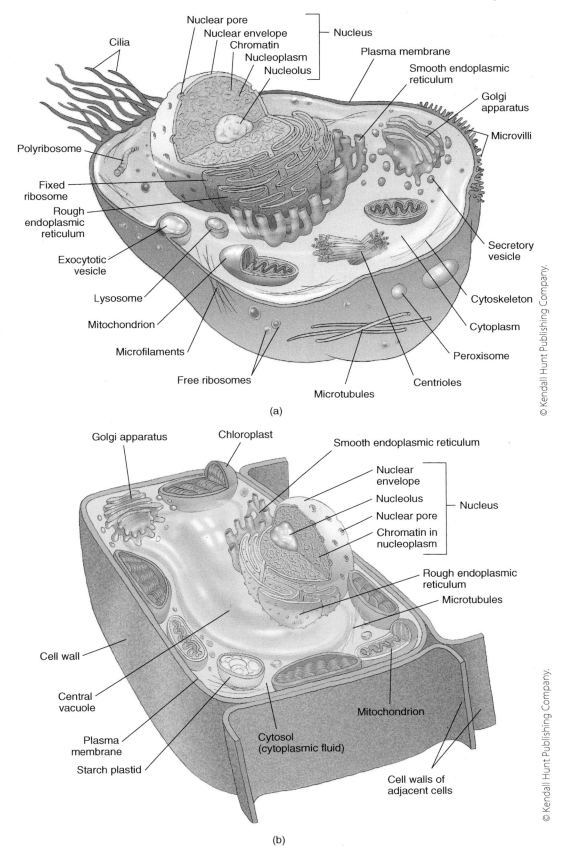

Figure 3.17a–b   (a) A typical animal cell and (b) a typical plant cell.

© Kendall Hunt Publishing Company.

division of labor
Cells in different parts of the plants become adapted for specific tasks and take on different shape and form, which allows the entire organism to become more efficient.

by photosynthesis. Some cells specialize in storage and others in structure support. This **division of labor** allows the entire organism to become more efficient. When cells come together to form a structural and functional unit, we call this unit *tissue*. There are two types of tissues: simple and complex. Simple tissues are made up of one type of cells whereas complex tissues are made up of two or more types of cells. When these tissues come together, they form a tissue system. Vascular plants have three types of tissue systems: ground, vascular, and dermal. The three tissue systems come together to form the four plant organs: roots, stems, leaves, and flowers (Figure 3.18).

Ground Tissue System: The Storage and Support System

The ground tissue system usually makes up the bulk of all herbaceous (soft, nonwoody) plants. This tissue system is composed of three simple tissues: parenchyma, collenchyma, and sclerenchyma. Each of these simple tissues consists of one type of cells and has unique functions. These three simple tissues can be easily distinguished from each other by the thickness and composition of their cell walls.

parenchyma tissue
Made up of parenchyma cells, which are the most abundant cells in the plant body; mainly responsible for storage.

Parenchyma tissue is made up of **parenchyma** cells, which are the most abundant cells in the plant body. Parenchyma cells have a thin primary cell wall but no secondary cell wall, which allows them to expand and assume various shapes. Parenchyma cells remain alive and metabolically active when mature. They also retain the ability to divide and change into other types of cells, which is important for tissue repair and propagation (Figure 3.19). The functions of parenchyma cells depend on what they contain and may range from storage, photosynthesis, and secretion to coloration.

Parenchyma cells of leaves contain numerous chloroplasts and are mainly for photosynthesis. Edible parts of fruits and vegetables contain parenchyma cells with many leucoplasts that are mainly for storage of starch, oil, and water. Coloration of fruits and flowers provided by parenchyma

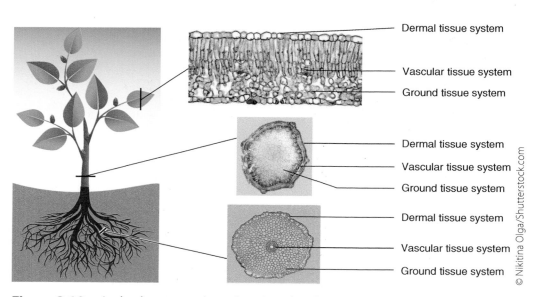

Figure 3.18   An herbaceous plant showing the three tissue systems in its root, stem, and leaf.

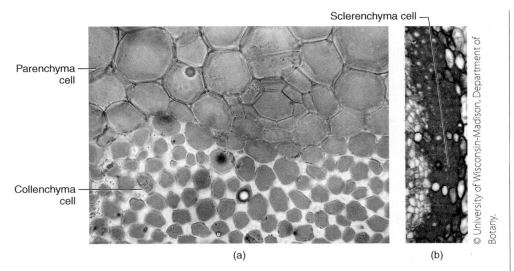

Sclerenchyma cell

Parenchyma cell

Collenchyma cell

© University of Wisconsin-Madison, Department of Botany.

(a)   (b)

Figure 3.19a–b   Simple tissues that make up the ground tissue system: (a) parenchyma, collenchymas, and (b) sclerenchyma.

cells contain many chromoplasts. Some parenchyma cells have extensive ridges and ingrowths on the inner surface of their cell wall to increase surface area and enhance secretion of nectar, hormones, enzymes, and resins.

Collenchyma tissue is made up of **collenchyma** cells that have thicker primary cell walls than parenchyma cells (Figure 3.19). The thickness of their cell walls is also more uneven than parenchyma cells and is especially noticeable in the corners. Like parenchyma cells, collenchyma cells remain alive and metabolically active when mature. These collenchyma cells are often found in long strands just beneath the epidermis or along leaf veins. The "strings" of celery stalk are mostly made up of collenchyma tissue. The main function of collenchyma tissue is to provide flexible support for herbaceous plants.

Sclerenchyma tissue is made up of **sclerenchyma** cells, which are specialized for structural support. Sclerenchyma cells have both primary and secondary cell walls but their secondary cell walls are extremely thick and tough because of the presence of lignin. *Sclero* in Greek means "hard." Unlike parenchyma and collenchyma cells, sclerenchyma cells are dead when mature and the thick cell walls occupy the entire cell (Figure 3.19). Sclerenchyma cells come in two forms: sclereids and fibers. **Sclereids** or stone cells are short and vary in shape. They are common in the shells of nuts and the pits of stone fruits such as peaches and cherries. The slightly gritty texture of pears is due to the presence of sclereid clusters. **Fibers** are long, tapered sclerenchyma cells often found in clumps. They are particularly common in wood, inner bark, and along leaf veins. Fibers from more than 40 families of plants are used to manufacture textile goods, ropes, string, and canvas.

Vascular Tissue System: The Transport System

The vascular tissue system is embedded in the ground tissue and responsible for transport of water and food materials throughout the plant. This transport system is composed of two complex tissues: **xylem** and **phloem**. Complex tissues are made up of two or more types of cells.

collenchyma tissue
Made up of collenchyma cells that have thicker primary cell walls than parenchyma cells; mainly responsible for providing flexible support for herbaceous plants.

sclerenchyma tissue
Made up of sclerenchyma cells that are specialized for rigid structural support.

sclereids
or stone cells Type of sclerenchyma cells that are short and vary in shape.

fibers
Long and tapered sclerenchyma cells often found in clumps.

xylem
Responsible for transporting water and dissolved minerals from the roots to the rest of the plant and also for providing structure support.

phloem
Responsible for transporting food (sugars) and other essential materials from the leaves to the rest of the plant.

Xylem is responsible for transporting water and dissolved minerals from the roots to the rest of the plant and also for providing structure support. Xylem is made up of four types of cells: tracheids, vessel elements, parenchyma cells, and fibers. Mature tracheids and vessel elements are dead cells that specialize in transporting water and dissolved minerals (Figure 3.20). Parenchyma cells and fibers are there to provide storage and support. **Tracheids** are long, tapered cells with thick secondary cell walls and are often found in patches or clumps. They have pits, which are thin areas of the cell wall with no secondary cell wall. These pits allow sideways or lateral movement of water from one tracheid to another. Tracheids are the only water-conducting cells found in gymnosperms (cone-bearing plants) and seedless vascular plants such as ferns. **Vessel elements** are much wider in diameter than tracheids and tend to be open or heavily perforated at both ends of the cell. Stacks of vessel elements function as water pipes and permit unhindered flow of water and minerals. Similar to tracheids, vessel elements have pits that allow lateral transport of water and minerals. Flowering plants are the only plant group that has both vessel elements and tracheids.

Phloem is responsible for transporting food (sugars) made by photosynthesis in the leaves to the rest of the plant. It is also responsible for transporting other essential materials such as amino acids, proteins, lipids, hormones, and signaling molecules. Phloem consists primarily of sieve-tube elements and companion cells but also utilizes parenchyma cells and fibers for storage and support. Links of these long, thin cells, called **sieve-tube elements**, form sieve tubes. Both ends of the sieve-tube element are heavily perforated, allowing cytoplasm to extend from one sieve-tube element into the next. These sieve-tube elements are responsible for transporting food and essential materials throughout the plant. Sieve-tube elements are among the most specialized and unusual living cells in nature. Although sieve-tube elements are living cells when mature, they have no nucleus

tracheids
Long, tapered cells with thick secondary cell walls that make up the xylem; often found in patches or clumps.

vessel elements
Much wider in diameter than tracheids and tend to be open or heavily perforated at both ends of the cell; make up the xylem.

sieve-tube elements
Links of long, thin cells that are heavily perforated, allowing cytoplasm to extend from one sieve-tube element into the next; responsible for transporting food and essential materials throughout the plant.

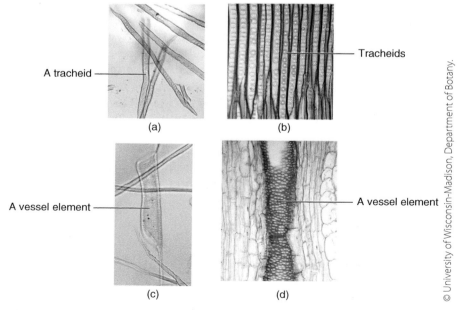

A tracheid — (a)

Tracheids — (b)

A vessel element — (c)

A vessel element — (d)

© University of Wisconsin-Madison, Department of Botany.

Figure 3.20a–d (a–b) Tracheid cells and (c–d) vessel elements, main components of xylem.

and much of the other cell structures such as mitochondria, vacuole, and ribosomes have also disappeared. Adjacent to each sieve-tube element is a **companion cell**, which is narrower and more tapered than a sieve-tube element. Companion cells are specialized parenchyma cells that have all the cell structures commonly found in living plant cells, including a nucleus. Their sole responsibility is to provide a life-support system for sieve-tube elements by synthesizing ATP, proteins, and RNAs for sieve-tube elements. Sieve-tube elements and their companion cells exchange materials through numerous plasmodesmata in their cell walls (Figure 3.21).

Dermal Tissue System: The Protective Covering

The dermal tissue system forms the outermost cell layer that covers and protects the whole plant. In herbaceous plants, the dermal tissue system is a layer of cells called the **epidermis** (Figure 3.22). Woody plants initially have an epidermis but it is eventually replaced with the **periderm** or bark.

The epidermis is made up mostly of unspecialized cells; however, depending on its location, other specialized cells such as guard cells, trichomes, and root hairs may be present. Most epidermal cells are both compactly arranged and have cell walls that are thicker toward the outside to provide considerable mechanical protection to the plant part. Above ground, the epidermis secretes a cuticle to minimize water loss. The **cuticle** consists mainly of waterproof materials called *cutin* and *wax*.

Although the cuticle greatly reduces water loss from plant surfaces, it also severely restricts gas exchange. Photosynthesis requires carbon dioxide and produces oxygen as a by-product. The movement of both gases is prevented by the cuticle. Openings or pores are needed on the epidermis for gas exchange. **Stomata** (singular: stoma) are tiny openings surrounded by two specialized epidermal cells called **guard cells** (Figure 3.23). Plant can open or close these stomata to regulate passage of carbon dioxide and oxygen, as well as water vapor. Stomata are discussed in greater detail in Chapter 6.

The epidermal cells in some stems and leaves may have specialized hairs called *trichomes*. **Trichomes** are used by some plants for protection. For example, leaf and stem trichomes of stinging nettle contain irritating chemicals

companion cell
Narrower and more tapered than a sieve-tube element; responsible for providing life support system for sieve-tube elements.

epidermis
Made up of mostly unspecialized cells forming a single layer covering the surface of herbaceous plants for protection.

periderm
Forms a thick protective layer around the roots and stems of woody plants.

cuticle
Found on the surface of stems and leaves; made up of cutin and wax, which make plant surfaces waterproof and reflective.

stomata
(singular *stoma*) Tiny openings surrounded by two guard cells; responsible for gas exchange and regulation of transpiration.

guard cells
Found beside the stoma; responsible for opening and closing the stomatal pore.

trichomes
Specialized epidermal cells that are used by some plants for protection.

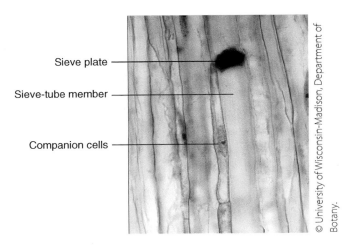

Sieve plate

Sieve-tube member

Companion cells

© University of Wisconsin-Madison, Department of Botany.

Figure 3.21 The two main components of phloem: sieve-tube members (without nucleus) and their companion cells (with nucleus).

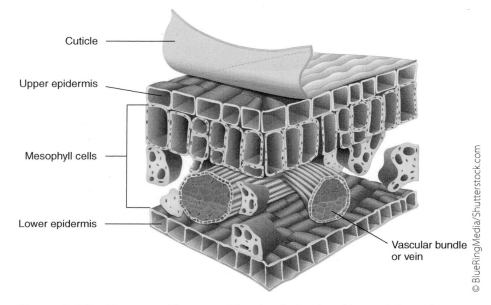

© BlueRingMedia/Shutterstock.com

Cuticle

Upper epidermis

Mesophyll cells

Lower epidermis

Vascular bundle
or vein

Figure 3.22 Upper and lower epidemis of a leaf and its cuticle and stoma.

root hairs
Simple extensions from individual epidermal cells in roots; responsible for increasing surface area of the root epidermis for absorption.

cork cells
Made up of the bark or periderm; cells that are dead when mature and their cell walls are heavily fortified with a waterproof substance called suberin.

suberin
Major component of the walls of cork cells in woody plants and root endodermal cells; helps to restrict water and mineral movement in the roots.

phelloderm
Living parenchyma tissue found in the bark; mainly responsible for storage.

cork cambium
Found in bark; responsible for producing cork cells and phelloderm.

that deter herbivorous animals from eating the plant. The saltbush (*Atriplex*) grows in salty soil and uses its leaf trichomes to remove excess salt. High trichome densities on the leaves of some desert plants can lower leaf temperature and rate of water loss by increasing reflection of light. Epiphytic bromeliads use leaf trichomes for absorption of water and minerals. Essential oil plants such as peppermint use glandular trichomes to make, store, and release essential oils. **Root hairs** are simple extensions from individual epidermal cells in roots. They are used to increase the surface area of the root epidermis so that water and minerals can be absorbed more rapidly.

The periderm or bark forms a thick protective layer around the roots and stems of woody plants. When a woody plant begins to increase in girth or width, its epidermis breaks up and is replaced by periderm. The periderm is a complex tissue made up of cork cells, phelloderm (cork parenchyma cells), and cork cambium. **Cork cells** are dead when mature

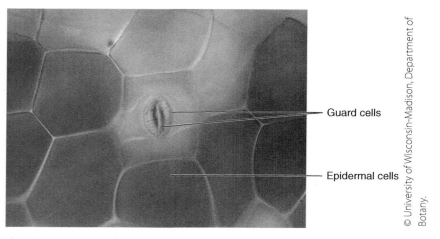

© University of Wisconsin-Madison, Department of Botany.

Guard cells

Epidermal cells

Figure 3.23 An open stoma with guard cells surrounded by epidermal cells.

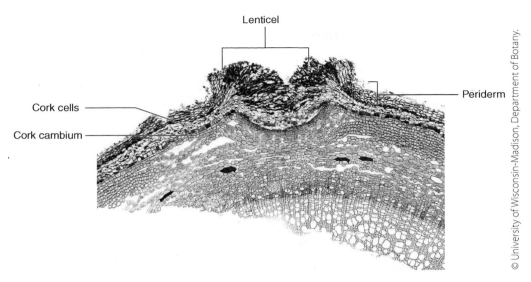

Figure 3.24 Various layers of the periderm and its lenticel.

© University of Wisconsin-Madison, Department of Botany.

Table 3.2 Components of Tissue Systems and Their Functions

Tissue System	Tissue Types	Cell Types	Primary Functions
Ground	Parenchyma tissue	Parenchyma cells	Storage, photosynthesis, coloration, and secretion
	Collenchyma tissue	Collenchyma cells	Flexible support
	Sclerenchyma tissue	Sclerenchyma cells (sclereids, fibers)	Rigid support
Vascular	Xylem	Tracheids, vessel elements, parenchyma cells, fibers	Water and mineral transport
	Phloem	Sieve-tube members, companion cells, parenchyma cells, fibers	Food (sugars) transport
Dermal	Epidermis	Unspecialized cells, guard cells, trichomes, root hairs	Mechanical protection, minimizes water loss, and regulates gas exchange via stomata
	Periderm	Cork cells, cork cambium, cork phelloderm	Replaces epidermis in some plants, mechanical protection, minimizes water loss, and regulates gas exchange via lenticels

and their cell walls are heavily fortified with a waterproof substance called **suberin**. **Phelloderm**, a living parenchyma tissue, is mainly responsible for storage. **Cork cambium** forms cork cells toward the surface and phelloderm toward the center of the plant. Although the cells of the periderm are compactly arranged for protection, **lenticels** are areas where the cells are loosely arranged to allow gas exchange to take place (Figure 3.24).

lenticels
Found on the surface of bark; areas where the cells are loosely arranged to allow gas exchange to take place.

Key Terms

Robert Hooke	smooth ER	secondary cell wall
cell biology	dictyosomes	lignin
cytology	Golgi apparatus	plasmodesmata
unicellular	central vacuole	division of labor
multicellular	turgor pressure	parenchyma
prokaryotic cell	mitochondria	collenchyma
eukaryotic cell	cristae	sclerenchyma
phospholipids	matrix	sclereids
proteins	theory of endosymbiosis	fibers
fluid-mosaic model	heterotrophs	xylem
cytosol	autotrophs	phloem
metabolism	Ribosomal RNA (rRNA)	tracheids
catabolic	plastids	vessel elements
anabolic	proplastids	sieve-tube elements
enzymes	chloroplasts	companion cell
cytoskeleton	grana	epidermis
microtubules	thylakoids	periderm
actin filaments	stroma	cuticle
cytoplasmic streaming	chromoplasts	stomata
nucleotides	leucoplasts	guard cells
nuclear envelope	amyloplasts	trichomes
nuclear pores	elaioplasts	root hairs
nucleoplasm	cell wall	cork cells
chromatin	cellulose	suberin
chromosome	pectin	phelloderm
endoplasmic reticulum (ER)	primary cell wall	cork cambium
rough ER	middle lamella	lenticels

Summary

- Discovery of cells is made possible with the invention and refinement of microscopes. In 1665, Robert Hooke discovered tiny pores in cork and coined the word *cells*. In the mid-1800s, Matthias Schleiden and Theodor Schwann are credited with developing the cell theory.

- Cells are small because they need a large surface area for efficient exchange of materials with the environment and a small volume for rapid transport of materials from the surface to the interior of the cell.

- Plants are made up of many cells and are called *multicellular*. Organisms that are made up of one cell are called *unicellular*.

- Cells are basically divided into two types: prokaryotic and eukaryotic. Eukaryotic cells have a nucleus and other organelles in their cells. Prokaryotic cells have neither a nucleus nor organelles in their cells. Eukaryotic plant cells are the focus of this chapter.

- All cells have these four common structures: the plasma membrane, the cytoplasm, DNA, and ribosomes. The plasma membrane provides a physical boundary separating the cell from its environment. It also regulates the movement of materials into and out of the cell. The cytoplasm is made up of cytosol, cytoskeleton, and all cellular structures found inside the plasma membrane. Deoxyribonucleic acid (DNA), the blueprint of life, is responsible for the storage, expression, and transmission of genetic information. Ribosomes are responsible for making all kinds of proteins.

- In addition to the four common cell structures, a typical eukaryotic cell also has a nucleus and many membrane-bounded structures or organelles. The nucleus is the most important organelle in eukaryotic cells because it contains and protects the genetic blueprints (DNA) of the cell.

- The endoplasmic reticulum (ER), a network of membranes, is responsible for manufacturing life-sustaining molecules and membrane components for the cell as well as channeling these molecules to different parts of the cell.

- Dictyosomes collect, process, and deliver proteins that are destined for use outside of the cell. In plant cells, dictyosomes also produce and deliver cellulose to the cell wall.

- As much as 90% of the volume of a mature plant cell may be taken up by a single large vacuole, often referred to as the central vacuole. The central vacuole helps plant cells to maintain their shape by taking up water and making the cells turgid.

- Mitochondria are tiny rod-shaped organelles that carry out aerobic respiration to break down fuel molecules into energy for the cell. Similar to the nucleus, each mitochondrion has two layers of membrane; its own DNA, which does not associate with histones; and its own ribosomes.

- Plastids are large organelles that are found in plants cells. The three main types are chloroplasts, chromoplasts, and leucoplasts. All plastids develop from proplastids and can change from one type to another with ease.

- Chloroplasts are large green disc-shaped plastids that carry out photosynthesis for plant cells. Similar to mitochondria, chloroplasts have two layers of membrane and also their own DNA and ribosomes, which are different from the cell.

- Chromoplasts are pigmented plastids that provide coloration for flowers, fruits, aging leaves, and some storage roots. Leucoplasts are colorless plastids and are mainly responsible for producing and storing starch, oils, or proteins.

- Plant cell wall is made up of layers of cellulose fibers that are held together by other complex carbohydrates, including pectin. The cell wall is responsible for protecting and supporting the plant cell. Plasmodesmata allow plant cells to communicate with one another.

- Animal cells do not have a cell wall, a central vacuole, or any plastids, which make them conspicuously different from plant cells. Animal cells have centrioles and lysosomes; plant cells do not. When animal cells divide, they pinch into two instead of forming a cell plate.

- Cells come together to form a structural and functional unit called *tissue*. There are two types of tissues: simple and complex. Simple tissues are made up of one type of cells whereas complex tissues are made up of two or more types of cells. When these tissues come together, they form a tissue system.

- The ground tissue system that makes up the bulk of all herbaceous plants is composed of three simple tissues: parenchyma, collenchyma, and sclerechyma. Parenchyma tissue is responsible for storage, photosynthesis, secretion, and coloration. Collenchyma tissue provides flexible support for herbaceous plants. Sclerenchyma tissue is specialized for structural support.

- The vascular tissue system is embedded in the ground tissue and responsible for transport of water and food materials throughout the plant. This transport system is composed of two complex tissues: xylem and phloem. Xylem is responsible for transporting water and dissolved minerals. Phloem is responsible for transporting food (sugars) made by photosynthesis and other essential materials such as amino acids, proteins, lipids, hormones, and signaling molecules.

- The dermal tissue system forms the outermost cell layer that covers and protects the whole plant. In herbaceous plants, the dermal tissue system is a layer of cells called the *epidermis*. Woody plants initially have an epidermis but it is eventually replaced with the periderm or bark.

Reflect

1. *Life on Mars?* Your uncle Andrew who works for NASA is having dinner with your family. He becomes excited when you tell him that you are studying cells and cell structures at the moment. Uncle Andrew told you that NASA's Mars exploration rover, Opportunity, just sent back an image of a microscopic cell-like structure found on a rock near Endeavour's rim. He asks you to help him identify whether it is a cell and what type of cell it is. If it is a cell, is it similar to a prokaryotic cell? Or is it closer to a eukaryotic cell? What will you tell him?

2. *Building a plant cell.* Draw a functional plant cell by including all the essential components. Label these components and explain why the plant cell needs them.

3. *Green sea slug.* You read an article about green sea slug from *Science News*. The slug *Elysia chlorotica*, shaped like a leaf, has a reputation for stealing chloroplasts and even some genes from algae. Is this slug still an animal? Or is it now a plant? Explain.

4. *A woody leaf?* All plant organs, including leaves, are made up of three tissue systems. Although you have not studied any of these organs yet, you can imagine what they might be like. Think about a plant leaf. Can you describe where you will find these tissue systems and explain why. Explain why plants will never produce woody leaves.

References

Gould, S. B., Waller, R. F., & McFadden, G. I. (2008). Plastid evolution. *Annual Review of Plant Biology, 59*, 491–517.

Harris, H. (1999). *The birth of the cell.* New Haven, CT: Yale University Press.

Hook, R. (1665). *Micrographia, or some physiological descriptions of minute bodies made by magnifying glasses with observations and inquiries thereupon.*

Keeling, P. (2010). The endosymbiotic origin, diversification, and fate of plastids. *Philosophical Transactions of the Royal Society B: Biological Sciences, 365*, 729–748.

Margulis, L. (1971). *Symbiosis and evolution. Scientific American, 225*, 48–57.

Schnepf, E., & Brown, R. M., Jr. (1971). On relationships between endosymbiosis and the origin of plastids and mitochondria. In H. Ursprung & J. Reinert (Eds.), *Origin and development of cell organelles.* New York: Springer-Verlag

Chapter 4

Roots

Source: Cynthia McKenney

Learning Objectives

- Identify the functions of roots
- Compare and contrast specialized root types
- Describe the steps of root development
- Identify the regions of the root
- Compare and contrast gymnosperm, monocot, and eudicot roots
- Identify the importance of roots to humans
- Describe the different types of symbiotic relationships
- Compare and contrast endotrophic and ectotrophic mycorrhizae

As you hike down a wooded path, you frequently see the large gnarling roots of nearby trees crossing your trail. These bulky woody structures seem to snake across the ground going above and below the surface almost at a whim. Compare these tree roots with the tenacious roots of a dandelion or clinging roots of an orchid and you begin to discover the diversity and function of these amazing plant organs hidden under the soil.

ROOT FUNCTION

As we have discovered previously in Chapter 3, plants are comprised of a variety of distinct organs that are in turn composed of an amazing array of tissues. The numerous combinations of these different tissues and organs accentuate the diversity found in plants. When we first contemplate what types of organs combine to make a functioning plant, we generally envision the exotic flowers, gnarled branches, or changing display of leaves, which are all easily taken in with a glance. Even though the roots are usually below ground and more difficult to observe, they are a vital portion of the plant and serve three primary functions. Roots provide strong **anchorage** to provide firm yet flexible support to the plant while aiding in knitting the soil together, reducing soil erosion.

In contrast to the image of a tree anchored in the soil are **epiphytes**, which are plants attached to the bark or branches of host plants and thrive without the benefit of soil; their roots anchor the epiphyte to the tree or other supporting structures and their moisture and nutrients are provided by rainfall and decomposing leaves, which fall in their proximity. Bromeliads and orchids are excellent examples of epiphytes, frequently attached to the branches of trees in garden settings to provide a striking focal point. The bromeliad (*Neoregelias spp.*) and orchid (*Dendrobiuum spp.*) shown in Figure 4.1 are two examples of this unique class of plants, which function entirely without soil. Plants require a strong foundation to establish and maintain their growth. The extensive system of firmly attached roots supplies the strength necessary to support a plant during periods of high winds, floods, and landslides.

anchorage
Support provided by the roots of a plant so it does not fall over.

epiphyte
Plant that attaches by its roots to another plant for support and does not need to have its roots in the soil.

(a)   (b)

Source: Cynthia McKenney.

Figure 4.1a–b (a) Bromeliads and (b) orchids are epiphytes, plants that are attached by their roots to another plant and do not need roots in soil.

The second primary function of a plant's roots is the **absorption** of water and dissolved minerals from the soil, providing the necessary components for plant growth. The moisture and essential plant nutrients are transported by the xylem up from the roots and throughout the plant. We look in more depth at this amazing process of water absorption through the root system later in this chapter. A third primary function of roots is to provide **storage** for surplus carbohydrates produced by the plant. Molecules of glucose, simple carbohydrates, are the sugars produced during the process of photosynthesis, which are transported by the phloem from the leaves throughout the plant and down to the roots where glucose molecules are converted to starch molecules, complex carbohydrates, to be held in reserve until needed. When the plant has a deficit of these materials, starch molecules are being broken down into simple carbohydrates. They are then transported back out of the roots and throughout the plant, providing the necessary sugars to maintain plant health.

TYPES OF ROOT SYSTEMS

Given the diversity of the functions of a root, it is surprising there are just two primary categories of root systems. A **taproot** system forms a single primary root from which many smaller roots branch off, such as in a carrot (*Daucus carota*) (Figure 4.2a). Frequently, eudicots and gymnosperms create a taproot system initially. In this case the **radicle**, which is the embryonic root, extends in length due to cell division and cell elongation, creating a primary root with smaller secondary roots forming from it (Figure 4.2b). Pecan trees, dandelions, and turnips are all examples of tap-rooted species. These roots grow deeper in the soil, allowing for access to water resources not available to more shallow-rooted species. In contrast, **fibrous root systems** have several embryonic roots called **seminal roots**; monocots such as grass plants usually produce fibrous root systems (Figure 4.3a). These roots develop into multiple main roots that branch into a mass of finer roots, creating a dense root system capable of absorbing water and nutrients for plants (Figure 4.3b). Many species of plants have fibrous root systems including begonias (*Begonia*

absorption
Ability of plant roots to take up water and absorb nutrients from the soil.

storage
Capacity of a plant to be able to put back and store surplus carbohydrates and other materials for use by the plant later.

taproot
Type of root structure that consists of a single primary root and much smaller secondary roots.

radicle
Initial embryonic root which protrudes through the seed coat and initiates the root system.

fibrous roots
Several initial embryonic roots that develop into multiple primary roots, creating a dense root system.

seminal roots
Adventitious seed roots arising from the base of the plant which grow laterally and help uptake soil moisture for the new seedling.

Tap root

© straga/Shutterstock.com

(a)

Primary root (tap root)

Lateral roots

© Mikus, Jo./Shutterstock.com

(b)

Figure 4.2a–b   (a) Carrots are an example of an edible taproot. (b) This seedling is an example of a developing taproot system.

Seminal roots

Primary roots
(Fibrous)

© Bogdan Wankowicz/Shutterstock.com

Mature fibrous
root system

© Krzystof Kostrubiec/Shutterstock.com

(a)

(b)

Figure 4.3a–b (a) The developing fibrous root system of a bean and (b) a more mature fibrous root system of a pepper.

semperflorens) and corn (*Zea mays*). After a year or two, tap-rooted tree species frequently develop a more fibrous secondary root system such as the sycamore tree (*Platanus occidentalis*) in Figure 4.4. Fibrous-rooted species generally maximize water utilization from resources closer to the soil surface. It is interesting to note that despite the massive size of mature trees, the vast majority of their roots are found within the top 18 inches of the soil!

SPECIALIZED ROOTS

Roots for Stabilization

prop roots
Adventitious roots that arise from the stem and grow vertically down to the soil where they root and supply support for the plant.

There are numerous diverse types of root modifications allowing for the specialized functions necessary for plant growth. Additional stabilization of the plant is one of these functions and may be provided by above-ground **prop roots**. These easily recognized adventitious roots originate from the stem and

Source: Cynthia McKenney

Figure 4.4 Sycamore trees develop a more fibrous root system after several years of growth.

Prop roots

Aerial roots

Source: Cynthia McKenney.

© David Davis/Shutterstock.com

(a)   (b)

Figure 4.5a–b   (a) The prop roots of this screw pine (*Pandanus utilis*) provide extra support. (b) The aerial roots of this strangler fig (*Ficus spp.*) have overgrown this temple and are rooted to the ground.

grow down to the soil where they root firmly and provide further support for the plant. The screw pine (*Pandanus utilis*) in Figure 4.5a exhibits prop roots that support this top-heavy plant during high winds, while aiding in the absorption of water and nutrients. Similarly, **aerial roots** are a second class of adventitious above-ground roots providing support and aiding in absorption of nutrients and moisture. These roots arise from branches, leaves, or other plant structures, rather than from the stem, and grow down to the soil where they root down then function as below-ground roots. Areas with high humidity, such as tropical locations, are generally more conducive to the development of aerial roots than more arid locations. The strangler fig or banyan tree (*Ficus benghalensis*) is an excellent example of this type of root system and have been known to consume an entire city lot with their prolific production of aerial roots (Figure 4.5b). Yet another root modification is massive flared roots at the base of the stem known as **buttress roots**. The name buttress roots is a result of their resemblance to the flying buttresses build on medieval cathedrals to hold the walls up without having internal support beams. True to their name, these convoluted vertical wings on the trunk provide extra support and anchorage during heavy storms. Research has shown that buttressed trees develop sinker roots that arise from the region of the buttresses themselves and provide almost twice the anchorage strength when compared to trees without this modification (Crook, Ennos, & Banks, 1997). These root flares may be 10–12 feet tall before they yield back into the tree trunk (Figure 4.6).

Roots for Extra Storage

Not all root modifications are related to plant support such as those variations that provide enhanced carbohydrate storage. **Food storage roots** are modified root systems with an increased capacity for retaining large amounts of stored carbohydrates. There are many important root crops we utilize as food that fall into this area of adaptation. Beets are swollen tap roots harvested for food (Figure 4.7a); sweet potatoes and yams are swollen fibrous root systems valued for their stored starch (Figure 4.7b). These modifications are botanically classified as **tuberous roots**. Another important type of root adaptation

aerial roots
Adventitious roots arising from branches and other above ground structures that grow vertically down to the soil where they root and provide additional support for the plant.

buttress roots
Massive roots at the trunk base that produce large flares to supply increased support of a plant. They are frequently found in tropical locations.

food storage roots
Roots which serve to store excess carbohydrates until they are needed at a later time.

tuberous roots
Root modification in which a swollen fibrous root system stores food for later distribution throughout the plant.

Figure 4.6   Buttress roots are common on tropical trees and flare to the ground, providing greater support.

(a)   (b)

Figure 4.7a–b   (a) Beets are swollen tap roots and (b) sweet potatoes are swollen fibrous roots.

water storage roots
Type of root system found primarily on arid and semiarid plants that retains water, allowing the plant to withstand periods of drought.

is **water storage roots**, which are often found on arid and semiarid plants exposed to prolonged periods of drought. The development of these roots allows the plants to survive extended periods of dry weather. The Cucurbitaceae or pumpkin family is known for this phenomenon (Figure 4.8). One member of the family is the buffalo stink gourd (*Cucuribita foetidissima*), in which roots over 8 feet in length have been routinely harvested!

Roots for Extra Anchorage

contractile roots
Roots found on bulbs and corns that shrink or contract during adverse conditions, pulling the storage structure lower in the soil and preventing it from heaving out during freezing weather.

The final root modification we explore is **contractile roots**. These are fleshy roots found on bulbs or corms that contract or shrink during winter months or prolonged drought, resulting in the bulb or corm being pulled lower into the soil for more protection. This continual tug of the roots on the bulb or corm maintains the specialized root at the appropriate depth, thus preventing damage from it being heaved out of the soil during growth. Crocus (*Crocus sativa*) corms are an excellent example of contractile roots (Figure 4.9).

Figure 4.8   The pumpkin family is known for having water storage roots.

Figure 4.9   Crocus (*Crocus sativus*) corms have contractile roots that help the plant over winter by maintaining the appropriate depth.

ROOT STRUCTURES

Examining the morphological features of roots provide us the opportunity to understand their unique functions. Even though there are many types of roots, there is a core of shared traits that are common to all of the root structures. When a seed germinates, primary roots create the initial root system that is composed of four primary regions, including the root cap and the regions of cell division, elongation, and maturation (Figure 4.10). The **root cap** plays a pivotal role for the root and is found at the very tip (Figure 4.11). This area of parenchyma cells is produced by the apical meristem and has immediate contact with the soil. As the root cap pushes through the soil, the peripheral cells are wiped off. These cells form a substance called **mucigel**, which lubricates the root tip and eases its journey through the soil. Mucigel consists of polysaccharides, which help to provide a suitable environment for beneficial microbes, assists in mineral absorption, and prevents excessive root drying (Graham, Graham, & Wilcox, 2006). The root cap is also responsible for perceiving gravity and water gradients, which impact root growth. Several theories exist regarding the nature of **gravitropism**, which is the movement of a plant in response to gravity. One theory suggests the movement of starch grains in the amyloplasts of the roots causes the

root cap
The tip of the root that have direct contact with the soil and ease the small root into the soil; also perceive gravity and water gradients.

mucigel
Mixture of poly-saccharides that provides a lubricant that eases the root tip through the soil.

gravitropism
Movement of a plant in response to gravity.

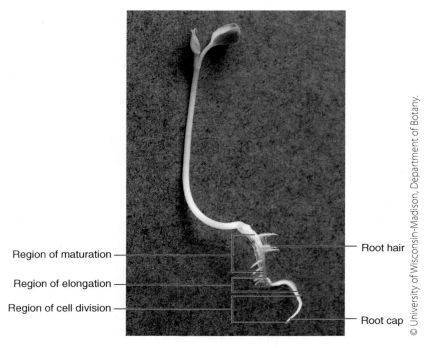

Region of maturation —

Region of elongation —

Region of cell division —

— Root hair

— Root cap

© University of Wisconsin-Madison, Department of Botany.

Figure 4.10   Radish (*Raphanus raphanistrum*) seedling demonstrating the regions of root growth.

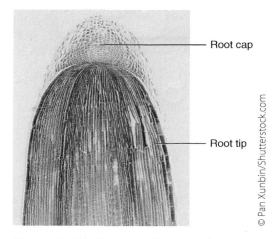

— Root cap

— Root tip

© Pan Xunbin/Shutterstock.com

Figure 4.11   Root tip showing the root cap and adjoining tissues.

region of cell division
see root apical meristem

root apical meristem (RAM)
Region synonymous with the region of cell division; area of rapidly dividing cells, resulting in root growth and production of the root cap.

procambium
Primary root tissue that gives rise to the xylem and phloem tissue.

ground meristem
Primary tissue that produces parenchyma cells, creating the cortex.

protoderm
Primary root tissue that gives rise to the epidermis of the plant.

gravity response in roots (Stern, 2006), whereas other experiments support the concept of protoplast cells in the root caps or specific proteins coating the plasma membrane exterior induce the root's response to gravity.

The Region of Cell Division

The root cap immediately precedes the **region of cell division** and is also known as the **root apical meristem** (Figure 4.10). In this region, cells rapidly grow and divide to produce the root cap and root growth (Figure 4.11). The root apical meristem creates the primary tissues of the root, which include the **procambium, ground meristem,** and **protoderm** (Figure 4.12). The procambium is in the central core of the root and gives rise to the

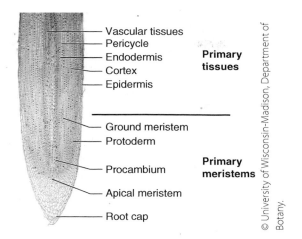

Vascular tissues
Pericycle
Endodermis
Cortex
Epidermis

Primary tissues

Ground meristem
Protoderm

Procambium

Apical meristem

Root cap

Primary meristems

© University of Wisconsin-Madison, Department of Botany.

Figure 4.12   Regions of the primary meristems and primary tissues of a pumpkin root.

primary phloem and primary xylem. The ground meristem is outside of the procambium and next to the outer protoderm and gives rise to the cortex. The outer surface of the root system is protected by a layer of epidermal tissues produced by the protoderm (Graham et al., 2006).

The Region of Cell Elongation

The **region of cell elongation**, located past the region of cell division, has the function for which it is named (Figure 4.10). In this region, the newly divided cells take on water, causing them to swell and expand and increasing the cells to their maximum size, thus increasing the length of the root system. The primary growth here is in length; however, there can be a slight increase in girth or width.

The Region of Maturation

As the cells stretch with absorbed water, they also begin to **differentiate** or mature and specialize. This is the **region of maturation** and it is adjacent to the region of cell elongation (Figure 4.10). As the cells mature, they differentiate into several different tissues including the epidermis, cortex, and vascular tissues. This region is the source of **root hairs**, which are small single-celled extensions of the epidermal cells and are the primary site for the absorption of water and minerals. Large numbers of fine root hairs develop, which vastly increases the surface area of water-absorbing roots. These delicate structures last only a few days, but they provide increased nutrient and water absorption for the plant. This massive absorptive surface created by the root hairs is an extremely valuable region of the root system as these structures perform the important function of selectively absorbing useful minerals while excluding harmful ones. Root hairs are vast in number; however, they generally last for only a few days. These structures are easily damaged when a plant is transplanted or by the abrasive nature of the soil.

Microscopic examination has determined there are specialized tissues found above the region of maturation. A cross section of a root from this area has led to the identification of the structure and function of these tissues. The protoderm cells differentiate to form the epidermis, which is usually one cell layer thick in roots and provides the protective external covering. The ground

region of elongation
Region of root system where newly divided cells expand in size; thus increasing the length of the root system.

differentiation
Process when cells take on different shape and form, allowing them to perform different functions.

region of cell maturity
Portion of the root system where cells differentiate and begin to specialize their tasks.

root hairs
Simple extensions from individual epidermal cells in roots; responsible for increasing surface area of the root epidermis for absorption.

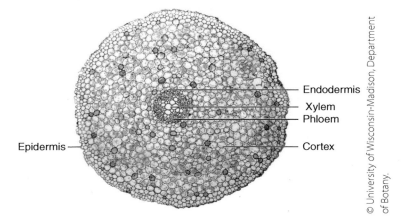

Epidermis

Endodermis
Xylem
Phloem
Cortex

© University of Wisconsin-Madison, Department of Botany.

Figure 4.13   Root cross section of a *Ranunculus* showing the vascular cylinder containing the xylem and phloem, the endodermis and the cortex.

cortex
Root tissue between the epidermis and vascular tissue; responsible for storage; allow oxygen to diffuse through.

plasmodesmata
(singular: *plasmodesma*) Tiny channels through the cell walls that connect the plasma membrane and cytoplasm of adjacent plant cells; responsible for cell communication.

endodermis
Innermost layer of cells in the cortex that regulate mineral absorption.

Casparian strip
Layer of waterproof tissue containing suberin, making it impermeable to the movement of water to and from the vascular cylinder.

meristem cells differentiate to form the **cortex** (Figure 4.13). The cortex is the plant tissue between the epidermis and vascular tissues and is composed primarily of storage parenchyma. The cortex generally stores starch and does not contain chloroplasts. However, this tissue does have intercellular spaces, which allow the necessary oxygen to diffuse through the roots. Without this oxygen, the root would not be able to survive. Once-mineral laden water is absorbed into the root hair cytoplasm, it may also be distributed from cell to cell through the cortex via small intercellular pores called **plasmodesmata** (Figure 4.14).

The interior layer of the cortex has a single wall of cells called the **endodermis**. This tissue helps to regulate the accumulation of minerals in the roots. The endodermis has a **Casparian strip**, which is a region of tightly packed cells impregnated with a waxy material called *suberin*. The presence of suberin results in this band of cells becoming impermeable to the movement of water, thus preventing its loss to the soil. The Casparian strip also serves to prevent sodium and other salts from invading the vascular cylinder during periods of stress (Karahara, Ikeda, Kondo, & Uetake, 2004). With the presence of the Casparian strip, all materials passing to and from the vascular tissues across the selective plasma membranes or through the connecting plasmodesmata are regulated, enhancing plant growth.

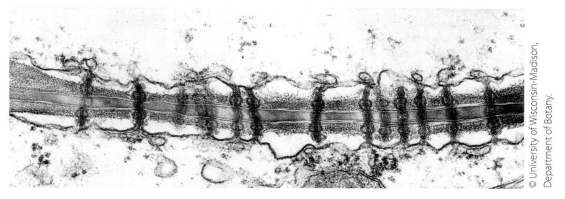

© University of Wisconsin-Madison, Department of Botany.

Figure 4.14   Water and minerals are able to move from cell to cell via small openings called *plasmodesmata*.

The procambium differentiates to form the vascular tissues, which are surrounded by a thin layer of parenchyma cells, called the **pericycle**. In angiosperms, the pericycle gives rise to lateral or secondary roots and cork cambium. Vascular cambium is responsible for girth increase in a plant and the tissues cambium produces are **secondary tissues**. The pericycle encloses a core of primary xylem and primary phloem known as the **vascular cylinder**. Recall from Chapter 3 that the primary xylem is the water-conducting tissue generally in the center of the vascular cylinder that forms ridge-like extensions. The function of the xylem is to carry water and dissolved minerals up from the roots throughout the plant. In contrast, the primary phloem is the vascular tissue responsible for transporting the food for the plant and is found in between the ridges of the primary xylem within the vascular cylinder (Raven, Evert, & Eichhorn, 2007).

pericycle
Layer of cells that surrounds the vascular tissues in the root gives rise to secondary roots.

secondary tissues
Produced by the vascular cambium of a plant and result in girth growth.

vascular cylinder
Composed of primary xylem and phloem; located in the center of the root.

BOX 4.1 The Importance of Roots to Humans

Before beginning this chapter you probably did not think much about the role roots play in your life. For just a moment, let's focus on the impact roots have on daily living. Food is generally the first thought that comes to mind. As mentioned earlier in the chapter, there are numerous food crops that are modified root structures, such as sweet potatoes, rutabagas, turnips, carrots, radishes, ginger, and beets. These foods are generally a good source of fiber as well as a storage site for carbohydrates. One root crop that is commercially important is the sugar beet (Figure Box 4.1a). This variety of beet has been bred to produce high concentrations of sugar, which is then extracted and processed to create granulated sugar. When purchasing your next bag of sugar, look to see if the sugar originates from beets or sugar cane. You might be surprised. Worldwide cassava (Figure Box 4.1b) is another important root crop that functions as one of the main sources of carbohydrates in tropical diets. It is important to note this root must be properly handled before consumption in order to remove a poisonous latex material inside. After appropriate processing to remove this juice, cassava becomes the source of tapioca.

© JP Chretien/Shutterstock.com

Figure Box 4.1a Sugar beets (*Beta vulgaris*) are important source of sugar.

▶▶▶

© My Life Graphic/Shutterstock.com

Figure Box 4.1b   Cassava, (*Manihot esculenta*), another important root crop for food.

Historically, many groups of peoples used roots for medicinal purposes. For example, Native Americans dug the roots of *Echinacea* (purple cone-flower; Figure Box 4.1c) and used a decoction to help fight off illness; today, tincture of *Echinacea* is readily available in drug stores to boost the immune system. *Oenothera biennis* (evening primrose; Figure Box 4.1d) was also a historically used herbal root for healing. The roots were dug and boiled and then consumed as German rampion. Today, the oil pressed from the seeds of *O. biennis* is sold in capsules as neutraceuticals to address a variety of skin and health disorders. *Panax quinquefolium* (Ginseng root; Figure Box 4.1e) is considered an all-purpose health plant in China, Japan and Korea where the roots are dug and sold in health food stores. This particular neutraceutical is attributed to having a calming effect on people.

© slowfish/Shutterstock.com

(c)

© Imageman/Shutterstock.com

(d)

Figure Box 4.1c and d   The roots of purple cone flower (*Echinacea purpurea*) and evening primrose (*Oenothera spp.*) are both harvested for use as nutraceuticals.

Figure Box 4.1e   Ginseng root (*Panax spp.*) is used as an all-purpose health plant.

Native Americans had many other root crops they depended on for uses other than food and medicine. Some roots were dug and used with a mordant as natural dyes. Other roots were used as a purgative in religious ceremonies. *Yucca spp.* (Yucca root; Figure Box 4.1f) was harvested and the juice extracted from the root to make soap. It is interesting to note yucca remains a constituent in botanically based shampoos and soaps today.

Figure Box 4.1f   The roots of yucca plant (*Yucca spp.*) are used for shampoo.

Roots are a primary defense against the erosion of soil that washes into our streams each year. Soil on barren slopes erodes during heavy storms. This is frequently seen on steep slopes and roadsides where the loss of topsoil confounds efforts to stabilize the area. Grass plants provide an excellent method of preventing erosion as the roots knit together and help bind the soil. Similarly, mature trees help hold soil in place with their extensive root systems. In the Western states, cover crops are used extensively to prevent wind erosion of topsoil, which leads to reduced agricultural productivity.

MONOCOT AND EUDICOT ROOTS

stele
Vascular cylinder of the root that contains the xylem, phloem, and, in monocots, the pith.

There are several differences in eudicot and monocot roots which may be used to categorize them. As mentioned previously, the vascular cylinder is located within the cortex of the root, which is also referred to as the **stele**. As discussed previously, this cylinder is surrounded by the pericycle, the endodermis, the cortex, and a protective epidermal layer. The stele itself is composed of the vascular tissues xylem and phloem. If the plant is a monocot, the stele will also include a central core of parenchyma called *pith* (Figure 4.15a). The xylem and phloem of the monocots is then located around the pith in alternating patches forming a ring. Contrastingly, when a transverse section of a eudicot root is observed, it is characterized by a cross-shaped central stele (Figure 4.15b). The lobes or arms of a eudicot stele have xylem lining the interior and phloem on their exterior. The phloem also extends between the lobes. The absence of pith in the center is notable.

ASSOCIATIONS AND SYMBIOTIC RELATIONSHIPS

symbiotic relationship
Close biological relationship between two different species.

mutualism
Symbiotic relationship in which both organisms benefit and neither are harmed.

commensalism
Relationship of two species where one benefits and the other is not harmed and does not benefit.

Frequently, two different species live in a close biological association that is called a **symbiotic relationship**. Symbiosis is important to many plants and may be observed in several different forms. **Mutualism** is a symbiotic relationship in which both organisms benefit from their association. A redbud tree (*Cercis canadensis*) is an example of this type of relationship as nitrogen-fixing bacteria live on the root system of the plant and help provide a useable source of nitrogen for the tree's growth. The bacteria themselves benefit from the rich source of nutrients diverted from the roots (Figure 4.16). A second example of mutualism is the pollination of a flower by a honeybee. The plant benefits from the association by the pollination of the flowers while the bee profits from the process by using the pollen as a food source.

 Commensalism is a second type of symbiotic relationship in which one organism benefits from the relationship and the other is not harmed or

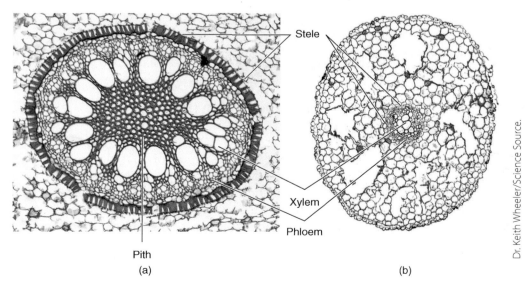

Figure 4.15a–b Transverse section of (a) a monocot and (b) a eudicot root.

Figure 4.16   This Texas redbud (*Cercis canadensis var. texensis*) is an example of a mutualistic relationship between it and its associated mycorrhizae (soil fungi).

Source: Cynthia McKenney.

negatively affected. A common example of a commensal relationship occurs when an animal brushes up next to a plant and a seed pod attaches to the animal's coat and is transported some distance before it detaches. The plant in this scenario benefits from the animal providing transport for seed dispersal while the animal is not harmed by the presence of the seed pod attached to its coat. The relationship between Texas bluebonnets (*Lupinus texensus*) and Indian paintbrush (*Castilleja indivisa*) is another example of commensalism. The Indian paintbrush lives on the roots of the Texas bluebonnet but does not harm the bluebonnet significantly (Figure 4.17a). This relationship is uniquely known as hemi-parasitism because the Indian paintbrush is capable of providing its own water and nutrients as well as attaching to a host plant via haustoria (Luna, 2005). Thus, hemi-parasitic plants can be both autonomous and live independent of a host plant or they can be parasitic.

The last symbiotic relationship is **parasitism** where one of the organisms benefits from the association of the two plants while the other organism is

parasitism
Symbiotic relationship in which one organism is harmed and the other organism benefits.

Bluebonnets

Indian
Paintbrush

Mistletoe

Source: Cynthia McKenney.

(a)   (b)

Figure 4.17a–b   (a) Bluebonnets (*Lupinus texensis*) and Indian Paintbrush (*Castilleja spp.*) are an example of a commensal relationship. (b) Mistletoe (*Phoradendron tomentosum*) is an example of a parasitic relationship.

harmed. The growth of mistletoe in trees is one of the classic examples of this type of symbiotic relationship. The mistletoe benefits from the nutrition and support provided by the tree (Figure 4.17b); however, the tree itself is damaged by the presence of the mistletoe's haustoria, which infest the wood of the tree and divert the food and nutrients toward itself. Heavy infestations of mistletoe can harm the growth of the host tree.

Association with Fungi

A very important type of symbiotic relationship exists between fungi and plants. **Mycorrhizae** are soil fungi that flourish in a mutualistic association with plant roots. Over evolutionary time, these associations have become so strong the organisms have become dependent on each other. In these mutualistic relationships, the plant benefits from the fungal mycelium functioning as an increased root system, providing more water, minerals, and fixed nitrogen for the plant to access while the fungi receive photosynthetic products from the plant as a food source. Mycorrhizae also assist in binding soils together, which reduces soil erosion and water loss. Mycorrhizal associations have been identified in the vast majority of plants, leading to the discovery that specific plants associate with different types of mycorrhizae. The positive impact on the growth of these plants has led to an increased usage of mycorrhizal inoculation in plant establishment.

Two common types of mycorrhizae are endomycorrhizae and ectomycorrhizae. The hyphae of endomycorrhizal fungi grow into the host plant's roots between the cells. The hyphae secrete enzymes allowing entrance into the cell walls of the host plant without piercing the cell membranes. The hyphae then divide, creating highly branched mats called **arbuscules**, which increase the functional surface area of the root system, facilitating the movement of water and minerals from the soil into the plant (Harrison, 1999). In return, the endomycorrhizal fungi benefits from photosynthate from the plant, providing a rich source of nutrients. Endomycorrhizae are known as **arbuscular mycorrhizae**.

A second type of mycorrhizae is ectomycorrhizae. **Ectomycorrhizae** differ from the endomycorrhizae in that fungi cover the outside root tissue but do not actually invade it. The roots and mycorrhizal fungi have a close association, allowing for the exchange of nutrients with each other, but the plant cells are not penetrated. It is of note that some tree species will not grow in the absence of their specific mycorrhizal partner, thus fostering an area of further research.

This mutualistic relationship between fungi and plant roots enhance plant tolerance to stresses and facilitate water and nutrient uptake. Mycorrhizae are found in all plant groups and may have been a key plant innovation for making the transition to land-based living. In Chapter 21, we discuss the challenges plants face living on dry land and their innovations for land-based living.

Association with Nitrogen-Fixing Bacteria

Another symbiotic relationship occurs between roots and **nitrogen-fixing bacteria** where these microbes convert nitrogen from the atmosphere into a plant-available form. This process is referred to as fixing nitrogen and the

mycorrhizae
Common soil fungi that form mutualistic relationships with the roots of plants.

arbuscules
Highly branched fungal mats of endomycorrhizal fungi formed within the tissues of a plant's roots.

arbuscular mycorrhizae
Endomycorrhizae that grow into the root cells of a plant and create mats of highly branched fungal hyphae. Hyphae of AM fungi grow in the spaces between root cell walls and plasma membranes, often forming highly branched and bushy arbuscules (from the word *arbor* meaning "tree").

ectomycorrhizae
Fungi that grow on the external surface of a plant's root in a mutualistic relationship.

nitrogen-fixing bacteria
Microbes that grow within the roots of plants and are capable of fixing nitrogen from the atmosphere.

nitrogen fixed in this manner is needed by the plant to create proteins and amino acids for growth. The close relationship between this bacteria and the host enhances the plant's growth. Of particular interest with this type of relationship are legumes and nitrogen-fixing bacteria. Legumes are an important source of protein for both humans and livestock. These plants secrete a chemical that the microbes respond to by releasing their own chemical materials into the soil. The root hairs then wrap around the bacteria and form nodules composed of infected and uninfected cells. The bacteria fix the nitrogen from the air and a portion of it is transported from the root nodules through the vascular system of the roots up through the plant. Just as with the mycorrhizae, there are specific bacteria associated with certain plants, precluding the one-bacteria-works-for-all concept of inoculation. In Chapter 13, we discuss how these symbioltic microbes provide nutrients for plants.

Key Terms

anchorage	root cap	Casparian strip
epiphytes	mucigel	pericycle
absorption	gravitropism	secondary tissues
storage	region of cell division	vascular cylinder
taproot	root apical meristem	stele
radicle	procambium	symbiotic relationship
fibrous root systems	ground meristem	mutualism
seminal roots	protoderm	commensalism
prop roots	region of cell elongation	parasitism
aerial roots	differentiate	mycorrhizae
buttress root	region of maturation	endomycorrhizae
food storage roots	root hairs	arbuscules
tuberous roots	cortex	arbuscular mycorrhizae
water storage roots	plasmodesmata	ectomycorrhizae
contractile roots	endodermis	nitrogen-fixing bacteria

Summary

- Roots provide anchorage, absorption and storage.
- Root systems may be divided into taproot systems and fibrous root systems.
- There are many types of diverse root modifications:
  - Prop roots, which provide additional support.
  - Aerial roots, which arise from branches and root to the ground, providing stability and nutrients for the plant.
  - Buttress roots, which are massive flare roots on trees that help prevent damage during storms.
  - Food storage roots such as carrots and sweet potatoes, which store photosynthate until needed at another time and provide a valuable source of food.
  - Contractile roots help to gently pull the bulb or corm lower in the soil, providing more protection from cold weather.
- The morphological features of a root include several structures:
  - The root cap, which is the very tip of the root and explores the soil as it moves forward.
  - The root apical meristem, which is the region of cell division and gives rise to the primary tissues of the root.

○ The region of cell elongation, which is where the newly divided root cells expand resulting in an increase in length.

○ The region of maturation is where root cells differentiate into different tissues such as the epidermis, cortex, and vascular tissues.

○ Root hairs are extensions of the epidermal cells and are the primary site for absorption of water and minerals.

○ Plasmodesmata provide access from the cytoplasm of one cell to another.

○ The Casparian strip is impregnated with a waxy material and prevents the movement of water.

○ The vascular cylinder houses the xylem and phloem.

• Roots are important to humans for food, medicinal purposes, a source of neutraceuticals, constituents of dyes and soaps, and as a defense against erosion.

• Symbiotic relationships include mutualism, commensalism, and parasitism.

• Mycorrhizal fungi have a mutualistic association with roots.

• Endomycorrhizal fungi enter the roots of a plant; ectomycorrhizal fungi coat the external surface of the plant.

• Nitrogen-fixing bacteria help to provide usable nitrogen for plant growth.

Reflect

1. ***Getting to the root of the problem.*** You have decided to open your own nursery production business specializing in native trees. You vaguely remember from your introductory plant science class there are good and bad fungi in the soil that can help or hurt your fledgling plants. How would you go about identifying which types of fungi are good for your plants and which are not? Would you make any special preparations in the soil or on the roots of your plants before you transplanted them into your production fields?

2. ***How unique are roots?*** Describe the major root functions and discuss some unique adaptions that allow roots to perform other functions.

3. ***Different zones, different functions.*** Identify the four major regions of roots and describe the importance of each region.

4. ***Monocots and eudicots.*** Explain how you can tell the two major plant groups apart based on their root structure and form.

5. ***Roots and their uses.*** Describe how plant roots are used by humans.

References

Crook, M. J., Ennos, A. R., & Banks, J.R. (1997). The function of buttress roots: A comparative study of the anchorage systems of buttressed (*Aglaia and Nephelium ramboutan* species) and non-buttressed (*Mallotus wrayi*) tropical trees. *Journal of Experimental Botany, 48*(9), 1703–1716.

Graham, L., Graham, H., & Wilcox, L. (2006). *Plant biology* (2nd ed.). Upper Saddle River, NJ: Pearson Education.

Harrison, M. (1999). Molecular and cellular aspects of the arbuscular mycorrhizal symbiosis. *Annual Review of Plant Physiology and Plant Molecular Biology, 50*(1), 361–389.

Karahara, I., Ikeda, A., Kondo, T., & Uetake, Y. (2004). Development of the Casparian strip in primary roots of maize under salt stress. *Planta, 219*(1), 41–47.

Luna, T. (2005). Propagation protocol for Indian Paintbrush Castilleja *species. Native Plant Journal, 6*(1), 62–68.

Raven, P., Evert, R. F., & Eichhorn, S. E. (2004). *Biology of plants* (7th ed.) New York: W. H. Freeman.

Stern, K. R. (2006). *Plant biology* (10th ed.). New York: McGraw-Hill.

Chapter 5

Stems

Source: Cynthia McKenney.

Learning Objectives

- Identify the functions of shoots and stems
- Explain the structural differences in stem development
- Describe the methods of stem classification
- Compare and contrast stem components
- Describe the steps of shoot development
- Clarify the importance of phyllotaxy
- Compare and contrast prickles, spines, and thorns
- Identify the importance of bark
- Describe the different types of specialized stems
- Explain how plants respond to the environment

The stems of plants vary almost as much as people do. There are herbaceous stems, woody stems, monocot and eudicot stems, stems specialized for food storage and other stems are underground and don't even look similar to their counterparts. Don't be fooled by these structures as they are not just empty halls connecting leaves to roots. In this chapter we explore the morphology and function of this diverse set of structures.

WHAT IS A STEM?

When we think of a stem, many times we envision a long, green plant part below a beautiful rose that has leaves and thorns scattered up and down its length. A stem is defined as the primary axis of a plant and it is a primary component of the **shoot**, which is the entire above ground portion of a young plant. Usually a stem attaches directly to the roots, is located above ground, and may be covered with leaves, flowers, fruits, and buds. As we continue thinking about the primary axis of a plant, we might think about the trunk of a tree and wonder if that is also a stem. Yes, it is also classified as a stem, along with a tulip bulb, the rhizome of an iris, or the fleshy pad of a cactus. Let's look back again at the definition of a stem and the function it serves to help determine how it is possible for such a diversity of plant parts to all be considered stems.

shoot
Above ground portion of a plant including the leaves and stems.

STEM FUNCTIONS

Stems have an extremely important function for plants as they provide the vascular system necessary to conduct water and dissolved minerals from the roots and the photosynthate from the leaves throughout the plant for utilization and storage (see Chapter 4). It is daunting to think of the great distance these materials are transported in giant coastal redwood trees (*Sequoia sempervirens*), which are thought to be the tallest trees in the world (Figure 5.1a); however, transportation of water and nutrients is just as important to lower-stature plants like these caladiums (*Caladium hortulanum*) to ensure they have sufficient water to sustain life (Figure 5.1b).

(a)   (b)

Figure 5.1a–b   (a) The vascular system of a redwood tree carries water and nutrients a great distance in comparison to (b) a lower-stature plant like the caladium.

In addition to the functions of housing the vascular system, stems support their attached leaves at an angle, allowing for the best light interception to facilitate the process of photosynthesis (Figure 5.2a.) Some stems such as the trunk of a date palm (*Phoenix dactylifera*) lift the foliage high, providing for greater access to sunlight and less chance of herbivory whereas other flexible stems, like those of groundcovers, support their leaves closer to the ground, reducing the plant's investment of resources in trunk development. The stem modification found in English ivy (*Hedera helix*) provides plant height and access to light by facilitating the rapidly growing shoot to climb and fasten on to another plant or structure (Figures 5.2b).

Just as roots serve to store carbohydrates, stems themselves also provide **storage** for water and the products of photosynthesis (see Chapter 4). The trunk of a giant sequoia tree (*Sequoiadendron giganteum*) stores photosynthate along with huge amounts of water, which has aided in the preservation of these ancient trees, currently estimated to be over 4,000 years old, during periods of fire (Figure 5.3a). The unusually shaped Grandidieri's baobab tree (*Adansonia grandidieri*) trunk has both water and food stored within, assisting with plant survival during periods of water stress (Figures 5.3b).

Of great importance to plant growth are the tissues produced throughout the life of the plant arising from the stem tissues. As we saw previously in Chapter 4, the dermal tissues cover and protect the whole plant. In an herbaceous plant, the dermal tissue is the epidermis and it functions to provide protection and reduce water loss; however, a woody plant's cells divide and mature, developing into the periderm or bark. The strength of a shoot or stem of a woody eudicot or gymnosperm is increased by the presence of the vascular and cork cambiums, which provide secondary growth for the plant resulting in an increase in girth not seen in monocots. Monocots such as palms seem to have a woody trunk. They do not have the cork cambium to create bark. They typically have an apical meristem, which is their only growing point and is enclosed in stiff overlapping leaf bases (Hodel, 2009).

storage
Capacity of a plant to be able to put back and store surplus carbohydrates and other materials for use by the plant later.

(a)   (b)

Source: Cynthia McKenney.

Figure 5.2a–b   (a) This palm trunk supplies support to the attached fronds so they are exposed to sunlight. (b) English ivy has stem modifications that help the plant cling to another structure for support.

Source: Cynthia McKenney

© Nazzu/Shutterstock.com

(a)   (b)

Figure 5.3a–b   (a) This giant sequoia has the ability to store water in its tissues, which has helped it to live through many many fires. (b) The trunk of the baobab tree stores photosynthates for periods of stress.

The "woody trunk" of palms is actually an accumulation of these overlapping leaf bases. Thus, secondary growth generally does not occur.

STRUCTURAL DIFFERENCES IN STEM DEVELOPMENT

Primary Growth

As the shoots mature, new tissues are created through **primary growth** resulting in an overall increase in height or length. The apical meristem at the tip of the shoot assists in the production of three primary meristemic tissues (Rost, Barbour, Stocking, & Murphy, 2006). These meristemic tissues differentiate into primary tissues, including the **protoderm**, evolve into the **epidermis**; in woody plants the epidermis develops into periderm or bark. The second primary meristemic tissue, called **ground meristem**, forms the **cortex** and **pith** with parenchyma cells for storage. When parenchyma cells are found in the central portion of the plant and surrounded by vascular tissues, they form the pith. The final of the three primary meristemic tissues is the **procambium**, which ultimately gives rise to the xylem and phloem, producing the vascular system of the plant.

Organization of Vascular Tissues

In eudicots, the vascular bundles containing the xylem and phloem are organized, creating a vascular cylinder; in contrast, the monocots develop their vascular bundles scattered throughout the ground meristem so no vascular cylinder is created and no pith is found. A cross section of the stems of a monocot and a eudicot are helpful in visualizing the difference in growth patterns of these two distinct plant groups. Monocots are characterized by the presence of the epidermis and the scattered vascular bundles within the ground meristem (Figure 5.4a). In contrast, young woody eudicot and herbaceous eudicot stems are unique given their vascular bundles

primary growth
Growth in height or length due to apical meristems.

protoderm
Primary tissue that gives rise to the epidermis of the plant.

epidermis
Made up of mostly unspecialized cells forming a single layer covering the surface of herbaceous plants for protection.

ground meristem
Primary tissue that produces parenchyma cells, creating the cortex.

cortex
Parenchyma tissues between the epidermis and vascular tissue; responsible for storage.

pith
Ground tissue comprised of parenchyma and located in the center of roots and stems.

procambium
Primary root tissue that gives rise to the xylem and phloem tissue.

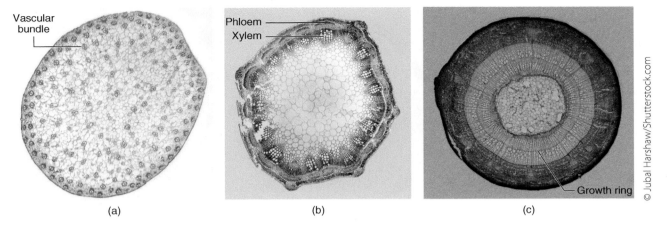

Figure 5.4a–c   (a) The cross section of a monocot stem discloses the orientation of scattered vascular bundles whereas (b) the herbaceous eudicot stem has the vascular bundles distinctly separate but not scattered. (c) The mature woody eudicot has the vascular bundles arranged in rings, allowing for girth growth.

are not merged together but are quite distinct (Figure 5.4b). As the woody eudicot matures, the stem structure evolves resulting in the presence of an epidermal layer of bark, the cork layer that may or may not be present, the phloem and xylem layers separated by a thin cambial layer, and the central pith (Figure 5.4c). The vascular bundles of the mature woody eudicot stem are oriented in a vascular cylinder, creating the appearance of rings and, in some species, may be dissected by xylem rays called **medullary rays**.

Secondary Growth of Eudicots and Gymnosperms

Frequently woody gymnosperms and eudicots have **secondary growth** resulting in an overall increase in girth. This radial growth creates distinctive rings as layers of secondary xylem are deposited directly next to one another, resulting in the "tree rings" frequently used to determine the age of a tree. The phloem, which conducts the photosynthate from the leaves throughout the plant, is oriented toward the epidermis; the xylem, tasked with transporting water and nutrients from the roots up through the plant, is oriented toward the center of the plant. The xylem portion of woody plants is commonly referred to as wood. The living xylem is known as **sapwood** and it forms the new layers (Figure 5.5a). Xylem may also create xylem rays composed of living tissue in the sapwood, which may appear to transverse the annual rings, thus allowing for the horizontal movement of water. Phloem rays, on the other hand, allow horizontal movement of photosynthate. With age, the sapwood becomes inactive and is more highly compressed. As this occurs, the tissue becomes classified as **heartwood**, which functions to supply support but does not contribute to available water or nutrients for the plant. With time, only the outer cambial layers of the plant are active in the trunk. If the heartwood rots out leaving only the cambial tissues present and functioning, a hollow tree is formed (Figure 5.5b). In some woody species there is an additional cambium known as cork cambium, which, along with the phloem, creates phellogen adding to the strength of the bark.

medullary rays
Rays of xylem which cut across the growth rings of a tree at a perpendicular angel to allow radial transportation of water.

secondary growth
Growth in the girth of a plant as a result of the cambium.

sapwood
Light-colored wood in a tree trunk that is the youngest tissue and conducts water and minerals.

heartwood
Wood at the center of a tree stem that is darkened and has become inactive.

Sapwood

Heartwood

(a)

(b)

Figure 5.5a–b (a) The darkened heartwood and lighter-colored sapwood are visible in this cross section of a tree trunk. (b) The internal heartwood has rotted out, leaving the living cambial tissues as a remnant of the original tree.

Monocots Lack Secondary Growth

In general, monocots do not produce secondary growth found in eudicots. Some monocots such as palms create a form of bark but do not have the same type of girth growth as eudicots and gymnosperms. Their "bark" is actually layers of leaf bases left behind when their leaves die, thus forming a pseudobark. It should be remembered their primary growth occurs at the apical meristem, which directs the development of stems and leaves. Thus, if you cut off the top of a monocot, it will not result in increased branching as you would find in a gymnosperm eudicot, but will kill the plant if the cut is below the apical meristem, e.g. the leaf base of palm fronds.

STEM CLASSIFICATION

Stems may be classified by several methods including whether the stem has been strengthened by lignification. A **woody plant** is one in which dense hard tissues composed of supportive tissues such as collenchyma, sclerenchyma, and phelloderm, along with the compound lignin, provide the ability of the plant to support itself. When thinking about woody stems, trees and shrubs easily come to mind such as the cherry tree (*Prunus serrulata*) in Figure 5.6a. These plants are generally winter hardy and do not die to the ground; however, when a woody stem is not able to support itself, such as the wisteria (*Wisteria sinensis*), it is termed a **liana** (Figure 5.6b). Inspection of a mature grape plant shows the presence of lignin and the ability to support itself from a trunk; therefore, it is appropriate to call the plants in a vineyard "grape lianas" rather than grapevines.

Herbaceous plants are composed of softer nonwoody tissues and stems also capable of supporting themselves. These plants may or may not be winter hardy and may in fact die to the ground to return the next year. Windflowers (*Anemone coronaria*) are an example of an herbaceous plant whose stem dies down in the winter and returns from the roots the following year (Figure 5.7a). Given this plant returns year after year, it

woody plant
A plant in which supportive tissues and lignin combine to provide a hard stem capable of supporting itself.

liana
Wood stem that is not able to support itself and resembles a vine.

herbaceous plants
A plant which is not woody and freezes down during the winter; perennial will return the following spring; annuals will not return.

Source: Cynthia McKenney.

(a) (b)

Figure 5.6a–b (a) A cherry tree with the ability to support itself compared to (b) a wisteria liana, which is not able to support itself.

Source: Cynthia McKenney.

(a) (b)

Figure 5.7a–b (a) Windflowers are herbaceous perennials (b) while buzzy Lizzy plants are herbaceous annuals.

would be considered an herbaceous perennial. In contrast, Buzzy Lizzy (*Impatiens wallerana*) is an example of an herbaceous plant whose stems freeze to the ground in the winter and does not return from the roots and is therefore considered to be an herbaceous annual (Figure 5.7b). When plants with an herbaceous stem are not able to be self-supportive, it is classified as a **vine**. Clematis (*Clematis spp.*) and morning glory (*Ipomoea purpurea*) are excellent examples of vines requiring support from another plant or structure given their soft stems. These plants generally entwine themselves around another plant, fence, or other structure to provide the height needed to access sufficient sunlight for photosynthesis (Figure 5.8a and b).

vine
Herbaceous stem that is not woody and cannot support itself.

(a)

Source: Cynthia McKenney.

(b)

© Charlene Bayerle/Shutterstock.com

Figure 5.8a–b (a) Clematis and (b) morning glory plants are excellent examples of herbaceous vines.

BRANCH AND STEM COMPONENTS

bud scales
Leafy structure protecting young buds or shoots.

lateral buds
see axillary buds.

axillary buds
Buds originating in the region between the stem and a petiole.

node
Region on a stem where leaves attach.

internode
Section of stem between two nodes.

As mentioned above, stems have a variety of characteristics that make them distinct; however, there are several components that are usually present. A woody twig provides a simplistic view of an average stem (Figure 5.9a and b). If we start at the tip of the branch, there is a terminal bud, which is a dormant apical meristem that may be covered by protective **bud scales**. When the terminal bud becomes active and opens, it provides the primary growing point exerting apical dominance over the branch and thus controls further growth. In addition to the terminal bud, there are **lateral buds**, also called **axillary buds**, due to their location in the axils of the attachments of the leaves to the stems. Lateral or axillary buds give rise to additional stems, flowers, and leaves. Leaves and flowers attach to stems in a region characterized by rapidly growing tissues called **nodes**. The area between two nodes is called an **internode** and may vary in length considerably. When new growth

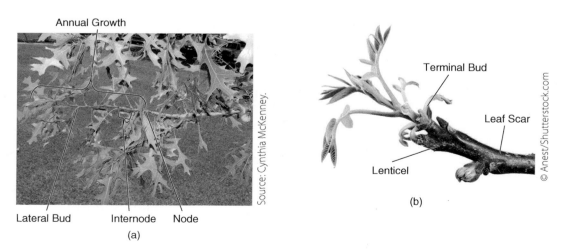

Annual Growth

Source: Cynthia McKenney.

Lateral Bud   Internode   Node

(a)

Terminal Bud

Leaf Scar

Lenticel

© Anest/Shutterstock.com

(b)

Figure 5.9a–b Woody twigs are a simple way to identify many of the parts of a stem.

arises from the terminal bud, the location where the terminal bud was located is scarred. The distance between two terminal bud scars is the amount of growth the plant has experienced during the growing season and is useful in determining the overall health of the plant. Additional scars on the stem are located where leaves formerly attached to the branch and, upon close examination, the size and shape of these **leaf scars** aid in plant identification as they are species specific. Leaf scars are composed of several small dots called **bundle scars** where the vascular tissues were previously attached to the leaf and are also characteristic by species. **Lenticels** are the final component of the stem we explore and they are areas where the outer cells of the stem are more loosely arranged, allowing for gas exchange with the atmosphere. Not all species have lenticels so their presence or absence on a mature stem is an additional aid to identification. Cherry tree bark is a classic example of obvious lenticels.

LEAF ATTACHMENT

Phyllotaxy is a term used to identify the way leaves are arranged on a plant stem and is of great benefit in identification of a specific plant family or genus. There are several common patterns of leaf attachment and the most common growth pattern is the alternate leaf arrangement. In this arrangement, leaves are attached staggered down the stem such that leaves are not directly across from one another (Figure 5.10a). In contrast, opposite leaf attachment is a growth pattern in which the leaves attach to the stem directly across from one another (Figure 5.10b) and, given its rarity, this phyllotaxy provides excellent information for plant identification. Occasionally, the phyllotaxy observed consists of leaves wrapping around the stem similar to the steps of a circular staircase and is referred to as a spiral leaf attachment (Figure 5.11a). Whorled leaf arrangement is the final phyllotaxy and is identified when three or more leaves attach in a circular pattern around each node on the stem and is extremely helpful in plant identification due to its scarcity (Figure 5.11b).

leaf scars
Markings left by the attachment of a petiole to a branch.

bundle scars
Small marks left on the leaf scar by the separation of the vascular bundles of a leaf.

lenticels
Found on the surface of bark; areas where the cells are loosely arranged to allow gas exchange to take place.

phyllotaxy
Classification of leaf arrangement on a stem.

(a)

(b)

Source: Cynthia McKenney.

Figure 5.10a–b   (a) Alternate leaf arrangement is characterized by one leaf attaching at a node on alternating sides of the stem and (b) opposite leaf arrangement had two leaves attached to a node.

(a)

Source: Cynthia McKenney.

(b)

© Pelevina Ksinia/Shutterstock.com

Figure 5.11a–b   (a) Spiral leaf attachment is characterized by the leaf attachment points rotating up the stem. (b) A whorled leaf attachment has three or more leaves attached at the same node.

BOX 5.1   Is Bark All the Same?

Prior to reading this chapter, you probably did not consider the appearance of bark; however, bark provides the vital function of protection for the plant. This protective coating of bark shields the plant from such mechanical damage as a sidewalk pressing against the bole of a tree (Figure Box 5.1a) or the prevention of infection from plant pests and diseases such as fungi (Figure Box 5.1b). Bark also has modifications that help arm them from animal herbivory and negative impacts from humans. Three of these common modifications are classified as **prickles**, **spines**, and **thorns**. Prickles are extension of the epidermis, which creates sharp spine-like structures as found on many types of roses. In some plants, the prickles are significant, as on the silk floss tree (*Ceiba speciosa*), and deter even the most inquisitive child from hanging on its branches (Figure Box 5.1c). Similarly, spines are modified leaves that form sharp projections as in leaf petioles of an ocotillo (Figure Box 5.1d). In contrast, thorns are sharp modified stems or branches located at the tip of the branch or at a node and may be single or compound. The thorns of the honey locust tree provide serious armor for the plant, preventing damage from animals grazing and stripping the bark off young plants (Figure Box 5.1e).

It is evident bark serves the function of aiding in plant identification due to its numerous variations including the peeling nature of a river birch (*Betula nigra*) or the patchy appearance of a lacebark elm (*Ulmus*

prickle
Extension of the epidermis making a sharp spine-like structure.

spine
Modified leaves forming a sharp projection frequently confused with a prickle.

thorn
Modified stems or branches creating a sharp projection.

Source: Cynthia McKenney.

Source: Cynthia McKenney.

© Melinda Fawver/Shutterstock.com

(a)

(b)

(c)

(d)

(e)

Figure Box 5.1a–e   (a) This sidewalk is causing mechanical damage to the tree. (b) An opening in the bark of the tree allowed a fungal mass to gain entrance and grow. (c) The prickles on the bark of the silk floss tree, (d) the spines on the stem of the ocotillo, and (e) the thorns of the honey locust tree are all examples of plant armor.

(f)   (g)

Source: Cynthia McKenney.

(h)   (i)

© Irina Mos/Shutterstock.com

Source: Cynthia McKenney.

Figure Box 5.1f–i   Peeling or patchy bark on this (f) river birch and (g) lacebark elm is helpful in the identification of both of these species. The unusual bark color of (h) the rainbow eucalyptus and (i) the Chinese parasol tree adds to their interest, as well as, helping with identification.

parvifolia) produced by exfoliation (Figure Box 5.1f-g). Unusual bark colors assist in rapid identification of the rainbow eucalyptus (*Eucalyptus deglupta*) and Chinese parasol (*Firmiana simplex*) trees (Figure Box 5.1h-i) as does the unique platy bark on the alligator juniper (*Juniperus deppeana*) or the presence of obvious lenticels on a cherry tree (*Prunus spp.*) (Figure Box 5.1jk). Historically, bark has also served a surprising number of human medicinal uses such as quinine extracted from the bark of the cinchona tree (*Cinchona officinalis*) and used as a fever reducer in the 17th century. Willow tree (*Salix babylonica*) bark (Figure Box 5.1l) was one of the original components used by Native Americans to treat aches and pains. Aspirin, a common over-the-counter medicine, is the synthetic form of the chemical compounds found in willow bark. (Figure Box 5.1m). The spice cinnamon is

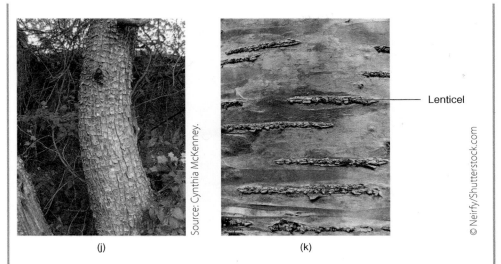

(j)

(k)

Lenticel

Figure Box 5.1j–k (j) The platy texture of the alligator juniper bark and (k) the prominent lenticels on cherry bark make them quite distinctive.

produced by grinding the inner bark of the cinnamon tree (*Cinnamomum verum*) (Figure Box 5.1n) and the inner bark of the cork oak (*Quercus suber*) is exploited to manufacture corks for wine bottles (Figure Box 5.1o). Moth orchids (*Phalaenopsis × hybrida*) are frequently found nestled in shredded fir bark rather than potting media to maximize drainage and aeration (Figure Box 5.1p) and redwood bark chips are frequently employed as an organic landscape mulch (Figure Box 5.1q).

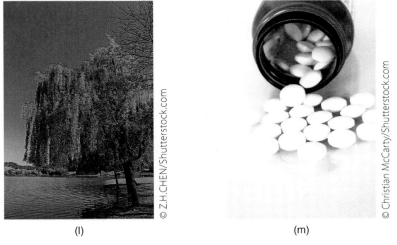

(l)

(m)

Figure Box 5.1l–m (l) Active compounds in the bark of the willow tree have similar properties to aspirin (m).

Figure Box 5.1n–q (n) The bark of the cinnamon tree which is ground and used as a spice and (o) the inner bark of the cork tree has been harvested to make cork. (p) Shredded fir bark is used fas a potting media for orchids and (q) redwood bark chips are a common type of landscape mulch.

SPECIALIZED STEMS

specialized storage structures
Modified compressed stems which are located under the ground and allow a plant to survive periods of adverse conditions by storing carbohydrates.

geophytes
Storage stems located below the soil surface.

When you go to a garden center or plant nursery, you generally don't ask for assistance with **specialized storage structures**. In the marketplace, these types of plant organs are loosely referred to as "bulbs"; however, not all specialized storage structures are bulbs, but all bulbs are a type of specialized storage structure! Botanists and taxonomists have a name for this group of specialized stems: **geophytes** (Dafni, Cohen, & Noy-Mier, 1981). Geophytes are storage structures composed of modified stems with their growing points located beneath the soil such as daffodils (*Narcissus spp.*) and hyacinths (*Hyacinthus orientalis*) (Figure 5.12). Generally these types of structures have a dormant period in which the foliage dies down to the ground as a method of stress avoidance and returns the following year utilizing the stored carbohydrates in the modified stem to initiate new growth.

Figure 5.12 Bulbs such as these yellow daffodils and pink and purple hyacinths are an example of geophytes with their growing points below the soil.

Bulbs

True bulbs are modified underground structures in which the stem is highly compressed to form a basal plate on to which their fleshy leaves are attached. The arrangement of these leaves and the presence or absence of a protective outer covering distinguishes these two bulb classifications known as laminate bulbs and scaly bulbs.

Laminate Bulbs are vertically compressed stems with nodes and internodes of which an onion is a familiar example (Figure 5.13a). The term *laminate* comes from the arrangement of the fleshy leaves in such a manner that they form layers of concentric circles around the stem. The internodal distance of the stem is extremely reduced and the entire structure is covered in a protective papery **tunic**, resulting in this type of bulb also

laminate bulbs
Vertical, compressed underground stem composed of fleshy storage leaves.

tunic
Dry, papery covering on a laminate bulb that provides protection for the bulb.

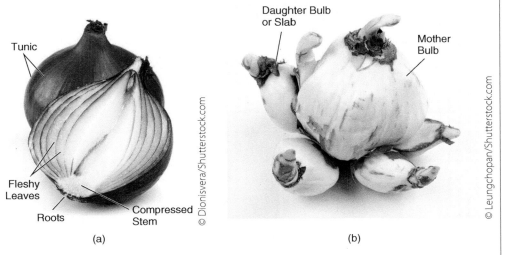

Figure 5.13a–b (a) This onion is a laminate bulb with the fleshy leaves and compressed stem all covered in a papery tunic. (b) Daffodil bulbs are called slab bulbs because of the flat side of the daughter bulb where it attaches to the mother bulb.

bulblets/bulbils
Small daughter bulbs forming at the base of the mother bulb.

scooping/scoring
Process of using cuttage to damage the basal plate of a bulb causing many smaller bulblets to form.

Scaly bulbs
Compressed, vertical underground stem in which the fleshy leaves are attached in a spiral pattern and are not enclosed in a tunic.

scaling
The process of gently removing each fleshy leaves from a scaly bulb for propagation.

being called a tunicate bulb. Flower spikes originate from the axial side of the compressed stem and penetrate between the fleshy leaves. Over time, the mother bulb generates small daughter bulbs called **bulblets** at the base of the compressed stem; however, these bulblets take several years to be mature enough to bloom themselves. One way to propagate bulbs is to gently pull bulblets off the mother bulb. Daffodils (*Narcissus spp.*) are referred to as slab bulb due to the side of the daughter bulb adjacent to the mother bulb having a flat appearance (Figure 5.13b). In addition to the formation of bulblets, there are two other methods of propagating bulbs known as **scooping** and **scoring**. Scooping a bulb is when one-third of the basal root plate is removed with a sharp knife or other implement, resulting in the formation of daughter bulbs on the callus produced at the wound. A second form of cuttage propagation is bulb scoring where a sharp knife is used to excise a small wedge of tissue from the root plate, resulting in the formation of small bulblets along the injured tissue.

Scaly bulbs have several similar features to laminate bulbs, such as a compressed stem and fleshy leaves; in contrast, the leaves of a scaly bulb are not tightly compressed or laminated like the tunicate bulbs and there is no tunic present. These leaves are spirally attached encircling the stem, giving an appearance similar to the leaves of an artichoke (Figure 5.14a). An Easter lily (*Lilium longiflorum*) is a classic example of a scaly bulb (Figure 5.14b). There are two common ways to propagate a scaly bulb, which are **scaling** and **bulbils**. Scaling is removing each of the bulb scales (fleshy leaves) from the underground stem and planting them with one-third of the scale base inserted in potting media. In a similar fashion, Easter lilies form aerial bulbs in the leaf axils where the leaf attaches to the stem and over time the aerial bulbs, called **bulbils**, mature. The bulbil with or without the attached leaf may be removed and placed onto damp potting media with one-third of the bulb submerged below the surface. The average lily has 90 leaves so this allows you to produce quite a crop of lilies from one plant. It should be noted when using either of these propagation techniques, it will take multiple growing seasons before the daughter bulbs are mature enough to flower.

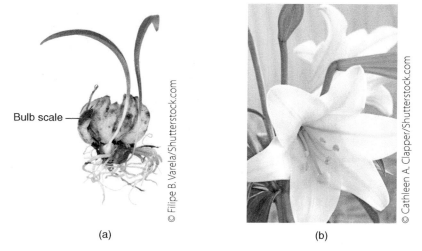

Bulb scale

© Filipe B. Varela/Shutterstock.com

© Cathleen A. Clapper/Shutterstock.com

(a)　　　　(b)

Figure 5.14a–b (a) Scaly bulbs have fleshy leaves, which are bulb scales surrounding the central axis. (b) Easter lilies are a classic example of scaly bulbs.

Corms are another category of compressed underground stems that resemble bulbs only superficially (Figure 5.15a). This unique structure is a vertical compressed stem with papery scales replacing the leaves and covering the exterior of the corm, much as the tunic of a tulip or onion. The external surface of the corm has both nodes and internodes where the papery scales attach; however, these are not the growing points as they are oriented at the top of this compressed stem. A cross section of a corm shows the interior of this structure is a solid mass of storage tissue without fleshy leaves and no evidence of nodes (Figure 5.15b). When the corm is severed into sections, these sections may resemble slices of potato; potatoes represent another category of specialized stems that we explore later in the chapter.

Spring flowering crocus (*Crocus vernus*) is an excellent example of a familiar plant that arises from this type of unique specialized stem (Figure 5.15c). Gladiolus (*Gladiolus spp.*), cyclamen (*Cyclamen persicum*), and freesia (*Freesia spp.*) are flowering crops originating from corms. Structurally, the base of the corm is a **basal plate** from which special **contractile roots**, named for their

corms
Specialized underground stem with nodes on the external surface and no fleshy leaves internally.

basal plate
Compressed stem of a corm where the roots attach.

contractile roots
Roots found on bulbs and corms that shrink or contract during adverse conditions, pulling the storage structure lower in the soil and preventing it from heaving out during freezing weather.

Figure 5.15a–d (a) Corms are characterized by the presence of the growing point at the top of the structure with the nodes visible on the body. (b) A slice through a corm shows the body is a mass of storage tissue. (c) Crocus are an example of a plant that produces corms. (d) The older mother corm, which is deteriorating, is visible below the new corm.

cormels/cormlets
Small daughter corms created near the basal plate of the mother corm.

rhizome
Horizontal below-ground stem with nodes and internodes; exhitbit monopodial growth.

monopodial
Growth of a plant from a single growing point in one direction.

tuber
Fleshy underground storage structure located at the end of a rhizome with nodes on the surface such as an Irish potato.

stolon
Horizontal stem found on or above the soil surface with nodes and internodes; not fleshy.

ability to shrink in response to soil temperature fluctuations, arise and serve to pull the corm lower in the soil as weather demands. Daughter corms, referred to as **cormlets** or **cormels**, form at the base of the stem and replace the mother corm when it dies, providing a continuity of the planting.

Rhizomes The German iris (*Iris germanica*) is an excellent example of a true **rhizome** (Figure 5.16a–b) that resembles a traditional stem a bit more than the bulbs and corms because this specialized storage structure is not as highly compressed as the other two we have explored. Rhizomes are horizontal stems that may or may not be fleshy depending on the amount of carbohydrates they have been able to store within. Unlike typical stems rhizomes are found below the soil surface, have both nodes and internodes, and frequently exhibit **monopodial** growth. With this distinctive growth pattern the rhizome grows in only one general direction, so when planning and planting a garden, you need to aim the rhizome growing points in the direction you would like the plant to spread. Frequently, rhizomes are planted in a circular pattern so the planting makes more of a complete display rather than a linear line. With age, the older portions of the rhizome become punky and then die, so lifting and dividing the rhizomes every few years is recommended so the senescing portions may be removed. The rhizome itself serves as a method of cuttage propagation with the understanding the nodes may be induced to form either shoots or roots. Simply remove a section of the rhizome with a minimum of three nodes and the associated growing tip (or fan) and let the cut heal for a few days in a warm, dry location before planting back in the soil.

Tubers The end of a rhizome swells and becomes an enlarged mass of storage parenchyma known as **tubers**. Irish potatoes (*Solanum tuberosum*) are tubers and the potato's "eyes" are actually small nodes on the structure (Figure 5.17a); this is why potatoes left in a dark, dry location may begin to sprout (Figure 5.17b). Tubers differ from the tuberous roots which we studied in Chapter 4 in that tubers are structures associated with stems and not roots, so there is no proximal or distal ends to the structures.

Stolons One of the most perplexing modified stems is the **stolon**. It is horizontal like a rhizome but is frequently not fleshy and is customarily

© Morphart Creation/Shutterstock.com

Source: Cynthia McKenney.

(a)   (b)

Figure 5.16a–b (a) A rhizome displaces is an example of monopodial growth such as this (b) iris plant.

(a)   (b)

Figure 5.17a–b   (a) Irish potatoes are probably the most easily recognized tuber. (b) When stored too long, the nodes, frequently called *eyes*, may sprout.

found on or above the soil surface (Figure 5.18a). Like other stems, they have nodes and internodes with adventitious roots and shoots forming at the nodes. Bermuda-grass is an example of a stolon and is able to root down and spread quickly due to its stoloniferous nature (Figure 5.18b).

Runners   Runners are similar to stolons as they are horizontal, nonfleshy, and aboveground, but they differ from stolons with the absence of nodes and the ability to produce roots only at the leading tip where a new plantlet is created. Strawberries are a familiar plant that spreads by runners. The mother strawberry plant produces the runner, which does not root down until a new plantlet forms at the tip.

Cladophylls   When the stem of a plant becomes flattened and is capable of functioning as a leaf, it is referred to as a **cladophyll**. Prickly pear cactus (*Opuntia spp.*) exemplifies the function of cladophylls as the leaves of this cactus are reduced to spines while the stems have modified into fleshy pads capable of conducting the functions of leaves (Figure 5.19). Note the fruit and flowers attach to the prickly pear cactus pads, confirming their function as stems.

cladophyll
Modified flattened stem functioning as a leaf.

Mother plant   Plantlet

(a)   (b)

Figure 5.18a–b   (a) Bermudagrass has actively growing stolons that root at the nodes and will knit down and produce a solid turf. (b) The strawberry mother plant has a small plantlet at the end of the runner.

Figure 5.19   Prickly pear cactus is an example of a clad-ophyll, which is a flattened stem that functions like a leaf.

BOX 5.2 Response to the Environment

Plants respond to their environment through a variety of methods. During periods of extreme heat and drought, a plant may roll its leaves, resulting in the reduction of leaf surface exposed to the sun and shedding a portion of the intense light. Similarly, herbaceous plants respond to similar environmental stress by wilting, which changes leaf surface angle and again limits sun exposure to leaf tissue, resulting in reduced water loss due to lower transpiration. In conjunction to these two drought and heat responses, stomata, which are small pores in the leaves, close, thus retaining moisture within the leaves while slowing photosynthesis. All of these responses to the environment are reacting toward heat and drought. However, there are other responses of plants to the environment, specifically involving stems and wind or soil slippage (Figure Box 5.2a–b).

(a)

(b)

Figure Box 5.2a–b   (a) The architecture of this conifer has been determined by the presence of strong winds. (b) The trunk of this palm has compensated for the slippage of the soil.

Reaction Wood

You have probably seen trees sway in the wind, but did you realize trees need to sway? The physiological response to this movement helps to strengthen trees by developing taper in the trunk. Mattheck (1997) identifies the engineering behind this movement and how it creates strength in the tree trunk by producing more wood at different stress points. This wood is known as **reaction wood** and it functions by allowing the tree to be structurally able to support the movement of the canopy during periods of high wind (Hellgren, Oloffson, & Sundberg, 2004; Wilson & Archer, 1977). Trees that are firmly tied to a stake or grown closely together are not able to sway enough and thus do not develop sufficient taper to support their branches, leading to more susceptibility to damage or tree failure. As the canopy becomes increasingly larger with age, the resistance it provides against the wind increases proportionally, causing mechanical leverage to occur at the base of the tree.

Tension Wood

Woody stems are induced to develop wood at other locations than just the base of the trunk. The point of attachment of the branches is also a stress point during periods of wind, resulting in wood depositing in different locations to match the plant architecture. **Tension wood** is created in broadleaf trees where there is an expansive canopy with no single apex. Arborists refer to this as a **decurrent** growth habit composed of **codominant** scaffold branches and no central leader (Figure Box 5.2c). The wood deposited on the upper side of leaning trunks or branch surfaces supports the branches by providing a type of internal contraction that aids in lifting the branches and stem up.

reaction wood
Secondary xylem created by a plant in response to mechanical stress.

tension wood
Reaction wood created by eudicots.

decurrent
Growth habit where codominant branches form the crown with no central leader.

codominant
When there are multiple trunks or branches on a tree which prevents the development of a central leader.

(c)

(d)

Courtesty Cynthia McKenney.

Figure Box 5.2c–d   (c) The codominant branches of this olive tree exhibit decurrent growth resulting in the creation of tension wood. (d) The strong central leader of this conifer exhibits excurrent growth, resulting in the creation of compression wood.

excurrent
Growth habit where a central leader forms the crown of a tree.

compression wood
Reaction wood present in gymnosperms.

Compression Wood

Conifers have more of a pyramidal shape called an **excurrent** growth pattern, where there is a prominent central leader (Figure Box 5.2b). The tension wood is deposited by the tree on the underside of leaning trunks and branch surfaces, referred to as **compression wood**. Compression wood acts to brace the plant by providing a compressive force against the trunk or branches, pushing them back toward the vertical where the tree will be stronger and less likely to fail.

Key Terms

shoot	lateral buds	scaling
conduct	axillary buds	bulbils
storage	nodes	corms
primary growth	internode	basal plate
protoderm	leaf scars	contractile roots
epidermis	bundle scars	cormlets (cormels)
ground meristem	lenticels	rhizomes
cortex	phyllotaxy	monopodial
pith	prickles	tubers
procambium	spines	stolon
medullary rays	thorns	cladophylls
secondary growth	specialized storage structures	reaction wood
sapwood	geophytes	tension wood
heartwood	laminate bulbs	decurrent
woody plant	tunic	codominant
liana	bulblets	excurrent
herbaceous plant	scooping	compression wood
vine	scoring	
terminal bud	scaly bulbs	

Summary

- Stems function in several ways:
  - Conduction
  - Support
  - Storage
  - Creation of new tissues
- Primary growth produces three primary meristemic tissues:
  - Protoderm
  - Ground meristem
  - Procambium

- The vascular system of monocots and eudicots differ in the following ways:
  - The vascular tissues of a monocot have scattered vascular bundles in the ground meristem.
  - Eudicots and gymnosperms have vascular bundles creating a vascular cylinder, which gives the appearance of rings. Eudicot stems have a central pith surrounded by a ring of vascular bundles.
- Eudicots and gymnosperms have secondary growth whereas monocots do not.
- Stems may be classified in several ways:
  - Woody versus herbaceous
  - Lianas versus vines
- Branch and stem components include:
  - Terminal buds
  - Bud scales
  - Lateral buds
  - Axillary buds
  - Nodes and internodes
  - Leaf and bundle scars
  - Lenticels
- Phyllotaxy, or leaf attachment to the stem, help with plant identification:
  - Alternate
  - Opposite
  - Whorled
  - Spiral
- Plant stems have adaptations to protect them from animals eating them:
  - Prickles
  - Spines
  - Thorns
- Bark provides economic uses for humans.
- Geophytes are specialized stems that include:
  - Bulbs including both laminate and scaly
  - Corms
  - Tubers
  - Rhizomes
  - Stolons
  - Runners
  - Cladophylls
- Plant stems respond to the environment in part by the production of reaction wood:
  - Tension wood in eudicots
  - Compression wood in gymnosperms

Reflect

1. *Specialized storage structures.* Many plants have storage structures under the soil that are quite diverse. How would you decide if one of these structures was a modified stem or a modified root? Explain.
2. *Stem characteristics.* Describe the major features that all stems have. What are their functions?
3. *Is this really a vineyard?* If grapes have woody vine-like structures called lianas, how can we have grape vineyards?

4. *Is this a monocot plant?* Compare and contrast monocot and eudicot stems.

5. *Cuttings.* If you are asked to make stem cuttings to propagate a plant and you are only allowed a small portion of the stem, what portion of the stem would you want and why?

6. *What is the difference?* Explain the relationship between shoots, stems, and trunks.

7. *Why underground?* Specialized stems are modifications serving to help stems adapt to their environment. What purpose does having these unusual stems under the ground serve?

References

Dafni, A., Cohen, D., & Noy-Mier, I. (1981). Life-cycle variations in geophytes. *Annals of the Missouri Botanical Garden, 68*(4), 652–660.

Hellgren, J. M., Olofsson, K., & Sundberg, B. (2004). Patterns of auxin distribution during gravitational induction of reaction wood in poplar and pine. *Journal of Plant Physiology, 135*(1), 212–220.

Hodel, D. R. (2009). Biology of palms and implications for management in the landscape. *HortTechnology, 19*(4), 676–681.

Mattheck, G. C. (1997). *Trees: The mechanical design.* New York: Springer Verlag.

Rost, T. L., Barbour, M., Stocking, C. R., & Murphy, T. M. (2006). *Plant biology* (2nd ed.). Belmont, CA: Thomson.

Wilson, B. F., & Archer, R. R. (1977). Reaction wood: Induction and mechanical action. *Annual Review of Plant Physiology, 28*, 23–43.

Chapter 6

Leaves

Source: Cynthia McKenney.

Learning Objectives

- Describe the external morphology of leaves
- Explain the internal morphology of leaves
- Identify the functions of leaves
- Describe the features used in leaf identification and classification
- Explain the economic uses of leaves
- Explain the response of leaves to the environment

Think of a windy day when you walked outside and saw colorful leaves falling to the ground. At first there may have been just a few leaves caught in a sudden gust; then perhaps there are more leaves accumulating in piles. A variety of thoughts might have crossed your mind as you see this change of season. Perhaps you thought how beautiful all the colorful leaves were or you may have thought you would have to rake the leaves up. Did it cross your mind these leaves are little photosynthetic factories and the plant has just shed them to decay away? Have you ever considered a pasture of grass as a photosynthetic factory creating food for cattle and wildlife? How about after a drought or a freeze; have you ever thought how this biological factory is now a lost food source to both the plants and the animals? In this chapter we investigate the many functions of leaves, from food and fiber to shade and environmental services. Hopefully, the next time you observe leaves, you will see them with a newfound appreciation.

LEAF MORPHOLOGY

In order to get started on our exploration of leaves, we need to identify the different structural components of these organs. As with our study of roots and stems, there will be differences once again in monocot and eudicot leaves, so we will concentrate on the eudicot leaves to identify the primary components and will then use these components as a contrast to the monocot structures.

External Morphology

lamina
Blade of a leaf that is the main surface for absorbing energy from the sun.

There are several primary components of the eudicot leaf and we now review each of these (Figure 6.1a). The leaf blade, called the **lamina**, is the large flat surface of the leaf, which provides the primary surface for absorbing radiant energy from the sun, similar in the monocot leaf. This structure has many variations in size, shape, color, and texture as exhibited by the color of the molten fire foliage (*Amaranthus tricolor*), the shape of the geranium plant leaf (*Pelargonium* × *hortorum*), and the waxy upper surface of the southern magnolia (*Magnolia grandiflora*) (Figure 6.2a–c). Variations such as

(a)   (b)

Figure 6.1a–b (a) The eudicot leaf has net venation and a petiole and (b) the monocot leaf has parallel venation and sheathing leaves.

(a)   (b)   (c)

Source: Cynthia McKenney.

Figure 6.2a–c   (a) The bright color of the molten fire foliage, (b) the shape and variation of the geranmium leaf, and (c) the waxy surface of the southern magnolia exhibits the diversity found in the leaves of plants.

these are useful in plant identification and are discussed in more depth later in the chapter. Close inspection of an oak leaf (*Quercus spp.*) reveals several other morphological features (Figure 6.1a). The **petiole** is a stem-like structure that may be large, small, or inconspicuous and serves to attach the lamina to the stem of the plant. The leaf attachment of the corn plant (*Zea mays*) has the characteristic monocot sheathing leaf that wraps around the stem (Figure 6.1b). The prominent vein running the length of the lamina is called the **midrib** and is found in both monocot and eudicot leaves. The venation pattern then diverges such that the eudicot oak leaf has net venation radiating out from the midrib and the monocot corn plant has parallel veins throughout the leaf. The **leaf margin** also varies significantly, with some leaves having smooth edges and others having toothed edges. The eudicot oak leaves have characteristically lobed margins; monocots such as corn plant usually have long and pointy leaves with smooth margins. The leaf tips, bases, shapes, and textures are all additional unique traits utilized in plant identification. An in-depth study of plants would necessitate familiarity with dozens of these leaf forms, surfaces, hair patterns, attachment methods, and other characteristics for a complete taxonomic classification.

Internal Morphology

As we focus on the internal structure of the leaf we find very distinct layers of tissues (Figure 6.3). On the upper surface is the **cuticle**, which secretes a **waxy bloom** over the leaf surface, reducing water loss to the environment. The tough **epidermis** is just below the cuticle and consists of several layers of cells, which serve as protection to the inner tissues of the leaf. The thicker the epidermis, the more adapted the plant is to tough environmental conditions such as heat and drought.

Photosynthesis is one of the primary functions of the leaf and the **mesophyll** is the chief photosynthetic tissue that facilitates this important plant process. Given this, the epidermal cells are generally translucent to allow light to penetrate efficiently. In eudicots, the mesophyll is

petiole
Structure attaching the leaf blade to the stem of the plant.

midrib
Primary vein that runs the length of a leaf.

leaf margin
Outer edge of a leaf that varies in shape and provides a method of identification.

cuticle
Found on the surface of stems and leaves; made up of cutin and wax, which make plant surfaces waterproof and reflective.

waxy bloom
Layer of material that reduces water loss from the leaf.

epidermis
Made up of mostly unspecialized cells forming a single layer covering the surface of herbaceous plants for protection.

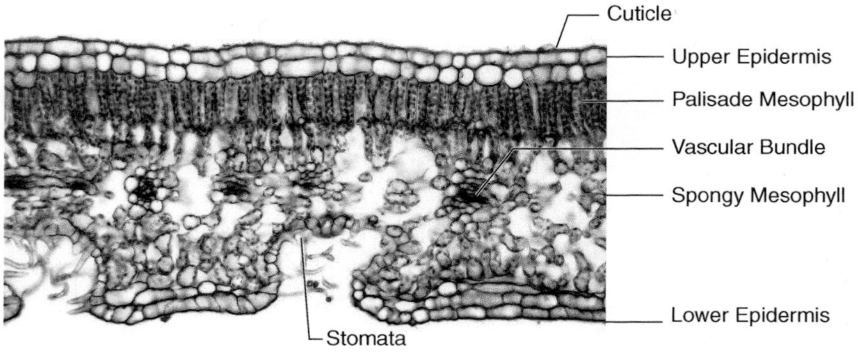

Figure labels: Cuticle, Upper Epidermis, Palisade Mesophyll, Vascular Bundle, Spongy Mesophyll, Lower Epidermis, Stomata

© University of Wisconsin-Madison, Department of Botany.

Figure 6.3   This cross section of a leaf shows the various tissues that compose a leaf. The stomata are clustered and recessed in a crypt to enhance their ability to retain moisture.

palisade mesophyll
Primary photosynthetic tissue of the leaf; tightly packed mesophyll cells found just below the upper epidermis.

spongy mesophyll
Loosely organized tissue below the palisade mesophyll with air spaces and vascular bundles passing through it.

vascular bundles
Xylem and phloem tissues that transport water and nutrients in the plant.

lower epidermis
Similar to the upper epidermis in providing protection to the inner tissues but differs in being the site for most of the stomata.

stomata
(singular stoma) Tiny openings surrounded by two guard cells; responsible for gas exchange and regulation of transpiration.

crypts
Depressions on the underside of some leaves where clustered of stomata are located.

differentiated into **palisade mesophyll** and **spongy mesophyll**. Palisade mesophyll cells are located on the upper side of the leaf and are tightly packed. The loosely organized mesophyll cells below the palisade mesophyll are the spongy mesophyll, which has open spaces allowing for air exchange. In monocots, the mesophyll cells are similar in shape and are not differentiated into palisade and spongy mesophyll cells. The vascular bundles containing the xylem and phloem necessary for the transport of water and sugars pass through the mesophyll. The xylem and phloem retain their orientation within the **vascular bundles**, with the xylem located toward the upper leaf surface and the phloem located below.

Finally, the **lower epidermis** is below the spongy mesophyll layer and is noticeably thinner than the upper epidermis as there is less need for protection on the underside of the leaf (Figure 6.4). The large depressions or pits in the lower epidermis are **stomata** (Figure 6.4), which function by allowing the exchange of gases within the leaf. In the cross section of the oleander leaf in Figure 6.3, the stomata are clustered together in depressions called **crypts** located on the underside of the leaf. They function to increase the relative humidity surrounding the stomata, thus allowing for better adaptation to drought (Graham, Graham, & Wilcox, 2006). In eudicots, the stomata are located on the lower surface of the leaf. In monocots, the stomata are located on both the upper and lower surfaces of the leaf.

Let us return our focus to the structure of a stoma. These small pores in the leaf allow for carbon dioxide to be absorbed from the air and water vapor and oxygen to be released in the atmosphere. This water loss from the stomata provides evaporative cooling but can lead to a water deficit in the plant. The primary method to control excess water loss is the presence of guard cells, which are two jelly bean–shaped cells surrounding the pore opening. When the plant is turgid or full of water, the guard cells swell and the pore opens. When the plant becomes flaccid or lacks sufficient water, the guard cells become lax, allowing the pore to close. This closure reduces the water loss by the plant through transpiration; however, it also limits the intake of carbon dioxide, which reduces the

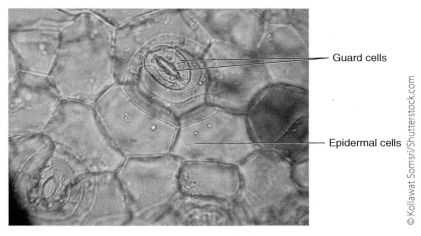

Guard cells

Epidermal cells

© Kollawat Somsri/Shutterstock.com

Figure 6.4   An open stoma surrounded by the two guard cells. Epidermal cells are identified around them.

production of carbohydrates. Over 90% of the water lost from a plant is transpired through the stomata (Graham et al., 2006). The control of the stomata plays an important role in the water status of the plant while also impacting the rate of photosynthesis by limiting the exchange of gases in the leaf. As mentioned earlier, the recessed crypts of the oleander are an additional modification that helps to retain some of the humidity; however, these structures are not common to many species.

LEAF FUNCTIONS

Leaves are amazing plant organs that act like miniature power generation plants, converting energy from the sun into stored chemical energy. All living organisms are energy dependent. Thus, the importance of plants and their leaves is apparent. In this section we explore several functions of leaves.

Processes

It is evident leaves serve an important function by converting sunlight into energy through the process of **photosynthesis**. These leaves convert sunlight and carbon dioxide from the air, providing important components for the process of photosynthesis while releasing oxygen and water vapor into the atmosphere helping to support life. Leaves are the primary location for photosynthesis in a plant; however, some plants also have other plant parts, such as stems, stipules, and bracts, capable of this process. The energy produced is available for later use by the plant, or it may be used indirectly by foraging animals. The process of photosynthesis is covered in more depth in Chapter 10. When water evaporates from the leaves, heat is removed from the leaf tissue. This process, which results in evaporative cooling by the leaf and movement of water through the plant, is called **transpiration**. Transpiration not only helps cool the plant but also moves water from the root to the rest of the plant body. The process of transpiration is covered in more depth in Chapter 10.

photosynthesis
Complex metabolic process that green organisms use to capture solar energy and convert it into chemical energy of organic molecules such as glucose or food.

transpiration
Loss of water vapor through the stomata in the leaves.

Water Retention

Leaves are quite diverse and have many adaptations, resulting in other functions than photosynthesis and transpiration. Leaf morphology modifications impact their overall function. Adaptations such as changes in the leaf surface help to retain precious water. Numerous plants native to arid or semiarid environs, including the artichoke agave (*Agave parryi*), have tough, fleshy leaves with a thick epidermis to reduce water loss to the environment (Figure 6.5a). Similarly, cacti have leaves that have modified into spines, which greatly reduces any water loss by the plant and also protects the plant by preventing animals feeding on the pads. Other plants such as lamb's ears (*Stachys byzantina*) have adapted to limiting water loss through the presence of fine hairs called **pubescence**, which reflect light from the leaf surface (Figure 6.6a). These bromeliad leaves have overlapping **trichomes**, which reduces exposure of the leaf to air movement, thus limiting transpirational water loss (Figure 6.6b). Similarly, succulent plants such as the painted

pubescence
Fine hairs covering the surface of a leaf.

trichomes
Specialized epidermal cells that are used by some plants for protection.

(a)

(b)

Source: Cynthia McKenney.

© thaikrit/Shutterstock.com

Figure 6.5a–b (a) The artichoke agave has tough fleshy foliage and (b) the cactus has leaves modified into spines. Both of these modifications help to conserve water during heat and drought.

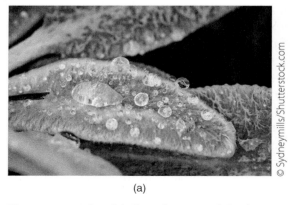

(a)

(b)

© Sydneymills/Shutterstock.com

Source: Cynthia McKenney.

Figure 6.6a–b (a) This closeup of the leaves of lamb's ears show fine hairs on the leaf surface to reflect sunlight, reducing transpirational water loss. (b) The trichomes on this bromeliad help reduce evaporative water loss.

beauty echeveria (*Echeveria nodulosa*) and the jade plant (*Crassula argentea*) have thick, succulent leaves coated with a waxy bloom, allowing the leaves to store water as well as function in photosynthesis (Figure 6.7a–b).

Another type of water adaptation is storage of free water by plants. Remember from Chapter 4 how epiphytes do not need to have roots in the soil and may grow on the branches of trees. These plants, such as guzmanias (*Guzmania sanquinea*), frequently have leaf attachments oriented to create a cup, which retains water for the plant and for small animals (Figure 6.8a). Similarly, the flower bracts and sheathing leaf bases of a traveler's palm (*Ravenala madagascariensis*) trap water and result in a source of free water accessible by small animals or thirsty travelers, thus creating its name (Figure 6.8b). All of these adaptations stress the importance of water to plants.

Source: Cynthia McKenney.

© Elena Elisseeva/Shutterstock.com

(a)   (b)

Figure 6.7a–b   (a) The painted beauty echeveria and (b) the common jade plant share the leaf adaptation of having a waxy cuticle, which aids in moisture retention.

© Todd Boland/Shutterstock.com

Source: Cynthia McKenney.

(a)   (b)

Figure 6.8a–b   (a) This guzmania stores water in the cup formed from its leaves; (b) the traveler's palm traps water in the space created by the sheathing leaf bases attached to the trunk.

Plant Protection

Leaf adaptations also provide protection from animal herbivory, which can set plants back substantially by losing the growth produced by the investment of resources. Leaf modifications similar to those on the stems help to reduce plant damage. The western stinging nettle (*Hesperocnide tenella*) has hollow trichomes containing a chemical compound similar to the substance found in a bee sting (Figure 6.9a); when this stinging hair is hooked onto skin, a strong irritation develops. Similarly, the leaf bases of a sago palm (*Cycas revoluta*) have some protective small barbs on the leaf bases; that, in conjunction with the toxic nature of the plant, renders it undesirable for foraging (Figure 6.9b).

Support

tendrils
Modified leaf that aids in plant support by wrapping around a structure or another plant.

Several leaf modifications function as a method of support for herbaceous vining plants. Some of these supple plants have modified leaves called **tendrils** that cling to structures, allowing the plant to attain greater height, which frequently translates into more available sunshine. The morning glory (*Ipomoea purpurea*) has tendrils forming at the tip of the vine, allowing the plant to scrabble up trellises or crawl along fences, positioning their flowers and foliage to best advantage for photosynthesis (Figure 6.10a). Another unique leaf structure providing support is the buoyant nature of the oversized leaves of the Amazon water platters (*Victoria amazonica*) found floating on the surface of still water (Figure 6.10b). Air-filled ribs in the platter itself distributes weight across the surface area of the leaf, providing the needed buoyancy to support the weight of a small child! The succulent nature of this large aquatic plant is protected by spiny projections on the undersides of the platters themselves, preventing damage by fish and small animals.

(a)

© Steve Mann/Shutterstock.com

(b)

Source: Cynthia McKenney.

Figure 6.9a–b (a) The stinging nettle leaf adaptation of stinging trichomes, which hook on the skin and prevent animals and people from coming too close. (b) This cycad has modified leaves that are sharp and leathery, making them less desirable for animal consumption.

Figure 6.10a–b (a) This morning glory is an example of a tendril providing support for the plant; (b) the Amazon water platter has air trapped in ribs on the underside of the pad to provide flotation.

Pollination

One of the most obvious characteristics of many plants is their attractive flowers; however, some plants have insignificant flowers that do not attract pollinators. **Bracts** are a type of leaf modification resulting in highly colored structures that frequently resemble an inflorescence attractive to the pollinators (Figure 6.11). The hot pink bracts of the paper flower (*Bougainvillea glabra*) and the golden yellow bracts of the lollipop plant (*Pachystachys lutea*) are especially attractive when compared to their insignificant white flowers in the center of the inflorescence. Frequently the attractive nature of the bract lasts for weeks or months after the flowers have faded, such as in lobster claw flowers (*Heliconia rostra*), which provide color for several weeks after the flowers are gone.

Bracts
Highly modified leaf frequently mistaken for a flower. It generally persists longer and attracts insect pollinators.

Figure 6.11a–c (a) The bougainvillea with its hot pink bracts, (b) the lollipop plant with its yellow bracts, and (c) the heliconia with the yellow and red bracts share the common trait of pollinator attracting bracts that are brightly colored while the flowers are more inconspicuous.

Trapping Food

Carnivorous plants exhibit amazing leaf modifications, resulting in the capacity of these plants to obtain at least a portion of their nutrition from the insects frequenting them. There are over 600 different species, which have fascinated observers for centuries and are frequently found in sunny bogs with acidic soils and limited nutrition (Ellison & Gotelli, 2001). The insectivorous modification provides an advantage for the plants; unfortunately, these intriguing carnivorous features have led to overcollection and species endangerment.

Hinged Leaf Traps There are several methods by which leaf modifications attract and trap prey. The Venus flytrap (*Dionaea muscipula*) is one of the best-known examples of carnivorous plants. It catches its prey through the development of a hinged leaf functioning as a trap (Figure 6.12a). The unsuspecting insect lands on one of the open traps and when the insect accidentally brushes against one of the trichomes, the trap is triggered and

(a)

(b)

(c)

(d)

© Thitirut Jenwanathankij/Shutterstock.com

© Marco Uliana/Shutterstock.com

© Matthijs Wetterauw/Shutterstock.com

© Sue McDonald/Shutterstock.com

Figure 6.12a–d (a) A Venus fly trap has modified leaves with small hairs triggered by insects, (b) causing the traps to close. (c) The sundew plant has sticky extensions that trap insects and (d) the pitcher plant lures insects into its pitfall trap where digestive juices dissolve them.

it springs closed (Figure 6.12b). Enzymes are then secreted and the trapped insect is then digested and the nutrients absorbed into the plant. It is important to realize the flytrap is not dependent on the insects to survive because it can make its own food by photosynthesis. In fact, these plants are easily "overfed" and eventually die when new owners place insects or bits of meat in the traps too often.

Sticky Traps The sundew plant (*Drosera rotundifolia*) is an example of a living sticky trap due to the long petioles with round leaves covered in tentacle-like hairs (Figure 6.12c). These innocent-looking hairs have a sticky mucilaginous liquid covering them. Insects are attracted to the colorful hairs and land on the lamina, becoming stuck. The leaf then curls around the stuck insect and the digestive enzymes are secreted. The sundew plant also produces its own food through photosynthesis; however, carnivorous evolutionary development allows these plants to have an advantage in low nutrient soils and wet bogs where there is sufficient sun to support the plant (Ellison & Gotelli, 2001).

Pitfall Traps The pitfall trap is a third carnivorous plant modification exemplified by the pitcher plant (*Sarracenia purpurea*). This plant sports a modified leaf resembling the shape of a pitcher (Figure 6.12d). A sweet secretion is discharged on the lip of the pitcher that attracts insects to land. As the insect crawls inside the pitcher looking for the source of the sweetness, sharp hairs lining the inside of the pitcher make it difficult for the insect to escape. The insect struggles to crawl out and becomes exhausted, ultimately falling into the digestive enzymes at the bottom of the structure where the nutrients are absorbed into the plant and only the exoskeletons are left behind.

Pollination Carnivorous plants depend on insects for pollination; however, there appears to be a conflict between pollination and capturing prey (Ellison & Gotelli, 2001). Some carnivorous plants have flowers not directly located by the traps that aid in allowing the pollinators to be separated from the prey. Other carnivorous plants are perennials that reproduce vegetatively, taking the pressure off the need for pollination (Figure 6.12c).

IDENTIFICATION AND CLASSIFICATION

The diversity found in leaves is amazing and these unique features serve to help with plant identification as well as plant classification. Entire courses are devoted to the study of plant identification and classification; however, for our purposes, we now look at just a few of the key concepts of plant identification based on leaf morphology.

Leaf Complexity

One way eudicot leaves may be categorized is by their leaf complexity. A simple leaf is composed of one lamina attached to a petiole (Figure 6.13a) in comparison to the **pinnately** and **palmately** compound leaves, which have multiple leaflets attached to the petiole (Figure 6.13b–e). By identifying the petiole by its slight swelling (axillary bud) and attachment to the stem, you

pinnately
Compound leaflet arrangement where leaflets are directly across from one another.

palmately
Compound leaflet arrangement where leaflets radiate from a central point like the fingers of a hand joining together at the palm.

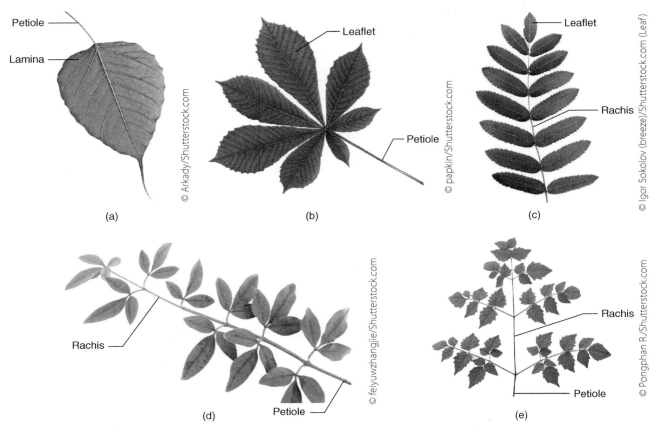

Petiole

Lamina

© Arkady/Shutterstock.com

(a)

Leaflet

Petiole

© papkin/Shutterstock.com

(b)

Leaflet

Rachis

© Igor Sokolov (breezel)/Shutterstock.com (Leaf)

(c)

Rachis

Petiole

© feiyuwzhangjie/Shutterstock.com

(d)

Rachis

Petiole

© Pongphan R./Shutterstock.com

(e)

Figure 6.13a–e These drawings show leaf complexity. Each of these images is of a single leaf. (a) is a simple leaf, (b) is a palmate compound leaf with 7 leaflets, (c) is a leaf with 15 leaflets pinnately arranged in an opposite pattern down the rachis (d) is a leaf with 10 clusters of leaflets, and (e) is one leaf with multiple clusters of leaflets.

rachis
Main axis of a compound leaf.

can determine whether a leaf is compound or simple. Thus every image in Figure 6.13 is a single leaf! Some of the leaves are characterized by the presence of multiple leaflets composing the leaf. Pinnately compound leaves have their leaflet arranged along the **rachis** (leaf axis) directly across from each other whereas a palmately compound leaf has multiple leaflets all arising from a central point, like the fingers of a hand joining together at the palm. Notice some leaves may be bipinnately or tripinnately compound depending on the number of divisions and type of attachments the leaflets have (Figure 6.13d and e). Every leaf only has one petiole, providing an excellent identifying characteristic when trying to determine the complexity of the leaf.

Leaf Venation

Leaf venation provides an additional method for plant identification and classification. There are three primary venation patterns: pinnately net veined, palmately net veined, and parallel veined (Figure 6.14a–c). Eudicot leaves have net venation patterns where intersecting veins originate from the midrib that then divide into finer veins throughout the leaf. Parallel veins are found in monocots and do not intersect, remaining parallel throughout the entire plant.

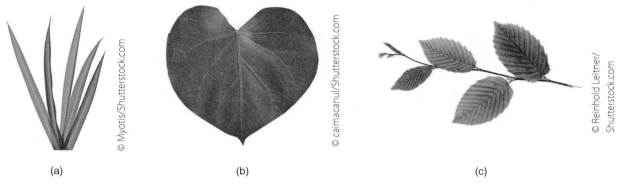

© Myotis/Shutterstock.com

© caimacanul/Shutterstock.com

© Reinhold Leitner/Shutterstock.com

(a)   (b)   (c)

Figure 6.14a–c   Leaf venation patterns: (a) monocot parallel venation, (b) palmate net venation, and (c) pinnate net venation.

Leaf Shape and Color

The shape and color of leaves emphasize the diversity found in these plant organs (Figure 6.15a–e). Leaves vary, from the ovate copper-colored leaves of a coleus plant (*Solenostemon scutellarioides*), to the variegated leaves of the zebra plant (*Aphelandra squarrosa*), to the trifoliate glossy leaf of Boston ivy

(a)   (b)   (c)

(d)   (e)

Source: Cynthia McKenney.

Figure 6.15a–e   (a) Ovate-shaped coleus leaves, (b) variegated leaves of the zebra plant, (c) trifoliate leaves of the Boston ivy, (d) linear leaves of Napier grass, and (e) compound leaves of the oxalis.

(*Parthenocissus tricuspidata*). The linear-shaped leaves of the maroon Napier grass (*Pennisetum purpureum*) contrasts with the trifoliate leaves of the oxalis foliage (*Oxalis sp.*). There are over 20 specific leaf shapes used for classification purposes and a similar number of leaf tip and leaf bases to distinguish specific plants. Leaf margins are also an aid in identification. A few of the margin types are included in Figure 6.16a–d.

(a)

(b)

(c)

(d)

© Teerasak/Shutterstock.com

© Volodymyr Burdiak/Shutterstock.com

© Brzostowska/Shutterstock.com

© Reinhold Leitner/Shutterstock.com

Figure 6.16a–d Leaf margins include (a) entire, (b) cleft, (c) dentate, and (d) serrate.

BOX 6.1 Economic Uses of Leaves

Structural Uses

Leaves have been very important to humans since the beginning of time and have been employed for functional uses such as making a palm frond roof on a Mayan home (Figure Box 6.1a) or serving as living fences, creating the feeling of a room (Figure Box 6.1b). Early Hawaiians used Ti plant (*Cordyline fruticosa*) leaves for ornamentation, as a living property boundary, to make leis or garlands, and to wrap food before placing it in a pit to cook (Figure Box 6.1c–d). The tough leaves of the sisal plant (*Agave sisalana*) are also harvested and the fibers removed to make sisal hemp used for rope.

(a)   (b)

Figure Box 6.1a–b   (a) Palm fronds are used to make a roof on this Mayan home. (b) Plant foliage is used to create a living fence around this estate.

(c)   (d)

Figure Box 6.1c–d   (c) The red-leaved Hawaiian Ti plant was used by early Pacific Islanders for ornamentation and property boundaries. (d) The sisal plant is an agave in which the fibers in the leaves are used to make rope.

Dyes and Medicinal Uses

For hundreds of years, plant extracts have been used for many purposes (Figure Box 6.1e–f). Red dyes were made from plant leaves, including red dye henna (*Lawsonia inermis*) used to dye skin and hair. When cotton became a common textile the blue dye indigo (*Indigo tinctoria*) was extracted from the foliage and ultimately became the traditional blue dye for denim jeans. Numerous plant leaves are exploited for use as a fabric dye, especially when a mordant such as salt or vinegar is utilized to fix the color. An example of a plant best used with a mordant is tea (*Camellia sinensis*). Brewing a strong pot of tea and using a fixative to help hold the color produces an excellent dye for natural fabrics.

Medicinal sedatives, remedies, and poisons have been concocted from the extracts of plants such as the poison hemlock plant (*Conium maculatum*), thus providing yet another historic use of plants (Figure Box 6.1g). Decoctions of the foxglove plant (*Digitalis purpureus*) have been used for

© Swapan/Shutterstock.com

© AKI's Palette/Shutterstock.com

(e)

(f)

© Bocman1973/Shutterstock.com

(g)

Figure Box 6.1e–g Plant foliage is also used to make dyes such as (e) henna and (f) indigo. (g) Extracts of poison hemlock were used as a medical sedatives and poisons.

treatment of cardiac arrhythmias (Figure Box 6.1h–i); Saint John's wort (*Hypericum perforatum*) is an herbal remedy that is thought to help with depression. Native peoples from around the globe have turned to plants for healing and a number of these traditional remedies are the basis for current medicinal treatments. The medicinal uses of plants is discussed in more depth in Chapter 29.

Leaves as a Food Source

Of all the uses of plant foliage provided here, harvesting them as a food source is probably the most important application globally. There is an amazing array of plant leaves suitable for human consumption. Some nutritious

Source: Cynthia McKenney.

© LianeM/Shutterstock.com

(h)

(i)

Figure Box 6.1h–i   (h) Extracts of foxglove have been used to treat heart arrhythmias; (i) Saint John's wort has been used as a homeopathic remedy for depression.

foliage includes spinach, lettuce, chard, greens, cabbage, and kale. The list increases dramatically when we think of all of the mixed greens available on the market now for salads. With increased interest in healthy eating, plant leaves supply a greater component of our daily diets.

Plant Reproduction by Leaves

Several plants have the unique ability to create plantlets from the leaves of a parent plant such as the mother-of-thousands plant (*Kalanchoe daigremontianum*), which produces small plantlets along the margins of their leaves (Figure Box 6.1j). By removing the leaf from the plant and placing it in direct contact with the soil, the leaf develops small plantlets along the edges which fall off automatically when mature. This type of propagation allows for a large number of propagules to be created from one mother plant. Rex begonias (Figure Box 6.1k) produce new plantlets but employ a slightly different process. A knife is used to cut across the large veins of the leaf and then it is pinned to the soil surface. Over time, the tissue that heals the wound along the cut veins gives rise to the new plantlets. African violets (Figure Box 6.1l) are easily propagated by leaf cuttings. By removing a leaf with the petiole attached, the lower half of the petiole may be placed in the soil without allowing the leaf blade to touch the soil surface, thus avoiding potential rot to occur. The leaf cutting will develop fibrous root systems and new plants will develop attached to the petiole of the leaf cutting. When done by micropropagation, large numbers of plants may be produced from a few stock plants.

Figure Box 6.1j–l (j) This edge of the mother-of-thousands plant has plantlets developing around the edge of the lamina. (k) Rex begonia plants may be propagated by laying a leaf on the soil surface. (l) African violets may be propagated by cutting on the leaf petiole.

RESPONSE OF LEAVES TO THE ENVIRONMENT

Heat and Drought

Plants respond to their environment in several ways. During periods of extreme heat or drought, plants may roll their leaves or raise them skyward, limiting the exposure of the leaf blade to the sun. This type of response reduces overall transpirational water loss and allows the plant to rehydrate after sunset. Similarly, leaves may respond to heat and drought by wilting, which again reduces exposure to the sun (Figure 6.17a). These responses are due to the change of water status in the plant, resulting in a change of leaf orientation.

Changing Seasons

As the weather changes, plants again respond to their environments. Cold weather triggers deciduous plants to shed leaves that will be replaced the following spring, which is advantageous as roots are less able to absorb water when the soil temperature is low. With the lower temperatures physiological changes take place and the hormone levels of the plants shift. In addition, photosynthate from the leaves is transported to the roots and stems for storage. These alternations prepare the plant to withstand low temperatures with less damage to the tissues.

The leaves of many species change color prior to **abscission**, the process of shedding leaves. The primary color of the leaf is due to the

abscission
Physiological process that causes leaves to shed from a plant.

photosynthetically active green pigment chlorophyll. As the temperatures decrease, the chlorophyll deteriorates, resulting in the visibility of the accessory pigments such as yellow xanthophylls and orange carotenes. These accessory pigments are masked by the presence of chlorophyll during most of the year, but become visible once the chlorophyll is denatured.

Prior to abscission, the leaves go through a period of **senescence** or a period of aging where ethylene is created, which enhances the leaf drop. When suberin is formed within the abscission zone, the leaf begins to separate from the stem (Figure 6.17b). Enzymes then work on the remaining vascular tissues, weakening their attachment. Eventually the weight of the leaf will pull it free from the stem.

senescence
Period of aging that eventually leads to the death of the leaves.

Air Pollution

Increasingly our environment is contaminated with a variety of pollutants. Leaves exhibit the effects of air pollution more readily given how thin the leaf blades are. Auto exhaust results in sulfur dioxide damage, which is usually indicated by brown tips and irregular spots on the foliage. Ozone damage, a form of industrial pollution, is usually manifested as surface flecks on the foliage with a reddish-brown color. The damage is also found on the upper surface with an irregular pattern similar to sulfur dioxide damage. Peroxyacetyl nitrate (PAN) is a photochemical smog that produces a silver or bronze color on the underside of the foliage. The color and location is very diagnostic, which is helpful for identification purposes.

Interior air quality has also become an issue as homes have become more energy efficient. A report by the U.S. Environmental Protection Agency (EPA) identified over 900 organic chemicals with elevated levels in public buildings such as office buildings and hospitals due in part to reduced ventilation (Wolverton & Wolverton, 1993). Research has determined plants are capable of removing significant amounts of contaminates such as ammonia, formaldehyde, and xylene from the air. The ability of foliage plants to improve air quality varies by species so further studies have continued to identify the best plants for specific situations.

(a)   Abscission layer   (b)

© Stephen VanHorn/Shutterstock.com

© University of Wisconsin-Madison, Department of Botany

Figure 6.17a–b   (a) This peace lily plant has wilted due to insufficient water. (b) During abscission, suberin is created, which forms a layer and separates the leaf from the stem.

Key Terms

lamina	spongy mesophyll	tendrils
petiole	vascular bundles	bract
midrib	lower epidermis	pinnately
leaf margin	stomata	palmately
cuticle	crypts	rachis
waxy bloom	photosynthesis	abscission
epidermis	transpiration	senescence
mesophyll	pubescence	
palisade mesophyll	trichomes	

Summary

- The primary components of a leaf include:
  - Lamina or leaf blade
  - Petiole
  - Midrib
  - Leaf margin, tips, and bases
- The internal leaf morphology of the leaf reveals layers including:
  - Cuticle
  - Epidermis
  - Palisade parenchyma
  - Spongy mesophyll with the presence of vascular bundles
  - Lower epidermis with stomata
- Differences between monocot and eudicot leaves:
  - Eudicot leaves have palisade and spongy mesophyll whereas monocot leaves have only a single type of mesophyll.
  - Eudicot leaves have netted venation and monocot leaves have parallel venation.
  - Stomata are found on the lower side of eudicot leaves whereas stomata are found on both the upper and lower side of monocot leaves.
- Stomata are small pores in the leaf allowing for the absorption of carbon dioxide and the release of water vapor and oxygen.
- The two guard cells surrounding each stoma open and close the pore in response to plant water status.
- Photosynthesis is the conversion of sunlight into energy.
- Transpiration is the loss of water vapor via stomata on the leaves.
- The process of transpiration evaporatively cools the leaf as well as creates a tension in the plant, which pulls water up through the vascular system.
- Leaf modifications assist in water retention:
  - Tough, fleshy leaves with a thick epidermis
  - Pubescence to reflect sunlight
  - Overlapping trichomes to shade the leaf surface
  - Waxy bloom on succulent leaves to limit water loss

- Development of a water-holding cup from the bases of the leaves
  - Sheathing leafs and flower bracts that trap water
- Leaf modifications create protection from herbivory by animals:
  - Hollow trichomes with chemicals inside
  - Leaf bases with protective barbs
- Leaf modifications called *tendrils* cling to structures and provide support for the plant.
- Bracts are modified leaves that are highly colored and appear to be petals of a flower. Bracts attract insect pollinators to the less noticeable flowers.
- Carnivorous plants have leaves modified into several kinds of traps:
  - Hinged-leaf traps
  - Sticky traps
  - Pitfall traps
  - Pollination
- Leaf identification and classification use the following characteristics:
  - Leaf complexity
    - Pinnately compound leaves
    - Palmately compound leaves
    - Simple leaves
  - Leaf venation
    - Pinnately net veined
    - Palmately net veined
    - Parallel veined
  - Leaf shape and color
- Economic uses of leaves:
  - Structural
  - Dyes and medicinal uses
  - Source of food
- Leaves are a method of plant propagation.
- Leaves respond to different factors in their environments:
  - Heat and drought result in wilting
  - Change of seasons result in leaf abscission
  - Air pollution creates foliar damage and stunted growth

Reflect

1. *Who's your housekeeper?* Plants are recognized as supplying environmental services for us such as purifying the air. What are other types of environmental services do they provide?
2. *How do carnivorous plants eat?* Since plants don't have mouth parts, how do these insectivores get the nutrients from the insects inside?
3. *Plant identification?* Which of the many characteristics of leaves do you feel provides the best method of plant identification? Support your answer.

4. *Winter plant identification.* In the winter eudicot plants lose their leaves. What other characteristics would allow you to identify a plant?

5. *Economic importance.* Plants have many facets of economic importance. Outline some of the major facets.

References

Ellison, A. M., & Gotelli, N. J. (2001). Evolutionary ecology of carnivorous plants. *Trends in Ecology and Evolution, 16*(11), 623–629.

Graham, L. E., Graham, J. M., & Wilcox, L. W. (2006). Plant biology (2nd ed.). Upper Saddle River, NJ: Pearson Education.

Wolverton, B. C., & Wolverton, J. D. (1993). Plants and soil microorganisms: Removal of formaldehyde, xylene, and ammonia from the indoor environment. *Journal of the Mississippi Academy of Sciences, 38*(2), 11–15.

REPRODUCTION IN THE GREEN WORLD

Chapter 7 **Flowers and Sexual Reproduction**

Chapter 8 **Fruits, Seeds, Dissemination, and Germinations**

Chapter 9 **Asexual Reproduction and Plant Propagation**

Flowers and Sexual Reproduction

Source: Cynthia McKenney.

Learning Objectives

- Discuss the process of sexual reproduction
- Identify the parts of a complete angiosperm flower and their function
- Define the terms *perfect*, *imperfect*, *monoecious*, and *dioecious* with regard to flowers
- Discuss methods of pollination
- Identify common corolla types
- Name all parts of an inflorescence
- Identify common types of inflorescences
- Explain what a floral formula is and how it is useful
- Explain the economic uses of flowers

sexual reproduction
Union of the gametes produced by two individuals produces offspring that are genetically different from their parents.

asexual reproduction
Does not involve gametes from two individuals and produces offspring that are genetically identical to their parent.

gametes
Sex cells (1n) that fuse together to create a zygote.

zygote
Single cell formed by the union of the male and female gametes.

sepals
Leaf-like structures that surround the petal and are collectively called the calyx.

petals
Showy component of a flower that attracts insects and are collectively called the corolla.

stamens
Male floral structure composed of the anther and filament.

pistils
Female flower component composed of the stigma, style, and ovary.

receptacle
Structure at the top of the stem to which the flower parts are attached.

complete flower
When a flower has all four primary parts including the sepals, petals, stamens, and pistils.

incomplete flower
When a flower is missing one or more of the four primary parts including the sepals, petals, stamens, and pistils.

Flowers have the ability to draw you close to them by both their fragrance and their exotic appearance. This attraction holds not only for people but for insects as well. In this chapter we focus primarily on angiosperm flowers as they are so abundant and provide an excellent example of many floral features and functions.

SEXUAL VERSUS ASEXUAL REPRODUCTION

Many plants depend on **sexual reproduction** (through the production of seeds) in order to continue their line. Other plants reproduce through **asexual reproduction** (cuttage or division of plant parts) only or can reproduce both sexually and asexually. There are advantages and disadvantages to both methods of reproduction. Primarily, asexual reproduction is faster and allows for the production of genetically identical plants known as *clones*. This is advantageous in the production of species that have poor germination of seed, limited seed production, or complex types of seed dormancy. In addition, with asexual reproduction, only one parent plant is required. We explore this method of plant continuance further in Chapter 9.

Sexual reproduction is basically the combination of genetic material from two parents, creating different sets of traits. The genetic material is contained in single-sex cells known as **gametes**. The fusion of a male and a female gamete results in the formation of a single-celled **zygote**, which will ultimately develop into an adult plant if the necessary conditions are present. This process is the foundation of sexual reproduction. The processes that direct these combinations are investigated in greater detail in Chapter 15.

As with asexual reproduction, there are pros and cons to sexual reproduction. The advantages include a greater potential for adaptation to the environment as a result of the many possible combinations of traits from the parents. If the resulting zygote has characteristics that make it more adaptable, it may allow the plant to populate a greater range. On the negative side, sexual reproduction takes a longer period of time to accomplish and requires the presence of two parents.

THE COMPLETE FLOWER

There are four primary floral parts in a typical flower: **sepals**, **petals**, **stamens**, and **pistils** (Figure 7.1). We look at each of these components separately next. The floral organs attach in whorls to a structure called the **receptacle**. The presence and absence of these four floral components provide a basis for classification of different flowers. When all four structures are present, the flower is referred to as a **complete flower**. It is said to be an **incomplete flower** if one or more of the structures are absent. For example, a flower without sepals would be incomplete.

Sepals and Tepals

Sepals are leaf-like structures that frequently enclose the maturing flower bud. As the bud develops, the sepals often separate and the corolla (the collective term for all the petals) expands as is evident in a rose (Figure 7.2a).

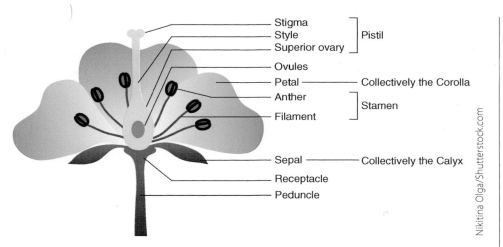

Figure 7.1   This stylized image of a perfect flower identifies the typical floral parts.

Collectively the sepals are called the **calyx**. In some flowers the calyx has ornamental value such as in a fuchsia (*Fuchsia spp.*; Figure 7.2b). In some flowers, the sepals and petals are indistinguishable, as in the daylily (*Hemerocallis spp.* Figure 7.2c). The showy upper three petals appear to be identical to the showy lower three sepals. In this situation, they are all referred to as **tepals**. The **perianth** is the collective term for the calyx and corolla or tepals.

Petals

Petals are usually the most obvious floral part as they frequently have eye-catching color and many different sizes and shapes (Figure 7.3a–c). The papery texture of the highly colored poppy flowers (*Papaver orientale*) contrasts with the pale waxy-textured wisteria (*Wisteria sinensis*) florets and the thin thread-like florets of the strawflower (*Helichrysum bracteatum*). Collectively the petals are referred to as the **corolla**. The color and fragrance of the corolla is important in attracting pollenating insects.

calyx
Refers to all sepals collectively.

tepals
Petals and sepals when they are indistinguishable from one another.

perianth
Refers to the calyx and corolla collectively.

corolla
Refers to all petals collectively.

Figure 7.2a–c   (a) The green leaf-like sepals of a rose contrast vividly with these red petals. (b) Similarly, the sepals of the fuchsia contrast with the petals but are also attractive components of the overall appearance of the flower. (c) The daylily blossom has sepals indistinguishable from the petals.

Source: Cynthia McKenney.

(a)

Source: Cynthia McKenney.

(b)

Source: Cynthia McKenney.

(c)

Figure 7.3a–c (a) The highly colored petals of this orange poppy attract insects; (b) the sweet fragrance of the wisteria flower accomplishes the same task; (c) the flowers of the strawflower are composed of many small florets that look like petals.

Stamens

stamens
Male floral structure composed of the anther and filament.

anther
Portion of the stemen that bears pollen.

filament
Stalk portion of a stamen that supports the anthers.

pollen
Male sex cells that originate in the anther.

self-pollination
When a flower is able to be pollinated by the pollen from the same plant; synonymous with the term *selfing*.

cross-pollination
When a plant is pollinated by the pollen of a different plant.

The male part of the flower is the **stamen**, which is composed of an **anther** and a **filament** (Figure 7.4). **Pollen** is the male sex cell and the anther is the structure that bears it. The brown pollen coats the flat anthers, which are attached to long, thin stalks called *filaments*. When the pollen is mature the anther splits or dehisces and the pollen is shed. The stamens surrounding the stigma of the tulip show mature pollen grains, which are able to be released in order to land on a stigma, resulting in the process of pollination. Some flowers are capable of **self-pollination** or "selfing", whereas other plants require the pollen from a different plant in order to successfully pollinate, called **cross-pollination**. Some flowers darken after pollination, such as the Big Bend bluebonnet (*Lupinus havardii*) or the black jewel orchid (*Ludisia discolor*), which provides an indicator to their status.

Given the importance of flowers to plant reproduction, another floral classification is based on the presence or absence of the stamens and pistils.

Pistil

Stamen

© Tatiana Makotra/Shutterstock.com

Figure 7.4 The yellow column in the center of this tulip is referred to as the pistil; the brown stalked structures surrounding the pistil are the stamens.

When both the male and female floral parts are present, the flower is said to be **perfect**. Likewise, if either of these reproductive organs is missing, the flower is **imperfect**. Imperfect flowers may be either male or female. If the pistil is missing and the flower only has the male components, it is referred to as **staminate** flower; female-only flowers are called **pistillate**. Some plants have both staminate and pistillate flowers, such as squash (*Cucurbita spp.*), corn (*Zea mays*), and pine (*Pinus spp.*). These plants are called **monoecious**, which means "one house" in Greek (Preece, 2005). Monoecious flowers are frequently wind pollinated. Contrastingly, when plants bear male and female flowers on separate plants, they are classed as **dioecious**, which means "two houses" in Greek. American holly (*Ilex opaca*) and buffalograss (*Buchloe dactyloides*) are examples of dioecious plants. If you are planting holly shrubs and you want to have lots of red berries for fall color, select one male plant as a pollinator for your yard. Then choose female plants for the remainder of your holly shrubs, increasing the number of plants producing the attractive fruits.

Pistils

If we return back to our concept of a perfect flower, the tulip (*Tulipa spp.*) in Figure 7.4 helps us to identify several other floral parts. The large yellow structure in the center of the flower is composed of several sections. This is called the **pistil** and it includes the **stigma**, **style**, and **ovary**. This female structure of the plant has a stigma at the very tip where the pollen grains are received. Immediately beneath is a thin style, which is the stalk joining the stigma to the ovary below. The ovary contains the ovules or female sex cells. Some plants have a single ovary chamber containing the ovules; however, other plants have a **compound pistil** in which the ovules are enclosed in multiple chambers called **locules** or **carpels** (Preece, 2005). The orientation and placement of the ovary and the ovules in the locules is useful in taxonomic identification, referred to as **placentation**.

FLOWERS AND THEIR POLLINATORS

Unlike animals, plants cannot move to find their mates for sexual reproduction. Most gymnosperms and some angiosperms rely on wind to move their pollen to the ovules. Wind pollination is passive and unreliable. To increase the chance of pollen reaching the ovules by wind, plants have to produce large amounts of pollen. Pollination by animals is more efficient and accurate than wind pollination and requires fewer pollen grains. Most angiosperms recruit animals to pollinate for them by having brightly colored blossoms such as the columbine (*Aquilegia × hybrida*), monkey flower (*Mimulus guttatus*), and impatiens (*Impatiens hawkeri*) (Figure 7.5a–c), in contrast to fragrant flowers attracting pollinators with their sweet-smelling nectar such as the lilac (*Syringa vulgaris*; Figure 7.6a). In some flowers, the nectar is not sweet-smelling but is more fetid. The corpse flower (*Amorphophallus titanium*) has such an intense stench that workers in the area near it must wear a breathing apparatus (Figure 7.6b). This flower may be up to 12 feet tall, thus it produces enough fumes to be noxious (Gandawijaja, Idris, Nasution, Nyman, & Arditti, 1982).

Nectar and pollen are provided as food for animals in return for their pollination service. Animal pollinators play a key role in the evolution and

---

perfect flower
When a flower has both male and female components.

imperfect flower
When a flower has only male or female components but not both.

staminate
Male flowers having no pistils.

pistillate
Flower having only female components.

monoecious
Male and female sex organs are found on the same individual.

dioecious
Male and female sex organs found on distinct individuals.

pistils
Female flower component composed of the stigma, style, and ovary.

stigma
Female floral structure at the tip of the style that receives pollen.

style
Female floral structure below the stigma where the pollen tube grows through to the ovary.

ovary
Female flower component that is at the base of the style and contains the eggs.

compound pistil
Plant that has ovules enclosed in multiple chambers or carpels.

locule
Ovary chamber.

carpel
Flower structure from a modified leaf bearing ovules; multiples may fuse together.

placentation
Placement of the ovules in the locules.

Source: Cynthia McKenney.

(a)

Source: Cynthia McKenney.

(b)

Source: Cynthia McKenney.

(c)

Figure 7.5a–c The brightly colored blossoms of (a) columbines, (b) monkey flowers, and (c) impatiens attract pollinators.

Source: Cynthia McKenney.

(a)

© Paul Marcus/Shutterstock.com

(b)

Figure 7.6a–b (a) The common lilac attracts pollinators with its sweet fragrance, in contrast to the (b) corpse flower, which attracts pollinators with the smell of rotted flesh.

diversity of flowers and are an important factor in the evolutionary success of flowering plants. When a plant species has few animal pollinators, the relationship between a flowering plant and its pollinators becomes highly specialized. To increase pollination efficacy, pollinator-specific floral traits began to emerge (Stebbins, 1970; Evert & Eichhorn, 2013) (Table 7.1).

In a few instances, the relationship between a flowering plant and its pollinator become so tightly linked or obligated that they completely depend on each other for survival. For example, *yucca* plants are pollinated exclusively by yucca moths in the genus *Tegeticula*. The moth larvae can only complete their development within yucca flowers (Powell, 1992). Yucca plants and yucca moths rely exclusively on one another for reproduction. The relationship between figs (*Ficus carica*) and *Pegoscapus* fig wasps is another classic example of an obligate mutualistic relationship between a flowering plant and its pollinator (Weibes, 1979).

Table 7.1   Characterization of Pollinator-Specific Floral Traits

Pollinator	Flower color	Flower structure	Nectar	Odor
Wind	Dull	Petals are small or absent, well-exposed anthers and stigmas, stigmas may be feathery	No	Odorless
Bees	Blue or yellow, rarely red	Petals are showy and have special markings or nectar guides visible in ultraviolet light	Yes	Sweet
Butterflies	Bright red, blue, yellow, or orange	Petals are used as landing platforms, have long floral tubes to hold nectar	Yes	Sweet
Moths	White or yellow	Petals are used as landing platforms, have long floral tubes to hold nectar, open at night	Yes	Heavy musky odor
Carrion flies	Dull red or brown	Flower parts produce heat	No	Smell like rotten meat
Birds	Bright red or yellow	Flowers are large and usually part of a large and sturdy inflorescence, have long floral tubes to hold nectar	Copious amount	Odorless
Bats	White	Flowers are large, have long floral tubes to hold nectar, open only at night	Copious amount	Fruit-like or musky odor

IDENTIFICATION AND CLASSIFICATION OF FLOWERS

The diversity of flower shape, symmetry, and complexity is astounding. Recognition of these features provides some of the primary tools used in plant identification. The shape of individual flowers provides an excellent place to begin with this taxonomic survey.

Symmetry

Actinomorphic flowers are also referred to as **radially symmetrical flowers** where the petals are evenly distributed around a central point and if a plane is passed through the flower in several directions, the halves would be matching. Windflowers (*Anemone × hybrida*) and Persian buttercups (*Ranunculus asiaticus*) are examples of radially symmetrical flowers (Figure 7.7a–b). The white lines dissecting the flowers indicate the planes passing through the flower, highlighting the symmetry. The flower parts on either side of the plane are mirror images no matter which direction the planes are rotated. By contrast, **zygomorphic flowers** are also said to have **bilateral symmetry**. In these flowers the petals are not uniform in shape or size and may not match in color. If a plane is passed through the flower, it would have mirror images in one direction but not if the plane is rotated in another direction. Snapdragons (*Antirrhinum majus*) and flamingo flowers (*Anthurium andraeanum*)

actinomorphic flowers
See *radial symmetry*.

radially symmetrical flowers
The petals are evenly distributed around a central point; mirror images created in multiple directions or planes.

zygomorphic flowers
See *bilateral symmetry*.

bilateral symmetry
Expressed by flowers when a plane passed through a flower creates mirror images in one direction but not in another.

Source: Cynthia McKenney.

Source: Cynthia McKenney.

(a) (b)

Figure 7.7a–b (a) These windflowers and (b) Persian buttercups are examples of actinomorphic flowers with radial symmetry. The white lines represent a plane passed through the flower itself.

are excellent examples of zygomorphic flowers with bilateral symmetry (Figure 7.8a–b). Dividing the flowers north and south, along the central plane, creates sides with mirror images. However, dividing the flowers east and west or other directions creates images that do not match.

Corolla Types

Corolla shapes provide an additional way to classify flowers. A familiarity with the taxonomic terms used to describe the shapes of corollas helps in communicating about flowers and in identifying an unknown plant. There are numerous types of corollas; however, Table 7.2 provides examples of only the most common forms. These corollas are classified by whether the

Source: Cynthia McKenney.

(a) (b)

Figure 7.8a–b (a) The two-lipped (bilabate) flowers of the snapdragon and (b) the heart-shaped flower of the flamingo flower demonstrate how the flowers cannot be dissected in all directions and have matching sides. This is zygomorphic symmetry.

Table 7.2   Classification, Definition, and Examples of Corolla Types

Corolla Shape	Plant Name	Description	Image
Bilabate	*Antirrhinum majus* Snapdragon	Composed of fused petals creating two lips surrounding a deep throat. Flowers exhibit bilateral symmetry.	© Iva Vagnerova/ Shutterstock.com
Campanulate	*Campanula × hybrida* Bellflower	Composed of fused petals creating a deep U-shaped vase. Flowers have radial symmetry.	Source: Cynthia McKenney.
Funnelform	*Brugmansia × hybrida* Angel's Trumpets	Composed of fused petals creating a funnel shape with a deep throat. Flowers exhibit radial symmetry.	Source: Cynthia McKenney.
Papilionaceous	*Lathyrus odoratus* Sweet Pea	Composed of a large petal at the top called a *standard* and two smaller overlapping lower petals called *keel*. Flowers have bilateral symmetry.	© Vilor/Shutterstock.com
Rotate	*Brunsfelsia australis* Yesterday, Today and Tomorrow	Composed of multiple petals fused together surrounding the center. Throat is shallow and the symmetry is radial.	Source: Cynthia McKenney.
Salverform	*Ipomoea purpurea* Morning Glory Vine	Composed of fused petals creating an elongated throat with a flat circular lip. Flowers exhibit radial symmetry.	Source: Cynthia McKenney.

inflorescence
Floral stalk bearing multiple flowers or a flower cluster.

peduncle
Flower stalk of a single flower or cluster of flowers.

sessile
When florets, flowers, fruits, leaflets, or leaves attach directly to a stem or axis.

pedicels
Small stalks which attach the individual florets to the main stalk of an inflorescence.

petals are fused, how many petals are present, the length of the throat if present, and the type of symmetry exhibited.

Inflorescence Morphology

When we think of a single flower, it is easy to understand how the flower attaches to the stem. An **inflorescence** is a structure composed of multiple flowers. The way these flowers attach to the stalk provides a method for classification. The main stalk of the inflorescence is called a **peduncle**. When flowers attach directly to the peduncle, they are said to be **sessile**. If there are smaller stalks attaching individual flowers to the peduncle, these are called **pedicels**. The pedicels attach to the **rachis**. There may be an increase in complexity if clusters of flowers attach to a pedicel. The composite head, called a **capitulum**, is so compact that there may be hundreds of flowers per head. Table 7.3 provides common inflorescence classifications, definitions, and examples.

Table 7.3   Classification, Definition, and Examples of Inflorescence Types

Corolla Shape	Plant Name	Description	Line Drawing	Image
Solitary	*Rosa × hybrida* Rose	A single flower born on the stem.		Source: Cynthia McKenney.
Spath and spadix	*Anthurium andraeanum* Flamingo Flower	Inflorescence includes a fleshy finger-like structure called a *spadix*, which is made up of many small flowers. A bract known as a spath surrounds the structure.		Source: Cynthia McKenney.
Spike	*Gladiolus × hybrida* Gladiolas	Inflorescence consists of sessile flowers on an unbranched peduncle. There is no secondary attachment.		Source: Cynthia McKenney.
Raceme	*Muscari spp.* Grape Hyacinths	Inflorescence is composed of multiple flowers attached to the rachis with small stalks called pedicels.		© Diana Taliun/ Shutterstock.com

Panicle	*Cortaderia sell-oana* Pampas Grass	Inflorescence is a compound raceme where the rachis branches into pedicels, which are in turned branched again.		
Corymb	*Spiraea japon-ica* Japanese spiraea	Multiple flowers forming a flat inflorescence in which the youngest flowers attach at the base.		
Umbel	*Allium gigan-teum* Giant Onion	The pedicels all attach to the main axis at the same point.		
Cyme	*Conium macu-latum* Poison Hemlock	Multiple flowers forming a flat-topped inflorescence where the oldest flowers attach at the base.		
Composite head (capitulum)	*Helianthus annuus* Sunflower	The capitulum is composed of ray flowers, disk flowers, or a combination of both.		

Constructing a Floral Formula

As you have seen, there are many modifications in flowers that allow for a variety of methods for classification. Taxonomists have found **floral formulae** to be a convenient form of shorthand to summarize key features of a flower. It records the type of symmetry, number of flower parts, whether the parts are fused within or between whorls, and the ovary position (Hickey & King 1997; Judd, Campbell, Kellogg, Stevens, & Donoghue, 2008). Floral formulae are used to determine whether a plant is a monocot or eudicot and to which plant family it belongs. They are also very useful for highlighting and contrasting floral features of different plant families. A typical floral formula

rachis
Main axis where pedicels attached to in an inflorescence."

capitulum
Inflorescence term synonymous with composite head.

floral formula
Shorthand summary of floral features for a specific flower.

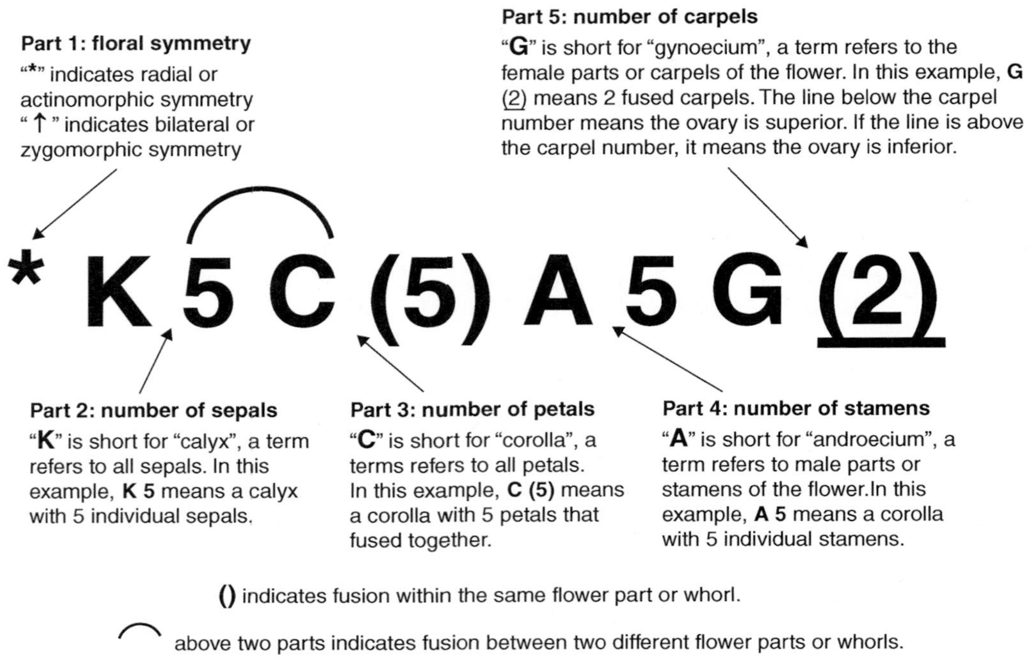

Part 1: floral symmetry

"★" indicates radial or actinomorphic symmetry

" ↑ " indicates bilateral or zygomorphic symmetry

Part 5: number of carpels

"**G**" is short for "gynoecium", a term refers to the female parts or carpels of the flower. In this example, **G** (2) means 2 fused carpels. The line below the carpel number means the ovary is superior. If the line is above the carpel number, it means the ovary is inferior.

$$* \; K\,5 \; C\,(5) \; A\,5 \; G\,\underline{(2)}$$

Part 2: number of sepals

"**K**" is short for "calyx", a term refers to all sepals. In this example, **K 5** means a calyx with 5 individual sepals.

Part 3: number of petals

"**C**" is short for "corolla", a terms refers to all petals. In this example, **C (5)** means a corolla with 5 petals that fused together.

Part 4: number of stamens

"**A**" is short for "androecium", a term refers to male parts or stamens of the flower. In this example, **A 5** means a corolla with 5 individual stamens.

() indicates fusion within the same flower part or whorl.

⌒ above two parts indicates fusion between two different flower parts or whorls.

Figure 7.9 A typical floral formula showing the five parts and their symbols with explanations.

consists of five symbols or parts (see Figure 7.9). Using Figure 7.9, you will be able to tell if the flower you are writing the floral formula for is a monocot or a eudicot flower by the number of flower parts it presents. In general, a monocot has flower parts in multiples of three whereas eudicots have flower parts in multiples of four or five.

Flowers vary greatly in their floral arrangement and deviations from a typical floral formula are often needed to properly describe a flower. Here are some common deviations:

A 5 + 5 The plus symbol (+) in this example is used to differentiate the stamens. There are 10 stamens in a flower but they are either in two whorls or different in height.

K 4 - 10 The dash symbol (-) in this example indicates a range. The number of petals in this family range from 4 to 10.

A ∞ The infinity symbol (∞) in this example indicates a large number of stamens.

T -6- In some flowers such as lilies, their sepals have the same color and are the same size as their petals and the two parts are difficult to differentiate. "T" is used instead of "K" and "C." "T" is short for "tepals," which is a term for sepals and petals that look alike. The hyphen (-) in front and after the number indicates the combined number of sepals and petals.

BOX 7.1   Economic Uses of Flowers

If you look at the energy invested by a plant to develop flowers, it should not be a surprise flowers are useful for more than attracting animals to aid in pollination. Each flower is composed of unique materials providing the fragrance, color, and texture needed. These elements result in a variety of uses for the floral components. An entire industry, the floral industry, revolves around cut flowers used for ornamentation. The landscape industry also incorporates colored plants in their designs to add interest. However, flowers are not just useful for their attractive appearance. They also have many other economic uses.

Flowers for Fragrances and Perfumes

Several flowers have been used through the years as a source of fragrance. Many perfumes are created with floral extracts to give them a sweet note to the fragrance. Lilacs (*Syringa vulgaris*), garden roses (*Rosa × hybrida*), and lavender (*Lavandula angustifolia*) are classic examples of flowers used for their delicate essence (Figure Box 7.1a–c).

(a)   Source: Cynthia McKenney

(b)   Source: Cynthia McKenney

(c)   Source: Cynthia McKenney

Figure Box 7.1a–c   Several flowering plants are used to make perfumes including (a) lilac, (b) rose, and (c) lavender.

(d)

(e)

(f)

Figure Box 7.1d–f (d) Broccoli, (e) cauliflower, and (f) globe artichoke are all common examples of edible flowers.

Flowers as a Food and Dye Source

Have you realized flowers are part of your diet? The flowers of several common plants such as broccoli and cauliflower (*Brassica oleracea*) and globe artichoke (*Cynara cardunculus* var. *scolymus*) are grown for their edible flowers (Figure Box 7.1d–f). More exotic flowers also find a place in a more adventurous diet (Figure Box 7.1g–l). Summer squash blossoms (*Cucurbita pepo*) can be stuffed and fried and dandelion petals (*Taraxacum officinale*) can be made into a wine. Nasturtium flowers (*Tropaeolum majus*) can be sprinkled in salads, made into pesto, or spread with a cream cheese filler to create appetizers. Chrysanthemum petals (*Chrysanthemum morifolium*) and daylily blossoms (*Hemerocallis spp.*) can be sprinkled into tossed salads for a pop of color. Violas (*Viola tricolor*) may be eaten raw or sugared to decorate desserts. One of the most drought-tolerant edible flowers is the soaptree yucca (*Yucca elata*) (Figure Box 7.1m). This semiarid to arid plant has panicles of cream-colored florets, which may be battered and deep fried or served fresh on a salad for a peppery taste. In contrast is the fall blooming saffron crocus (*Crocus sativus*), which blooms close to the ground in the cool of the fall season (Figure Box 7.1n). Saffron is the most expensive edible flower. The floral parts consumed are the stigma and style, which are harvested, dried, and sold as the spice saffron. Saffron is used for both its yellow food coloring and flavoring in rice, soups, and breads. Another plant that imparts a yellow color is yellow chamomile (*Anthemis tinctoria*; Figure Box 7.1o). This plant is a relative of the medicinal chamomile

Figure Box 7.1g–l   Exotic edible flowers include (g) squash blossoms, (h) dandelion flowers, (i) nasturtium flowers, (j) chrysanthemum petals, (k) daylily petals, and (l) viola flowers.

used to make tea. Yellow chamomile is used to produce a bright yellow dye, which may be used on both textiles and skins.

Flowers as Medicinals

Flowers have long provided a source of alternative medicines and herbal teas (Figure Box 7.1p). Probably the most easily recognized medicinal herb is purple coneflower (*Echinacea purpurea*). This native plant has desirable

Figure Box 7.1m–o (m) The soaptree yucca has blossoms that are delicious battered and deep fried; (n) the stigmas and styles of the saffron crocus provide both food coloring and flavor; (o) yellow chamomile is used an herbal dye.

Figure Box 7.1p–r Flowers providing alternative medicines include (p) purple coneflower, (q) St. John's wort, and (r) dead nettle flowers.

medicinal compounds found in all parts of the plant. A capsule of ground plant tissue or a few drops of a tincture is used to enhance the immune system (Barrett, 2003). There is conflicting opinions on the effectiveness and safety of using this alternative medicinal. Another alternative medicinal plant is St. John's wort (*Hypericum perforatum*) for the treatment of mild depression (Gaster & Holroyd, 2000). This is a common weed found in ditches and open areas across the country. As with the purple coneflower, use of this plant as an alternative medicine is debated in the literature. A tea of dead nettle flowers (*Lamium album*) is considered a moderate source for antioxidants (Buřičova & Reblova, 2008).

Flowers as Insecticides

The last use of flowers we now explore is as an insecticide. With the concern over toxicity and decomposition of insecticides in the environment, plants and plant extracts show some promise as an environmentally softer alternative. Marigolds (*Tagetes patula*) have long been considered a companion plant (Figure Box 7.1s). They have a rather strong smell that is recognized as a repellant to some insects, resulting in less damage to the primary crop (Latheef & Irwin, 1980). Another floral crop, pyrethrum daisies (*Chrysanthemum coccineum*) is such an effective insecticide; it is sold commercially (Figure Box 7.1t). The pyrethrum flowers are dried, crushed, and made into an insecticide. Next time you are at a garden center, check the insecticide display and you will probably see a pyrethrum insecticide. Remember, just because this is a plant-based product does not mean it is nontoxic!

(s)   Source: Cynthia McKenney.   (t)   © Birute Vijeikiene/Shutterstock.com

Figure Box 7.1s–t   (s) French marigolds are used as an insect repellant and (t) pyrethrum flowers are dried and made into a pesticide.

Key Terms

sexual reproduction	stamen	tepals
asexual reproduction	pistils	perianth
gametes	receptacle	corolla
zygote	complete flower	stamen
sepals	incomplete flower	anther
petals	calyx	filament

pollen
self-pollination
cross-pollination
perfect flower
imperfect flower
staminate
pistillate
monoecious
dioecious
pistil

stigma
style
ovary
compound pistil
locules
carpels
placentation
actinomorphic flowers
radially symmetrical flowers
zygomorphic flowers

bilateral symmetry
inflorescence
peduncle
sessile
pedicels
rachis
capitulum
floral formulae

Summary

- The primary components and functions of a flower:
  - Reproduction
    - Sexual by fusion of gametes producing seed
    - Asexual by cuttage, division, or separation
  - Complete flower
    - Petals collectively the corolla
    - Sepals collectively the calyx
    - Stamens including the anther and filament
    - Pistils including the stigma, style, and ovary
  - Classification of flowers by morphology:
    - Perfect with both male and female flower parts
    - Imperfect with either male or female flowers
    - Monoecious with both types of flowers on the same plant
    - Dioecious with only one type of flower on a plant
  - Plants rely on wind and attracting animals to help with pollination
    - Brightly colored corollas attract insects to a flower
    - Fragrance attracts pollinators
  - Some plants have highly developed relationships with specific pollinators
    - Yucca moth to yucca flowers
    - Fig wasp to figs
  - Identification and classification of flowers:
    - Symmetry
      - Actinomorphic
      - Zygomorphic
    - Corolla type
      - Bilabate
      - Campanulate
      - Funnelform
      - Papilionaceous

- Rotate
- Salverform
  - Inflorescence morphology
    - Parts include the peduncle, pedicel, and rachis
    - Classification includes:
      - Solitary
      - Spath and spadix
      - Spike
      - Raceme
      - Panicle
      - Corymb
      - Umbel
      - Cyme
      - Composite head
  - Floral formulae
    - Shorthand to summarize key floral features
    - Consists of five symbols
  - Economic uses of flowers:
    - Fragrances and perfumes
    - Food and dye source
    - Medicinal use
    - Insecticides

Reflect

1. *Floral formulae. How can you read this?* Try your hand at writing floral formulae by selecting three common flowers and identifying the five symbols that would comprise each of their formulae. Be sure to include an image of each of the plants.
2. *Plant identification?* Flowers have many different characteristics that may be used to aid in plant identification. If you are only allowed to use one characteristic, which would you choose? Explain your answer.
3. *Economic importance.* Flowers have many levels of economic importance. What are some other uses for flowers that are not listed in this discussion?

References

Barrett, B. (2003). Medicinal properties of echinacea: A critical review. *Phytomedicine, 10*(1), 66–86.

Buřičova, L., & Reblova, Z. (2008). Czech medicinal plants as possible sources of antioxidants. *Czechoslavakian Journal of Food Science, 26,* 132–138.

Evert, R. F., & Eichhorn, S. E. (2013). *Raven biology of plants* (8th ed.). New York: W.H. Freeman.

Gandawijaja, D., Idris, S., Nasution, R., Nyman, L. P., & Arditti, J. (1982). *Amorphophallis titanium* Becc.: A historical review and some recent observations. *Annals of Botany, 51*(3), 269–278.

Gaster, B., & Holroyd, J. (2000). St. John's wort for depression: A systematic review. *Archives of Internal Medicine, 160*(2), 152–156.

Hickey, M., & King, C. (1997). *Common families of flowering plants.* Cambridge, UK: Cambridge University Press.

Judd, W. S., Campbell, C. S., Kellogg, E. A., Stevens P. F., & Donoghue, M. J. (2008). *Plant systematics* (3rd ed.). Sunderland, MA: Sinauer Associates.

Latheef, M. A., & Irwin, R. D. (1980). Effects of companionate planting on snap bean insects, *Epilachna varivestia* and *Heliothis zed. Environmental Entomology, 9*(2), 195–198.

Preece, J. E. (2005). *The biology of horticulture: An introductory textbook* (2nd ed.). Hoboken, NJ: Wiley.

Powell, J. A. (1992). Interrelationships of yuccas and yucca moths. *Trends in Ecology and Evolution, 7,* 10–14.

Stebbins, G. L. (1970). Adaptive radiation of reproductive characteristics in angiosperms, I: pollination mechanisms. *Annual Review of Ecology and Systematics, 1,* 307–326.

Weibes, J. T. (1979). Co-evolution of figs and their insect pollinators. *Annual Review of Ecology and Systematics, 10,* 1–12.

Chapter 8

Fruits, Seeds, Dissemination, and Germination

© Adisa/Shutterstock.com

Learning Objectives

- Explain the steps in fruit development
- Discuss the structural components of fruits
- Identify six common methods of fruit dispersal
- Explain the identification and classification systems of common fruits
- Discuss the economic uses of seeds
- Explain the steps in the process of seed germination
- Discuss different types of dormancy
- Identify the different levels of seed certification
- Explain the concepts used to enhance seed preservation

Have you ever taken a bite of an apple and noticed several small seeds are located in the center, but a peach just has one large seed? Has it crossed your mind how you can raise seedless watermelons from seed? Once again, we are able to see the great diversity in plants and how they function. In Chapter 7 we investigated how flowers are pollinated resulting in the development of a ripened ovary called a **fruit**. Inside the fruit are seeds for the perpetuation of the plant. In this chapter we look more in depth at the characteristics of fruits, seeds, methods of seed dissemination, and the process of germination.

FRUIT DEVELOPMENT AND STRUCTURE

Fruit Development

Usually a fruit is defined as a ripened ovary. If a flower is not pollinated, it will fall from the plant and not set fruit. As we learned in Chapter 7, there are many different methods assisting in pollination such as insects, wind, or even your Aunt Polly hitting the tomato plants with a broom! Cans of blossom set sprays are available at garden centers to help aid in the pollination process, especially for tomatoes. These sprays have plant hormones that are designed to stimulate pollination of the flowers, increasing fruit production. However, weather and plant water status play important roles in fruit set. If a plant flowers too early, it may be hit by a cold snap and the leaves and fruit are not able to develop properly. Plants flowering too early may not have the necessary insects available to assist in their pollination, resulting in a low yield. Heat may also be a problem. Many plants like tomatoes stall during extremely hot conditions and will not set fruit until the night temperatures are lower.

Structural Components

When all the conditions are right and the flower is pollinated, the ovary begins to transform. As the ovary ripens and becomes recognized as a developing fruit, the ovary wall is referred to as a **pericarp**. In general, there are several different layers in the pericarp. The outer layer is called the **exocarp**. It may be thin, succulent, hard, or leathery. The next layer, which is the middle layer, is termed the **mesocarp**. The mesocarp varies with different types of fruits but frequently it is one of the fleshy portions of the fruit that is consumed. The **endocarp** is the innermost layer, which may be soft and succulent like a tomato or hard and stony like the pit of a peach. Identification of these different layers of the pericarp provides a useful method of botanical classification and identification, as seen in Figure 8.1. Here the cross section of the orange identifies the peel, which is composed of the exocarp and mesocarp, with the endocarp being the luscious fruit you consume.

Generally, seeds result from pollination of the flower; however, in some species application of the plant hormone auxin can induce fruit set. When hormones induce fruit growth without benefit of pollination, the fruit is called **parthenocarpic**. Genetically, some parthenocarpic fruit is formed when a plant is a triploid (when a plant has three complete sets of chromosomes) or has other odd ploidy levels. Some common examples of fruits without seeds include navel oranges, bananas, seedless watermelons, and

Fruit
Ripened ovary.

pericarp
Ovary wall of a developing fruit.

exocarp
Outer layer of the pericarp.

mesocarp
Middle layer of the pericarp.

endocarp
Innermost layer of the pericarp of a fleshy fruit.

parthenocarpic
When fruit is developed without the process of pollination.

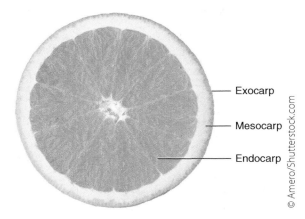

Exocarp

Mesocarp

Endocarp

© Amero/Shutterstock.com

Figure 8.1   The layers composing the pericarp of an orange include the exocarp, mesocarp, and endocarp.

some types of grapes. When a seedless fruit is cut open, rudimentary seeds called **vestigial seeds** may be seen.

vestigial seeds
Rudimentary seeds that develop in a fruit that is considered seedless.

FRUITS AND THEIR SEED DISPERSAL MECHANISMS

After the fruit ripens, the seeds are developed and capable of germination. But how do plants distribute their fruits/seeds? There are a variety of ways seeds are dispersed. Some fruits are carried by the wind due to feathery attachments to their fruits; others float on water to be washed up on a distant shore. Some fruits have hooked barbs, allowing them to catch on clothing or the coat of an animal, thus transporting them greater distances before they eventually come loose. With the introduction of differing methods of transportation such as trains, planes, and trucks, seeds manage to hitch a ride across the country effortlessly. However, some seeds stay closer to home and simply drop to the ground when the seed pod splits open along its sutures. Table 8.1 illustrates some of the more common methods of seed dispersal exhibited by plants.

IDENTIFICATION AND CLASSIFICATION OF FRUITING STRUCTURES

The diversity of fruit morphology provides an excellent characteristic by which plant materials may be identified. Recognition of these features may serve in determining the plant family and ultimately the identification of the plant bearing the fruit. This is a good thing to know before you decide to pop it in your mouth or plant it in your garden! Along with your new familiarity on how seeds are dispersed, other botanical characteristics are helpful in understanding these diverse fruiting structures. Fruit complexity is determined by whether the fruit is composed of a single carpel ovary known as a **simple fruit**, or from a multiple carpel ovary resulting in a **compound fruit**.

simple fruit
Fruit that is developed from an ovary with a single carpel.

compound fruit
Fruit that develops from an ovary with multiple carpels.

Table 8.1   Common Seed Dispersal Mechanisms of Plants

Dispersal Agent	Fruit	Description	Image
Wind	Dandelion *Taraxacum officinale*	Fruits have feathery attachments that are caught by the wind	© Alexey U/Shutterstock.com
Water	Coconut *Cocos nucifera*	Fruits drop and are carried by water to the next location where it is buried by wave action	© OlegD/Shutterstock.com
Animal coat	Cocklebur *Xanthium strumarium*	Hooks on the outer surface of the fruit attach to the coat of the animal and travel some distance before coming loose	© chinahbzyg/Shutterstock.com
Animal digestive tract	Possum-Haw *Ilex decidua*	Birds, cattle, and wild animals eat berries and pass the seeds through their digestive tracts where they then are excreted in a new location	Source: Cynthia McKenney.
Cars, planes, and trains	Wheat *Triticum aestivum*	Crop being transported intentionally in trucks, in the cuff of a pair of slacks, on a train, or on the tread of a tire	© Robert Lucian Crusitu/Shutterstock.com
Gravity	Green Bean *Phaseolus vulgaris*	Pod splits down the suture and the seeds drop to the ground, rolling a short distance	© Kelvin Wong/Shutterstock.com

fleshy fruit
Fruit in which the pericarp is thick and succulent when developed.

Further diversity is found in whether the fruit is dry or fleshy. Many of the fruits we purchase at the grocery store in the produce section are examples of **fleshy fruits**. Pears, grapes, apples, oranges, lemons, bananas, blueberries, apricots, and cherries are all fleshy fruits. Generally, when you think about eating a piece of fruit, you are referring to a fleshy fruit. Table 8.2 provides a review of many of the common forms of fleshy fruits.

There are several other unique fleshy fruit types. These fruits have distinctive modifications, making them easy to identify. Aggregate fruits result when several ovaries are produced by a single flower and share a receptacle. Raspberries are an example of aggregate fruits. In contrast, multiple fruits are formed when several flowers fuse together during the maturation process, resulting in a fruit with a common receptacle creating a core. Mulberries, pineapples, and figs are examples of multiple fruits. The final classification of fleshy fruits is known as an accessory fruit. These fruits develop when the tissues close to the ovary become fleshy. A strawberry not only is a aggregate fruit but is also an accessory fruit in which the receptacle becomes fleshy and the seeds are achenes attached to the surface of the fruit. Pomes such as apples and pears are also examples of accessory

Table 8.2   Types of Fleshy Fruits

Fruit Type	Plant Name	Description	Image
Berry	Tomato *Lycopersicum esculentum*	Fleshy fruit in which the entire pericarp is fleshy and the seeds are distributed throughout the endocarp.	© Thomas Klee/Shutterstock.com
Pepo	Squash *Cucurbita maxima*	Fleshy fruit in which the exocarp thickens to form a rind and the mesocarp is the portion of the fruit consumed.	© Peter Zijlstra/Shutterstock.com
Pome	Apple *Malus domestica*	Fruit in which the pericarp is enclosed by a fleshy hypanthium (floral tube) and the endocarp develops into a core containing the seeds.	© Ekkapon/Shutterstock.com
Drupe/stone fruit	Plum *Prunus domestica*	Fruit in which the exocarp and mesocarp are fleshy while the endocarp forms a stony pit enclosing a single seed.	© Dionisvera/Shutterstock.com
Hesperidium	Orange *Citrus sinensis*	Fruit in which the exocarp and mesocarp form a leathery rind containing essential oils.	© kaband/Shutterstock.com

fruits. Table 8.3 provides descriptions and examples of these three types of fruits.

When a fruit is not succulent and fleshy, it is considered to be a **dry fruit**. Dry fruits are characterized by the number of sutures they have and whether or not the sutures split and the seeds dehisce. Table 8.4 reviews characteristics and examples of dry **dehiscent fruits**. These structures use a wide variety of dispersal mechanisms explored previously.

Dry **indehiscent fruits** are dry fruits characterized by the sutures not splitting during the ripening process. Like dehiscent fruits, these fruiting structures exhibit a variety of dispersal mechanisms. Table 8.5 provides examples and descriptions of many of the common types of dry indehiscent fruits.

PROPAGATION BY SEED

Propagating plants by seed is one of the primary ways of producing quality plants. There are a few basic practices to follow when planting seeds. First, the planting medium should have a good water-holding capacity while still providing adequate drainage and oxygen. There also needs to be adequate temperature, moisture, and, for some species, light. The seed should be planted at a depth of two to three times the diameter of the seed. Most seeds will germinate in 3–21 days.

dry fruit
Fruit in which the pericarp is not fleshy or succulent when developed.

dehiscent
Fruiting structure that splits along a suture at maturity, releasing the seeds within.

indehiscent fruits
When a dry fruit has sutures that do not split and the seed is distributed by other mechanisms.

Table 8.3 Types of Unique Fleshy Fruits

Fruit Type	Plant Name	Description	Image
Aggregate fruit	Blackberry *Rubus fruticosus*	A fruit in which multiple ovaries are produced by a single flower and are attached to a common receptacle.	© BMJ/Shutterstock.com
Multiple Fruit	Pineapple *Ananas comosus*	A fruit formed from several flowers that fuse together during ripening.	© naluwan/Shutterstock.com
Accessory Fruit	Strawberry *Fragaria virginiana*	A fruit in which tissues adjacent to the ovary become fleshy components.	© Swetlana Wall/Shutterstock.com

Table 8.4 Types of Dry Dehiscent Fruits

Fruit Type	Plant Name	Description	Image
Legume	Garden pea *Pisum sativum*	Has one carpel in which two sutures split and disperse the seed by dropping them due to gravity.	© Nattika/Shutterstock.com
Silique	Canola *Brassica napus*	Has two carpels in which the length is longer than it is wide and the two sutures split dispersing the seed, leaving behind a type of septum that divides the chambers.	© Crepesoles/Shutterstock.com
Capsule	Thorn Apple *Datura stramonium*	Has multiple carpels and the sutures split open at the top and the seeds fall or may be thrown due to the movement of the sutures.	© Dr. Morley Read/Shutterstock.com
Follicle	Milkweed *Asclepias syriaca*	Has a single carpel that contains multiple seeds and dehisces by a single suture that opens.	© Le Do/Shutterstock.com

germination
When a seed goes through multiple physical and chemical changes that results in growth of the enclosed embryo.

imbibition
When a seed absorbs moisture resulting in initiation of growth activities.

GERMINATION

The process of **germination** is a series of complex processes that initiates growth activity in the seed. The first step is the **imbibition** of water. In this process the dry seed absorbs moisture, which results in a swelling of the seed and a breakdown of the stored foods. Respiration occurs, which burns the food producing energy that allows for the synthesis of DNA, RNA, proteins, and enzymes, used to develop the embryo of the seed. New cells and tissues

Table 8.5   Types of Dry Indehiscent Fruits

Fruit Type	Plant Name	Description	Image
Nut	Pecan *Carya illinoinensis*	Exocarp surrounds a stony mesocarp, protecting the desirable endocarp.	 © Jiri Hera/ Shutterstock.com
Caryopsis	Corn *Zea mays*	A simple dry indehiscent fruit in which the fruit wall is fused to the seed, creating the appearance of a single unit.	 © Maks Narodenko/ Shutterstock.com
Schizocarp	Carrot *Daucus carota*	A simple dry fruit that is composed of two single-seeded achene-like seeds that separate at maturity.	 © Claudio Zaccherini/ Shutterstock.com
Achene	Sunflower *Helianthus annuus*	Simple dry indehiscent fruit with one carpel containing one seed that may be easily separated.	 © Brzostowska/Shutterstock.com
Samara	Maple *Acer saccharum*	Simple dry indehiscent winged fruit with one carpel that contains a single seed.	 © marekuliasz/ Shutterstock.com

BOX 8.1 Economic Uses of Seeds

Plants invest considerable energy in the development of fruits and seeds. This stored energy has many economic uses. Seeds may be used as a source of culinary spices, cooking oil, renewable energy, and confectionary ingredients. Increasingly, new economic uses for seeds are found, enriching our quality of life.

Seeds as Spices and Flavorings

Seeds have been used throughout the ages as sources of spice. The ground seed of mustard, dill, and nutmeg are just a few examples of seed used as culinary flavoring and for preservation (Figure Box 8.1-1a–e). The seed of cacao, coffee, and vanilla are also prepared for use in beverages and as extracts for cooking (Figure Box 8.1-2a–c). Spices and extracts command a high price, historically leading to the trade routes developed to secure these items. In our economy, coffee and chocolate have become expected daily pleasures rather than occasional extravagances. Internationally, efforts are currently being made to try and rebuild the shrinking cacao industry.

Figure Box 8.1-1a–e Seeds of (a) anise, (b) dill, (c) mustard, (d) nutmeg, and (e) pepper are ground and used as culinary spices.

Seeds as a Source of Oil

Several plants are cropped for the oil produced from their seeds. Bladder pod, canola, olive, evening primrose, coconut, and palm are just a few of the examples of crops used to produce renewable sources of oil with specific beneficial characteristics (Figure Box 8.1-3a–d). Canola and olive are healthful sources of cooking oils; bladder pod oil is used in making plastics, cosmetics, and lubricating oils. Evening primrose oil has been identified as useful to treat skin disorders (Balch, McKenney, & Auld, 2003). Coconut and palm oils are also used as sources of cooking oil. Next time you buy a snack item such as crackers or chips, be sure to look on the label and you will find some of these oils listed.

(a)

(b)

(c)

Figure Box 8.1-2a–c   Seeds of (a) cacao, (b) coffee, and (c) vanilla are used to create beverages and extracts for cooking.

Seeds as an Energy Source

As the price of fuel continues to increase, renewable sources of energy are in high demand. Globally, the oil extracted from the seed of castor, corn, and palm is used for the production of these alternative energy sources (Figure Box 8.1-4a–c). Increasingly in the United States, biodiesel is being manufactured, especially from corn given it yields 93% more energy than used in its production, while ethanol yields only 25% (Hill, Nelson, Tilman, Polasky, & Tiffany, 2006).

The many uses of corn, ranging from human consumption to cattle feed, creates concerns regarding the competition between food sources and energy. In addition, the massive global production of palm oil is altering ecosystems as large areas of rainforests, savannas, and grasslands are converted for its production. Serious environmental concerns have been

▶▶▶

Figure Box 8.1-3a–d Seeds of (a) canola, (b) coconut, (c) olive, and (d) palm are used as sources of culinary oils.

raised regarding the "biofuel carbon debt" (Fargione, Hill, Tilman, Polasky, & Hawthorne, 2008). Estimates of the carbon debt created by the production of palm oil vary between 17 and 420 times the amount of greenhouse gas reduction achieved through using biofuels rather than fossil fuels. A balance will have to be achieved between the many environmental, health, and food security impacts resulting from alternative fuel production.

Seeds as Sources of Protein

For health-conscious individuals, sources of protein are an important component of any diet. Nuts provide a healthy snack rich in protein. Likewise, pulse crops such as beans and peas provide an excellent source of protein (Figure Box 8.1-5a–b). As more individuals reduce their dependence on red meat, nuts and pulse crops are increasingly turned to for their delicious and nutritious attributes.

(a)

(b)

(c)

Figure Box 8.1-4a–c   Seeds of (a) castor, (b) corn, and (c) palm are used to manufacture renewable sources of energy including biodiesel.

(a)

(b)

Figure Box 8.1-5a–b   (a) Dried beans and (b) Nuts are examples of seeds that provide a rich source of protein to our diets.

plumule
Rudimentary portion of a plant embryo giving rise to the shoot.

epigeous
Type of germination in which eudicot seeds develop a hypocotyl arch that emerges first through the soil before the cotyledons.

hypogeous
Type of germination in which monocot seeds have the plumule emerge through the soil first and the cotyledon remains below the soil surface.

testa
Hard outer seed coat frequently called the integument or seed coat.

epicotyls
Shoot that develops above the cotyledons.

hypocotyls
Shoot that develops below the cotyledons.

are formed, resulting in the emergence of the radicle (see Chapter 4 for a review of this process) and **plumule**.

There are two types of germination: **epigeous** and **hypogeous**. Epigeous germination is associated with eudicot species. Figure 8.2 is an image of a eudicot seed in which the stored food that is used to produce energy is the cotyledon. The outer seed coat is the **testa**. This provides protection and may have to be altered physically to allow for imbibition in hard-seeded species. The **epicotyl** and **hypocotyl** form the shoot along with the radicle. These mature and break through the testa to create the components of the seedling. Figure 8.3 shows epigeous germination. The hypocotyl grows and bends to create an arch, dragging the cotyledons out of the soil as it elongates. After the cotyledons emerge, the true leaves form and the cotyledons wither and fall.

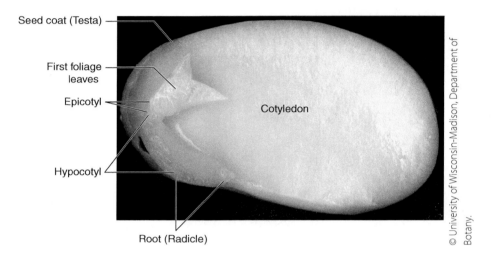

© University of Wisconsin-Madison, Department of Botany.

Figure 8.2 The cross section of the bean seed shows one of two eudicot cotyledons. The primordial components have been labeled.

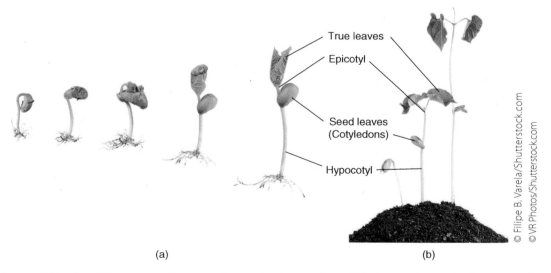

© Filipe B. Varela/Shutterstock.com
© VR Photos/Shutterstock.com

(a)   (b)

Figure 8.3a–b When a eudicot seed germinates such as this bean plant, the hypocotyl elongates and bends to form an arch, which breaks through the soil first and pulls the two seed leaves out after it. (b) After the two seed leaves open, the seedling produces the first true leaves.

Hypogeous germination is represented by monocots and some eudicots. In this type of germination, the cotyledon(s) remain below the soil surface. In Figure 8.4 a kernel of corn provides an example of a monocot seed undergoing hypogeous germination. There is only one cotyledon. The embryo includes the **coleorhiza**, which precedes the radicle emerging through the seed coat; the radicle, which will grow and initiate the root system; and the **coleoptile**, which will push up through the soil, followed by the plumule forming true leaves. Figure 8.5 provides a visual example of hypogeous germination of a monocot. Notice in hypogeous germination the plumule or epicotyl rather than the hypocotyl elongates.

coleorhizae
Portion of the germinating monocot seed that precedes the emerging radicle.

coleoptiles
First structure of a monocot seedling that emerges through the soil.

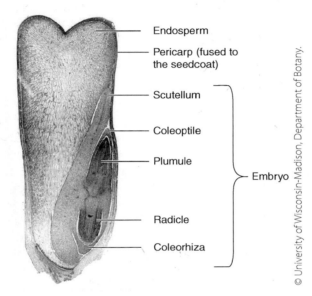

Endosperm

Pericarp (fused to the seedcoat)

Scutellum

Coleoptile

Plumule

Embryo

Radicle

Coleorhiza

© University of Wisconsin-Madison, Department of Botany.

Figure 8.4   Cross section of a monocot seed showing the primordial tissues which develop into the seedling.

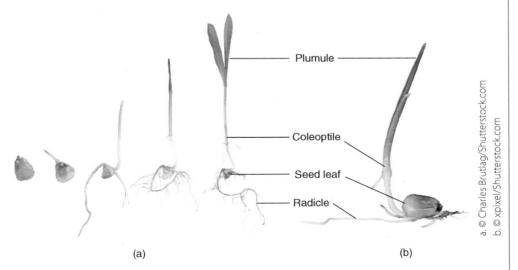

Plumule

Coleoptile

Seed leaf

Radicle

(a)   (b)

a. © Charles Brutlag/Shutterstock.com
b. © xpixel/Shutterstock.com

Figure 8.5   When a monocot like corn germinates, the radicle grows down and then the plumule grows up. The coleoptile protects the young stem as it pushes through the soil. There is only one seed leaf in monocots.

DORMANCY

Germination is not always immediately possible for seeds. This has both advantages and disadvantages for the plant itself. This delay in germination is called **dormancy**. Dormancy, also known as rest, is caused by physiological inhibition, anatomical constraints, or a combination of both. Failure of the seed to germinate even when the required environmental conditions of appropriate temperature, available oxygen, acceptable pH, and sufficient water are all present constitutes dormancy.

dormancy
Form of rest for the seed where germination does not occur due to physiological or anatomical issues.

Quiescence

If seed does not have the appropriate environmental conditions above and does not germinate, it is called **quiescence**. Dormancy is favorable when seed does not germinate until ideal environmental conditions are likely. This prevents the seed from being wasted during poor climatic conditions. In contrast, dormancy is not favorable when seed has all of the environmental conditions met but the seed will not germinate and the crop is needed. These situations have resulted in an area of study called *seed science*. Seed scientists work in part to determine the best set of environmental conditions to facilitate early, uniform germination.

quiescence
When a seed does not develop even until the appropriate environmental conditions are provided.

There are several classifications of seed dormancy. *Innate* or **primary dormancy** occurs when seed is not immediately able to germinate after ripening or harvest. Contrastingly, **secondary dormancy**, also called *induced dormancy*, occurs when the seed is exposed to extreme stress such as high temperatures, drought, or a lack of oxygen. For this reason, it is important to store seed in dry, cool conditions to prevent loss of viability.

primary dormancy
When a seed is not capable of germination immediately after harvest.

secondary dormancy
Induced by extreme stress or inappropriate storage.

Physiological Dormancy

Physiological dormancy is also termed *embryo dormancy* and it occurs when a specific treatment must be given to the embryo in order to initiate growth. Seed of temperate zone woody plants frequently require stratification while tropical seed generally does not. **Stratification** is the process of chilling seed in a moist medium such as sand, perlite, or potting media from 6 to 12 weeks between 35 and 45°F (1.7 and 7.2°C). Physiologically immature seeds also present physiological dormancy and require specific treatment to allow the necessary biochemical and enzymatic changes to occur. This process is referred to as **after ripening** and where the seed is stored providing time for the embryos to completely mature. Other types of seed treatments that enhance germination include soaking in hot water, cold water, dilute bleach, or smoke water. These can aid in leaching chemicals within the seed, inhibiting active growth of the embryo. All of these treatments are specific to individual plant species.

physiological dormancy
Internal dormancy in which a specific treatment must occur in order for the embryo to germinate.

stratification
Seed treatment in which seed is chilled in a moist media.

after ripening
When seed is held in appropriate storage conditions to provide time for the embryo to develop.

Physical Dormancy

In contrast, structural or **physical dormancy** also impacts the germination of seed, frequently called *seed coat dormancy*. This name refers to a seed coat that is so hard it is impervious to moisture, thus preventing it from imbibing water and initiating the germination process. The honey locust tree (*Gleditsia triacanthos*) is an example of this type of dormancy.

physical dormancy
Type of dormancy where the hard seed coat prevents imbibitions of water to initiate germination.

There are several methods of overcoming physical dormancy, including damaging the seed coat, called **scarification**. This may be accomplished using a triangular file or tumbling the seed with gravel. An alternate method used by professional propagators is to soak the seed in concentrated sulfuric acid for a specified period of time.

Double Dormancy

Double dormancy refers to seed that has a combination of both physical and physiological dormancy. Seed from the green ash tree (*Fraxinus pennsylvanica*) requires multiple cycles of both cold and warm treatments, which mimic the changing seasons in the environment. Providing this stratification and after-ripening treatment allows the seed to be germinated in a shorter span of time as compared to seed left in the environment.

SEED PRODUCTION AND STORAGE

Seed Certification

Seed production is an important industry that is regulated through very strict quality-control measures. Seed certification agencies serve to inspect and certify seeds for governmental agencies and for international partners. The initial seed that is first produced is termed **breeder's seed**. There must be enough of the seed to allow for further production and the seed needs to be maintained to retain its unique characteristics. Isolation of the field is important to prevent cross pollination by other sources. **Foundation seed** is used to maintain the cultivar to provide a very pure, high-quality seed for production. Foundation plantings are planted from breeder's seed or other foundation seed. In order to maintain this high level of genetic purity, the seed is subjected to testing and inspection. The third level of seed is **registered seed**. This seed is established from foundation seed or another source of registered seed. This seed is kept at a high level of purity but does not have as much testing and inspection as the foundation seed. The final level of seed purity is the seed generally sold to growers. This is **certified seed** and may be established from foundation or registered seed. This seed has even less regulation but is more genetically pure and generally has better uniformity for cropping than **noncertified seed**.

Quality and purity of the seed is important such that certain information is required on a seed label. Name of the crop and cultivar, date, percent germination, percent pure seed, percent other crop seed, percent inert material, percent weed seed, and percent noxious weed seed must be listed on the label so the grower is aware of the quality of the seed being purchased. Federal, state, and international laws regulate the sale and import of seeds.

Seed Preservation

Seed storage is important to retain the viability of the seed. Globally, genetic diversity has become a huge concern, resulting in the creation of the International Board for Plant Genetics Resources (Vertucci & Roos, 1990). This organization has facilitated the development of germplasm

scarification
Method of damaging an impervious seed coat to allow for imbibitions of water.

double dormancy
When a seed experiences both physical and physiological dormancy issues.

breeder's seed
Intial seed resulting from a desired cross used to establish the line.

foundation seed
Seed from breeder's seed that is used to maintain a very pure seed line.

registered seed
Seed established from foundation seed and is kept at a level of purity that is high, but not at the level of foundation seed.

certified seed
Seed line with less regulation and is established from registered seed, which maintains uniform cropping.

noncertified seed
Seed that has the least regulation and is used for general cropping.

storage facilities around the world such as the U.S. National Seed Storage Laboratory in Fort Collins, Colorado. Seed are stored here indefinitely via cryopreservation to ensure high viability levels. This repository is maintained to maximize the genetic diversity available in crops. Small companies and individuals are also able to store seed using less sophisticated but very practical techniques. In general, seed should be stored in a cool, dry place. Most species double the life of the seed for each 10% decrease in seed moisture. Similarly, the life of the seed is also doubled for each 10°F (5.6°C) temperature reduction. Storing seed using liquid nitrogen is very successful for most seeds. This method of seed storage is referred to as **cryopreservation** and there are seed banks across the country and around the world dedicated to preserving **germplasm** samples.

cryopreservation
Seed held under liquid nitrogen for long-term storage.

germplasm
Collection of seed that provides specific genetic resources.

Key Terms

fruit	imbibition	stratification
pericarp	plumule	after ripening
exocarp	epigeous	physical dormancy
mesocarp	hypogeous	scarification
endocarp	testa	double dormancy
parthenocarpic	epicotyls	breeder's seed
vestigial seed	hypocotyl	foundation seed
simple fruit	coleorhizae	registered seed
compound fruit	coleoptile	certified seed
fleshy fruit	dormancy	noncertified seed
dry fruit	quiescence	cryopreservation
dehiscent fruit	primary dormancy	germplasm
indehiscent fruit	secondary dormancy	
germination	physiological dormancy	

Summary

- Fruit is defined as a ripened ovary.
- Fruit set is dependent on pollination.
- The wall of the ripened ovary is referred to as the pericarp.
  - Exocarp is the outer layer.
  - Mesocarp is the middle layer.
  - Endocarp is the innermost layer.
- Seeds that develop without benefit of pollination are called parthenocarpic.
- Diverse methods for seed dispersal:
  - Wind
  - Water
  - Barbs on the seed, which attach to passing objects
  - Animal digestive tracts
  - Vehicles
  - Gravity

- Fruiting structures are classified by morphological features:
  - Simple fruit
  - Compound fruit
  - Flesh fruit
    - Berry
    - Pepo
    - Pome
    - Drupe
    - Hesperidium
    - Aggregate fruit
    - Multiple fruit
    - Accessory fruit
  - Dry dehiscent fruit
    - Legume
    - Silique
    - Capsule
    - Follicle
  - Dry indehiscent fruit
    - Nut
    - Caryopsis
    - Schizocarp
    - Achene
    - Samara
- Economic uses for seed:
  - Spices and flavorings
  - Oils
  - Energy source
  - Protein source
- Propagation by seed
- Germination
  - Epigeous growth of eudicots
  - Hypogeous growth of monocots
- Dormancy
  - Quiescence
  - Primary and secondary dormancy
  - Physiological dormancy
  - Physical dormancy
  - Seed treatments
  - Double dormancy
- Seed production and storage
  - Certification
    - Breeder's seed
    - Foundation seed
    - Registered seed

◦ Certified seed

◦ Noncertified seed

◦ Seed preservation

Reflect

1. *What layer of the pericarp do you eat?* Try your hand at identifying which layers of the pericarp are eaten in the following fruits: orange, cucumber, apricot, and apple.

2. *How is this seed disseminated?* As you are walking through your neighborhood you notice fruit on several landscape plants. There is a pecan tree in fruit, a holly shrub with berries, and grass with seed heads poking above the foliage. How are each of these disseminated?

3. *Economic importance.* As mentioned, seeds and fruits have a huge economic importance. What are some other uses for each not listed in this discussion?

4. *Describe that seed.* There are many causes for dormancy. One of the most obvious types of dormancy is physical dormancy. Describe how you think a physically dormant seed would look.

5. *Monocot versus eudicot.* Describe the difference in germination modes between a corn kernel and a bean seed.

6. *Harvesting seed.* While working in a friend's garden you find several plants you would like to propagate for your own yard. How would you go about harvesting and storing the seed? What time of the year would you do this?

7. *Energy balance.* What are the pros and cons of growing our own fuel?

References

Balch, S. A., McKenney, C. B., & Auld, D. L. (2003). Evaluation of gamma-linolenic acid composition of evening primrose *Oenothera* species native to Texas. *HortScience, 38*, 595–598.

Fargione, J., Hill, J., Tilman, D., Polasky, S., & Hawthorne, P. (2008). Land clearing and biofuel carbon debt. *Science, 319*(5867), 1235–1238.

Hill, J., Nelson, E., Tilman, D., Polasky, S., & Tiffany, D. (2006). Environmental, economic and energetic costs and benefits of biodiesel and ethanol biofuels. *Proceedings of the National Academy of Sciences of the United States of America, 103*(30), 11206–11210.

Vertucci, C. W., & Roos, E. E. (1990). Theoretical basis of protocols for seed storage. *Plant Physiology, 94*, 1019–1023.

Chapter 9

Asexual Reproduction and Plant Propagation

Source: Cynthia McKenney.

Learning Objectives

- Discuss the pros and cons of using asexual propagation over sexual propagation
- Identify the primary methods of asexual propagation
- Explain the differences between budding, grafting, layering, and vegetative cuttings
- Discuss the different portions of a plant that may be used to make a cutting and how the process of root development differs in each
- Explain how underground storage structures may be propagated vegetatively
- Outline the factors impacting the success of vegetative cuttings
- Identify the basic elements of a good cutting
- Discuss the differences between grafting and budding
- Explain the process of layering
- Outline the steps of propagating geophytes via division and separation

Have you ever moved into a new home and found a plant in your landscape you would like to replace but you can't find a match to your existing planting at your local garden center? When you mention this dilemma to one of your neighbors they tell you to take a "start" off your plant. What are they talking about? The solution to this puzzle is based on asexual reproduction. In contrast to sexual reproduction, **asexual reproduction** produces offspring without the fusion of gametes and these offspring are genetically identical to the parent (see Chapter 7). Asexual reproduction, also known as *vegetative propagation* because vegetative tissues are the **propagules**, is the process of taking a portion of the original desired plant and using it to create new plantlets. This is basically plant propagation without using seed or spores. The type of vegetative tissue(s) needed and the environmental conditions required to successfully propagate plants are very species specific. Woody plants may require a certain age of wood, time of year to propagate, or specific size of cutting such as these variegated corn plants (*Dracaena massangeana*; Figure 9.1a–b). They may also require a rooting hormone or bottom heat to initiate roots. In contrast, the leaf of a succulent may fall and touch the potting media and immediately begin rooting without any other help, as seen with the jade (*Crassula argentea*) leaf and the mother-of-thousands (*Kalanchoe daigremontiana*) leaf Figure 9.2a–b. This is just an example of the extravagant complexity of plants. Once you get a few basic rules down, it will be easy for you to "make little ones out of big ones!"

ADVANTAGES AND DISADVANTAGES OF ASEXUAL REPRODUCTION

Asexual propagation is the process where a specific portion of a plant is excised and placed in an environment where it will be able to initiate roots and create a new plant. The resulting plants are referred to as **clones** given they have genetic uniformity. Clones have many practical uses such as

asexual reproduction
Does not involve gametes from two individuals and produces offspring that are genetically different from their parent.

propagules
Structures that develop into individual plants.

clones
Genetically identical plants arising asexually from the same mother plant.

(a)   (b)

Source: Cynthia McKenney.

Figure 9.1a–b (a) Mature cane cuttings are made from variegated corn plant in preparation for them to be planted in large nursery containers. (b) Large stock plants grown in field plots in Hawaii serve to provide the cuttings.

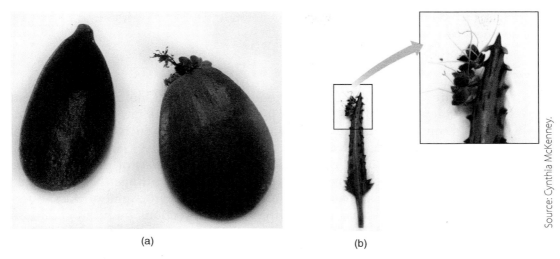

Source: Cynthia McKenney.

(a)   (b)

Figure 9.2a–b   (a) Leaves of a jade plant that have fallen off the mother plant. Note the one on the right has started rooting without benefit of potting media. (b) The mother-of-thousands plant gets its name from the small plantlets that form on the edges of the leaves.

solving the problem posed in the introduction to this chapter. All of the plants will look identical because both their genotype and phenotype will be the same (Figure 9.3).

Advantages of Asexual Propagation

Plants propagated asexually have several distinct advantages over sexually propagated plants. They are characterized as having very uniform growth, allow a unique characteristic to be maintained, and reach a mature size more quickly. Fewer days in production equals a savings in money for commercial producers. In addition, some plants must be propagated vegetatively because they do not produce seed or the seed is very difficult to germinate such as

Source: Cynthia McKenney.

Figure 9.3   The clones of these bromeliads are valued because they have identical genotypes and phenotypes.

in orchids (Figure 9.4a–b). Using vegetative tissues allows you to propagate these difficult plants.

Disadvantages of Asexual Propagation

There are disadvantages to asexual propagation. One example is plant propagules may only be held for just a period of days in specific environmental conditions before use in comparison to seed, which may be kept cool and dry for months or years before they lose viability. Plant- or soil-borne diseases such as Southern wilt (*Ralstonia solanacearum*) of geraniums may also be transmitted by systemically contaminated cuttings (Norman et al., 2009). This bacterium has been found in imported propagation stock entering North America with devastating results. Infections found in geranium growing stock produced offshore resulted in quarantines in the United States as well as eventual closure of one of the companies due to the economic impact caused by the infection. A final disadvantage is the number of cuttings or plant propagules obtained from one stock plant may be significantly lower than the amount of seed able to be collected from a plant. Knowledge of the crop, the economics in the marketplace, and evaluation of your growing facility all play a part in making an informed and profitable decision between asexual and sexual propagation.

TYPES OF ASEXUAL PROPAGATION

Vegetative Cuttings

As with many other plant characteristics, the choice of asexual propagation method is dependent on plant characteristics. One of the most common methods of plant propagation is vegetative cuttings. Cuttings are made when a portion of a stem, root, or leaf is excised from the plant and used as a propagule (Figure 9.5a–b). In general, the plant part is placed in damp

Source: Cynthia McKenney.

(a)   (b)

Figure 9.4a–b   (a) Asexual propagation is important in orchids because they are very difficult to germinate. (b) Containers of tissue-cultured orchid cuttings are imported for commercial propagation.

(a)   (b)

Figure 9.5a–b   (a) Coleus tip cuttings are an example of the most common type of asexual propagation. It is important to pinch out the tip of the cutting to direct stored energy to the roots. Note how the roots form primarily at the node. (b) During growth poinsettia cuttings will also have to be pinched back.

media or other growing material and allowed to root. The tip is generally pinched out to direct stored energy to the developing root system. Some plants such as poinsettias (*Euphorbia pulcherrima*) may require further pinching to produce a more dense plant. Several factors impact the success rate of this type of propagation. These include age of wood, time of year, light source and intensity, water, temperature, **exogenous** or **endogenous** hormones, humidity, bottom heat, removal of any flowers, pinching out the tip to alter the source:sink ratio, and the portion of the plant propagated.

It is important to remove the flowers from cuttings as they will use an excessive amount of water; the cutting does not have the root system to support it (Figure 9.6a). For example, when making cuttings of germaniums (*Pelargonium × hortorum*), you may take **tip cuttings** or **stem cuttings**. Each cutting needs a minimum of one node above the soil and one node

exogenous
When a material is found or applied on the outside of a plant.

Endogenous
When a material is found within a plant.

tip cuttings
A method of plant propagatino in which the tip or apical growing portion is excised and used as propagule.

stem cuttings
A method of plant propagation where portion of the stem is excised and used as propagule.

(a)   (b)

Figure 9.6a–b   (a) It is important to remove the flowers from a cutting to reduce water loss. (b) Crop covers are used to retain humidity around new cuttings.

softwood cuttings
Portion of the stem that is composed of early spring growth, which is less mature and more flexible.

semihardwood cutting
Portions of a stem that is the current season's growth but more mature and generally more stiff.

hardwood cutting
When a cutting is made from stiff, inflexible wood from the previous season's growth.

leaf cuttings
When a leaf is used as the primary plant propagule for production of new plants.

meristematic tissue
Region of rapidly dividing tissue that allows for the formation of roots.

below the soil. The standard size for a cutting is 4–6 cm depending on the internodal distance. When propagating by tip cuttings, the terminal bud remains intact and the lower ½ to ⅔ of the foliage is removed before the stem is placed in damp media. These lower leaves are removed to prevent excessive transpirational water loss given there are no roots to absorb water. Generally, it is important for foliage to not touch the soil surface, preventing several types of rot. In some instances, the cutting may be dipped in a rooting hormone powder or solution to accelerate rooting. In addition, mist or bench covers may be used in the propagation area to raise humidity and reduce transpiration (Figure 9.6b). It is important for a few leaves to remain on the cutting so hormone movement and photosynthesis continue to occur within the cutting.

Stem cutting, taken just below the tip, may also be used to produce new plants. In this situation, it is important to ensure you have nodes above and below the media level. In addition, this type of cutting will usually take longer to produce a viable plant. Cuttings from soft, flexible spring growth are referred to as **softwood cuttings**. Cuttings of herbaceous or interior plants are also considered to be softwood cuttings. If the plant material originates from spring growth but is more mature and stiff, it is classified as a **semihardwood cutting**. **Hardwood cuttings** are from previous season's growth and may be deciduous or narrow-leaved evergreens. These cuttings are generally longer, and may need specific media temperature and more humidity to root. Narrow-leaved evergreens in particular require strong hormones and additional light in order to produce roots.

Leaf cuttings are prevalent primarily in herbaceous plants, especially succulents. These plants have leaves in which the leaf itself or the petiole has rapidly dividing tissue known as **meristematic tissue**. The leaf of a succulent such as the hen and chicks (*Echeveria elegans*) or the felt bush (*Kalanchoe beharensis*) must be placed in potting media one-quarter of the way up the leaf and then it is allowed to root (Figure 9.7a–b). Another form of leaf propagation is accomplished by placing the entire leaf on the soil such as with a Rieger begonia (Begonia × *hiemalis*).

(a)   (b)

Source: Cynthia McKenney.

Figure 9.7a–b Some plants such as the (a) hen and chicks or the (b) felt bush have leaves that root without a stem or node present.

Root cuttings are used to propagate some types of woody and tropical plants. In this type of cutting, 2–6 inches (5–15 cm) of root tissue is excised and placed in damp medium (Figure 9.8a–b). The **distal** end of the cutting is placed into the soil where the roots will primarily form at the nodes. It is important to propagate some plants from root cuttings in order to retain their appearance. The golden snake plant (*Sansevieria trifasciata* var. *laurentii*) root system has a **chimera** (Figure 9.9), when two different tissues are directly in contact with each other. In the golden snake plant, if a leaf cutting is made, the resulting plantlet will not have the golden edge because the chimeric tissue is not included. A piece of the root system with the chimera must be propagated to maintain the attractive golden edge on the foliage.

Root cuttings
A method of plant propagation where a portion of the root system is excised and used as propagule.

distal
Portion of the plant furthest from the central axis.

chimera
When a plant is composed of two genetically different tissues growing directly next to each other.

(a)

(b)

Source: Cynthia McKenney.

Figure 9.8a–b   (a) Geranium tip cuttings root easily in damp media. (b) Geraniums have large leaves requiring the majority of the foliage to be removed to reduce transpiration.

Source: Cynthia McKenney.

Figure 9.9   The snake plant has two genetically different tissues growing from the same meristem, which is referred to as a *chimera*.

Factors Impacting Success of Vegetative Cuttings There are many environmental and physiological conditions that need to be considered when making vegetative cuttings. The type of environment in which the cuttings will be placed plays a huge role. Having good media contact with the base of the cutting is important to supply the necessary moisture for the root system to develop. Rooting media also provides support for the stem, allowing the cutting to stay upright. In addition, the media contains the nutrients needed for plant growth once the roots have started to form. Be sure to use well-drained weed- and disease-free media for best results.

Temperature is also important for the development of vegetative cuttings. Semihardwood cuttings and hardwood cuttings may need increased bottom heat in order to develop the new roots needed for plant growth. In some cases, the overall temperature of the growing area needs to be addressed. If the temperature is too high, the plant may respond with greater transpirational water loss, resulting in desiccation and death. Conversely, if the temperature of the growing area is too low, the plant metabolic processes may be slowed, inhibiting adventitious rooting.

The age of wood, time of year, and time of day cuttings are made also impacts vegetative cutting success. Many types of woody and herbaceous plants root more easily when new growth is harvested in the spring. This is because the ratio of carbon to nitrogen in the tissues is optimal for root generation. Older plants have a greater amount of carbon than nitrogen in their tissues and this slows rooting down when compared to younger plant materials. Selecting new growth in the spring off of these older plants frequently helps overcome this difficulty. The importance of this timing is species specific so taking a few moments to consult a plant production resource may save time, energy, and plant material. The time of day is just as important as the time of year. This is because early in the morning, the plant's water status is generally higher given it has rehydrated after the evaporative demand of the day. Cuttings that are turgid (having a high water status) are more capable of producing roots before they desiccate.

Moisture is important in more than one form. It is important that the rooting media is sufficiently moist to allow generation of the new root hairs. Roots will not grow into dry media effectively. Relative humidity, the percent of moisture in the air at a certain temperature, is also important in preventing desiccation. When there is sufficient humidity, the evaporative demand on the cutting is lowered and there is less water lost before the roots are able to replenish the water status of the cutting. Professional growers have misting benches with fog systems where they place their cuttings to root. The periodic misting improves rooting success by reducing evapotranspirational water loss.

Rooting hormones also improve rooting success. These hormones are available as a powder or as a liquid dip. In either case, the hormones help induce rooting, especially in difficult-to-root woody plants. Once again, response to these materials is species specific so you will want to again check your propagation resource to determine which plants will most benefit from this type of treatment.

wounding
When a cutting is intentionally damaged, resulting in a concentration of hormones in the affected region.

Similar to the use of rooting hormones is the practice of **wounding** a cutting. When the cambium is damaged, the flow of hormones up and down the stem of the plant is interrupted. This results in an accumulation of hormones around the wound. Plant propagators take advantage of this plant response to help stimulate the development of adventitious roots on difficult-to-root species such as narrow-leaved evergreens. The practice of wounding cuttings is generally reserved for woody plants rather than herbaceous plants.

Basics for Producing a Good Cutting There are several common features good cuttings have. First, there needs to be a minimum of at least one node below the soil. If you are able to put two or more in the soil, your success is improved. Nodes are regions of actively dividing plant cells. The production of roots will generally result from the undifferentiated tissues located at the nodes. Some species will only root at the nodes, limiting root production to that area only, such as the wandering traveler plant (*Zebrina pendula*; Figure 9.10a). Other plants, such as the wax begonia (*Begonia semperflorens*), will root up and down the entire stem (Figure 9.10b).

At least one node and some foliage must be above the soil. Having leaves on the top portion of the cutting is important as they are the photosynthesizing tissues producing the necessary photosynthates (see Chapter 10) and are referred to as the "source." The tissues or organs which require the import of these photosynthates are called the "sink." In the absence of fruit, the root system becomes the primary sink. The sink increases the transpiration rate, which also enhances photosynthesis and the movement of nutrients and hormones throughout the cutting. Thus this source–sink relationship supports the development of the root system on the new propagule. It is important that about two-thirds of the foliage be removed from the lower portion of the cutting to reduce the amount of transpirational water loss. Too many leaves may contribute to the loss of water and ultimate failure of the cutting. Similarly, it is necessary to remove the flowers and developing flower buds from the cutting as they will contribute extensively to water and nutrient loss.

A good tip or stem cutting is sufficiently long such that the leaves will not touch the media. This helps prevent the development of fungal rot. Obviously, plants that are able to be propagated by leaf cuttings do not follow this practice. Using a media that is weed and disease free is also very

Source: Cynthia McKenney.

(a)   (b)

Figure 9.10a–b   (a) Wandering traveler roots only at the very obvious nodes; (b) wax begonia roots easily at the nodes and also along the internodes.

important for propagation success. A good stem cutting is also sufficiently sturdy to stay upright in the pot or flat on its own strength. Be sure the orientation on the cutting is correct so the stem is not placed in the media upside down. Also, gently pat the media to help stabilize the cutting.

Grafting

grafting
Process of creating an improved plant by placing the rootstock of one plant in direct contact with the scion of another plant and allowing them to heal together.

stock
Roots and stem portion of a grafted plant that is selected for the tolerance to soils or disease or may be selected for its ability to dwarf the newly created plant.

scion
Top portion of a grafted plant that is the desirable cultivar selected for the best flowers or fruit.

graft junction
Point of contact between the scion and the rootstock.

graft compatibility
Degree of success attained when two plant portions are placed in direct contact and allowed to heal together.

There are situations in which the attributes of a single plant are not as successful as desired. For instance, heavy soils and temperature extremes may preclude raising a very desirable cultivar of a crop. In these circumstances, **grafting** may be used to provide a solution to the problem. Grafting is accomplished by joining two different plants and allowing the parts to heal together to form a more desirable new plant. Frequently, the cultivar of the plant, which provides tolerant roots to the environmental problem, does not produce desirable fruit or flowers. In this situation, the tolerant plant serves as the **stock**, which is the lower part of the plant including the roots and stem (Figure 9.11). The **scion** is the upper portion of this new conjoined plant, which is not tolerant to the environmental problem but has the desirable fruit or flowers. The two components are placed next to each other such that the cambium lines up and then they are held fast together and allowed to heal. The result is a new plant with a root system and shoot from two different plants. The point where the two parts of the grafted plant join together and heal is called the **graft junction** (Figure 9.11).

Grafting Methods Grafting techniques are varied but rely on a few basic concepts. First, it is imperative the cambium lines up. These tissues may be held in place by a variety of options such as grafting wax, rubber wraps, or plastic ties, which are designed for this process. The two plants must also be compatible for a successful union to occur. **Graft compatibility** is generally improved when several factors are taken into consideration (Acquaah, 2009). Success improves the more genetically related the scion and stock are. The scion and stock must be of similar diameter. Usually 1-year-old wood about the diameter of a pencil or your little finger is selected. Larger-diameter materials take longer to heal and require special techniques. An additional factor

© Petar Ivanov Ishmiriev/Shutterstock.com

Figure 9.11 The graft on this fruit tree is composed of the scion and the stock.

relating to successful grafting is the physiological state of both components. Dormant plants are utilized extensively for woody species. As with any propagation technique, it is important the moisture is controlled and the blades are sharp, allowing for a clean cut. Given the diversity of plant materials, the age of the materials, and the conditions surrounding them, several different methods of grafting have been developed (Figure 9.12a–c).

Grafting Vegetables In the past, grafting was primarily used on trees such as apple trees, shrubs such as rose bushes, and vines such as grapes. Increasingly, grafting is also being utilized by the vegetable industry, allowing heirloom varieties to be produced on rootstocks which have disease resistance in them. Some of the vegetables that are now being propagated by grafting include tomatoes, peppers, and eggplants. Specific equipment has been developed to graft these seedlings when they are very small. Tomatoes have become increasingly popular given a trendy heirloom cultivar may be attached to a tolerant rootstock (Figure 9.13). Utilization of this technique has allowed the diversity of many of our flavorful varieties to be grown once again without having crop loss due to disease.

Budding

There are times when it is appropriate to use a very small portion of scion material to provide the desirable cultivar of fruit. This is referred to as **budding**. When budding is used, the rudimentary bud along with a small portion of the stem is excised and placed in contact with the vascular tissue of the stock plant. There are several methods of budding but by far the most common type is the t-bud Figure 9.14. All types of budding have the

budding
The process of placing a bud from a desirable cultivar on the stock plant of another cultivar.

(a) © vallefrias/Shutterstock.com
(a)

(b) © Bork/Shutterstock.com
(b)

(c) © Bork/Shutterstock.com
(c)

Figure 9.12a–c   (a) The cleft graft is used to change over or rework an orchard to a new cultivar. (b) The whip graft is used on smaller-diameter wood. (c) The whip and tongue graft is also used on smaller-diameter wood but provides increased contact between the vascular cambium of the two components.

Source: Cynthia McKenney.

Figure 9.13 This grafted tomato is an excellent example of an heirloom variety being grafted on a tolerant rootstock using new vegetable grafting technology.

common attributes of having a tolerant rootstock, desirable cultivar, and direct contact between the two sets of vascular tissues. Different species respond to differing types of budding so a brief investigation as to the most desirable method is worth the time and effort.

Layering

Like budding, layering is another method of vegetative plant propagation where a portion of the plant is placed in contact with the soil while it is still attached to the mother plant. The section of the plant touching the soil develops roots and after they are sufficiently grown, the rooted section is cut

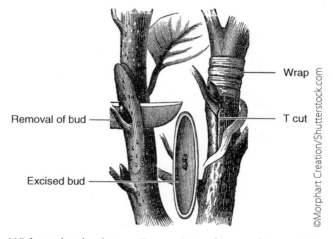

©Morphart Creation/Shutterstock.com

Figure 9.14 With a t-bud, a letter *T* is cut into the stock just deep enough to open a slit and peel the bark back to expose the cambium. The bud of the desired cultivar is then excised and inserted into the T pouch. The wound is then wrapped closed.

off and the new plantlet is transplanted as a rooted cutting. This practice is helpful for difficult-to-root species or plants with supple stems, which easily lend themselves to the process of layering. Tip layering, serpentine layering, and mound layering are several common methods of layering Figure 9.15a–c. Once again, the method of layering is species specific.

Underground Plant Parts

As previously discussed in Chapter 5, underground storage structures, commonly referred to as *geophytes*, may serve as a method of plant propagation. These compressed stems lend themselves to two primary methods of asexual propagation: **division** and **separation**. Division is the process of using a knife or blade to cut different structures apart, resulting in portions capable of starting a new plant; separation is the process of pulling the structure apart by hand to create the plant propagules.

Rhizomes, tubers, and stolons are usually propagated by division (Figure 9.16a). It is important to ensure there are viable nodes present for each excised piece. If we use the Irish potato as an example, the nodes of the potato are frequently called the "eyes." When you cut the potato in portions, there must be an eye present on each piece or the piece will not root. Iris rhizomes are similar as they must have viable nodes present for each "fan" propagated.

division
Practice of using a knife to cut an underground storage structure into multiple propagules.

separation
Process of physically pulling and removing daughter bulbs from the mother bulb.

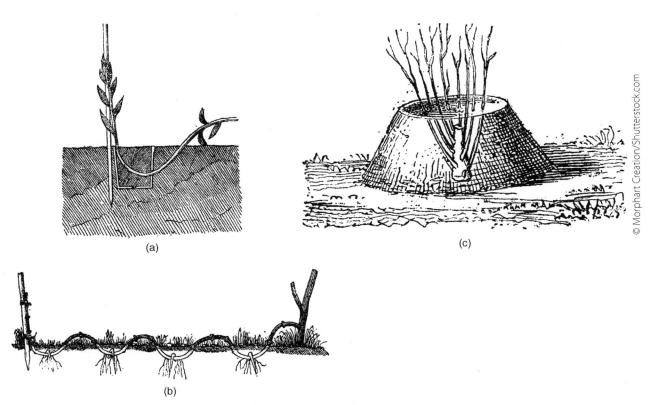

(a)

(b)

(c)

© Morphart Creation/Shutterstock.com

Figure 9.15a–c   There are several types of layering: (a) tip layering, (b) serpentine layering, and (c) mound layering. The sections produce roots while still attached to the mother plant and then are removed, creating rooted cuttings.

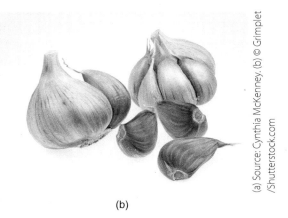

(a) Source: Cynthia McKenney. (b) © Grimplet /Shutterstock.com

(a)

(b)

Figure 9.16a–b   (a) Caladiums are propagated by division where a knife is used to cut the tuber. (b) Garlic is an example of separation where the individual cloves are physically pulled away from the bulb.

Bulbs and corms are frequently propagated by separation (Figure 9.16b). In this process, the mother bulb or corms has the smaller daughter bulbs or corms pulled loose and planted separately. These smaller bulbs are referred to as *bulblets* and the immature corms are called *cormels* or *cormlets*. Generally there will need to be several more years of growth before the new propagule is the size needed to flower. Tulips and gladiolas are examples of bulbs and corms, respectively.

Micropropagation

Orchid, ferns, and other high-value crops are produced by micropropagation. This is the process where very small portions of a plant are removed and placed on a nutrient-enriched agar media in sealed containers. The closed containers allow for greater control of humidity and prevent contamination. Roots and shoots are induced by the presence of hormones in the agar and new plantlets develop. Orchids and ferns are propagated in this fashion (Figure 9.17).

Apomixis

apomixis
Asexual reproduction where seeds develop without fertilization.

One last method of asexual propagation is apomixis. **Apomixis** is seed production from maternal tissues without the benefit of fertilization (Hofmann, 2010). This unique development is quite useful in plant breeding, allowing for superior cultivars to be developed more efficiently as the progeny is genetically identical to the parent. Breeders desire to transfer the trait of apomixis to other crops, allowing for the production of true-breeding hybrids with much less breeding expense (Koltunow & Grossniklaus, 2003). It is also desirable for resource-poor farmers as they would be able to replant their own seed (Frisvold, Bicknell, & Bicknell, 2005). Naturally apomicts are frequently found in disturbed sites or areas with short growing seasons, limiting the opportunity for cross-pollination (Bicknell, 2004). A protocol for inducing apomixis has not been fully identified. Apomixis appears to occur more frequently in

© nongpimmie/Shutterstock.com

Figure 9.17 Micropropagation produces plant propagules by excising small portions of a plant and growing them on a nutrient-enriched agar within a closed container. Sanitation is very important.

polyploids, which are plants with more than twice the usual number of chromosomes. Mutagens are also being researched as a potential method of introducing apomixis.

Key Terms

asexual reproduction	hardwood cuttings	scion
propagules	leaf cuttings	graft junction
clones	meristematic tissue	graft compatibility
exogenous	root cuttings	budding
endogenous	distal	division
tip cuttings	chimera	separation
stem cuttings	wounding	apomixis
softwood cuttings	grafting	
semihardwood cuttings	stock	

Summary

- Asexual propagation is the process of taking a portion of a plant and using it to create new plantlets.
- Advantages of asexual propagation:
  - Crop uniformity
  - Capability of maintaining a unique characteristic
  - Fewer days to mature size
  - Capable of producing more plants if the seeds are difficult to germinate
- Disadvantages of asexual propagation:
  - Plant propagules may only be held for a few days before they expire
  - Difficulties with spreading disease
  - Fewer plantlets yielded from each plant when compared to seed production

- Types of asexual propagation:
  - Vegetative cuttings
    - Tip cuttings
    - Stem cuttings
      - Softwood cuttings
      - Semihardwood cuttings
      - Hardwood cuttings
    - Leaf cuttings
    - Root cuttings
  - Grafting
    - Composed of a scion and rootstock
    - Join together at the graft junction
    - Graft compatibility is necessary
    - Grafted vegetables are a popular new product
  - Budding
    - When a very small plant part is used as the scion
    - Bud is placed on a tolerant rootstock
  - Layering
    - Supple plant stems are placed in contact with the soil to initiate roots
    - Pieces of the stem are then cut from the stem, creating rooted cuttings
  - Underground plant parts
    - Geophytes are underground storage structures
    - May be propagated by asexual propagations
      - Division is cutting the geophyte into pieces with a sharp knife
      - Separation is pulling the geophyte apart to remove small propagules forming on the mother geophyte
- Factors impacting the success of vegetative cuttings:
  - A good planting media that is well drained and weed-, insect-, and disease-free
  - Temperature of the rooting area is acceptable and bottom heat is provided for difficult-to-root species
  - The age of wood is fairly young, collected in the spring and during the early morning hours to maximize the correct carbon:nitrogen ratio and maintaining good water status in the cutting
  - Moisture in the soil as well as good humidity is available
  - Plant hormone powders or dips are used as needed
  - Wounding the cutting is implemented on evergreen
- Basics for a good cutting
  - Nodes below and above the media
  - Reduce but do not eliminate all of the foliage
  - Remove flowers

- ○ Keep the foliage of stem cuttings from touching the soil
- ○ Firm the soil around the cutting
- Grafting
  - ○ Union of a desirable scion with a tolerant rootstock
  - ○ Vascular tissues must line up
  - ○ The scion must be compatible with the rootstock
- Grafting vegetables
  - ○ Increasingly popular in the industry
  - ○ Allows heirloom and other varieties to be produced on tolerant rootstocks
- Budding is when a bud, acting as a very small scion, is placed in contact with a rootstock
- Layering is when plant tissues still attached to the plant are placed in contact with the soil and allowed to root before being detached.
- Underground geophytes may be asexually propagated
  - ○ Division is when a knife is used to cut pieces apart
  - ○ Separation is when the parts are simply pulled apart
- Apomixis is the development of seeds without fertilization
  - ○ Provides an inexpensive way to produce hybrid seed
  - ○ Research is being conducted to transfer the trait into other crops

Reflect

1. **Which method of propagation?** How can you determine what is the best method of propagating a plant? Defend your answer.
2. **What is your best guess?** You receive a houseplant for a gift. You decide you would like to try your hand at propagating it. What steps would you follow?
3. **Parts are parts!** Compare the differences between making a tip cutting, a stem cutting, and a leaf cutting.
4. **Humidity and cuttings.** You live in an arid environment and you would like to make some cuttings of some of your houseplants. How could you go about raising the humidity for your cuttings to improve your success?
5. **Manufacturing plants.** You have discovered the secret of how to transform plants so they produce seeds via apomixis. Which plant will you work with first and why?

References

Acquaah, G. (2009). *Horticulture: Principles and practices* (4th ed.). Upper Saddle River, NJ: Pearson Education.

Bicknell, R. A. (2004). Understanding apomixis: Recent advances and remaining conundrums. *Plant Cell*, 16, S228–S245.

Frisvold, G., Bicknell, K., & Bicknell, R. (2005, August). *A preliminary analysis of the benefits of introducing apomixis into rice.* Proceedings of the 2005 New Zealand Agricultural and Resource Economics Society Conference, Nelson, New Zealand.

Hofmann, N. (2010). Apomixis and gene expression in *Boechera. Plant Cell*, 22, 539.

Koltunow, A. M., & Grossniklaus, U. (2003). Apomixis: A developmental perspective. *Annual Review of Plant Biology*, 54, 547–574.

Norman, D. J., Zapata, M., Gabriel, D.W., Duan, Y. P., Yuen, J. M. F., Manfravita-Novo, A., (2009). Genetic diversity and host range variation of *Ralstonia solanacearum* strains entering North America. *Phytopathology*, 99, 1070–1077.

Chapter 10   Photosynthesis

Chapter 11   Respiration

Chapter 12   Plant Responses to Hormonal and Environmental Stimuli

Chapter 13   Soils, Plant Nutrition, and Transport in Plants

Photosynthesis

© Elenamiv, 2014. Used under license with Shutterstock, Inc..

Learning Objectives

- Discuss the central role of photosynthesis for life on earth
- Describe the electromagnetic spectrum of light and the wavelengths important for photosynthesis
- Discuss the general role and function of chlorophyll, other pigments, and photosystem I and II in the process of the light reactions
- Sketch the reactions of carbon fixation and the location where they occur
- Distinguish C_3, C_4, and CAM photosynthesis and compare the similarities and differences

Most organisms on earth depend exclusively on green plant tissue to live. When looking at a plant, we may admire its beautiful flowers, relish fruit or other edible parts, or enjoy the shade from a tree canopy. While all of these features are important, the most essential process occurring in the green plant tissue is photosynthesis. Life on earth depends on the conversion of solar energy into energy sources that nourish the planet's inhabitants. **Photosynthesis** converts carbon dioxide and water with light energy from the sun into carbohydrates and oxygen. In this chapter, we examine the basic principles of photosynthesis, the cell components involved in photosynthesis, and adaptations some plants have developed to cope with extreme environmental conditions.

A simplified summary of photosynthesis in the following equation shows that carbon dioxide and water plus light energy and chlorophyll are needed to produce carbohydrates such as glucose, oxygen, and water. Carbohydrates are actually produced outside the carbon fixation reaction and outside the chloroplast from products of the photosynthesis reaction.

photosynthesis
Complex metabolic process that green organisms use to capture solar energy and convert it into chemical energy of organic molecules such as glucose or food.

$$6CO_2 + 12\,H_2O + \text{Light energy} \xrightarrow{\text{chlorophyll}} C_6H_{12}O_6 + 6O_2 + 6H_2O$$

Carbon dioxide   Water   Glucose   Oxygen Water

Photosynthesis is a metabolic chemical process occurring in plant, algal, and some bacterial cells. Photosynthesis is an **anabolic** and **endergonic** process building more complex molecules from smaller molecules in several steps that use energy. Electrons are carriers of energy from one molecule to another. When a molecule is oxidized it loses one or more electrons. When a molecule is reduced, it gains one or more electrons (Figure 10.1). In many reactions as in photosynthesis, oxidation and reduction are connected as one molecule loses electrons and the other one is adding those electrons.

Photosynthesis involves two reactions linked through electron transfer. The first reaction requires light and produces chemical energy in the form of NADPH and ATP (Figure 10.2). During the first reaction, water is oxidized to release oxygen. The second reaction of photosynthesis is the Calvin–Benson cycle, which relies on the energy produced from the first reaction. In

anabolic
Process resulting in the biosynthesis or construction of larger molecules; opposite of catabolic.

endergonic
Chemical process that requires energy input; the opposite of exergonic.

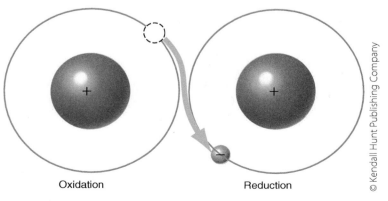

Oxidation   Reduction

© Kendall Hunt Publishing Company

Figure 10.1   Oxidation and reduction, loss and gain of electrons.

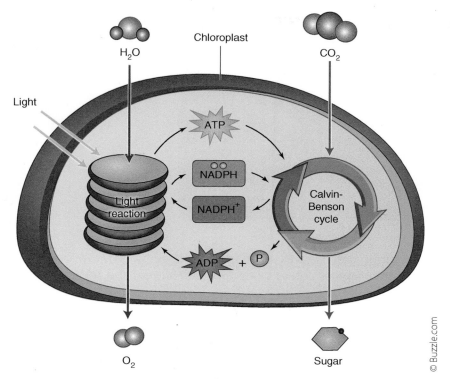

Figure 10.2   The two processes in photosynthesis, the light reactions and the Calvin–Benson cycle.

this step, NADPH and ATP are utilized to reduce carbon dioxide to glucose and other carbohydrates. The process of respiration, which is covered in Chapter 11, oxidizes carbohydrates to produce the energy used in metabolic plant processes such as the production of starch, lipids, and synthesis of nucleic acids.

THE ROLE OF PHOTOSYNTHESIS FOR LIFE ON EARTH

Photosynthesis fixes carbon in organic molecules, the basis for all life. All living systems including humans, animals, and anything produced from living system such as food, fiber, or fuel are organic. Photosynthesis is central for life on earth, undisputable the most important process supporting life as we know it. Plant growth and development depend on the sun's energy and the process of photosynthesis while humans, animals, and other organisms rely on plants for energy and oxygen. Plants are indispensable for humans as they generate food, fiber, energy, medicines, and building supplies. The oxygen we breathe is a byproduct of photosynthesis and vital for most life. If all animals would disappear from earth today, plants will continue to thrive. If all plants would disappear from earth today, animals will soon deplete their food supply and perish.

Plants, algae, and bacteria carrying out photosynthesis sustain themselves by this process and are called **autotroph** (self-feeder) or photoautotroph (Figure 10.3). The chlorophyll in their leaves or bark enables them to capture light energy and with carbon dioxide produce their own food.

autotrophs
Organisms that are able to make their own food using photosynthesis.

(a)   (b)   (c)

Source: Ursula Schuch.   Source: Ursula Schuch.   Source: Ursula Schuch.

Figure10.3a–c   Photosynthesis occurs in green leaves and bark of autotrophs: (a) Mexican redbud (*Cercis mexicana*), (b) the red and green foliage of potato vines (*Ipomoea batata*) and, (c) and the bark of a paloverde tree (*Parkinsonia microphylla*).

heterotrophs
Organisms that cannot produce their own food but have to depend on other organisms for food.

chemoautotrophs
Autotrophic organisms that use chemical energy for food production; compare with photoautotrophs.

Contrary, herbivores, carnivores, and some bacteria are **heterotrophs** (different feeders) because they acquire their energy and oxygen directly and indirectly from plants. Most heterotrophs depend for their nutrition on organic molecules produced by plants. Examples of plants that are heterotrophic and lack chlorophyll are found in more than 400 species of vascular plants (Leake, 1994). Most of these plants inhabit forest floors in dense shade and obtain their energy from mycorrhizal fungi associated with the roots of trees (see Chapter 4). The brown-colored bird's-nest orchid (*Neottia nidus-avis*) and the white-colored Indian pipe (*Monotropa uniflora*) are examples of heterotrophs. **Chemoautotrophs** are organisms that use a third venue to obtain energy independent of photosynthesis or the byproducts of photosynthesis. Chemoautotrophs use carbon dioxide and sources such as sulfur or ammonia to produce organic compounds. Examples are bacteria living in undersea vents or similar inhospitable conditions.

Photosynthesis sequesters large amounts of carbon from the atmosphere and produces organic molecules. The earth's atmosphere contains 78% nitrogen, 21% oxygen, and 0.04% or 400 ppm carbon dioxide. This carbon is incorporated into plants through photosynthesis. Rates of incorporation and net productivity of photosynthesis measure the amount of biomass or oxygen production in the ocean and terrestrial ecosystems. Net photosynthetic production on earth, which is the biomass produced by photosynthesis minus losses through respiration, is estimated at 224×10^9 metric tons carbon per year (Vitousek, Ehrlich, Ehrlich, & Matson, 1986). Terrestrial photoautotroph organisms produce 59% while marine and other aquatic organisms contribute 41% of the biomass resulting from photosynthesis. Humans consume the majority of the total organic material produced, an estimated 35% from terrestrial and 2% from marine photosynthetic net production for their primary use including food, animal feed, and wood products. An even larger percentage is used by humans if consequences from their activities such as growing crops, clearing forests, and decreasing plant growth through pollution are counted. No other species on earth uses this disproportionately large amount of resources for their size population. One of the great challenges in years to come is how to feed, clothe, and shelter the rapidly growing world population. Changes in climate and other effects from human activities add uncertainty about how much life can be sustained by the planet's photoautotroph organisms. Some scientists look to the oceans to develop this so far scarcely tapped biomass. Considering the oceans cover 71% of the Earth's surface, marine ecosystems will likely play a greater role for humans in the future.

BOX 10.1. The Carbon Cycle and Global Warming

Carbon is stored in the atmosphere, the biosphere (all living organisms), the hydrosphere (oceans and lakes), and the geosphere (rocks). Carbon is continuously exchanged between those reservoirs through biological, chemical, and physical processes. Processes releasing carbon include respiration in plankton, animals, and plants, decomposing organic material, or burning of fossil fuels (red arrows in Figure Box 10.1). In respiration, glucose is oxidized

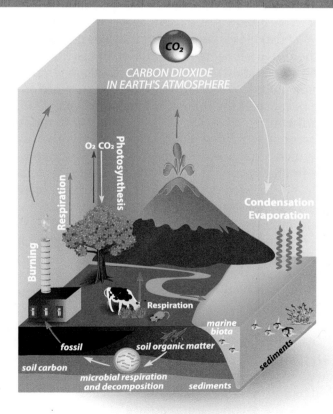

GLOBAL CARBON CYCLE

Carbon stored in

- Atmosphere
- Vegetation
- Surface water
- Deep water
- Soil
- Old carbon (fossil)

© Designua/Shutterstock.com

Figure Box 10.1   The carbon cycle.

and energy in the form of ATP is produced. Carbon is stored in plants through photosynthesis and in fossil fuels (yellow arrows in Figure Box 10.1). The ocean harbors the largest global carbon storage reservoir, primarily in the form of dissolved inorganic carbon (Houghton, 2007). Terrestrial carbon storage is largest in the soil, followed by vegetation, and decomposing plant biomass. Forests are important carbon reservoirs because trees store more carbon than any other type of plants (Houghton, 2007).

Carbon continually cycles between the different pools. CO_2 is fixed by photosynthesis in plant tissue, can be broken down in respiration, and released into the atmosphere. Another possible fate of a carbon atom in a plant might be that it becomes an integral part of the plant structure such as cellulose or lignin and once the plant dies and decomposes, the carbon atom is released again into the atmosphere. An alternative scenario is that the carbon molecule was fixed 300 million years ago in a plant subsequently turned into fossil fuel and still remains fixed in the geosphere.

By estimations, a carbon atom in atmospheric CO_2 passes through soil organic matter somewhere on the globe about every 12 years (Amundson, 2001). The dynamics of carbon flux between soil and atmosphere are affected by human activities, climate, and some natural fluctuations. Some carbon remains fixed in soils through managed grazing, no tillage of agricultural fields, and other land management strategies. The tropics are currently considered a net source of carbon release caused by deforestation and other soil disturbances like agriculture or reforestation (Amundson, 2001). The fluxes caused by these activities pale compared to the carbon release caused by global warming.

Global warming is the rise in global temperature of almost 1.5°F (0.83°C) over the last century at an accelerating rate. Greatest increases in temperature occurred in the last three decades and are projected to increase further. Temperatures are rising because of the release of large amounts of CO_2 and other gases including methane, nitrous oxide, and water vapor into the atmosphere. These gases, also referred to as greenhouse gases, are found naturally in the atmosphere and maintain a stable earth temperature by absorbing some of the infrared radiation emitted back into space. Greenhouse gases reradiate the energy back to earth and the lower atmosphere, causing global warming. From the middle of the 18th century until 2010, atmospheric CO_2 concentrations have increased from 280 ppm to 390 ppm (Houghton, 2007). Some plants can produce more biomass with the higher CO_2 concentration in the atmosphere but not all vegetation benefits from this. Changes in temperature and moisture availability due to climate change affect plant growth worldwide and influence carbon absorption by the different reservoirs. Global warming causes the melting of the Earth's glaciers and a slow rise of the sea level. Potential submersion of coastal agricultural land and habitat pose great challenges.

Stabilization of CO_2 in the atmosphere is desired to control climate change (NRC, 2010). This requires a severe reduction of CO_2 emissions. Currently, the primary source of CO_2 emission is the burning of fossil fuels, coal, and oil, which increased by 29% between 2000 and 2008 (Le Quéré, Raupach, Canadell, & Marland, 2009). The second largest source of human caused rise in CO_2 emissions are changes in land use, including deforestation. Almost half (43%) of the CO_2 emissions each year stayed in the atmosphere in the last 50 years instead of being absorbed by terrestrial or hydrospheric carbon sinks.

THE NATURE AND FUNCTION OF LIGHT IN PHOTOSYNTHESIS

Light provides the energy for the process of photosynthesis, although only 5% of the total solar energy reaching the earth is used to produce carbohydrates (Taiz, Zaiger, Møller, & Murphy, 2018). When plants are deprived of light, they cannot carry out photosynthesis. A potted plant in a dark room or grass covered with an object for even a few days will result in yellow foliage and eventually the plants will die for lack of light. Light controls many other physiological and morphological processes in plants, but this chapter is focused on how light energy drives photosynthesis.

An exploration of the properties of light reveals the central role this resource plays for life on earth. The physicist Sir Isaac Newton (1642–1727) discovered that light can be separated into the visible spectrum from violet to red. Observing light through a prism bends colors at different angles and shows how white light is composed of all the colors in the visible spectrum. A hundred years later, James Clerk Maxwell (1831–1879) discovered the **electromagnetic spectrum** and the fact that visible light perceived by the human eye represents only a small portion of the electromagnetic spectrum (Figure 10.4). The electromagnetic spectrum is radiation moving in waves and is characterized by different wavelengths. Wavelengths are measured by the distance from the top of one wave to the top of the next wave in units of nanometers (one nanometer equals 1.0×10^{-9} meter).

Light is composed of particles of energy or quanta called **photons**. The amount of energy carried by a photon depends on the wavelengths. Wavelengths carry a specific amount of energy and shorter wavelengths such as UV or X rays are associated with greater energy while longer wavelengths such as radio waves carry less energy. When a photon strikes a biological molecule, the energy of the photon is transferred to the molecule, which enters

electromagnetic spectrum
Entire range of radiation of different wavelengths from shortest wavelength with highest energy (gamma rays) to longest wavelength with lowest energy (radio waves).

photon
Elementary light particle.

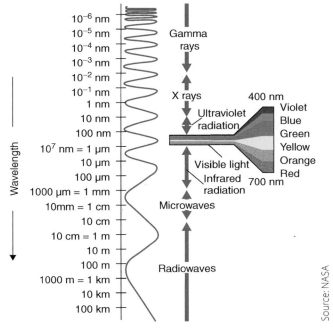

Figure 10.4   The electromagnetic spectrum.

an excited state, a higher energy level because electrons in the molecule are moved to an orbit farther away from the nucleus. If the electron is accepted by an electron acceptor molecule, then it moves along the electron transport chain and is used for synthesis into chemical energy in photosynthesis. This does not always occur, and electrons may lose their higher energy by releasing heat and fluorescence. Another possibility is that they may transfer the energy to a neighboring pigment from where it can move to an electron acceptor and become part of photosynthesis. This energy transfer can be from one chlorophyll to another or from a phycobilin to a chlorophyll molecule.

PIGMENTS AND THEIR ROLE IN PHOTOSYNTHESIS

pigment
Molecules that absorb light of certain wavelengths.

Light can only be used by organisms containing **pigments**, which are light absorbing substances. Pigments absorb different wavelengths of light and reflect the wavelengths they do not absorb. Pigments that look black absorb all wavelengths of light. Chlorophyll a, chlorophyll b, carotenoids, and phycobilins are the pigments that absorb the spectrum of light used in photosynthesis. Wavelengths absorbed by some of these pigments are shown in Figure 10.5. The absorption spectrum of these pigments matches closely the **action spectrum** of photosynthesis, which is the range of wavelengths used in this process. **Chlorophyll** pigments absorb light primarily in the blue and red wavelengths. Chlorophylls do not absorb light in the green range but reflect and transmit these wavelengths, which give plants their green color. Carotenoids absorb visible light in the blue and green range. They get their characteristic yellow, orange, and red color by transmitting these wavelengths.

action spectrum
Range of the light spectrum that causes particular responses in plants.

chlorophyll
Green photosynthetic pigments that are essential for photosynthesis.

Chlorophyll

Chlorophyll *a* is the essential pigment for the light reaction process of photosynthesis. It accounts for three quarters of the pigments involved in photosynthesis in green leaves and is present in all eukaryotic autotrophs and cyanobacteria. The molecular structure of chlorophyll contains a porphyrin ring with magnesium at the center and a long tail composed of hydrocarbons, anchoring the molecule in the thylakoid membrane (Figure 10.6a). Chlorophyll *b* differs from chlorophyll *a* only by a CHO group instead of a CH_3 group in the porphyrin ring. The molecular formula for chlorophyll *a* is $C_{55}H_{72}O_5N_4Mg$ and for chlorophyll *b* is $C_{55}H_{70}O_6N_4Mg$. Chlorophyll *b* is an accessory pigment and occurs in green algae and plants. It absorbs light in a different range than chlorophyll *a* (Figure 10.5) and transfers the gathered energy to chlorophyll *a*; however, chlorophyll *b* is not involved directly in energy transfer.

Accessory Pigments

Two other accessory pigments, carotenoids and phycobilins, operate similarly to chlorophyll *b* by increasing the range of wavelengths absorbed and transferring the energy to chlorophyll *a*. Phycobilins are water-soluble and occur in cyanobacteria and the chloroplasts of red algae.

Carotenoids are lipid-soluble chains of hydrocarbons with a ring at each end (Figure 10.6.b). They do not absorb wavelengths in the yellow, orange,

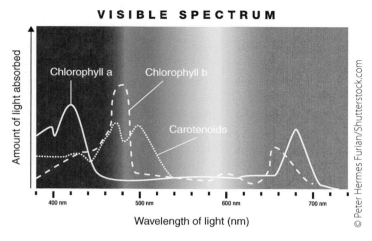

Figure 10.5   Absorption spectrum of pigments important in photosynthesis.

and red spectrum, which gives them their distinct colors. Carotenoids are located in the thylakoid membrane of chloroplasts and in cyanobacteria. They are not visible in green leaves where chlorophyll masks their presence. They are the dominant pigment in vegetables such as carrots and sweet potato. The yellow and red fall color of deciduous tree leaves originates from the degradation of chlorophyll and the emerging dominance of carotenoids and xanthophylls, yellow-colored pigments. Carotenoids and flavonoids, other pigments which absorb UV rays, have antioxidant properties and protect the chlorophyll molecule from oxidation damage. Photooxidation damages chlorophyll molecules when oxygen molecules become reactive under high light intensity due to UV exposure. Although carotenoids absorb light for photosynthesis, their primary function is to protect chlorophyll from photooxidation damage. Carotenoids are important for humans and other animals relying on vitamin A, which is synthesized from beta-carotene.

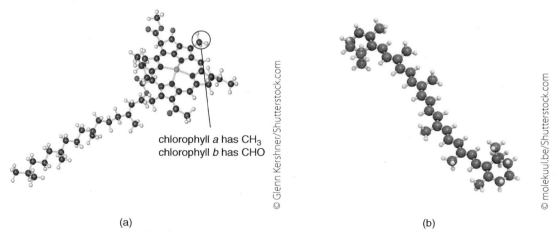

chlorophyll *a* has CH₃
chlorophyll *b* has CHO

(a)   (b)

Figure 10.6a–b   Molecular structure of the pigments: (a) chlorophyll and (b) β-carotene. Atoms are shown as balls and are color coded: magnesium (green), nitrogen (blue), oxygen (red), carbon (grey), and hydrogen (white).

chloroplasts
Most common plastids found mostly in leaf and stem cells; responsible for carrying out photosynthesis to produce food and oxygen.

photosystems
Pigments organized in membrane proteins having a reaction center with chlorophyll at the center surrounded by antennae pigments; they absorb light energy for photosynthesis.

reaction center
Complex of chlorophyll molecule and proteins in the center of the photosystem that converts light energy into chemical energy.

CHLOROPLASTS AND PHOTOSYSTEMS

Photosynthesis occurs in **chloroplasts**, which are organelles located in the mesophyll cells of leaves (Figure 10.7). Within the chloroplasts are **thylakoid** membranes where all chlorophyll molecules are located and where the light reactions of photosynthesis occur. The space inside the thylakoid membranes is the thylakoid lumen. Thylakoid membranes tightly stacked into flattened stacks are **grana** (granum is singular) and are surrounded by the stroma. **Stroma lamella** are unstacked thylakoid membranes and connect grana lamella (Taiz et al., 2018). Chlorophyll and accessory pigments are embedded in the thylakoid membranes. One mesophyll leaf cell can have as many as 40–50 chloroplasts, and one square millimeter of leaf can have about half a million chloroplasts.

Photosystems

Photosystems consist of several hundred molecules of light absorbing pigments harvesting light. They are arranged in clusters in the thylakoid membranes of chloroplasts. Photosystems have two essential components, the reaction center and the antennae complex (Figure 10.8). The **reaction center**

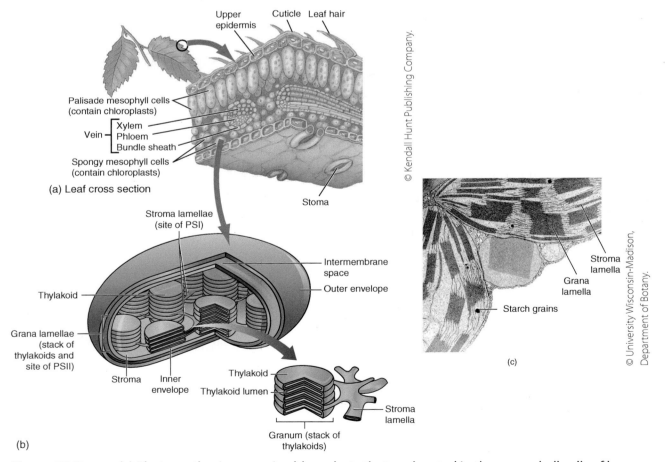

(a) Leaf cross section

(b)

(c)

Figure 10.7a–c   (a) Photosynthesis occurs in chloroplasts that are located in the mesophyll cells of leaves. (b) Photosystems are on the thylakoid membranes that are stacked into grana. (c) Chloroplasts with grana and stroma thylakoids connecting grana. Starch grains (dark circles) store photosynthesis products.

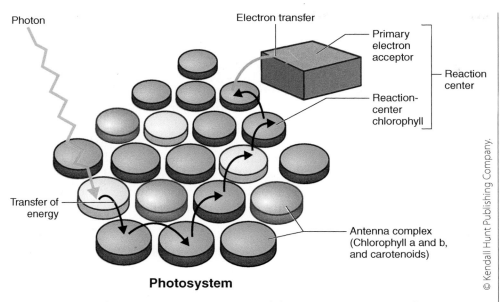

Photon

Electron transfer

Primary electron acceptor

Reaction center

Reaction-center chlorophyll

Transfer of energy

Antenna complex (Chlorophyll a and b, and carotenoids)

Photosystem

© Kendall Hunt Publishing Company.

Figure 10.8   The photosystem consists of the reaction center and antenna complex.

consists of a special chlorophyll *a* molecule solely responsible for absorbing light energy, and a primary electron acceptor. The numerous **antennae pigments**, chlorophyll *a*, chlorophyll *b*, and carotenoids, absorb light over a broader spectrum (Figure 10.5) than single pigments. When a photon strikes a pigment, the energy is transferred within the antennae complex from molecule to molecule until it reaches the reaction center chlorophyll molecule and then is passed on to the primary electron acceptor.

There are two photosystems named in order of their discovery, photosystem I and **photosystem II**. The reaction center molecules of photosystem II are called P_{680}, which refers to the pigment with a peak absorption spectrum at 680 nm. This is where the first step in photosynthesis takes place. The reaction center molecule of **photosystem I** is known as P_{700}, referring to the pigments peak absorption spectrum at 700 nm. Photosystem I and II are connected by an electron transport chain and work continuously together in the first part of photosynthesis where electromagnetic energy is converted into chemical energy in the light reactions.

The two photosystems are located in separate parts of the thylakoid membrane (Taiz et al., 2018). Photosystem II and the proteins carrying out electron transport to photosystem I are located in the grana lamella, the stacked thylakoid membranes. All components of photosystem I, the enzyme that produces ATP (ATP synthase) and the electron transport chain, are located in the stroma lamella, thylakoid membranes that are not stacked and are always in direct contact with the stroma. The cytochrome b_6f complex connecting the photosystems through the electron chain is found throughout the thylakoid membranes in both the stroma and grana lamella. The diffusible electron carriers plastocyanin and plastoquinone are also components of the electron transport chain between the two photosystems. Plastocyanin is located in the thylakoid lumen, plastoquinone in the thylakoid membrane.

antennae complex
Pigment molecules in photosystem I and II that absorb light and direct it to the reaction center.

photosystem II (PS II)
Protein and pigment complex that is used to harness energy from light to make ATP; an important component of light reactions of photosynthesis.

P_{680}
Reaction center molecules of photosystem II with peak absorption at 680 nm.

photosystem I (PS I)
Protein and pigment complex that is used to harness energy from light to make NADPH and occasionally ATP; an important component of light reactions of photosynthesis.

P_{700}
Reaction center molecules of photosystem I with peak absorption at 700 nm.

THE LIGHT REACTIONS AND THE CALVIN–BENSON CYCLE

Photosynthesis occurs in two steps, the light reactions and the Calvin–Benson cycle. In the light reactions, the energy from sunlight is converted into chemical energy, ATP (adenosine triphosphate) and NADPH (reduced nicotinamide adenine dinucleotide phosphate). ATP is the universal energy currency and supplies the energy for many biochemical processes within the plant (Figure 10.9). This reaction splits water molecules and releases oxygen. In the Calvin–Benson cycle, the energy produced from the light reactions reduces carbon dioxide to chemical energy in the form of glucose. Electrons are the carriers of energy in both processes.

The Light Reactions

The light reactions, also referred to as energy-transduction reactions, occur in and across the thylakoid membrane within the chloroplast. Light energy in photosynthesis is transferred through electrons. When photons strike a chloroplast, there are three outcomes for electrons. Photons have different amounts of energy depending on their wavelengths. Blue photons have a higher energy state than red photons; infrared wavelengths have even less energy. Thus, if a pigment absorbs energy from a photon with little energy, the electron falls back to the ground state, releasing the absorbed energy as heat and fluorescence. Photosynthesis cannot start. The second option is that energy from the photon is passed to a nearby chlorophyll molecule in the antennae complex and then from one pigment molecule to another until the electrons reach the reaction center. The third option is that an excited electron in a higher energy state is transferred to an electron acceptor molecule in the reaction center. In both cases, the molecule is reduced and the electromagnetic energy from sunlight has initiated the process of photosynthesis.

Photosynthesis starts when light energy strikes a chloroplast and is absorbed by photosystem II (Figure 10.10). The photons excite the special chlorophyll a molecule P_{680} and an energized electron is eventually passed to a primary electron acceptor molecule pheophytin. The energized electron

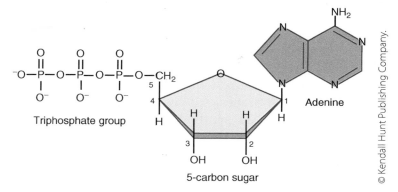

Figure 10.9　ATP (adenosine triphosphate) is a coenzyme and the major transport form of energy inside cells. It is synthesized from ADP (adenosine diphosphate) and inorganic phosphate during photosynthesis and cellular respiration.

travels from one to the next electron acceptor molecule along the electron transport system. These molecules include plastoquinone and plastocyanin and the cytochrome b_6f complex. As P_{680} is losing one electron, it is positively charged and attracts an electron from water molecules. **Photolysis** is the process where enzymes split water into two electrons, two protons (H^+), and one oxygen atom. Photosynthesis depends on the oxidation of water as the source of electrons for this first step. The oxygen we breathe is this atom released from water. Most organisms on earth depend on oxygen, a byproduct of this reaction.

The movement of electrons along the electron transport chain involves a sequence of oxidation–reduction reactions called **chemiosmosis** (Figure 10.11) where energy is released and used in the synthesis of ATP.

photolysis
Splitting of water molecules in photosystem II during the light reaction of photosynthesis

chemiosmosis
Synthesis of ATP that is coupled with an electrochemical gradient across a membrane producing the required energy; occurs in photosynthesis and aerobic respiration.

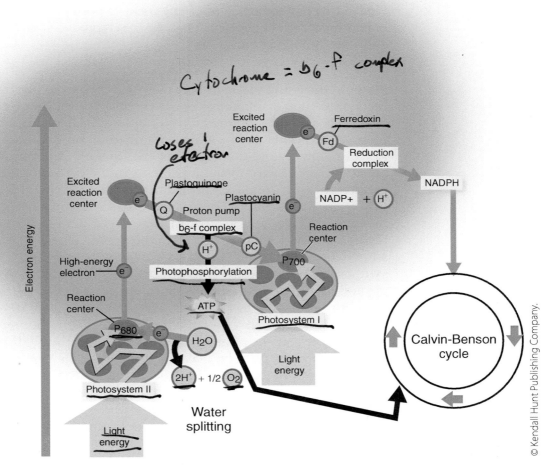

© Kendall Hunt Publishing Company.

Figure 10.10   The blue-colored zigzag or Z scheme shows the connection between photosystem II and I with the electron transport chain also known as noncyclic electron flow. The Z-scheme illustrates the energy state of electrons as they move through the two photosystems. ATP and NADPH from the light-dependent reaction are used in the Calvin-Benson cycle to reduce CO_2 to glucose.

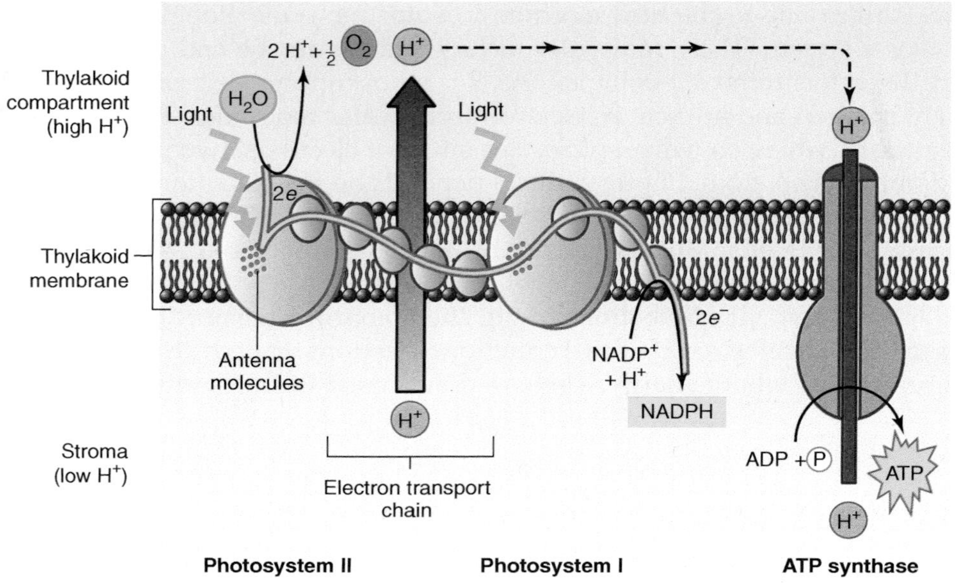

Figure 10.11 Production of ATP by chemiosmosis during the light reaction of photosynthesis in photosystem II. NADPH is produced with energy from electrons through photosystem I.

Pheophytin is the electron acceptor and starts the electron chain in the reaction center of photosystem II (Figure 10.10). The electron transport chain has several quinones. Plastoquinone is a central molecule because it moves electrons and protons from pheophytin in the stroma to the interior of the thylakoid membrane or the lumen to the cytochrome b_6f complex. Electrons move through proteins in the cytochrome b_6f complex via oxidation–reduction. This process is facilitated when H^+ ions from photolysis are released into the thylakoid space or lumen (Figure. 10.11). The movement of protons results in a gradient with pH 5 in the thylakoid membrane and lumen, and pH 8 in the stroma. This pH gradient results in a force moving protons across the thylakoid membrane. The enzyme ATP synthase uses the energy from protons flowing from the thylakoid interior through the ATP synthase complex embedded in the thylakoid membrane and into the stroma to attach an inorganic phosphate (P_i) to ADP for the final product ATP. This process of adding P_i to another molecule is called phosphorylation. In photosynthesis, this phosphorylation is referred to as **photophosphorylation** because it relies on the energy harvested from light through photosystem II.

It is noteworthy here to mention the essential role of phosphorus in the regulation of photosynthesis. Inorganic phosphate (P_i) along with CO_2 and water is one of the primary substrates in photosynthesis and is required continuously. Lack of phosphate limits the rate of photosynthesis both short-term and long-term (Rychter and Rao, 2005).

Plastocyanin, the protein at the end of the electron chain in photosystem II moves electrons to photosystem I to produce NADPH (Figure 10.10). Electrons passing along the electron transport chain from photosystem II to photosystem I are boosted to a higher energy state when pigments in

photophosphorylation
Addition of an inorganic phosphate (P_i) to a molecule such as addition of P_i to ADP synthesizing ATP; using energy from the light.

photosystem I harvest photons that excite the P_{700} molecule. Electrons from the photosystem II reaction center move to a primary electron acceptor and along another electron transport chain. Electrons move in the electron transport chain through several carriers including phylloquinone, iron-sulfur proteins, on to ferredoxin, and finally ferredoxin NADP$^+$ reductase where NADPH is synthesized from NADP$^+$ and two electrons. This is the final step in the light reactions, the noncyclic electron transport flow that starts with oxidation of water and culminates in the production of NADPH.

The process of photosynthesis through photosystem II and photosystem I is also known as the Z scheme or **noncyclic electron flow**. Following the blue line in Figure 10.10 shows the Z-shaped path of electrons with their energy state indicated in height along the electron energy axis on the left. Electrons flow in one direction (noncyclic), starting from photosystem II with the splitting of water and production of ATP to photosystem I, and concluding with the reduction of NADP$^+$ to NADPH. The energy state of electrons is raised first in P_{680}, slowly decreases through the electron transport chain as ATP is produced. Upon arriving in photosystem I (P_{700}), electrons are excited again by light to a higher energy state, enter the electron transport chain, and move through protein complexes and ferredoxin NADP$^+$ reductase to produce NADPH.

The two photosystems operate at optimal levels when both are working together through an additive or enhancement effect. Exposure of chloroplast cells to only one wavelength results in a low rate of photosynthesis and subsequent exposure to another wavelength elicits a similar rate of photosynthesis. Exposing chloroplast cells to both wavelengths at the same time more than doubles the rate of photosynthesis. This enhancement effect was discovered when algae cells were exposed to far-red light (700 nm) and red light (680 nm) first separately and then together and photosynthesis was boosted to rates multiple times higher with the combination of both wavelengths.

Photosystem I can also work independently and create ATP in a process called cyclic electron flow or **cyclic photophosphorylation** (Figure 10.12). In contrast to noncyclic photophosphorylation, this process occurs only in photosystem I where neither NADPH nor oxygen is produced. Energized electrons in P_{700} are moved along the electron transport chain to the b_6f cytochrome complex as in the step of the light reaction when electrons flow from photosystem II to photosystem I. Electrons then move back into the reaction center of P_{700}. This electron flow stimulates proton pumping and yields ATP.

Cyclic photophosphorylation is necessary because the Calvin–Benson cycle uses more ATP than NADPH. The light reaction produces similar quantities of ATP and NADPH and eventually not enough ATP is available for the Calvin–Benson cycle. When NADPH accumulates in chloroplasts, cyclic photophosphorylation is stimulated to ensure sufficient supply of ATP for the operation of the Calvin–Benson cycle.

The light reactions of photosynthesis require electrons in photosystems II and I to be excited to a higher energy state, have a steady supply of sunlight and water, have chlorophyll, accessory pigments, enzymes, and electron carriers (plastoquinones, b_6f cytochrome complexes, plastocyanins, ferredoxins) to produce the energy storing molecules ATP and NADPH. The two photosystems produce the highest level of energy when they operate

noncyclic electron flow
Flow of electrons from water to PSII, ETC, PSI, and ETC producing ATP and NADPH; also known as Z-scheme.

cyclic photophosphorylation
Electrons from PSI are diverted to cytochrome complex by ferredoxin then back to PSI producing ATP instead of NADPH.

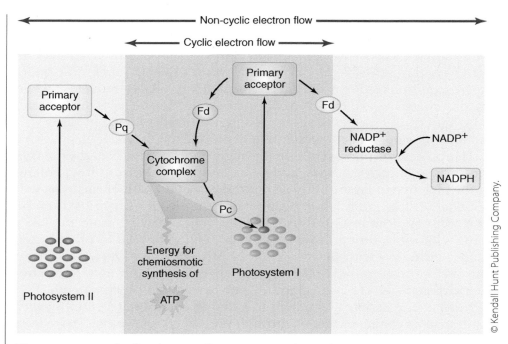

Non-cyclic electron flow

Cyclic electron flow

Primary acceptor

Primary acceptor

Fd

Pq

Fd

NADP+ reductase

NADP+

Cytochrome complex

NADPH

Pc

Energy for chemiosmotic synthesis of

Photosystem I

Photosystem II

ATP

© Kendall Hunt Publishing Company.

Figure 10.12 Cyclic electron flow occurs only in photosystem I and produces ATP.

both at the same time. The light reactions take place in and across the thylakoid membranes, with photosystem II in the grana lamella and photosystem I in the stroma lamella within the chloroplast. The energy stored in ATP and NADPH is used in the subsequent Calvin–Benson cycle, also known as the carbon fixing and reducing reaction.

The Calvin–Benson Cycle

The light energy captured into ATP and NADPH during the light reactions is used to synthesize simple sugars in the endergonic reaction called the Calvin–Benson cycle or the Calvin–Benson–Bassham cycle. This process is also known as the carbon reactions of photosynthesis. The cycle occurs in the stroma of chloroplasts and consists of a series of redox reactions that reduce carbon dioxide to glucose. Dr. Melvin Calvin, James Bassham, and Andrew Benson studied how carbon dioxide was converted into simple sugars through photosynthesis in algae. They added radioactive CO_2 to algae and analyzed the progress of radioactive compounds, which eventually led to the final product, the three carbon sugars. This discovery earned Dr. Calvin the Nobel Prize in 1961 and the pathway was named after him. Plants carrying out the Calvin–Benson cycle are called **C_3 plants** after the first stable product of the pathway, the three-carbon molecule **3-phosphoglycerate (PGA)**.

The Calvin–Benson cycle occurs in the stroma of the chloroplasts where CO_2 diffuses after entering through the stomata into the mesophyll cells. The Calvin–Benson cycle consists of a chain of reactions where CO_2 molecules are linked, adding one molecule with each completion of the Calvin–Benson cycle, until the final product **glyceraldehyde 3-phosphate (G3P)** is

C_3 plants
Plants that use the Calvin-Benson cycle to fix CO_2 with the enzyme rubisco and produce 3-phosphoglycerate, a three-carbon compound, as the first stable molecule.

3-phosphoglycerate (PGA)
Molecule containing three carbon atoms, first stable compound in the Calvin-Benson cycle.

glyceraldehyde 3-phosphate (G3P)
Product of the reduction phase of the Calvin-Benson cycle; one molecule of G3P is produced after three completed cycles.

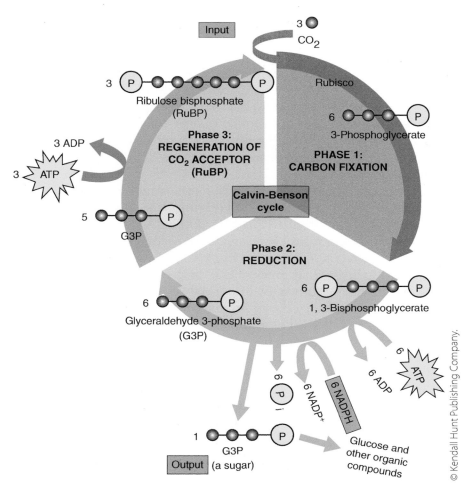

Figure 10.13   Carbon dioxide is reduced to glyceraldehyde 3-phosphate in the Calvin–Benson cycle. It takes three completed cycles to produce one molecule of glyceraldehyde-3-phosphate. The Calvin–Benson cycle starts and ends with ribulose bisphophate (RuBP).

synthesized (Figure 10.13). Although the reactions in the Calvin–Benson cycle are not directly dependent on light, the enzymes mediating the process require light, and the ATP and NADPH produced in the light-dependent reaction need to be synthesized continuously for use in the Calvin–Benson cycle. ATP and NADPH are not stored in large quantities in the chloroplasts; rather the light reactions and Calvin–Benson cycle occur simultaneously as the proton gradient across the thylakoid membrane supplies the energy for ATP production. ADP, NADP$^+$, and phosphate are released during the Calvin–Benson cycle and are used again in the light reactions.

The Calvin–Benson cycle starts and ends with **ribulose 1,5-bisphosphate (RuBP)**, a five-carbon sugar with two phosphate groups. The cycle adds one carbon of one CO_2 molecule each time and it takes three completed cycles to obtain one molecule of G3P. Two molecules of the three-carbon molecule G3P are combined as the six-carbon sugar molecule glucose. There are three phases in the Calvin–Benson cycle, first the fixation of CO_2, second the phosphorylation or reduction of 3-phosphoglycerate to G3P, and third the rearrangement of G3P into RuBP molecules to start the cycle again.

RuBP (ribulose 1,5-bisphosphate) Five-carbon sugar with which the Calvin-Benson cycle starts and ends.

RuBP carboxylase/oxygenase or rubisco
see rubisco

rubisco
Enzyme RuBP carboxylase/oxygenase mediates the first step in the Calvin-Benson cycle and is the most common enzyme on Earth.

In the first phase, one molecule of CO_2, one molecule of water, and one molecule of RuBP are combined. The product is an unstable six-carbon molecule, which immediately splits into two three-carbon PGA molecules (3-phosphoglycerate). The enzyme **RuBP carboxylase/oxygenase** often abbreviated as **rubisco** catalyzes this reaction. Rubisco is the most abundant enzyme on earth and the most common protein in leaf tissue. This first phase is exergonic and does not require energy for the generation of PGA. To produce one molecule of G3P in the second phase of the Calvin–Benson cycle, this process has to be repeated three times, fixing carbon from three CO_2 molecules and yielding 6 PGA molecules.

The second phase of the Calvin–Benson cycle involves reduction of PGA into glyceraldehyde 3-phosphate (G3P) (Figure 10.13). In this two-step reaction, first one phosphate from ATP is added to PGA forming the intermediate compound 1,3-bisphosphoglycerate. In the second step, the intermediate compound is reduced using electrons from NADPH, and phosphate (PO_4) is removed to produce the final product G3P of the Calvin–Benson cycle. Two molecules of G3P can be converted into fructose-6-phosphate in subsequent reactions and then synthesized into sucrose and more complex sugars in the cytosol and into starch in the chloroplasts. Both steps, phosphorylation and reduction, require enzymes and the energy of six ATP and six NADPH molecules. The enzymes required are 3-phosphoglycerate kinase and NADP-glyceraldehyde-3-phosphate dehydrogenase.

In the third phase of the Calvin–Benson cycle, five of the six molecules of G3P produced in phase two are regenerated in a series of reactions, using three ATP to yield three molecules of RuBP. These RuBP molecules feed again into phase one, carbon fixation of the Calvin–Benson cycle. The third phase includes nine steps and requires eight different enzymes.

The three phases of the Calvin–Benson cycle can be summarized as follows:

1. Carboxylation of CO_2: $3\,RuBP + 3\,H_2O + 3\,CO_2 \rightarrow 6\,\text{3-phosphoglycerates}$
2. Reduction of 3-phosphoglycerate to G3P: 6 3-phosphoglycerates + 6 ATP + 6 NADPH $\rightarrow$ 6 G3P
3. Regeneration of RuBP from G3P: 5 G3P + 3 ATP $\rightarrow$ 3 RuBP

C_4 Photosynthesis

The C_4 pathway carries out photosynthesis with little energy loss to **photorespiration**. Energy is lost by photorespiration in C_3 plants because the enzyme rubisco binds oxygen instead of carbon dioxide when CO_2 concentration in chloroplasts drops to low levels. Rubisco can act as both a carboxylase binding carbon dioxide or as an oxygenase binding oxygen. When carbon dioxide concentrations in the chloroplasts are high, rubisco performs as carboxylase. C_3 plants growing in dry, hot climates and exposed to water stress close their stomata to prevent further water loss but at the same time also restrict CO_2 uptake. As CO_2 concentration decreases and the ratio of carbon dioxide to oxygen concentration in the chloroplasts favors oxygen, the rate of photosynthesis decreases and rubisco acts as an oxygenase. Instead of producing two 3-phosphoglycerates in the first phase of the Calvin–Benson cycle during photosynthesis, rubisco binding to oxygen in the process of photorespiration results in one 3-phosphoglycerate and

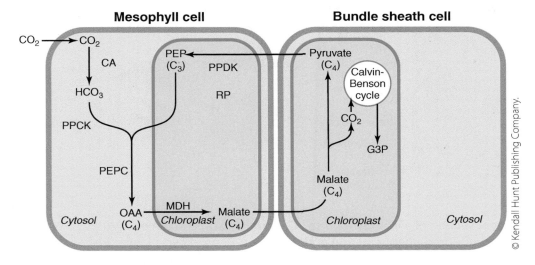

Figure 10.14   The C_4 pathway of photosynthesis is carried out in two locations, the mesophyll cells and the bundle sheath cells. Carbon is concentrated in the bundle sheath cells to prevent photorespiration under hot, dry conditions (OAA, oxaloacetate; PEP, phosphoenolpyruvate; G3P, glyceraldehyde 3-phosphate).

one 2-phosphoglycolate. When 2-phosphoglycolate is processed, CO_2 is released, and ATP is used. Thus, this process is the opposite of photosynthesis, consuming energy and releasing carbon dioxide.

Plant species adapted to tropical areas and growing predominantly under high temperatures, high light intensities, and dry conditions use C_4 photosynthesis. Angiosperms from 19 different families optimize photosynthesis with the C_4 pathway. Many tropical grasses such as sugar cane and corn are highly productive because they are C_4 plants.

C_4 photosynthesis starts by concentrating carbon and maintaining the carboxylase activity of rubisco. This pathway produces a four-carbon molecule as the first compound of CO_2 fixation compared to the three-carbon molecule produced in C_3 photosynthesis of the Calvin–Benson cycle. C_4 photosynthesis is carried out in two locations: the mesophyll cells and the bundle sheath cells (Figure. 10.14).

The four-carbon compound oxaloacetate is produced in the cytosol of mesophyll cells by adding one molecule of hydrated CO_2, which is the bicarbonate ion (HCO_3^-) to **phosphoenolpyruvate (PEP)** by the enzyme **PEP carboxylase**. Oxaloacetate is converted to either malate or the amino acid aspartate depending on the plant species. This process takes place in the chloroplast of the mesophyll cells. Once malate is produced, it is moved through a diffusion barrier to **bundle sheath cell**. A decarboxylation enzyme releases CO_2 from malate and CO_2 then enters the Calvin–Benson cycle. This accumulation of CO_2 in the bundle sheath cells ensures carboxylase activity from rubisco and high efficiency of photosynthesis. The remaining three-carbon acid pyruvate moves back into the mesophyll cell. Pyruvate is converted back to PEP by the enzyme pyruvate-phosphate dikinase and the carbon accumulation cycle starts again. This conversion to regenerate PEP from pyruvate requires two molecules of ATP.

The bundle sheath cells in C_4 plants are arranged in a circle referred to as Kranz anatomy (after the German word wreath) around the vascular

phosphoenolpyruvate (PEP)
Molecule used in the CO_2 fixation step of the C_4 pathway.

PEP carboxylase
Enzyme in C_4 photosynthesis that fixes CO_2 to phosphoenolpyruvate to form oxaloacetate, a four-carbon organic acid.

bundle sheath cells
Specialized cells in C_4 plants surrounding the vascular bundles where the Calvin-Benson cycle occurs.

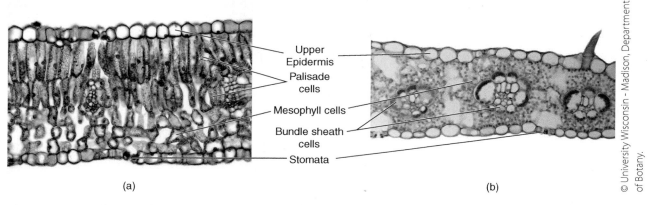

Upper
Epidermis

Palisade
cells

Mesophyll cells

Bundle sheath
cells

Stomata

(a)

(b)

Figure 10.15a–b Leaf cross section of a lilac (*Syringa* sp.) C_3 plant (a), and corn (*Zea mays*) a C_4 plant (b). Photosynthesis with the light dependent reaction and the Calvin-Benson cycle in lilac leaves takes place in the chloroplasts located in the palisade and mesophyll cells. In corn plants, vascular bundles of phloem and xylem are surrounded by bundle sheath cells outside of which mesophyll cells are located. This morphology is unique to C_4 plants. It enables them to start the C_4 pathway in the mesophyll cells and then transports malate to the bundle sheath cells where the Calvin-Benson cycle occurs.

bundles of phloem and xylem (Figure 10.15). Mesophyll cells where the CO_2 is first fixed are surrounding the bundle sheath cells. This spatial compartmentalization allows for the Calvin–Benson cycle to operate efficiently in the bundle sheath cells where the concentration of CO_2 is always high enough for rubisco to bind to CO_2, thus preventing photorespiration. Recently, a few plant species without the Kranz anatomy as well as eukaryotic algae were found to perform C_4 photosynthesis in single cells (Taiz et al., 2018). The spatial compartmentalization of the process occurs in different regions of the cytosol of chlorenchyma cells.

COMPARING THE C_3 AND C_4 PATHWAYS

The energy cost to fix one molecule of CO_2 in the C_3 pathway requires three ATP molecules. This process takes five ATP molecules in the C_4 pathway because of the need to synthesize one molecule of PEP in the final regeneration step at the cost of two ATP. This higher energy cost is compensated for with the improved CO_2 fixation of PEP carboxylase which is not inhibited by oxygen, and the fact that CO_2 molecules lost to photorespiration in the bundle sheath cells move to the surrounding mesophyll cells where they can be fixed again in the C_4 pathway.

A comparison of the efficiency of plants using either pathway shows that efficiency in C_4 plants is greater, meaning that more sugars are produced by C_4 plants than C_3 plants under the same environmental conditions because photorespiration is minimized in C_4 plants (Table 10.1). The temperature range for optimum growth of C_4 plants is higher than for C_3 plants. This leads to greater yields of C_4 plants in environments with higher temperatures. Smaller size stomata in many C_4 plants compared to C_3 plant causes less water loss when stomata are open and supports greater photosynthesis efficiency. C_4 plants rely on the Kranz anatomy of the mesophyll and bundle sheath cells and the transport pathways with the associated enzymes to allow for the carbon concentration prior to CO_2 fixation in the Calvin–Benson cycle.

Table 10.1.   Comparing Physiological, Anatomical, and Morphological Characteristics of the Different Mechanisms of Photosynthesis

	C_3	C_4	CAM
Types of pants	Many higher plants, including grasses, algae, bacteria	Angiosperms including monocots and eudicots	Angiosperms and gymnosperms
Examples	Wheat, oats, rye, rice, soybeans, potatoes	Corn, sugarcane, sorghum, Bermuda grass, crabgrass	Pineapple, agave, bromeliads, orchids, ice plant
Typical environments and optimum temperatures	Cool, moderate temperature habitats, 59–77°F (15–25°C)	Warm, tropical or subtropical areas, 86–117°F (30–47°C)	Arid or semi-arid environments, about 95°F (35°C)
Leaf anatomy	Palisade mesophyll and spongy mesophyll cells	Mesophyll cells and bundle sheath cells	Mesophyll cells with large vacuoles
Enzymes	Rubisco	Mesophyll cells: PEP carboxylase, Bundle sheath cells: rubisco	Day: Rubisco, Night: PEP carboxylase
CO_2 compensation points (ppm)	30–70	0–10	0–5 at night
Advantage	High efficiency of photosynthesis under cooler conditions. Medium dry matter production (22 tons/ha/year)	High efficiency of photosynthesis under warm temperatures. Highest dry matter production (39 tons/ha/year) No loss of CO_2 or energy to photorespiration	Grows under extreme arid conditions where C_3 and C_4 plants cannot survive.
Disadvantage	Low efficiency under warmer conditions due to photorespiration	Low photosynthesis efficiency under cool temperatures	Slow growth and small plants due to low biomass production.

CRASSULACEAN ACID METABOLISM (CAM)

Crassulacean acid metabolism or CAM photosynthesis is used by about 7% of plants in over 30 families. It was named after the plant family where it was first discovered, the Crassulaceae family. CAM plants have evolved in extremely arid areas such as deserts, and semiarid environments with seasonal or intermittent water availability such as the Mediterranean or tropical epiphytic environment (Cushman, 2001). CAM photosynthesis occurs exclusively in almost all the species of the Crassulaceae and cactus families. Most recently, CAM plants were identified in aquatic habitats with limited

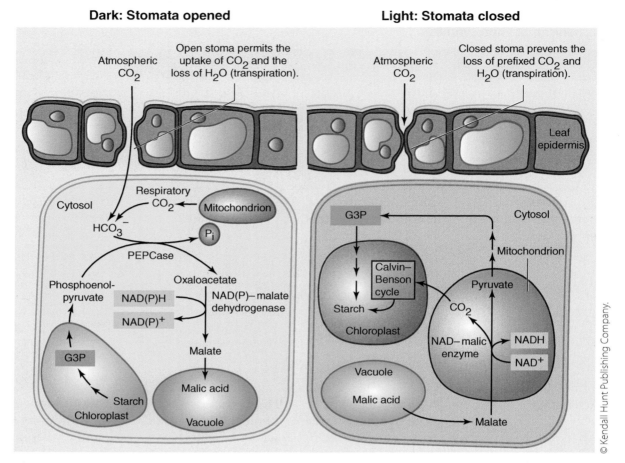

Figure 10.16   CAM photosynthesis separates CO_2 uptake at night to conserve water from CO_2 fixation during the day.

CO_2 availability. Commercially important CAM plants include pineapple, orchids, cacti, agave, and floriculture plants such as kalanchoe.

CAM plants open their stomata at night to take up carbon dioxide and conserve water. The enzyme PEP carboxylase converts hydrated CO_2 (HCO_3^-) and PEP to the four-carbon organic acid oxaloacetate (Figure 10.16). The enzyme NADP-malate dehydrogenase converts oxaloacetate and NADPH and one proton to malate and $NADP^+$. Malate diffuses into the large vacuole for storage until daytime (Figure 10.16). This leads to a low pH or acidification of the leaf. During the day, stomata remain closed to minimize water loss. Malate moves into the mitochondrion and NAD malic enzyme decarboxylates the malate into pyruvate and the CO_2, which is then used in the Calvin–Benson cycle. The resulting glyceraldehyde 3-phosphate (G3P) is converted to starch which accumulates in the chloroplast or to sucrose which moves from the cytosol into the vascular tissue.

The concentration of CO_2 during the day promotes the carboxylase and suppresses the oxygenase activity of rubisco, thus minimizing photorespiration. This process occurs in the mesophyll cells. CAM photosynthesis uses a temporal separation of carbon fixation, nocturnal CO_2 uptake, and daytime CO_2 fixation compared to the spatial separation of the C_4 metabolism in mesophyll and bundle sheath cells (Figures 10.16 and 10.17).

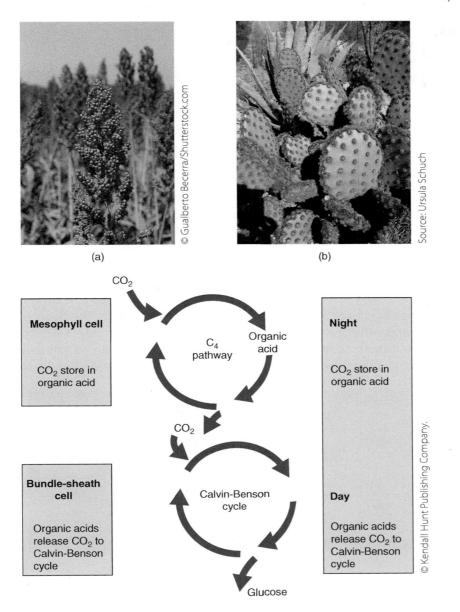

(a)

(b)

Figure 10.17   Comparison of C_4 (a) and CAM (b). photosynthesis. Both types of photosynthesis use the C_4 and the C_3 pathways. Photosynthesis is separated by location in different cells in C_4 plants and by time in CAM plants.

CAM plants have several anatomical or morphological adaptations minimizing water loss. Many succulents exhibit these features such as smaller or fewer stomata, thick cuticles, small surface area, and large cells and vacuoles for water storage (Cushman, 2001). CAM plants have much higher water use efficiency than C_3 or C_4 plants (Table 10.1). However, because they grow primarily in water limiting environments, their growth rates are slow.

BOX 10.2. Shifty Photosynthesis

Some plants employ different photosynthetic metabolisms depending on the stage of development or environmental factors such as seasonal or temporary moisture availability (Figure Box 10.2). The most important CAM food plant is pineapple (*Ananas comosus*) in the bromeliad family. This plant gathers energy prevalently through CAM, but shifts to C_3 photosynthesis when temperature, light, and water conditions are favorable. Conversely, the common ice plant (*Mesembryanthemum crystallinum*) changes from C_3 to CAM photosynthesis after exposure to water or salinity stress (Cushman, 2001). Purslane (*Portulaca oleracea*), an annual succulent weed and super food because of its high content of omega-3 fatty acids, shifts from facultative C_4 metabolism to CAM photosynthesis during short days or periods of drought. In many plants, these changes are reversible when moisture and other environmental conditions favor the more efficient C_4 or C_3 metabolism. This flexibility in photosynthetic metabolism confers advantages so plants can optimize their growth during periods of abundant moisture but can still conduct photosynthesis when water is scarce or under environmental stress.

(a)

(b)

© Tami Freed, 2014. Used under license with Shutterstock, Inc.

© Lagui, 2014. Used under license with Shutterstock, Inc.

(c)

© Pu Su Lan, 2014. Used under license with Shutterstock, Inc.

Figure Box 10.2a–c Purslane (*Portulaca oleracea*) (a), shifts from C_4 to CAM under stress, the common ice plant (*Mesembryanthemum crystallinum*) (b) changes from C_3 to CAM under stress, and (c) pineapple (*Ananas comosus*) switches from CAM to C_3 photosynthesis when environmental condition are favorable.

ENVIRONMENTAL FACTORS AFFECTING PHOTOSYNTHESIS

Light Intensity

The amount of light leaves receive affects net photosynthesis. Light intensity is measured as photosynthetic active radiation, which measures only the wavelengths plants use in photosynthesis, not all visible light. Leaves need a minimum light intensity to start absorbing and fixing CO_2. This light intensity is referred to as **light compensation point** where net photosynthesis is zero and photosynthesis and respiration are at equilibrium (Figure 10.18). At that point, loss of CO_2 to respiration is equal to gain of CO_2 through photosynthesis. At light intensities below the light compensation point the rate of respiration is higher than photosynthesis and results in a loss of CO_2 from the plant. With increasing light intensity net photosynthesis increases. **The light saturation point** is where net photosynthesis remains steady and is not increasing any more in response to increasing light intensity.

Plant species have different light compensation and light saturation points. Most plants have light saturation points between 500 and 1,000 μmol m^{-2} s^{-1}, far less than full sunlight which is usually about 2,000 μmol m^{-2} s^{-1} (Taiz et al., 2018). Plants thriving in full sun will have a higher light saturation point than plants adapted to shade. Optimum plant growth can be achieved when a plant receives enough light to reach the light saturation provided other factors are also within the optimum range. Greenhouse growers do this by supplementing light or shading plants, depending on the plants' light requirements.

light compensation point
When photosynthesis and respiration are at an equilibrium and net photosynthesis is zero.

light saturation point
Net photosynthesis remains steady even if light intensity is further increased.

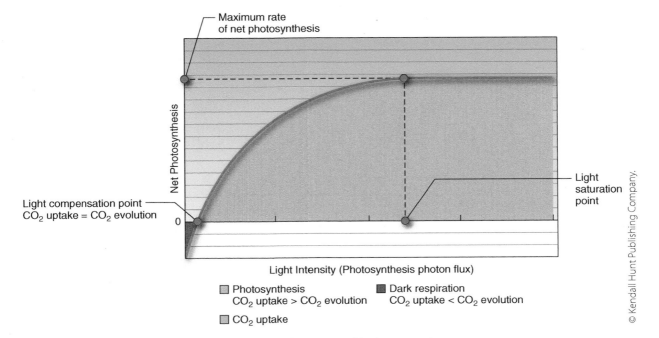

Figure 10.18 Rate of photosynthesis as a function of light intensity.

Temperature

Photosynthetic CO_2 assimilation as a function of temperature is typically a bell shaped curve. At the lower end, increasing temperatures will result in higher rates of photosynthesis until an optimum temperature range where photosynthesis has reached its highest rates. At temperatures above the optimum range CO_2 assimilation will decrease with further increases in temperature, generally because photorespiration increases. Typically, C_3 plants have their optimum range at lower temperatures while C_4 and CAM plants have their highest rates of photosynthesis at higher temperatures (Bhagwat, 2005) (Table 10.1).

Carbon Dioxide

Photosynthesis rates can be increased by raising the CO_2 concentration above the ambient concentration of about 400 ppm. Many C_3 plants can double their growth when CO_2 concentration is raised to twice the ambient concentration because photorespiration is decreasing (Taiz et al., 2018). Greenhouse producers can increase productivity of vegetables and other plants by increasing CO_2 provided water, nutrition, and light intensity and duration are optimum. When one of the inputs becomes limiting, net production of photosynthesis and consequently plant production will drop to the level supported by the most limiting factor.

Key Terms

photosynthesis	reaction center	CAM plants
anabolic	antennae pigments	3-phosphoglycerate (PGA)
endergonic	photosystem II	glyceraldehyde 3-phosphate
autotrophs	P_{680}	(G3P)
heterotrophs	photosystem I	ribulose, 1, 5-bisphosphate
chemoautotrophs	P_{700}	(RuBP)
electromagnetic spectrum	photolysis	RuBP carboxylase/oxygenase
photons	chemiosmosis	rubisco
pigments	photophosphorylation	phosphoenolpyruvate (PEP)
action spectrum	noncyclic electron flow	PEP carboxylase
chlorophyll	cyclic photophosphorylation	bundle sheath cells
chloroplasts	C_3 plants	light compensation point
photosystems	C_4 plants	light saturation point

Summary

- Photosynthesis occurs in green tissue in the chloroplasts. Photosynthesis uses radiant energy from light converting it into chemical energy. The process requires light, water, and carbon dioxide to produce sugars. Oxygen is a byproduct. Life on earth depends on photosynthesis.

- Autotroph organisms carry out photosynthesis and sustain themselves through this process. Heterotroph organisms depend on autotrophs for energy and oxygen. Chemoautotrophs rely on inorganic molecules and carbon dioxide.

- Light has the property of a wave and particles that carry energy. Chlorophyll absorbs light for photosynthesis in the red and blue region of the visible spectrum, but transmits green.

- Photosynthesis consists of two reactions, the light reaction in the thylakoid membranes of chloroplast and the Calvin–Benson cycle in the stroma of chloroplasts.

- In the light reaction, chlorophyll and other pigments harness energy first in photosystem II and subsequently in photosystem I where electrons are boosted to a higher energy state. Electrons flow along a chain of electron transport carriers. Water is split (photolysis), releasing oxygen, protons and electrons. The released protons result in a gradient across the thylakoid membrane which generates energy in the form of ATP. This noncyclic photophosphorylation is a chemiosmotic process. The final product is energy as NADPH and ATP necessary for subsequent production of glucose in the Calvin–Benson cycle.

- Cyclic electron flow produces ATP when electrons move down the electron transport chain through photosystem I. Energized electrons in P_{700} are moved along the electron transport chain to the $b_6 f$ cytochrome complex and then move back into the reaction center of P_{700}. This electron flow stimulates proton pumping and yields ATP, but no production of NADPH.

- In the Calvin–Benson cycle, carbon dioxide is reduced to carbohydrates. NADPH and ATP are necessary. Carbon fixation occurs in the first phase where the enzyme rubisco catalyzes the fixation of CO_2 to a ribulose 1,5-bisphosphate (RuBP), a 5-carbon compound. RuBP is the starting and ending molecule in the Calvin–Benson cycle. Then 3-phosphoglycerate is reduced to G3P. Finally G3P is rearranged into RuBP to start the cycle again. It takes three turn of the Calvin–Benson cycle to yield one molecule of RuBP which can be used to synthesize sugars, amino acids, or lipids.

- The enzyme rubisco can react with oxygen or carbon dioxide. When CO_2 concentrations in the leaf drop because stomata have closed during dry conditions, rubisco will bind with oxygen and initiate photorespiration, which uses sugars previously produced through photosynthesis.

- C_3 photosynthesis is the process where CO_2 is fixed in the Calvin–Benson cycle and the first product is 3-phosphoglycerate. Two other pathways of photosynthesis are C_4 and CAM.

- In C_4 photosynthesis carbon dioxide is fixed with PEP carboxylase to yield oxaloacetate in the mesophyll cells of the leaf. Oxaloacetate is converted to malic acid and moves to the bundle sheath cells. CO_2 is released and enters the Calvin–Benson cycle for carbon fixation. C_4 plants are more efficient under high temperatures and in the use of CO_2 than C_3 plants. C_4 plants benefit from PEP carboxylase, which is not inhibited by oxygen compared to rubisco.

- CAM (crassulacean acid metabolism) plants fix CO_2 to phosphoenolpyruvate (PEP) at night when they open their stomata. PEP is then stored as malic acid in the vacuole. During the day when stomata are closed malic acid is decarboxylated, the released CO_2 is used in the Calvin–Benson cycle. Most succulents are CAM plants and grow successful in arid environments.

- Photosynthesis and respiration are interdependent; photosynthesis is an anabolic and endergonic process; and respiration is a catabolic and exergonic process.

- Photosynthesis is affected by environmental conditions and CO_2 concentration, light quality, light intensity, and temperature affect net photosynthesis.

- The light compensation point is the light intensity or photon flux where net photosynthesis is zero and photosynthesis and respiration are at equilibrium. The light saturation point is where net photosynthesis remains steady and is not increasing any more in response to increasing light intensity.

Reflect

1. *Photosynthesis and life on earth.* Why is it possible and why can't we live without it?

2. *Rubisco and photosynthesis.* Explain the role of rubisco in photosynthesis, why it is such an important protein in nature, and how plants try to optimize its function.

3. *The nature and function of light.* You are a greenhouse vegetable producer and want to maximize your production. How will you manipulate light, temperature, and carbon dioxide in your greenhouses?

4. *Chemiosmosis in cyclic and noncyclic electron flow.* Describe the differences and similarities between the two processes.

5. *Is it a C_3, a C_4, or a CAM plant?* What do you examine or measure to find out? What are some obvious hints looking at a plant that it may be one or the other?

6. *Engineering photosynthesis.* If you could change one process or factor in photosynthesis to increase efficiency, what would that be?

References

Amundson, R. (2001). The carbon budget in soils. *Annual Review of Earth and Planetary Sciencs, 29,* 535–562.

Bhagwat, A. (2005). Photosynthetic carbon assimilation of C3, C4, and CAM pathways. In M. Pessarakli (Ed.). *Handbook of photosynthesis* (2nd ed., pp. 367–389). Boca Raton, Florida: Taylor and Francis.

Cushman, J. C. (2001). Crassulacean acid metabolism. A plastic photosynthetic adaptation to arid environments. *Plant Physiology, 2001*(127), 1439–1448.

Houghton, R. A. (2007). Balancing the Global Carbon Budget. *Annual Review of Earth and Planetary Sciencs, 35,* 313–347.

Leake, J. R. (1994). The biology of myco-heterotrophic ('saprophytic') plants. *The New Phytologist,* 127, 171–216.

Le Quéré, C., Raupach, M. R., Canadell, J. G., & Marland, G. (2009). Trends in the sources and sinks of carbon dioxide. *Nature Geoscience, 2,* 831–836.

NRC. (2010). *Advancing the science of climate change.* Washington, DC, USA: National Research Council. The National Academies Press.

Rychter, A. M., & Rao, I. M. (2005). Role of phosphorus in photosynthetic carbon metabolism. In M. Pessarakli (Ed.). *Handbook of photosynthesis* (2nd ed., pp. 123–148). Boca Raton, Florida: Taylor and Francis.

Taiz, L., Zaiger, E., Møller, I. M., & Murphy, A. (2018). *Fundamentals of plant physiology* (1st ed.). New York: Oxford University Press.

Vitousek, P. M., Ehrlich, P. R., Ehrlich, A. H., & Matson, P.A. (1986). Human appropriations of the products of photosynthesis. *BioScience, 36*(6), 368–373.

Chapter 11

Respiration

Source: Ursula Schuch

Learning Objectives

- Describe the four major steps involved in cellular respiration
- Describe the fate of one molecule of glucose in glycolysis
- Describe where and how CO_2 evolves during respiration
- Explain the role of oxidative phosphorylation in the process of respiration
- Discuss the similarities and differences between aerobic respiration, anaerobic respiration, and fermentation
- Explain why respiration is controlled during the shipment of fresh produce and what steps are taken

respiration
Breakdown of organic molecules to yield energy (ATP) through oxidation of glucose by enzyme-mediated steps.

Respiration is the reverse process of photosynthesis. Carbon products generated in photosynthesis are broken down in small controlled reactions, and the released energy is conserved in the energy storing molecule ATP for use in cellular processes. Respiration occurs in all living cells of both plants and animals at all times and is independent of light or photosynthesis. Respiration oxidizes glucose molecules in a multistep process converting the energy released during each step into ATP. The many steps in respiration ensure maximum extraction of energy from each glucose molecule and prevent damage to cellular structures that might result if energy is released too fast. Aerobic respiration requires the presence of oxygen while anaerobic respiration and fermentation occur in the absence of oxygen. The three pathways differ in their final electron acceptors. Organisms carrying out aerobic respiration depend on oxygen as electron acceptor, those performing anaerobic respiration use nitrate or sulfate, and organisms carrying out alcohol fermentation use acetaldehyde as electron acceptor.

AEROBIC RESPIRATION

Aerobic respiration is the most common way in plants to release stored energy. An overview of aerobic respiration is shown in Figure 11.1. The process can be divided into four steps: glycolysis, oxidative pentose phosphate pathway or pyruvate processing, tricarboxylic acid (TCA) or Krebs cycle, and the oxidative phosphorylation (Taiz, Zaiger, Møller, & Murphy, 2018).

Glycolysis and the oxidative pentose phosphate pathway occur in the cytosol and plastids, the remaining processes take place in the mitochondria (Figure 11.1). The mitochondria are referred to as the cellular power generator because this is where most of the energy is generated. Recall from Chapter 3 that the mitochondria contain an outer membrane and inner membrane, which is folded and called cristae. Inside the mitochondria is a liquid matrix holding DNA, RNA, ribosomes, and many enzymes needed for the remaining respiration processes (TCA cycle, oxidative phosphorylation) and other metabolic steps. The inner membrane has a very large surface area due to the folds and contains complex protein structures, which are highly selective, allowing only few molecules to pass through it. The structure of the mitochondria is essential for the measured release of energy in the process of respiration.

During the majority of the following discussion glucose, the product of photosynthesis is considered the main substrate for aerobic respiration. However, sucrose, other sugars, and other organic acids are also used as substrates. Proteins are broken down into amino acids, and lipids are converted to fatty acids and finally into acetyl coenzyme A where they too enter the Krebs cycle.

ATP is the main product of respiration. There are two ways in which ATP is produced. During oxidative phosphorylation, ATP is produced from ADP and inorganic phosphate in the electron transport chain of the mitochondria. This process requires a proton gradient, the enzyme ATP synthase, and NADH and $FADH_2$ which are oxidized. Substrate-level phosphorylation, the production of ATP from ADP and inorganic phosphate takes place during glycolysis. The main difference to oxidative phosphorylation is the energy to synthesize ATP that comes from a phosphorylated compound. With the aid of an enzyme, the phosphate group is transferred from the phosphorylated molecule to ADP, resulting in ATP.

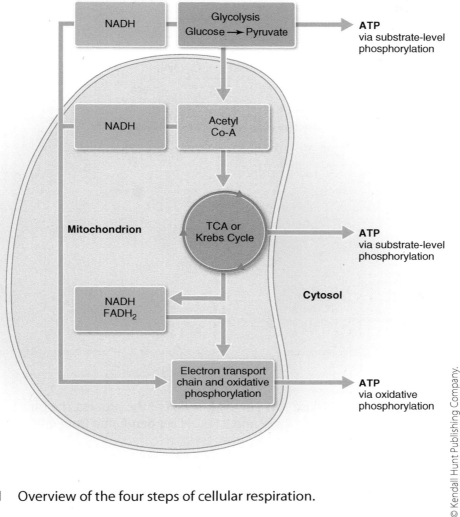

Figure 11.1   Overview of the four steps of cellular respiration.

The general equation for respiration is:

$$C_6H_{12}O_6 + 6\ O_2 \rightarrow 6\ CO_2 + 6\ H_2O + \text{energy (ATP)}$$

Glycolysis

Glycolysis (*glyco* meaning sugar and *lysis* meaning splitting) is the process where the 6-carbon sugar glucose is broken down to two 3-carbon molecules of pyruvate, 2 ATP, and 1 NADH (Figure 11.2). The enzyme-mediated reactions occur in the cytosol and start with glucose, use 2 ATP for phosphorylation, and after five steps produce two 3-carbon molecules, dihydroxyacetone phosphate, and glyceraldehyde 3-phosphate. Glycolysis was named after this cleavage step. After these energy consuming steps, glycolysis continues with reactions that conserve energy in ATP and NADH. In the next step, the oxidation of glyceraldehyde 3-phosphate to 1,3-bisphosphoglycerate yields 2 NADH molecules. NADH is an electron carrier and is produced by adding one proton and two electrons to NAD^+. These electrons are later available to reduce other molecules. The following reaction of 1,3-bisphosphoglycerate to 3-phosphoglycerate yields 2 ATP.

glycolysis
The phosphorylation of glucose in the first step of aerobic respiration resulting in energy and the final product glyceraldehyde-3-phosphate; occurs in the cytosol.

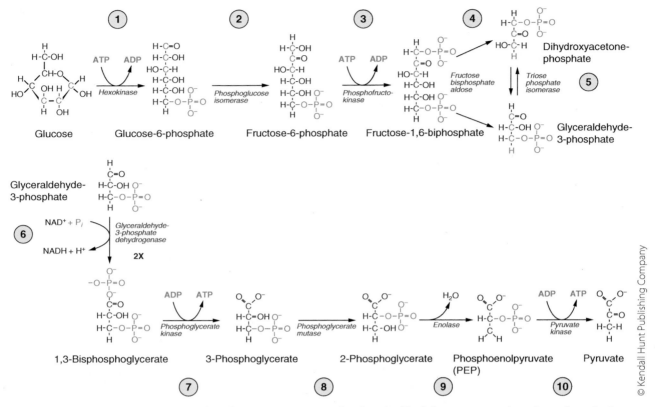

Figure 11.2 The reactions in glycolysis use 2 ATP in the first half of the process, the phosphorylation of glucose. In the second half of the process 4 ATP and 2 NADH are generated from one molecule of glucose.

The final reaction in glycolysis is a substrate-level phosphorylation where phosphoenolpyruvate is converted to pyruvate, resulting in the gain of two more ATP. Glycolysis releases only two ATP and two NADH, a relatively small gain; however, about 80% of the energy from one glucose molecule still remains in the two pyruvate molecules produced (Taiz et al., 2018).

As seen in these reactions, glycolysis does not require oxygen and is considered the most ancient form of energy release. Glycolysis occurs in prokaryotic and eukaryotic cells. Intermediate products of glycolysis are used for many biochemical processes in plants. They supply the building blocks for amino acids, nucleic acids, fatty acids, and important polysaccharides such as cellulose and hemicellulose, which are the major constituents of cell walls.

Pyruvate Conversion to Acetyl Coenzyme A

This pathway is between glycolysis and the TCA or Krebs cycle, and the reactions take place in the mitochondria matrix. This space is the area inside the inner mitochondria membrane. In the first step, carbon is removed as CO_2 from pyruvate and electrons are transferred to NAD^+ to form NADH (Figure 11.3). In the last part of this conversion, the acetyl group combines with coenzyme A to acetyl coenzyme A. This reaction uses pyruvate, NAD^+,

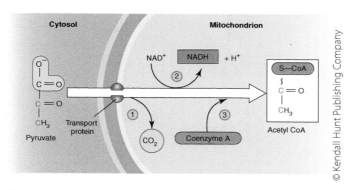

© Kendall Hunt Publishing Company

Figure 11.3   The oxidation of pyruvate to acetyl coenzyme A. Pyruvate is the product of glycolysis and acetyl coenzyme A feeds into the TCA or Krebs cycle.

coenzyme A, and the enzyme pyruvate dehydrogenase. It results in acetyl coenzyme A, NADH plus one proton, and CO_2. This is the first release of CO_2 in the process of respiration.

The TCA or Krebs Cycle

The third step in aerobic respiration is the **tricarboxylic acid (TCA) cycle or Krebs cycle**, also known as the citric acid cycle (Figure 11.4). It is named after the British biochemist Sir Hans Krebs who elucidated this multistep pathway in the 1930s and in 1953 received the Nobel Prize for this discovery. The cycle is known under the other two names because the first two products in the cycle, citric acid and isocitrate, are both tricarboxylic acids.

The series of eight reactions of the TCA cycle take place in the mitochondrial matrix. The reactions start and end with oxaloacetate and occur in a continuous circle. Although the TCA cycle is part of aerobic respiration, no oxygen is used.

First, acetyl coenzyme A from the previous pathway and oxaloacetic acid, a 4-carbon molecule, combine to produce citric acid, a 6-carbon molecule, and the first tricarboxylic acid in the cycle. Following, citric acid is converted to isocitrate, another tricarboxylic acid. In each of the next two steps, one carbon as CO_2 is removed and one NADH is produced. The two reactions are oxidative decarboxylations and yield the 4-carbon product succinyl-coenzyme A. The following four steps of the TCA cycle oxidize succinyl-coenzyme A to the final and beginning product of the cycle, oxaloacetate, via the intermediates succinate, fumarate, and malate. Energy produced in these steps is one ATP, one NADH and one $FADH_2$, both electron carriers. These electron carriers will be used in the production of ATP in the last part of the respiration process, the process of oxidative phosphorylation.

Release of carbon dioxide during aerobic respiration is limited to the intermediate production of acetyl coenzyme A and the TCA cycle. Energy gained in the TCA cycle includes one molecule ATP and electron carrying compounds, one molecule $FADH_2$, and three molecules NADH.

TCA or Krebs cycle
This part of the aerobic respiration includes several reactions that releases energy starting with acetyl Co-A as substrate and occurs in the matrix of mitochondria.

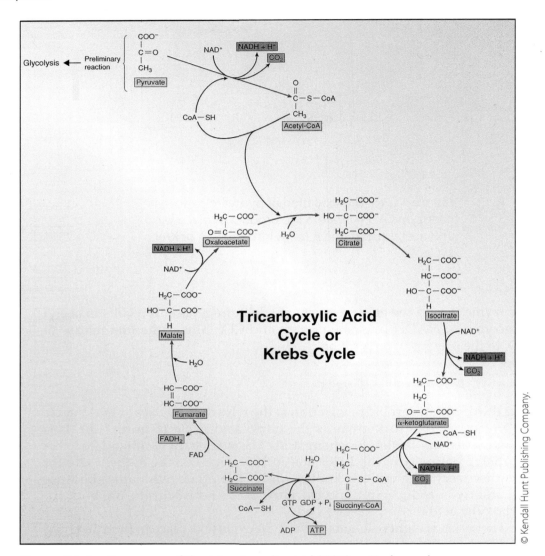

Figure 11.4 Summary of the tricarboxylic acid (TCA) or Krebs cycle.

The Electron Transport Chain and Oxidative Phosphorylation

electron transport chain
This final step in aerobic respiration occurs in the mitochondrial membrane and results in energy as electrons are moving through protein complexes.

The final and highest energy-releasing step in respiration is the **electron transport chain** followed by ATP production (Figure 11.5). In this process, electrons are moved along a chain of electron carriers to molecular oxygen; with each passage, the electrons release energy, leaving them at a slightly lower energy level. The electron carriers are molecules referred to as multiprotein complex. They accept electrons from the previous carrier and donate electrons to the subsequent carrier, while producing NADH in the process. Electrons donated by NADH and FADH$_2$ gradually lose energy as they move down the electron transport chain, the small loss in energy at each step allows for production of ATP or other energy storing molecules. The electron carriers are embedded in the mitochondrial membrane and regulate the passage of electrons.

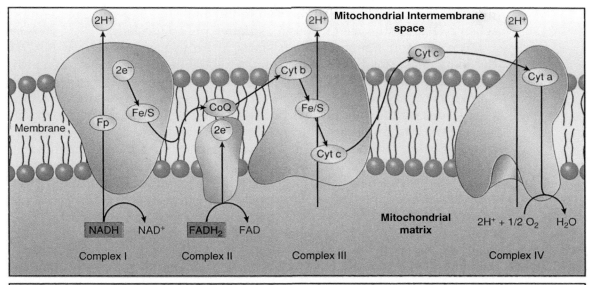

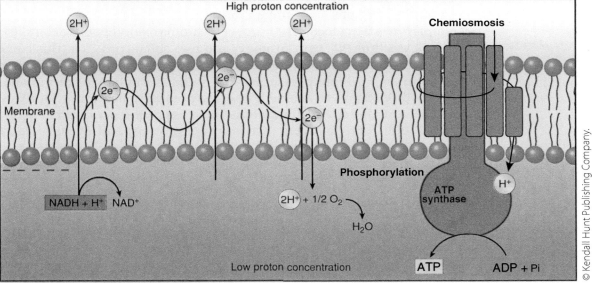

Figure 11.5   The electron transport chain and oxidative phosphorylation produce energy and oxidize water in the final step of aerobic respiration.

The energy for transport of the electron transport chain is derived from a chemiosmotic coupling mechanism. This mechanism was first proposed by Peter Mitchell in 1961 but was not accepted until later experimental evidence confirmed his hypothesis. In 1978, he received the Nobel Prize for this work. The mechanism relies on establishing an electrochemical gradient through the electron transport chain and subsequent ATP production from the energy of this gradient. The electrochemical gradient is created when some of the energy from the passage of electrons through the electron transfer chain is used by three of the four protein complexes to pump protons from the matrix into the intermembrane space between cristae and outer membrane (Figure 11.5). Since the cristae is impermeable to protons, the high concentration of protons in the intermembrane space leads to a greater positive charge compared to the matrix. The potential energy from this electrochemical gradient is used in the final step of aerobic respiration when ATP is synthesized through oxidative phosphorylation.

The electron transport chain starts in protein complex I, an NADH dehydrogenase, oxidizes electrons from NADH in the mitochondria matrix (Figure 11.5). The electrons are passed from complex I to ubiquinone, also known as coenzyme Q (coQ). For every electron pair moving through complex I, four protons are pumped across the mitochondria membrane from the matrix into the intermembrane space. Ubiquinone is an electron and proton carrier, is lipid-soluble, and freely moves in the inner membrane. Ubiquinone is the oxidized form, ubiquinol the reduced form of the molecule. Ubiquinone passes the electrons either to complex III or to alternative oxidase. Ubiquinone further accepts electrons from complex II. Additional electrons from NADH in the matrix or cytosol can be passed on to ubiquinone by NAD (by dehydrogenases located on the matrix and intermembrane side of the cristae (Taiz et al., 2018)).

Complex II, succinate dehydrogenase, oxidizes succinate to fumarate. The resulting energy from the reaction of $FADH_2$ to FAD is transferred to ubiquinone. No protons are pumped through complex II.

Complex III is the cytochrome b, c or cytochrome oxidoreductase complex. It contains cytochrome b, cytochrome c, and iron sulfate proteins. This complex moves electrons received from ubiquinone via cytochrome c to complex IV and pumps four protons per electron pair received through the membrane.

Complex IV, the cytochrome c oxidase complex, has three proteins, cytochrome c, a, and a_3. In this terminal process of the electron chain, O_2 is reduced with four electrons, takes up two protons, and produces two molecules of H_2O. Two protons are pumped from the matrix to the intermembrane space.

Chemiosmosis and Oxidative Phosphorylation

The electron transport chain establishes an electrochemical gradient across the inner mitochondria membrane as protons are pumped by complex I, III, and IV from the matrix into the intermembrane space. This results in a greater accumulation of hydrogen ions in the intermembrane space. In order to move through the membrane back into the matrix, hydrogen ions cannot freely diffuse through the phospholipid membrane and need to pass through the protein ATP synthase. As they move through this complex, the addition of an inorganic phosphate to ADP results in ATP in a process called **chemiosmosis**. Chemiosmosis is the process where the electrochemical gradient across a membrane results in ATP production.

Oxidative phosphorylation is the process when ATP is generated through chemiosmosis in the mitochondria in aerobic respiration as described earlier. ATP synthesis results from the movement of protons through ATP synthase back into the matrix. In comparison, photophosphorylation (Chapter 10) uses the same process of ATP production across the thylakoid membrane inside the chloroplasts. Protons are pumped from the stroma inside chloroplasts across the membrane into the lumen of the membrane to establish the gradient; subsequent proton movement through ATP synthase across the thylakoid membrane to the stroma produces ATP in the chloroplast.

In contrast to oxidative phosphorylation, substrate-level phosphorylation occurs when an enzyme catalyzes the movement of a phosphate group

chemiosmosis
The process where the electrochemical gradient across a membrane results in ATP production.

oxidative phosphorylation
The process when ATP is generated through chemiosmosis in the mitochondria in aerobic respiration.

from a phosphorylated substrate to ADP, producing ATP. The energy for this process comes from the phosphorylated substrate, not from an electrochemical gradient. Substrate-level phosphorylation occurs in glycolysis and in the TCA cycle.

The energy yield from oxidation of one molecule of glucose is up to 30 ATP, with the majority of ATP synthesized in the electron transport chain and oxidative phosphorylation step. Overall energy yield is calculated in ATP, one NADH molecule generates 1.5 ATP in the glycolysis cycle and 2.5 ATP in the TCA cycle and electron transfer chain, and one $FADH_2$ molecule generates 1.5 ATP (Taiz et al., 2018). Thus, glycolysis and the TCA cycle yield four ATP from substrate phosphorylation, two NADH in the cytosol, and eight NADH plus two $FADH_2$ in the matrix of the mitochondria. The ATP yield from aerobic respiration is an estimate as different factors influence the total energy conversion into ATP. Intermediate products of the aerobic respiration are used in many biosynthesis pathways such as nitrogen assimilation, amino acid, lipid, and nucleotide synthesis. The type of sugar used in respiration also affects ATP efficacy.

Aerobic respiration converts about 38% of the energy contained in each glucose molecule; the remaining energy is lost as heat. Aerobic respiration is more efficient than another form of respiration, the anaerobic process.

Control of Respiration Through Feedback Mechanisms

Respiration controls biosynthesis of substances by a mechanism called feedback inhibition. This ensures the efficient use of resources and the appropriate interaction between the multiple endergonic and exergonic metabolic processes of cellular respiration and other metabolic reactions coupled to this pathway. A process can be stimulated or slowed down depending on the access to a substrate through transport proteins. Specific enzymes are a major control factor in respiration. For example, if excess substrate for a process is available, that substrate inhibits the enzyme that mediates the first step of the process. Similarly, if a compound is in short supply, the process will be stimulated. The breakdown of compounds is regulated similarly. The availability of energy currency in the cell provides a strong feedback mechanism; if ATP in a cell is getting low, respiration accelerates, if ATP supply is high, ATP inhibits the enzyme, phosphofructokinase, which is responsible for phosphorylate glucose in step 3 of glycolysis (see Figure 11.2) to slow down respiration.

ANAEROBIC RESPIRATION

Aerobic respiration cannot function in environments lacking oxygen. Without oxygen to accept electrons and no other electron acceptor available, electrons in NADH cannot move to an electron acceptor, NAD+ cannot be regenerated, and the electron transport chain stops. Pyruvate cannot be converted to acetyl coenzyme A and move to the TCA cycle and the electron transfer chain under anaerobic conditions.

Two processes of respiration are possible under anaerobic conditions: Anaerobic cellular respiration and fermentation. **Anaerobic cellular respiration** is similar to aerobic cellular respiration in that electrons extracted

anaerobic respiration
Respiration without the use of oxygen as an electron acceptor. Energy from organic compounds is released through lactic acid or alcohol fermentation.

from a molecule are moving through an electron transport chain. Sulfate or nitrate are common final electron acceptors, but other molecules can serve in this function as well. Organisms using anaerobic cellular respiration are prokaryotes including some bacteria and archaea that live under low oxygen conditions. In these organisms, the complexes of the electron transport chain and ATP synthase are found in the plasma membrane. The energy gain from anaerobic respiration is low with only two ATP or 7% of the total energy from one glucose molecule versus up to 30 ATP in aerobic cellular respiration. It is considered an inefficient process; however, it is important because it allows for regeneration of NAD^+. Without the ability to oxidize NADH to NAD^+, glycolysis would stop.

Some bacteria and archaea are obligate anaerobes meaning that they will die in the presence of oxygen. However, many others are facultative anaerobes where they can switch to fermentation or anaerobic respiration in the absence of oxygen, but carry out the more energy efficient aerobic respiration when oxygen is available. Anaerobic respiration is an important pathway for survival of organisms when oxygen is lacking.

Fermentation is a type of anaerobic respiration where pyruvate from glycolysis can be converted and result in regeneration of NAD^+ to continue glycolysis. This method of releasing energy from organic molecules is believed to have evolved early in earth's evolution when oxygen was absent in the atmosphere. Fermentation occurs in the cytosol compared to aerobic respiration which takes place in the mitochondria. There are two types of fermentation, lactic acid and alcohol fermentation (Figure 11.6).

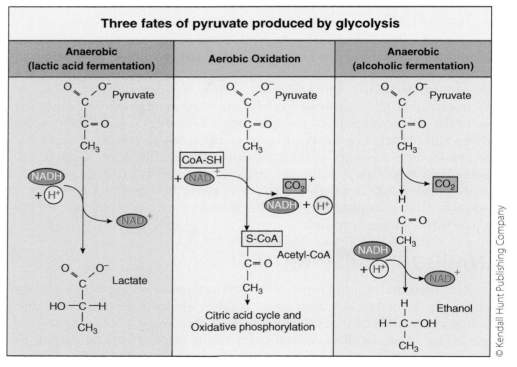

Figure 11.6　Comparison of three types of cellular respiration starting with pyruvate produced by glycolysis: aerobic respiration, anaerobic lactic acid fermentation, and anaerobic alcohol fermentation.

Lactic acid fermentation is common in bacteria, fungi, and animal cells. The process of lactic acid fermentation is essential for the production of food such as sour milk, yoghurt, and sauerkraut. Pyruvate becomes the electron acceptor and is converted to lactic acid while producing NAD^+.

Alcohol fermentation is common in most plant cells and in yeasts. This process also releases carbon dioxide and produces ethanol and energy. Pyruvate is decarboxylated to acetaldehyde, a two-carbon intermediate, which becomes the electron acceptor to convert $NADH + H^+$ to NAD^+. Acetaldehyde is then converted to ethanol.

Controlled alcohol fermentation also has many commercial applications. Baking bread and brewing wine or beer rely on fermentation of yeasts with a sugar substrate. The mixture of yeast, starch (flour), and sugar starts to ferment, producing CO_2 that makes dough rise. Similarly, grape juice with yeast will ferment and produce CO_2 and alcohol. Originally, wine fermentation used only the crushed grapes and depended on the natural presence of yeast on the grapes. Modern wine makers use the grape juice and add yeast to control the fermentation process. When the alcohol concentration reaches 12%, the yeast cells die and fermentation stops. This is why wines usually have no higher alcohol concentration.

Lack of oxygen in the root zone of plants through waterlogging or compaction of soil causes problems for plant health. When plant roots are waterlogged, they lack oxygen and have to switch from aerobic respiration to fermentation. Plants differ widely in their flood tolerance, sometimes even within the same genus (Taiz et al., 2018). Plants living in permanently flooded conditions such as wetlands or those tolerating temporary flooding have to rely on anaerobic respiration for energy production. They also have developed morphological adaptations to obtain oxygen. Under prolonged exposure to anaerobic conditions, plants can die because fermentation is not sufficient to provide enough energy a plant needs to maintain metabolic processes. Extreme acidification of the cytosol is often the cause of cell death because the pH gradient between cytosol and vacuole cannot be maintained.

RESPIRATION AND FRESH PRODUCE

Fresh produce, cut flowers, and potted plants are often transported many miles from where they are grown to the location where they are sold. Maintaining optimum quality and reducing losses from harvest to consumers is the goal of postharvest practices. Once harvested, the water and nutrient supply from the roots is cut off and fruit, vegetables, and cut flowers respire, using up their stored carbohydrates. Live plants shipped inside dark boxes or trucks also respire without light. Low temperatures are key for minimizing respiration rates, which generally increase two to three fold for every 10°C rise in temperature. That is why fresh produce and cut flowers are usually shipped in refrigerated containers. Shipping and storage temperatures differ for each type of plant because of their tolerance to low temperatures and the metabolic processes that will be inhibited or stimulated. Optimal storage temperatures of subtropical plants such as basil (52–59°F or 11–15°C), ripe or bananas (56–60°F or 13–16°C) are much higher than for apples (30–40°F or −1–4°C).

lactic acid fermentation
Process of fermentation where pyruvate is converted to lactic acid because of the lack of oxygen.

alcohol fermentation
Process of fermentation where pyruvate is converted to ethanol.

Respiration of fruit and vegetables depends on the degree of ripeness, the normal loss of quality due to ageing, and the extent of physical damage. Many cut and processed fruit and vegetables have respiration rates several times higher than the whole fruit or vegetable. To extend the shelf life of fresh plant material, the atmosphere in the packaging is sometimes changed to reduce the concentration of oxygen and to slow down aerobic respiration. Oxygen concentrations below 5% for whole fruit or vegetables and below 2–3% for cut fruit or vegetables lowers respiration rates (Taiz et al., 2018). Special packing materials have been developed to allow a limited gas exchange, which will prevent the buildup of excess oxygen.

Key Terms

respiration	electron transport chain	lactic acid fermentation
aerobic respiration	oxidative phosphorylation	alcohol fermentation
glycolysis	chemiosmosis	
TCA or Krebs cycle	anaerobic respiration	

Summary

- Respiration is the breakdown of glucose to obtain energy. All living cells perform respiration. Aerobic respiration requires the presence of oxygen, anaerobic respiration or fermentation operate without oxygen.

- Aerobic respiration can be divided into four steps: glycolysis, conversion of pyruvate to acetyl coenzyme A, the TCA or Krebs cycle, and the electron transport chain with oxidative phosphorylation.

- In glycolysis, one molecule of glucose (6-carbon molecule) is broken down to two pyruvate (3-carbon) molecules. In the process, two ATP are used and two NADPH, and four ATP are produced through substrate-level phosphorylation. This process occurs in the cytosol.

- In the following step, pyruvate is oxidized to acetyl coenzyme A, producing one molecule of carbon dioxide and one molecule of NADH. This process takes place in the mitochondria.

- In the TCA or Krebs cycle, acetyl coenzyme A reacts with oxaloacetate and the following reactions regenerate oxaloacetate. Products of this cycle are two molecules of carbon dioxide, one molecule ATP though substrate-level phosphorylation, one molecule $FADH_2$, and three molecules NADH.

- The electron transport chain uses the energy from the electron donors $FADH_2$ and NADH to pump protons across the inner mitochondrial membrane. ATP synthase uses the energy to synthesize ATP through oxidative phosphorylation. Products are water and ATP, so the total process of aerobic respiration can yield up to 30 ATP from one glucose molecule.

- Anaerobic respiration occurs in some bacteria and archaea when oxygen is not available. Sulfate or nitrate are common final electron acceptors for lack of oxygen, but other molecules can serve in this function as well.

- Fermentation is the process of anaerobic respiration when oxygen is not present as the final electron acceptor in the electron transport chain. Fermentation produces NAD+ from NADH so glycolysis can continue. Fermentation can yield lactic acid or alcohol. The energy gain from fermentation is very low, only two ATP molecules from one molecule of glucose.

- Respiration controls biosynthesis of substances by a mechanism called feedback inhibition. If there is enough end product of a process available, that end product inhibits the enzyme that mediates the first step of that process.

Reflect

1. *Aerobic respiration and photorespiration.* Examine the two processes and where they occur.
2. *Engineering respiration.* If you could change one process or factor in aerobic respiration to increase efficiency, what would that be?
3. *Aerobic respiration and fermentation.* Contrast the two processes and point out the similarities and differences.

References

Taiz, L., Zaiger, E., Møller, I. M., & Murphy, A. (2018). *Fundamentals of plant physiology* (1st ed.). New York: Oxford University Press.

Chapter 12

Plant Responses to Hormonal and Environmental Stimuli

© Elenamiv/Shutterstock.com

Learning Objectives

- Describe the information processing within a plant to external or internal stimuli
- Outline the steps it takes to stop the actions of a hormone
- Explain the nature of a plant hormone and give examples of hormones stimulating or inhibiting growth
- Compare the functions of auxin and ethylene in plant growth and development
- Compare the mode of transport for different hormones
- Explain the concept of totipotency and the role it plays in plant production
- Explain how plants sense light and give examples of some responses
- Outline how plants protect themselves from growing during unfavorable environmental conditions
- Compare how plants perceive and respond to gravity and touch
- Describe how plants defend themselves against intruding organisms
- Discuss how humans rely on plant hormones to produce fruits and vegetables

Plants are rooted in one place for their entire lifetime. They have developed sophisticated systems of sensing and responding to interact with their environment and to other factors (Figure 12.1). Plants detect with their shoot systems and with their root systems. Roots sense the supply of nutrients and water in the soil and send signals to the leaves about opening or closing the stomata. Leaves sense the presence of light and know when to begin and end photosynthesis each day. They sense the daily duration of light and know when it is time to start initiating the flower process. The ability to perceive ambient temperatures protects plants from growing when it is too cold or during a short unseasonably warm spell in winter. Plants activate metabolic pathways to respond to wounding or an attack by invasive organisms. Plants even know when they are upside down because they sense gravity and will change the direction of new shoot growth toward the sun. Plants produce their own hormones to detect different stimuli and signal to other plant parts and even to neighboring plants how to respond appropriately. Plants are indeed intriguing organisms!

SIGNAL RECEPTION AND TRANSDUCTION

Plants receive information about their environment on a continuous basis (Figure 12.2). Their ability to thrive, survive, and reproduce depends on appropriate responses to the different cues. There are three steps in how information is transmitted in a plant (Taiz et al., 2018). First, a receptor cell senses an external stimulus and translates it into an internal signal. Then an enzyme or other cell-to-cell signal sends information through the plant, often with the aid of second messengers. Finally, a receptor cell receives the plant-to-plant signal and translates it into a signal that causes change of metabolic activity. The plant now responds to the signal it received.

Some proteins are specialized to receive signals. Most **signal receptors** are located in the plasma membrane. When a receptor protein receives a signal it binds to it and the protein changes shape. The signal is converted from an exterior signal to a signal inside the cell. This process is called **signal transduction**. The inside signal is then amplified and causes a response in the cell.

signal receptor
Specialized protein in plasma membrane receiving signals from outside the cell and binding to a signal molecule.

signal transduction
Conversion of a signal from outside the cell to within the cell.

Exogenous signals	Endogenous signals
Light	Hormones
Quality	Auxin
Quantity	Cytokinins
Duration	Gibberellins
Direction	Abscisic acid
Mechanical	Ethylene
Wind	Brassinosteroids
Touch	Mechanical
Structures	Growth induced
Herbivores	Tissue compression
Soil	and tension
Nutrients	Defense signals
Water	Jasmonic acids
Atmosphere	Salicylic acid
Temperature	Secondary compounds
Relative humidity	Maturity/Juvenility
Neighboring plants	
Gravity	
Air pollution	
Pathogens	

© Nikitina Olga/Shutterstock.com

Figure 12.1   Internal and external factors influencing growth and development of plants.

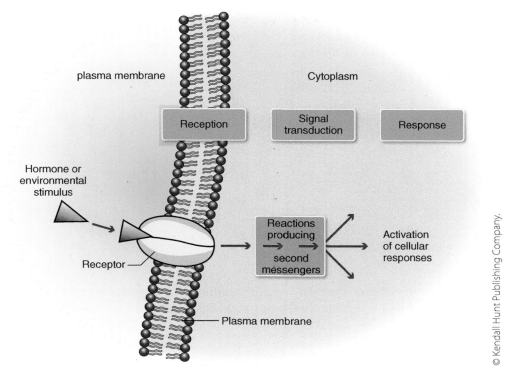

Figure 12.2   Mechanism of how plant cells receive information, transduce the signal, and activate a response.

Signal transduction in the cell can occur via G-proteins, ion channels, and enzyme-linked receptors. The production of **second messengers** allows for the amplification of a signal and rapid dissemination. **G-proteins** are located inside the cell and are closely associated with protein receptors. They are named G-proteins because they bind guanosine triphosphate (GTP) and guanosine diphosphate (GDP). GTP has high potential energy similar to ATP and causes a protein to change structure and function when binding to it. G-proteins cause the amplification of the signal through the production of large amounts of second messengers inside the cell, which trigger the appropriate response to the signal within the cell. The process works as follows (Figure 12.3.a). When inactive, the G-protein binds to GDP. Upon the arrival of a signal molecule, the first messenger which can be a hormone or environmental signal, GDP is replaced by GTP and binds to the now activated receptor protein. The active G-protein then binds to an enzyme, catalyzing the production of second messengers, which cause a response in the cell.

Second messenger molecules are small and can be produced rapidly in great numbers, which allows amplification and swift spreading of the original signal. Calcium ions (Ca^{2+}) can act as second messengers. Second messengers can trigger different responses in different cell types. Often more than one second messenger is involved in triggering the same cellular response. An example of a cellular response is increased or decreased gene expression.

Ion channels are another way of signal transduction (Figure 12.3.b). Ion channels allow the passage of specific ions through the plasma membrane

second messengers
Small molecules produced rapidly to amplify signal transduction and response.

G-proteins
Amplify signal in signal transduction and produce large amounts of second messengers.

ion channels
Channel protein allowing passage of specific ions through the plasma membrane down an electrochemical gradient.

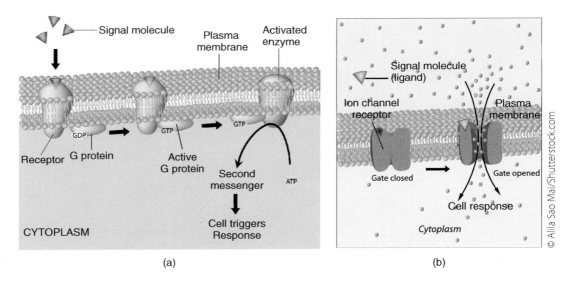

Figure 12.3a–b Signal transduction and second messenger production that cause cellular response with (a) G-proteins or (b) ion channel receptors.

enzyme-linked receptors
Transmembrane protein activated in signal transduction and responsible for starting the phosphorylation cascade, which amplifies signals received by the cell.

phosphorylation cascade
Number of phosphorylation (a process that adds phosphate to another molecule) events triggered by enzymes as a result of signal transduction to amplify a signal through the plasma membrane.

signal deactivation
When no more signal is received by the signal receptor and signal transduction stops.

when a signal molecule or ligand binds to a ligand-gated ion channel receptor. The gate opens and allows specific ions to pass into the cell and trigger a response to the signal. Transport through ion channels is always passive and in the direction of the electrochemical gradient.

Enzyme-linked receptors cause a **phosphorylation cascade** when they receive a signal. The signal binds to a receptor located in the plasma membrane. One ATP molecule binds to the receptor and activates it through phosphorylation. The activated receptor is now an active enzyme that links to a nearby membrane protein. This membrane protein exchanges GTP to GDP, resulting in phosphorylation of a protein. This protein in turn phosphorylates another protein and so on, resulting in the phosphorylation cascade that finally triggers the cell response. This process can amplify a signal rapidly as one activated receptor can result in thousands of proteins through the cascade.

Signal transduction processes intersect and are not strictly separated. They form a network to respond to the many signals perceived by plant cells (Taiz et al., 2018). Signal response varies from cell to cell and depends on the signal that was received. Second messengers and the phosphorylation cascade activate or deactivate proteins in the cell. When root cells require glucose, they send a signal to the source that eventually leads to phloem loading and the movement of glucose to the root tips. When sufficient glucose has been translocated to the sink, the signal will be deactivated to stop the transport.

Signal deactivation occurs through intracellular signals; some are the reversal of the activation steps. When GTP is converted to GDP the enzyme is deactivated and no more second messengers are produced. Ion channel gates are closed when the ligand does not bind anymore to the ion channel receptor. Enzyme-mediated phosphorylation cascades are slowed down by phosphatases, enzymes that remove phosphate groups from proteins. This eventually stops the process of phosphorylation of proteins. Second messengers are short lived and allow for a quick deactivation of signal transduction. In plants with defective genes, the failure of turning off a process when the signal is diminished or has disappeared can result in permanent damage.

Signal transduction responds to very small changes in hormones or the presence of few or many signal receptors. Signal transduction can be rapidly turned on and off in the plant and serves to coordinate intercellular signals throughout the whole organism.

PLANT HORMONES

Plant hormones are defined as naturally occurring, organic compounds that affect physiological processes at low concentrations (Davies, 2010). Previous definitions often added that hormones send signals over long distances in the plant. This notion comes from the fact that plant hormones were considered similar to mammalian hormones, which are produced in one tissue, and are sent through the bloodstream to a target tissue area where they cause a response. Auxin, the first plant hormone discovered, fits this scheme because it senses a stimulus in one tissue and causes a response in a distant tissue. It is now known transport of plant hormones is not essential for their action. Plant hormones can be produced in different tissues and cells, they may be transported to different parts of the plant, but they also can act within the cell where they are produced.

Plant hormones are also known as phytohormones, a term coined by early investigators but less often used now. Synthetic or manmade chemicals similar to plant hormones are called *plant growth regulators* (PGRs). They are used to influence plant growth of crop plants, weeds, and plants or plant cells cultured *in vitro*, an artificial environment such as a test tube. One advantage of PGRs over natural hormones is they are not degraded by enzymes when signal transduction ceases, providing longer-lasting effects.

Plant hormones regulate vegetative growth, the formation of flowers and fruits, timing of flowering, and dropping of leaves. Plant hormones determine when seeds germinate and direct root growth downward and shoot growth upward. They signal stress in one part of the plant to another part to elicit the appropriate response.

Plant hormones occur in very low concentrations in cells and tissues. Generally they are present at a few micrograms per kilogram of plant tissue (Davies, 2010). Exact amounts and locations within cells or tissue are still not known for many plant hormones. The detection and analysis of minute amounts of these substances still presents a challenge for current methods of analysis. Instead, many genes responsible for the synthesis of hormones have been identified and localized. Hormones work concurrently with each other and different responses can result from the action of two hormones when they are present in different concentrations or ratios. Different hormones can work in addition or opposition to each other and it is the balance that determines the growth or developmental response.

There are five major hormones naturally produced by the plant: auxins, cytokinins, gibberellins, abscisic acid, and ethylene. Other hormones have been identified and include brassinosteroids, jasmonates, salicylic acid, and signal peptides. Plant hormones differ widely in their chemical structures (Figure 12.4).

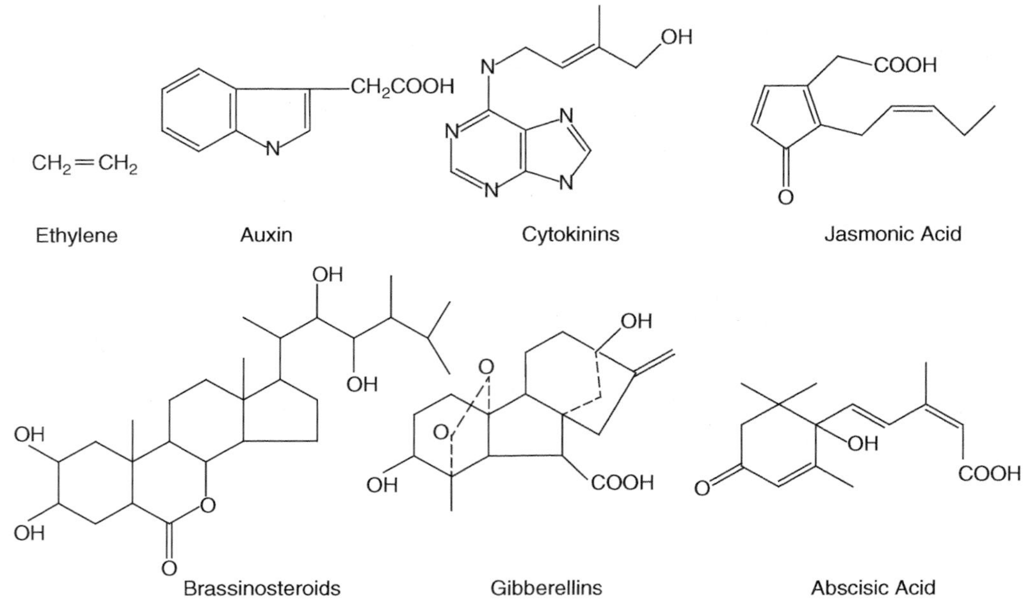

Figure 12.4 Chemical structures of plant hormones.

auxin
growth stimulating hormone involved in phototropism, apical dominance, root formation, and other processes.

Auxin

The plant hormone auxin was the first one discovered. In 1880, Charles Darwin and his son Francis observed grass seedlings bend toward the light, if the tip of the coleoptile, the first leaf that forms a protective sheath around the stem tip, is present. When they removed the tip or covered it with an opaque cap, the stem did not bend toward the light (Figure 12.5). They concluded from this observation that the shoot tip senses the light and sends a signal to the region of bending in the stem. Further experiments by Danish plant physiologist Peter Boysen-Jensen about 30 years later determined the signal can move through agar, but not through a layer of impermeable mica. Final evidence of a mobile growth-stimulating substance in coleoptile tips was provided with Fritz Went's classic experiments with oat coleoptiles in 1926 (Went, 1927). He placed severed coleoptile tips on blocks of agar for a period of time until the substance had diffused into the agar. Then he placed agar blocks exposed to oat tips on coleoptiles with the tips removed where they stimulated growth. When he placed just agar blocks not previously exposed to oat tip coleoptiles without tips they did not grow, confirming the substance was a growth-stimulating hormone. Later he discovered when the agar block containing the substance from a coleoptile tip was placed on half of the cut surface, the coleoptile tips grew only on that side. When this experiment was conducted in the dark, the coleoptiles also grew and bent, indicating light was not required for the movement. Went called the hormone *auxin*, based on the Greek *auxein*, which means "to increase." A few years later the structure of auxin was discovered as **indoleacetic acid (IAA)**, the most common auxin in plants (Figure 12.4).

indoleacetic acid (IAA)
See *auxin*.

acid growth hypothesis
Based on auxin causing an increase in proton pumps in the cell wall, acidification of the cell wall leading to loosening, and subsequent expansion of the cell through incoming water.

Auxin causes cell elongation through a mechanism called the **acid growth hypothesis**. When the cell receives a signal and auxin binds to the auxin-binding protein, signal transduction is triggered (Figure 12.6). First, the number of proton pumps in the cell wall is increased. Protons are pumped outside the cell wall, acidifying the outside. This causes cations

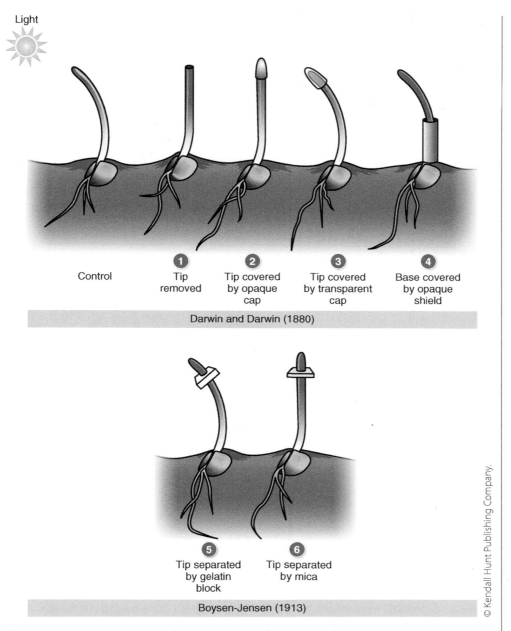

Light

Control

1 Tip removed

2 Tip covered by opaque cap

3 Tip covered by transparent cap

4 Base covered by opaque shield

Darwin and Darwin (1880)

5 Tip separated by gelatin block

6 Tip separated by mica

Boysen-Jensen (1913)

© Kendall Hunt Publishing Company.

Figure 12.5   Experiments leading to the discovery of phototropism and the role of auxin in this response.

such as K⁺ and sugars to move into the cell and water following via osmosis. Turgor inside the expanding cell is increasing. The acidity of the cell wall activates **expansin**, a protein in the cell wall to loosen cellulose microfibrils. Expansins allow for cell elongation through new cellulose microfibrils.

Auxin is produced in young leaf tissue and developing seeds. The hormone moves in phloem parenchyma and parenchyma cells surrounding the vascular tissue toward the base in shoots and toward tips in the roots. This unidirectional transport is called **polar** and is unique to auxin. Concentration of auxin in shoot tips is much greater than in roots. Auxin enters cells through the top with a cotransporter (for details on active transport, see Chapter 13) and two protons. Cotransporters are membrane proteins moving protons

expansin
Protein in the cell wall involved in loosening cellulose microfibrils.

polar
Movement of auxin in one direction, with gravity, from shoot tips to the base and towards root tips.

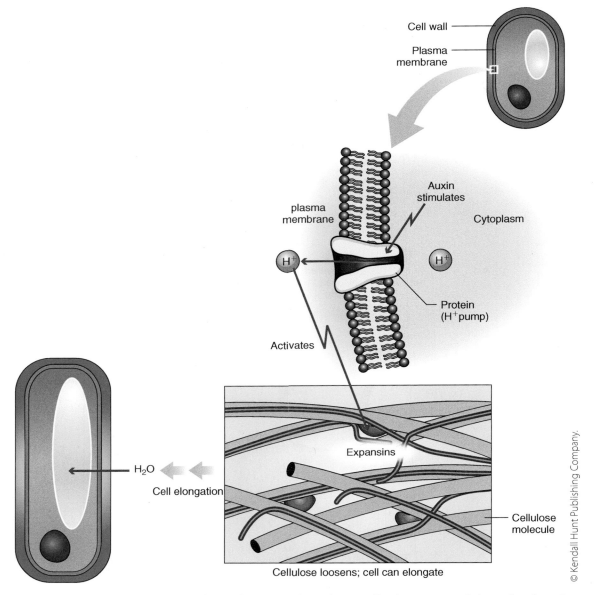

Figure 12.6   The acid growth hypothesis explains how cells elongate and the role of auxin in cell elongation.

along their electrochemical gradient across the membrane while facilitating the movement of molecules such as auxin across the membrane. During diffusion through the cytoplasm some auxin is destroyed by enzymes. Once at the bottom of the cell, auxin molecules are transported out of the cell with carrier proteins. The movement of auxin in parenchyma cells at 10 centimeters per hour is much slower than movement of molecules in the vascular tissue.

Auxin causes a number of different responses in plants in addition to cell elongation (Table 12.1). Auxin initiates roots on stem cuttings and in tissue culture. The synthetic auxins naphthaleneacetic acid (NAA) and indolebutyric acid (IBA) are used in plant propagation to stimulate root formation. The basal end of a cutting is dipped into a powder or solution containing auxin and is then stuck into a growing medium and kept under mist until roots have formed (Figure 12.7a).

Table 12.1   Plant Hormone Characteristics and Their Functions

Plant Hormone	Effects	Characteristics
Auxin	• Promotes cell enlargement and cell division • Initiates roots on stem cuttings and in tissue culture • Mediates response to gravity and light (tropism) • Promotes apical dominance • Stimulates phloem and xylem differentiation • Delays leaf senescence and fruit ripening and abscission • Promotes growth of flower parts and femaleness in dioecious flowers • High auxin concentrations can stimulate ethylene production	• Indole-acetic acid (IAA) primary auxin • First hormone identified • Produced in young leaves, leaf primordial, and developing seeds • Transported from cell to cell in cambium and phloem
Gibberellins (GA)	• Stem elongation; GA_1 causes cell division and cell elongation • Bolting (flowering) in long-day plants • Induces seed germination • Induces fruit set and growth when applied to plants • Can substitute for dormancy requirements • Induces maleness in dioecious flowers	• More than 125 gibberellins known • GA_1 most important for stem elongation • GA_3 most widely available • Produced in apical meristems and developing seeds
Cytokinin	• Stimulates cell division in the presence of auxin • Promotes shoot initiation in tissue culture • Promotes growth of lateral buds • Promotes chloroplast development • Delays leaf senescence	• Most common cytokinin is zeatin • Produced in root tips and developing seeds • Transported in xylem from roots to shoots
Ethylene	• Produced in stressed plants and stimulates defense responses • Stimulates stem thickening, stem elongation, and horizontal bending of stems • Promotes senescence of leaves, flowers, and fruit • Stimulates fruit ripening • Induces flowering in some plants and femaleness in dioecious flowers	• Ethylene is a gas and hydrocarbon (C_2H_4) • Not essential for mature plant growth • Produced by most tissues in response to stress • Moves by diffusion

(continued)

Table 12.1 Plant Hormone Characteristics and Their Functions (Continued)

Plant Hormone	Effects	Characteristics
Abscisic acid (ABA)	• Causes stomata to close in water-stressed plants • Inhibits shoot growth and seed germination • Induces storage protein synthesis in seeds	• Single compound • Produced in roots and mature leaves • Transported in phloem and xylem
Brassinos-teroids	• Promote cell division and elongation • Stimulate phloem and xylem development and fertility • Promote ethylene production • Inhibit root growth	• First isolated from *Brassica* pollen • More than 60 steroidal compounds

phototropism
Movement or growth of a plant in response to light.

gravitropism
Movement of a plant in response to gravity.

Auxin promotes apical dominance, the trait allowing a terminal bud to suppress growth of axillary buds. When a terminal bud is removed or damaged, the auxin-producing signal in that meristem stops and the subtending axillary buds will start to grow through stimulation by the growth hormone cytokinin. The top one or two new shoots will take over the role of the previous dominant shoot and eventually produce enough auxin to suppress growth of lower buds again. The practice of pinching plants by removing the shoot tips manipulates the natural hormone concentration and causes bushy plants with a greater number of side shoots.

Auxin causes bending of the stems by being distributed asymmetrically on the shaded side of the stem (Figure 12.7b). This initiates greater elongation of the cells on the shaded versus the illuminated side and the curving of the stem toward the light. This response is also known as **phototropism**. Auxin regulates growth in response to gravity, a phenomenon known as **gravitropism**.

Low concentrations of auxin delay leaf aging, fruit ripening, and abscission. However, high auxin concentrations stimulate ethylene production and accelerate these processes. The synthetic auxin 2,4,-dichlorophenoxyacetic acid (2, 4-D), a widely used broadleaf herbicide not affecting most grasses, induces uncontrolled growth, the production of high concentrations of ethylene, and finally plant death. Other synthetic auxins also have efficacy to control weeds (Figure 12.7c).

Auxin stimulates development of vascular tissue by differentiating xylem, phloem, and cambium tissue. The hormone promotes growth of flower parts and femaleness in dioecious flowers. When auxin levels decrease in leaf tissue and fruit in fall, the aging process and subsequent abscission begin. It is likely auxins are involved in further developmental processes in plants.

Cytokinins

The hormones cytokinins regulate both cell division and differentiation. They are named after their important role in cell division or cytokinesis. They stimulate the growth of lateral buds, delay leaf senescence or aging, and promote

Figure 12.7a–c   Plant responses to auxin: (a) Cuttings of Coleus (*Solenostemon spp*) treated with low and higher dose of auxin (left to right). (b) Phototropism, bending of stems towards the light where auxin is moved to the shaded side of the stem, causing greater cell elongation on the shaded side and stem bending. (c) High auxin concentrations are used in herbicides and are effective in killing broadleaf weeds.

chloroplast development (Table 12.1). Cytokinins along with auxins are necessary to activate genes that maintain the cell cycle and regulate shoot and root growth (Taiz et al., 2018). Cytokinins are produced in the tips of roots, in growing buds, young fruits, and other developing tissues. They move in the xylem to different parts of the plant.

Cytokinins were discovered as components in the liquid endosperm of coconuts (*Cocos nucifera*) called coconut water. Researchers observed that coconut water maintained cells in culture by promoting cell division and growth. Kinetin was the first cytokinin discovered. Zeatin, isolated first from corn endosperm, is the most common cytokinin and is found in most plants.

The concept of producing a whole plant from one parenchyma cell is called **totipotency**. The research team of Folke Skoog achieved this in the 1950s using tobacco cells in culture and adding different hormones and vitamins to the culture (Miller & Skoog, 1953). Cytokinins initiate shoot differentiation and promote cell division in tissue culture. The ratio of auxin to cytokinin and their concentration determines how a cell culture grows (Figure 12.8a). Auxin alone results in cell growth but no division. Equal parts of auxin and cytokinin cause the growth of undifferentiated cell masses called **callus**. When cytokinins are supplied in higher concentrations than auxin, cells divide, grow, and develop into shoots. Roots initiate when the ratio is reversed and high auxin and low amounts of cytokinin are supplied (Figure 12.8b).

totipotency
Capability of cells to divide and grow into a complete mature organism.

callus
Mass of undifferentiated cells capable of developing into root and shoot tissue.

Gibberellins (GA)

The hormone activity of gibberellins was first described in 1926 by Japanese researcher Eiichi Kurosawa. He documented that rice (*Oryza sativa*) plants infected with a fungus grew twice as tall but with a weak stem, which eventually failed and resulted in plant death. Applied extracts of the fungus

Figure 12.8a–b (a) Alfalfa (*Medicago sativa*) tissue culture in a petri dish with callus on the right and shoots differentiating on the left. Higher cytokinin to auxin ratio promotes shoot growth. (b) These seedlings have fully developed shoots and roots and are ready to be transferred from tissue culture to growing medium.

caused a similar response in previously uninfected plants. Several years later the growth-stimulating substance was isolated and named gibberellin after the fungus causing the disease (*Gibberella fujikuroi*). The chemical structure (Figure 12.4) of the hormone was not determined until the 1950s.

More than 125 different gibberellins have been isolated from different plants and fungi. Gibberellin is commonly abbreviated as GA and different GAs are identified by their subscripts. GA is produced in apical meristems and developing seeds.

GA plays an important role in seed germination. Dormancy is a metabolic state with little or no activity to protect plants or seeds from starting to grow before the onset of winter or during mild winter periods. To overcome dormancy and resume or start growth, buds or seeds generally require a period of cold temperatures, drought, or a combination of cold and wet conditions. GA stimulates growth in seeds by activating the enzyme α-amylase, a digestive enzyme breaking down stored starch. The enzyme is located in the aleurone layer, a tissue inside the seed. Sugars released from the starch support the growing embryo. The addition of GA to the aleurone layer will stimulate the production of α-amylase, hastening seed germination.

GAs have growth stimulating responses especially in shoot elongation (Table 12.1; Figure 12.9a). GA_1 is essential for stem elongation. GAs are often applied to plants to stimulate shoot elongation and flowering, or to break dormancy. GA stimulates stem growth in dwarf plants and in some grasses (Taiz et al., 2018). GA regulates the development of plants from the juvenile to the mature form. In some conifers the application of $GA_4 + GA_7$ can hasten the beginning of the reproductive phase, an advantage in seed orchards. Other commercial applications primarily of GA_3 include accelerating the malting of barley (*Hordeum vulgare*) seed, increasing the yield of sugar from sugarcane (*Saccharum officinarum*) through internode elongation, and regulating fruit production. Seedless grapes naturally form tight bunches with short internodes and tightly packed grapes. Grapes treated with GA during fruit development grow longer internodes and larger grapes. The elongated shape of commercially available Delicous-type apples (*Malus* sp.) is often a result of a spray containing $GA_4 + GA_7$ and benzyladenine, a cytokinin (Figure 12.9b).

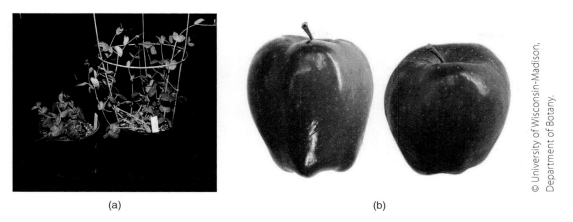

(a)   (b)

Figure 12.9a–b (a) Gibberellic acid stimulates internode elongation (right treated), and (b) causes fruit elongation and a more typical shape of apples (left treated).

Inhibitors of gibberellin biosynthesis are also commonly used in plant production, particularly in floriculture to decrease stem length of flowers. Many greenhouse-grown potted flowers such as chrysanthemum (*Dendranthema morifolium*), poinsettia (*Euphorbia pulcherrima*), and bedding plant species are generally treated with GA biosynthesis inhibitors to produce shorter, compact plants. Cereal species in cooler, wetter climates are sometimes sprayed with these plant growth regulators to prevent lodging, or falling over, before they are ready for harvest.

Abscisic Acid

Inhibitory effects of the hormone abscisic acid (ABA) were first observed in the 1940s by botanist Torsten Hemberg. He documented an inhibitory substance in dormant buds, preventing the growth-promoting effects of auxin (Hemberg, 1947). ABA was identified and named by the late 1960s.

Abscisic acid is produced in developing seeds where it promotes synthesis of seed storage proteins and prevents the premature germination of seeds on the plant. When ABA in the seed decreases, dormancy is released and seeds can start germinating.

In water-stressed plants elevated levels of ABA are produced in the leaves and cause stomatal closure to protect plants from further water loss. In drought-stressed plants ABA increases in both the root and leaf tissue and can lead to greater root growth. When sufficient water becomes available, ABA breaks down and stomata open again.

Ethylene

Ethylene is a simple hydrocarbon (Figure 12.4) and the only hormone occurring as a gas (Table 12.1). Ethylene is a component of illuminating gas used in the 19th century. Leaking gas was found to defoliate shade trees located nearby and in 1901 ethylene was identified as the compound causing the defoliation. Pea (*Pisum sativum*) seedlings exposed to ethylene respond with a triple response in epicotyl growth: decreasing shoot elongation, thickening of the shoot and root, and horizontal bending (Figure 12.10). In addition to this growth response, ethylene affects fruit maturation, abscission of fruit and leaves, and senescence. Ethylene is stimulated by wounding, air

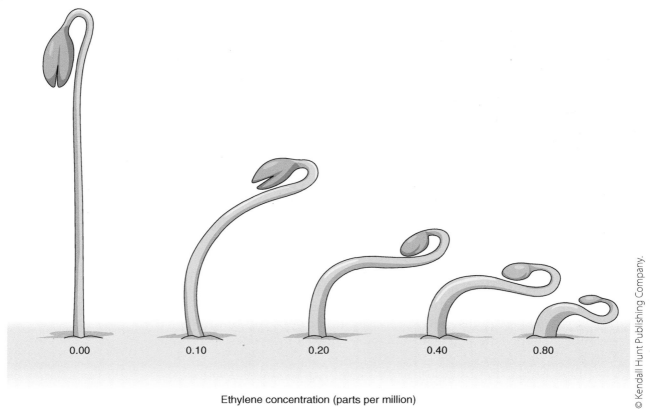

Ethylene concentration (parts per million)

© Kendall Hunt Publishing Company.

Figure 12.10 Increasing ethylene concentration results in shorter stems, thicker stem diameter, and increasingly horizontal growth.

climacteric
Fruits that have high respiration rates during ripening and concurrent high ethylene production.

pollution, and by high ABA concentrations. Ethylene also stimulates shoot elongation in semiaquatic species such as rice in response to flooding and initiates flowering in pineapple and related species in the bromeliad family.

The role of ethylene in fruit ripening is well established. Ethylene stimulates several of the processes involved in fruit ripening such as change in fruit color, enzyme activity to soften the fleshy part of the fruit, and the conversion of starches into sugars. Some fruit are called **climacteric**, referring to their increased cellular respiration rates just before ripening, which is preceded by increased ethylene synthesis (Figure 12.11). Examples of climacteric fruit are apples, pears, peaches, avocado, bananas, cantaloupes, mango, tomatoes, and plums. Nonclimacteric fruit such as strawberries, grapes, bell peppers, watermelon, or citrus ripen gradually without the surge of ethylene followed by a surge of CO_2. Ethylene is also responsible for spoiling fruit caused by one excessively ripe or wounded fruit among others. The high rates of ethylene release from the ripe or wounded fruit will hasten ripening of the surrounding fruit.

Genetic changes to plants have altered ethylene synthesis and the sensitivity of plants to ethylene. This allows fruit to be picked green and stored in the absence of ethylene without the risk of quick ripening. Just before shipping fruit are exposed to ethylene and begin to ripen in time for marketing. Another approach to extend the storage capacity of climacteric fruit is

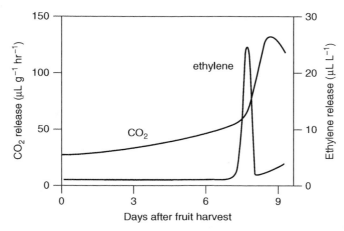

Figure 12.11   Climacteric fruit ripening has a spike of ethylene production before the climacteric increase in respiration rate (CO_2 release), indicating that ethylene triggers the ripening process.

to select for slow-ripening fruit or store fruit at low temperatures, high CO_2, and low oxygen. This will prevent ethylene synthesis and allow storage of fruit like apples long after they would have otherwise spoiled.

Ethylene stimulates abscission of leaves, flowers, and fruit in many plants by activating enzymes that degrade the cell walls in the abscission layer. Auxin counteracts ethylene activity and can prevent abscission. Very high concentrations of auxin, however, can stimulate ethylene synthesis and subsequent abscission. Ethylene-releasing growth regulators are used to stimulate abscission of peaches for fruit thinning, of cotton to defoliate before harvest, and of berries to loosen fruit before mechanical harvest.

BOX 12.1 The FLAVR SAVR Tomato

The FLAVR SAVR tomato was introduced to supermarkets in 1994. Its claim to fame was that it was the first genetically engineered food plant (Bruening & Lyons, 2000). The common term "genetic-modified organism" (GMO) was not coined yet. These tomatoes were genetically engineered by Calgene Inc. to delay the ripening process with the characteristic spike in ethylene, and keep ripe fruit firm for a longer time. Tomatoes ripened on the vine usually have a bright red color, slightly softened flesh, and delicious flavor. Their greatest downside is the short shelf-life and easy bruising during transport. At the time, most tomatoes were picked green and artificially ripened by treating them with ethylene. Although these tomatoes develop a red color by the time they arrive at the store, they do not develop the flavor associated with natural ripening.

Ripening tomatoes develop the enzyme polygalacturonase (PG), which softens the fruit by dissolving cell wall pectins. Calgene researchers constructed a reverse-orientation antisense gene that almost eliminated synthesis of PG in developing fruit. It took several years from the time of constructing the PG antisense gene until tomato plants with the gene were produced.

Subsequent restricted field trials satisfied the U.S. Department of Agriculture that the plants were not a plant-pest risk. Animal feeding studies fulfilled requirements by the U.S. Food and Drug Administration that the genetically modified tomatoes were unlikely to pose a health risk for humans or animals. The tomatoes were similar to nontransformed fruit except for the delay in pectin degradation in fruit cell walls and higher viscosity of tomato paste.

FLAVR SAVR fresh tomatoes were well received by consumers. However, despite steady high demand they were never profitable because of the high cost associated with production and distribution and the sale of the tomatoes was eventually discontinued. Tomato paste produced from the modified fruit and labeled accordingly was sold successfully in the United Kingdom for about three years. The product's demise started when a researcher working on genetically modified potatoes claimed biological effects on rats fed with the modified potatoes could be due to the GMO potatoes. Although the researchers' conclusions were proven incorrect later, sales of the tomato paste produced with FLAVR SAVR tomatoes was discontinued.

More than a dozen crop plants have been genetically modified and introduced to the world market since the FLAVR SAVR tomato. The majority of GMO crops are targeted to improve resistance against pests, diseases, or herbicides. Further discussion of genetic modification is included in Chapter 17.

Brassinosteroids and Other Plant Hormones

Brassinosteroids are among the more recently identified hormones and were first isolated from plants in the *Brassica* or cabbage genus. Steroid hormones were previously thought to be synthesized only in animals. They promote cell division and elongation in stems and xylem development and inhibit root growth and leaf senescence (Table 12.1). Brassinosteroids cause responses similar to auxin in plants.

Jasmonates, salicylic acid, and signal peptides are also considered hormones because they initiate responses at very low concentrations (Davies, 2010). Jasmonates are named after the jasmine plants from which they were isolated. Jasmonates promote plant defense mechanisms, senescence, fruit ripening, pigment formation, and tendril coiling. They inhibit growth and seed germination.

system-acquired resistance
Development of resistance to pathogen invasion of an entire plant in response to a localized pathogen attack.

Salicylic acid, found in willow bark, induces proteins in response to pathogen attacks on older leaves. The response is called **system-acquired resistance (SAR)** and transmits pathogen resistance to younger leaves. Salicylic acid causes thermogenesis, the warming around *Arum* flowers that melts snow and allows for pollination. Salicylic acid also blocks wound response, enhances flower longevity, and inhibits ethylene production and seed germination.

Signal peptides have been discovered to promote plant defense, cell division, growth, and development. More than 10 different compounds have been identified.

Strigolactones are plant hormones that occur in more than three quarter of all plants. They inhibit the branching of shoots, stimulate the germination of some parasitic plants, and the growth of arbuscular mycorrhizae. In roots they promote the growth of root hairs and reduce the formation of adventitous and lateral roots (Taiz et al., 2018).

RESPONSES TO LIGHT

Light affects many growth and developmental functions in plants. Receptors in plants sense certain wavelengths and trigger a range of responses including growth toward or away from light, flower initiation, seed germination, and opening of stomata, to name a few. Plant responses to light are called **photomorphogenesis**.

Blue Light

Plants respond to a light source by growing toward the light, a phenomenon called **phototropism**. Receptors in the tissue respond to the blue wavelengths and plants will not bend toward a light source lacking wavelengths in the blue spectrum. Blue light receptors causing phototropic growth of plants are called **phototropins**. Recall from Chapter 10 that chlorophyll *a* and *b* absorb significant amounts of light in the blue range. The genes PHOT1 and PHOT2 respond to blue light as it triggers the phosphorylation of phototropins that subsequently cause the blue light response (Taiz et al., 2018). Auxin moves to the shaded side of the stem and causes greater elongation of cells compared to cells on the illuminated side. This results in the bending movement of stems (Figure 12.7b).

Blue light receptors also trigger a number of other responses to light. Chloroplasts in leaves orient themselves to optimize light interception for photosynthesis, or shade each other for protection from damage under high light intensity. This movement is initiated by blue light receptors. Phototropins and other pigments called **zeaxanthins** cause stomata to open and gas exchange and photosynthesis begins. Chryptochromes are photoreceptors that play a role in flowering, sensing day length, and stimulating stem growth in shaded plants. Blue light causes roots to grow away from it, a response observed when germinating seedling roots grow into the soil.

Red and Far-Red Light

Red and far-red light controls seed germination, stem elongation, and flowering. Red light is important for photosynthesis while far-red light is not critical for this process. **Phytochrome** is the photoreceptor absorbing red and far-red light. Phytochrome exposure to red light (620–700 nm) causes it to convert to a form that absorbs far-red light (710–850 nm) (Figure 12.12). When this phytochrome is exposed to far-red light it converts to phytochrome-absorbing red light. P_{fr} and P_r are abbreviations for

photomorphogenesis
Growth and developmental responses of plants to light.

phototropism
Growth of a plant shoot towards a light source.

phototropins
Blue light receptors in plants that cause growth responses.

zeaxanthins
Carotenoid pigment responding to blue light and regulating stomatal opening.

phytochrome
Photoreceptor that absorbs red or far-red light; exists in two shapes and reverses shape in response to the illuminating wavelengths; affects flowering and seed germination.

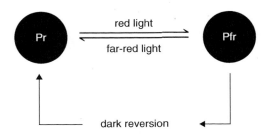

Figure 12.12   The pigment phytochrome exists in two forms: P_r absorbs red light (660 nm) and P_{fr} absorbs far-red light (730 nm). When P_r absorbs red light it changes to P_{fr}. When P_{fr} absorbs far-red light it converts to P_r. During the day red light is prevalent and P_{fr} accumulates, it reverts to P_r during the night.

photoreversibility
One pigment exists in two forms; phytochrome absorbs red light and changes to the far-red conformation, in the far-red conformation it absorbs far-red light and reverts to the red conformation.

photoperiodism
Growth and development of a plant in response to the hours of day and night in a 24-hour period.

photoperiod
Number of hours of daylight in a 24-hour period.

long-day plant
Plant that flowers in response to short night periods. See also *day-neutral plant* and *long-day plant*.

short-day plant
Flowering of this plant is initiated by exposure to long nights. See also *day-neutral plant* and *long-day plant*.

day-neutral plant
Flowering of these plants is not affected by the relative length of day and night. See also *short-day plant* and *long-day plant*.

the phytochrome-absorbing far-red and red light, respectively. The fact the same pigment exists in two shapes, the P_{fr} or P_r conformation, is called **photoreversibility**.

The P_{fr} conformation stimulates seed germination; the P_r conformation inhibits germination. The promoting or inhibiting effect of the two wavelengths was studied on seed germination of lettuce (Borthwick, Hendricks, Toole, & Toole, 1954). Lettuce seed was illuminated with light alternating between red and far-red wavelengths. Lettuce seeds are known to grow best in bright light. Most seeds germinated when red light was applied last; very few seeds germinated when far-red light was applied last. Therefore the red light that induces the P_{fr} conformation is responsible for the stimulating effect. Far-red light passes through leaves and indicates shaded conditions to plants, whereas red light indicates high light conditions.

Phytochrome responses differ based on whether they occur at high or low light levels. Plants growing outdoors are exposed to a wide spectrum of light. Leaves in full sunlight can absorb the abundant red light, leaving more far-red light for leaves within the shaded canopy. Physiological responses will vary for those leaves and are regulated by phytochrome (Taiz et al., 2018).

Photoperiodism is the growth and development response of an organism to photoperiod. **Photoperiod**, the number of hours of daylight and night during a 24-hour period, affects different processes, flowering being the major one. Plants are grouped into **long-day**, **short-day**, or **day-neutral plants** with regard to how they respond to different photoperiods (Figure 12.13). Long-day plants bloom in summer when days are longest and nights are shorter. The short nights rather than the long days are the critical component that is sensed by plants and initiates flowering. Lettuce, spinach, wheat, corn, and

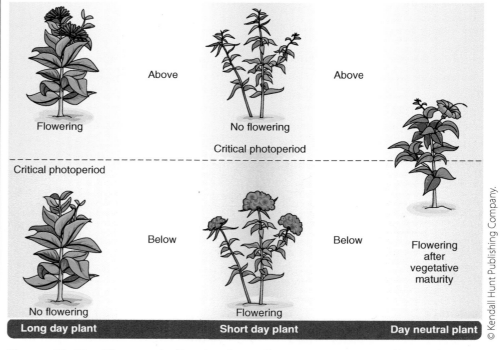

Figure 12.13   Photoperiodism in plants, flowering response of long-day, short-day, and day-neutral plants.

iris are examples of long-day plants. Short-day plants will flower when the daylight hours are below a critical number and an uninterrupted period of long nights will initiate flowering. Chrysanthemum, poinsettia, dahlia, sorghum, strawberries, and ragweed are examples of short-day plants. Those plants naturally flower in fall or spring when days are shorter. Day-neutral plants will flower regardless of photoperiod and include many plants originating in the tropics where day length changes very little annually. Examples of day-neutral plants are sunflower, tomato, carnation, cotton, and roses.

The discovery of photoperiod on flowering was investigated with a short-day plant. When the plant was exposed to long days it failed to flower. Upon receiving less than the critical number of hours of daylight the plant eventually flowered. When the light period was interrupted with short dark intervals, flowering was not affected. In contrast, when the dark period was interrupted with short bursts of light, flowering again failed, demonstrating the importance of an uninterrupted long night.

Further examination of the effect of photoperiod on flowering investigated what happened when long nights were interrupted with short bursts of red light, far-red light, or alternating light sources (Figure 12.14). Interrupting long nights with red light inhibited flowering of short-day plants, but promoted flowering in long-day plants. Following the red light night interruption with far-red light illumination cancelled the effect of the red light and resulted in flowering of the short-day plants and no flowering of the long-day plants. Alternating the light sources showed that the last burst of light determined whether plants perceived a long night (with far-red) or a short night (with red).

Phytochrome affects stem elongation. Plants exposed to red light sense higher light intensity and will elongate less than plants exposed to far-red light, indicating shaded conditions. Greater elongation under shaded conditions such as the understory in a forest gives plants the possible advantage of intercepting more light for photosynthesis. When germinating seedlings are grown in the dark they rapidly grow out of the soil and grow upward with little leaf expansion. Those seedlings are yellow or white because light is required for chlorophyll synthesis. The elongated stems and small leaves allow plants to grow taller, hypothetically toward the light, and use less stored starch for leaf expansion.

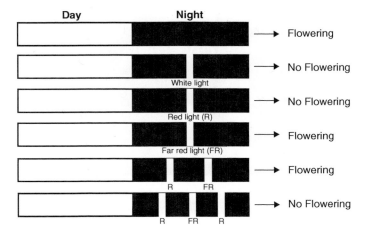

Figure 12.14   Effect of interrupted night with red or far-red light on flowering of a short-day plant.

Growers of floriculture crops use the concept of photoperiodism to produce flowering plants at different times of the year regardless of their natural response to day length. This practice is called "forcing" flowering. Poinsettias (*Euphorbia pulcherrima*) are short-day plants and growers need the plants in flower for the winter holidays. Flowering is initiated by exposing plants to several weeks of 14-hour-long nights with the aid of light-blocking curtains. Some chrysanthemum cultivars are short-day plants and will tend to flower naturally in the winter. Later flowering can be desirable to market plants at a different time. This can be accomplished by interrupting the long natural nights with light and delaying the onset of flowering. Some plants need additional stimuli such as cold temperatures in addition to certain photoperiod requirements to flower. Breeders have also developed cultivars from the same species that flower in response to different photoperiods.

RESPONSES TO GRAVITY

statolith hypothesis
Role of statoliths in sensing gravity by sinking to the bottom of cells, triggering a gravity receptor, and causing reorientation of the cell.

amyloplasts
Leucoplasts that synthesize and store starch.

Plants sense gravity and respond to it, a phenomenon known as gravitropism. The response to gravity guides plants in how to position their roots and shoots (Figure 12.15). The **statolith hypothesis** explains how plant cells sense gravity. Statoliths are special gravity-sensing **amyloplasts**, organelles filled with starch granules. The statolith hypothesis is based on the idea that heavy amyloplasts will sink to the bottom of a statolith cell, triggering a gravity receptor that may cause changes in cell orientation. A gravity receptor has not been identified, although the movement of amyloplasts in response to gravity has been confirmed.

Cells in the center of the root cap sense gravity in roots. In shoots and coleoptiles, statoliths are found in the starch sheath, a layer of cells around the vascular tissue of the shoot. The starch sheath extends to the endodermis of the roots, but contains only amyloplasts and no statoliths in the endodermis of the root (Taiz et al., 2018).

© University of Wisconsin-Madison, Department of Botany.

Figure 12.15 Regardless of the orientation of the germinating seeds, upside down or sideways, the roots grow downward (positive gravitropism) and the stems grow upward (negative gravitropism). The tomato plant (*Lycopersicon esculentum*) was growing upright at first and once laid on its side reoriented itself to grow against gravity (negative gravitropism).

Auxin signals a change in direction when it is redistributed to the bottom of root cells growing in a horizontal direction. This results in elongation of the cells on top and inhibition of elongation on the bottom cells, causing a downward bending of the root. When roots are growing downward auxin is distributed in the center of the root.

RESPONSES TO MECHANICAL STIMULI

Plants respond to touch and wind by stiffening cell walls, growing shorter, and increasing in stem diameter. This can be observed on trees in windy locations growing shorter than the same tree growing taller in less exposed conditions. Plants brushed a few times each day will grow shorter and may have a delay in flowering, similar to plants being treated with a plant growth regulator decreasing internode length. These morphological changes are slower and are common among all higher plants (Braam, 2005). The response is called **thigmomorphogenesis**, or touch-induced reaction of plants not specialized in specific responses to mechanical stimuli. Other responses are increased rigidity or increased flexibility, effects on dormancy, drought, low temperature, and pathogen resistance. Thigmomorphogenetic changes may also be induced by a plant's increased weight due to growth.

thigmomorphogenesis Changes in growth and development of a plant in response to touch or mechanical stimuli.

Hormones and other compounds are involved in signaling changes due to touch. Calcium ions (Ca^{2+}) within the cell are important second messengers (Braam, 2005). Ethylene's role in increasing radial stem diameter in response to touch has been confirmed. However, reduced elongation is not dependent on ethylene.

Thigmotropism is the response to touch. Tendrils, the coiling modified leaves or stems allowing plants to adhere to support structures, begin to curve around an object they are touching continuously (Figure 12.16). This occurs when the opposite side of the tendril touching the object grows more rapidly, thereby wrapping itself around the support. The same response enables roots to grow around obstacles in the soil.

thigmotropism Movement or growth of a plant in response to touch.

The most rapid response to touch can be seen in the Venus flytrap (*Dionaea muscipula*) (Braam, 2005). Sensory hairs on the leaf margin have receptor

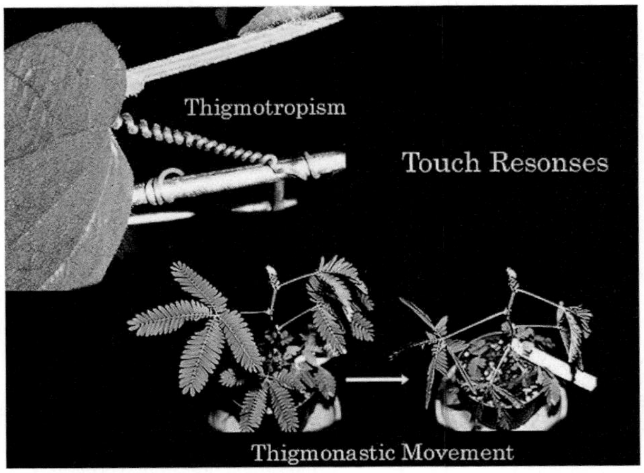

Figure 12.16 Thigmotropism responds to touch with tendrils coiling around the object they touch. Thigmonastic movement of mimosa leaves occurs immediately after the touch.

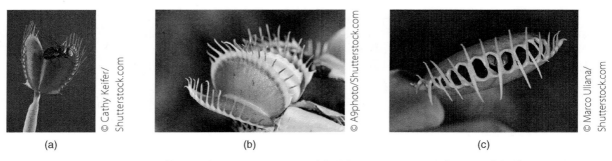

(a) (b) (c)

© Cathy Keifer/Shutterstock.com

© A9photo/Shutterstock.com

© Marco Uliana/Shutterstock.com

Figure 12.17a–c Venus flytrap (*Dionaea muscipula*). (a) Insect in open leaves. (b) Closeup with trigger hairs. (c) Insect trapped in leaves.

cells at the leaf base (Figure 12.17). When stimulated they produce an intercellular electrical signal that generates an action potential. This causes swelling of the cells and rapid closure of the trap in less than one second.

The leaf movements of the mimosa or the sensitive plant (*Mimosa pudica*) in response to touch are also rapid and can affect not only the touched leaf but neighboring leaves as well (Braam, 2005). The leaves fold due to turgor loss in special cells located at the base of the leaflet and petiole (Figure 12.18). Potassium (K⁺) and chloride (Cl⁻) leave the cells and are followed by water.

(a) (b)

© LianeM, 2014. Used under license with Shutterstock, Inc.

© Jafaris Mustafa/Shutterstock.com

Leaflet

Pulvinus

Vascular tissue

Cross-section of open leaf
(c)

Parenchyma cells retaining turgor

Decrease of turgor in parenchyma cells

Cross-section of closed leaf
(d)

© Kendall Hunt Publishing Company.

Figure 12.18a–d Thigmotropism in *Mimosa pudica*.

Up to 25% of the water volume can exit the cell within one second, likely through **aquaporins** and solute-water **cotransporters** (Braam, 2005). Signaling to remote leaves is still being investigated.

RESPONSES TO ENVIRONMENTAL SIGNALS

Plants respond to annual changes in the growing season to protect themselves against being killed during predictably unfavorable conditions such as freezing temperatures or drought. In temperate and cold climates plants prepare for winter by responding to shorter days and cooler daytime temperatures in the fall. In tropical and desert areas the onset of the dry season triggers dormancy. Plants also begin osmotic adjustment to prepare for frost or drought. Hormones regulate the movement of nutrients and other elements out of the leaves of deciduous plants to conserve them in the stem. Chlorophyll starts to break down in leaves and the development of the abscission zone of leaves in preparation for leaf drop begins. In healthy, growing leaves, auxin is produced and leaves are not affected by ethylene, which can be produced in response to environmental stress or decreasing photoperiod. As leaves age they produce less auxin and sensitivity to ethylene increases in the abscission layer, where the petiole is attached to the branch. Cells in the abscission zone closest to the stem store suberin to seal off the opening once the leaves drop. Cells at the base of the leaf blade degrade and weaken the attachment until the leaf drops. This process can be delayed with application of cytokinins, maintaining green leaves well beyond the natural time.

Acclimation of plants to inclement environmental conditions begins when meristem growth stops and buds enter dormancy, a temporary cessation of growth. Scales form on the outside of buds to protect them from harmful temperatures or desiccation. As buds are exposed to cold temperatures or to drought, dormancy requirements are fulfilled and once enough hours of chilling or drought have been accumulated, meristems will grow again as soon as favorable conditions begin. This mechanism prevents plants from growing during a brief period of warm weather in winter or a brief spell of rain during the dry period.

RESPONSES TO PATHOGENS, INSECTS, AND HERBIVORES

Pathogens, viruses, insects, and large herbivores feed on almost every part of plants. Humans rely on plants for food and have great interest in minimizing damage to crops caused by these organisms. The first line of defense is the outer layer of the plant, the cuticle. Some cuticles are thick, waxy, and difficult to penetrate; others have trichomes armed with poisonous compounds to deter intruders. Some plants have spines or thorns to prevent herbivores from consuming them.

Defense mechanisms are an important survival strategy of plants. Many plants produce compounds such as phenolics, terpenes, and alkaloids poisonous to organisms trying to consume or actually consuming them. Plants need to spend energy to produce these defense compounds continuously whether they are needed or not. Therefore another strategy of protection is to rapidly mobilize defenses only when attacked. This response is called

aquaporins
Channel proteins in the plasma membrane to facilitate water transport.

cotransporters
Membrane protein transporting two ions, one in the direction of its electrochemical gradient, while simultaneously transporting another ion *in the same or opposite direction* but against its electrochemical gradient. See also *symporter* and *antiporter*.

inducible defensive response and can be activated at the site of injury or systemically throughout the plant.

Hypersensitive Response

Plants intruded by bacteria, virus, or fungi will react with a **hypersensitive response (HR)**. Affected cells will kill themselves rapidly, thereby starving or killing the intruder. When HR is activated, the toxic compounds produced include reactive oxygen intermediates, nitric oxide, and superoxide ions (O_2^-). Cells synthesize different compounds to strengthen cell walls as a physical barrier. This induced defense activated after a disease or other organism entered a plant is an effective protection mechanism for many plants (Taiz et al., 2018).

The **gene-for-gene hypothesis** explains how plants sense intruding organisms and mount the HR (Figure 12.19). Plants have **disease-resistant genes (R-genes)**, which produce receptors able to sense pathogens in the cell. Disease resistance is a genetic trait and these genes are inheritable. Pathogens have a gene known as the **avirulence gene (avr)** that encode for products called **elicitors**, which are recognized only by plants with the corresponding R-gene (Taiz et al., 2018).

Scientist Harold Henry Flor studied genetics of rust (*Malampsora lini*), a fungal disease, and the host plant flax (*Linum usitatissimum*). He reported in the 1950s plants encoding receptors from a specific R gene were resistant to the corresponding pathogen elicitors from an avr gene. When flax is attacked by rust, the fungus produces numerous compounds, including elicitors from the avr gene. If the elicitors are not recognized by a receptor from the R-gene in flax, the fungus can continue to spread and colonize the

hypersensitive response
Plant defense mechanism where cells intruded by a pathogen will kill themselves and surrounding cells to starve the pathogen.

gene-for-gene hypothesis
Explains how plants sense pathogen infection and how they activate the hypersensitive response. Resistant genes in plants produce receptors that will recognize and neutralize corresponding avirulence gene products from the pathogen.

disease-resistant genes (R-genes)
Disease-resistant genes in plants with the ability to recognize corresponding avirulence genes in pathogens.

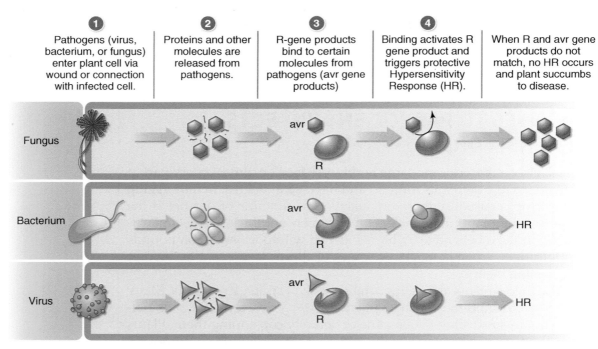

❶	❷	❸	❹	
Pathogens (virus, bacterium, or fungus) enter plant cell via wound or connection with infected cell.	Proteins and other molecules are released from pathogens.	R-gene products bind to certain molecules from pathogens (avr gene products)	Binding activates R gene product and triggers protective Hypersensitivity Response (HR).	When R and avr gene products do not match, no HR occurs and plant succumbs to disease.

© Kendall Hunt Publishing Company.

Figure 12.19   Hypersensitive reactions are produced when the R-gene in a host plant matches the avr gene of the intruding organism.

plant tissue. If the receptor recognizes the elicitors, HR is activated and cells and pathogens are killed immediately. As long as the pathogen elicitor is unable to escape recognition by the receptor, disease resistance will prevail.

Gene-for-gene relationships are common in plants and the pathogens attacking them. The corresponding genes in both organisms are important for disease resistance of plants. Based on this relationship, breeders screen for plants in new selections carrying R-genes against certain pathogens. The greater the number of R genes a plant contains, the greater the possibility to detect many different pathogens and activate HR.

System-Acquired Resistance

Salicylic acid is a hormone that induces system-acquired resistance (SAR) as a follow-up precaution to the entire plant once HR has been activated. Salicylic acid is produced at the site of the infection and moves slowly throughout the plant, triggering resistance in cells to the intruding organism (Figure 12.20). The hormone activates the resistance through expression

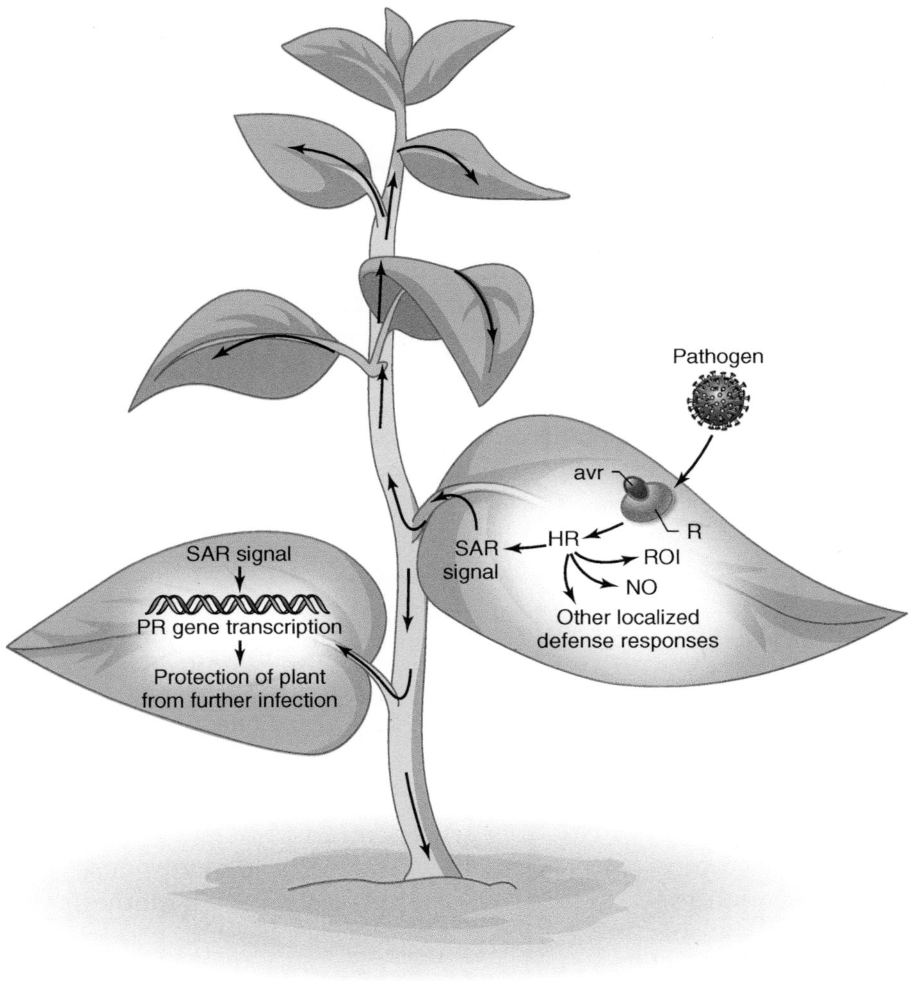

avirulence gene
Gene in a pathogen that produces elicitors that can be recognized by products from the corresponding R-gene in plants.

elicitors
Compound produced in response to a stimulus and only recognized by plants with a specific gene.

© Kendall Hunt Publishing Company.

Figure 12.20   System-acquired resistance transfers protection against future attacks from pathogens from the attached leaves to the parts of the plant.

of pathogenesis-related genes (PR genes) and confers protection against future attack to the entire plant.

Defenses against Herbivores

Plants are food for many species of herbivores. Plants have developed mechanisms to render leaves or other parts unpalatable or toxic through chemical compounds. These secondary metabolites include phenolic compounds (flavonoids and lignin), terpenes, nitrogen-containing compounds such as alkaloids, and protein inhibitors. Proteinase inhibitors in plant tissue are enzymes blocking digestion of proteins in herbivores. Large amounts of proteinase inhibitors can cause herbivores to get sick so they will not consume the leaf if the compound is present in high enough amounts.

The hormone **systemin**, a polypeptide, is synthesized when plant cells are wounded. The hormone triggers a wound response in other parts of the plant and proteinase inhibitors are produced in undamaged cells and tissues, making a follow-up attack less likely. Systemin binds to a receptor protein in the membrane of healthy cells. This triggers a series of events that eventually produces jasmonic acid, which in turn results in the production of protease inhibitors.

Plants can respond to chemical signals sent from a distance by plants of other species attacked by herbivores or pathogens. This extraordinary communication between plants induced production of proteinase inhibitors in leaves of undamaged plants (Farmer & Ryan, 1990). Methyl jasmonate, a secondary metabolite, induced proteinase inhibitors in plants treated with the compound as well as neighboring plants that were untreated. The volatile signal was also effective when methyl jasmonate was extracted from leaves and supplied in the atmosphere.

Plant secondary metabolites are used for many purposes, including insecticides, fungicides, pharmaceuticals, and other purposes. Further discussion of secondary metabolites are included in Chapter 29.

systemin
Hormone that is produced after wounding of a plant; results in the production of proteinase inhibitors, making plants unpalatable to herbivores.

Key Terms

signal receptors
signal transduction
G-proteins
enzyme-linked receptors
second messengers
ion channels
phosphorylation cascade
signal deactivation
auxin
indoleacetic acid (IAA)
acid growth hypothesis
expansin
polar
phototropism
gravitropism

totipotency
callus
climacteric
system-acquired resistance
   (SAR)
photomorphogenesis
phototropism
phototropins
zeaxanthins
phytochrome
photoreversibility
photoperiodism
photoperiod
long-day plant
short-day plant

day-neutral plant
statolith hypothesis
amyloplasts
thigmomorphogenesis
thigmotropism
aquaporins
cotransporters
hypersensitive response (HR)
gene-for-gene hypothesis
disease-resistant genes (R
genes)
elicitors
avirulence gene (avr)
systemin

Summary

- Plants sense their environment, process the information, and trigger appropriate responses to different signals to ensure they can thrive, survive, and reproduce. Signals are deactivated when a response is no longer necessary.

- Signal reception and transduction occurs in three steps: (1) Receptor cells sense a stimulus to the cell and translates the extracellular signal to an intracellular signal. (2) Signal transduction proceeds with either a phosphorylation cascade or second messengers amplifying and sending the signal to the affected location in the plant. (3) Responder cells initiate an appropriate change in cellular metabolism.

- Signal receptors are generally specialized proteins located in the plasma membrane. They transduce the signal via ion channels, G-proteins, or enzyme-linked receptors.

- Plant hormones are naturally occurring organic compounds that affect plant growth and development at low concentrations. Auxin, cytokinin, and gibberellin are growth-promoting hormones; abscisic acid and ethylene retard growth and play a role in senescing tissue. Most growth responses depend on the interaction of several hormones.

- Synthetic plant growth regulators are important in crop production to stimulate flowering and fruiting, increase fruit size, and control plant elongation.

- Phototropism is initiated by photoreceptors in shoot tips sensing blue light. Auxin is moved to the shaded side of a stem, which curve and grow toward the light.

- Red and far-red light are perceived by plants with phytochrome, a pigment existing in two forms and changing shape in response to red and far-red light. Red light signals day or sunlight to plants; far-red light signals shade or night. Red and far-red light affect seed germination and flowering.

- Photoperiodism is the response of plants to bloom based on their exposure to the relative day length in a 24-hour period. Plants are classified as short-day, long-day, or day-neutral plants.

- Gravitropism is the response of plants to gravity. Specialized cells sense gravity with statoliths and can reorient their growth in the direction of gravity typical for roots or in the opposite direction typical for shoots.

- Thigmotropism is the plant response to touch or wind. Slow response is seen when repeated mechanical stimuli can result in shorter plants with increased stem diameter. Rapid response results from an insect triggering the closure of a Venus flytrap or an object touching leaves that quickly fold.

- Plants defend themselves against pathogens, insects, and herbivores in different ways. Preventative strategies include mechanical deterrents like thorns, or thick cuticles and chemical deterrents such as secondary compounds toxic to intruders.

- Plant-induced defenses are mobilized after infection by a pathogen or attack by an herbivore. The hypersensitive response kills host cells and pathogens. Systemic acquired resistance confers whole plant resistance to a future attack in response to a localized intrusion. Communication between plants can result in increased resistance of undamaged plants due to signals from damaged plants.

Reflect

1. *Hormones affect physiological processes in the plant in low concentrations.* Describe the role of auxins and cytokinins in tissue culture and the role of gibberellins and ABA in seed germination.

2. *Plants respond to environmental conditions in their growth and development.* Plants from a cool climate in northern latitudes do not flower and do not produce fruit in the tropics. What is the possible problem?

3. ***Plants respond to gravity and grow toward the light.*** Explain how plants perceive these stimuli and why appropriate plant response is important.

4. ***Phytochrome is a pigment important for flowering and germination of seeds.*** Describe how it works and how it affects flowering and seed germination.

5. ***Ethylene is important in flowering and fruit ripening.*** Describe the roles of ethylene and some uses in agriculture and horticulture.

6. ***Plants use different defense systems against intruding organisms.*** What is the role of secondary metabolites?

References

Borthwick, H. A., Hendricks, S. B., Toole, E. H., & Toole, V. K. (1954). Action of light on lettuce seed germination. *Botanical Gazette, 115*(3), 205–225.

Braam, J. (2005). In touch: Plant responses to mechanical stimuli. *New Phytologist, 165*, 373–389.

Bruening, G., & Lyons, J. M. (2000). The case of the FLAVR SAVR tomato. *California Agriculture, 54*(4), 6–7.

Davies, P. J. (2010). The plant hormones: their nature, occurrence, and functions. In P. J. Davies, *Plant hormones: Physiology, biochemistry and molecular biology* (pp. 1–15). Dordrecht, Netherlands: Kluwer Academic.

Farmer, E. E., & Ryan, C. A. (1990). Interplant communication: Airborne methyl jasmonate induces synthesis of proteinase inhibitors in plant leaves. *Proceedings of the National Academy of Sciences USA, 87*, 7713–7716.

Hemberg, T. (1947). Studies of auxins and growth-inhibiting substances in the potato tuber and their significance with regard to its rest-period. *Acta Horticulturae, 14*, 133–220.

Kurosawa, E. (1926). Experimental studies on the nature of the substance secreted by the "bakanae" fungus. *Natural History Society of Formosa, 16*, 213–227.

Miller, C., & Skoog, F. (1953). Chemical control of bud formation in tobacco stem segments. *American Journal of Botany, 40*, 768–773.

Taiz, L., E. Zaiger, I. M. Møller, and A. Murphy. 2018. Fundamentals of Plant Physiology. 1st Ed., Oxford University Press, New York.

Went, F. W. (1927). *Wuchsstoff und Wachstum*. PhD thesis, Amsterdam, de Bussy.

Chapter 13

Soils, Plant Nutrition, and Transport in Plants

© Denis and Yulia Pogostins/Shutterstock.com

Learning Objectives

- Examine the role of soils for plant growth and how humans have substituted soil for plant production
- Discuss the role of essential elements for plant growth
- Describe the role microorganisms in the soil play for plant growth
- Describe and compare the movement of molecules across membranes with different mechanisms
- Discuss the movement of organic molecules from source to sink
- Describe the flow of water from the soil through the plant and the loss of water to the atmosphere

Soil is essential for plants in our environment because it supplies the physical support and the mineral nutrients necessary for plant growth. Soil is sometimes referred to as dirt, implying something that requires cleaning or a potential nuisance. Nothing could be further from the truth. Soil is an important natural resource, and just like water and air warrants careful stewardship.

Soils develop from weathering rock over long periods of time. The parent material, climate, topography, biological factors, and time will affect how soils evolve and how suitable they are as a substrate for plant growth. Both the physical and chemical properties of soil affect plant growth. Plant roots and other subterranean plant parts share the soil environment with many different organisms including microorganisms, earthworms, and insects that can affect plant growth. Plant nutrition is focused on the inorganic mineral nutrients in soil and how they affect plant growth. Nutrients taken up by plants are used and recycled in complex systems involving different organisms and soil. The science of plant nutrition, especially how much of the different nutrients are required for different crops, is vital for the efficient production of food for humans and animals.

SOILS

Soil is the product of weathered rock. Rain, wind, and freezing and thawing cycles break up the solid bedrock on the surface of the earth's crust. The physical disintegration and chemical breakdown of large rocks takes place over thousands of years. Weathered material might stay at the site where the parent rock is disintegrating or can be carried substantial distances by wind, water, or glaciers. The abrasive forces during transport break particles further apart. Dilute acids formed by water, carbon dioxide, and sulfur adds another element assisting in decomposition. Bacteria, fungi, lichen, and other organisms settle in the rock particles if temperature and light conditions are favorable and contribute to the production of soil eventually colonized by larger plants.

Soil Horizons

Soil is organized in horizons (Figure 13.1), which are layers parallel to the surface differing in physical, chemical, and biological properties. The layers can differ in color, texture, structure, the presence of biological organisms, pH, and mineral content. The horizon at the surface is called A horizon or topsoil and is best suited to support plant growth. It generally has sand, silt, and clay particles and contains organic matter and biological organisms like earthworms, insects, and microorganisms. Topsoil ranges in depth from a few centimeters to one or more meters; in general, the deeper the topsoil, the greater the fertility level. In forest soil that has not been plowed or cultivated there may be an O horizon on top of the A horizon consisting of organic material like decomposing leaves, branches, or dead vegetation in various stages of decomposition. This layer is called *humus* if the organic matter is fully decomposed and physically stable. The layer below the A horizon is called B horizon and has larger soil particles like gravel and rock and generally has very little or no organic matter. The C horizon is located below the B horizon, rocky and barely affected by weathering. Bedrock is located below the C horizon and is the earth's crust.

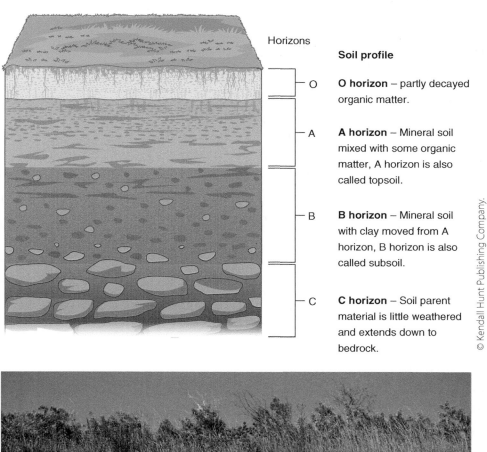

Horizons

Soil profile

O   **O horizon** – partly decayed organic matter.

A   **A horizon** – Mineral soil mixed with some organic matter, A horizon is also called topsoil.

B   **B horizon** – Mineral soil with clay moved from A horizon, B horizon is also called subsoil.

C   **C horizon** – Soil parent material is little weathered and extends down to bedrock.

© Kendall Hunt Publishing Company.

Kenneth W. Fink/Science Source.

Figure 13.1   Horizons in a soil profile.

Soil Texture and Structure

Soil texture is classified by particle size. The diameter of sand particles is 2.0–0.02 millimeters (mm), of silt particles is 0.02–0.002 mm, and clay particles are smaller than 0.002 mm (Fig. 13.2). **Loam** refers to a soil consisting of similar amounts of sand, silt, and clay. When one of the three particle sizes dominates, a loam soil can be classified as sandy loam or silty clay loam. Loam soils have beneficial structural properties supporting good water infiltration, water-holding capacity, and drainage. They

soil texture
Proportion of different-sized soil particles used to classify physical soil characteristics.

loam
Soil containing approximately one-third sand, one-third silt, and one-third clay particles.

Clay <0.002 mm

Silt 0.002–0.02 mm

Sand 0.02–2.0 mm

Sheila Terry/Science Source.

Figure 13.2 Soil texture is characterized by soil particle size. When soil is suspended in water the larger sand particles settle first followed by silt and finally clay. Organic matter stays on top.

organic matter
In soils, organic matter refers to once living material now decaying and recycling nutrients back into the soil.

infiltration rate
Measure of how much water per unit of time a soil can absorb.

water-holding capacity
Amount of water a soil can hold based on the soil texture.

field capacity
Percentage of water held in a soil saturated with water and drained.

are generally easy to till and are desirable for gardening and agricultural production, particularly if they contain more **organic matter** than sandy or clay soils.

Soil texture influences several physical soil properties affecting plant growth: water infiltration, water-holding capacity, and compaction. About half of the soil volume consists of solid matter, the soil particles, and the other half is pore space. In ideal conditions half of the pores are filled with water and the other half with air. The proportion of pores filled with water fluctuates and will be higher after rain or irrigation and will decrease as water drains or is taken up by roots. A small percentage of the soil volume of topsoil, up to 6%, is occupied by organic matter. Roots grow and extend in between the soil particles. A coarse-textured soil dominated by sand particles has larger-size pores for water and air and for roots to penetrate compared to a fine-textured soil with more clay and silt particles. While water will infiltrate fast into the coarse-textured soil, it will also drain fast from the large pores. These characteristics of a coarse-textured soil are called high **infiltration rate** and a low **water-holding capacity**. Conversely, fine-textured soils with very small pores have a low infiltration rate and a high water-holding capacity. The water droplets stick to the clay and silt particles, but much less to sand and larger rocks. A clay soil can hold three to six times more water as the same amount of sand.

Fine-textured soils are subject to compaction, which can further decrease the pore sizes and the volume of soil occupied by air. Compacted soils are problematic for roots unable to find pores large enough to grow into and lack oxygen in the root zone. Oxygen in the air pockets around roots is vital for cellular respiration.

The amount of water a soil can hold is also characterized by the term **field capacity** (Figure 13.3). This is the percentage of water soil can hold after rain or irrigation have saturated the soil and excess moisture has drained. Water is stored in capillaries but has drained from the largest pore spaces. Plants growing in a soil at field capacity can extract water until it is

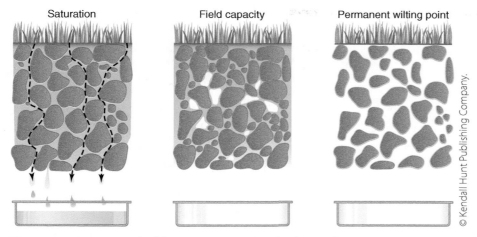

Figure 13.3   Water-holding capacity of soils depends on pore size. Water is available for uptake by roots from field capacity to the permanent wilting point.

depleted to the percentage of soil moisture where they cannot absorb any more and start to wilt. This is called the **permanent wilting point**, when the capillaries in the soil are depleted of water. Although individual soil particles are surrounded by a shell of water, the molecules adhere so tight that roots cannot take it up. Soils vary in their ability to hold water. Soils containing a great percentage of sand hold less water at field capacity and reach the permanent wilting point sooner. Soils with greater amounts of silt and clay hold more water at field capacity, thus providing water longer to plants than sandy soils.

permanent wilting point
Percentage of soil moisture below which plants cannot absorb more water and begin to wilt.

ESSENTIAL ELEMENTS FOR PLANT GROWTH

There are 17 elements identified as essential for plant growth (Table 13.1). The need for essential elements becomes evident when one or more elements are not present in the right amount. Plants will show specific deficiency symptoms and will start to decline in growth. If the lack of even one essential element persists, the plant will eventually die. This knowledge has led to the development of criteria that define essential elements: (1) an element is essential if the plant cannot complete its life cycle and (2) an element is essential if it is part of an essential molecule or metabolic function.

Nutrients are classified as macronutrients and micronutrients based on their concentration in plants (Table 13.1). The macronutrients carbon, oxygen, and hydrogen obtained from air or water make up 96% of a plant's dry weight. The average dry weight of macronutrients from the soil ranges from 1.5% for nitrogen to 0.1% for sulfur. Micronutrients are present in even smaller concentrations, ranging from less than 0.01% chlorine to 0.00001% molybdenum. Macronutrients are needed in greater amounts because they are part of compounds that are the major building blocks of the plant system. Micronutrients are required in much smaller amounts because they are not part of the building blocks.

Table 13.1 Essential Elements for Plant Growth, Their Function, and Deficiency Symptoms

Element	Form Available to Plants	Average % Dry Weight	Function	Deficiency Symptoms
Oxygen	O_2, H_2O	45	Major component in organic compounds, cellular respiration	Roots suffocate and rot, plants wilt and eventually die
Carbon	CO_2	45	Photosynthesis, major component in organic compounds	Slow growth due to starvation
Hydrogen	H_2O	6	Major component of organic compounds, electrochemical gradients	Slow growth due to cell dehydration
Macronutrients				
Nitrogen	NO_3^- (nitrate) NH_4^+ (ammonium)	1.5	Component in nucleic acids, amino acids, proteins, coenzymes, and hormones	General chlorosis especially in older leaves, plants don't thrive
Potassium	K^+	1.0	Cofactor for enzymes, balances osmotic adjustment, stomatal opening	Mottled or chlorotic leaves at tips and margins mostly on older leaves, weak stems, short internodes
Calcium	Ca^{2+}	0.5	Regulates membrane and enzyme activities, component of cell wall structure, role in signal transduction	Death of young shoot and root tips, young leaves first deformed, root system stunted
Magnesium	Mg^{2+}	0.2	Chlorophyll component, enzyme activator	Mottled or chlorotic leaves first on older leaves, thin stems, premature leaf drop
Phosphorus	$H_2PO_4^-$ (dihydrogen phosphate) $H_2PO_4^{2-}$ (hydrogen phosphate)	0.2	Component of ATP nucleic acids, coenzymes, and phospholipids	Dark green leaves, stunted growth

Element	Form Available to Plants	Average % Dry Weight	Function	Deficiency Symptoms
Sulfur	SO_4^{2-} (sulfate)	0.1	Component of amino acids, proteins, and coenzymes	Chlorosis of leaves, young leaves with light green veins, stunted growth
Micronutrients				
Chlorine	Cl^-	0.01	Essential for oxygen-releasing step in photosynthesis, regulates water balance	Wilted leaves with chlorotic and necrotic spots, leaves can turn bronze color, thickened root tips
Iron	Fe^{3+} (ferric iron) Fe^{2+} (ferrous iron)	0.01	Essential for chlorophyll synthesis, part of cytochrome, enzyme cofactor	Interveinal chlorosis of young leaves; short, thin stems
Manganese	Mn^{2+}	0.005	Enzyme activator, chloroplast integrity, required for oxygen release in photosynthesis	Interveinal chlorosis followed by necrosis, starts on older or younger leaves depending on species
Zinc	Zn^{2+}	0.002	Enzyme activator, essential in auxin biosynthesis	Smaller leaf and shorter internodes, distorted leaf margins, interveinal chlorosis
Boron	$H_2BO_3^-$ (borate)	0.002	Cofactor in chlorophyll synthesis, nucleic acid synthesis	Bud necrosis and young leaf necrosis, shoot dieback
Copper	Cu^+ (cuprous ion) Cu^{2+} (cupric ion)	0.0006	Cofactor of enzyme function	Dark green young leaves, twisted, misshapen, with necrotic spots
Nickel	Ni^{2+}	Not known	Essential cofactor in nitrogen metabolism	Necrosis of leaf tips
Molybdenum	MoO_4^{2-} (molybdate ion)	0.00001	Required for nitrogen fixation and nitrate reduction	Interveinal chlorosis starting in older leaves, followed by necrosis

Each essential nutrient serves important functions in the plant (Table 13.1). Carbon, oxygen, and hydrogen are the major constituents of plants and a major component of all organic compounds. They are also involved in major metabolic functions like photosynthesis. Macronutrients are major parts of proteins, carbohydrates, nucleic acids, and lipids. They also have major regulatory functions in plant metabolism. Micronutrients are often enzyme activators or components of enzymes.

Nitrogen was the first element discovered as essential for plant growth in the early 1800s (Barker & Pilbeam, 2007). Essentiality for the other macronutrients and iron was established within the next 60 years. The essential nature of the other micronutrients was determined in the 1900s, with the last element nickel classified as essential in 1987. Until that time nickel had been considered beneficial. Beneficial elements such as selenium, sodium, aluminum, and vanadium stimulate plant growth and may be required for some species of plants. Sodium appears to be required for some saltbush species (*Atriplex* sp.) and some C_4 grasses (*Distichlis spicata, Panicum miliaceum*) (Gorham, 2007). It stimulates growth of Joseph's coat (*Amaranthus tricolor*) and marsh grass (*Sporobolus virginicus*).

Nutrient Deficiency

Lack of essential elements results in foliar deficiency symptoms and plant disorders that can be quite distinctive (Table 13.1). The general symptoms vary by the severity of the deficiency and sometimes by species (Figure 13.4). Studies using hydroponic systems (Box 13.1) have identified the symptoms of nutrient deficiencies in different species. Hydroponic systems use solutions containing all the essential elements instead of soil to grow a healthy plant. When one element is withheld, plants will develop the corresponding deficiency symptoms.

Mobility of elements within a plant can help with the identification of a deficiency. Elements mobile in the plant such as nitrogen, phosphorus, potassium, zinc, or molybdenum will translocate the minerals to the young tissue where it is needed most, thus developing deficiency symptoms in older leaves first. Deficiency of immobile elements will show first in young leaves and is characteristic for calcium, sulfur, iron, boron, and copper.

In a field situation, deficiency symptoms are sometimes more difficult to identify because they can be based on deficiency of multiple elements, be chronic or acute, and be combined with other stress factors such as drought. Interactions between essential elements can occur when they are present in a deficient or excessive amount and induce deficiency or toxicity in another element. Plant tissue analysis is the best way to determine which elements are missing in a plant. It requires knowledge of the range of optimum tissue concentrations of each element for a particular species. For most crop plants these values have been established even for different stages of plant growth to ensure optimum production. Before fertilizer is added to a field or orchard, a soil analysis provides vital information to decide what type of fertilizer and how much to add to the soil to correct or prevent a deficiency.

Plants tolerate a range of nutrient levels before symptoms are expressed through plant growth and appearance. This applies to deficiency and excess concentrations of mineral nutrients. General symptoms that can be expected under different levels of fertility are summarized in Table 13.2.

(a)   (b)   (c)

(d)   (e)

Nigel Cattlin/Science Source.

Figure 13.4a–e   Examples of nutrient deficiencies: (a) copper deficiency in wheat (*Triticum aestivum*); (b) potassium deficiency in potato (*Solanum tuberosum*); (c) magnesium deficiency in potato; (d) boron deficiency in cucumber (*Cucumis sativus*); and (e) Phosphorus deficiency in barley (*Hordeum vulgare*).

BOX 13.1 Hydroponic Growing Systems

Hydroponics is the cultivation of plants with their roots in a nutrient solution but without soil. This soilless form of cultivation was developed in the 19th century to study nutrient requirements and deficiency symptoms. Several systems have been developed since and today hydroponic production of high-value vegetables and leafy greens in greenhouses is an important industry worldwide (Jensen, 2002). Hydroponic soilless systems are capital and technology intensive, but they can conserve land and water resources and produce food without the problems of soil-borne diseases.

The basic growth systems are liquid deep flow hydroponic, nutrient film technique, and aeroponic. In a liquid deep flow hydroponic system plant roots are immersed in a nutrient solution and air is circulated through the water to supply oxygen (Figure Box 13.1). A support ensures the plants are held upright. Plants growing in the nutrient film system have their roots in a shallow trough through which nutrient solution is pumped continuously. The liquid deep flow hydroponic system and the nutrient film system are generally closed and recirculate the nutrient solution. A major advantage of the nutrient film technique is it uses relatively smaller amounts of solution than the liquid deep flow system and solutions can be cooled in tropical environments or heated in cooler environments to meet the crop environmental needs. The ability to cool the nutrient solution has led to the successful production of

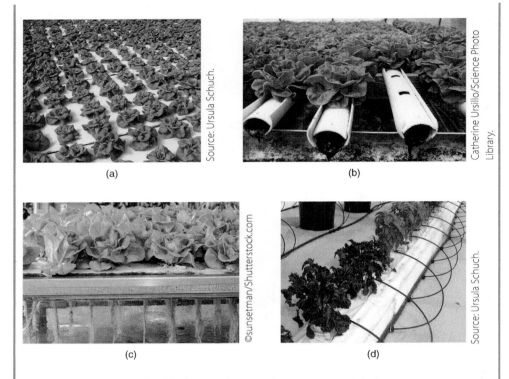

Figure Box 13.1a–d Hydroponic growing systems: (a) deep water culture, (b) nutrient film technique, (c) aeroponic system for lettuce, and (d) aggregate system using rockwool blocks for basil cultivation.

lettuce in the Caribbean, which was previously prevented because of bolting, the premature development of a flower stalk accompanied by bitter-tasting leaves. The deep flow hydroponic system and the nutrient film technique are most widely used in commercial production. The aeroponic system suspends plants over a chamber with nutrient solution that is applied as a mist to the roots.

Aggregate hydroponic systems are a variation of hydroponic systems and use a solid inert medium to support the crop (Jensen, 2002). Rockwool, perlite, volcanic rock, or coconut coir are popular media to anchor the root system (Figure Box 13.1). The nutrient solution is applied through frequent applications of drip irrigation.

Nutrient Availability and Cation Exchange

Plant growth depends on water, oxygen, and nutrients in the soil solution surrounding the roots. The elements plants absorb from the soil solution occur as ions (Table 13.1). The ions are charged either positive or negative. Negatively charged **anions** readily dissolve in water and are easily absorbed by roots. The most common anions in soil are NO_3^-, SO_4^{2-}, HCO_3^-, and OH^-. Phosphate anions are an exception and do not dissolve easily in water. Instead, they form insoluble complexes with cations such as iron, aluminum, and calcium. Anions are subject to **leaching**, the loss of nutrients through movement in the soil solution. Rain or irrigation can wash anions

anion
Negatively charged ion. See also *cation*.

leaching
Loss of ions in the soil when they move with water below the root zone.

Table 13.2   Plant Symptoms for a Range of Nutritional Levels

Nutritional Level	Symptom
Toxic	Visible symptoms to the plants due to excess fertility
Superoptimal	Excess fertility but does not cause visible symptoms
Optimal	Perfect amount of fertility, maximum growth, no symptoms
Suboptimal	No visible symptoms but reduced plant growth due to limited fertility
Deficient	Visible symptoms due to very limited fertility

below the root zone where they become unavailable for plant uptake. Leaching, the application of large amounts of water to soil, is a management practice to rid soils with high salinity of high concentrations of ions such as chloride and sodium. The leaching process, however, does not distinguish between desirable or undesirable ions and also results in the loss of essential nutrients.

Positively charged **cations** are attracted and bound to the negatively charged clay particles or negatively charged organic matter particles such as humus. This makes cations more difficult for plants to take up but prevents them from leaching. Nitrogen in the form of ammonia (NH_4^+), potassium, calcium, and magnesium are macronutrients that occur in anion form.

Soil pH is the concentration of hydrogen ions in a solution. Soil pH influences whether nutrients can be absorbed by plant roots through a process called **cation exchange**. Cations bound to negatively charged soil particles will be released when protons (hydrogen ions) or other cations bind in their place and release them into the soil solution. They can be absorbed by roots or washed away by rain or irrigation. Respiring roots release carbon dioxide, which dissolves in the soil solution and forms carbonic acid (H_2CO_3). Protons (H^+) available for cation exchange and bicarbonates (HCO_3^-) are produced when carbonic acid ionizes. Plants affect this process by releasing protons into the soil solution to release ions of essential nutrients for uptake. Soils are characterized by their ability to exchange cations with the term *cation exchange capacity*. Soils rich in clay and organic matter have a high cation exchange capacity; sandy soils have low cation exchange capacity.

Soil pH also affects the availability of some inorganic nutrients. In alkaline soils with pH greater than 7 some cations such as iron, zinc, manganese, and copper are precipitated into insoluble complexes and are not available for plant uptake. Mycorrhizal fungi (see Chapter 4) are important to aid plants in the uptake of phosphorus, copper, and zinc (Marschner & Dell, 1994).

cation
Positively charged ion. See also *anion*.

cation exchange
Cations such as potassium or magnesium become available for uptake by plants after they are released from negatively charged soil particles and are replaced by protons or other cations.

NUTRIENT AND WATER UPTAKE

Nutrients and water are taken up through the roots of most plants and are transported in the xylem (see Chapter 4). Root hairs are the site of nutrient uptake and are located in the zone of maturation just behind the root tip. Root hairs have a large surface area and efficiently absorb the ions in their vicinity. They create a zone where nutrients become scarce. Root hairs are constantly regenerated near the root tips to explore new nutrient-rich soil solutions.

MOVEMENT OF MOLECULES ACROSS MEMBRANES

Molecules essential for plant growth are taken up into the plant, transported within the plant, and moved out of the roots. These include water, minerals, and organic molecules. Molecules and ions are either moved across membranes by passive or active transport.

Passive Transport: Diffusion, Channels, and Carriers

Passive transport across membranes occurs when molecules or ions move by **diffusion** with the electrochemical or concentration gradient and without energy being spent by the plant. Small, noncharged molecules can directly move through the membrane, but transport of charged or larger molecules is not feasible in this manner. They move across the phospholipid bilayer through membrane proteins that are either channels or carriers (Figure 13.5).

Channels or channel proteins extend across the entire membrane and allow passage of certain ions in a controlled way (Taiz et al., 2018). Channels specific for water molecules are called aquaporins and allow water transport many times faster than through diffusion. Other channels are specific for elements such as sodium, potassium, or larger molecules.

Carrier proteins or transporters allow facilitated diffusion by changing shape and moving a substrate across the membrane (Taiz et al., 2018). For example, when glucose needs to be transported into the cell, the carrier protein for glucose has a binding site ready to accept glucose on the outside of the membrane. Glucose binds to the carrier protein, which changes conformation and transports glucose to the inside of the cell where it is released. This process still relies on diffusion and a gradient with higher glucose concentration outside the cell than inside.

passive transport
Movement of a molecule across a membrane from higher to lower concentration through channels, channel proteins, or by simple diffusion; requires no energy.

diffusion
Movement of particles from an area of high concentration to low concentration. See also *facilitated diffusion*.

channels or channel proteins
Proteins embedded in membranes allowing specific ions to pass through.

carrier proteins or transporters
Membrane proteins binding to molecules and undergoing a reversible change when moving the molecule across a membrane.

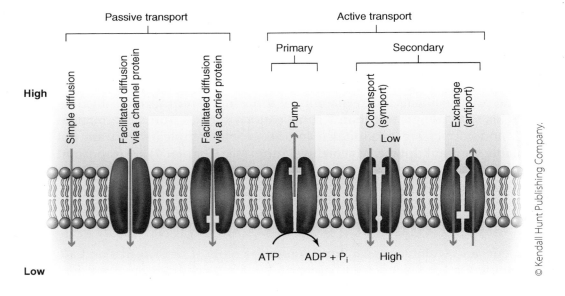

Figure 13.5 Active and passive movement of ions and molecules across plasma membranes.

Active Transport: Proton Pumps

Active transport of ions and molecules against an electrochemical or concentration gradient requires membrane proteins and energy in the form of ATP or energized electrons. The concentration of sucrose molecules inside a cell is higher than outside, yet more sucrose needs to move through the membrane toward the higher sucrose concentration in the cytoplasm. **Proton pumps** in membrane proteins move H^+ from inside the cell to the outside against a gradient with the aid of ATP (Figure 13.5, left). This gradient allows movement of sucrose molecules through membrane proteins known as **cotransporters** as sucrose along with the H^+ diffuses into the cell cytoplasm through the cotransporter (Figure 13.5). The transport gradient across the membrane is electrochemical and allows the transport of sucrose toward a higher concentration and the flow of protons from a high to a low concentration. The proton pump maintains the electrochemical gradient with greater concentration of protons outside the cytoplasm. Transport of ions through cotransporters and proton pumps is an active process and requires energy.

Passive Ion Exclusion

Minerals such as sodium, chloride, or heavy metals can become toxic to plants in higher concentrations. Recall from Chapter 4 that the Casparian strip forces all solutes to move across the endodermis before entering the xylem. Ions or molecules can move into the plant via the apoplast, the space outside the plasma membrane between cell walls of adjoining cells until they reach the Casparian strip. At the Casparian strip, this selective membrane facilitates movement across the membrane only for those ions for which a channel or transporter exists. A second area of exclusion is at the root hairs when solutes try to enter through the membrane. Only those ions for which membrane proteins exist can enter.

Active Ion Exclusion

In some cases plants cannot exclude ions, particularly when the concentration of the ions in the soil solution is very high. Once inside the plant, metal ions can be inactivated by small proteins called **metallothioneins**. These proteins bind to the metal ion and the metal cannot become toxic anymore.

The second active exclusion can occur when transport proteins in the tonoplast act as **antiporters** and remove toxic ions from the cytoplasm and into the vacuole. The antiporter moves the toxic ion against the electrochemical gradient into the vacuole while at the same time moving protons out of the vacuole. This ability is used in some plants to sequester or take up large quantities of toxic ions to clean contaminated soils or water, a process known as **phytoremediation**. Plants such as *Brassica juncea* (mustard greens) extract lead and selenium from soil (Salt, Smith, & Raskind, 1998). Almost 400 taxa of metal-accumulating plants are known and genes encoding antiporters or protein transporters of heavy metals into the vacuole have been identified. Some plants accumulate sodium or chloride, which is toxic to many plants above certain concentrations. They assist in the remediation of saline soils or in the use of saline water. Mature plants used for phytoremediation are harvested and disposed of.

proton pumps
Membrane proteins moving protons across a cell membrane against an electrochemical gradient.

cotransporters
Membrane protein transporting two ions, one in the direction of its electrochemical gradient, while simultaneously transporting another ion *in the same or opposite direction* but against its electrochemical gradient. See also *symporter* and *antiporter*.

metallothionein
Protein binding to a metal ion and for storage or transport, preventing the metal from becoming toxic in cells.

antiporter
Membrane protein transporting two molecules or ions across a membrane in opposite directions. One ion or molecule moves along its electrochemical gradient as a different ion or molecule moves *in the opposite direction* against its own electrochemical gradient. See also *symporter*.

phytoremediation
Using plants for the reduction or removal of contaminants from soil, water, or air; also known as bioremediation.

Nutrient Uptake through Leaves

Mineral nutrients are generally applied to the soil, but in some cases are sprayed on the leaves in a solution, a process known as **foliar application** (Taiz et al., 2018). Foliar nutrient application is more effective than soil application for several micronutrients such as copper, zinc, iron, and manganese. These nutrients are often bound in the soil and not available for plant uptake. Foliar applied nutrients are also immediately usable by the plant and can correct deficiencies in a timely manner.

Ions are taken up through the cuticle and diffuse into plant cells. Foliar sprays need to contain a surfactant to lower the surface tension, otherwise the spray just rolls off the leaf instead of forming a thin film over the leaf surface. Timing of foliar spray applications is critical to prevent leaf damage, for example, if sprays were applied during a hot day. Foliar application of nutrients is successfully used in many tree crops and grapes and in some cereals.

SYMBIOTIC MICROORGANISMS PROVIDING NUTRIENTS

Many plants rely on mycorrhizal fungi and on *Rhizobium* bacteria to help in the absorption of essential nutrients (see chapter 4). Mycorrhizae are involved in several different aspects of nutrient cycling and even protect plants from stress such as drought, salinity, heat, heavy metal toxicity, and plant pathogens (Garg & Chadel, 2010). In addition to some macronutrients, mycorrhizae assist plants in the uptake of zinc, copper, and iron. Ectomycorrhizal fungi colonize the outside of the root and hyphae penetrate only in between the cells of the outer root surface. They increase the surface area of trees growing in temperate and northern climates. The fungi release enzymes into the slowly decaying organic matter speeding up the decomposition process to make nitrogen available for plant uptake. The fungi are also very effective at taking up phosphorus from the organic or clay soil particles.

In grasslands and in the tropics arbuscular mycorrhizae or endomycorrhizae dominate. More than 80% of plants and more than 92% of plant families form a symbiotic relationship between their roots and arbuscular mycorrhizal fungi (Garg & Chadel, 2010). They grow into the cells of the roots but also extend out into the soil pores. Their primary function is to absorb phosphorus for the plant.

Rhizobium bacteria colonize the roots of plants in the bean or legume family (*Fabaceae*). They fix atmospheric nitrogen (N_2) and reduce it to two molecules of NH_3 in the roots for plant use. This multistep process requires up to 24 ATP molecules and several enzymes, including nitrogenase. In return the plant supplies the bacteria with carbohydrates. There is great interest to understand how genes trigger nitrogen fixation in order to insert these genes into important food crops. If successful, these nitrogen-fixing crops then would require less synthetic fertilizer input for production.

WATER AND SOLUTE TRANSPORT IN THE PLANT

Water and minerals move from the roots to the leaves in the xylem (see Chapter 5). Water and solutes move considerable distances in tall plants. When stomata in the leaves are open, water constantly moves through the plant in the process of transpiration. Water is lost to the drier atmosphere surrounding the leaf and is replenished by water taken up from the soil through the root hairs. When nutrients move into the plant they are always in an aquaeous solution. Water can move through aquaporins across membranes and into the plant. There are, however, additional mechanisms important to understand how water moves into the root hairs, against gravity up the stem, and sometimes 50 meters or more against gravity into the canopy of some trees.

Sugars are transported from the source where they are produced by photosynthesis to the sink where they are used for growth or other metabolic functions. This transport occurs in the phloem (see Chapter 5) and along a gradient of pressure. The process follows a pressure-flow mechanism that is maintained by osmosis.

WATER POTENTIAL AND WATER TRANSPORT

The loss of water vapor through stomata to the surrounding atmosphere is called **transpiration**. Plants open stomata to take up carbon dioxide for photosynthesis and at the same time they lose water to transpiration (see Chapters 6 and 10). Stomata regulate the amount of water lost through transpiration. As long as the plant is taking up sufficient water through the roots, water will be lost through the open stomata because of the lower humidity in the atmosphere compared to the water-saturated atmosphere in the leaf. Water loss also cools the leaves to maintain favorable temperatures for photosynthesis.

Plants take up different amounts of water, almost 99% of which is lost through transpiration. Different species of trees with an average height of 21 meters use between 10 and 200 kilograms of water per day (Wullschleger, Meinzer, & Vertessy, 1998). In comparison, an overstory tree in the rainforest of the Amazon uses 1,180 kilograms per day. Plant water use depends on many factors, on the plant species and where it evolved, the climate where it grows, the time of year, temperature, relative humidity, and whether the plant is solitary or surrounded by other plants.

Water movement from the roots to the leaves is a passive process requiring no energy from the plant. **Water potential** is the energy of water in an environment. It is represented by the Greek letter Ψ (psi) and is measured in megapascals (MPa) or atmospheres (atm), which measures pressure per unit area. The atmospheric pressure at sea level is 1 atm or 0.01 MPa. Water always moves from a place of high water potential to low water potential. Within a plant there is a water potential gradient that is higher in the soil and roots and lower in the leaves, thereby determining the direction of water movement.

The water potential (Ψ_w) within a plant is the sum of the solute (Ψ_s) or osmotic potential and the pressure potential (Ψ_p). The **pressure potential** in the plant is the force applied by the cell wall from the inside in response to pressure from water entering the cell. The pressure potential protects the plant cell from bursting and presses the plasma membrane against the cell wall when it is fully hydrated. This pressure is also known as **turgor pressure**. Plants with turgid cells are stiff and hydrated and have a positive turgor pressure. The opposite are flaccid cells without turgor pressure ($\Psi_p = 0$) and wilted leaves (Figure 13.6).

transpiration
Loss of water vapor through the stomata in the leaves.

water potential
Energy of water in an environment; in plants is the sum of the solute or osmotic potential and pressure potential.

pressure potential
See *turgor pressure*.

turgor pressure
When water enters the central vacuole and causes it to swell and exert pressure against the cell wall; responsible for providing strength to keep nonwoody plants upright.

solute or osmotic potential
Also known as solute potential, osmotic potential is part of the water potential and refers to the difference in energy of water based on the solute concentration; values are negative.

osmosis
Diffusion of water across a semipermeable membrane; water will move from an area of higher water concentration to an area of lower water concentration.

hypotonic
Solution with a lower solute concentration or potential, and a higher water concentration than a comparison solution. See also *isotonic* and *hypertonic*.

Isotonic
Solution with the same solute concentration and the same water concentration than a comparison solution. See also *hypotonic* and *hypertonic*.

hypertonic
Solution with a higher solute concentration or potential and a lower water concentration than a comparison solution. See also *isotonic* and *hypotonic*.

symplast
Continuous pathway in the roots leading through cells and plasmodesmata.

apoplast
Pathway through cell walls and intercellular spaces.

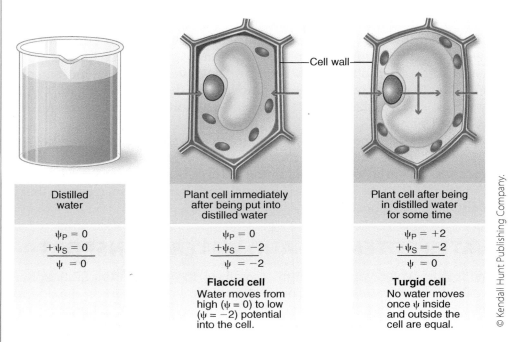

Distilled water

$$\psi_P = 0$$
$$+\psi_S = 0$$
$$\overline{\psi = 0}$$

Plant cell immediately after being put into distilled water

$$\psi_P = 0$$
$$+\psi_S = -2$$
$$\overline{\psi = -2}$$

Flaccid cell
Water moves from high ($\psi = 0$) to low ($\psi = -2$) potential into the cell.

—Cell wall—

Plant cell after being in distilled water for some time

$$\psi_P = +2$$
$$+\psi_S = -2$$
$$\overline{\psi = 0}$$

Turgid cell
No water moves once ψ inside and outside the cell are equal.

© Kendall Hunt Publishing Company.

Figure 13.6 Water potential (Ψ) is affected by pressure potential (Ψ_p) and solute potential (Ψ_s).

The **solute or osmotic potential** is the concentration of solutes compared to pure water. The Ψs in a cell is always negative because pure water, which by definition contains no solutes, has a solute potential of zero. A cell always has solutes, therefore the Ψs is negative. The greater the negative number, the more solutes or the more concentrated the solution.

Osmosis occurs when water moves across a semipermeable membrane in the direction from high water potential (low solute potential) to low water potential (high solute potential) until equilibrium between the two solutions is reached (Figure 13.7). Water will move into the cell when it is surrounded by a **hypotonic** solution, characterized by lower solute potential and higher water potential than inside the cell. No water will move across the membrane when the solute potential inside a cell and outside is equal; this occurs when the cell is in an **isotonic** solution. This happens when the concentrations of the solutions within and outside the cell are the same. Cells in a **hypertonic** solution will lose water to the outside solution because it has a higher solute concentration and a lower water potential than inside the cell. This water loss can lead to plasmolysis, death of the cell, a characteristic condition of cells damaged by salinity.

When water moves from the root hairs to the xylem it can take two different pathways: the apoplast or the symplast (Figure 13.8). The **symplast** is the pathway that leads through cells and plasmodesmata that connect adjacent cells. The **apoplast** is the pathway through the porous cell walls and spaces in between cells. Water can move by this route through the root epidermis, the cortex, and into the endodermis. There the apoplast pathway is blocked by the Casparian strip and water and solutes pass through the cell, the symplast way.

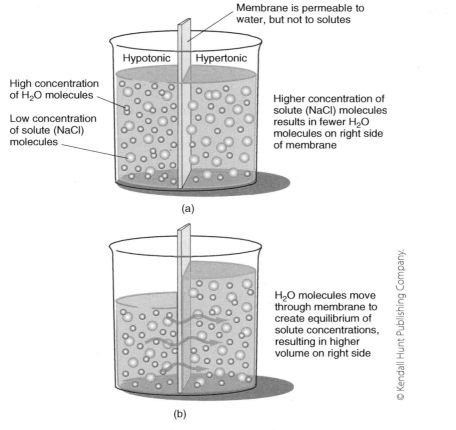

Membrane is permeable to water, but not to solutes

Hypotonic   Hypertonic

High concentration of H₂O molecules

Low concentration of solute (NaCl) molecules

Higher concentration of solute (NaCl) molecules results in fewer H₂O molecules on right side of membrane

(a)

H₂O molecules move through membrane to create equilibrium of solute concentrations, resulting in higher volume on right side

(b)

© Kendall Hunt Publishing Company.

Figure 13.7   Osmosis is the movement of water across a semipermeable membrane.

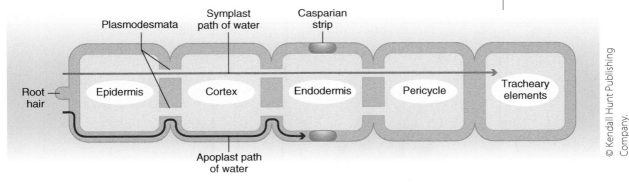

Plasmodesmata

Symplast path of water

Casparian strip

Root hair

Epidermis   Cortex   Endodermis   Pericycle   Tracheary elements

Apoplast path of water

© Kendall Hunt Publishing Company.

Figure 13.8   Symplast and apoplast path of water into the roots.

Water loss is regulated through the stomata and depends on water availability within the plant. Loss of turgor results in the guard cells becoming flaccid and closing. While this conserves water it also prevents the uptake of carbon dioxide and photosynthesis. When water is again available the guard cells become turgid and open. Potassium and the hormone abscisic acid are involved in the regulation of stomata opening. Well-watered plants will open their stomata during the daylight hours to take advantage of photosynthesis.

Water potential becomes greater or more negative from the soil solution into the roots, up to the leaves, and finally in the atmosphere (Figure 13.9). Water potentials in the soil, plant, and atmosphere change constantly depending on the availability of soil water, the concentration of solutes in the soil, stomata opening, and the atmospheric relative humidity. During periods of fog or rain transpiration is low as the water potential of the atmosphere, also known as *evaporative demand*, will not be as high or negative than during sunny weather and high temperatures. In hot, windy conditions, transpiration will be high and more water has to move from the soil into the plant and through the stomata.

Generally there is a diurnal cycle for plant water potential. Values are low or least negative the morning just before sunrise and become more negative as stomata open and photosynthesis begins in the morning. Water potential increases, becoming more negative as transpiration rates increase and depleted water needs to be replenished within the plant. Highest transpiration demand occurs in the early afternoon and water potential generally peaks midafternoon. Transpiration demand decreases as temperatures decrease later in the day. The water potential starts to recover after the sun sets and transpiration stops.

Root Pressure

Root pressure builds when water and ions move from the soil into the root, although stomata are closed. Without transpiration removing water, ions moving into the xylem lower the water potential and cause more water from surrounding root cells to flow into the xylem. This causes pressure to build up and water can move into the leaves in low-growing plants. In some plants water is forced out as droplets through the leaves, an event called **guttation** (Figure 13.10). It is visible in the early morning when transpiration rates are very low. Root pressure can aid the movement of water over short distances, but cannot accomplish long-distance transport.

guttation
Droplets of water are forced out of leaves as a result of root pressure.

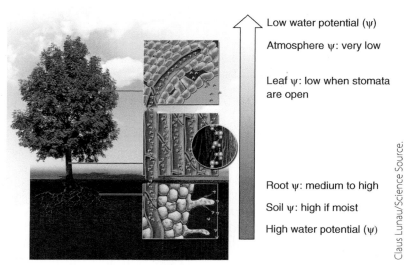

Low water potential (ψ)

Atmosphere ψ: very low

Leaf ψ: low when stomata are open

Root ψ: medium to high

Soil ψ: high if moist

High water potential (ψ)

Claus Lunau/Science Source.

Figure 13.9 Water potential (Ψ) from the soil through the tree into the atmosphere. Water moves from high water potential to low water potential.

Source: Ursula Schuch.

Figure 13.10   Leaf of watermelon showing guttation from root pressure.

Capillary Action

Capillary action is one component of long-distance movement of water in plants. Xylem tissue acts like a small capillary tube where water rises (Figure 13.11). This is possible because water has the properties of cohesion, adhesion, and surface tension. Water molecules are characterized by their polarity with a positive charge near the oxygen and a negative charge near the hydrogen atoms (see Appendix). They attract each other and form a hydrogen bond. **Cohesion** is the property of water molecules binding to each other. This allows the formation of columns of water in the xylem.

cohesion
Ability of similar molecules to adhere or stick to each other.

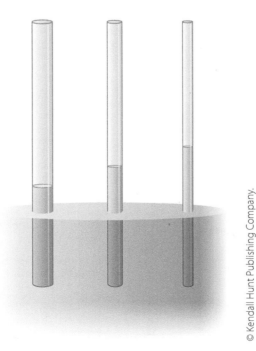

© Kendall Hunt Publishing Company.

Figure 13.11   Capillary action occurs because of the adhesion and cohesion properties of water.

adhesion
Ability of different molecules to stick to each other.

Adhesion is the attraction of water molecules to other polar molecules. In the xylem, water is attracted to the cell walls, facilitating movement into the canopy. Surface tension exists at the water–air interface where water molecules attract each other and are attracted to the surface wall, forming a curve called *meniscus*. Capillary action can lift water in xylem vessels about 1 meter high.

The Cohesion-Tension Theory

cohesion-tension theory
Tension or negative pressure from transpiration pulls water from the roots through the xylem into the leaves and stomata.

The movement of water through the xylem is often compared to a continuous column of water from the soil solution into the root hairs and into the xylem cells. From there water travels through the tracheids and vessels into the mesophyll of the leaf where it will leave as water vapor through the stomata and into the atmosphere. This continuum has led to the **cohesion-tension theory**, stating that tension or negative pressure caused by the pull of transpiration moves water from the soil into the roots and all the way up into the canopy of tall trees (Figure 13.12). Of course, this same theory applies to shorter plants. Cohesion of water molecules and the small diameter of xylem tissue elements are also necessary for maintaining the column of water. This remarkable process does not require energy from the plant.

Tension within the plant comes from the menisci in the spongy mesophyll where water changes from liquid to gas when it evaporates from the stomata into the atmosphere. Tensions created by those menisci can amount up to –2.0 MPa, a force to lift water up to 100 meters. In comparison, a column of water can only be lifted to a height of 10 meters when the tension comes from the top; at greater height the column will break. Vessel elements and tracheids in the xylem are exposed to great negative tension, which might cause them to collapse. However, lignin in the cell walls of tracheids and vessels strengthens them sufficiently.

Evaporation in leaves create pulling force
- Atmosphere has a very low water potential which causes water to evaporate from the leaves
- Low water potential pulls water columns up

Cohesion in xylem
- Water molecules stick together (cohesion) create water columns within the xylem tissues

Water uptake by roots
- Low water potential in roots draws water in from soil
- Root hairs greatly increase surface area for absorption
- Water moves through root cells via osmosis

Claus Lunau/Science Source.

Figure 13.12   The cohesion-tension theory of water movement in xylem.

Breaking of the water column results in interruption of water transport and the formation of air bubbles. **Cavitation** is the rupture of the water column and **embolism** is when air or water vapor fills the void (Taiz et al., 2018). If an element is embolized the plant contains it and water moves around it. Bordered pits on vessel elements prevent the air from moving into adjacent vessel elements. Embolisms in the xylem can be repaired. Air bubbles can dissolve into the solution when pressure in the xylem increases due to root pressure or low transpiration demand. Cavitation occurs in drought-stressed trees and frozen tissue or due to physical damage.

PLANT WATER UPTAKE IN DRY OR SALINE SOILS

Plants living in permanently dry places have adapted in several ways. From previous chapters we know that there are anatomical adaptations in leaf size and morphology (Chapter 6) and physiological adaptations in photosynthesis (Chapter 10). Another way to ensure that water is taken up from relatively dry soil is to increase the solute potential in cells so that it is more negative than the solute potential of the soil solution. Leaf water potential of plants living in dry places is also very low or negative during periods of drought. This ensures that water will still move into the plant from the soil instead of the other way around where plants would lose water to the soil.

Plants living in soil with high salinity, which is soil with a high solute potential because of high concentration of minerals, follow a similar strategy as plants in dry places. They lower their solute potential by increasing the amount of ions in the vacuole and the concentrations of organic molecules in the cytoplasm. Plants unable to cope in this way will perish in saline soil because water will be drawn out of their roots, flowing from an area of low solute potential to an area of high solute potential. Flushing the soil with pure or good-quality water and moving most of the solutes in the soil solution below the root zone is the only way to save plants when a soil has become too saline.

TRANSLOCATION

The transport of organic molecules including sugars occurs in the phloem and is called **translocation**. Products of photosynthesis move from a source to a sink. A **source** is the location where the sugars are entering the phloem (Taiz et al., 2018). A **sink** is where sugars are transported to and exit the phloem to be used in metabolic processes or storage. Sinks are shoots, roots, and fruit where energy is needed for growth. Leaves can be both a source because they produce sugars and a sink because they might use sugars to complete growth. Mature leaves and stems conducting photosynthesis are sources of sugar. In early spring before leaves emerge in deciduous plants, roots are a source because they transport stored carbohydrates to the shoots to use for new growth of leaves. Mature leaves can retain about 10% of the carbon fixed in photosynthesis while growing leaves retain up to two-thirds of the fixed carbon.

cavitation
Rupture of the water column in the xylem.

embolism
Air or water vapor filling a void in the xylem tissue.

translocation
Transport of organic molecules and sugars in the phloem.

source
Location where sugars are produced by photosynthesis and enter the phloem.

sink
Location where sugars exit the phloem and are used in metabolic processes.

The Pressure-Flow Hypothesis

The movement of sugars is facilitated through water uptake by osmosis and can move by the symplast or apoplast pathway. Sugars have to move from the spongy mesophyll cells into the phloem for long-distance transport. About 90 years ago German plant physiologist Ernst Münch established the pressure flow hypothesis to explain translocation (Figure 13.13). It states that high turgor pressure caused by high sucrose concentration near a source causes the phloem solution to move by bulk flow to a sink, an area of low turgor pressure. Phloem sap moves along the water potential gradient established by the difference in pressure potential. Water moves between the xylem and the phloem, moving into sieve-tube members close to a source and back into the xylem near a sink. The transport of sugars in the phloem requires energy in the form of ATP to establish the difference in turgor pressure.

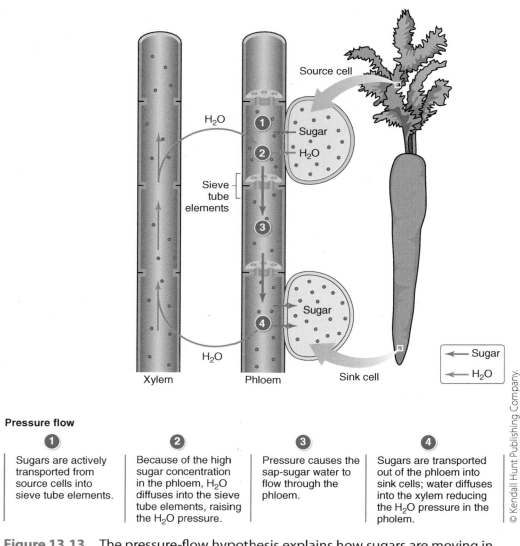

Pressure flow

1	**2**	**3**	**4**
Sugars are actively transported from source cells into sieve tube elements.	Because of the high sugar concentration in the phloem, H_2O diffuses into the sieve tube elements, raising the H_2O pressure.	Pressure causes the sap-sugar water to flow through the phloem.	Sugars are transported out of the phloem into sink cells; water diffuses into the xylem reducing the H_2O pressure in the pholem.

Figure 13.13   The pressure-flow hypothesis explains how sugars are moving in the phloem and how they get from a source cell where they are produced to a sink cell where they are used.

Phloem Loading and Unloading

Large amounts of sugars in the phloem sap move into the phloem at the source location. In the first step, sugars move over a short distance from the mesophyll source cells to the closest sieve cells in the nearest leaf veins. Once in the phloem, sugars can be transported over long distances within the plant to various sinks. Unloading from the phloem at the sink involves movement over a short distance into the cells where the sugars will be used or stored.

Phloem sap moves through sieve tubes, which are formed by sieve tube elements found in angiosperms, the individual cells of the phloem vascular system. Recall from Chapter 3 that these cells are perforated at both ends with sieve plate pores, which are open channels. These cells also have large pores in their lateral structure for easy translocation of phloem sap between sieve elements. Sieve tube elements have one or more companion cells right next to them. The plasmodesmata between these two cell types allow fast exchange of solutes and companion cells perform several metabolic functions, including production of ATP, for the sieve tube elements, which are specialized in transport and have lost some of the other functions. Because of their close proximity and functional relationship, the two cell types are also known as the sieve element–companion cell complex (Taiz et al., 2018). In case of damage, sieve tube elements produce callose, glucose polymers within their cells to form a plug and reduce or stop the loss of phloem sap. P-proteins, which are readily found in angiosperms, but not in gymnosperms, serve the same purpose.

Phloem loading differs by species and can occur through the symplast or the apoplast. Three mechanisms of phloem loading are known: (1) passive diffusion via the symplast, (2) polymer trapping via the symplast, and (3) sucrose movement through the apoplast with sucrose protein transporters (Zhang and Turgeon, 2018). Phloem loading by diffusion, a passive movement along the concentration gradient, occurs in most herbaceous species and also in many tree species. This mechanism is attributed to numerous plasmodesmata between the sieve element–companion cell complex and surrounding cells. The diffusion gradient from mesophyll cells to the sieve elements is maintained by high levels of sucrose from photosynthesis in the source cells; sucrose is present in high concentrations throughout the leaf (Figure 13.14a).

The second phloem loading mechanism is the polymer-trapping model. For short distance transport via the symplast, sugars move through the plasmodesmata from mesophyll cells to bundle sheath cells to companion cells and eventually into the sieve elements. In this case, sucrose moves against a concentration gradient, with sugars more concentrated in the sieve element–companion cell complex than in the mesophyll cells. How do the sugars stay in the sieve elements and do not diffuse back into the bundle sheath cells? As sucrose moves toward the sieve elements and into the bundle sheath cells, larger sugar molecules such as raffinose and stachyose are formed in the companion cells (sometimes referred to as intermediate cells), allowing only translocation through the numerous large plasmodesmata into the sieve elements (Figure 13.14b). Sucrose keeps diffusing from the source cells into the companion cells where it is used for the synthesis of oligosaccharides and subsequently moved into the sieve elements.

The third phloem loading mechanism transports sucrose via the apoplast and requires energy and sucrose transporter proteins. Many species perform the short distance transport of sugars through cell walls via the apoplast, especially into the companion cells. Often this pathway leads first through the symplast from mesophyll cells to companion cells. Phloem loading through the apoplast against a concentration gradient requires energy supplied by a proton pump. The movement of sucrose from companion cells via the apoplast into sieve elements is facilitated by a sucrose-H^+ symporter and sucrose transporter proteins (Figure 13.14c). It is thought that other molecules such as amino acids and sugar alcohols are also transported via this apoplastic loading process. However, other compounds such as hormones may enter the phloem passively through diffusion.

Export of sugars from a source requires maturity and the presence of infrastructure in the leaf. The leaf has to produce more sucrose than it requires for its own metabolism. The small veins in the vicinity of the mesophyll cells producing excess photosynthates have to be mature and functional for transport. Finally, for those plants using the apoplast for translocation, the sucrose-H^+ symporter has to work. Young, immature leaves are a sink and

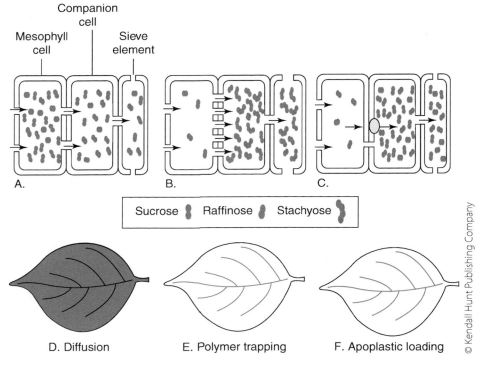

© Kendall Hunt Publishing Company

Figure 13.14a–c Three mechanisms of phloem loading: (a) Sucrose passively diffuses along a concentration gradient via the symplast through plasmodesmata from mesophyll source cells through companion cells to sieve elements in the leaf veins. The entire leaf is high in sucrose concentration. (b) Polymer trapping forms larger carbohydrate molecules that move through plasmodesmata and move into the sieve elements. Concentrations of raffinose and stachyose are high in leaf veins. (c) Sugars move partially via symplast and through the apoplast facilitated by a sucrose-H^+ symporter and protein transporter (yellow circle). Leaf veins are high in sucrose concentration.

import sugars from surrounding source leaves. They gradually transition to a source as they mature, starting at the leaf tip and moving toward the petiole. Other parts of the plant such as seeds, inflorescences, and roots also compete as sinks. The sugars are allocated to sinks based on their size and metabolic rates.

Phloem unloading can occur via the symplast or apoplast. This journey to the sink cells involves again short distance transport from the sieve elements to their final destination followed by either storage or metabolism. Phloem unloading mechanisms vary widely depending on the type of sink such as flowers, fruits, roots, or leaves. Translocation through the symplast or apoplast can vary depending on the stage of development in a sink and it may change over time. Phloem unloading through the symplast occurs by diffusion, and the use of sugars for respiration in sinks maintains the concentration gradient facilitating the flow. Apoplast transport requires energy because two membranes, those of the source and sink cells are crossed. Transporters for this movement have yet to be identified.

Key Terms

soil texture	carrier proteins	isotonic
loam	transporters	hypertonic
organic matter	proton pumps	symplast
infiltration rate	cotransporters	apoplast
water-holding capacity	metallothioneins	guttation
field capacity	antiporters	cohesion
permanent wilting point	phytoremediation	adhesion
anions	foliar application	cohesion-tension theory
leaching	transpiration	cavitation
cations	water potential	embolism
cation exchange	pressure potential	translocation
passive transport	turgor pressure	source
diffusion	solute or osmotic potential	sink
channels	osmosis	
channel proteins	hypotonic	

Summary

- Soil evolves from weathering of rock and the accumulation of organic material. It consists of sand, silt, clay and organic matter.

- Plants need 17 essential elements to complete their life cycle. Macronutrients are required in larger amounts and are obtained from the atmosphere and the soil; micronutrients are needed in smaller amounts and are obtained from the soil.

- Mineral nutrients are in some cases applied as foliar spray and ions enter the leaf through the cuticle.

- Negatively charged soil particles like clay and organic matter can bind cations; excessive water application can leach nutrients into lower soil horizons.

- Hydroponic growing systems allow the cultivation of plants in nutrient solutions without soil.

- Mycorrhizae and *Rhizobium* bacteria form symbiotic relationships with plant roots and provide nutrients to the plant, providing water and energy to the microorganisms.

- Nutrients and water are taken up through root hairs located in the root maturation zone behind the root cap.
- Passive movement of molecules across membranes occurs via diffusion, channels or channel proteins, and carrier proteins or transporters.
- Diffusion moves molecules along a concentration gradient from an area of higher concentration to an area of lower concentration.
- Proton pumps facilitate active transport and are proteins located in the plasma membrane, which facilitate the transport of H^+ from inside the cell to the outside. This establishes an electrochemical or concentration gradient and cations enter through ion channels into the cell toward a higher ion concentration.
- Passive ion exclusion can occur in two ways: at the endodermis for solutes entering through the apoplast, and at the plasma membrane from solutes entering through the symplast. In both locations ions are excluded if the correct membrane proteins for those ions are not present.
- Metallothioneins are proteins binding metal ions that entered a plant cell and render them nontoxic.
- Antiporters exclude potentially toxic ions by moving them from the cytoplasm across the tonoplast into the vacuole where they are concentrated.
- Antiporters are membrane proteins transporting two molecules or ions across a membrane in opposite directions. One ion or molecule moves along its electrochemical gradient as a different ion or molecule moves in the opposite direction against its own electrochemical gradient.
- Symporters are membrane proteins carrying a small molecule down an electrochemical gradient while taking a second different ion or molecule along in the same direction but against its electrochemical gradient.
- Osmosis is the process when water moves across a selectively permeable membrane from low solute concentration or high water concentration to high solute concentration or low water concentration. Water movement will cease when concentrations are the same on both sides of the membrane. Osmosis is a passive process and requires no energy from the plant.
- Water potential is the energy in water and is defined as the sum of the pressure or turgor potential and the solute or osmotic potential. Pressure potential is the force acting upon the cell wall from inside the cell and solute potential is the concentration of a cell's solutes. Water potential is negative in growing plants.
- Transpiration is the loss of water from the plant through stomata. It is a driving force in moving water from the roots to the leaves and is controlled by the stomata.
- The cohesion-tension theory describes how water moves from the soil solution through the plant into the leaves and employs osmosis, capillary action, and tension to transport water against gravity.
- Plants growing in dry or saline soils increase their solute or osmotic potential in their cells to ensure water uptake.
- Translocation is the transport of organic molecules, including sugars, in the phloem from a source tissue where they are produced to a sink tissue where they are used.
- The pressure-flow hypothesis explains how sugars are loaded into the phloem because of high concentration near a source. Water enters the phloem by osmosis and pushes the solutes through the phloem toward a sink because of a difference in turgor pressure.
- Phloem loading differs by species and can occur through the symplast or the apoplast.
- Phloem loading can occur through three mechanisms: 1. passive diffusion via the symplast, 2. polymer trapping via the symplast, and 3. sucrose movement through the apoplast with sucrose protein transporters.

Reflect

1. *Nitrogen is most important for plant growth.* Describe different ways the plant can obtain nitrogen, how it is used in the cell, what forms are available, and how it moves into the plant.

2. *Poor plant growth indicates a possible nutrient deficiency.* Explain how to diagnose and remedy the problem.

3. *Soil texture affects soil properties and plant growth.* Compare a sandy soil and a loam regarding nutrient storage and availability, water management, and biomass production.

4. *Describe the mechanisms of how ions and molecules move across membranes.* Explain diffusion, facilitated diffusion, and active transport.

5. *Water potential of the soil, the plant, and the atmosphere affect water and mineral transport in plants.* Explain the challenges plants growing in hot, dry climates or saline soils experience.

6. *Translocation in plants is central for the distribution of photosynthates.* Describe the typical movement of a glucose molecule from the leaf to a sink in summer and from the root (source) to a sink in early spring.

References

Barker, A. V., & Pilbeam, D. J. (2007). Introduction. In A. V. Barker & D. J. Pilbeam (Eds.), *Handbook of plant nutrition* (pp. 3–18). Boca Raton, FL: Taylor & Francis.

Garg, N., & Chandel, S. (2010). Arbuscular mycorrhizal networks: Process and functions. A review. *Agronomy for Sustainable Development, 30*(3), 581–599.

Gorham, J. (2007). Sodium. In A. V. Barker & D. J. Pilbeam (Eds.), *Handbook of plant nutrition* (pp. 569–583). Boca Raton, FL: Taylor & Francis.

Jensen, M. H. (2002). Controlled environment agriculture in deserts, tropics and temperate regions: A world review. *Acta Horticulturae, 578*, 19–25

Marschner, H., & Dell, B. (1994). Nutrient uptake in mycorrhizal symbiosis. *Plant and Soil, 159*, 89–102.

Salt, D. E., Smith, R. D., & Raskin, I. (1998). Phytoremediation. *Annual Review of Plant Physiology and Plant Molecular Biology, 49*, 643–668.

Taiz, L., Zaiger, E., Møller, I. M., & Murphy, A. (2018). *Fundamentals of plant physiology* (1st ed.). New York: Oxford University Press.

Wullschleger, S. D., Meinzer, F. C., & Vertessy, R. A. (1998). A review of whole-plant water use studies in trees. *Tree Physiology, 18*, 499–512.

Zhang, C., & Turgeon, R. (2018). Mechanisms of phloem loading. *Current Opinion in Plant Biology, 43*, 71–75.

SECTION IV

CONTINUITY OF THE GREEN WORLD

Chapter 14   Cell Cycle and Plant Life Cycle

Chapter 15   Patterns of Inheritance

Chapter 16   Molecular Basis of Inheritance

Chapter 17   Biotechnology and Genetically Modified Plants

Chapter 18   Evolution

Cell Cycle and Plant Life Cycle

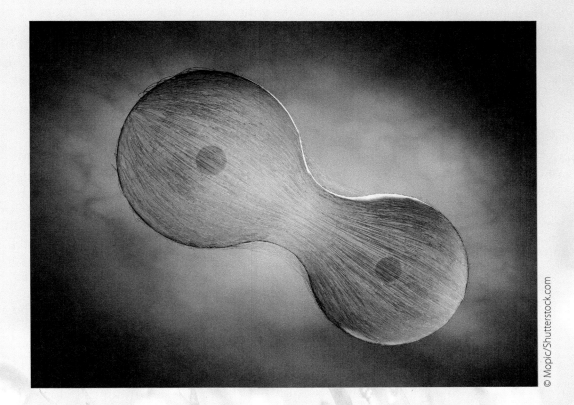

© Mopic/Shutterstock.com

Learning Objectives

- Distinguish between cell cycle and cell division
- Discuss the purposes of mitosis and meiosis
- Outline the phases of a mitotic cell cycle and describe what is happening to the cell during each phase
- Outline the phases of a meiotic cell cycle and describe what is happening to the cell during each phase
- Compare and contrast mitosis and meiosis
- Discuss how meiosis and sexual reproduction produce genetic variation
- Explain the role of mitosis and meiosis in plant life cycle

cell division
Divided into two stages: mitosis and cytokinesis; also referred to as the M phase

interphase
A preparatory stage or period in the cell cycle (*inter* in Latin means "between")

cell cycle
Composed of interphase and cell division

binary fission
Asexual reproduction of a prokaryotic cell by "splitting in half"

prokaryote
Organism that made up of a prokaryotic cell (cell without a nucleus)

chromosome
The condensed and visible form of DNA/histone complex

Plants are complex organisms and consist of many cells. Where do all these cells come from? The cell theory states cells come from preexisting cells (see Chapter 3). A cell divides and produces two new cells. These two cells, in turn, divide and produce four new cells. This process of cell multiplication is called **cell division**. In unicellular organisms, such as bacteria and some algae, cell division increases the number of individuals in a population. In multicellular organisms, such as plants and animals, cell division and enlargement enable the organisms to grow in size or repair tissues. When cells divide, they first go through a preparatory stage or period called **interphase** (*inter* in Latin means "between"). The **cell cycle** consists of two main stages: interphase and cell division. In an onion cell, the entire cell cycle lasts about 16 hours. However, the length of the cell cycle varies widely from one species to another and from one cell type to another. Prokaryotic cells divide by a process called **binary fission**. In eukaryotic cells, there are two main types of cell division, mitosis and meiosis. For this chapter, we explore the essential role of cell division in growth and reproduction of living organisms, especially in plants.

PROKARYOTIC CELL CYCLE

Prokaryotes, such as bacteria and archaea, reproduce by a relatively simple cell cycle. Recall from Chapter 3, prokaryotic cells are all unicellular, have no nucleus or organelles, and are relatively simple in structure compared to eukaryotic cells. During interphase, a prokaryotic cell replicates its circular DNA or **chromosome** while growing around twice its size (Figure 14.1). It then divides into two identical halves or cells. This process of "cell division" is called binary fission (Figure 14.1). Although unicellular eukaryotes also divide in half, the stages of interphase and cell division are much more complex than for prokaryotes.

In prokaryotes, such as the bacterium *Escherichia coli*, most of the genes are found on a single circular chromosome. The process of cell division begins when the circular chromosome replicates at a specific region on the DNA called the "origin of replication" and produces two origins. DNA replication continues from the origin of replication and moves bidirectionally toward the opposite side of the chromosome. During cell elongation, the

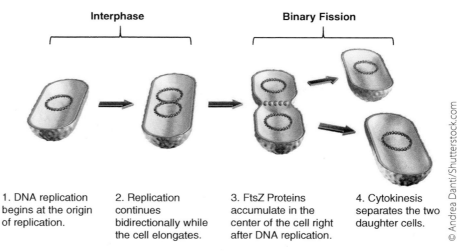

Interphase   **Binary Fission**

1. DNA replication begins at the origin of replication.

2. Replication continues bidirectionally while the cell elongates.

3. FtsZ Proteins accumulate in the center of the cell right after DNA replication.

4. Cytokinesis separates the two daughter cells.

© Andrea Danti/Shutterstock.com

Figure 14.1   Prokaryotic cell cycle.

two origins move to opposite ends of the cell and separate the two circular chromosomes. How chromosomes move in prokaryotic cells is still not fully understood. However, proteins resembling eukaryotic actins apparently play a role in chromosome movement during binary fission. DNA replication occupies the entire time between cellular division.

Upon completion of DNA replication, special proteins called FtsZ proteins accumulate at the center of the cell to form a ring of fibers (Bramhill & Thompson, 1994). These fibers, which resemble tubulins, help the cell membrane to pinch in and eventually separate the cytoplasm of the two daughter cells (Figure 14.1). This process of separating the cytoplasm of the two daughter cells is called **cytokinesis**.

EUKARYOTIC CELL CYCLE

The eukaryotic cell cycle also includes interphase and cell division, but the process has more complex and distinct stages than the prokaryotic cell cycle. There are two different types of cell division: mitosis and meiosis. Cell division, also referred to as the M phase of the cell cycle, is divided into two stages: nuclear division and cytoplasmic division. **Nuclear division** refers to the sorting and separation of duplicated chromosomes during **mitosis** and **meiosis**. **Cytoplasmic division** refers to the physical separation of the original cell into two new cells during cytokinesis. In the following section, we examine stages of eukaryotic cell cycle and contrast mitosis and meiosis.

Interphase: Preparation between Cell Division

During interphase, cells are actively preparing for cell division and go through three distinct phases: G_1, S, and G_2 (Figure 14.2).

- **G_1 phase** (first growth or gap phase) begins right after a nucleus has divided and ends before DNA replication. G_1 is the longest period in the cell cycle and the cell increases in size during this time.
- During the **S phase** (synthesis), DNA replication takes place and a duplicate copy of the genetic blueprint (DNA) is made. Now each chromosome consists of two sister chromatids or duplicated chromosomes (Figure 14.3).
- Right after the S phase, the cell enters **G_2 phase** (second growth or gap phase) to produce more proteins, replicate organelles, and accumulate energy in the form of ATP. The cell is now ready for cell division.

Up to this point, the chromosomes in the cell are not visible. To recognize interphase under the light microscope at 400× total magnification, look for cells with a nucleus that has one or more distinct nucleoli inside. It is not possible to distinguish among G_1, S, and G_2 under the light microscope (Figure 14.4).

Mitosis: Cell Division That Produces Growth

Mitosis is the cell division that produces two genetically identical cells. This cell division is essential for growth, repair, renewal, and cell maintenance. Mitosis is generally divided into five phases: prophase, prometaphase, metaphase, anaphase, and telophase (Figure 14.5). After mitosis is complete, cytokinesis follows.

cytokinesis
Division of the cytoplasm follows mitosis

nuclear division
Refers to the sorting and separation of duplicated chromosomes during mitosis and meiosis.

mitosis
A type of cell division; divided into five stages: prophase, prometaphase, metaphase, anaphase, and telophase; responsible for asexual reproduction, growth, and repair

meiosis
A type of cell division that undergoes two rounds of divisions, meiosis I and meiosis II. Each division contains five stages; responsible for sexual reproduction

cytoplasmic division
Refers to the physical separation of the original cell into two new cells during cytokinesis.

G_1 phase
Part of the cell cycle where cells increase in size; begins right after a nucleus has divided and ends before DNA replication

S phase
Part of the cell cycle where DNA replication takes place and a duplicate copy of the blueprint is made

G_2 phase
Part of the cell cycle where cells produce more proteins, replicate organelles, and accumulate energy in form of ATP; begins right after DNA replication is completed

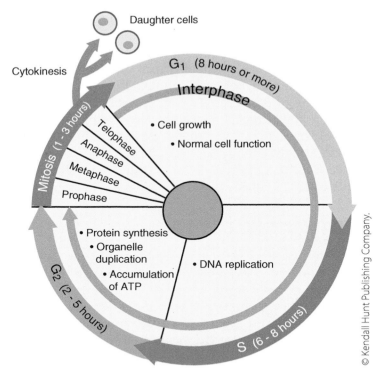

Figure 14.2 Eukaryotic cell cycle.

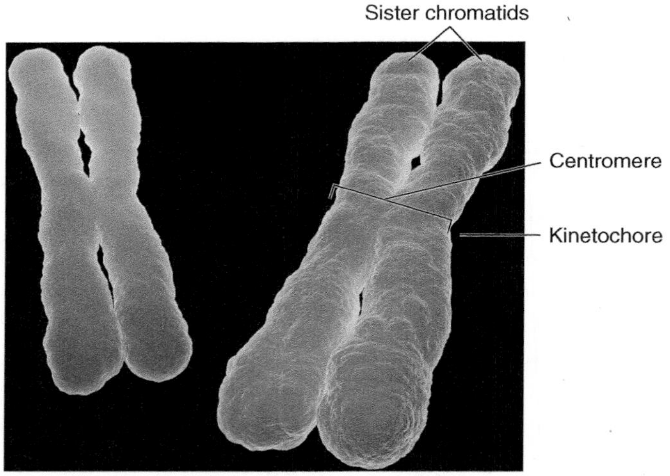

prophase
First stage of mitosis where chromosomes become visible, nuclear envelope and nucleolus start to disappear

centromere
A structure or region that connects the two sister chromatids

prometaphase
Mitotic spindles begin to attach to each sister chromatid. The nuclear envelope is now completely fragmented. Chromosomes become more condensed.

mitotic spindles
Thick protein fibers called microtubules that are involved in the movement of chromosomes during nuclear division

kinetochore
A structure of proteins attached to the centromere that connects each sister chromatid to the mitotic spindle

Figure 14.3 A duplicated chromosome and its structures.

- During **Prophase**, individual chromosomes begin to condense and become visible. The nuclear envelope begins to break apart and the nucleolus inside the nucleus also disappears. The two sister chromatids are connected by a structure or region called the **centromere** (Figure 14.3). Protein fibers called mitotic spindles begin to form. To recognize prophase under the light microscope (at 400×), look for cells with thick distinctive lines (condensed chromosomes) inside their nuclei (Figure 14.4).

- **Prometaphase** is when individual chromosomes become more condensed. **Mitotic spindles** begin to attach to **kinetochore** around the

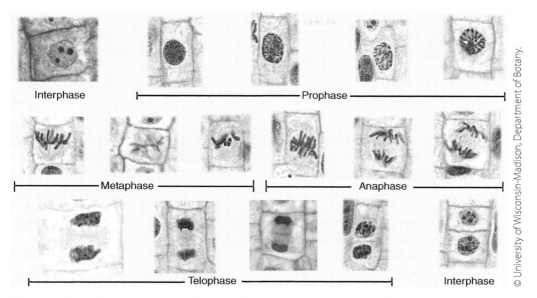

Figure 14.4   Various stages of the cell cycle: interphase, prophase, metaphase, anaphase, and telophase.

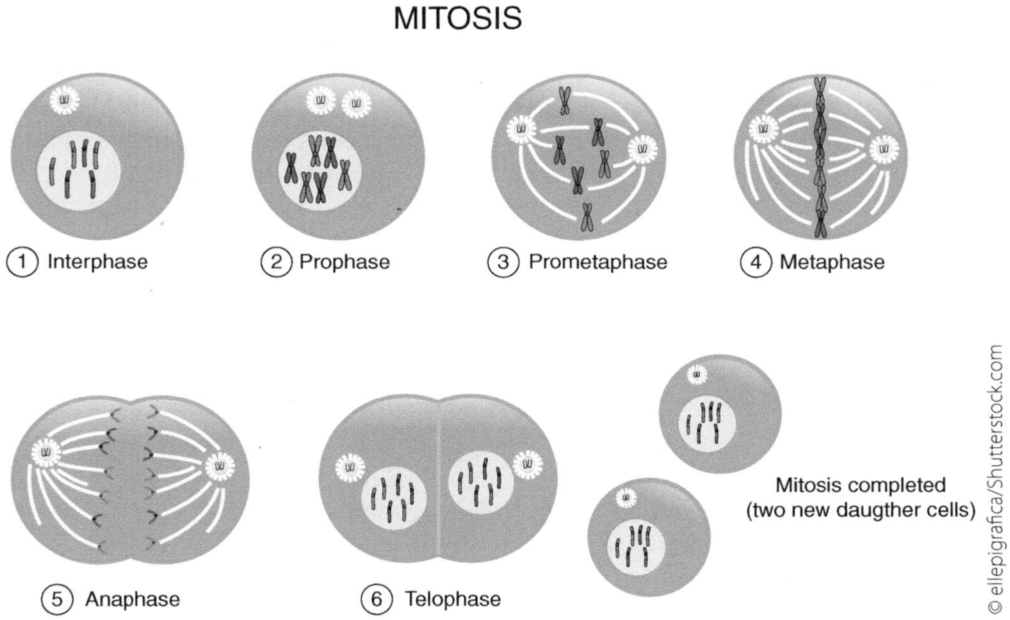

MITOSIS

(1) Interphase   (2) Prophase   (3) Prometaphase   (4) Metaphase

(5) Anaphase   (6) Telophase

Mitosis completed
(two new daugther cells)

Figure 14.5   The five phases of mitosis showing how chromosomes change shape.

centromere region on each sister chromatid. The nuclear envelope is now completely fragmented.

- **Metaphase** is when the chromosomes line up in the middle of the cell. Each sister chromatid connects to spindles from an opposite pole. The chromosomes are most condensed at this stage and appear thick and distinct. To recognize metaphase under the light microscope (at 400×), look for cells with condensed chromosomes lining up along the center of the cell or mitotic plane (Figure 14.4).

metaphase
Where chromosomes begin to line up in the middle of the cell

anaphase
Where duplicated chromosomes begin to separate and move to opposite ends of the cell

telophase
Last stage of mitosis where chromosomes begin to decondense and become invisible, nuclear envelope and nucleolus start to reform

cell plate
Separates the cell into two compartments during cytokinesis

phragmoplast
A barrel-shaped system of short microtubules found between the two newly form nuclei; responsible for the formation of the cell plate

somatic cells
Body cells contain two sets of chromosomes (diploid) and are not specialized for sexual reproduction

diploid
Having two sets (2n) of chromosomes in the cell(s)

gametes
Haploid reproductive cells such as sperms or eggs. Gametes unite during sexual reproduction to produce a diploid zygote

haploid
Having one set (n) of chromosomes in the cell(s)

- **Anaphase** is when the two sister chromatids separate and become two individual chromosomes. The spindles then pull these chromosomes to opposite poles of the cell. Anaphase ends when the chromosomes reach the poles. It is the shortest stage of mitosis. To recognize anaphase under the light microscope (at 400×), look for cells with two sets of v-shaped condensed chromosomes moving to opposite ends of the cell (Figure 14.4).

- **Telophase** is the exact opposite to prophase. During this stage, the chromosomes begin to de-condense and become invisible. The nuclear envelope begins to reform around the chromosomes. The nucleolus begins to reorganize and becomes visible again. Spindles begin to disappear. Mitosis is now complete. To recognize telophase under the light microscope (at 400×), look for cells with two rectangular clumps of chromosomes separated by a think line called the **cell plate** (Figure 14.4).

Cytokinesis: Physical Separation of Cells

Cytokinesis always follows mitosis and is the division of the cytoplasm. In plants, cytokinesis begins during early telophase by forming the **phragmoplast**, which is a barrel-shaped system of short microtubules found between the two newly formed nuclei. The phragmoplast facilitates the formation of the cell plate, which starts in the middle of the cell and extends toward the edges of the cell. The cell plate eventually separates the cell into two compartments, and a middle lamella forms between the two cells and cements them together. The two cells produce new plasma membranes and primary cell walls. The cell wall of the original cell stretches and ruptures as the two new cells grow in size. In dividing animal cells, the phragmoplast and cell plate are absent and the cell simply pinches into two.

Meiosis: Essential Cell Division for Sexual Reproduction

The body of most multicellular organisms is made up of **somatic cells**, cells that are not specialized for sexual reproduction. These **diploid** somatic cells contain two sets of chromosomes (2n) and divide through the process of mitosis to produce two new diploid cells. When we consider the production of sex cells or **gametes** in eukaryotes, the cells go through a more extensive process called meiosis. This process is similar to and different from mitosis (see Table 14.1). The process of meiosis results in the formation of four sex cells, which are comprised of unpaired chromosomes; thus, the chromosome number is reduced from diploid (2n) to **haploid** (n). When two of these haploid gamete cells fuse together, they create a zygote which conserves the original diploid number of chromosomes in this original cell. Similar to mitosis, interphase precedes meiosis and the cell goes through G_1, S, and G_2 to prepare for meiosis. The process of meiosis has two distinct series of cell divisions, meiosis I and meiosis II (Figure 14.6). Let's examine the different phases of meiosis.

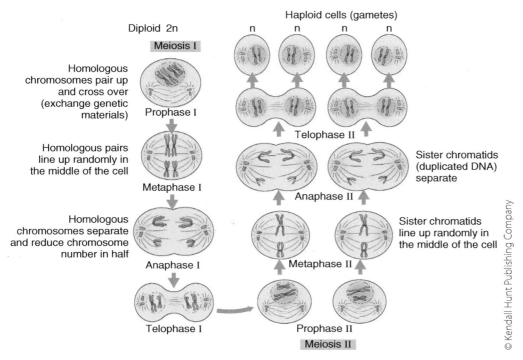

Figure 14.6   Overview of meiosis I and II.

Table 14.1   Differences between Mitosis and Meiosis

Characteristics	Mitosis	Meiosis
Purposes	Growth, repair, renewal, cell maintenance, and asexual reproduction	Sexual reproduction
Interphase	The preparatory stages include G_1, S, G_2	
Number of cell division and stages	One; prophase, prometaphase, metaphase, anaphase, and telophase	Two; prophase I, prometaphase I, metaphase I, anaphase I, telophase I, prophase II, prometaphase II, metaphase II, anaphase II, and telophase II
Homologous chromosomes pair up and cross-over	Does not take place	Takes place during prophase I
Cytokinesis	Follows cell division. In some organisms, the cells may skip the first cytokinesis after meiosis I and proceed directly to prophase II after telophase I.	
Number and type of daughter cells	Two genetically identical cells	Four genetically different cells
Sets of chromosomes in daughter cells (Figure 14.7)	Identical to the mother cell. • If the parent cell is diploid (2n), the two daughter cells will be diploid (2n). • If the parent cell is haploid (n), the two daughter cells will be haploid (n).	Half of the mother cell. • If the parent cell is diploid (2n), the four daughter cells will be haploid (n). • If the parent cell is haploid, no daughter cells will be produced.

© Kendall Hunt Publishing Company

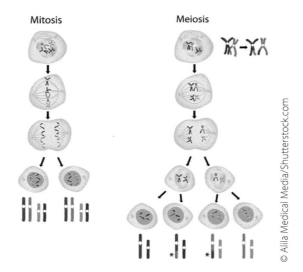

prophase I
Has all the elements of prophase of mitosis. The key features found only in this stage of meiosis I are homologous chromosomes pair up and crossing over occurs between the homologous chromosomes.

Figure 14.7 Contrast mitosis and meiosis.

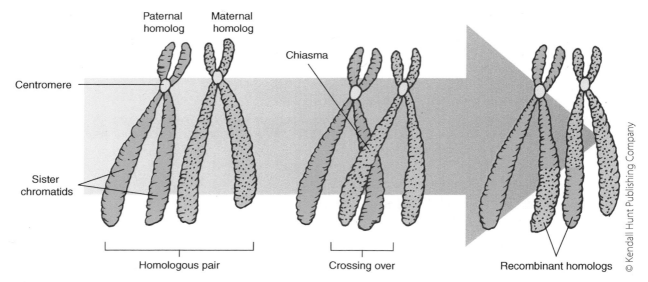

Figure 14.8 Crossing over of homologous chromosomes.

homologous chromosomes
A pair of chromosomes of the same length, centromere position, staining pattern, and genes in corresponding loci. One homologous chromosome is inherited from the paternal side and the other from the maternal side. Also called homologs, or a homologous pair

In the first meiotic division or meiosis I, the homologous chromosomes are separated and result in two daughter cells with one set (n) of chromosomes. Meiosis I is often referred to as the reduction division and is generally divided into five phases: prophase I, prometaphase I, metaphase I, anaphase I, and telophase I.

- During **Prophase I**, individual chromosomes condense and become visible, the nuclear envelope and nucleoli begin to disappear, and mitotic spindles begin to form. The key features found only in prophase I of meiosis are (1) **homologous chromosomes** pair up, and (2) **crossing over** occurs when the homologous chromosomes exchange sections of their chromosomes (Figure 14.8). This process of crossing over produces genetic variation.

- **Prometaphase I** is similar to prometaphase of mitosis except the homologous chromosomes paired up and mitotic spindles begin to attach to the homologous chromosomes of each pair.

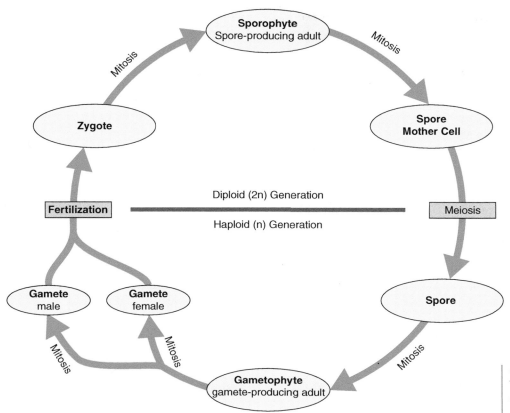

Figure 14.9   A typical plant life cycle: alternation of generations.

- During **Metaphase I**, the homologous pairs line up in the middle of the cell. Each homologous chromosome connects to spindles from an opposite pole. The alignment of the homologous pairs further produces genetic variation.

- During **Anaphase I**, the two homologous chromosomes separate and are being pulled to opposite poles of the cell. Anaphase ends when the chromosomes have reached the poles. The sister chromatids of each chromosome are still connected at this point.

- **Telophase I** is very similar to telophase of mitosis. The only difference is that the sister chromatids are still connected.

In many organisms, the chromosomes proceed directly to prophase II, the next meiotic phase. During the second meiotic division, the sister chromatids in the daughter cells separate into four new haploid cells. Meiosis II is often referred to as the equational division (Figure 14.6). The five phases of meiosis II are essentially identical to mitosis; they are referred to as prophase II, prometaphase II, metaphase II, anaphase II, and telophase II. When meiosis is complete, cytokinesis follows.

Meiosis and Sexual Reproduction Generates Genetic Variation

Unlike animals, plants can reproduce both sexually and asexually. In Chapter 9, you learned that many of our crops are propagated asexually to

crossing over
The reciprocal exchange of genetic material between homologous (non-sister) chromosomes during prophase I of meiosis

prometaphase I
Similar to prometaphase of mitosis except the homologous chromosomes paired up and mitotic spindles begin to attach to the homologous chromosomes of each pair

metaphase I
The homologous pairs line up in the middle of the cell

anaphase I
The two homologous chromosomes separate and are being pulled to opposite poles of the cell; the sister chromatids of each chromosome are still connected at this point.

telophase I
Is very similar to telophase of mitosis; the only difference is that the sister chromatids are still connected

produce plants with certain colors or characteristics. Sexual reproduction, however, is essential to the survival of a species because it produces genetic variation and enables the species to adapt to changes.

Genetic variation comes from four different sources: mutations, crossing over of homologous chromosomes, independent assortment of chromosomes, and random fusion of gametes during fertilization. We will discuss mutations in a later chapter. For species that reproduce sexually, the behavior of chromosomes during meiosis and fertilization is responsible for the variation that arises in each generation. As we learned in this chapter, homologous chromosomes pair up and exchange their genetic materials (crossing over) during prophase I of meiosis (Figure 14.8). These homologous pairs then randomly orientate in metaphase I of meiosis. In metaphase II, the nonidentical sister chromatids again independently orientate. The behavior of chromosomes during meiosis increases the number of genetic types of daughter cells or gametes. When the gametes from two individuals come together during fertilization, they introduce even more genetic variation to the next generation. In the next chapter, we will learn more about how characteristics are inherited from one generation to the next.

KEY ROLE OF CELL DIVISION IN THE UNIQUE LIFE CYCLE OF PLANTS

sporophytes
Adult diploid plants produce spores by meiosis in their sporangia

gametophytes
Adult haploid plants produce gametes by mitosis in their gametangia

alternation of generations
An unique life cycle of plants; plants spend part of their lives as multicellular haploid gametophytes and part as multicellular diploid sporophytes

A unique characteristic of plants is that they produce two adult forms, called **sporophyte** and **gametophyte**, and the two forms alternate in the life cycle. The sporophyte is diploid whereas the gametophyte is haploid. Plants spend half of their lives in a haploid stage and the other half in a diploid stage. This unique life cycle of plants is called **alternation of generations** (Figure 14.9). Meiosis is essential to sexual reproduction in plants but the production of gametes (sex cells) is indirect. Instead of gametes, meiosis in plants produces haploid spores. These spores germinate and mature into haploid adults called gametophytes. These adult plants produce gametes using mitosis. In plants, mitosis is essential not only to the growth of both sporophyte and gametophyte but also for sexual reproduction. Unlike animals, plant gametes are produced by both meiosis and mitosis.

Key Terms

alternation of generations
anaphase
anaphase I
binary fission
cell cycle
cell division
cell plate
centromere
chromosome
crossing over
cytokinesis
cytoplasmic division
diploid

G_1 phase
G_2 phase
gametes
gametophytes
haploid
homologous chromosomes
interphase
kinetochore
meiosis
metaphase
metaphase I
mitosis
mitotic spindles

nuclear division
phragmoplast
prometaphase
prometaphase I
prophase
prophase I
prokaryote
S phase
somatic cells
sporophytes
telophase
telophase I

Summary

- When cells divide, they need to first go through a preparatory stage or period called interphase (inter in Latin means "between"). Interphase and cell division together form the cell cycle.
- Prokaryotic cells divide by a process called binary fission.
  - DNA replication continues from the origin of replication and moves bidirectionally toward the opposite sides of the chromosome in prokaryotes. DNA replication occupies the entire time between cell divisions.
  - Special proteins called FtsZ proteins are essential for cytokinesis in prokaryotes.
- Cell division in eukaryotes is divided into two stages: nuclear division and cytoplasmic division.
  - Nuclear division refers to the sorting and separation of duplicated chromosomes during mitosis and meiosis.
  - Cytoplasmic division refers to the physical separation of the original cell into two new cells during cytokinesis.
- Summary of a mitotic cell cycle

Stages	Cellular Events
G_1 phase	Begins right after a nucleus has divided and ends before DNA replication. G_1 is the longest period in the cell cycle and the cell increases in size during this period of time.
S phase	DNA replication takes place and a duplicate copy of the genetic blueprint is made. Now each chromosome consists of two sister chromatids or duplicated chromosomes.
G_2 phase	Begins to produce more proteins, replicates organelles, and accumulates energy in form of ATP. The cell is now ready for cell division.
Prophase	Is when individual chromosomes begin to condense and become visible. The nuclear envelope begins to break apart and the nucleolus inside the nucleus also disappears. The two sister chromatids are connected by a structure or region called the centromere. Protein fibers called mitotic spindles begin to form.
Prometaphase	Is when individual chromosomes become more condensed. Mitotic spindles begin to attach to kinetochore around the centromere region on each sister chromatid. The nuclear envelope is completely fragmented.
Metaphase	Begins when the chromosomes line up in the middle of the cell. Each sister chromatid connects to spindles from an opposite pole. The chromosomes are more condensed at this stage.
Anaphase	Is when the two sister chromatids separate and become two individual chromosomes. The spindles then pull these chromosomes to opposite poles of the cell.
Telophase	Begins when the chromosomes begin to de-condense and become invisible. The nuclear envelope begins to reform around the chromosomes. The nucleolus begins to reorganize and become visible again. Spindles begin to disappear. Mitosis is now complete.
Cytokinesis	Begins during early telophase by forming the phragmoplast, which facilitates the formation of the cell plate. The cell plate eventually separates the cell into two separate cells.

- Summary of a meiotic cell cycle

Stages	Cellular Events
G_1 phase	Begins right after a nucleus has divided and ends before DNA replication. G_1 is the longest period in the cell cycle and the cell increases in size during this time.
S phase	DNA replication takes place and a duplicate copy of the genetic blueprint is made. Now each chromosome consists of two sister chromatids or duplicated chromosomes.
G_2 phase	Begins to produce more proteins, replicates organelles, and accumulates energy in form of ATP. The cell is ready for cell division.
Prophase I	Has all elements of prophase of mitosis. The key features found only in prophase I of meiosis are: (1) homologous chromosomes pair up, and (2) crossing over occurs when the homologous chromosomes exchange sections of their chromosomes.
Prometaphase I	Is similar to prometaphase of mitosis except the homologous chromosomes pair up and mitotic spindles begin to attach to the homologous chromosomes of each pair.
Metaphase I	Begins when homologous pairs line up in the middle of the cell. Each homologous chromosome connects to spindles from an opposite pole.
Anaphase I	The two homologous chromosomes separate and are being pulled to opposite poles of the cell. The sister chromatids of each chromosome are still connected at this point.
Telophase I	Is very similar to telophase of mitosis. The only difference is that the sister chromatids are still connected.
Cytokinesis I	May or may not take place.
Prophase II to Telophase II	The same as prophase to telophase of mitosis
Cytokinesis II	Begins during early telophase by forming the phragmoplast which facilitates the formation of the cell plate. The cell plate eventually separates the cell into two separate cells.

- Differences between mitotic and meiotic cell cycle

Characteristics	Mitosis	Meiosis
Purposes	Growth, repair, renewal, cell maintenance, and asexual reproduction	Sexual reproduction
Interphase	The preparatory stages include G_1, S, G_2	
Number of cell division and stages	One; prophase I, prometaphase I, metaphase I, anaphase I, and telophase I	Two; prophase I, prometaphase I, metaphase I, anaphase I, telophase I, prophase II, prometaphase II, metaphase II, anaphase II, and telophase II
Homologous chromosomes pair up and cross-over	Does not take place	Take place during prophase I

Cytokinesis	Follows cell division. In some organisms, the cells may skip the first cytokinesis after meiosis I and proceed directly to prophase II after telophase I.	
Number and type of daughter cells	Two genetically identical cells	Four genetically different cells
Sets of chromosomes in daughter cells	The same as the mother cell. • If the parent cell is diploid (2n), the two daughter cells will be diploid (2n). • If the parent cell is haploid (n), the two daughter cells will be haploid (n).	Half of the mother cell. • If the parent cell is diploid (2n), the four daughter cells will be haploid (n). • If the parent cell is haploid, no daughter cells will be produced.

- Plants have a unique life cycle called alternation of generations in which they spend half of their lives in a haploid stage and the other half in a diploid stage.
  - Sporophyte and gametophyte are two multicellular adult forms of plants, but the sporophyte is diploid whereas the gametophyte is haploid.
  - Meiosis is essential to sexual reproduction but the production of gametes (sex cells) is indirect.
  - Meiosis in plants produces haploid spores, which germinate and mature into haploid gametophytes.
  - In plants, a haploid gametophyte produces gametes by mitosis.
  - In plants, mitosis is essential not only to the growth of both sporophyte and gametophyte but also sexual reproduction.

Reflect

1. *A cycle of life*. Cell division is an important process for all cells. Use a diagram to map out the stages that a cell will go through in order to produce new cells. Explain what is happening to the cell in each stage.

2. *A Mad Scientist*. You want to create a multicellular organism from a single cell. What do you need to do? What is required for a single cell to become a functioning multicellular organism?

3. *Alternation of generations*. Plants have a unique life cycle called alternation of generations. Use a diagram to show this unique life cycle and make sure you include the process by which plant cells change from haploid (n) to diploid (2n) and also the other process that enables plant cells to change from 2n to n. Make sure you start with the first and finish with the last cells of each generation and include its adult form.

Reference

Bramhill, D., & Thompson, C. M. (1994). GTP-dependent polymerization of *Escherichia coli* FtsZ protein to form tubules. *Proceedings of the National Academy of Sciences of the USA, 91*, 5813–5817.

Chapter 15

Patterns of Inheritance

David Cavagaro/Visuals Unlimited.

Learning Objectives

- Define the following genetics terms: *genotype, phenotype, dominant, recessive, homozygous,* and *heterozygous*
- Solve genetics problems involving single-character or two-character inheritance
- Discuss the limitations of Mendel's principles
- Distinguish between simple and complex patterns of inheritance
- Compare and contrast the three main plant breeding strategies
- Discuss the importance and impact of plant breeding

You love Gala apples and decided to grow your own Gala apple trees. You took the seeds from your favorite apple variety and planted them in the backyard. You finally have your own apple trees but the apples produced by your trees tasted so different from what you expected. What is going on? Did something change in the seeds of your favorite Gala apple? How do growers produce the same variety of fruits or same color ornamental crops every time? To answer these questions, we need to understand genetics, the branch of biology studying how characteristics are passed from parents to offspring. Although the study of genetics began with Gregor Mendel in 1865, his rules and principles of inheritance were not taken seriously or widely accepted until very recently. Genetics is a relatively young field, barely 150 year old. This field of study explores how many genes organisms have, how genes direct cellular activities, how organisms develop, how species evolve, and how defective genes cause disease. Our understanding of plant genetics may help us improve crop quality, enhance crop production, and mass-produce pharmaceutical products.

PATTERNS OF INHERITANCE

Genetics has traditionally been divided into three major areas: transmission, molecular, and population (Brooker, 2012). Transmission genetics explores the patterns of inheritance as traits are passed from parents to offspring. Molecular genetics studies the biochemical and molecular structure of the genetic blueprint or hereditary material. Population genetics focuses on genetic variation and its role in evolution. In this section, we take a closer look at transmission genetics, how traits are passed from one generation to the next.

Mendelian Inheritance

hybridization
When two distinct individuals with different characteristics are crossed or mated.

The origin of transmission genetics can be traced back to Gregor Johann Mendel (Figure 15.1), a Moravian monk and a botanist, and his path-breaking studies of pea plants in the mid-to late 1800s. Mendel was particularly interested in how traits were inherited. He conducted **hybridization**

Figure 15.1 Gregor Johann Mendel (1822–1884), the father of genetics, deduced the rules governing how traits are inherited in organisms.

experiments that crossed or mated two distinct pea plants with different characteristics to produce offspring or **hybrids**. Mendel's meticulous record keeping and careful analyses enabled him to deduct patterns of trait inheritance in pea plants. Gregor Mendel is recognized as the father of genetics because his principles of inheritance laid the basic foundation for genetics.

Mendel chose the garden pea, *Pisum sativum*, as his experimental organisms for three main reasons. First, a number of pea varieties differing in height, flowers, seeds, and pods were available at that time. Mendel studied seven characters of pea plants: stem length, flower color, flower position, pod color, pod shape, seed color, and seed shape. Each of these seven characters has two traits or variants (Figure 15.2). **Characters** refer to the

hybrids
Offspring of two distinct individuals.

characters
Morphological characteristics of an organism such as flower color or plant height.

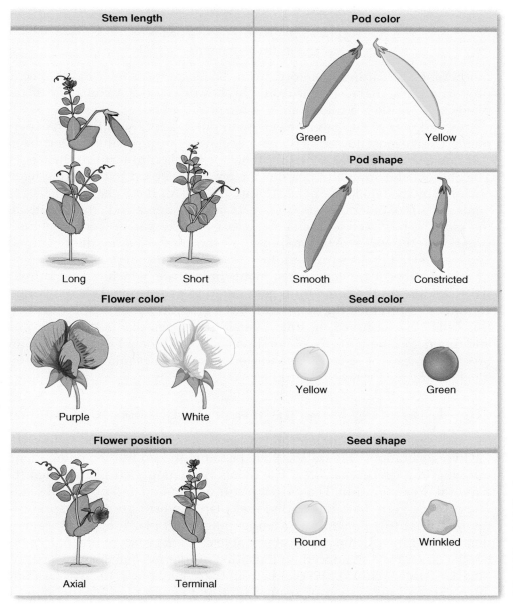

© Kendall Hunt Publishing Company.

Figure 15.2 An illustration of the seven characters studied by Mendel. Each character has two distinct traits.

Figure 15.3 A closeup of a pea flower and its floral structure.

traits
Specific properties of a character.

morphological features of an organism. The term **traits** refer to the specific properties of a character. For example, flower color is a character of pea plants and purple is a color trait of some pea flowers.

Second, pea plants can be easily manipulated for hybridization experiments. Unlike most flowering plants, pea flowers are normally self-fertilized. Pea flowers self-fertilize by using their own pollen to fertilize their eggs. The anthers and stigma of pea flowers are not exposed but are completely enclosed by the two lower petals (Figure 15.3). This floral characteristic prevents pollens of other flowers from reaching the stigma and cross-fertilizing the plant. When Mendel wanted to cross-fertilize pea flowers, he simply pried open the lower petals, removed all the anthers before they matured, and transferred pollens from another pea plant to the stigma.

Third, most characters of pea plants breed true, which means individual traits remain the same from generation to generation. When a variety continues to produce the same trait after generations of self-fertilization, it is called a **true-breeding line**. The ability of pea plants to self-fertilize and retain specific traits for generations allowed Mendel to maintain pure breeding lines. These unique characteristics of pea plants make them the ideal organisms for heredity studies.

true-breeding line
Variety that continues to produce the same trait generations after generations of self-fertilization.

parental (P) generation
Two parent plants that begin a cross.

first filial (F$_1$) generation
Offspring of the P generation.

phenotype
Visible characteristics of an individual.

second filial (F$_2$) generation
Offspring of the F$_1$ generation.

dominant
Trait or effect is always expressed or manifested.

One-Character Inheritance or Single-Factor Crosses

Mendel began with a single character such as flower color and followed it for two generations (Figure 15.4). He began with two true-breeding plants differing in a single character. The plants beginning a cross are called the **parental (P) generation**. The parent plants are crossed and produced seeds. These seeds germinate into individual plants, which are the offspring or the **first filial (F1) generation**. Mendel noticed all plants of the F$_1$ generation showed the **phenotype** or observable characteristics of one parent but not the other. He also noticed no transition or intermediate forms were produced. Mendel allowed two plants from the F$_1$ generation to cross and produce a second generation or **second filial (F$_2$) generation**. Mendel noticed three-quarters or 75% of the F$_2$ plants produced the phenotype of one P parent, which he referred to as the **dominant** trait, and the remaining 25% produced the phenotype of the other P parent, which he referred to as the

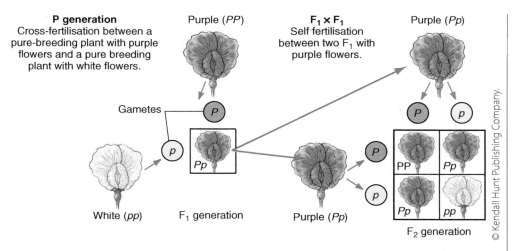

Figure 15.4   The outcomes of a single-character inheritance, flower color, as it passed through two generations.

recessive trait (Figure 15.5). Mendel demonstrated all seven characters produced a 3:1 phenotypic ratio (3 dominant to 1 recessive phenotype) when two hybrids were crossed. A cross between two heterozygous individuals is called **monohybrid cross** (*mono* means "one character," *hybrid* means "heterozygous") and it always produces a 3:1 phenotypic ratio.

Mendel suggested a parental factor, which is now known as a **gene**, was passed unchanged to the offspring and the traits, which are now known as **alleles**, always occur in pairs. See Table 15.1 for other key genetic terms.

Conceptualizing the Genetic Composition   Mendel began using letters or signs to describe the genetic composition or **genotype** of the pea plants (Mendel, 1866). Letters are still used today to denote various genotypes. For one-character inheritance, the rules for writing out genotypes are:

1. Two letters are used for each genotype: PP, Pp, or pp.
2. Uppercase letter (P) represents the dominant allele and always comes first. Lowercase of the same letter (p) represents the recessive allele.

P generation	F1 generation	F2 generation	Ratio
Long × Short	All Long	787 Long, 277 Short	2.84:1
Purple × White Flowers	All Purple	705 Purple, 224 White	3.15:1
Axial × Terminal flowers	All Axial	651 Axial, 207 Terminal	3.14:1
Green × yellow pods	All Green	428 Green, 152 Yellow	2.82:1
Smooth × constricted pods	All Smooth	882 Smooth, 299 Constricted	2.95:1
Yellow × Green seeds	All Yellow	6,022 Yellow, 2,001 Green	3.01:1
Round × Wrinkled seeds	All Round	5,474 Round, 1,850 Wrinkled	2.96:1
Total	All Dominant	14,949 Dominant, 5,010 Recessive	2.98:1

Figure 15.5   Original data from Gregor Mendel (1866).

recessive
Trait or effect is repressed or masked by the presence of a dominate allele.

monohybrid cross
When two hybrids with one different character are crossed (*mono* means "one," *hybrid* means "heterozygous"), producing a 3:1 phenotypic ratio.

gene
Organization unit or segment of DNA that produces a functional product.

alleles
Different forms of the same gene.

genotype
Genetic composition of an individual.

Table 15.1 Key Genetic Terminology

Term	Definition
Gene	An organization unit or segment of DNA that produces a functional product.
Alleles	Different forms of the same gene.
Characters	The morphological characteristics of an organism.
Traits	Specific properties of a character.
Dominant	Trait or effect is always expressed or manifested.
Recessive	Trait or effect is repressed or masked by the presence of a dominant allele.
Homozygous	Two identical alleles of a gene.
Heterozygous	Two different alleles of a gene.
Genotype	The genetic composition of an individual.
Phenotype	The visible characteristics of an individual.
Haploid (n)	One set of chromosomes.
Diploid (2n)	Two sets of chromosomes.

3. When a genotype has two identical alleles such as PP or pp, they represent two identical copies of the same gene and are referred to as homozygous. Individuals that are either homozygous dominant (PP) or homozygous recessive (pp) are referred to as pure-breeding lines.

4. **Heterozygous** refers to the genotype that has two different alleles such as Pp. Individuals with this type of genotype are referred to as hybrids.

heterozygous
Two different alleles of a gene.

Let us go back to the example of flower color in pea plants (Figure 15.4). In the P generation, Mendel crossed a pure-breeding plant with purple flowers (PP) and a pure-breeding plant with white flowers (pp) to produce hybrid F_1 plants (Pp) with purple flowers. The hybrids (Pp) have the same phenotype as their homozygous dominant parent (PP) because the dominant allele (P) always expresses itself. When the two hybrids were crossed, they produced three kinds of genotypes, PP, Pp, and pp, in a 1:2:1 ratio. Although PP and Pp produced the same phenotype, purple flowers in this case, they were different genetically. Mendel's studies have shown plants with the same phenotype may not have the same genotype.

Predicting the Outcome of Crosses with Punnett Squares Once you know the genotype of the parents, you can easily predict the phenotype and genotype of all their offspring using a Punnett square. The **Punnett square** was invented by British geneticist Reginald C. Punnett and is a diagrammatic presentation of all the possible offspring genotypes. It is important you learn how to determine the kind(s) of gametes produced by a given genotype. Meiosis is the cell division responsible for gamete production (see Chapter 14). It produces haploid gametes (eggs or sperms) from diploid cells. For example, an individual with a genotype PP will produce gametes carrying only one allele, in this case P. An individual with a genotype Pp will

Punnett square
Graphic representation that shows all the possible combinations of gametes to form offspring; developed by Reginald Punnett.

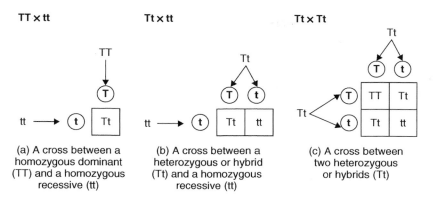

TT x tt   Tt x tt   Tt x Tt

(a) A cross between a
homozygous dominant
(TT) and a homozygous
recessive (tt)

(b) A cross between a
heterozygous or hybrid
(Tt) and a homozygous
recessive (tt)

(c) A cross between
two heterozygous
or hybrids (Tt)

Figure 15.6a–c   The size of a Punnett square is determined by the kind(s) of gametes (in circles) produced by each parent. There is no one-size-fits-all. a) a cross between a homozygous dominant (TT) and a homozygous recessive (tt), b) a cross between a heterozygous or hybrid (Tt) and a homozygous recessive (tt), and c) a cross between two heterozygous or hybrid (Tt).

produce two kinds of gametes and each carries only one allele, either P or p. To take this one step further, an individual with a genotype AABBCCDD will produce one kind of gametes with an allele combination of ABCD. An individual with a genotype AABbCCDD will produce two kinds of gametes with an allele combination of either ABCD or AbCD.

The size and shape of a Punnett square is determined by the kind(s) of gametes produced by the two parents (Figure 15.6). For example, when you cross two pure-breeding or homozygous parents (TT × tt), each parent produces only one kind of gametes and the Punnett square will have only one cell or combination. All the offspring will have the same phenotype and genotype (Tt) (Figure 15.6a). When you cross two heterozygous or hybrids (Tt × Tt), each parent produces two kinds of gametes, either T or t, and the Punnett square will have four cells or combinations producing one offspring with genotype TT, two with genotype Tt, and one with genotype tt (Figure 15.6c). It is important to determine the kind(s) of gametes produced by each parent and use these gametes to construct your Punnett square. Remember, there is no one-size-fits-all.

Using a Test Cross   As you might recall, a pea plant with purple flowers, a dominant phenotype, can be either a homozygous dominant (PP) or a heterozygous (Pp). One of the easiest ways to determine the genotype of a plant with the dominant phenotype is to cross it with a homozygous recessive plant. This is called a **testcross**. In our example, the plant with purple flowers is crossed with a homozygous recessive plant (pp) with white flowers (Figure 15.7). If the unknown plant is a homozygous dominant (PP), all the offspring will produce purple flowers. If the unknown plant is a heterozygous, half of the offspring will produce purple flowers and the other half white flowers.

Two-Character Inheritance or Two-Factor Crosses

Mendel also crossed plants with two different characters and investigated their pattern of inheritance. He began with two true-breeding plants differing in two characters: seed color and seed shape. For two-character inheritance,

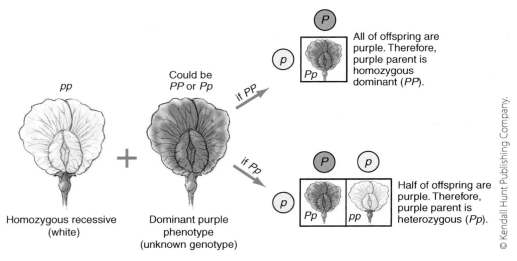

Figure 15.7 A testcross between a known homozygous recessive parent and an unknown dominant parent to determine its genotype.

the rules for writing out genotypes and gametes are similar to the rules used for single-character inheritance with the following modifications:

1. Four letters are used for each genotype (i.e., YYRR, YYRr, YyRr).
2. Uppercase letter for the dominant allele and lowercase for the recessive allele. The alleles of the same gene stay together (i.e. YyRr not YRyr).
3. Homozygous dominant refers to YYRR; homozygous recessive refers to yyrr.
4. Heterozygous refers to YyRr.
5. Gametes have two letters or alleles, one allele from each character or gene. A genotype YYRR will produce only one type of gametes with an allele combination of YR. A genotype YyRr will produce four types of gametes and the allele combinations are YR, Yr, yR, and yr.

Let's get back to the example of Mendel's two-character inheritance. Mendel crossed the two true-breeding parents YYRR and yyrr. Their offspring were all yellow and round YyRr (Figure 15.8). When he crossed the two F$_1$ plants, their offspring could be divided into four groups based on their distinct phenotypes or nine groups based on their different genotypes. Mendel demonstrated all seven characters produced a 9:3:3:1 phenotypic ratio (nine homozygous dominant, six heterozygous, and one homozygous recessive) when two F$_1$ hybrids with the genotype YyRr were crossed. This is called **dihybrid cross** (*di* means "two characters," *hybrid* means "heterozygous") and it always produces a 9:3:3:1 phenotypic ratio. Similar to one-character inheritance, the size and shape of a Punnett square used for two-character inheritance is also determined by the kind(s) of gametes produced by the parents (Figure 15.9).

dihybrid cross
When two hybrids with two different characters are crossed (*di* means "two," *hybrid* means "heterozygous"), producing a 9:3:3:1 phenotypic ratio.

Mendel's Principles of Inheritance

Mendel's hybridization experiments were aimed at understanding the principles governing how characters are inherited. At that time, no one knew anything about the molecular structure of the genetic material or how it is transferred

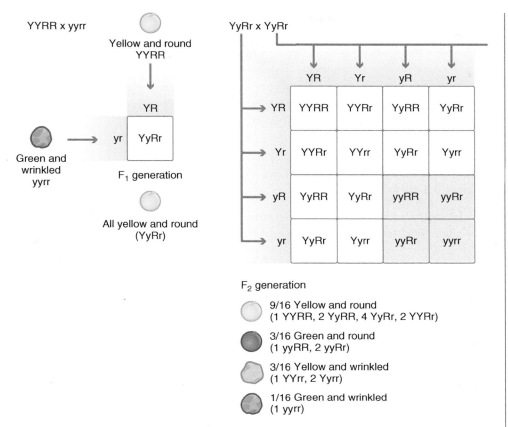

Figure 15.8   The outcomes of a two-character inheritance, seed color and seed shape, as they passed through two generations.

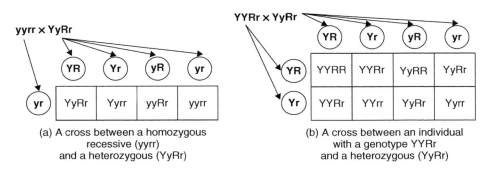

(a) A cross between a homozygous recessive (yyrr) and a heterozygous (YyRr)

(b) A cross between an individual with a genotype YYRr and a heterozygous (YyRr)

Figure 15.9a–b   The kind(s) of gametes (in circles) produced by the parents determine the size of a Punnett square. a) a cross between a homozygous recessive (yyrr) and a heterozygoud (YyRr), b) a cross between an individual with a genotype YYRr and a heterozygous (YyRr).

during gamete formation and fertilization. We now know the genetic material is made up of deoxyribonucleic acid (DNA). Each of the seven characters studied by Mendel is governed by a different gene and each gene has two alleles. The principles of inheritance Mendel deduced from his experiments are:

1. The genetic determinants of characters are passed along as "unit factors" from generation to generation. These unit factors are now called genes. This is referred to as the **particulate theory of inheritance**.

particulate theory of inheritance
Genetic determinants of characters passed along as "unit factors" or genes from generation to generation.

BOX 15.1 Solving Genetics Problems

While some of you might find it easy to solve genetics problems, many of you might find it difficult. The following instructions will provide a step-by-step approach to handle genetics problems.

Step 1: **Read** the genetics problem carefully.

Step 2: **Identify** whether it is a single-character or two-character inheritance and which allele is dominant.

- **Identify** which allele is dominant and designate a capital letter for this allele. For the recessive allele, use the same letter but lowercase.

  Examples: **T** for tall allele and **t** for short allele (**T** and **t** are two versions of the same gene that controls plant height and tall is dominant).

- **Single-character inheritance**: all genotypes (parent and offspring) are expressed using two letters (two alleles of the same gene in a diploid organism).

  Examples: **TT** for homozygous dominant tall plant, **Tt** for heterozygous tall plant, or **tt** for homozygous recessive short plant.

  Special note: **monohybrid cross** refers to the cross between two heterozygous individuals (Tt x Tt).

- **Two-character inheritance**: all genotypes (parent and offspring) are expressed using four letters [two letters (alleles) per character (gene)].

  Examples: **TTPP** for homozygous dominant, **TtPp** for heterozygous, or **ttpp** for homozygous recessive

  Special note: **dihybrid cross** refers to the cross between two heterozygous individuals (TtPp × TtPp).

Step 3: **Give** the genotype and phenotype of each parent!

Step 4: **Determine** how many kinds of gametes each parent produces and what allele(s) each gamete is carrying (gamete carries only half of the genetic makeup of the parent).

  Examples for one-character inheritance: A parent with a genotype TT produces only one kind of gametes, **T** (see Figure 15.6a). A parent with a genotype Tt produces two kinds of gametes, **T** and **t** (see Figure 15.6b).

  Examples for two-character inheritance: A parent with a genotype TTPP produces only one kind of gametes, **TP** (see Figure 15.9a). A parent with a genotype TtPp produces four kinds of gametes, **TP**, **Tp**, **tP**, and **tp** (see Figure 15.9b).

Step 5: **Use** the different kinds of gametes produced by each parent to determine the size of a Punnett square. Punnett square is used to show all the possible combinations from the different kinds of gametes.

Examples: If one parent produces two different kinds of gametes and the other parent produces one kind of gametes, then the Punnett square is 2 × 1 or has two cells or combinations (see Figure 15.6b)

Step 6: **Give** phenotypic or genotypic ratio for the offspring. When the number of each combination is divided by the total combinations, it gives the proportion or ratio of that phenotype or genotype.

Examples: There are 3 tall plants and 1 short in the F_2 generation (see Figure 15.6c) and the phenotypic ratio will be 3/4:1/4, which can be simplified to 3:1. The F_2 generation produces 1 TT, two Tt, and 1 tt. The genotypic ratio will be 1/4:2/4:1/4, which can be simplified to 1:2:1.

Step 7: **Check** to make sure you provide the genotypes and phenotypes for all parents and offspring, including genotypic and phenotypic ratios.

Let's practice what you learn and solve the genetics questions at the end of this chapter.

2. When two different alleles are present, the dominant allele masks the recessive allele and is the only one being expressed. This is called **complete dominance** or the **principle of dominance**. For example, Pp has the same phenotype as PP because of the dominant allele, P.

3. The two alleles of the same gene segregate or separate from each other during meiosis to form gametes. Each gamete receives only one allele for each gene. This is called the principle of segregation or **Mendel's law of segregation**. For example, a parent with a genotype Pp will produce two kinds of gametes, P or p.

4. When two or more characters or genes are involved, they will be inherited independently and their alleles will be randomly assorted. This is called the principle of independent assortment or **Mendel's law of independent assortment**. For example, a parent with a genotype PpTt will produce four kinds of gametes because the alleles of P gene and the alleles of T gene will assort independently and produce four possible combinations: PT, Pt, pT, and pt.

Complex Pattern of Inheritance

Mendel's principles of inheritance form the basic foundation of modern genetics. However, we now know not all patterns of inheritance can be explained by Mendel's principles. Mendel chose a relative simple genetic system for his studies and his principles apply only when all the following four criteria are met:

- The character is controlled by one gene.
- The gene has only two alleles.
- One allele is completely dominant over the other, a condition known as **complete dominance**.
- When two or more genes are involved, these genes are not linked but independently inherited.

complete dominance
When one allele completely masks the expression of the other allele.

principle of dominance
When two different alleles are present, the dominant allele masks the recessive allele and is the only one being expressed.

Mendel's law of segregation
When two alleles of the same gene segregate or separate from each other during meiosis to form gametes and each gamete receives only one allele for each gene; also known as principle of segregation.

Mendel's law of independent assortment
When two or more characters or genes are involved, they will be inherited independently.

complete dominance
When one allele completely masks the expression of the other allele.

We will highlight some of the complex patterns of inheritance that do not follow Mendel's principles. The complex patterns of inheritance we briefly cover here are polygenic inheritance, pleiotropy, incomplete dominance, linkage, and environmental influence.

Polygenic Inheritance Many characters of an organism are controlled by more than one gene and exhibit a **polygenic inheritance**. Each gene makes a contribution to the phenotype of the organism. A character controlled by multiple genes or polygenes shows a much wider range of traits or variability. Herman Nilsson-Ehle (1909), a Swedish geneticist, was the first to experimentally demonstrate hull color of wheat grain, *Triticum aestivum*, exhibited a polygenic inheritance. Hull color is controlled by two genes, each with two alleles. Although the dark red hull color is dominant to white, the hull color of the offspring varies from dark red, medium red, light red, to white. The intensity of the red color is determined by the numbers and types of alleles and genes involved.

Pleiotropy Many genes control only one character of an organism, but some genes control multiple characters. **Pleiotropy** (*pleio* means "more" and *trope* means "turnings") refers to the situation when a single gene controls multiple characters of an organism. In garden pea, flower color, seed color, and leaf axil spot are controlled by a single gene with dominant and recessive alleles. These characters are inherited as a single unit.

Incomplete Dominance In some plants such as cyclamens, snapdragons, and four o'clock plants, a cross between two pure-breeding parents produces offspring with an intermediate character. When a cyclamen with red flowers is crossed with one with white flowers, the F_1 cyclamens all have pink flowers instead of red flowers (Figure 15.10). When two F_1 cyclamens are crossed, they produce offspring with three phenotypes in a ratio of 1

polygenic inheritance
Character of an organism controlled by more than one gene and each gene contributes to the phenotype of the organism; character controlled by multiple genes or polygenes show a much wider range of traits or variability.

pleiotropy
When a single gene controls multiple characters of an organism; *pleio* means "more" and *trope* means "turnings."

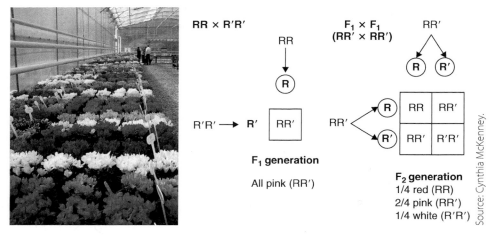

Figure 15.10 Incomplete dominance in cyclamens. When a cyclamen with red flowers (RR) is crossed with one with white flowers (R'R'), all F_1 plants (RR') will have flowers with an intermediate or transitional color, pink. When two F_1 cyclamens are crossed, the F_2 generation will produce three phenotypes: 1 red, 2 pink, and 1 white.

red:2 pink:1 white. Flower color in cyclamens is a character that exhibits **incomplete dominance**. The R allele cannot completely mask the expression of the other allele, which we now call R′ instead of r.

Linkage When two or more genes are inherited, they may or may not be linked. If the genes are found on the same chromosome and very close to each other, they will be linked and inherited together. The pattern of inheritance exhibited by linked genes cannot be explained by Mendel's principles. **Gene linkage** or coupling was first discovered in the early 1900s by William Bateson, an English geneticist. He studied sweet peas, *Lathyrus odoratus*, and showed two characters, petal color and pollen shape, were linked and inherited together. When Bateson crossed a homozygous dominant (purple with long pollen) with a homozygous recessive (red with round pollen), all F_1 offspring have purple flowers and long pollen. When he crossed two F_1 individuals, they produced offspring with four different phenotypes (Bateson, 1912) (see Table 15.2). Most of the F_2 plants (see the top and the bottom rows) were the same as the original parental plants and their numbers were much higher than what Mendel predicted. The other two phenotypes were new combinations and their numbers were much lower than expected. Bateson's results confirmed linked genes were inherited together and had a lower probability of assorting independently.

Environmental Influence Characteristics of an organism are influenced by not only its genetic composition but also the environment in which the organism lives. For example, *Hydrangea macrophylla*, a common ornamental plant, normally produces large clusters of blue flowers but under certain conditions it may also produce pink flowers (Figure 15.11). The flower color of *Hydrangea* is influenced by soil acidity and availability of aluminum ions for uptake (Allen, 1943). *Hydrangea* produces blue flowers when the soil is acidic and has more aluminum ions. Pink flowers are produced when the soil becomes more basic and has less aluminum ions. In some snapdragons, flower color is influenced by temperature and light level. In certain aquatic plants, leaf shape is determined by whether the leaves are submerged or in the air. These are just some of the examples showing how environmental factors influence phenotypic expression of an organism.

incomplete dominance
When one allele cannot completely mask the expression of the other allele.

gene linkage
When genes are very close to each other on the same chromosome, they are linked and will be inherited together.

Table 15.2   Comparison Between Bateson's (1912) Original Data and Mendel's Prediction

F2 Phenotype	Bateson's Observation	Predicted Number Based on Mendel's Dihybrid Ratio
Purple petals and long pollen	177	144
Purple petals and round pollen	15	48
Red petals and long pollen	15	48
Red petals and round pollen	49	16

© Martina Ebel/Shutterstock.com

Figure 15.11   Environmental influence on flower color. Flower color of *Hydrangea* changes with soil pH. When the soil is acidic, *Hydrangea* produces blue flowers. When the soil becomes basic, the flowers turn pink. In soil with neutral pH, the *Hydrangea* produces lavender flowers.

IMPORTANCE AND IMPACT OF PLANT BREEDING

Recall from Chapter 1 that humans began domesticating plants some 13,000 years ago. Plant domestication not only altered the course of human history (see Box 1.1) but also altered the genetic makeup of domesticated plants. As soon as we began sowing seeds of selected plants instead of simply harvesting wild plants, we altered the course of crop plant evolution. Domesticated species became very different morphologically from their wild ancestor and relatives (see Figure Box 1.1).

plant breeding
Branch of agriculture that focuses on manipulating plant genetics to produce more desirable crops.

Plant breeding is a branch of agriculture focusing on manipulating plant genetics to produce more desirable crops. Plant breeders rely on the principles of genetics and use genetic analysis to develop plant lines to meet consumer needs. Some plant breeders specialize on field crops such as soybean, corn, and cotton. Others focus on horticultural crops (such as vegetables), ornamentals (such as orchids, bedding plants, and shade trees), forage crops (such as alfalfa), or turf. The goals of plant breeding are to improve crop yield, crop quality, pest resistance, stress tolerance, and ease of harvesting (Welsh, 1981). In ornamental horticulture, the goal of plant breeding is to develop varieties or cultivars with new colors and forms. With the advances in plant biotechnology, plants can now be used as bioreactors to produce certain pharmaceuticals. Plant breeding is now applicable in fields beyond agriculture such as **biopharming**. The goal of biopharming is to use genetically modified (GM) plants to mass-produce pharmaceuticals efficiently and economically. Plant breeders achieve these goals by manipulating crop genetics, which alters crop characteristics. We highlight three strategies plant breeders use to manipulate crop characteristics.

biopharming
Uses genetically modified plants to mass-produce pharmaceuticals efficiently and economically.

Traditional Breeding and Hybridization

Crop plants are basically divided into two groups: those that self-fertilize and those that cross-fertilize. Crop plants such as wheat, rice, peas, tomatoes, eggplants, and peaches are primarily self-fertilized (Stoskopf et al., 1993). **Pure-line selection** is the oldest method to breed these self-fertilized crops. It involves growing seeds from several plants in individual rows and selecting the most desirable row as the new pure-line variety. Crosses can be made between self-fertilized plants to produce hybrids. Hybrid plants have the desirable features of both parents, are usually more vigorous, and produce a greater yield. The most famous hybrids, high-yield wheat varieties, created by Norman E. Borlaug in the 1950s, revolutionized agriculture in developing countries such as Mexico, India, and Pakistan (Borlaug, 1983). These wheat varieties have shorter and stronger stems and are much better adapted to the environments of the countries for which they were bred. Using these varieties, Mexico greatly increased its wheat production from 750 Kg/ha to about 4,500 Kg/ha and transformed from being an importer of wheat to a large exporter (Hanson, Borlaug, & Anderson, 1982). These varieties boosted India's wheat production as much as seven times. The **Green Revolution** refers to the transformation of crop plants and agricultural practices, which dramatically increased food production in many developing countries. Borlaug is known as the father of the "Green Revolution" and was awarded the Nobel Peace Prize in 1970 for increasing the world's food supply (Figure 15.12).

Corn, rye, alfalfa, most fruits, nuts, and vegetables are primarily cross-fertilized. **Mass selection** is the simplest method to breed these crops. Seeds from many plants from a population are selected and used to produce the next generation. Seeds from the best next-generation plants are selected and grown. Over time, breeders mold and shape the genetic makeup of their crops to fit their needs. Crosses between genetically different plants in cross-fertilized crops also results in vigorous hybrids.

pure-line selection
Most primitive method to breed self-fertilized crops by growing seeds from several plants in individual rows and selecting the most desirable row as the new pure-line variety.

green revolution
Transformations of crop plants and agricultural practices that dramatically increase food production in many developing countries.

mass selection
Simplest method to breed cross-fertilized crops; breeders slowly mold and shape the genetic makeup of their crops to fit their preference.

Sylvan Wittwer/Visuals Unlimited.

Figure 15.12   Norman E. Borlaug (1914–2009) (*center*), the father of the "Green Revolution" and the Nobel Prize Laureate for Peace (1970).

Self-fertilization of these cross-fertilized plants, on the other hand, produce inferior plants due to inbreeding depression. Plant breeders often force the cross-fertilized plants to self-fertilize for several generations in order to eliminate any deleterious genes. Selected inbred lines are crossed to produce hybrids. For example, Shull (1910) crossed inbred lines of corn to produce high-yield hybrids. Most of the corn in the United States is grown from these hybrids.

Somatic Fusion and Tissue Culture

Traditional breeding can only be done by crossing sexually compatible plants. However, protoplast fusion or somatic fusion allows breeders to produce hybrids from two different varieties of the same species (intraspecific), two different species (interspecific), and even two plants with different ploidy levels. For example, somatic fusion has been used to produce higher-quality and/or disease-resistant hybrids. The domestic potato, *Solanum tuberosum*, is susceptible to pathogens such as the potato leaf roll virus. However, crossing the domestic potato with a wild, non-tuber-forming potato, *Solanum brevidens*, produces hybrids that are resistant to the virus (Helgeson, Hunt, Haberlach, & Austin, 1986). Carlson, Smith, and Dearing (1972) developed the most successful commercial hybrid of tobacco by somatically fusing the cells from two tobacco plants: *Nicotiana glauca* and *N. langsdorffi*.

Somatic fusion usually involves four steps: (1) cells from each plant are treated with enzymes to remove their cell wall, (2) the two cell types are induced to fuse together by electric shock or chemical treatment, (3) hybrid cells are induced with hormones to form their own cell wall, and (4) hybrid cells are grown in tissue culture until they are ready to be planted. Tissue culture is a technique for growing cells, tissues, and whole plants on artificial nutrients under sterile conditions. Hybrids resulting from the somatic fusion of two different plants are called **somatic hybrids**. Hybrids produced by traditional breeding, on the other hand, are called **sexual hybrids** because they are formed by combining gametes or sex cells (Bidlack & Jansky, 2011).

Genetic Modification or Engineering

Although humans have been manipulating plant genetics for a long time, plant breeding has taken a quantum leap forward with genetic modification or engineering. Genetic engineering has a number of advantages over traditional breeding. Traditional breeding requires a long period of time to develop new varieties. Months to years are required for plants to reach reproductive maturity, allowing crosses to be made. Hundreds or thousands of crosses have to be made before desirable traits are found. Often a significant period of time is needed before the characteristics of the offspring can be evaluated. Genetic engineering, on the other hand, produces new varieties much faster. New genes can be inserted into cell genomes within a matter of minutes. Transformed cells are grown in tissue culture and identified in a matter of days. Within weeks, the characteristics of genetically modified plants can be evaluated. Plants expressing the new traits can then be grown. Seeds of genetically modified plants can be brought to market much faster than those from traditional breeding. We will discuss biotechnology and genetically modified plants in Chapter 17.

somatic hybrids
Hybrids produced by protoplast fusion are the results of combining somatic or body cells from two different plants together.

sexual hybrids
Hybrids produced by conventional breeding as the result of gamete or sex cell fusion.

Key Terms

hybridization
hybrids
characters
traits
true-breeding line
parental (P) generation
first filial (F_1) generation
phenotype
second filial (F_2) generation
dominant
recessive
monohybrid cross
gene

alleles
genotype
heterozygous
Punnett square
testcross
dihybrid cross
particulate theory of
   inheritance
complete dominance
principle of dominance
Mendel's law of segregation
Mendel's law of independent
   assortment

complete dominance
polygenic inheritance
pleiotropy
incomplete dominance
gene linkage
biopharming
plant breeding
pure-line selection
Green Revolution
mass selection
somatic hybrids
sexual hybrids

Summary

- Genetics has been traditionally divided into three major areas: transmission, molecular, and population.

- Gregor Mendel, the father of genetics, was particularly interested in how traits were inherited from one generation to the next. His hybridization experiments enabled him to formulate basic principles and laws of inheritance, which laid the foundation of genetics.

- One-character inheritance focuses on a single character and how it passes from one generation to the next. Two-character inheritance focuses on two characters and how they pass from one generation to the next independently.

- Monohybrid cross is a special case of a one-character inheritance and always produces a 3:1 phenotypic ratio. Dihybrid cross is a special case of a two-character inheritance and always produces a 9:3:3:1 phenotypic ratio.

- It is essential to know all the genetic terms and their definitions.

- Letters are still used today to denote various genotypes. The rules for writing out genotypes are slightly different between one-character and two-character inheritances.

- A Punnett square is a diagram of all the possible offspring genotypes. Its size and shape are determined by the kind(s) of gametes produced by the two parents.

- A testcross is used to determine the genotype of a plant expressing a dominant phenotype. It is done by crossing this plant with a homozygous recessive plant.

- Genetics problems can be solved using seven logical steps.

- The four principles of inheritance Mendel deduced from his experiments are the particulate theory of inheritance, the principle of dominance, the principle of segregation, and the principle of independent assortment.

- Mendel chose a relatively simple genetic system for his studies and his principles of inheritance only apply when specific criteria are met.

- The complex patterns of inheritance we briefly covered are polygenic inheritance, pleiotropy, incomplete dominance, linkage, and environmental influence. These patterns of inheritance cannot be explained by Mendel's principles.

- Plant breeding is a branch of agriculture focusing on manipulating plant genetics to produce more desirable crops. The three breeding strategies are traditional breeding and hybridization, somatic fusion and tissue culture, and genetic engineering.

Reflect

1. *Mendel's principles have limits and can only explain certain patterns of inheritance.* Discuss the limitations of Mendel's principles. Give examples of these non-Mendelian inheritance patterns and explain why they do not follow Mendel's principles.

2. *Meeting consumer demands.* What is plant breeding? Discuss the pros and cons of the three strategies used by plant breeders.

3. *Mutants in the produce aisle!* Explain why crops such as bananas, seedless watermelons, and potatoes are mutants. Why are their genomes so unusual? Explain why these mutant plants are desirable.

Genetics Problems

1. *Making gametes.* Complete the table by listing all possible gametes produced by each parental or diploid cell.

2. In peppers, round fruit (R) is dominant to square fruit. A homozygous dominant plant is crossed with a heterozygous plant. Give the genotype and phenotype of each parent. Use a Punnett square to show all possible genotypes of the resulting offspring. Give the offspring phenotypes and their phenotypic ratio.

3. In peas, purple flower is dominant to white flower. Two heterozygous pea plants are crossed to produce 480 offspring. How many of these offspring will have purple flowers? How many of them will have a genotype Pp? How many of them will have white flowers? How many of them will have a genotype of PP with purple flowers?

4. In corn plants, yellow kernels are dominant to white kernels and smooth kernels are dominant to wrinkled kernels. Two corn plants that are heterozygous for both traits are crossed. Give the genotypes and phenotypes of these two parents. Use a Punnett square to show the genotypes of resulting offspring. Give the phenotype(s) of their offspring and the phenotype ratio.

Parental Phenotype	Parental Genotype (Diploid)	Possible Gametes (Haploid)
Purple flower	PP	all P
Purple flower	Pp	P, p
White flower	pp	
Tall plant with purple flower	TTPP	
Tall plant with purple flower	TTPp	
Tall plant with purple flower	TtPp	
Tall plant with white flower	Ttpp	
Dwarf plant with white flower	ttpp	
Long stem with wrinkled seed	LLrr	
Long stem with round seed	LlRr	
Short stem with round seed	llRR	
Long stem, round seed, and purple flower	LLRRPp	
—	AABbCC	
—	AABBCCDd	
—	AABBCCDDEEFFGG	

5. Half of the thorn apple trees in the area produce white flowers and the other half produce purple flowers. It is known that purple flowers are dominant to white flowers. What parental genotypes are required to produce this 1:1 phenotypic ratio.

6. In sesames, one-chamber seed pod is dominant to three-chamber seed pod and normal leaves are dominant to wrinkled leaves. Determine the genotypes and phenotypes of the two parents that produced the following offspring: 301 with one-chamber seed pods and normal leaves, 98 one-chamber seed pods and wrinkled leaves, 297 three-chamber seed pods and normal leaves, and 102 three-chamber seed pods and wrinkled leaves.

References

Allen, R. C. (1943). Influence of aluminum on the flower color of *Hydrangea macrophylla* DC. *Boyce Thompson Institute, 13,* 221–242.

Bateson, W. (1912). Facts limiting the theory of heredity. *Proceedings of the Seventh International Zoological Congress,* pp. 306–319.

Bidlack, J. E., & Jansky, S. H. (2011). *Stern's introductory plant biology* (12th ed.). New York: McGraw-Hill.

Borlaug, N. E. (1983). Contributions of conventional plant breeding to food production. *Science, 219,* 689–693.

Brooker, R. J. (2012). *Genetics: Analysis and principles* (4th ed.). New York: McGraw-Hill.

Carlson, P. S., Smith, H. H., & Dearing, R. D. (1972). Parasexual interspecific plant hybridization. *Proceedings of the National Academy of Sciences of the USA, 69,* 2292–2294.

Davis, K. M. (2009) Modifying anthocyanin production in flowers. In K. Gould, K. Davies, & C. Winefield (Eds.), *Anthocyanins biosynthesis, functions and applications* (pp. 49–84). New York: Springer.

Hanson, H., Borlaug, N. E., & Anderson, R. G. (1982). *Wheat in the third world.* Boulder, CO: Westview Press.

Helgeson, J. P., Hunt, G. J., Haberlach, G. T., & Austin S. (1986). Somatic hybrids between *Solanum brevidens* and *Solanum tuberosum*: Expression of a late blight resistance gene and potato leaf roll resistance. *Plant Cell Reports, 5,* 212–214.

Mendel, G. (1866) Versuche Über Plflanzenhybriden [Experiments in plant hybridization]. *Verhandlungen des naturforschenden Vereines in Brünn, Bd. IV für das Jahr 1865, Abhandlungen,* 3–47. Retrieved June 22, 2012, from www.esp.org/timeline/.

Nilsson-Ehle, H. (1909). Kreuzungsuntersuchungen an Hafer und Weizen. *Academic Dissertation, Lund.*

Shull, G. H. (1910). Hybridization methods in corn breeding. *Journal of Heredity, 1,* 98–107.

Stoskopf, N. C., Tomes, D. T., & Christie, B. R. (1993). *Plant breeding: theory and practice.* Boulder, CO: Westview Press.

Welsh, J. R. (1981). *Fundamentals of plant genetics and breeding.* New York: Wiley.

Chapter 16

Molecular Basis of Inheritance

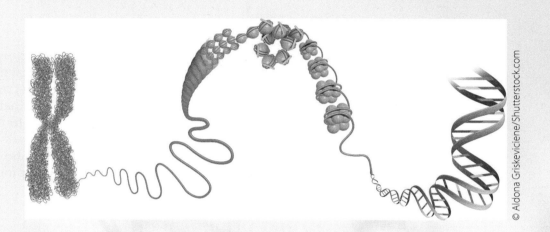

© Aldona Griskeviciene/Shutterstock.com

Learning Objectives

- Distinguish the different levels of DNA structure and their unique characteristics
- Outline the process of DNA replication and explain the term *semiconservative*
- Compare and contrast DNA and RNA molecules and describe their functions
- Explain gene expression and summarize its two main processes
- Distinguish between polyploidy and aneuploidy and discuss its role in crop improvement

As you might recall from Chapter 1, all living organisms reproduce. Reproduction is the most distinctive characteristic of living organisms and enables the genetic blueprint, DNA, to pass from one generation to the next. We already examined the mechanisms of plant reproduction in Chapters 7 and 9. In Chapter 15, you learned how characteristics are passed from parents to offspring. All characteristics of an organism, whether they are flower color, fruit shape, and seed color, are encoded in the DNA genetic blueprint of the organism. In this chapter, we take a closer look at the genetic blueprint, its unique structure and organization, how the genetic blueprint is replicated, and finally how information on the blueprint is used to make functional products.

MOLECULAR BASIS OF INHERITANCE

Molecular genetics studies the biochemical and molecular structure of the genetic blueprint or hereditary material. Hereditary material must meet the following three requirements. First, it has to encode information governing the organization, development, characteristics, and functions of that organism. Second, it has to be duplicated so copies can be passed onto the next generation. Third, it has to account for the differences among individuals. We take a closer look at the structure and organization of DNA, how information is encoded in DNA, how DNA replicates, and how the information on DNA is expressed.

DNA Structure and Organization

The hereditary material of all living organisms is made up of DNA (deoxyribonucleic acid) molecules and can be studied at five different levels of organization (Figure 16.1):

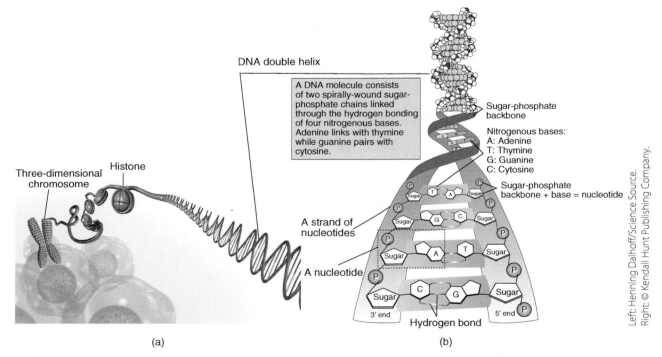

DNA double helix

A DNA molecule consists of two spirally-wound sugar-phosphate chains linked through the hydrogen bonding of four nitrogenous bases. Adenine links with thymine while guanine pairs with cytosine.

Sugar-phosphate backbone

Nitrogenous bases:
A: Adenine
T: Thymine
G: Guanine
C: Cytosine

Sugar-phosphate backbone + base = nucleotide

Three-dimensional chromosome

Histone

A strand of nucleotides

A nucleotide

3' end

5' end

Hydrogen bond

(a)

(b)

Left: Henning Dalhoff/Science Source.
Right: © Kendall Hunt Publishing Company.

Figure 16.1   Chromosome structure and organization.

1. **Nucleotides** are the building blocks of nucleic acids such as DNA and RNA (recall from Chapter 3). Each nucleotide has a phosphate on one side, a five-carbon sugar in the middle, and a base on the other side (Figure 16.1). The four types of nucleotides or bases making up a DNA molecule are adenine (A), guanine (G), cytosine (C), and thymine (T).

2. Nucleotides are linked together to form a long chain or **strand** of DNA or RNA. For example, a strand of DNA will look like –ATTCCTTAC–. The sequence of nucleotides determines the genetic information it encodes. For example, the genetic information encoded in the sequence –TGCTTATTT– will be different from the one in the sequence –GCTTCCATA–.

3. Two strands of DNA come together (= double) and coil into a helical shape (= helix), hence the name "double helix." The two strands are **complementary** to each other and are **antiparallel**, which means they run in opposite directions. For example, if you know the sequence of one DNA strand being 5′–ATTCCTTAC–3′, the complementary strand will have to be 3′–TAAGGAATG–5′. The first strand runs in a 5′ to 3′ direction and the second or complementary stand runs in the opposite direction from 3′ to 5′. The 5′ end of a DNA strand has a phosphate group (Figure 16.1) whereas the 3′ end does not. When you look at the orientation of the sugar molecules within each strand, you will notice the sugar molecules run in opposite directions as well. Three key advances in molecular biology and biochemistry contribute to our understanding of the DNA structure and configuration.

   • James Watson and Francis Crick deduced the basic structure of DNA and came up with the double helix model in 1953 (Figure 16.2a). Watson and Crick relied heavily on previous work done by Erwin Chargaff, Rosalind Franklin, and Maurice Wilkins.

nucleotides
Basic units of building blocks that make up nucleic acids.

strand
Long chain of nucleotides that make up either DNA or RNA.

double helix
Two strands of DNA come together (= double) and coil into a helical shape (= helix).

complimentary
Two strands of DNA are complimentary; if you know the sequence of one strand, you will know the sequence of the complimentary strand. See *AT/GC rule.*

antiparallel
Two strands of DNA making up a DNA molecule or double helix run in opposite directions.

(a)   (b)

a: Barrington Brown/Science Source.
b: SPL/Science Source.

Figure 16.2a–b   (a) James D. Watson (*left*) and Francis H. C. Crick (*right*) with their DNA model, which they constructed. They deduced the structure and organization of DNA based on previous work done by (b) Erwin Chargaff, Rosalind Franklin and Maurice Wilkins (see Figure 16.3a–b).

AT/GC rule
Adenine of one strand of DNA always pairs up with thymine of the other strand; guanine always pairs up with cytosine.

histones
DNA-binding proteins that enable a DNA molecule to twist and fold around them.

chromosome
Condensed and visible form of DNA/histone complex.

genome
All chromosomes of an organism.

- During the late 1940s and early 1950s, Erwin Chargaff determined the amount of adenine is always equal to the amount of thymine and the amount of guanine is always equal to the amount of cytosine. His finding led to the concept of complementary base pairing or the **AT/GC rule**. For example, adenine of one strand of DNA always pairs up with thymine of the complementary strand; guanine always pairs up with cytosine.

- During the early 1950s, Rosalind Franklin and Maurice Wilkins used x-ray diffraction to study DNA molecules and were the first to show the helical shape of DNA molecules (Figure 16.3c).

4. A double helix folds and twists around numerous DNA-binding proteins such as **histones** to form a three-dimensional structure called a **chromosome** (Figure 16.1a). Each chromosome contains hundreds or thousands of genes, which are segments of DNA encoded for specific functional products. The number of chromosomes varies greatly among living organisms. For example, humans have 46 chromosomes in each of their cells. The adderstongue fern, *Ophioglossum reticulatum*, has 1,260 chromosomes in their cells. The jack jumper ant, *Myrmecia pilosula*, has only two chromosomes in their cells.

5. The term **genome** refers to all chromosomes of an organism. Recall from Chapter 3 that plant cells have three different sets of DNA: one in the nucleus, one in the mitochondria, and one in the chloroplasts. As a result, plants have three distinct genomes: nuclear, mitochondrial, and chloroplast.

DNA Replication

Another key feature of hereditary material is that it must be duplicated so copies can be passed onto the next generation. This duplication process or DNA replication must be precise and controlled by the cell. Recall from Chapter 3 that DNA replication takes place during the S phase of

(a)

National Library of Medicine/Science Source.

(b)

SPL/Science Source.

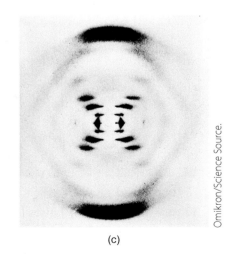

(c)

Omikron/Science Source.

Figure 16.3a–c (a) Rosalind Franklin and (b) Maurice Wilkins subjected DNA molecules to x-rays and (c) were the first to show the helical shape of DNA molecules.

the cell cycle before the cell undergoes either mitosis or meiosis. On each double helix, DNA replication begins at specific locations called *replication origins* (Figure 16.4). **Helicase**, an enzyme, recognizes a replication origin and begins to unzip or open up the two strands of DNA. The two original strands are used as templates to build new DNA strands.

An enzyme complex called **DNA polymerase** uses each template to match the nucleotides and produces its own complementary strand. Each double helix has one original strand and a new strand. This process of keeping one original strand in each newly synthesized double helix is called **semiconservative replication**. When DNA polymerase reaches a termination signal, it detaches from the template and stops the replication process. The result of DNA replication is two identical double helix molecules.

Gene Expression

The hereditary material has to not only encode genetic information but also express it. Gene expression is the process that uses the genetic information encoded in the genes to synthesize functional proteins. **Proteins** are large and complex molecules made up of smaller units called **amino acids. Polypeptide** refers to a long chain of amino acids. Some proteins are made up of a single polypeptide but others are made up of two or more polypeptides. These proteins not only govern the organization, development, characteristics, and functions of an organism but also produce its phenotypic makeup. Gene expression occurs throughout the life of the cell and is divided into two main processes: transcription and translation.

helicase
Enzyme that recognizes the replication origin and begins to unzip or open up the two strands of DNA at the beginning of DNA replication.

DNA polymerase
Enzyme uses each template to match the nucleotides and produce its own complimentary strand during DNA replication.

semiconservative replication
Process of using one original strand to synthesize a new strand and incorporating it as part of the newly synthesized double helix.

protein
One of four organic molecules that are essential to life.

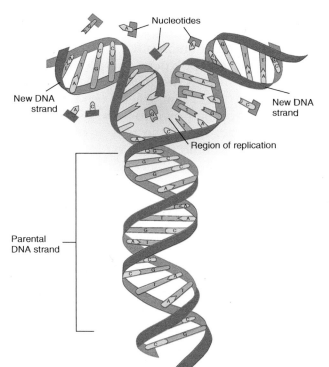

1. Initiating replication:
 • Enzymes such as helicase, DNA polymerases are needed
 • Replication begins at the replication origin
 • Semiconservative replication

2. Synthesizing DNA:
 • DNA polymerases match and join the nucleotides together

3. Terminating replication:
 • Termination signal is used to end replication
 • One double helix molecule becomes two double helix molecules

© Kendall Hunt Publishing Company.

Figure 16.4   DNA replication.

amino acids
Small units that make up a polypeptide or protein.

polypeptide
Long chain of amino acids.

transcription
First process of gene expression, which is all about synthesizing RNA from DNA.

messenger RNA (mRNA)
Delivers the genetic information or protein recipe to the ribosome, the protein factory of the cell.

transfer RNA (tRNA)
Transfers specific amino acids to the ribosome during protein synthesis.

ribosomal RNA (rRNA)
Ribonucleic acids that make up ribosomes.

promoter region
Segment of DNA in front of or upstream to a gene that signals RNA polymerase to begin transcription.

RNA polymerase
Enzyme complex that opens up the DNA double helix and begins transcribing or copying the genetic information of a gene to a RNA transcript.

codons
Sequences of three nucleotides found on mRNA.

genetic code
Contains all the codons found in mRNA and their corresponding amino acids.

anticodon
Found on tRNA and complimentary to the codon. It helps each tRNA in matching the right amino acid to the right codon.

Types of RNA **Transcription** is the first process of gene expression and is all about synthesizing RNA from DNA. Many types of RNA molecules are made in the nucleus of plant cells. In this chapter, we focus only on three types: messenger RNA, transfer RNA, and ribosomal RNA. **Messenger RNA (mRNA)** delivers the genetic information or protein recipe to the ribosome, the protein factory of the cell (see Chapter 3). **Transfer RNA (tRNA)** transfers specific amino acids to the ribosome during protein synthesis. **Ribosomal RNA (rRNA)** is part of the ribosome (see Chapter 3). Although DNA and RNA are both nucleic acids, they are distinct from each other (see Table 16.1).

Transcription Transcription begins at the **promoter region**, which is a segment of DNA in front of or upstream to a gene that signals **RNA polymerase** to begin transcription. RNA polymerase attaches itself to the promoter region, opens up the DNA double helix, and begins transcribing or copying the genetic information of that gene (Figure 16.5). RNA polymerase makes a complementary RNA strand or transcript from the DNA template. RNA polymerase matches guanine with cytosine and pairs adenine with uracil but not thymine. For example, if the sequence of a gene is 3'–AATCGATAC–5', its complementary RNA strand will be 5'–UUAGCUAUG–3'. The RNA strand is always made in one direction, from 5' to 3'. Transcription stops when RNA polymerase reaches the termination signal at the end of a gene and detaches itself from the DNA. Unlike DNA replication, transcription copies a gene, a small segment of DNA, and not the entire DNA double helix. Transcription produces a single RNA strand instead of a double helix. The newly synthesized mRNA or RNA transcript leaves the nucleus and attaches itself to a ribosome in the cytosol to begin protein synthesis or translation.

Decoding the Genetic Information for Translation Translation is the synthesis of a polypeptide specified by mRNA. This process involves tRNA and ribosomes to decode the genetic information on mRNA and produce a specific polypeptide. As we discussed earlier, a polypeptide is a long chain of amino acids. The information carried by mRNA specifies the sequence and type of amino acids to put together. There are 22 amino acids used by all living organisms and these amino acids have their own **codons**, sequences of three nucleotides (Figure 16.6). These codons form the **genetic code** of life. For example, the codon UUA specifies an amino acid called *leucine*, or *Leu* for short. The codon GUU specifies an amino acid called *valine*, or *Val*. The codons are found on the mRNA and must be interpreted or translated by tRNA in order to transfer the right amino acid to the ribosome. Each tRNA contains a sequence of three nucleotides called **anticodon**. The anticodon is complementary to the codon and helps each tRNA match the right amino

Table 16.1 Comparison Between DNA and RNA

Characteristics	Deoxyribonucleic acid	Ribonucleic acid
Nucleotides	A,G,C,T	A,G,C,U
Shape	Double helix	Single strand
Type(s)	One	Many Examples: mRNA, tRNA, rRNA

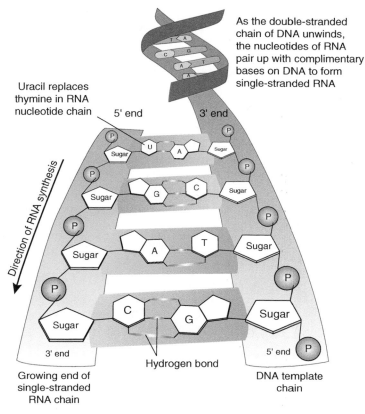

As the double-stranded chain of DNA unwinds, the nucleotides of RNA pair up with complimentary bases on DNA to form single-stranded RNA

Uracil replaces thymine in RNA nucleotide chain

5' end

3' end

Direction of RNA synthesis

Growing end of single-stranded RNA chain

3' end

Hydrogen bond

5' end

DNA template chain

Key Features of Transcription:
1. Only the genetic information of a gene is copied not the whole DNA.

2. Transcription begins when RNA polymerase attaches to the promoter region at the beginning of a gene.

3. RNA transcript is synthesized from 5′ to 3′ direction as RNA polymerase move down the DNA template.

4. Transcription stops when RNA polymerase detaches itself from the DNA template at the termination signal, at the end of a gene.

© Kendall Hunt Publishing Company.

Figure 16.5   Transcription: RNA synthesis.

Second Base

First Base		U	C	A	G		Third Base
U		UUU phenylalanine	UCU serine	UAU tyrosine	UGU cysteine	**U**	
		UUC phenylalanine	UCC serine	UAC tyrosine	UGC cysteine	**C**	
		UUA leucine	UCA serine	UAA stop	UGA stop	**A**	
		UUG leucine	UCG serine	UAG stop	UGG tryptophan	**G**	
C		CUU leucine	CCU proline	CAU histidine	CGU arginine	**U**	
		CUC leucine	CCC proline	CAC histidine	CGC arginine	**C**	
		CUA leucine	CCA proline	CAA glutamine	CGA arginine	**A**	
		CUG leucine	CCG proline	CAG glutamine	CGG arginine	**G**	
A		AUU isoleucine	ACU threonine	AAU asparagine	AGU serine	**U**	
		AUC isoleucine	ACC threonine	AAC asparagine	AGC serine	**C**	
		AUA isoleucine	ACA threonine	AAA lysine	AGA arginine	**A**	
		AUG(start) methionine	ACG threonine	AAG lysine	AGG arginine	**G**	
G		GUU valine	GCU alanine	GAU aspartate	GGU glycine	**U**	
		GUC valine	GCC alanine	GAC aspartate	GGC glycine	**C**	
		GUA valine	GCA alanine	GAA glutamate	GGA glycine	**A**	
		GUG valine	GCG alanine	GAG glutamate	GGG glycine	**G**	

Information carried by a gene on the DNA:

3′ – TTTACAACTCAACT – 5′   Template strand
5′ – AAATGTTGAGTTGA – 3′   Coding strand

Transcription

mRNA made from the template DNA strand:

5′ – AAAUGUUGAGUUGA – 3′

Translation

A polypeptide specified by mRNA:

Met – Leu – Ser

© Kendall Hunt Publishing Company.

Figure 16.6   The genetic code and gene expression.

acid to the right codon. For example, the codon GCU specifies an amino acid called *alanine*, or *Ala*. The tRNA that carries Ala has an anticodon CGA to match the codon GCU.

Translation Translation begins at the 5′ end of the mRNA and moves toward the 3′ end. The first tRNA will search for the **start codon (AUG)** and attach itself to the mRNA and the ribosome to form the initiation complex. This first tRNA carries a specific amino acid called *methionine*, or *Met*. Another tRNA will come along, match up with the next codon, and transfer another amino acid to the ribosome (Figure 16.7). The matching process continues until the mRNA–tRNA–ribosome complex reaches one of the **stop codons, UAA, UAG,** or **UGA**. The complex falls apart and translation is complete. The newly synthesized polypeptide is ready for transport or further processing.

start codon (AUG)
Found on the mRNA; signals the start of translation.

stop codon (UAA, UAG, or UGA)
Found on the mRNA; signals the stop of translation.

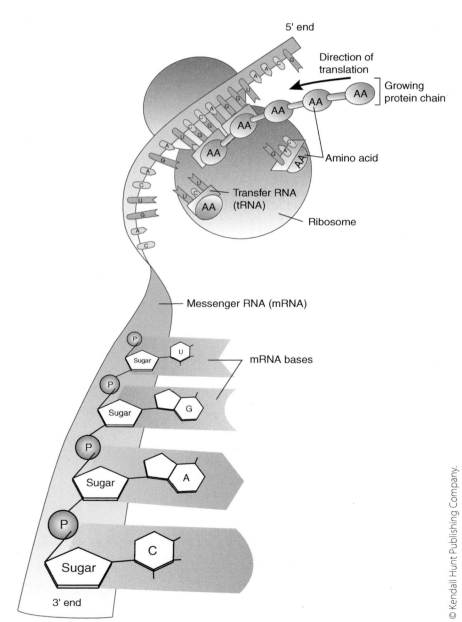

Figure 16.7 Translation: protein synthesis.

Mutations

The genetic information of life is encoded in DNA sequences. What would happen if the DNA sequence is changed accidentally? What if a piece of a chromosome is broken off and attached itself to a different chromosome? What if there are more or less chromosomes than normal? Changes in the genetic information result in **mutations**. Some mutations are harmless as long as they do not change the structure and function of the protein products. However, most mutations are harmful because they either halt the production of certain proteins or alter the structure and function of those proteins. If mutations occur during gamete production, they can be passed to the next generation. Many believe mutations may provide genetic variation in a population that allows changes or evolution to take place over long spans of time.

Changes in Chromosome Number   Many organisms such as plants and humans are diploid (2n). If a diploid organism has an extra set of chromosomes, they become **triploid** (3n). If it has two extra sets of chromosomes, they become **tetraploid** (4n). Organisms with three or more sets of chromosomes are also called **polyploid**. When an organism has one or two extra chromosomes or one or two fewer chromosomes, they are called **aneuploid**. Animals including humans are very sensitive to changes in chromosome numbers. For example, individuals with Down syndrome have 47 chromosomes, instead of the normal 46. Individuals with Turner syndrome have 45 chromosomes, one less than normal. Polyploid animals normally cannot even complete their development and are extremely rare in nature. Unlike animals, polyploid plants occur naturally and are very common. Polyploidy may play an important role in plant evolution. About 47–70% of flowering plants and 95% of ferns originate from polyploid ancestors (see review by Ranney, 2006).

Polyploidy and New Plant Development   Polyploidy also plays an important role in crop improvement. Polyploid crop plants are ideal because they are often larger in size, more vigorous, and yield more than their diploid counterparts (Figure 16.8). About 40% of crop plants such as potato, cotton,

mutation
Random change of genetic composition of an organism.

triploid
(3n) Organisms with three sets of chromosomes.

tetraploid
(4n) Organisms with four sets of chromosomes.

polyploid
Organisms with three or more sets of chromosomes.

aneuploid
Organisms with one or two extra chromosomes or one or two fewer chromosomes.

Peggy Greb/U.S. Department of Agriculture/Science Source.

Figure 16.8   Polyploid potato plants are larger in size than their diploid relatives.

wheat, oat, tobacco, banana, sugarcane, and strawberries are polyploid (Stoskopf, Tomes, & Christie, 1993). For example, the bread wheat, *T. aestivum*, is a hexaploid (6n) developed by fusing three diploid relatives. The upland cotton *Gossypium hirsutum* is a tetraploid (4n) developed by fusing Asiatic cottons with American cottons. Different species of strawberries have different chromosome numbers ranging from diploid, tetraploid, to even octaploid (8n).

Plants with an even number of chromosome sets (4n, 6n, 8n) may be fertile or sterile. Polyploid plants with an odd number of chromosome sets (3n, 5n, 7n) are usually sterile (Stoskopf et al., 1993). An odd number of chromosome sets cannot be divided evenly during gamete production, which results in the production of aneuploid or unviable gametes. Although sterility is generally considered to be detrimental, it can be desirable in agriculture because it produces seedless fruits. For example, domestic bananas and some varieties of watermelons are seedless because they are triploid. Triploid marigolds developed by Burpee can divert more energy into flower production instead of seed production because they are sterile.

Key Terms

nucleotides	semiconservative replication	genetic code
strand	proteins	anticodon
double helix	amino acids	start codon (AUG)
complementary	polypeptide	stop codon
antiparallel	transcription	mutations
AT/GC rule	messenger RNA (mRNA)	triploid
histones	transfer RNA (tRNA)	tetraploid
chromosomes	ribosomoal RNA (rRNA)	polyploidy
genome	promoter region	aneuploid
helicase	RNA polymerase	
DNA polymerase	codons	

Summary

- The hereditary material of all living organisms is made up of DNA (deoxyribonucleic acid) molecules and can be studied at five different levels of organization.

- Hereditary material must be duplicated so copies can be passed onto the next generation. This duplication process is called *DNA replication*.

- Hereditary material must encode genetic information and also express it. Gene expression is the process of decoding genetic information into functional proteins.

- Gene expression occurs throughout the life of the cell and is divided into two main processes: transcription and translation.

- Transcription is all about synthesizing RNA from DNA. It is important to know the differences between DNA and RNA and how cells synthesize RNA.

- Translation is the synthesis of a polypeptide specified by mRNA. This process also involves tRNA and ribosomes. It is important to know the genetic code and how to decode mRNA.

- Although polyploidy is a type of mutation, it plays an important role in plant evolution and crop improvement.

Reflect

1. *Are chromosomes and DNA the same thing?* Distinguish between chromosomes and DNA. Explain how terms such as *chromosomes, genes, DNA, alleles, genomes,* and *nucleotides* are related.

2. *DNA replication and gene expression.* Compare and contrast DNA replication and gene expression.

3. *Can you decipher this genetic information?* Using what you learned from this chapter, decode the information on the DNA below. Use the top DNA strand as the template to produce a polypeptide.

3′ G G T C T A C C C T C A T A C T G T A T C A G 5′

5′ C C A G A T G G G A G T A T G A C A U A G T C 3′

4. *Mutants in the produce aisle!* Explain why crops such as bananas, seedless watermelons, and potatoes are mutants. Why are their genomes so unusual? Explain why these mutant plants are desirable.

References

Ranney, T. G. (2006). Polyploidy: from evolution to new plant development. *Combined Proceedings of the International Plant Propagator's Society, 56,* 137–142.

Stoskopf, N. C., Tomes, D. T., & Christie, B. R. (1993). *Plant breeding: theory and practice.* Boulder, CO: Westview Press.

Chapter 17

Biotechnology and Genetically Modified Plants

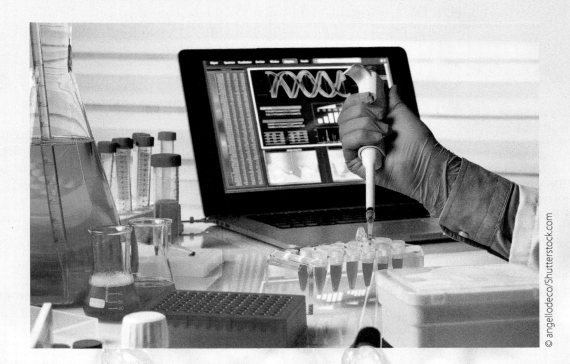

© angellodeco/Shutterstock.com

Learning Objectives

- Discuss the role of restriction enzymes and DNA ligase in creating recombinant DNA.
- Outline the steps of recombinant DNA technology.
- Compare and contrast plasmids and viruses as vectors.
- Explain the role of reporter genes in recombinant DNA technology.
- Evaluate the various methods used to deliver foreign DNA molecules into cells.
- Identify the two main methods used to genetically modify plants.
- Give examples of GM crops and foods and explain the purpose of these GM varieties.
- Evaluate the benefits and risks of GM crops and foods.

Are you eating genetically modified food? What are genetically modified organisms (GMO)? How do we genetically modify organisms? What are the reasons for genetically modifying plants? To answer these questions, we need to understand genetics, DNA structure, and techniques for genetic modification. In Chapter 15, we discussed how characteristics are passed from parents to offspring. In Chapter 16, we explored DNA structure, discussed how information is encoded on DNA molecules and eventually used to make functioning products such as proteins. In this chapter, we will examine restriction enzymes and DNA ligases, the two essential tools of recombinant DNA technology. We will also explore ways to introduce foreign DNA into cells and screen for the transformed cells. Finally, we will discuss practical applications and concerns regarding recombinant DNA technology in plant breeding.

COMBINING DNA FROM DIFFERENT SOURCES WITH RECOMBINANT DNA TECHNOLOGY

World Health Organization (2014) defines **genetically modified organisms (GMO)** as organisms (i.e., plants, animals, or microorganisms) in which the genetic material (DNA) has been altered in a way that does not occur naturally by mating and natural recombination. "GM foods" refer to foods produced from genetically modified plants or animals. Genetic modification, in particular, recombinant DNA Technology, is a biotechnology that alters the genetic composition of living organisms.

Recombinant DNA Technology

The discoveries of **restriction enzymes** and **DNA ligases** more than half a century ago revolutionized the study of DNA and accelerated our ability to both sequence and modify genomes. These enzymes become the most powerful tools for combining DNA from different sources to generate recombinant DNA molecules. **Recombinant DNA (rDNA) molecules** are DNA molecules created artificially by combining at least two or more sources of DNA. The source of these DNA molecules can be from the same organism, different members of the same species, or entirely different organisms. Recall in Chapter 14, the process of genetic recombination occurs naturally during meiosis when the homologous pairs cross over and exchange bits of their DNA materials. The process of rDNA is artificial genetic recombination.

Are rDNA molecules functional when inserted into a living cell? In 1973, Stanley Cohen, Herbert Boyer, and their laboratories gave a resounding yes to this question experimentally (Cohen, Chang, Boyer, & Helling, 1973). They isolated two different plasmids from *Escherichia coli* cells, each containing a different antibiotic resistance gene. They cut the plasmids with a restriction enzyme, *Eco*RI, and used a DNA ligase to rejoin them. The recombinant plasmids were then inserted into new *E. coli* cells, and these transformed cells were able to grow in the presence of both antibiotics. Their experiment was a watershed event in molecular biology, giving birth to recombinant DNA technology.

Recombinant DNA technology is widely used in research to identify, map, clone, and sequence genes. To determine gene function, rDNA probes are used to analyze gene expression. Recombinant proteins are extensively

---

genetically modified organisms (GMO)
Organisms (i.e., plants, animals, or microorganisms) in which the genetic material (DNA) has been altered in a way that does not occur naturally by mating and natural recombination.

GM foods
Refer to foods produced from genetically modified plants or animals.

restriction enzymes
Enzymes used by bacteriophage-resistant bacteria as self-defense mechanism to restrict bacteriophage's growth in the cell. These enzymes are used to cut DNA into fragments.

DNA ligases
Enzymes used to join DNA fragments together by catalyzing formation of a phosphodiester bond between 5'-phosphoryl (−PO4) and 3'-hydroxyl (−OH) ends of the two sugar–phosphate backbones.

recombinant DNA (rDNA) molecules
DNA molecules that are created artificially by combining at least two or more sources of DNA.

used as reagents in experiments and also for generating antibody probes to examine protein synthesis within cells and organisms (Bruce et al., 2015). There are many additional practical applications of recombinant DNA in industry, food production, human and veterinary medicine, agriculture, and bioengineering. Let's first take a closer look at restriction enzymes and DNA ligases, the two essential tools for recombinant DNA technology.

Restriction Enzymes Cut DNA into Fragments

Restriction enzymes or **restriction endonucleases** were first noticed in early 1950s by Salvador Luria, Giuseppe Bertani, Jean Weigle, and their colleagues (Chial, 2014). These scientists showed that some strains of bacteria were more resistant to viral infections than others. A group of enzymes called restriction enzymes were used by these bacteriophage-resistant bacteria as a self-defense mechanism to restrict bacteriophage growth in the cell. However, it took more than 15 years before the restriction mechanism was fully understood by Werner Arber, Dan Nathans, Hamilton Smith, and their laboratories (Chial, 2014). The use of a restriction enzyme to cut a DNA molecule into smaller fragments is called a **restriction enzyme digest**. Restriction enzymes quickly became powerful tools for generating physical maps of genomes in the early days of genome sequencing. For this groundbreaking set of discoveries, Arber, Nathans, and Smith were jointly awarded the Nobel Prize in Physiology or Medicine in 1978.

Since the initial discovery of restriction enzymes, we have now identified more than 4,000 restriction enzymes and isolated more than 360 different recognition sequences (Roberts, Vincze, Posfai, & Macelis, 2014). Over 600 restriction enzymes are available commercially. New restriction enzymes are continuously being discovered. Current information on restriction enzymes and their commercial availability can be found on REBASE website (http://rebase.neb.com), a database maintained by Richard J. Roberts and Dana Macelis (REBASE, 2017). Restriction enzymes are found primarily in bacterial genomes and plasmids. However, some restriction enzymes are found in archaea, viruses, and even in eukaryotes. The recognition sequences of these enzymes are typically four to eight base pairs long, and usually **palindromic**, which means that their recognition sequence reads the same in the 5' to 3' direction on both DNA strands (Figure 17.1). Restriction enzymes are classified into four major types (I, II, III, and IV) based on their structure, recognition sequences, cleavage sites, cofactors, and activators (see Williams, 2003 for a detailed review). Type I restriction enzymes cut DNA at random locations far from their recognition sequence. Type II restriction enzymes cut close to or within their recognition sequence

restriction endonucleases
see restriction enzymes

restriction enzyme digest
Refers to the process using restriction enzymes to cut a DNA molecule into smaller fragments.

Palindromic
Refers to a sequence, which reads the same backward as forward, such as GATTAG.

Recognition sequence (inside the red box):

5' AACCTTACTAC GAATTC GGGATTA 3'
3' TTGGAATGATG CTTAAG CCCTAAT 5'

Cleavage pattern:

5' AACCTTACTACG AATTC GGGATTA 3'
3' TTGGAATGATGCTTAA GCCCTAAT 5'

Source: Amanda Chau

Figure 17.1   Recognition sequence and cleavage pattern for the restriction enzyme, *EcoRI*.

whereas Type III cut outside of their recognition sequence. Type IV restriction enzymes typically recognize a modified recognition sequence.

Type II restriction enzymes are the most useful for laboratory experiments because they cut DNA close to or within their recognition sequence (Table 17.1). Some of these enzymes such as *ALU*I and *Eco*RV make a straight-cut to DNA strands, producing blunt ends (Table 17.1). Others such as *Eco*RI, *Bam*HI, and *Hind*III make a stagger-cut that produce fragments with overhangs or sticky ends (Figure 17.2). Sticky ends of different fragments that are complimentary can form hydrogen bonds between the base pairs.

Table 17.1 Examples of Restriction Enzymes or Endonucleases and Their Recognition Sequence

Enzyme Name	Microbial Source	Recognition Sequence[a]
*ALU*I	*ARTHROBACTER LUTEUS*	↓ 5′—A—G—C—T—3′ 3′—T—C—G—A—5′ 　　　　↑
*BAM*HI	*BACILLUS AMYLOLIQUEFACIENS* H	↓ 5′—G—G—A—T—C—C—3′ 3′—C—C—T—A—G—G—5′ 　　　　　　　↑
*Eco*RI	*ESCHERICHIA COLI*	↓ 5′—G—A—A—T—T—C—3′ 3′—C—T—T—A—A—G—5′ 　　　　　　　↑
*Eco*RV	*ESCHERICHIA COLI*	↓ 5′—G—A—T—A—T—C—3′ 3′—C—T—A—T—A—G—5′ 　　　　　↑
*HAE*III	*HAEMOPHILUS AEGYPTIUS*	↓ 5′—G—G—C—C—3′ 3′—C—C—G—G—5′ 　　　　↑
*HIND*III	*HAEMOPHILUS INFLUENZAE*	↓ 5′—A—A—G—C—T—T—3′ 3′—T—T—C—G—A—A—5′ 　　　　　　　↑
*PST*I	*PROVIDENCIA STUARTII*	↓ 5′—C—T—G—C—A—G—3′ 3′—G—A—C—G—T—C—5′ 　　↑
*SAL*I	*STREPTOMYCES ALBUS*	↓ 5′—G—T—C—G—A—C—3′ 3′—C—A—G—C—T—G—5′ 　　　　　　　↑

[a]The arrows indicate the cleavage sites on each strand.

Source: From Lansing M. Prescott, John P. Harley, and Donald A. Klein, Microbiology, 3rd edition.

(1) Using a specific restriction enzyme, *Eco*RI, to cut the first DNA molecule into two fragments. Each fragment has a sticky end.

5′ AACCTTACTACG·AATTCGGGATTA 3′
3′ TTGGAATGATGCTTAA·GCCCTAAT 5′

5′ AACCTTACTACG
3′ TTGGAATGATGCTTAA   AATTCGGGATTA 3′
                                     GCCCTAAT 5′

(2) Using the same *Eco*RI to cut the second DNA molecule into two fragments. Again each fragment has a sticky end.

5′ GCTTTAGGACG·AATTCAATGAT 3′
3′ CGAAATCCTGCTTAA·GTTACTA 5′

5′ GCTTTAGGACG
3′ CGAAATCCTGCTTAA   AATTCAATGAT 3′
                                     GTTACTA 5′

(3) Using DNA ligase to join the different fragments to form two recombinant DNA molecules.

5′ AACCTTACTACGAATTCAATGAT 3′
3′ TTGGAATGATGCTTAAGTTACTA 5′

    5′ GCTTTAGGACGAATTCGGGATTA 3′
    3′ CGAAATCCTGCTTAAGCCCTAAT 5′

Source: Amanda Chau

Figure 17.2   Recombinant DNA molecules are produced using the restriction enzyme, *Eco*RI, to cut the two DNA molecules and DNA ligase to join the fragments.

DNA ligase then joins the two different DNA fragments by forming phosphodiester bonds between the sugar-phosphate backbones (Figure 17.2).

In addition to these traditional restriction enzymes, emergence of engineered restriction enzymes during the past decade makes it feasible to target specific DNA sequence for genome editing. **Chimeric restriction enzymes** such as Zinc-finger nucleases (ZFNs) and **transcription activator-like effector nucleases (TALENs)** are a novel class of engineered restriction enzymes in which a nonspecific endonuclease is linked to sequence-specific DNA-binding modules. These site-specific endonucleases can target a specific DNA sequence to inactivate or replace it, or even add other sequences to it. **Clustered regulatory interspaced short palindromic repeat (CRISPR)/Cas-based RNA-guided DNA endonucleases** are an example of programmable endonucleases. By redesigning the CRISPR RNA, CRISPR/Cas-based endonucleases can be targeted to edit virtually any DNA sequence. For detailed reviews, please see Gaj, Gersbach, and Barbas (2013) and Doudna and Charpentier (2014).

DNA Ligases Glue DNA Fragments Together

Restriction enzymes are used to cut DNA into fragments. In order to produce rDNA molecules, DNA ligases are needed to join the fragments together. DNA ligases were discovered in 1967 by Martin Gellert, I. R. Lehman, Charles C. Richardson, Jerard Hurwitz, and their laboratories (Shuman, 2009). DNA ligases catalyze formation of a phosphodiester bond between 5′-phosphoryl (–PO4) and 3′-hydroxyl (–OH) ends of the two sugar–phosphate backbones

phosphodiester bond
The covalent chemical bond found between two nucleotides.

chimeric restriction enzymes
A novel class of engineered restriction enzymes in which a non-specific endonuclease is linked to sequence-specific DNA-binding modules.

transcription activator-like effector nucleases (TALENs)
A novel class of engineered restriction enzymes in which a non-specific endonuclease is linked to sequence-specific DNA-binding modules.

clustered regulatory interspaced short palindromic repeat (CRISPR)/Cas-based RNA-guided DNA endonucleases
Examples of programmable endonucleases.

Okazaki fragments
Short, newly synthesized DNA fragments that are formed on the lagging template strand during DNA replication. These fragments named after the molecular biologist, Reiji Okazaki, who discovered them.

transgenes
Refers to the foreign DNA that is introduced into a host cell.

transgenic
A host cell or organism contains transgenes.

transformation
When introduced DNA molecules change the genetic composition of the host cells.

transfection
When introduced DNA molecules change the genetic composition of the animal cells

vector
A carrier DNA sequence which can replicate on its own or integrate into a host chromosome.

totipotent
Cells that are capable of developing into a complete organism or differentiating into any of its cells or tissues.

plasmids
Small circular DNA molecules found in many prokaryotic cells and are distinct from the main circular DNA molecule or chromosome. Plasmids contain genes that confer resistance to antibiotics or unusual metabolic capacities.

(Lehnman, 1974). Most organisms have multiple DNA ligases that either function in DNA replication (by joining Okazaki fragments in the lagging strand) or are dedicated to particular DNA repair pathways. These enzymes play an important role in genome integrity (see Chapter 16).

DNA ligases are divided into two types, ATP-dependent ligases and NAD+-dependent ligases, based on the substrate required for part of the ligation process (Shuman, 2009). NAD+-dependent DNA ligases are found in all bacteria whereas ATP-dependent DNA ligases are found in all known eukaryotes. For example, humans have four ATP-dependent DNA ligases and plants have three whereas fungi have two. ATP-dependent ligases are also found in all known archaea, which suggest a common ancestry for DNA replication machinery in archaea and eukaryotes. Purified DNA ligases have proved to be useful reagents in the construction of rDNA molecules *in vitro* (Lehnman, 1974).

Ways to Deliver Foreign DNA Molecules into Cells

Transgenes refers to foreign DNA that is introduced into a host cell. Recombinant DNA molecules have no biological significant unless they are inserted into living cells, which then replicate and transcribe the information on these transgenes. A host cell or organism containing transgenes is called **transgenic** or genetically modified (GM). When foreign DNA molecules are introduced into host cells, the host cells undergo **transformation** or **transfection** if host cells are derived from an animal.

Using Prokaryotic and Eukaryotic Cells as Host Cells

Foreign DNA can be inserted into vectors to facilitate the introduction of rDNA into prokaryotic or eukaryotic cells. Bacterial cells such as *E. coli* are excellent host cells because they are easily grown and manipulated in the laboratory. Bacteria have plasmids that can be used as vectors to introduce foreign DNA into the cells. A **vector** is a carrier DNA sequence that can replicate on its own or integrate into a host chromosome. Yeasts such as *Saccharomyces cerevisiae* are commonly used as eukaryotic hosts for rDNA research. Like bacteria, yeasts are easily grown and manipulated in the laboratory. Unlike bacteria, yeasts are more compatible hosts for expression of certain eukaryotic genes. Plant cells are great hosts because of their ability to produce **totipotent** stem cells from mature plant tissues. These stem cells can be transformed and then studied in culture, or grown into mature plants. Animal cells grown in culture can be used to study gene expression in human or animal genes for medical purposes. Transgenic animals can be made by introducing foreign DNA into egg cells.

Using Plasmids as Vectors to Deliver DNA

Plasmids are small circular DNA molecules found in many prokaryotic cells. Plasmids are distinct from the main circular DNA molecule or chromosome and contain genes that confer resistance to antibiotics or unusual metabolic capacities. They replicate independently from the main chromosome. Plasmids are excellent vectors because they are relatively small (2,000–6,000 base pairs), making them easy to manipulate in the laboratory. A plasmid often has one or more restriction-enzyme recognition sequences, which facilitates insertion of new DNA into it. Many plasmids contain

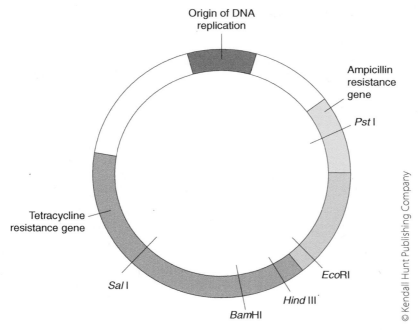

Figure 17.3   Plasmid pBR322 from *Escherichia coli* showing its origin of replication, restriction enzyme recognition sites, antibiotic resistance genes.

genes that confer resistance to antibiotics and can be used as reporter genes (Figure 17.3). Plasmids have their own replication origin, which allows them to replicate independently from the host chromosome.

Using Viruses as Vectors to Deliver DNA

Plasmids are good vectors for introducing short DNA strands of less than 10,000 base pairs in length. However eukaryotic genes are usually much larger and need vectors that can accommodate larger genes. Viruses such as Bacteriophage λ make good vectors for eukaryotic genes. Bacteriophage λ, which infects *E. coli*, has a genome of about 49,000 base pairs of which half are essential for the bacteriophage to complete its lytic cycle. The nonessential half of the genome can be deleted and replaced with DNA from another organism (Glover, 1980). Viruses are better vectors than plasmids because they infect cells naturally and can accommodate bigger genes. Host-range of the viruses may be a limiting factor for this delivery method.

Using Direct Gene Transfer to Deliver DNA

Protoplasts are plant, fungi, or bacterial cells without a cell wall, and these cells can be induced to take up DNA directly. This process is called direct gene transfer or DNA-mediated gene transfer. Electroporation and PEG-mediated transformation are the two main methods of direct gene transfer (Narusaka, Narusaka, Yamasaki, & Iwabuchi, 2012). **Electroporation** refers to the application of a short high-voltage pulse of electricity to protoplasts in order to form micropores in the cell membrane (Figure 17.4).

electroporation
Refers to the application of a short high-voltage pulse of electricity to protoplasts, which cause micropores to form in the cell membrane, allowing foreign DNA to enter the cell and then the nucleus.

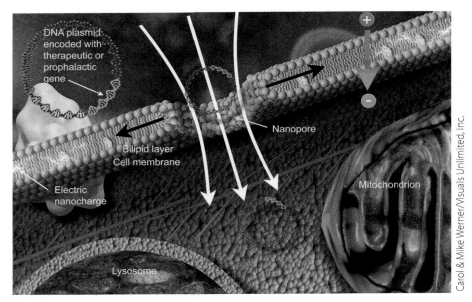

Labels in figure: DNA plasmid encoded with therapeutic or prophalactic gene; Nanopore; Bilipid layer Cell membrane; Electric nanocharge; Mitochondrion; Lysosome; Carol & Mike Werner/Visuals Unlimited, Inc.

Figure 17.4 Electroporation: the use of a short high-voltage pulse of electricity to create transient nanopores in the cell membranes to allow foreign DNA to enter the cell.

PEG-mediated transformation
Refers to the treatment of protoplasts with polyethylene glycol (PEG), which changes the permeability of the membrane, making it more readily to take up foreign DNA.

biolistic transformation or biolistics
A physical means of directly introducing foreign DNA into cells.

direct microinjection
A physical method for introducing foreign DNA by glass micropipettes or microneedles into animal and plant cells.

This allows foreign DNA to enter the cell and then the nucleus. **PEG-mediated transformation** refers to the treatment of protoplasts with polyethylene glycol (PEG) or a similar polyvalent cation in order to change the permeability of the membrane so that it more readily takes up foreign DNA. Protoplasts are then cultured under conditions to regrow their cell walls. Unlike using vectors, these methods are host-range independent.

Using Physical Methods to Deliver DNA

Biolistic transformation or particle bombardment is a physical means of directly introducing foreign DNA into cells (Sanford, Klein, Wolf, & Allen, 1987). The term biolistics stands for biological ballistics. DNA molecules are applied onto high-density gold or tungsten microparticles (1–2 μm) that are delivered at high velocity by a helium pulse generated with a gene gun (Figure 17.5). The microparticles are driven through cell walls and membranes into the target. Direct gene transfer can be used on a wide range of targets that include cell cultures, tissues, organs, plants, animals, bacteria, and organelles. **Direct microinjection** is another physical method of delivery that introduces foreign DNA by glass micropipettes or microneedles into animal cells (Capecchi, 1980) and plant cells (Schnorf et al., 1991). These direct physical methods penetrate intact cell walls and do not require a protoplast-cell-wall regeneration system. Like direct gene transfer, host-range is not an issue with biolistic transformation and direct microinjection.

Use of Reporter Genes to Identify Transformed Cells

Regardless of the methods of DNA delivery, not all cells or tissues are transformed or transfected successfully. How do we identify cells or regenerating plants that contain the desired DNA sequence? The use of reporter genes

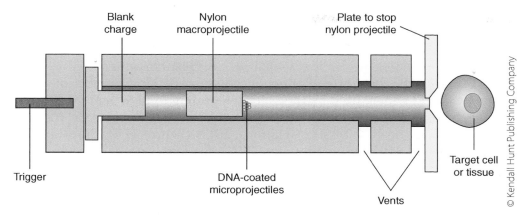

Figure 17.5   Diagram of a gene gun loaded with DNA-coated high-density gold microprojectiles or microparticles.

make it possible to distinguish between transformed and untransformed cells. **Reporter genes** are genes that have observable expression and are divided into two types: selectable marker genes and visual marker genes. Genes that confer resistance to antibiotics or tolerance to herbicides are good selectable marker genes. Only transformed cells or plants will survive after exposure to the selected antibiotic or herbicide. Green fluorescent protein (GFP), which normally occurs in the jellyfish, *Aequorea victoria*, emits visible green light when exposed to ultraviolet light (Tsien, 1998). The gene of this protein has been isolated and incorporated into vectors. The GFP gene is a good visual marker for identifying transformed cells (Figure 17.6).

reporter genes
Genes that have observable expression are used to distinguish between transformed and untransformed cells.

TECHNIQUES FOR GENETIC MODIFICATION OF PLANTS

All plant breeding involves the alteration of plant genes, whether it is through the crossing of different varieties or related species, selection of a naturally occurring mutant, or the artificial induction of random mutations through chemical or radiation mutagenesis. Nevertheless, the term

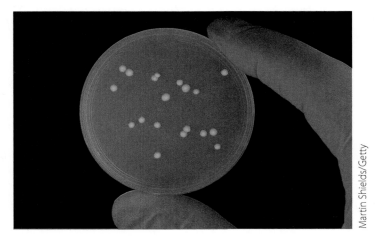

Figure 17.6   Transformed bacteria colonies emit green light when exposed to UV because of the Green fluorescent protein (GFP).

"genetically modified (GM) plants" is now used specifically to describe plants produced by the artificial insertion of a single gene or small group of genes into its DNA. Genetic modification has been an extremely valuable tool in plant genetic research as well as in crop plant breeding.

Plants can be transformed or genetically modified in a number of ways. The two most widely used methods for integrating foreign genes into plant cells are *Agrobacterium*-mediated transformation and direct gene transfer using a gene gun or **biolistics** (Halford, 2012) (Figure 17.7). Biolistics has been a method for delivering foreign DNA into plant genome since the early 1990s and is commonly used in plant science. The *Agrobacterium*-mediated method was first used to successfully transform eudicotyledonous plants (eudicots). During the mid-1990s, biolistics was the main method used to transform monocotyledonous plants (monocots) such as maize and rice (Gao & Nielsen, 2013).

Using *Agrobacterium tumefaciens* to Transform Plant Cells

The choice of transformation method is dependent on the *in vitro* regeneration capability of the given plant species. *Agrobacterium*-mediated transformation is commonly used for plants such as potato, tobacco, or tomato that

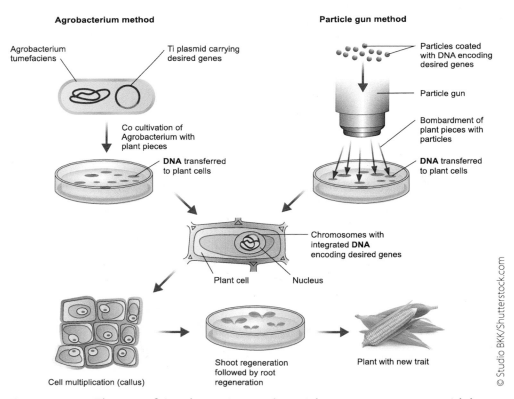

Figure 17.7 The use of *Agrobacterium* and particle gun are two most widely used methods to introduce foreign DNA into plant cells.

have good regeneration capability (Galun & Galun, 2001). *Agrobacterium tumefaciens* is a common soil bacterium that infects a broad range of dicots and gymnosperms. This bacterium invades injured plant tissue and causes the disease known as crown gall tumor (Figure 17.8). In 1977, Chilton et al. showed that only a small fragment of the tumor-inducing (Ti) plasmid carried by *A. tumefaciens* was inserted into the DNA of host plant cells during the infection process. Once integrated, the genes in this small "transferred" fragment, called transferred DNA (T-DNA) become active and encode proteins that cause plant cells to grow and divide continuously into a tumor-like structure called the crown gall. The cells of a crown gall or **callus** are not differentiated and can be removed from the plant and cultured as long as they are supplied with light and nutrients. All these cells contain T-DNA from the Ti plasmid and can be inherited stably through callus culture or regenerations of genetically modified (GM) plants from the callus. Some of the genes in T-DNA induce the host cell to produce novel amino acid derivatives such as octopine and nopaline which *A. tumefaciens* feeds on.

callus
A growing mass of undifferentiated plant parenchyma cells.

The Ti plasmid of *A. tumefaciens* can be modified to vector and introduce any foreign DNA into plant cells. In 1983, Zambryski et al. showed that only parts of the T-DNA, a short region of 25 base pairs at each end or border, are required for the transfer process. Anything between these border regions will be transferred into the DNA of the host plant cell. Caplan et al. (1983) confirmed that the tumor-causing genes can be removed so that they no longer interfere with normal plant growth and differentiation. Also Fraley et al. (1983) showed that antibiotic resistance genes can be inserted into the Ti plasmid of *A. tumefaciens* as reporter genes (Figure 17.9). Alternatively, the lucficerase gene from the firefly, *Photinus pyralis*, can be inserted into the Ti plasmid and used as a reporter gene (Ow et al., 1986). The transformed plants or cells will emit light in the presence of luciferin and ATP (Figure 17.10). Bevan (1984) developed so-called binary vectors, plasmids that would replicate in both *E. coli* and *A. tumefaciens*. Binary vectors contain the left and right T-DNA borders but none of the genes present in "wild-type" T-DNA. On their own, they are unable to induce transfer of the T-DNA into a plant cell because the required virulence (*VIR*) genes are missing. However when the binary vectors are present in *A. tumefaciens* with a

virulence
Refers to a pathogen's ability to infect or damage a host.

Figure 17.8 *Agrobacterium tumefaciens*, a common soil bacterium that causes the disease known as crown gall tumor.

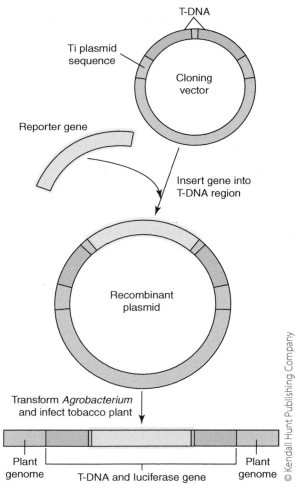

Figure 17.9   Reporter gene is incorporated into T-DNA of the Ti plasmid to help identify transformed cells or plants.

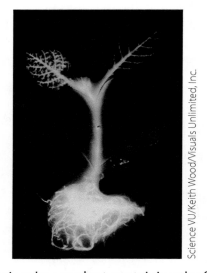

Figure 17.10   Transgenic tobacco plant containing the firefly luciferase gene, a luminescent reporter gene. The plant was watered with luciferin, resulting in the firefly glow.

"helper" plasmid containing the *VIR* genes, the region of DNA between the T-DNA borders is transferred with any genes placed there (Hansen, Shillito, & Chilton, 1997). The transformed callus is then transferred to culture medium in order to regenerate into a mature plant. *Agrobacterium*-mediated transformation is one of the most reliable and widely used means of transferring foreign DNA into plants.

Using *Agrobacterium*-mediated Transformation Without Tissue Culture

Although *Agrobacterium*-mediated transformation is relatively straightforward with many plant species such as soybean and potato, the two main drawbacks are (1) relatively expensive laboratory facilities required to carry out tissue culture after transformation and (2) mutations occur when plants are regenerated from single cells. To overcome these drawbacks, methods of plant transformation have been developed that do not require tissue culture. The most successful of these methods is germ-line transformation, which was first developed in 1993 by Georges Pelletier, Nicole Bechtold, and Jeff Ellis using the model plant, *Arabidopsis thaliana* and vacuum infiltration (Bechtold, Ellis, & Pelletier, 1993). Flowering plants are placed in a suspension of *A. tumefaciens*, and a vacuum is used to replace the air around the plant tissue with bacteria suspension. Treated plants are then grown for a few more weeks and seeds are collected. Transformed seedlings are identified by germinating them on selective media. Because intact plants are used rather than cultured callus, no tissue culture or plant regeneration procedures are necessary. Typically, around 1% of the seeds are genetically modified. This method is now widely used in basic research with *Arabidopsis* and has been adapted with some success for use with other plant species, including soybean and rice (Halford, 2012).

Using Biolistics to Transform Plant Cells

Biolistic transformation is the method of choice to transform plant species that are difficult to regenerate into whole plants from transformed callus. Biolistic transformation has two advantages over *Agrobacterium*-mediated transformation. Unlike *Agrobacterium*-mediated transformation, biolistic transformation is not limited by the bacterium's host range so this method can be used with a much wider range of plant species. Another advantage of biolistic transformation is that it is relative simple to engineer the rDNA. However, the two major drawbacks of biolistic transformation are that the integration of DNA into plant DNA and expression of transgenes are unpredictable. Unlike biolistics, the T-DNA of *Agrobacteria* is rather precisely integrated into plant DNA and expression of transgenes is not problematic.

Using Biolistics to Transform Plant Plastids

Biolistic transformation remains the only method of choice for transforming chloroplasts and mitochondria (Bock & Khan, 2004). Plasmid transformation is useful when large amounts of the transgene product are needed.

Pal Maliga and colleagues pioneered the transformation of chloroplasts in plants using biolistic transformation (Svab, Hajdukiewicz, & Maliga, 1990). The integration of foreign DNA into plastid DNA is simpler compared to integration into plant nuclear DNA. For plastid transformation, the foreign DNA is precisely constructed so that it combines with similar sequences in the target plastid DNA, using homologous recombination. Another advantage of plastid transformation is that transgene escape via pollen is avoided since plastids are only maternally inherited in most plant species.

GENETICALLY MODIFIED PLANTS FOR HUMAN BENEFIT

Traditional breeding relies on genes present in domesticated plants or their wild relatives. If the desirable traits are absent, they cannot be added. With advances in biotechnology, we can now transfer foreign genes into plant genomes. These GM plants will then produce novel characteristics. In the following sections, we will take a closer look at the growth and development of GM crops in the United States.

Genetically Modified (GM) Crops in the United States

Transgenic or GM crops were first introduced for commercial production in 1996 and planted on 4.2 million acres worldwide. By 2016, global area of GM crops has grown to 457.4 million acres. A total of 26 countries, 19 developing and 7 industrial countries, planted GM crops in 2016. The top 10 countries, each of which grew over 2.5 million acres in 2016, include the United States, Brazil, Argentina, Canada, India, Paraguay, Pakistan, China, South Africa, and Uruguay (ISAAA, 2016). The United States is by far the largest producer of GM crops and accounts for 39% or 180.1 million acres of the global area of GM crops. In 2015 alone, global economic gains from GM crops were $15.4 billion (USD) and the United States gained the most, $6.9 billion (USD), from GM crops (ISAAA, 2016). GM crops most commonly grown in the United States are maize or corn, soybean, cotton, alfalfa, canola, squash, and papaya. The acreage for GM crops has grown steadily over the past decade (Figure 17.11). By 2017, 94% of soybeans, 96% of cotton, and 82% of corn planted in the United States were genetically modified (USDA/NASS, 2017). The two most common GM traits for these GM crops are herbicide tolerance and insect resistance.

GM Traits that Confer Tolerance to Herbicides

Weed control is an essential part of all types of agriculture and is achieved mainly by spraying fields with chemical herbicides. Most herbicides are selective in the types of plant that they kill but a grower has to use an herbicide that is tolerated by his/her crop. Herbicide use presents a number of problems for growers. For example, some herbicides may have to go into the ground before planting and cannot be used to control weeds after crop seeds have germinated. Some herbicides may be toxic to humans and

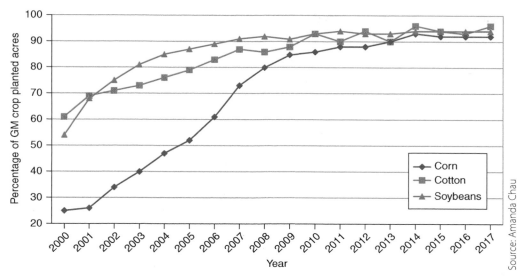

Figure 17.11   Percentages of genetically modified (GM) corn, upland cotton, and soybeans planted acres in the United States from 2000 to 2017. Based on United States Department of Agriculture's National Agricultural Statistics Service (NASS) in the June Agricultural Survey for 2000 through 2017, available at https://www.ers.usda.gov/data-products/adoption-of-genetically-engineered-crops-in-the-us.aspx.

dangerous to handle. Protective clothing and equipment are needed which adds cost to the production. Some herbicides persist in the soil from one season to the next, making crop rotation difficult. Many of these problems have been overcome by the introduction of GM crops that tolerate broad range herbicides.

Since 1996 Roundup-Ready™ soybeans, the first GM crop with broad-range-herbicide tolerance, were produced and marketed by Monsanto, a U.S. agrochemical company based in Missouri. Roundup™ is Monsanto's trade name for glyphosate, a broad-range herbicide introduced as a commercial product by Monsanto in 1974. It is now marketed under many different trade names in various agricultural and garden products. Glyphosate is taken up through plant foliage, so it is effective after weeds have become established. Glyphosate targets a pathway that is not present in animals so it is considered to have low toxicity for insects, birds, fish, or mammals including humans. For these reasons, growers have always been very comfortable in using this herbicide. Two decades after the introduction of Roundup-Ready™ soybeans, their use rose to 94% of all the soybeans planted in the United States (Figure 17.11). In addition to soybeans, other agricultural crops such as alfalfa, canola, cotton, corn, creeping bentgrass, and potato are transformed to be Roundup-Ready as well (ISAAA, 2017). GM trait that tolerates other herbicides such as 2,4-dichlorophenoxyacetic acid (2,4-D), dicamba, glufosinate, mesotrione, and oxynil are being developed for soybean, cotton, and corn. GM traits that tolerate herbicides have not only been developed for agricultural crops but also for ornamental crops. Moon™ carnations have GM traits that give them their unique blue color and also tolerance to sulfonylurea herbicide (Table 17.2).

Table17.2 Examples of GM Crops and Their Modified Traits (Agrawal & Sharma, 2011; ISAAA, 2017)

Crop	GM trait(s)	Modification(s)
Apple	• Produce non-browning phenotype.	Double-stranded RNA (dsRNA) was used to suppresses PPO (polyphenol oxidase) resulting in apples with a non-browning phenotype.
Carnations	• Produce novel blue, mauve, to violet colors for carnation flowers. • Resistant to sulfonylurea herbicides.	Genes from *Petunia* and *Viola* that catalyze the production of blue-colored pigment were added to the plant. A gene from *Nicotiana* confers tolerance to sulfonylurea herbicides was added.
Corn (field)	• Resistant to glyphosate or glufosinate herbicides. • Resistant to insects using Bt proteins as pesticides. • Increase Vitamin A and C, folate content.	New genes from bacteria and other organisms were transferred into plant genome.
Corn (sweet)	• Resistant to insects using Bt proteins as pesticides.	Gene from the bacterium *Bacillus thuringiensis* (Bt) was added to the plant.
Cotton (cottonseed oil)	• Resistant to insects using Bt proteins as pesticides.	Gene from the bacterium *Bacillus thuringiensis* (Bt) was added to the plant.
Papaya	• Resistant to the papaya ringspot virus.	New genes were transferred into plant genome.
Potatoes	• Resistant to glyphosate or glufosinate herbicides.	New genes added/transferred into plant genome.
Potatoes (Innate Acclimate)	• Block brown spot bruise development. • Suppress asparagine formation. • Resistance to potato late blight.	Genes derived from potato were used to transform the potato genome.
Potatoes (Amflor)	• Produce starch composed almost exclusively of the amylopectin, making paper and yarn glossier and stronger.	An antisense gene or reverse copy of the gene for granule bound starch synthase (GBSS) was added to switch of the synthesis of amylose. Amflora will be produced solely under contract farming conditions and not made available on the general market.
Rapeseed (Canola)	• Resistant to glyphosate or glufosinate herbicides. • Increase laurate canola.	New genes added/transferred into plant genome.

Crop	GM trait(s)	Modification(s)
Rice	• Increase Vitamin A (beta-carotene) content.	Two genes from daffodils and one gene from a bacterium were added to rice genome.
Soybeans	• Resistant to glyphosate or glufosinate herbicides.	Herbicide resistant gene taken from bacteria was inserted into soybean.
Tomatoes (see Box 12.1)	• Suppression of enzyme polygalacturonase (PG) production, retarding fruit softening after harvesting.	An antisense gene or reverse copy of the gene responsible for the production of PG enzyme was added into plant genome. Taken off the market due to commercial failure.

Stacking GM Traits

The current trend in genetic modification is **gene stacking**, which refers to the process of transferring two or more genes of interest into a single plant. Compared to mono-trait crop varieties, stacked varieties offer broader agronomic enhancements to meet grower needs under complex production conditions. Growers can increase productivity because stacked varieties are better equipped to overcome the myriad of problems in the field such as insect pests, diseases, weeds, and environmental stresses. For example, Moon™ carnations that were discussed previously are examples of stacked varieties that produce a unique blue color and also tolerate sulfonylurea herbicide. Other examples of stacked varieties include transgenic cotton, corn, and soybean that express both insect resistance and herbicide tolerance.

The first stacked variety that gained regulatory approval in 1995 was a dual stacked cotton produced by crossing Bollgard™ cotton that produces Bt toxin and Roundup Ready™ cotton that tolerates glyphosate (ISAAA, 2013). Adoption of stacked varieties has accelerated in recent years. Approximately 80% of cotton acreage and 77% of corn acreage were planted with stacked seeds in 2017 (Figure 17.12). Following commercial success of this dual stacked cotton, developers sought to stack up more GM traits into their crops to create multistacked varieties. In 2008, the triple stacked cotton that combines two Bt genes with the glyphosate resistance gene was grown on more than 54% of U.S. cotton acreage. The eight-stacked corn known by its trade name SmartStax™ was released commercially in 2010 and is the result of combining two herbicide tolerance genes with six Bt genes (ISAAA, 2013). SmartStax™ corn features dual modes of control for weeds, lepidopteran insects and coleopteran insects.

GM Traits that Confer Resistance to Insects

Insect damage can cause substantial loss to growers. Chemical insecticides and, to a lesser extent, biological control agents are used by growers to prevent insect damage. Organic growers have been using a biological pesticide called Bt to control insects for several decades. *Bacillus thuringiensis* (Bt) is a soil bacterium which produces a protein called Cry (crystal) protein that is toxic to some insects. Bt pesticides are applied as powders, granules, and aqueous or oil based liquids. They have no toxicity to mammals, birds, or

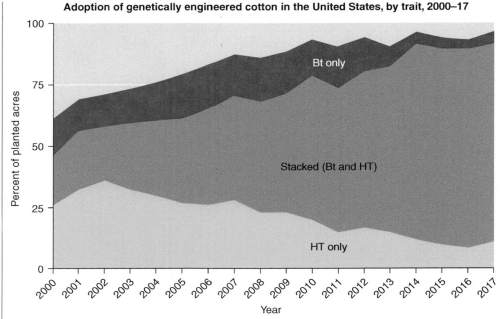

Source: USDA, Economic Research Service using data from USDA, National Agricultural Statistics Service, *June Agricultural Survey*.

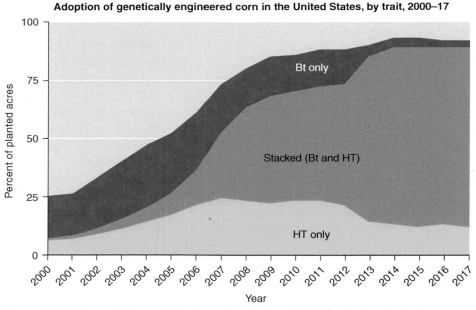

Source: USDA, Economic Research Service using data from USDA, National Agricultural Statistics Service, *June Agricultural Survey*.

Figure 17.12 Percentages of stacked varieties of cotton (left) and corn (right) [with both herbicide tolerance (HT) and insect resistant (Bt)] planted acres in the United States from 2000 to 2017. Based on United States Department of Agriculture's National Agricultural Statistics Service (NASS) in the June Agricultural Survey for 2000 through 2017, available at https://www.ers.usda.gov/web-docs/charts/55237/biotechcorn.png?v=42565.

fish and also have an extremely good safety record (Halford, 2012). The two main disadvantages are that Bt pesticides are insect-specific and do not remain effective for long after application.

Different strains of *B. thuringiensis* produce different versions of the Cry protein, CryI–CryIV. Each version is further divided into subgroups A, B, C, etc. The different Cry proteins are effective against different types of insects. CryI proteins, for example, are effective against the larvae of lepidopterans (butterflies and moths) while CryIII proteins are effective against coleopterans (beetles) (Halford, 2012). Various Cry genes have been introduced to cotton, eggplant, corn, potato, rice, and soybean. These GM varieties are generally referred to as Bt varieties and are marketed under different trade names. The use of Bt varieties has dramatically reduced chemical pesticide use in agriculture. The use and rapid adoption of Bt cotton is a good case in point. One-quarter of U.S. insecticide production was used on cotton, which is very susceptible to insect damage (Halford, 2012). In areas under severe pressure from tobacco budworm, cotton bollworm, and pink bollworm, the three major pests of cotton, Bt cotton on average requires only 15–20% of the insecticide used on conventional cotton (Halford, 2012). It is no surprise that the use of Bt cotton rose rapidly during the last two decades. In 2017, 96% of all the cotton planted in the United States were Bt cotton seeds (Figure 17.11).

GM Traits that Confer Resistance to Viruses

Plant diseases can cause major economic loss as well as loss in human lives. Viruses such as cassava mosaic virus and the feathery mottle virus of sweet potato, for example, are responsible for the deaths of millions of people in developing countries every year through the destruction of vital food crops (Halford, 2012). Farmers attempt to prevent viral plant diseases by controlling the insect vectors that carry the disease or applying virucidal chemicals.

Two different types of viral resistance are introduced to plants. The first type uses cross protection, in which infection by a mild strain of a virus induces resistance to subsequent infection by a more virulent strain. Cross protection appears to involve the coat protein of the virus. The coat protein of the papaya ringspot virus (PRSV) was introduced to papaya by scientists from Cornell University and University of Hawaii (ISAAA, 2017). PRSV is by far the most widespread and damaging virus that infects papaya and there is no other known solution to an epidemic of PRSV. This virus almost destroyed the papaya industry in the Puna district of Hawaii (Gonsalves, 1998). When growers switched to GM papaya varieties (trade names Rainbow and SunUp) in 1998, the industry was saved from being devastated by PRSV. Another type of viral resistance uses gene suppression techniques to block the activity of viral genes. Monsanto, for example, targeted a replicase gene from potato leaf roll virus (PLRV) to induce resistance to PLRV in potato. A GM potato variety containing this trait and the Bt insect-resistance trait was marketed under the trade name NewLeaf Plus™ in 1998 but the variety was withdrawn from market in 2001 due to poor sales (Halford, 2012).

cross protection
A phenomenon in which infection of a plant with a mild virus strain protects it from subsequent infection by a more virulent strain.

GM Traits that Modified Product Quality

Herbicide tolerance and insect resistance were the first traits to come on to the market in the major commodity crops such as cotton, corn, and soybean. These traits remain by far the most successful of those introduced into crop plants by genetic modification. Other traits that have been success-fully introduced into plants include traits that allow fruits to ripen without softening, increase nutritional quality, produce unique flower color, reduce browning of fruit, and increase oil content. For a more current and comprehensive list of GM traits, please visit ISAAA Data Base (http://www.isaaa.org/gmapprovaldatabase/gmtraitslist/default.asp).

GM Trait that Allows Fruits to Ripen without Softening

The first commercially grown GM crop was a tomato (called Flavr Savr™), which was modified to ripen without softening, by a California company called Calgene, later a subsidiary of Monsanto (Table 17.2 and Box 12.1) (Kramer & Redenbaugh, 1994). Calgene obtained FDA approval for its release in 1994 without any special labeling. Although consumers were willing to pay a substantial premium for Flavr Savr™ tomatoes, production problems and competition from a conventionally bred, longer shelf-life variety prevented the product from becoming profitable (Agrawal & Sharma, 2011). A variant of the Flavr Savr™ was used by Zeneca to produce tomato paste, which was sold in United Kingdom from 1996 through early 1999. The labeling and pricing were designed as a marketing experiment, which proved, at the time, that European consumers would accept genetically engineered foods (Bruening & Lyons, 2000).

GM Trait that Enriches Nutritional Quality of Rice

Plant breeders have been trying to improve the nutritional quality of rice by adding beta-carotene to its grains. Beta-carotene is the precursor to Vitamin A which is essential for eye health and bone development. Vitamin A deficiency is particularly problematic in developing countries where the staple food is rice, as white rice is almost totally carbohydrate and contains no carotenoids. Rice provides around 80% of the carbohydrate daily for half the world, about 3.5 billion people, and is the staple crop in most of Asia. Even in Africa, rice is becoming more and more important. However, none of the domesticated varieties of rice or its wild relatives produce beta-carotene in their grains. Genetic modification, on the other hand, can incorporate the genes from a wide range of organisms into crop plants. For example, genes from daffodils that produce beta-carotene were successfully added to the rice genome (Table 17.2). This genetically modified or transgenic rice produces beta-carotene in its grains, which gives them an orange hue, hence the name "Golden Rice" (Ye et al., 2000) (Figure 17.13). In 2001, the inventors of Golden Rice licensed their technology to Syngenta Seeds AG for commercial uses. In exchange Syngenta agreed to support the inventors' humanitarian project and Syngenta scientists made subsequent improvements to the technology. For more information on the history and current development of Golden Rice, please visit the Golden Rice Project website (http://www.goldenrice.org/index.php).

GM Trait that Produces Novel Flower Color

Genetic modification is being applied not only to agricultural crops but also ornamental crops such as carnations and roses to produce novel color (ISAAA, 2017). Popular cut flowers do not normally produce blue, deep violet, or purple colors. Scientists are able to insert genes from a blue flowering petunia into carnation genomes (Davies, 2009). These transgenic

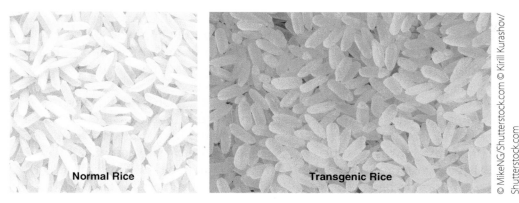

Normal Rice

Transgenic Rice

Figure 17.13 Grains from transgenic rice or golden rice are easily recognizable by their orange color; the deeper the orange the more beta carotene (pro-Vitamin A).

carnations produce novel blue, violet to purple colors and are marketed as Moon™ carnations by Florigene, a biotechnology company based in Australia (Table 17.2). These petunia genes can also be used to transform chrysanthemum, gerbera, and rose to produce varieties with blue flowers.

GM Trait that Produces Non-Browning Apples and Potatoes

The first GM non-browning apples went on sale in the United States in 2017. This GM apple, trade name Arctic™ apple, was developed by Okanagan Specialty Fruits Inc., a Canadian company, and was FDA approved in 2015. When apple cells are ruptured by bruising, biting, or cutting, the browning reaction begins when the enzyme polyphenol oxidase (PPO), found in the plastids, mix with polyphenolics, found in the vacuoles of the cell. PPO catalyzes the oxidation of polyphenolics to melanins, causing browning. Scientists constructed transgenes derived from apple genes to silence the PPO gene by the process of RNA interference (Rommens, Haring, Swords, Davies, & Belknap, 2007). This GM apple produces no PPO and, therefore, does not turn brown. The process of RNA interference was used to transform potato by J.R. Simplot Co., a U.S. company based in Idaho, to produce Innate™ potato. Like apples, browning in potatoes is due to PPO. However, browning in potatoes also occurs nonenzymatically when starch is partially degraded into sucrose and fructose, causing brown spots to accumulate. Scientists at Simplot silenced both the PPO and invertase gene in potato to reduce the expression of PPO and conversion of starch to sucrose and fructose, respectively (Rommens, Ye, Richael, & Swords, 2006).

GM Trait Increases Oil Content

The main components of plant seed oils are fatty acids and the combination of different fatty acids gives different plant oils their unique properties. Fatty acids are used not only to store energy for plants but also to form cell membranes. Some fatty acids are converted into hormones and other compounds with vital functions in the human body. Lauric acid and palmitic acid are well-known plant fatty acids found in coconut and palm kernel oil. Lauric acid is used in manufacture of cosmetics and detergents. For example, the sodium lauryl sulphate that is commonly used in shampoos is derived from lauric acid. Palmitic acid is used in both detergents and foods. Some fatty acids such as oleic acid and linoleic acid are unsaturated fatty acids because they have one or more double bonds in their structure. Oleic acid is found in olive and oilseed rape or canola oil while linoleic acid is found in safflower, sunflower, and corn oil, and makes up about 20% of canola oil.

Canola oil has long been the target for genetic modification, partly because the oil is one of the cheapest edible oils on the market, so growers and processors are interested in anything that adds value to it. The first GM canola variety was modified by Calgene and released commercially in 1995. This GM variety produces high levels of lauric acid in its oil. A gene from the Californian bay plant was introduced into canola. This gene encodes an enzyme called thioesterase that increases the level of lauric acid to approximately 40% of the total oil content, compared with 0.1% in non-GM varieties, making the oil an alternative to palm and coconut oil as a source of lauric acid for detergents. However, palm oil production in particular is very efficient and oil from the GM canola did not gain a foothold in the market (Halford, 2012).

Soybean is the other major crop that has been genetically modified to increase its oil level. The GM variety was produced by Pioneer Hi-Bred International (PBI), a subsidiary of DuPont. It accumulates oleic acid to approximately 80% of its total oil content, compared with approximately 20% in non-GM varieties. In non-GM soybean, much of the oleic acid is converted to linoleic acid by an enzyme called delta-12 desaturase. In the GM varieties, the activity of the gene that encodes the delta-12 desaturase is reduced to increase the level of oleic acid levels while decreasing linoleic acid level. Although linoleic acid is an essential amino acid and is nutritionally important, oleic acid is very stable during frying and cooking and is less prone to oxidation than polyunsaturated oils. The traditional method of preventing polyunsaturated oil oxidation involves hydrogenation which runs the risk of creating *trans* fatty acids. The oil produced by high-oleic-acid GM soybean requires less hydrogenation, which lowers the risk of *trans* fatty acid formation (Halford, 2012).

Omega-3 polyunsaturated fatty acids, especially eicosapentaenoic acid (EPA) and docosahexaenoic acid (DHA), play important roles in many aspects of human health. Marine fish and organisms that are rich in these beneficial fatty acids are our primary dietary source. However, over-fishing and concerns about pollution of the marine environment indicate a need to develop alternative, sustainable sources of very long chain polyunsaturated fatty acids such as EPA and DHA. One of the most promising strategies is to develop transgenic plants to produce these so-called fish oils. Although considerable progress has been made towards this goal, there is presently no commercial GM varieties of this kind (Napier, 2007; Venegas-Calerón et al., 2009).

GM Traits that Confer Tolerance to Abiotic Stress

One of the major limiting factors for plant production in many parts of the world is water availability. If the predictions of climate change are correct, growers face the challenge of developing crop varieties for an environment that is going to change dramatically over the coming decades. There is an urgent need for new, high-yielding varieties with enhanced drought tolerance. However, developing high-yielding drought tolerant varieties is a major challenge and progress has been limited by the complexity of the underlying traits, which are often determined by multiple genes, and the seasonal and yearly variation of water availability. Drought is often, although not always, accompanied by high temperatures and although these two stresses provoke different responses in plants the dividing line is often blurred as one stress exacerbates the effects of the other.

Currently, there is only one GM corn variety being marketed commercially in the United States for improved drought or heat tolerance, but many major biotechnology companies claim to have such varieties in development in other parts of the world (Waltz, 2014). This GM drought-tolerant corn, trade name DroughtGuard™, was developed by Monsanto and received regulatory approval in 2011. A gene from a common soil bacterium, *Bacillus subtilis*, was introduced to corn. The gene produces a cold shock protein (CSP), which acts as RNA chaperone to preserve RNA stability and translation, maintaining normal cellular functions under water stress (Castiglioni et al., 2008). Although, the gain in yield under drought conditions seems to be relatively modest, 6.7–13.4%, these varieties show promise for long-term improvement.

GM Traits that Produce Therapeutic Products

Biopharming refers to the use of GM plants to produce pharmaceuticals, vaccines, antibodies, and even proteins for industrial uses. Plants have been proposed as an attractive alternative for the production of pharmaceutical proteins to current mammalian or microbial cell-based systems (Fischer, Stoger, Schillberg, Christou, & Twyman, 2004). GM microbes have long been used to produce recombinant proteins. For example, insulin was first produced in GM bacteria in 1981, enabling people with diabetes to be treated with human insulin instead of livestock insulin. The demand for insulin is increasing rapidly with the rise in cases of diabetes and the development of new delivery methods such as inhalation and oral ingestion which require more insulin to compensate for reduced bioavailability. In 2007, SemBioSys Genetics, a Canadian company, developed a GM safflower to produce human insulin in its seeds (Halford, 2012).

Plant-based hepatitis B vaccine was the first vaccine produced successfully using plants like potato, cherry tomatillo, and corn (Streatfield, 2005). Hepatitis B causes acute and chronic liver disease and is associated with liver failure and liver cancer. It causes high mortality throughout the developing world. The first hepatitis B vaccine was a surface antigen extracted from the blood of people infected with the disease and was produced in the 1970s. As HIV spread in the 1980s, producing vaccine extracted from human blood was viewed as too dangerous. The gene that encoded the surface antigen protein was introduced into yeast to produce the vaccine, which has been made this way ever since. Unfortunately, it is too expensive for many developing countries to afford. Introducing this surface antigen into plants to produce edible plant-based vaccines seems to provide an excellent tool for mass prevention, an alternative to injection as well as lower production cost (Pniewski, 2013). However, the concept of edible vaccines is now recognized to be difficult to implement because of problems with dosage control and the requirement for an adjuvant, a protein that stimulates the immune response and increases the effectiveness of a vaccine (Halford, 2012).

biopharming
Refers to the use of GM plants to produce pharmaceuticals, vaccines, antibodies, and even proteins for industrial uses.

The advantages of using plant-based systems are lower risk of mammalian pathogen contamination, reduced production cost, and larger scale of production. These advantages have led many to predict that plant-based systems will offer the next wave for pharmaceutical product manufacturing. However, for this to become a reality, the quality of the products at a relevant scale must equal or exceed the predetermined release criteria set by pharmaceutical regulatory agencies. Pogue et al. (2010) reviewed two different methods for transforming tobacco plants to produce pharmaceutical-grade recombinant aprotinin or monoclonal antibody (mAb). They concluded that plant-based systems have moved beyond the proof-of-concept stage in development and offer a legitimate cost-competitive alternative for recombinant protein production.

BENEFITS AND RISKS OF GENETICALLY MODIFIED PLANTS

GM crops offer many potential benefits such as economic gain, improved food nutrition, reduced pesticide use and exposure, enhanced yields, and resistance to insects and diseases. However, like other technologies, genetic modification also poses known and unknown risks.

Controversies and public concern surrounding GM foods and crops commonly focus on human and environmental safety, labelling and consumer choice, intellectual property rights, ethics, food security, poverty reduction, and environmental conservation. In the following section, we will discuss three benefits and three risks of selected GM foods and crops.

Increased Economic Benefits to Growers

Analysis of the global economic impact of GM crops showed that there were continuing and significant net economic benefits at the farm level. In 2014, GM crops were used by 18 million farmers and the net economic benefits amounted to $17.7 billion. Over the 19-year period of 1996–2014, the net economic benefits amounted to $150.3 billion (Brookes & Barfoot, 2016). About 65% of the gains were derived from yield and production gains with the remaining 35% coming from cost savings. The GM traits that confer insect resistance have mostly delivered higher incomes through improved yields in all countries. Many growers, especially in developed countries, have also benefited from lower costs of production (less spending on insecticides). The gains from GM traits that confer herbicide tolerance have mostly come from reduced costs of production, notably on weed control. Genetic modification has also made important contributions to increasing global production levels of soybean, corn, cotton, and canola. For example, since the introduction of GM crops, an additional 174.2 million tons (158 million metric tons) of soybean and 355 million tons (322 million metric tons) of corn were produced globally (Brookes & Barfoot, 2016). If genetic modification had not been available, it would have required an increment of 11% of the arable land in the United States, or 32% of the cereal area in the EU, to maintain equivalent production levels. Reports on global status of biotech/GM crops produced by ISAAA (2016) also came to the same conclusions and showed that GM crops increased crop productivity and alleviated poverty by helping small farmers in developing countries from 1996 to 2015. For full analyses, please see Brookes and Barfoot (2016) and ISAAA (2016).

Reduced Environmental Impacts from Pesticides

Based on the global economic analyses discussed previously, GM crops not only increased yields for growers but also lowered spending on and use of agrochemicals. On a global basis, GM crops have reduced pesticide use but the degree of reduction varies between crops and the introduced traits. Phipps and Park (2002) estimated that the use of Bt cotton and herbicide-tolerant (HT) soybean, canola, cotton, and corn reduced pesticide use by a total of 22.3 million kg of formulated product (4.4 million kg of active ingredient) in the year 2000. In addition to the reduction of pesticide use, there would be a reduction of 18.5 million acres (7.5 million hectares) sprayed which would save 5.4 million gallons (20.5 million liters) of diesel and result in a reduction of approximately 80,469 tons (73,000 metric tons) of carbon dioxide being released into the atmosphere. Furthermore, a review by Carpenter (2011) showed that GM crops have reduced the impacts of agriculture on biodiversity, not only through reduction of pesticide use, but also through enhanced adoption of conservation tillage practices and increasing yields to alleviate pressure to convert additional land into agricultural use. For full review, please see Carpenter (2011) and Phipps and Park (2002).

Enhanced Product Quality and Improved Post-Harvest Processing

Genetic modification that increases nutrients in food crops would provide a sustainable and long-term strategy for delivering micronutrients such as vitamins to populations in developing countries. Vitamin A deficiency (VAD) continues to be a major public health problem affecting developing countries where people eat mostly rice as a staple food. Between 1995 and 2005, World Health Organization estimated that preschool-age children (<5 years old) in 45 countries had VAD based on the prevalence of night blindness and in 122 countries based on low serum retinol concentration (<0.70 µmol/L), a biochemical VAD (WHO, 2009). Night blindness was estimated to affect 5.2 million preschool-age children and 9.8 million pregnant women. Low serum retinol concentration affected an estimated 190 million preschool-age children and 19.1 million pregnant women. This corresponds to 33.3% of the preschool-age population and 15.3% of pregnant women in populations at risk of VAD, globally (WHO, 2009). The WHO Regions of Africa and South-East Asia were found to be the most affected by vitamin A deficiency for both population groups. Golden rice, discussed previously, was designed to alleviate VAD in developing countries. De Moura et al. (2016) assessed the possible impact of replacing regular white rice with golden rice in regions of Asia where rice is a major staple food. Their study showed that rice enriched with β-carotene, such as golden rice, can substantially increase vitamin A intake and consequently reduce the prevalence of vitamin A deficiency. Increasing vitamin A intake through golden rice, in combination with programs that increase adoption of golden rice by growers, could be an effective method to combat VAD.

Genetic modification is also used to improve post-harvest processing. GM crops such as Flavr Savr™ tomatoes and non-browning apples would improve post-harvest processing. These GM crops are more suitable for mechanical harvesting and will have less bruising from packing line handling. Fresh-cut processors will find that non-browning apples dramatically reduce their costs and improve product quality. Foodservice operators will save preparation time and labor costs. Consumers will have more convenient access to sliced apples.

Potential Health Risks

The debates over GM foods focus mostly on uncertainties concerning the potential adverse effects of GM foods and crops on human health and the environment. Three major health risks potentially associated with GM foods are: toxicity, allergenicity, and unintended hazards. These risks may result directly from the inserted genes and their expressed proteins, indirectly from interactions of genes or modifications of protein products, and from the spread of transgene to consumers. So far, the transgene proteins from two GM crops have been reported to potentially cause allergic reactions in humans. A GM soybean variety was transformed with a gene from the Brazil nut to improve its methionine content. The extracts of this GM soybean and a non-GM soybean were used in skin-prick tests on subjects that have Brazil nut allergy. Subjects developed positive reactions to the extracts of the GM soybean but negative reactions to extracts of the non-GM soybean (Nordlee, Taylor, Townsend, Thomas, & Bush, 1996). This GM soybean never reached the market because of its allergenicity. A GM corn variety, trade name StarLink, was transformed to express Cry9C protein from a Bt gene. StarLink corn was only approved for animal feed in the United States in 1998 but was later found in tacos and tortillas in 2000. The Food and Drug Administration (FDA) received 51 reports of adverse reaction to these corn products and 28 of these individuals appeared to develop allergic reactions. However, the Centers for Disease Control and Prevention (CDC) found no conclusive evidence of hypersensitivity to the Cry9C protein in these subjects (CDC, 2001). Nonetheless, the developer Aventis voluntarily withdrew the registration for StarLink corn in the same year. The unpredictable and hidden effects of transgenes to the GM organisms and consumers are one underlying reasons for public concerns regarding GM foods.

Potential Impact on Nontarget Organisms

Adoption of GM crops may endanger nontarget species in the environment. Losey, Rayor, & Carter, (1999) conducted a laboratory study and found that caterpillars of the monarch butterfly suffered higher mortality levels when forced fed large quantities of pollen from GM insect-resistant corn than caterpillars that were not fed the pollen. However, experts were extremely skeptical that corn pollen would ever accumulate in such a large amount on milkweed in the wild. Field-based studies published subsequently bore this out. Similar laboratory-based experiments have shown that the survival rate of predators such as lacewings and ladybirds can be reduced if they are fed exclusively prey species reared on GM insect-resistant varieties. However, none of these results have been replicated in the field (Halford, 2012).

The use of GM insect-resistant crops can actually lead to an increase in beneficial insects due to changes in tillage practices and reduction of insecticide use (Carpenter, 2011). With GM corn, for example, the grower does not have to use an early spray against the corn borer. The grower may need to spray against other pests but, if this can be avoided, predatory insects and other beneficial insects can thrive. If this happens, the predatory insects may prevent a late infestation of red spider mites, further reducing pesticide use (Halford, 2012). Furthermore, the review by Nicolia, Manzo, Veronesi, & Rosellini, (2014) concluded that there is no consolidated scientific evidence showing the negative impacts of GM crops on the biodiversity of nontarget species such as birds, snakes, and macro- or micro-soil-fauna.

Potential Contamination of Foreign Genes into Wild Species

Another key concern associated with genetic modification is that transgenes may migrate out of the GM organism to wild species and even to unrelated species. As discussed in the previous section, the use of reporter genes such as antibiotic resistance genes has been extremely valuable in the development of GM technology. The potential risk associated with these genes is that they may pass to bacteria present in the environment or in the consumers' gut. However, these antibiotic resistance genes are designed to work in plants and would not be active in bacteria. Furthermore, the occurrence is unlikely given the extremely low frequency of uptake for exogenous DNA by bacteria (Nicolia et al., 2014). Many scientific and regulatory bodies have also considered the safety of antibiotic resistance genes in food and have concluded that those used in GM technology do not pose a health threat (Halford, 2016).

The risk associated with the use of GM herbicide-tolerant crops is that weeds become resistant to herbicides and develop into uncontrollable "superweeds". The main concern is that the transgenes that confer herbicide tolerance may pass into related wild species through cross-pollination and the resulting hybrids would become uncontrollable weeds (Bawa & Anilakumar, 2013). The potential risk of cross-pollination of GM crops with wild species has to be assessed case by case. Guan et al. (2015) showed that the transgenes that confer herbicide tolerant in GM soybean could be transferred to wild soybeans. The escaped transgenes did not appear to adversely affect the growth of "hybrid" soybeans and could persist in nature in the absence of herbicide use. Although canola can cross with other cultivated and wild Brassicas, it does not necessarily mean that GM canola represents a problem, just that the potential risk represented by GM canola is higher than other GM varieties. The extent of such hybridization in agricultural systems is the subject of continuing research. Studies carried out in Canada, where there are millions of hectares of GM herbicide-tolerant canola, have not found any hybrids between the GM crop and wild relatives or glyphosate- or glufosinate-resistant weeds (Beckie et al., 2011). Other studies have found no viable hybrids between GM corn, potato, and wheat and their wild relatives (Halford, 2012).

Another concern is that the prevalent use of GM herbicide-tolerant crops would further encourage the use of herbicides by growers and promote resistance in weeds. Many GM crops are modified for glyphosate resistance and growers are now more likely to use glyphosate-containing herbicides on their crops. As a result, glyphosate resistant weeds and Bt-toxin resistant pests have been reported (Nicolia et al., 2014). The repetitive and high-volume use of herbicides has created a vicious cycle and ironically made weed control more difficult.

GM crops have made a number of important contributions to sustainability. GM crops lower production costs, make food more affordable, conserve biodiversity by changing tillage practices, reduce agricultural impact on the environment by reducing pesticide input, and bring economic benefits to growers. However, it is important to evaluate their potential risks against potential benefits or gains.

Key Terms

biolistic transformation or biolistics
biopharming
callus
chimeric restriction enzymes
clustered regulatory interspaced short palindromic repeat (CRISPR)/Cas-based RNA-guided DNA endonucleases
cross protection
direct microinjection
DNA ligases

electroporation
GM foods
genetically modified organisms (GMO)
Okazaki fragments
Palindromic
PEG-mediated transformation
plasmids
phosphodiester bond
recombinant DNA (rDNA) molecules
reporter genes
restriction endonucleases

restriction enzymes
restriction enzyme digest
transcription activator-like effector nucleases (TALENs)
transformation
transfection
transgenes
transgenic
totipotent
vector
virulence

Summary

- Restriction enzymes and DNA ligase are the most powerful tools for combining DNA from different sources to generate recombinant DNA molecules. These molecular tools revolutionized the study of DNA and accelerated our ability to both sequence and modify genomes.
- Recombinant DNA (rDNA) molecules are DNA molecules created artificially by combining at least two or more sources of DNA.
- The experiments carried out by Stanley Cohen, Herbert Boyer, and their laboratories in 1973 showed that rDNA molecules can be functional when inserted into a living cell.
- Restriction enzymes or restriction endonucleases that cut DNA into fragments were first discovered during the early 1950s by Salvador Luria, Giuseppe Bertani, Jean Weigle and their colleagues.
- Type II restriction enzymes are the most useful molecular tools for recombinant DNA experiments because they cut DNA close to or within their recognition sequence.
- DNA ligases catalyze formation of a phosphodiester bond between 5'-phosphoryl (–PO4) and 3'-hydroxyl (–OH) ends of the two sugar–phosphate backbones. These ligases were discovered in 1967 by Martin Gellert, I.R. Lehman, Charles C. Richardson, Jerard Hurwitz, and their laboratories.
- All known eukaryotic DNA ligases are ATP-dependent. For example, humans have four ATP-dependent DNA ligases, plants have three whereas fungi have two. NAD^+-dependent DNA ligases are found in all bacteria.

- Foreign DNA can be inserted into vectors and the rDNA can then be introduced into prokaryotic or eukaryotic cells.

- Plasmids are excellent vectors because they are relatively small, making them easy to manipulate in the laboratory. They contain genes that confer resistance to antibiotics or unusual metabolic capacities, and they replicate independently from the main chromosome.

- Viruses are better vectors than plasmids because they infect cells naturally and can accommodate bigger genes.

- Electroporation and PEG-mediated transformation are the two main methods of direct gene transfer, which are used to induce protoplasts to take up DNA directly.

- Biolistic transformation or particle bombardment is a direct physical method for introducing foreign DNA into cells.

- The use of reporter gene makes it possible to distinguish between transformed and untransformed cells.

- *Agrobacterium*-mediated transformation is commonly used for plants such as potato, tobacco, or tomato that have a good regeneration capability. The Ti plasmid of *Agrobacterium tumefaciens* can be modified to vector and introduce any foreign DNA into plant cells.

- Vacuum infiltration is one of the methods of plant transformation that require less time and skill and do not require tissue culture.

- Biolistic transformation is the method of choice to transform plant species that are difficulty to regenerate into whole plants from transformed callus. It is the preferred method for transforming chloroplasts and mitochondria.

- The United States is by far the largest GM crops producer and accounts for 39% or 180.1 million acres of the global area of GM crops. In 2015 alone, economic gains from GM crops globally were $15.4 billion (USD) and the United States gained the most at $6.9 billion (USD).

- Herbicide tolerance and insect resistance were the first traits to come to the market in the major commodity crops such as cotton, corn, and soybean. These traits remain by far the most successful traits introduced into crop plants by genetic modification.

- Weed control is an essential part of all types of agriculture and is achieved mainly by spraying fields with chemical herbicides. Many problems caused by spraying herbicides have been overcome by the introduction of GM crops that tolerate broad range herbicides.

- The current trend in genetic modification is gene stacking, which refers to the process of transferring two or more genes of interest into a single plant. Stacked varieties are engineered to have better chances of overcoming the myriad of problems in the field such as insect pests, diseases, weeds, and environmental stresses so that growers can increase crop productivity.

- Insect damage can cause substantial loss to growers. Chemical insecticides and, to a lesser extent, biological control agents are used by growers to prevent insect damage. Bt cotton on average requires only 15–20% of the insecticide used on conventional cotton.

- Other traits that have been successfully introduced into plants include traits that allow fruits to ripen without softening, increase nutritional quality, produce unique flower color, reduce browning of fruit, and increase oil content.

- One of the major limiting factors for plant production in many parts of the world is water availability. Although, the gain in yield of the GM varieties under drought conditions seems to be relatively modest, 6.7–13.4%, these varieties show promise for long-term improvement.

- Biopharming refers to the use of GM plants to produce pharmaceuticals, vaccines, antibodies, and even proteins for industrial uses. Plants have been proposed as an attractive alternative to current mammalian or microbial cell-based systems for the production of pharmaceutical proteins.

- GM crops offer many potential benefits such as economic gain, improved food nutrition, reduced pesticides use and exposure, enhanced yields, and resistance to insects and diseases. However, like other technologies, genetic modification also poses some risks, both known and unknown.

- The debates over GM foods focus mostly on uncertainties concerning the potential adverse effects of GM foods and crops on human health and the environment. Three major health risks potentially associated with GM foods are: toxicity, allergenicity, and unintended hazards.

- Adoption of GM crops may endanger nontarget species in the environment.

- Another key concern associated with genetic modification is that transgenes may migrate out of the GM organism to wild species and even to unrelated species.

- It is important to assess these risks so they can be evaluated against the potential benefits or gains of using GM crops.

Reflect

1. *Strengths and Weaknesses* Compare and contrast the use of *Agrobacterium*-mediated transformation and biolistic transformation to modify plants.

2. *Dive Deeper.* Pick one of the commercially available GM crops and research its history, benefits, and risks.

3. *Frontiers of Genetic Modification.* Conduct a literature review and see what genetic modifications have been done to plants recently and discuss the rationale for developing these GM crops.

4. *Worlds Apart.* Compare and contrast the reception of GM crops and foods in Europe and the United States.

References

Agrawal, A., & Sharma, B. (2011). Genetically modified food: Prospects and retrospects. In M. P. Singh, A. Agrawal, and B. Sharma (Eds.). *Recent trends in biotechnology* (Vol. 2). New York: Nova Science Publishers.

Bawa, A. S., & Anilakumar, K. R. (2013). Genetically modified foods: Safety, risks and public concerns—A review. *Journal of Food Science and Technolology, 50*, 1035–1046.

Bechtold, N., Ellis, J., & Pelletier, G. (1993). In planta *Agrobacterium*-mediated gene transfer by infiltration of adult *Arabidopsis thaliana* plants. *Comptes Rendus de l Academie des Sciences Paris, , Sciences de la vie/Life Sciences, 316*, 1194–1199.

Beckie, H. J., Harker, K. N., Légère, A., Morrison, M. J., Séguin-Swartz, G., & Falk, K. C. (2011). GM canola: The Canadian experience. *Farm Policy Journal, 8*, 43–49.

Bevan, M. (1984). Binary Agrobacterium vectors for plant transformation. *Nucleic Acids Research, 26*, 8711–8721.

Bock, R., & Khan, M. S. (2004). Taming plasmids for a green future. *Trends in Biotechnology, 22*, 311–318.

Brookes, G., & Barfoot, P. (2016). Global income and production impacts of using GM crop technology 1996–2014. *GM Crops & Food, 7*, 38–77.

Bruce, A., Johnson, A., Lewis, J., Raff, M., Roberts, K., & Walter, P. (2015). Molecular biology of the cell (6th ed.). New York: Garland Science.

Bruening, G., & Lyons, J. M. (2000). The case of the Flavr Savr tomato. *California Agriculture, 54*, 6–7.

Capecchi, M. R. (1980). High efficiency transformation by direct microinjection of DNA into cultured mammalian cells. *Cell, 22*, 479–488.

Caplan, A., Herrera-Estrella, L., Inzé, D., Van Haute, E., Van Montagu, M., Schell, J., et al. (1983). Introduction of genetic material into plan cells. *Science, 222,* 815–821.

Carpenter, J. E., (2011). Impact of GM crops on biodiversity. *GM Crops, 2,* 7–23.

Castiglioni, P., Warner, D., Bensen, R. J., Anstrom, D. C., Harrison, J. Stoecker, M., et al. (2008). Bacterial RNA chaperones confer abiotic stress tolerance in plants and improved grain yield in maize under water-limited conditions. *Plant Physiology, 147,* 446–455.

CDC. (2001). CDC report to FDA: Investigation of human health effects associated with potential exposure to genetically modified corn. Retrieved December 28, 2018, from the Centers for Disease Control and Prevention website, https://www.cdc.gov/nceh/ehhe/Cry9cReport/pdfs/cry9creport.pdf

Chial, H. (2014). Restriction enzymes. Retrieved December 28, 2018, from Scitable by Nature Education Spotlights website, https://www.nature.com/scitable/spotlight/restriction-enzymes-18458113

Chilton, M.-D., Drummond, M. H., Merlo, D. J., Sciaky, D., Montoya, A. L., Gordon, M. P., et al. (1977). Stable incorporation of plasmid DNA into higher plant cells: The molecular basis of crown gall tumorigenesis. *Cell, 11,* 263–271.

Cohen, S. N., Chang, A. C. Y., Boyer, H. W., & Helling, R. B. (1973). Construction of biologically functional bacterial plasmids in vitro. *Proceedings of the National Academy of Sciences of the United States of America, 70,* 3240–3244.

Davis, K. M. (2009). Modifying anthocyanin production in flowers. In K. Gould, K. Davies, & C. Winefield (Eds.). *Anthocyanins biosynthesis, functions and applications* (pp. 49–84). New York: Springer.

De Moura, F. F., Moursi, M., Angel, M. D., Angeles-Agdeppa, I., Atmarita, A., Gironella, G. M., et al. (2016). Biofortified β-carotene rice improves vitamin A intake and reduces the prevalence of inadequacy among women and young children in a simulated analysis in Bangladesh, Indonesia, and the Philippines. *The American Journal of Clinical Nutrition, 104,* 769–775.

Doudna, J. A., & Charpentier, E. (2014). The new frontier of genome engineering with CRISPR-Cas9. *Science, 346,* 1258096-1–1258096-9.

Fraley, R., Roger, S. G., Horsch, R. B., Sanders, P. R., Flick, J. S., Adams, S. P., et al. (1983). Expression of bacterial genes in plant cells. *Proceedings of the National Academy of Sciences of the United States of America, 80,* 4803–4807.

Fischer, R., Stoger, E., Schillberg, S., Christou, P., & Twyman, R. M. (2004). Plant-based production of biopharmaceuticals. *Current Opinion in Plant Biology, 7,* 152–158.

Gaj, T., Gersbach, C. A., & Barbas III, C. F. (2013). ZFN, TALEN, and CRISPR/Cas-based methods for genome engineering. *Trends in Biotechnology, 31,* 397–405.

Galun, E., & Galun, E. (2001). The manufacture of medical and health products by transgenic plants. London: Imperial College Press.

Gao, C., & Nielsen, K. K. (2013). Comparision Between *Agrobacterium*-mediated and direct gene transfer using the gene gun. In S. Sudowe & A. B. Reske-Kunz (Eds.). Biolistic DNA delivery: Methods and protocols. New York: Humana Press/Springer.

Glover D. M. (1980). Bacteriophage λ vectors. In Genetic engineering cloning DNA. Genetic engineering: Principles and methods. Dordrecht: Springer.

Gonsalves, D., (1998). Control of papaya ringspot virus in papaya: A case study. *Annual Review of Phytopathology, 36,* 415–437.

Guan, Z. J, Zhang, P. F., Wei, W., Mi, X. C., Kang, D. M., & Liu, B. (2015). Performance of hybrid progeny formed between genetically modified herbicide-tolerant soybean and its wild ancestor. *AoB Plants, 7,* plv121. Retrieved December 28, 2018, from National Center for Biotechnology Information, U.S. National Library of Medicine website https://www.ncbi.nlm.nih.gov/pmc/articles/PMC4670487/pdf/plv121.pdf

Halford, N. G. (2012). Genetically modified crops (2nd ed.). London: Imperial College Press.

Hansen, G., Shillito, R. D., & Chilton, M. D. (1997). T-strand integration in maize protoplasts after co-delivery of a T-DNA substrate and virulence genes. *Proceedings of the National Academy of Sciences of the United States of America*, 94, 11726–11730.

ISAAA. (2013). *Stacked traits in biotech crops*. ISAAA Pocket K No. 42. ISAAA: Ithaca, NY.

ISAAA. (2016). *Global status of commercialized biotech/GM crops: 2016*. ISAAA Brief No. 52. ISAAA: Ithaca, NY.

ISAAA. (2017). *GM approval database*. Retrieved December 28, 2018, from the International Service for the Acquisition of Agri-Biotech Applications (ISAAA) website, http://www.isaaa.org/gmapprovaldatabase/commercialtraitlist/default.asp

Kramer, M. G., & Redenbaugh, K. (1994). Commercialization of a tomato with an antisense polygalacturonase gene: The Flavr Savr™ tomato story. *Euphytica*, 79, 293–297.

Lehnman, I. R. (1974). DNA ligase: Structure, mechanism, and function. *Science*, 186, 790–797.

Losey, J. E., Rayor, L. S., & Carter, M. E. (1999). Transgenic pollen harms monarch larvae. *Nature*, 399, 214.

Napier, J. A. (2007). The production of unusual fatty acids in transgenic plants. *Annual Review of Plant Biology*, 58, 295–319.

Narusaka, Y., Narusaka, M., Yamasaki, S., & Iwabuchi, M. (2012). Methods to transfer foreign genes to plants. In Yelda Ozden Çiftçi (Ed.). Transgenic plants—Advances and limitations. Retrieved December 28, 2018 from InTech: https://www.intechopen.com/books/transgenic-plants-advances-and-limitations/methods-to-transfer-foreign-genes-to-plants

Nicolia, A., Manzo, A., Veronesi, F., & Rosellini, D., (2014). An overview of the last 10 years of genetically engineered crops safety research. *Critical Review in Biotechnology*, 34, 77–88.

Nordlee, J. A., Taylor, S. L., Townsend, J. A., Thomas, L. A., & Bush, R. K. (1996). Identification of a Brazil-nut allergen in transgenic soybeans. *The New England Journal of Medicine*, 334, 688–692.

Ow, D. W., Wood, K. V., DeLuca, M., De Wet, J. R., Helinski, D. R., & Howell, S. H. (1986). Transient and stable expression of the firefly luciferase gene in plant cells and transgenic plants. *Science*, 234, 856–859.

Phipps, R. H., & Park, J. R. (2002). Environmental benefits of genetically modified crops: Global and european perspectives on their ability to reduce pesticide use. *Journal of Animal and Feed Sciences*, 11, 1–18.

Pniewski, T. (2013). The twenty-year story of a plant-based vaccine against hepatitis B: stagnation or promising prospects? *International Journal of Molecular Science*, 14, 1978–1998.

Pogue, G. P., Vojdani, F., Palmer, K. E., Hiatt, E., Hume, S., Phelps, J., et al. (2010). Production of pharmaceutical-grade recombinant aprotinin and a monoclonal antibody product using plant-based transient expression systems. *Plant Biotechnology Journal*, 8, 638–654.

REBASE. (2017). *The restriction enzyme database*. Retrieved December 28, 2018, http://rebase.neb.com/rebase/rebase.html

Roberts, R. J., Vincze, T., Posfai, J., & Macelis, D. (2015). REBASE-A database for DNA restriction and modification: Enzymes, genes and genomes. *Nucleic Acids Research*, 43, D298–D299.

Rommens, C. M., Ye, J., Richael, C., & Swords, K. (2006). Improving potato storage and processing characteristics through all-native DNA transformation. *Journal of Agricultural and Food Chemistry*, 54, 9882–9887.

Rommens, C. M., Haring, M. A., Swords, K., Davies, H. V., & Belknap, W. R. (2007). The intragenic approach as a new extension to traditional plant breeding. *Trends in Plant Science*, 12, 397–403.

Sanford, J. C., Klein, T. M., Wolf, E. D., & Allen, N. (1987). Delivery of substances into cells and tissues using a particle bombardment process. *Particulate Science and Technology*, 5, 27–37 .

Schnorf, M., Heuhaus-Url, G., Galli, A., Iida, S., Potrykus, I., & Neuhaus, G. (1991). An improved approach for transformation of plant cells by microinjection: Molecular and genetic analysis. *Transgenic Research*, 1, 23–30.

Shuman, S. (2009). DNA ligases: Progress and prospects. *Journal of Biological Chemistry, 284*, 17365–17369.

Streatfield, S. J., (2005). Oral hepatitis B vaccine candidates produced and delivered in plant material. *Immunology and Cell Biology, 83*, 257–262.

Svab, Z., Hajdukiewicz, P., & Maliga, P. (1990). Stable transformation of plastids in higher plants. *Proceedings of the National Academy of Sciences of the United States of America, 87*, 8526–8530.

Tsien, R. Y. (1998). The green fluorescent protein. *Annual Review in Biochemistry, 67*, 509–544.

USDA/NASS. (2017). *June agricultural survey: Adoption of genetically engineered crops in the U.S., Data Set 2000–2017.* Retrieved December 28, 2018, from the United States Department of Agriculture, Economic Research Service website, https://www.ers.usda.gov/data-products/adoption-of-genetically-engineered-crops-in-the-us/

Waltz, E., (2014). Beating the heat. *Nature Biotechnology, 32*, 610–613.

Williams, R. J. (2003). Restriction endonucleases—Classification, properties, and applications. *Molecular Biotechnology, 23*, 225–243.

WHO. (2009). *Global prevalence of vitamin A deficiency in populations at risk 1995–2005. WHO global database on vitamin A deficiency.* Geneva: World Health Organization.

WHO. (2014). Retrieved December 28, 2018, from the World Health Organization website, http://www.who.int/foodsafety/areas_work/food-technology/faq-genetically-modified-food/en/

Ye, X., Al-Babili, S., Klöti, A., Zhang, J., Lucca, P., Beyer, P., et al. (2000). Engineering the provitamin A (β-carotene) biosynthetic pathway into (carotenoid-free) rice endosperm. *Science, 287*, 303–305.

Zambryski, P., Joost, H., Genetellol, C., Leemans, J., Van Montagu, M., & Schell, J. (1983). Ti plasmid vector for the introduction of DNA into plant cells without alteration of their normal regeneration capacity. *The EMBO Journal, 2*, 2143–2150.

Chapter 18

Evolution

© Stanislav Bokach/Shutterstock.com

Learning Objectives

- Describe how fossils and radioactive dating provided evidence of evolution
- Explain how Darwin and Wallace arrived at similar theories of natural selection
- Understand the four concepts underlying the process of natural selection
- Describe the Hardy Weinberg principle and its assumptions
- Describe contemporary examples of evolution
- Be able to define microevolution, macroevolution, homology, coevolution, stabilizing and disruptive natural selection, allopatry, and sympatry
- Explain different mechanisms of speciation

Evolution describes how life forms changed over millennia and explains how organisms continue to change. In this chapter we examine the mechanisms of evolution, how different forces shape evolution, and supporting evidence for the evolution of different species. When we look at a natural landscape we see rock formations, soil, plants, and animals. Rocks and soil have evolved over time as have all living organisms. Now we look at the living organisms, plants, and what caused them to change over many generations and millions of years.

HISTORY OF EVOLUTION

The modern idea of evolution is a cornerstone of science, but was not formulated until the middle of the 19th century. Early attempts to explain the nature of living things and how they are related were documented by Greek philosophers starting in the 4th century B.C. Plato theorized that all living beings were brought to life by a Creator, an idea developed further by several religions in the following centuries. Aristotle developed the scale of nature, also called the *ladder of life*, that started with nonliving things at the bottom followed by organisms of increasing size and complexity. Humans took the top rank of the ladder. He believed that all organisms were created for a purpose.

Islamic biologists and philosophers interpreted their observations of the natural world from the 9th to the 11th century such that transformations of animals can be caused by their environment in order to ensure survival. In the Middle Ages the prevalent view of Western civilizations was that all organisms were created in their current form according to the biblical narrative in the Book of Genesis and had survived unchanged. Leonardo da Vinci recorded his observations about fossils and rocks at the end of the 15th century and formed hypotheses that Earth was much older than creation, a fact not readily accepted during his time. By the 17th and 18th century the theory of spiritual creation of life was still prevalent. Carl Linnaeus, the founder of modern taxonomy and the binomial nomenclature, authored his work on nomenclature in the 1750s assuming he was cataloguing the work of divine creation.

Subsequent developments of natural history and other sciences, including the collection of fossils, led to different interpretations and more solid ideas about evolution. Erasmus Darwin, grandfather of Charles Darwin, suggested that all warm-blooded animals can be traced back to one original organism, or "filament" as named by him. French naturalist Comte de Buffon proposed that the earth was more than 70,000 years old, more than 10 times the age of what was believed at the time. His other important theory, although later refuted just like his estimated age of the earth, stated that species change over time and during migrations and that species can become extinct. George Cuvier, a French expert on animal anatomy, firmly established the concept of extinction of species by the beginning of the 1800s. Fossil records of elephants in Italy and the fact that the animals were not found in this country anymore, but were still living in Africa, supported the idea that they had become extinct. Cuvier also demonstrated anatomical differences between fossil elephant records from different geographical areas as further evidence of extinction.

In 1809 Jean-Baptiste Lamarck, a French naturalist and biologist, published the first theory of evolution or "transmutation." He thought that simple organisms appeared through spontaneous generation, following Buffon's idea, and over time increased in complexity. His theory is known as the inheritance of acquired characteristics and was based on the beliefs that the changes in organisms were guided by a goal-oriented force organizing life. Lamarck's theory stated that organisms adapted to their environment based on how often certain organs or senses were used or not used. Examples to support his theory were that moles became blind because they did not use their eyes underground and that the neck of giraffes lengthened because they had to stretch it further and further to reach the higher leaves on trees. He proposed that these adaptive changes were then passed on to the following generations. Lamarck believed species kept evolving and would not become extinct. His theories of evolution were most developed, but lacked accuracy because it was not known at the time how traits were passed on to the next generation.

Evolution by Natural Selection According to Darwin and Wallace

Charles Darwin and Alfred Russel Wallace proposed the principle of evolution by natural selection in the 1800s. Their claim to fame is based on the recognition that variation among individuals of a population will lead to changes of a species over time. Desirable traits helping a species to produce more offspring will be passed on to subsequent generations and other traits that prove favorable for survival and procreation will also persist in future offspring. Therefore the individuals best adapted to local conditions will have the best chance for survival. Darwin and Wallace, both British naturalists, worked on this evolution theory independently. Darwin had collected evidence for his theory for over 20 years when Wallace corresponded with him and proposed his own theory of evolution, which was the same as Darwin's, in a paper he planned to publish. This proposition spurred Darwin to promptly publish his own paper and both authors' papers on evolution by natural selection were read at the Linnean Society in London at the same time. Darwin then finished writing the book *On the Origin of Species*, which was published a year later in 1859 while Wallace continued his studies on wildlife biogeography in the Southern hemisphere. The success of Darwin's book was phenomenal and although it took time for the scientific community to fully accept his theory of evolution by natural selection, he became famous as the first naturalist to publish this idea.

Darwin's passion for natural science developed further when he traveled as a geologist on a five-year journey on the *HMS Beagle* (Figure 18.1). The voyage gave Darwin ample opportunities to study the natural world on the trip along the South American coastline, to the Galapagos Islands, Tahiti, around South Africa, and back to England. He spent most of the voyage on land and based on his observations and study of specimens in different geographic locations he started to formulate his ideas on how individuals in a population change over time. His detailed observations of the different beak sizes and shapes in finches found on the Galapagos Islands led him to the conclusion that the many variations of beaks resulted from birds adapting to different foods, but having evolved from a common

(a)

(b)

SPL/Science Source.

SPL/Photo Researchers, Inc.

Figure 18.1 Charles Darwin authored *On the Origin of Species*, the first account of how species evolved. His ideas on evolution were supported by observations he made while working as a naturalist on the *HMS Beagle*, which circumnavigated the world over a period of five years. Studies of plants and animals along the South American coast and on the Galapagos Islands were instrumental for his theory of natural selection.

ancestor finch (Figure 18.2). Slight changes in appearance of tortoises on the different Galapagos Islands led him to the same conclusion as with the finches.

Darwin's theory of evolution was based on the assumptions species change over time, different species have common ancestors, and changes in evolution happen over long periods of time. Previous scholars and contemporary naturalists of Darwin's time had discussed some of these ideas. Fossil evidence and observations in support of the theory began to accumulate. Darwin, however, was the first to publish these ideas as evolution theory with comprehensive evidence in his book *On the Origin of Species*.

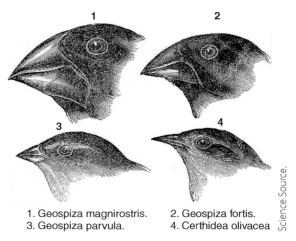

1. Geospiza magnirostris. 2. Geospiza fortis.
3. Geospiza parvula. 4. Certhidea olivacea

Science Source.

Figure 18.2 Variation in beaks of finches from the Galapagos Islands as observed by Charles Darwin.

EVIDENCE IN SUPPORT OF EVOLUTION

The change of populations over long periods of time is documented through **fossil records**, the evidence of history of life, such as remains of bones, plants, or imprints of organisms living in earlier times (Figure 18.3a–b). Fossil records are generally preserved in sedimentary rock and in aquatic habitats and are grouped based on their age. Fossils found in upper layers of sediment are younger and more complex, while with increasing depth they are older and simpler. Fossil records show that organisms in a particular layer of sediment differ from organisms found in layers above or below and deposited thousands or millions of years later or earlier. These records show changes in organisms that include great differences in size, shape, or special features. Some of the species found as fossils, however, resemble current living species, which led to the conclusion that species change over time and previously living species are related to contemporary species. Fossil records gave evidence to support the idea of **extinction** because many species preserved as fossils could not be found living anywhere on Earth. Extinction is now believed to occur both through catastrophic events and gradually over time.

Common ancestry of species living at different times has been supported by fossil evidence. When related fossils are arranged in chronological order and are compared to similar currently living species, evolutionary lines can be traced back based on changes in morphological features. Fossils of extinct ancestors are often discovered in the same geographic region where currently living descendants are found. Fossil records are sporadic and most often preserved when the original material was hard, such as shells, bone, or wood, and decay of organic material was prevented through the absence of oxygen. Fossils are most likely to develop when organisms are covered in fine-textured soil and immersed in water, buried with sand in an arid climate, or preserved in arctic ice. Some fossils are conserved in tar pits, volcanic ash, or lava. Fossil development is uncommon in environments where decomposition of organic material is rapid such as in terrestrial ecosystems with a warm, moist climate.

Methods of dating fossils confirmed the slow changes in evolution over thousands and millions of years. A common method of dating rocks or fossils

fossil records
Evidence of organisms living in the past on earth and recorded in the scientific literature.

extinction
Permanent disappearance of a species from earth due to death of all individuals of the species.

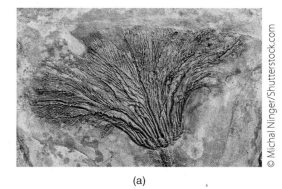

(a)

(b)

Figure 18.3a–b   The imprint of a fossil plant (a) and petrified wood (b) are used to date when different species of plants lived on Earth. The petrified wood, more than 200 million years old, developed when logs buried in sediments were infiltrated with groundwater. Dissolved minerals in the water substitute for the wood and turn the specimen into stone.

radiocarbon dating
Estimates the age of organic remains by measuring the amount of the isotope ^{14}C and calculating the ratio between ^{14}C and ^{12}C.

measures the half-life of natural radioactive elements in these materials. **Radiocarbon dating** uses the naturally occurring ^{14}C to date plant material. ^{14}C is a radioactive isotope of carbon formed naturally in the atmosphere and occurring in trace amounts at a known ratio to ^{12}C. When plants take up carbon dioxide for photosynthesis they take up the radioactive isotope and ^{12}C in the proportion found in the air at that time. ^{14}C has a half-life of 5,730 years and begins to decay upon the death of an organism. The amount of ^{14}C in a fossil sample can be determined and the ratio of ^{12}C to ^{14}C can be calculated. Comparisons to samples of known age based on other dating techniques such as tree rings or mineral deposits in caves increase the accuracy of radioactive carbon dating. Radiocarbon dating is limited to about 60,000 years because of the short half-life of the isotope. Other naturally occurring isotopes with longer periods of decay are used to date older organisms or geological specimens. Radioisotopes of uranium-235 have a half-life of 704 million years, radioisotopes of potassium-40 and uranium-238 have 1.3 and 4.5 billion years, respectively. These methods supported Darwin's claims of the slow changes of evolution, although they were not discovered until the last century.

molecular dating
Uses genomic sequence and the rate of change of nucleotide sequences to estimate when different species evolved.

Molecular dating uses the analysis of genomic sequence data to compare DNA, RNA, or amino acid sequence of proteins to estimate the relationship of organisms. If the sequences are similar, the organisms are closely related and evolved fairly recently into different species. Organisms with less similar DNA, RNA, and proteins are not closely related and likely diverged in their development a long time ago.

homology
Different organisms inherited the same characteristic from a common ancestor. See *analogy*.

Another evidence for evolution is **homology**. This is the case when different species inherited the same trait from a common ancestor. Plants have inherited leaves from a common ancestor. Although the leaves from a palm, a cypress, a sycamore, and a cactus spine look different, they all share the same basic structure. Homology is also evident in the many vegetables selected from the common ancestor wild mustard (*Brassica oleracea*). Breeders have capitalized on different traits of wild mustard and through artificial selection produced the different cultivar groups cauliflower, cabbage, kale and collard greens, broccoli, kohlrabi, and brussel sprouts.

Basis of Natural Selection

Darwin's theory of evolution by natural selection, the fact that modifications could be passed on to the next generation, was based on the following four concepts:

1. Individuals in a population vary in size and form, known as their *traits*. Darwin recognized that variability within a population where certain traits could be passed on to the next generation was the basis for change. For example, drought-tolerant plants will pass on this trait to their offspring.
2. Overpopulation produces many more offspring for each species than can be sustained. Only a small percentage will survive and reproduce.
3. Individuals compete for limited resources in their environment. Although many acorns are produced by plants such as oaks, very few will survive to the age of reproduction and even fewer will successfully produce offspring. If seeds escape herbivores and land in a favorable place to germinate they have to compete for light, water, and minerals to grow successfully to the age where they can produce seeds themselves.

4. Individuals best adapted to an environment will survive and reproduce in greatest numbers. Natural selection favors certain traits that organisms will pass on to the next generation. This increases the survival and reproduction of individuals with these traits versus organisms that do not possess them. This concept of survival to reproduce is known as survival of the fittest.

Evolution is a change in genetic characteristics of individuals in a population. Differences advantageous to individuals may be passed on to the next generation. Over time and with changing environment, individuals within a population change enough that eventually two groups with different traits can be distinguished within that original population. These gradual changes resulted in different species and over time the great diversity of organisms on Earth. Darwin observed this on the Galapagos Islands where closely related species have changed over time to adapt to certain food sources, like the beaks of finches.

GENETIC COMPOSITION AND EVOLUTION

Evolution is the change in genetic traits of a population where certain traits are inherited and increased in frequency by following generations. Modern genetics defines evolution as the change in allele frequencies from one generation to the next. An allele is an alternative form of the same gene that produces a certain effect. Remember from Chapter 15 that diploid organisms have two sets of alleles that are homozygous if they are identical and heterozygous if they are different. In a population, a group of the same species is able to interbreed and each individual has genes. Population geneticists study the gene pool, which consists of all the alleles of all genes occurring in all individuals of a population. They study change in a population by measuring how often certain alleles occur, how they change over time, and what might be the cause for the change in allele frequency.

Populations naturally have genetic variability and some alleles are inherited in greater numbers than others. Individual organisms possessing alleles for favorable characteristics in a certain environment—for example, the ability to survive drought—are more likely to pass on the alleles for this trait under dry conditions whereas individuals without these alleles most likely will not be able to reproduce. Over time the genotype of the successful individuals in the dry environment will change and alleles conferring this trait will be more frequently present in individuals of the next generations. Alleles found in individuals that do not reproduce will eventually decrease or disappear from the population. This example illustrates an evolutionary change in the gene pool of a population.

Fitness as defined by population genetics describes the potential of an individual to reproduce and pass on its alleles to the next generation. A plant with small leaves and a large root system will have greater fitness to survive in a dry environment than a plant with large leaves and a small root system. **Adaptation** in this context describes a genetic trait that is passed on to the next generation and improves the survival and reproduction of an individual under certain conditions. Improved adaptation increases the fitness of an individual organism and if most organisms in a population possess this trait it increases the fitness of the population.

evolution
Theory that organisms on Earth developed over a long time period from common ancestors and changed by natural selection. On a population level evolution is the change in genetic composition or allele frequency.

fitness
In biology, the level of an individual's capacity to survive and reproduce.

adaptation
Inheritance of genetic traits that improve the fitness of an organism or population.

Modern genetics added the last piece of the puzzle to the Darwinian theory of natural selection by explaining how traits that result in differences in survival and number of offspring are inherited from one generation to the next. One important aspect of natural selection is that it affects individuals, but evolutionary changes happen in a population. For example, plants with large leaves and a small root system that germinates in a dry environment will not suddenly grow small leaves and a large root system. The individual plant with large leaves and a small root system, however, is less likely to survive and produce seeds than the plants with small leaves and large root systems. This example demonstrates that individuals in the population are not changing morphologically; rather, the alleles conferring drought tolerance will become more frequent in the population as long as the arid conditions persist. After several generations of this natural selection there will be many more plants with small leaves and large root systems and very few with large leaves and small root systems.

Acclimation occurs when an organism changes in response to slow changes in the environmental conditions. Acclimation can optimize the performance of the organism under new, different conditions. When plant species slowly expand their range, some plants may become acclimated to a higher elevation whereas others of the same species may acclimate to a particular site with shallow soil and dry conditions. This is possible because individuals within a species differ in genetic and morphological characteristics. A plant growing at low elevation may grow larger than a plant of the same species at a higher elevation or at the drier location. The favorable environment at low elevation might provide a longer growing season while colder temperatures in spring and fall prevent the plant at higher elevation to grow to the larger size. Similarly, plants at the drier location will be smaller because less moisture is available. The change in morphology affects only the phenotype but not the genes. Acclimation causes several morphological changes, but these changes are not inherited to the next generation and will not affect evolution.

acclimation
Gradual change of an organism in response to slow changes in the environment.

THE HARDY-WEINBERG LAW

In the early 20th century Godfrey Harold Hardy, a mathematician, and G. Weinberg, a physician, independently proposed an approach to answer the question of why both dominant and recessive traits are conserved in individuals of a population. One can hypothesize that over time the dominant alleles will lead to the extinction of recessive alleles from the gene pool of a population; however, this is not what population biologists observe in nature. Hardy and Weinberg demonstrated mathematically that the proportion or frequency of genotypes and alleles in large populations remain constant from one generation to the next. Equilibrium in the gene pool of a population is maintained and no evolutionary changes occur in the population if the following five assumptions are met:

1. There are no mutations.
2. Populations are isolated from other populations so no alleles will leave or join the gene pool.
3. The population is large enough so the probability is high that the frequency of alleles is not changed by chance alone.

BOX 18.1 Calculating Genotype Frequency According to the Hardy-Weinberg Principle

The Hardy-Weinberg equation predicts the genetic makeup of a new generation assuming the random combination of two alleles B and b in a gene pool from a large population. For this example we assume allele B occurs with a frequency p of 0.7 (p = 0.7 of genotype B) and allele b occurs with a frequency q of 0.3 (q = 0.3 of genotype b). With two alleles in the gene pool, their frequency must total 1. This means that 70% of the genotypes are made up of allele B and 30% by allele b, which adds up to 100% of genotypes. The Hardy-Weinberg equation predicts that each generation will have the following allele frequencies of allele B and b in the population:

$$p^2 + 2pq + q^2 = 1$$

This means the probabilities for genotypes in the next generation are as follows. The three possible genotype combinations in the next generation are BB, Bb, and bb. Their frequencies are:

BB (B, p = 0.7)	offspring genotype frequency	0.7 × 0.7 = 0.49
Bb and bB (B, p = 0.7; b, q = 0.3)	offspring genotype frequency	2(0.7 × 0.3) = 0.42
bb (b, q = 0.3)	offspring genotype frequency	0.3 × 0.3 = 0.09

From the parent generation with 70% alleles B and 30% alleles b the genotypes of the offspring will be 49% BB, 42% Bb, and 9% bb. The allele frequency of B of this new generation can be calculated by taking the frequency of BB plus half the frequency of Bb (because we only count the frequency of B here). Similarly, allele frequency of b equals half of Bb plus bb.

p (allele frequency of B) = 0.49 + 0.5 (0.42) = 0.7 or 70%

q (allele frequency of b) = 0.5 (0.42) + 0.09 = 0.3 or 30%

The original ratio of 0.7 to 0.3 in allele frequency of B and b has been preserved in the next generation according to the Hardy-Weinberg principle. All assumptions of the equation have been fulfilled and no change in the gene pool of the population has occurred.

4. Individuals in the population mate at random.
5. There is no natural selection in the population.

The Hardy-Weinberg principle uses an idealized population and the assumptions that alleles in the gene pool remain in a steady state from one generation to the next. This artificial set of circumstances provides a framework for defining an equilibrium where no change occurs and also a baseline to compare changes to. In nature changes in alleles occur spontaneously or slowly through different mechanisms and the Hardy-Weinberg equation can quantify this change from one generation to the next.

PROCESSES OF EVOLUTION

Changes in populations and species are caused by mutation, gene flow, genetic drift, nonrandom mating, and natural selection.

Mutation

mutation
Random change of genetic composition of an organism.

Random changes in the genetic composition of an organism are **mutations**. Mutations change the overall DNA sequence and expression of proteins and result from base changes, deletion, or insertion of DNA segments. They can alter entire chromosomes, alleles, or some nucleotides and may or may not affect the phenotype (Figure 18.4). Changes in chromosome number also result in mutations. When chromosomes fail to separate during meiosis, the offspring become either polypoids or aneuploids. We discuss mutation in detail in Chapter 15. Mutations occur randomly at low rates in all organisms, and the cause is generally not known. Mutations are not always apparent in a phenotype because they can be recessive. Mutations are not affected by their environment and are not directed toward being beneficial or detrimental to an individual organism and their descendants. Mutations introduce new genetic material, which later can be influenced by other factors and eventually contribute to evolutionary changes.

Gene Flow

gene flow
Movement of alleles between populations; occurs when individuals join or leave one population and produce offspring with an individual from a different population.

When individuals leave or when they join a population, allele frequency of the population changes. This migration into or out of a population is called **gene flow**. This migration is frequent for pollen or seeds carried from one population to another, often over large distances. Gene flow is usually slow when plants depend on special conditions, such as soils high in salt, or extreme pH conditions. For larger populations inhabiting larger areas, gene flow can be faster than for small populations in limited locations.

Genetic Drift

genetic drift
Random effects influencing genotypes of a population.

Random events affecting genotypes of a population are called **genetic drift**. The chance for genetic drift is higher in smaller than in larger populations. Two circumstances favor genetic drift by reducing a population

Dr. Jeremy Burgess/Science Source.

Figure 18.4 Flowers of the African violet have a transposable DNA element (jumping gene), which results in a single purple flower instead of the double pink flower. This mutation is an example how rapid evolutionary change can occur.

size: the bottleneck effect and the founder effect. The **bottleneck effect** occurs when a population is decreased nonselectively through a natural disaster such as volcanic eruption, flood, or fire. The new, smaller population likely has a different frequency of alleles than the original, larger population and thus evolutionary change has affected the population. The **founder effect** causes a change in allele frequency when a small number of individuals split off the larger parent population and establish a new population. Allele frequency in the new population, now separated from the parent population, differs from the original one. A typical example illustrating the founder effect is when a few individuals colonize a new area such as an island.

Nonrandom Mating

When an individual from a population chooses any other individual from that population for mating, then **random mating** occurs. Mating is not random when individuals with certain phenotypic characteristics are preferred over other individuals without those characteristics. Another case of nonrandom mating occurs when individuals in close proximity mate more often than those further apart. An example of this type of nonrandom mating is self-fertilization of plants. This promotes inbreeding and reduces the number of heterozygotes in a population. When mating is not random, allele frequency in the population changes and some genotypes might have less of a chance to reproduce.

Natural Selection

Natural selection favors individuals in a population that are better adapted to their environment than others. Not all genotypes will reproduce at the same rate and the ones better adapted generally will produce a greater number of offspring, as discussed earlier. Natural selection is common in nature and promotes evolutionary change. Phenotypes with great genetic fitness still carry different genotypes and environmental factors further affect natural selection.

Studying finches on the Galapagos Islands since 1973, Peter and Rosemary Grant measured the bird's beak size, body size, and the food they ate (Weiner, 1994). They found that extreme weather could quickly change the composition of the finch population. A severe drought in 1977 killed many smaller finches with smaller beaks because seed production was severely reduced and the smaller, softer seeds were soon consumed. Larger birds with longer beaks survived on the larger, tough seeds usually not consumed by finches and not available to finches with small beaks. Offspring from the survivors were larger with stronger beaks. Later, in a wet year, many small, soft seeds were produced by the plants. Smaller finches with smaller beaks were best adapted for this food and produced more offspring with these characteristics (Boag & Grant, 1981). In this case evolution happened quickly in response to the weather, which affects the food source for birds occupying a different niche.

bottleneck effect
Change in allele frequency of a population compared to their ancestors due to nonselective reduction of a population, for example by a volcanic eruption.

founder effect
Change in allele frequency when a small number of individuals split off the larger parent population and establish a new population that differs genetically from the original one.

random mating
When an individual from a population is equally likely to mate with any other individual in the population.

natural selection
Mechanism of evolution by which organisms with favorable traits will produce more offspring than those without the traits.

stabilizing selection
Type of natural selection where the extreme phenotypes are eliminated while the average of the population is maintained. See *directional selection* and *disruptive selection*.

directional selection
Type of natural selection where one extreme of the population is favored, resulting in a shift of the average toward that extreme. See *disruptive selection* and *stabilizing selection*.

disruptive or diversifying selection
Type of natural selection that eliminates the average phenotype, favoring phenotypes at both ends of the spectrum. See *directional selection* and *stabilizing selection*.

microevolution
Changes in the gene pool from one generation to the next within a species.

There are different types of natural selection based on how a population changes in phenotype. The three types of natural selection are stabilizing, directional, and disruptive. They change the frequency of individual organisms with particular phenotypic characteristics (e.g., flower size). The **stabilizing selection** purges plants with extreme flower sizes, the very small and the very large ones. The population still has a normal distribution with the same average flower size as the original population, but the smallest- and largest-size flowers are gone (Figure 18.5a), thus resulting in a population with greater numbers of intermediate-sized flowers. **Directional selection** is characterized by the removal of one extreme, the largest flower size. This shifts the new distribution curve to the left with a lower average-size flower and largest-size flowers missing (Figure 18.5b). The **disruptive** or **diversifying selection** eliminates the average size flowers and increases the number of smaller and larger flowers. The new distribution curve is bimodal with two peaks and a greater range of flower sizes in the new population than the original one (Figure 18.5c). If disruptive selection continues over time, two species can eventually develop. The different types of natural selection can be a result of competition for limited resources, changes in environmental conditions, or other issues affecting the fitness of traits in a population.

Microevolution and Macroevolution

Microevolution characterizes changes in the gene pool of a population such as the ones commonly seen from one generation to the next. Microevolution is affected by mutation, gene flow or migration, genetic drift, nonrandom mating, and natural selection. The rapid change in finch population characteristics observed by Peter and Rosemary Grant is an example of microevolution. Flowers of a species might show a different color or larger flowers compared to the parent population from which they descended. Another example is the development of herbicide resistance of a weed.

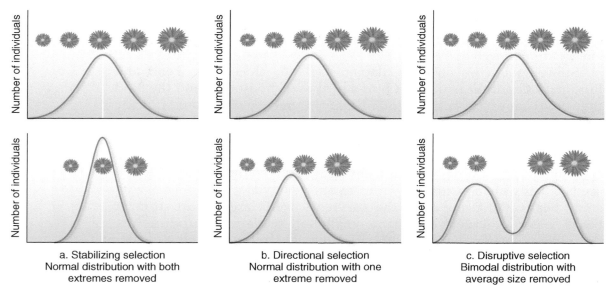

a. Stabilizing selection
Normal distribution with both extremes removed

b. Directional selection
Normal distribution with one extreme removed

c. Disruptive selection
Bimodal distribution with average size removed

Figure 18.5   Natural selection changes the flower size from the original, normal distribution in the top row to the new frequency distributions below. The new flower size frequency in the evolving population is shown in the curves below the original population.

Macroevolution describes large changes in the gene pool that occur as small changes add up over many generations and results in formation of new species. Major evolutionary changes can develop over thousands or millions of years and can span the development of different life forms. These changes are documented in how allele frequencies in the gene pool changed. These changes are documented in the tree of life and in diagrams showing the evolution of different species over time (Figure 18.6). They show many species with a common ancestor as well as ancestral lines that stopped or became extinct.

macroevolution
Changes in the gene pool over many generations and thousands or millions of years results in speciation.

The Pace of Evolution

Early theories of evolution including Darwin's hypotheses assumed a slow and gradual pace of evolution. Evolution can occur over thousands or millions of years, but it can also happen at a faster pace. The **punctuated equilibrium** model by Eldredge and Gould (1993) suggests a stair-step progression in evolution. Steps or evolutionary changes where new species emerge are interspersed with long periods where no changes are detected. The rate of evolution in this model can be very rapid such as a few hundred or thousands of years.

Plant biologists have observed examples of rapid evolution. Samples of *Agrostis tenuis*, bent grass found in meadows but also on mine tailings with soil high in heavy metals, were tested for tolerance to heavy metals (Gregory & Bradshaw, 1965). Most plants are unable to grow on soil high in heavy metals, but bent grass was observed growing on mine tailings. Researchers collected bent grass from the mine tailings and from meadows and grew the plants

punctuated equilibrium
Model of evolutionary change proposing a stair-step progression in evolution where no change occurs for a long time followed by times when many new species emerge.

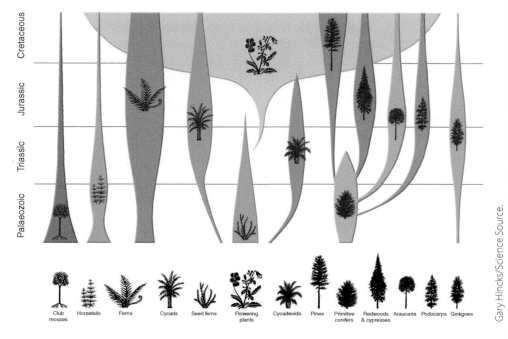

Gary Hincks/Science Source.

Figure 18.6   Plant evolution of major plant groups over four geologic time periods (542 million years ago to 65 million years ago). All plants evolved from a common ancestor but diverged over time into different species. Currently, flowering plants dominate the plant kingdom.

together in soil from the mines and in uncontaminated soil. Grass from the mine populations were tolerant to the soils high in heavy metals, but plants from the meadow did not survive or barely grew at all. Up to 2% of meadow grass survived on the contaminated soil, indicating a small amount of natural resistance. Meadow grass grew normally in the uncontaminated soil, while the heavy metal tolerant grass from the mine tailings did poorly on the clean soil. The plants tolerant to the heavy metals growing on the mine tailing likely evolved from the small number of heavy metal tolerant meadow grass plants. It took only about 100 years from the first seed that became established on the mine tailings until tolerant bent grass plants produced viable seed, expanding the growing area on the mine tailings.

Another contemporary example of rapid evolution is the emergence of herbicide-resistant weeds, such as glyphosate-resistant weeds (Powles, 2008). Glyphosate is the most widely used herbicide in the world. It is nonselective and affects many different plant species, making it a popular broad-spectrum choice in agriculture, natural, and urban settings. Glyphosate resistance in weed species was not reported for 20 years since the introduction of the herbicide in 1974, but now several glyphosate-resistant weeds are reported worldwide. Resistance developed over a period of more than 20 years, most rapidly in genetically diverse species, and primarily when glyphosate alone was the only weed control method.

In 1996, genetic engineering introduced agricultural crops such as cotton, soybean, corn, and canola resistant to glyphosate. This allowed farmers to apply glyphosate over the crop canopy, selectively controlling all plants except the glyphosate-resistant crops. Conversely to the general weed populations, glyphosate-resistant weeds in transgenic glyphosate-resistant crops were first reported in 2001, only five years after release of the transgenic glyphosate-resistant crops (Powles, 2008). Mutation of the EPSPS gene and the reduction of glyphosate translocation to meristems have been identified as the cause for resistance and can be inherited as single gene trait (Powles & Preston, 2006). This example shows how quick plant populations like weeds can adapt to selection pressure from overuse of a single herbicide.

Adaptive Radiation

adaptive radiation
Evolutionary event when several new species are evolving simultaneously from one species and adapting to diverse environments.

Adaptive radiation refers to an evolutionary event when a species rapidly produces several new species with diverse adaptations. Typical examples are when plants or animals colonize an island or an area where no competitors of this species have settled. The colonizing species will often rapidly develop a number of new species with diverse adaptations. Adaptive radiation differs from natural selection by requiring two criteria: emergence of many new species from a common ancestor in a short period of time and adaptation by natural selection (Glor, 2010). A short period can be a few million years or less. Adaptation alone without many new species from one ancestor and only new species from one ancestor without adaptation are not considered adaptive radiation.

The Hawaiian silversword (*Argyroxiphium sandwicense*) and 27 other species from a tarweed subtribe have evolved through adaptive radiation in the isolation of the Hawaiian islands. The ancestor to these species came from California and nearby regions as shown through DNA sequencing and through successful hybridization of modern California and Hawaiian

Noble Proctor/Science Source.

© Steven Maltby/Shutterstock.com

(a)   (b)

Figure 18.7a–b (a) The mountain dubautia or Haleakala naenae plant (*Dubautia menziesii*) is a relative of (b) the silversword (*Argyroxiphium sandwicense macrocephalum*). Both species have a common tarweed ancestor from which they diverged through adaptive radiation on the Hawaiian Islands.

species in these genera. The 28 species can all hybridize, although they range from trees, climbing vines, shrubs, and short, mat-forming shrubs (Figure 18.7). They grow in extremely diverse habitats ranging from lava beds at high elevation, dry scrubland or woodland, wet forest, bogs, and from desert to rainforest. Leaf size and morphology vary from very large leaves on plants in wet ecosystems to small leaves on plants in dry ecosystems. Leaves of the silversword growing on the lava slopes of the Haleakala Crater on Maui are covered with a thick mat of silvery hairs believed to protect the leaves from the extreme solar radiation. Plants in moist or wet habitats lack any hairs on the leaf surface. The plants growing in wet environments have little tolerance to drought stress compared to those species in the arid habitats. Tarweed species of Hawaii and the West Coast of North America are an example of adaptive radiation confirmed by different scientific methods. It is likely that more species have evolved through adaptive radiation over time, but fossil records are not conducive to investigate these relationships.

Convergent Evolution

Organisms with similar appearance that developed independently have evolved by **convergent evolution**. A prominent example of convergent evolution are cacti and Euphorbia species, which share many similarities because they both evolved in arid climates. However, they are not related but adapted independently to the environmental pressures of high temperatures and scarce water. Recall from Chapter 5 that cacti native to South and North America have a large stem for water storage and spines that are modified leaves. Succulent Euphorbia species native to Africa have thorns that are modified shoots. Euphorbia such as the Canary Island spurge (Figure 18.8a) and cacti such as the organ pipe cactus (Figure 18.8b) have developed through convergent evolution but have unrelated ancestors.

convergent evolution
Unrelated organisms evolve independently but have a similar appearance.

(a)

(b)

© Christian Musat/Shutterstock.com

© Martha Marks/Shutterstock.com

Figure 18.8a–b Convergent evolution, the development of similar morphological form in unrelated species, is evident in (a) Canary Island spurge (*Euphoribia canariensis*) and (b) the organ pipe cactus (*Stenocereus thurberi*). The species evolved on different continents but similar climates.

Coevolution

coevolution
Evolution of two species that interact with each other and continuously shape each other's adaptations.

When two or more species are going through **coevolution** they influence each other's adaptations and become interdependent. Recall from Chapter 7 that many plants have coevolved with their pollinators by adapting to special features of birds, bats, or insects depending on their pollen and the plants depending on pollination. The long beaks of hummingbirds are especially well suited to visit tubular flowers (Figure 18.9). Plants have evolved special flower structures to accommodate hummingbirds and preventing other pollinators from reaching the pollen and sometimes nectar that serves as food for pollinators. The yucca plant (*Yucca whipplei*) relies exclusively on the yucca moth (*Tegeticula maculata*) for pollination (see Chapter 7). Larvae of the moth rely on the yucca seeds for food. The moth has special mouth parts to gather and transport the sticky pollen from plant to plant.

SPECIATION

speciation
Process when new species develop from a common ancestor species.

What is **speciation**? Species are populations of interbreeding individuals. Most species do not interbreed with other species. In the rare cases when they do, a hybrid offspring is the result that carries genes from both parents. It occurs when the lineage of a species is split and two species develop. Tracing back the lines on a tree of life diagram, also known as a **phylogenetic tree**, of two closely related species, there is a common ancestor. The point where the common ancestor line is split in two or more species is where the process of speciation happened. We discuss the phylogenetic tree in detail in Chapter 19.

A primary reason for speciation is lack of gene flow between populations, eventually leading to species that are independent from each other. The species have allele frequencies distinctly different from each other and are evolutionarily independent. Three criteria were developed to distinguish evolutionarily independent species both in living organisms and in

(a)

(b)

(c)

© Florian Andronache/Shutterstock.com
© Glenn Price/Shutterstock.com
© Mariusz S. Jurgielewicz/Shutterstock.com

Figure 18.9a–c   Examples of coevolution. The carrion flower (*Stapelia* sp.) produces a pungent odor and relies on flies and carrion beetles for pollination (*a*). The yucca (*Yucca whipplei*) relies exclusively on the yucca moth (*Tegeticula* sp.) for pollination (*b*). Many flowers have evolved to accommodate special pollinators like hummingbirds (c).

fossils. The criteria are the biological species concept, the morphospecies concept, and the phylogenetic species concept.

Reproductive Isolation

The **biological species concept** uses **reproductive isolation** of populations as an indicator that gene flow between populations has stopped and that they do not interbreed and produce viable offspring. This model is useful when reproductive isolation can be confirmed but is not relevant for asexually reproducing species or fossil species.

Reproductive isolation can be based on prezygotic and postzygotic isolation, whether the isolation occurs before or after breeding. **Prezygotic isolation** relates to factors that prevent mating for various reasons. They can

biological species concept
Classifies members of a population that can freely interbreed as a species.

reproductive isolation
When gene flow between population stops and no interbreeding occurs.

prezygotic isolation
Reproductive isolation before pollination where no viable zygote is produced.

be related to time of mating, habitat, behavior, and genetic incompatibility. Different species might be receptive for pollination at different times, therefore they cannot produce an offspring. Several species of a genus might occupy the same geographic area, but are specialized in different habitats with little or no opportunity to interbreed. Scarlet oak (*Quercus coccinea*) and black oak (*Quercus velutina*) share a similar geographic region in the eastern United States. In nature they rarely produce hybrids because the scarlet oak is located on moist, acidic, wet soils while the black oak inhabits well-drained soils. The two species easily generate viable hybrids in cultivation, but are reproductively isolated in nature. Behavior can be a barrier when a flower is open only during a certain time of day, favoring specific pollinators that visit only this species and none of the related ones. Genetic incompatibility prevents successful germination of pollen, successful growth into the pollen tube, and subsequent fertilization of the egg. Genes control whether all steps in this process are successful or whether the match is not compatible.

postzygotic isolation
Reproductive isolation after pollination where the offspring is not viable or sterile.

Postzygotic isolation occurs when hybrids are produced through successful fertilization and zygote develops, but the zygote is either not viable or is sterile if it reaches maturity. Zygotes may die when the endosperm fails to develop and supply nutrients to the developing zygote. Hybrids between species or genera are often sterile or have low fitness.

morphospecies concept
Uses distinct and heritable morphological features to distinguish species.

The **morphospecies concept** relies on distinct and heritable morphological features to distinguish species. It is an older method applicable to both sexually and asexually reproducing organisms (Niklas, 1997). The **phylogenetic species concept** is based on cladistics and defines a species as the smallest group of populations (monophyletic group) of sexual organisms or lineages of asexual organisms, which can be distinguished by a distinct combination of traits in all individuals of the group (Niklas, 1997). This method would distinguish the different selections of vegetables such as broccoli, kale, cabbage, and others from wild mustard as the smallest distinct group. This approach uses genetically fixed traits to distinguish between lineages; the drawback is that few phylogenies are currently established.

phylogenetic species concept
Uses the species as the smallest monophyletic group; see also morphospecies concept and biological concept.

Allopatry

New species develop when gene flow between populations is disrupted. This disruption of gene flow can be due to physical separation and is called **allopatric** (different homeland) speciation. Individuals of a population start to form two populations because they disperse and colonize a different geographic area such as an island. The second mechanism for allopatry is through **vicariance** or a physical barrier separating two populations that previously were one population (Freeman, 2008). This can happen when changes in major barriers such as rivers, glaciers, or mountain ranges separate individuals from the original population.

allopatry
Populations living in separate geographic areas or divided by a physical barrier. See *sympatry*.

vicariance
Division of a population into smaller fragments due to a physical barrier.

Ernst Mayr, a pioneer in evolutionary biology in the 20th century, suggested that speciation happens first because physical separation interrupts gene flow and second because genetic drift will cause further deviations from the gene pool of the original population. As discussed earlier, genetic bottlenecks or the founder effect can cause rapid genetic changes of a smaller population that left the original population and colonized a new area. Environmental pressures in the new habitat can cause natural selection to further change the allele frequency in the population until a new species has developed.

Sympatry

Species living in **sympatry** (the same homeland) live close enough in a geographic area to interbreed. The belief was that sympatric populations would not fragment into different species because of their physically close proximity to allow gene flow. The assumption was that gene flow would homogenize the allele frequency in the gene pool and counteract any shift toward speciation.

Speciation in sympatric populations where no physical or geographic barriers exist occurs through **polyploidy.** Organisms with more than two sets of chromosomes are called *polyploid*. They originate when chromosomes during cell division do not separate and each original chromosome from the diploid individual (2n) is copied, resulting in four copies of a tetraploid organism (4n). The polyploid individual is now reproductively isolated from the parent individuals, which have only half the number of chromosomes and cannot interbreed anymore. Polyploid individuals can successfully interbreed. In nature more than 50% of flowering plants are polyploid and more than 80% of members of the grass family are polyploid. Many important food crops are polyploid, oat and bread wheat are hexaploid (6n), durum wheat is tetraploid (4n), and citrus is triploid (3n). Some polyploid plants are very successful invasive weeds. We discuss polyploidy and new plant development further in Chapter 15.

Allopolyploid organisms are offspring from two different species, where the unpaired chromosomes of the hybrid cannot complete meiosis and the chromosomes are spontaneously doubled. They can now pair and produce diploid gametes, enabling sexual reproduction. Hybrid populations are sometimes sterile, but can reproduce asexually. This can occur vegetatively or through **apomixis,** when embryos in seeds develop without fertilization. The allopolyploid individuals are reproductively isolated from the parent population because of their genetic difference stemming from the two parent species, providing the first step for speciation.

Rejoining of Isolated Populations

What happens when two populations that originated from the same mother population have the opportunity to interbreed again? If the populations developed significantly differently and prezygotic isolation exists, interbreeding will be rare. If, however, no prezygotic isolation exists, the two populations can interbreed and the two gene pools can be joined again. The outcome of such a union can result in reinforcement, hybrid zones, or new species via hybridization (Freeman, 2008). **Reinforcement** occurs when the two populations produce hybrids that are either infertile or have less fitness than their parent populations. This process increases the reproductive isolation and results in further speciation. **Hybrid zones** refer to a geographic area where the hybrid offspring reproduce successfully and show traits of both parent populations. **Hybrid speciation** is when individuals from the two populations mate and a new species with a distinct phenotype evolves. The sunflower species *Helianthus anomalus* is a cross between *H. petiolaris* and *H. annuus*. The hybrid is found in the area of the American West where the ranges of the two parent species overlap. The hybrid has morphological traits and the same number of

---

sympatry
Two or more populations living close enough to allow interbreeding. See *allopatry*.

polyploid
Organisms with three or more sets of chromosomes.

allopolyploid
Characteristic of an individual with more than two sets of chromosomes due to hybridization between two species.

apomixis
Asexual reproduction where seeds develop without fertilization.

reinforcement
Production of hybrids with less fitness or sterility, resulting in further speciation from the parent population.

hybrid zones
Geographic area where hybrid offspring reproduce successfully and show traits of both parent populations.

hybrid speciation
When individuals from two populations mate and a new species with a distinct phenotype evolves.

chromosomes of both parents, so polyploidy is not the mechanism by which it was generated (Freeman, 2008). Gene sequencing confirmed the close relationship between the two parent species and the hybrid. Hybridization in nature is another mechanism how new species evolve and breeders have used this knowledge to create hybrids with new traits artificially.

Key Terms

fossil record	random mating	reproductive isolation
extinction	natural selection	prezygotic isolation
radiocarbon dating	stabilizing selection	postzygotic isolation
molecular dating	directional selection	morphospecies concept
homology	disruptive selection	phylogenetic species concept
evolution	diversifying selection	allopatry
fitness	microevolution	vicariance
adaptation	macroevolution	sympatry
acclimation	punctuated equilibrium	polyploidy
mutation	adaptive radiation	allopolyploid
gene flow	convergent evolution	apomixis
genetic drift	coevolution	reinforcement
bottleneck effect	speciation	hybrid zones
founder effect	biological species concept	hybrid speciation

Summary

- Evolution explains the great diversity of organisms living on Earth and how they developed from a single life form over millions of years. Evolution is also the change in genetic composition of a population over time.

- Naturalists and philosophers started to speculate on how different living organisms developed and how they might be related by the 4th century B.C. attributing living organisms to a creator. The first ideas about evolution and an ancient earth were documented in the 15th century.

- Charles Darwin and Alfred Wallace proposed the idea of evolution by natural selection independently about 1850. Darwin expanded the theory and stated that (1) individuals in a population have inheritable variation and (2) natural selection causes individuals with better-adapted characteristics to have more offspring than others.

- Fossil records are used to estimate the age of organisms living in earlier times based on where they are found and the geological age of the formation. The youngest fossils are found in upper layers of sediments and the older ones are deposited at greater depths. Fossil records support the idea of extinction, the permanent loss of a species, and the idea of slow change in evolution.

- Radioisotope dating of fossils is a precise method to estimate age based on the decay of naturally occurring isotopes of elements such as carbon, uranium, and potassium. Molecular dating uses the sequence of DNA, RNA, or nucleotide sequences to estimate how closely related different organisms are.

- Species inheriting the same trait from a common ancestor are homologous. In contrast, unrelated species developing similar appearance evolved by convergence.

- The theory of natural selection has four basic ideas: (1) Individuals in a population vary in their traits. (2) Most populations will overproduce the number of offspring and only a few will survive. (3) Individuals compete for limited resources. (4) Individuals in a population best adapted to an environment will survive and reproduce in the greatest numbers.

- Evolution of a population is defined as the change in genetic characteristics or the change in allele frequency. Individuals are affected by natural selection, but populations change through evolution. Individual organisms cannot be affected by evolution.

- The Hardy-Weinberg principle states that the proportion of alleles in a population remains constant from one generation to the next if the following five assumptions are met: (1) There are no mutations, (2) populations are isolated from other populations, (3) populations are large enough so frequency of alleles will not change (4) individuals in the population mate at random, and (5) there is no natural selection in the population.

- Evolution is caused by mutation, gene flow, genetic drift, nonrandom mating, and natural selection. Mutations are random changes in the genetic composition of an organism and introduce new alleles into the gene pool. Gene flow occurs when individuals leave or join a population. Genetic drift affect genotypes, especially in smaller populations separated from a larger population through founder effect or bottleneck effect. Nonrandom mating occurs when individuals favor mating with other individuals that possess a certain trait.

- Natural selection favors organisms with the best adapted traits. Selection can be stabilizing by excluding the two extremes, directional by favoring one extreme, or disruptive or diversifying by excluding the average and favoring organisms with traits that are not average.

- Macroevolution describes large changes in the gene pool over long periods of time and results in formation of new species, whereas microevolution describes changes from one generation to the next within a species.

- Evolution can occur over long periods of time or gradually, but can also happen rapidly. Punctuated equilibrium is the rapid evolution of organisms to adapt to certain conditions.

- Adaptive radiation is the rapid emergence of new species from a common ancestor in a short period of time and adaptation by natural selection.

- Coevolution indicates two species evolving together and becoming interdependent. Pollinators adapted to special flowers with special morphology are an example of coevolution.

- Speciation is the development of new species. Species are populations of interbreeding organisms. New species can evolve if they are separated geographically or by natural selection.

- Speciation can occur because of reproductive isolation either before pollination, prezygotic, or after pollination, postzygotic. Prezygotic isolation can be caused by unsuccessful pollination or improper pollen tube growth. Postzygotic isolation indicates the zygote is not viable or the offspring plant is sterile.

- Allopatry contributes to speciation by physically separating part of a population from the original population either through dispersal or vicariance (fragmentation). The new population changes due to genetic drift, mutations, and natural selection until a new species has evolved.

- Speciation in sympatric populations, which are not separated geographically, occurs through polyploidy or hybridization. If hybrid offspring are sterile, they can reproduce asexually or through apomixis.

Reflect

1. *Darwin's theory of evolution.* How did Darwin's theory of evolution differ from previous ideas and how did his voyage on the *HMS Beagle* contribute to the formulation of his hypothesis? What limiting factors did Darwin encounter?

2. *Fossil records.* Your team just discovered a large number of fossil animal bone and residue fragments and fossilized plants and imprints of plants in a particular location. Explain the steps you will take to estimate the age of the fossils and what species they might have been.

3. *The Hardy-Weinberg principle.* Explain the principle and how it can be used to quantify evolution.

4. *Mechanisms of evolution.* What are the major mechanisms of evolution? Which one contributes most to evolution and why?

5. *Reproductive isolation.* Three species with similar morphology have been found in the same geographic area. How can you determine if they are reproductively isolated?

6. *Speciation.* How are small populations that colonize a new area affected by genetic drift and natural selection?

References

Boag, P. T., & Grant, P.R. (1981). Intense natural selection in a population of Darwin's finches (*Geospizinae*) in the Galapagos. *Science, 214,* 82–85.

Freeman, S. (2008). *Biological science* (3rd ed.). San Francisco: Pearson Benjamin Cummings.

Glor, R. E. (2010). Phylogenetic insights on adaptive radiation. *Annual Review of Ecology, Evolution, and Systematics, 41,* 251–270.

Gould, S. J., & Eldredge, N. (1993). Punctuated equilibrium comes of age. *Nature, 366,* 223–227.

Gregory R. P. D., & Bradshaw, A. D. (1965). Heavy metal tolerance in populations of *Agrostis tenuis* Sibth. and other grasses. *New Phytologist, 64,* 131–143.

Niklas, K. (1997). *The evolutionary biology of plants.* Chicago: University of Chicago Press.

Powles, S. B. (2008). Evolved glyphosate-resistant weeds around the world: lessons to be learnt. *Pest Management Science, 64,* 360–365.

Powles, S. B., & Preston, C. (2006). Evolved glyphosate resistance in plants: Biochemical and genetic basis of resistance. *Weed Technology, 20*(2), 282–289.

Weiner, J. (1994). *The beak of the finch: A story of evolution in our time.* New York: Knopf.

SECTION V

DIVERSITY OF THE GREEN WORLD

Chapter 19 **Phylogeny and Taxonomy**

Chapter 20 **Cyanobacteria and Algae**

Chapter 21 **The Plant Kingdom**

Chapter 22 **Bryophytes**

Chapter 23 **Lycophytes and Ferns**

Chapter 24 **Gymnosperms**

Chapter 25 **Angiosperms**

Chapter 26 **Fungi: Friends or Foes of the Green**

Phylogeny and Taxonomy

Source: Cynthia McKenney.

Learning Objectives

- Discuss five plant classification systems
- Identify the primary levels of hierarchical classification
- Define the terms *hydrophyte*, *mesophyte*, and *xeriphyte*
- Explain the cladistics approach to systematics
- List the domains and kingdoms recognized by modern taxonomists
- Complete identification of a plant using a dichotomous key

animals
Kingdom of multicellular eukaryotes that are heterotrophic. They are composed of different tissues and have cell membranes but no cell walls.

archaea
Kingdom of single-celled microbes without any membrane-bound organelles or a nucleus; have no peptidoglycan in their cell walls.

bacteria
Kingdom of prokaryotic organisms without nuclei and lack membrane-bound organelles. Usually they are single celled, but may be found in chains or clusters; have peptidoglycan in their cell walls.

fungi
Kingdom of eukaryotic organisms with chitin in the cell walls and classified by their fruiting structures. They have membrane-bound nuclei and organelles but do not have chloroplasts.

plants
Kingdom of multicellular eukaryotes that have chloroplasts and are autotrophs. They have membrane-bound organelles and cell walls.

protists
Kingdom of eukaryotic single-celled or multicellular organisms that do not have specialized tissues. They may be either heterotrophic or autotrophic.

autotrophs
Organisms that are able to make their own food using photosynthesis.

heterotrophs
Organisms that cannot produce their own food but have to depend on other organisms for food.

CLASSIFICATION SYSTEMS

The process of evolution has progressed for 400 million years. As time has advanced the resulting diversity of organisms has resulted in over 280,000 plant species. This vast number does not include the thousands of extinct species known only through the fossil record. The investigation and usage of plant materials requires the ability to communicate effectively about specific plants, eliminating any question as to which plant is being referred to. One of the ways to help communicate about specific plants is by placing them in a classification system. There are numerous ways to classify plant materials including plant growth requirements, usage, morphology, physiology, and hardiness (Table 19.1). These methods of classification are used more frequently in daily life than other scientific classification methods.

HIERARCHICAL CLASSIFICATION AND TAXONOMY

One natural way to organize and classify living organisms is through their hereditary relationships. As you remember from Chapter 1, taxonomy is the aspect of systematics in which organisms are categorized, named, and described using hierarchical groupings. Carolus Linnaeus, a Swedish botanist, provided the basis for this system of classification, which is based on progressively smaller groupings called **taxa**. A testament to the flexibility of this traditional system is its value and use, with only minor revisions even after thousands more organisms have been identified. This system has the most inclusive grouping, referred to as a *kingdom*. There were six identified kingdoms providing taxa to include all living organisms whose characteristics are described in Table 19.1. The kingdoms include **animals**, **archaea**, **bacteria**, **fungi**, **plants**, and **protists**. Recent phylogenetic analyses reveal that protists do not form a monophyletic group. Kingdom Protista has now been demolished and its members are now classified within several eukaryotic supergroups. The taxon of supergroup lies between domain and kingdom. You may wish to review these kingdoms and the major plant groupings in the kingdom Plantae before you continue with this chapter (see Chapters 1 and 21). The kingdoms are divided by the presence or absence of membrane-bound organelles, cell walls, chloroplasts, and nuclei. In addition, whether they are able to produce energy (**autotrophs**) or must consume their energy (**heterotrophs**) and if the organism is a single cell or multicellular also contributes to the classification system. A summary of these characteristics is provided in Table 19.2.

Each subsequent taxa is more exclusive and narrows from phylum to class, order, family, genus and finally to species or specific epithet. The red line in Figure 19.1 presents an illustration of the **hierarchical classification** of the Japanese rose (*Rosa rugosa*).

BINOMIAL NOMENCLATURE

If we focus back on the 280,000-plus identified species, it becomes evident a system to name plants is necessary. Common names used casually by individuals are colloquial and vary from location to location and individual to individual. For example, Creeping Jenny is a common name that is used for

Table 19.1   Classification Systems of Plants Based on Plant Requirements, Usage, and Performance in Variable Environments

Classification Criteria	Classification Categories	Example
Plant requirements	• Water	• Hydrophyte (water loving) • Mesophyte (moderate water) • Xeriphyte (drought loving)
	• Light	• Shade tolerant • Sun tolerant
	• Nutrients	• Heavy fertilization • Light fertilization
	• Soil type	• Acidic soil (low pH) • Alkaline soil (high pH)
	• Habitat	• Epiphytic (on another plant) • Terrestrial (on land) • Aquatic (in water)
Plant usage	• Food	• Fruit • Vegetables
	• Recreation	• Football field • Tennis court
	• Aesthetic	• Shade tree • Foundation shrub • Perennial border • Color bed • Ground cover
Plant morphology	• Herbaceous (dies to the ground and returns)	• Herb (self-supporting) • Vine (requires support)
	• Woody (does not die to the ground)	• Shrub (multi-axis) • Tree (single axis) • Liana (requires support)
Plant physiology	• Foliage persistence	• Deciduous (leafless part of the year) • Evergreen (leaves persist) • Semi-evergreen (variable with climate)
	• Lifespan	• Annual (life cycle in one growing season) • Biennial (life cycle in two growing seasons) • Perennial (life cycle in three or more growing seasons)
Hardiness	• Tender	• Annuals (freeze in current climate) • Perennials (freeze in current climate)
	• Hardy	• Annuals (do not freeze in current climate) • Perennials (do not freeze in current climate)

Table 19.2 Characteristics, Examples, and Images of the Six Taxonomic Kingdoms

Kingdom	Characteristics	Example	Example Image
Animalia (Animals)	Multicellular **eukaryotes**, which do not have the ability to carry out photosynthesis and so are heterotrophic. They are composed of different tissues and have only cell membranes but no cell walls.	Turtle	Source: Cynthia McKenney.
Archaea	Single-celled microbes without any membrane-bound organelles or a nucleus (prokaryotes). Do not have peptidoglycan in their cell walls.	Geothermal pool where archaea are found	©James Mattil/Shutterstock.com
Bacteria	Prokaryotic organisms that do not have a nucleus and usually do not have membrane-bound organelles. Frequently are single celled, but may be found in chain-like or cluster-like colonies. Have peptidoglycan in their cell walls.	Bacillus bacteria	©Fedorov Oleksy/Shutterstock.com
Fungi	Eukaryotic organisms with chitin in the cell walls and frequently classified by their fruiting structures. They have membrane-bound nuclei and organelles but do not have chloroplasts.	Shelf fungi	©Aleksander Bolbot/Shutterstock.com
Plantae (Plants)	Multicellular eukaryotes that have chloroplasts and are autotrophs able to create their own energy. They have membrane-bound organelles and cell walls made of cellulose.	Nasturtiums	Source: Cynthia McKenney.
Protista (Protists)	Eukaryotic single-celled or multicellular organisms that do not have specialized tissues. They may be either heterotrophic or autotrophic depending on the presence or absence of chloroplast.	Diatoms	©Jubal Harshaw/Shutterstock.com

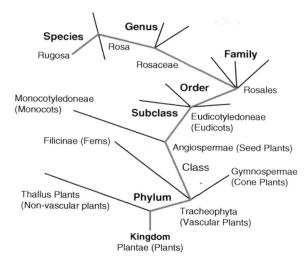

Figure 19.1   Hierarchical classification of the Japanese rose, *Rosa rugosa*.

over a dozen trailing plants ranging from hanging baskets to groundcovers. When someone asks a question about Creeping Jenny, you do not know which plant is being discussed. Alternately, there are dozens of different names historically used for the white water lily. Binomial nomenclature is an efficient two-part naming system developed to reflect the phylogenetic relationships of plants. In this system plants are assigned a **binomial** or a two-part name that consists of two terms. The first term identifies the genus or generic name. A genus is a grouping of closely related species. The second term of the binomial name is the specific epithet, which is the species within the genus. A species is a grouping of very similar organisms that may cross and produce viable offspring. The binomial name, which results from the union of these two terms, is referred to as the scientific name and it is unique to a specific plant. Scientific names are Latin based, are used universally, and are frequently descriptive of the plant. For example, *Morus alba* is the white mulberry (*alba* is Latin for "white").

Rules for Writing Scientific Names

There are several rules for writing scientific names. First, the genus is always capitalized and the specific epithet is not. In print, both parts of the name are italicized; however, when written by hand, the two terms are both underlined. *Tagetes patula* is an example of the two-part scientific name for French marigolds. Occasionally plant names are even more specific and a third term is included, which results in a trinomial. *Gleditsia triacanthos* var. *inermis* is the thornless Honeylocust. The third term *inermis* is included to identify the naturally occurring thornless variety of the tree and this third term is also written in italics. When plants are intentionally hybridized, the resulting plants are said to be cultivars. *Cultivar* is a term that refers to the concept of a cultivated variety. When cultivars are included in a scientific name, the cultivar is placed in single quotes and is not italicized. For example, *Lobularia maritima* 'Snow Crystals' is an intentional cross bred to create a compact and floriferous sweet alyssum.

hierarchical classification
System of grouping organisms into progressively more limiting taxa or groupings based on their common and divergent traits.

binomial
Two-part naming system that is composed of the genus and the specific epithet.

(a)   (b)

Figure 19.2a–b (a) English ivy and (b) Japanese fatsia have been successfully crossed resulting in the intergeneric hybrid commonly called botanical wonder.

How to Handle Hybrids

Some plants, such as petunias, are crossed repeatedly over many generations until the parentage is more obscure. In these situations, the scientific name reflects the extensive breeding programs. Note the convention when writing this type of name such that the "x" in *Petunia* x *hybrida* 'Silver Wave' is lowercase and not italicized. The specific epithet is *hybrida* and the cultivar is written as normal. Hybridization is rare between genera. With the advent of new technology, there have been a few intergeneric hybrids such as X*Fatshedera lizei*, also known as botanical wonder or tree ivy (Figure 19.2a–c). This plant is a cross between *Hedera helix* (English Ivy) and *Fatsia japonica* (Japanese fatsia). Note an upper case "X" representing the hybrid is placed before the first term in the binomial to represent the intergeneric hybridization and it would be read, "the hybrid *Fatshedera lizei*." Do not consider the "x" when alphabetizing plants in a list as genera are always listed first.

SYSTEMATICS AND CLADISTICS

Systematics

systematics
Study of biological diversity and evolutionary relationships among organisms.

phylogeny
Evolutionary history of an organism.

monophyletic group
Natural group that includes all of the taxa sharing a common ancestor.

paraphyletic group
Includes all of the taxa sharing a common ancestor but missing one or more species.

polyphyletic group
Includes diverse taxa representing several different evolutionary lines.

Technology has also provided the opportunity to obtain more information about plants at the molecular level including DNA and RNA sequences. This enhanced information has led to the development of multiple classification systems based not only on the morphology of the organisms but also on molecular information. The study of these evolutionary relationships is known as **systematics**. Systematists attempt to determine the evolutionary relationships between plants, thus creating a **phylogeny** or evolutionary history. As is the want of systematists, taxa are grouped together in several ways. If there is a true evolutionary relationship between all of the taxa and they have a common ancestor, then they comprise a natural grouping called **monophyletic** taxa (Figure 19.3). When all of the taxa have a common ancestor but one or more of the group is not included, this is referred to as a **paraphyletic** group (Figure 19.3). However, if there is no common ancestor, the evolutionary relationships are more diverse and may represent several different evolutionary lines, then these groupings are referred to as **polyphyletic** (Figure 19.3) and taxonomists work to determine more accurate phylogeny for them.

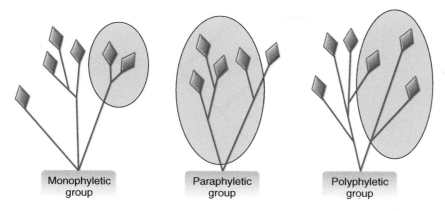

Figure 19.3   The three types of groupings: monophyletic, paraphyletic, and polyphyletic.

Cladistics

Classification based on evolutionary relationships has increasingly gained in popularity as the ability to research evolutionary lines has improved. **Cladistics** is classification based on the phylogeny of an organism, focusing on how the lineage branches in different directions. The identification of branching points helps to determine the common ancestral lines and results in a visual representation of the evolutionary relationship in the form of a diagram called a **cladogram**. Cladograms consist of branches and nodes. A node represents when an evolutionary change in a single characteristic divides related organisms. The characteristic is held by two lines that have a shared common descent and evolve in independent directions from one another. The point on the cladogram where the division occurs is called a node. Everything prior to the node has common ancestry. Everything past the node has common ancestry only within those segregated plants on that path from the *node*. Each node farther up the branch represents a divergence more recent than the one below it. Figure 19.4 is a simplistic cladogram showing the node division points for four branches of divergence. In an actual cladogram, there would be many more nodes providing the genetic divergence of this shrub rose.

When there are multiple species with several different characteristics to consider, it becomes apparent there are numerous ways a cladogram may be constructed. The **principle of maximum parsimony** helps to determine the most appropriate cladogram. The principle is also called "Occam's razor," named after William of Occam, a 14th-century English philosopher. William of Occam advocated this minimalist problem-solving approach. According to this principle, we should construct a cladogram or phylogentic tree that requires the fewest evolutionary events or shared derived morphological characters if it is based on morphology. For phylogenies based on DNA sequence, the cladogram or phylogentic tree should require the fewest base changes.

DOMAINS

For many years the taxonomic system functioned fairly well and was flexible enough to accommodate the addition of new plant families. However,

cladistics
Classification system based on the phylogeny of an organism focusing on how the lineage branches.

cladogram
Diagram providing a visual representation of the evolutionary relationship of an organism.

maximum parsimony
Minimalist problem-solving approach. When considering multiple explanations for an observation, one should investigate first the simplest explanation that is consistent with the evidence.

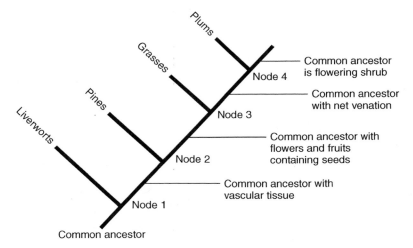

Figure 19.4   Simplistic cladogram of *Rosa*.

as the complexities of living organisms became more apparent, the division into kingdoms failed to provide adequate levels of hierarchy. With the advent of electron microscopy, it was determined the members of the plant, animal, fungi, and protist kingdoms are all eukaryotes. Their DNA is enclosed in a membrane creating a true nucleus. In contrast, bacteria do not have a membrane around their DNA so there is no true nucleus classifying them as prokaryotes. This distinction provided a very distinctive characteristic for classification. Carl Woese, an American bacteriologist, discovered prokaryotes had a natural division in their evolutionary decent and have followed divergent evolutionary paths since that time. Woese and his team determined the DNA coding for ribosomal RNA varied more among prokaryotes than between plants, animals, and fungi. Given the entrenched concept of the plant, animal, and fungi kingdoms, another taxa was created to provide an appropriate level of grouping. This higher level of organization is called a **domain** (Figure 19.5) Currently

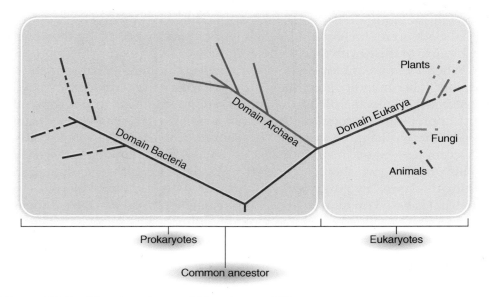

Figure 19.5   The domains and kingdoms of life.

there are two recognized prokaryotic domains that include the domain Bacteria and the domain Archaea. There is a third domain, which is the domain Eukarya. The plants, animals, and fungi are all included in the domain Eukarya along with the components of the protist kingdom. As more information has shed light on the structure of protists, there is a debate over whether they should be segregated into several supergroups. Similarly, the domains Bacteria and Archaea may be divided in the future. Classification is as dynamic as the living organisms it organizes. As the knowledge base continues to grow with the advent of more technology, it is natural to expect the classification system to continue to grow concurrently.

BOX 19.1   Dichotomous Key

Dichotomous Key of Common Plants

1. a. The plant is a monocot (parallel venation) 11
   b. The plant is a eudicot (net venation).. 2
2. a. The plant has simple leaves (leaves have a single blade or lamina) ... 3
   b. The plant has compound leaves (there are several leaflets that make up the blade) 9
3. a. The plant has entire margins (edges of the leaves are smooth) 4
   b. The plant has margins that are not entire.. 10
4. a. The leaf is succulent (stores water, covered in wax)............................. 5
   b. The leaf is not succulent (does not store water, not covered in wax).. 6
5. a. The leaf is gray ....*Echeveria cante* 'White Shadow'
   b. The leaf is not gray ....*Echeveria Compton* 'Carousel'
6. a. Leaf is cordate (heart shaped).. 7
   b. Leaf is not cordate (not heart shaped) 8
7. a. Leaf is dark blue-green .... *Anthurium andraeanum*
   b. Leaf is shiny spring green.....................................*Colocasias × hybrida*
8. a. Leaf has olive and gray variegation...................*Aglaonema commutatum*
   b. Leaf has olive and maroon variegation ....*Maranta leuconeura*
9. a. Leaves are trifoliate (three leaflets) ....*Oxalis triangularis* 'Allure Burgundy'
   b. Leaves are pinnate (leaflets directly opposite each other along an axis) ....*Tagetes patula* 'African Discovery Orange'
10. a. Leaves have dissected margins (cuts in the edges)....*Monstera deliciosa*
    b. Leaves have serrate margins (edges are toothed) ......*Rubus occidentalis*
11. a. Leaves are gray-green and grass-like........... *Festuca glauca* 'Elijah Blue'
    b. Leaves are palmate, gray-green blades ....*Bismarckia nobilis*

BOX 19.2 Dichotomous Key

dichotomous key
Series of couplets guiding the stepwise identification of an organism.

couplets
Question with results in two divergent selections, allowing for the determination of specific characteristics for identification.

A **dichotomous key** is helpful in identifying plants correctly. Through a series of questions called **couplets**, the participant is guided stepwise to the correct identity of the plant. In this box, there is a simple dichotomous key (Table 19.1) that initiates the process using very general questions about a plant's characteristics. Each question has two options to select from to proceed. Select the couplet that best matches the plant you are trying to identify and go to the question number provided at the end of the couplet. Each successive couplet progresses to more narrow traits until ultimately the plant is identified. Try your hand at using a dichotomous key by selecting a plant from Figure Box 19.1. After looking at the plant image carefully, determine the leaf characteristics of the plant from Figure Box 19.2. Finally, look at Table Box 19.1 and read the first couplet. Select the couplet which is most correct for your plant and go to the next couplet indicated at the end of the description. Continue to do this until your discover the name of the plant. Compare your results with the name of the plant in Figure 19.1. Did you correctly identify your plant? Taxonomic terms are defined in the question to aid in your process. A taxonomic image dictionary of selected terms found in the Dichotomous Key is also provided (Figure 19.2). Normally there will not be names with the images to be identified; however, they are provided in this exercise to aid in your learning the process of using a dichotomous key.

Figure Box 19.1 Sample Plant Materials Included in the Dichotomous Key of Common Plants

Aglaonema commutatum

©Pekka Nikonen/Shutterstock.com

Anthurium andraeanum

Source: Cynthia McKenney.

Bismarckia nobilis

©Madlen/Shutterstock.com

Colocasias × hybrida

©pixelsnap/Shutterstock.com

Echeveria cante 'White Shadow'

Source: Cynthia McKenney.

Echeveria Compton 'Carousel'

Source: Cynthia McKenney.

Festuca glauca

©Kelly MacDonald/Shutterstock.com

Maranta leuconeura

©Pekka Nikonen/Shutterstock.com

Monstera deliciosa

©Robynrg,/Shutterstock.com

Oxalis triangularis 'Allure Burgundy'

Source: Cynthia McKenney.

Rubus occidentalis

©Richard Peterson/Shutterstock.com

Tagetes erecta

©Quang Ho/Shutterstock.com

Figure Box 19.2   Taxonomic Image Dictionary of Selected Terms Found in the Dichotomous Key of Common Plants

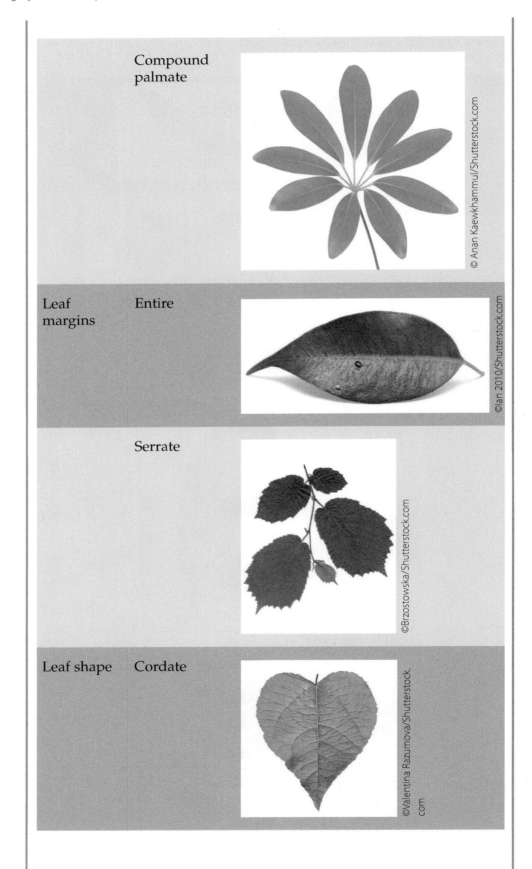

Compound palmate

© Anan Kaewkhammul/Shutterstock.com

Leaf margins   Entire

©Ian 2010/Shutterstock.com

Serrate

©Brzostowska/Shutterstock.com

Leaf shape   Cordate

©Valentina Razumova/Shutterstock.com

Ovate

Linear

Leaf tips   Acute

Obtuse

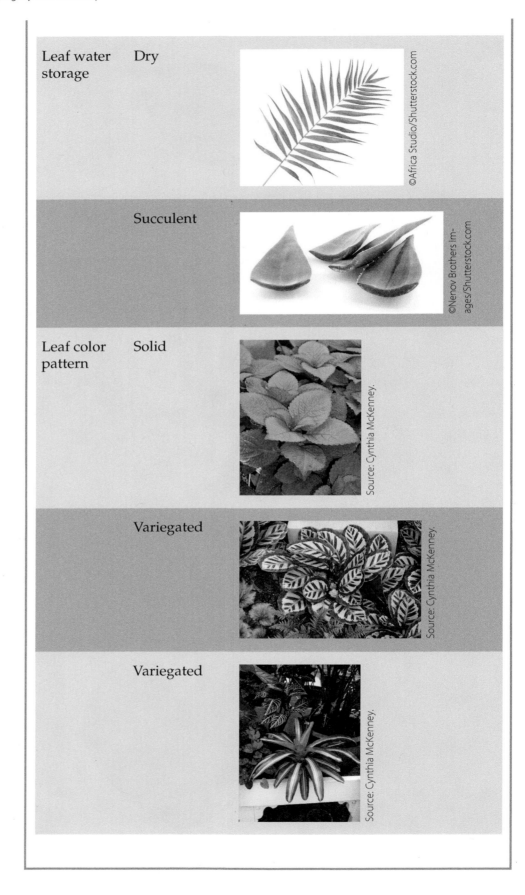

Leaf water storage	Dry	©Africa Studio/Shutterstock.com
	Succulent	©Nenov Brothers Images/Shutterstock.com
Leaf color pattern	Solid	Source: Cynthia McKenney.
	Variegated	Source: Cynthia McKenney.
	Variegated	Source: Cynthia McKenney.

Key Terms

animals	heterotrophs	polyphyletic
archaea	hierarchical classification	cladistics
bacteria	binomial	cladogram
fungi	systematics	principle of maximum
plants	phylogeny	parsimony
protists	monophyletic	dichotomous key
autotrophs	paraphyletic	couplets

Summary

- There are many different types of classifications systems:
  - Plant requirements
  - Plant usage
  - Plant morphology
  - Plant physiology
  - Plant hardiness
- Hierarchical classification and taxonomy
  - Based on evolutionary relationships
  - Organisms may be classified by energy production
    - Energy-producing organisms are called autotrophs
    - Energy-consuming organisms are heterotrophs
  - Organisms are divided into six kingdoms:
    - Animals are multicellular eukaryotes that do not have the ability to carry out photosynthesis. They are composed of different tissues and have only cell membranes but no cell walls.
    - Archaea that are single-celled microbes without a membrane-bound nucleus. Do not have peptidoglycan in their cell walls.
    - Bacteria are prokaryotic organisms that do not have a nucleus and usually don't have membrane-bound organelles. Have peptidoglycan in their cell walls.
    - Fungi are eukaryotic organisms with chitin in the cell walls and have membrane-bound nuclei and organelles but do not have chloroplasts.
    - Plants are multicellular eukaryotes that have chloroplasts and are autotrophs able to create their own energy. They have membrane-bound organelles and cell walls made of cellulose.
    - Protists are eukaryotic single-celled or multicellular organisms that do not have specialized tissues. They may be either heterotrophic or autotrophic depending on the presence or absence of chloroplasts.
- Binomial nomenclature
  - System of naming plants in which a plant has a unique name consisting of two terms, the genus and the specific epithet.
  - Reflects phylogenetic relationships of plants
  - There are several rules for writing the species names of plants:
    - Capitalize the genus but not the specific epithet
    - "x" between terms means it is an intraspecific cross

- "X" before the genus means it is a rare interspecific cross
- Occasionally there is a third term to identify naturally occurring varieties
- Intentionally hybridized crosses produce cultivated varieties known as cultivars. These are capitalized and in single quotes.
  - Systematics is a classification system based on evolutionary relationships
  - Cladistics classifies organisms by their evolutionary relationship by focusing on how the lineage branches into divergent paths. The diagrams are called cladograms.
  - Nodes are the points that the paths diverge from a common ancestor
  - Domains are a taxonomic grouping that is greater than a kingdom
    - There are three domains:
      - Bacteria
      - Archaea
      - Eukarya
  - Dichotomous Key
    - Used to identify plants
    - Composed of couplets leading to increasing levels of complexity

Reflect

1. As time progresses, technology will advance and our understanding of hereditary relationships will deepen. In which domain do you think there will be the most changes and why?

2. Taxonomists have continued to divide large groupings into increasingly smaller groupings based on DNA and cellular-level characteristics. Do you think there will ever be a movement to group organisms back into larger groupings rather than parsing them into smaller groups and, if so, to what advantage would this be?

3. What are the general advantages and disadvantages of using common names and scientific names?

4. If you were developing your own dichotomous key, what would be your first set of couplets?

Chapter 20

Cyanobacteria and Algae

© Elizabeth C. Doerner/Shutterstock.com

Learning Objectives

- Name the key features that are shared by all inhabitants of the Green World
- Compare and contrast prokaryotic cyanobacteria and eukaryotic algae
- Distinguish between cyanobacteria and prochlorophytes
- Compare and contrast the six algal groups
- Explain the evolutionary importance of cyanobacteria and algae to plants
- Describe the importance of cyanobacteria and algae to humans

Although the primary focus of this book is on plants, cyanobacteria, formerly known as blue-green algae, and algae are important members of the Green World. Like plants, cyanobacteria and algae are autotrophic organisms that carry out photosynthesis to produce their own food and generate oxygen as a by-product. They are also important contributors to the global carbon and nitrogen cycles. Plants are the dominant primary producers on land and cyanobacteria and algae are their counterparts in freshwater and marine environments. Cyanobacteria and algae might even be considered the "pioneers" of this Green World. According to paleontological, geological, and geochemical evidence, cyanobacterial ancestors evolved approximately 3 to 3.5 billion years ago (Schopf, 2002). These photosynthetic bacteria enriched the atmosphere with oxygen to a level that transformed the geochemistry of the planet and allowed oxygen-dependent life forms to flourish. The oxygen in the atmosphere eventually formed the protective ozone layer, which protects organic life forms against harmful ultraviolet (UV) radiation from the sun.

From an evolutionary perspective, primeval cyanobacteria may have been the ancestor of chloroplasts and thus played an important role in the evolution of photosynthetic eukaryotes such as algae and plants. Recall from Chapter 3 that eukaryotes are organisms that are made up of cells with a nucleus. In this chapter we investigate unique characteristics of cyanobacteria and algae, the "pioneers" of the Green World, and their ecological and evolutionary importance.

CHARACTERISTICS OF THE INHABITANTS OF THE GREEN WORLD

Plants, cyanobacteria, and algae are photoautotrophs that harness energy from the sun and convert it to chemical energy by photosynthesis, an important process that was explored in great detail in Chapter 10. There are a number of similarities in how these organisms carry out photosynthesis. A variety of photosynthetic pigments such as chlorophylls (a, b, c, d, and f), carotenoids, and phycobilins are used by different organisms to capture light energy. Plants, cyanobacteria, and algae all use **chlorophyll** as the main photosynthetic pigments, thus giving them a green color. They all possess two photosystems, **photosystem II (PS II)** and **photosystem I (PS I)**, and these photosystems work together to produce ATP and NADPH for the Calvin cycle. They all use **water as the electron donor** for their PS II and release oxygen when water is being split apart during the light reactions. In addition to these similarities, molecular and genetic evidence suggest that primeval cyanobacteria were the ancestors of the chloroplasts found in plants and most algae (see Chapter 3, Box 3.1).

TAKING A CLOSER LOOK AT CYANOBACTERIA

Prior to 1970, cyanobacteria were thought to be primitive eukaryotic algae or even a transition between prokaryotes and eukaryotes and hence they were called blue-green algae (Box 20.1). In contrast to eukaryotes, prokaryotes do not have a nucleus in their cells (see Chapter 3). In addition to chlorophyll a, a green pigment, cyanobacteria also use **phycobilins**, blue or red pigments, for photosynthesis. As a result, they look bluish-green, hence the

chlorophyll
Green photosynthetic pigments that are essential for photosynthesis.

photosystem II (PS II)
Protein and pigment complex that is used to harness energy from light to make ATP; an important component of light reactions of photosynthesis.

photosystem I (PS I)
Protein and pigment complex that is used to harness energy from light to make NADPH and occasionally ATP; an important component of light reactions of photosynthesis.

water as the electron donor
Water molecules are used as source of electrons.

phycobilins
Blue or red pigments that are used by photosynthetic bacteria to carry out photosynthesis.

BOX 20.1   Algae or Bacteria?: The Case of Mistaken Identity

Until the late 1960s, cyanobacteria were thought to be closely related to eukaryotic algae and called blue-green algae. At that time, the prevailing thought was that prokaryotes had gradually evolved into eukaryotes through the enclosure of the nucleoid by the membrane system and cell enlargement. Prokaryotic cells (about 1 µm in diameter) are generally 10 times smaller than eukaryotic cells. Most cyanobacteria are large in comparison. For example, the cell of a marine cyanobacterium, *Lyngbya majuscula*, is about 80 µm in diameter and 8 µm in height, one of the largest prokaryotic cells (Schulz & Jørgensen, 2001).

Many cyanobacterial cells aggregate to form large colonies or filaments and some of the cells even undergo differentiation. Many cyanobacteria form heterocysts, cells specialized for nitrogen fixation (Figure Box 20.1). These colonies or filaments with more than one cell type are analogous to the bodies of many eukaryotic algae. In addition, cyanobacteria carry out the same type of photosynthesis as algae. As we mentioned earlier, cyanobacteria, like algae, use chlorophyll a as the main photosynthetic pigment, have two photosystems, and use water as the electron donor. The evidence suggested that cyanobacteria had all the characteristics of an intermediate stage between prokaryotes and eukaryotes.

However, the theory of endosymbiosis proposed by Lynn Margulis in 1971 radically changed our views on the origin and evolution of eukaryotic cells (see Chapter 2). Since the 1970s, evidence began to accumulate in support of the eukaryotic cell origin as explained by the theory of endosymbiosis.

Electron microscopy has advanced our understanding of prokaryotic ultrastructure (fine details of cell structures that can only be seen with an electron microscope) and enabled us to correct this case of mistaken identity. Cyanobacteria are not eukaryotic algae. They are prokaryotes and play an important role in the evolution of autotrophic eukaryotes. This case of mistaken identity also illustrates an important principle in how science works. When new evidence arises, a hypothesis should be reevaluated, modified, or

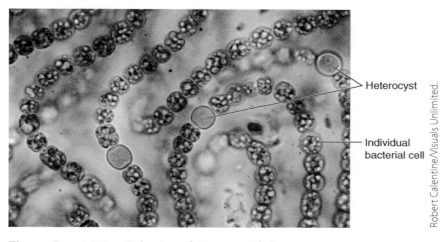

Heterocyst

Individual
bacterial cell

Robert Calentine/Visuals Unlimited.

Figure Box 20.1   Colonies of *Nostoc* with heterocysts.

even abandoned. It is important to realize that we do not know absolutely everything about a subject and data that we gathered on a given subject may not be complete or even correct. By "riding the scientific method cycle" (Chapter 1), we learn more from being wrong than we do from being right. A wrong hypothesis often opens up even more questions and experiments as we attempt to understand why our predictions were incorrect.

name blue-green algae. Cyanobacteria are actually prokaryotes and belong to the phylum Cyanobacteria in the domain Bacteria. The phylum Cyanobacteria is the only bacteria phylum capable of photosynthesis and oxygen production. These bacteria live in a wide range of habitats (Whitton & Potts, 2002). Some are found in or on the soil and also in marine, fresh, and brackish waters. Some are found in extremely inhospitable conditions such as hypersaline (water with extremely high salt content) environments, the mineral water of hot springs, or the frigid lakes of Antarctica (Figure 20.1). However, they are usually absent in acidic waters where eukaryotic algae are common. Cyanobacteria are known to form microbial mats with other heterotrophic bacteria and even algae. These mats may calcify into laminated rocks called *stromatolites* (Figure 20.2). Chemical analyses of stromatolites in Western Australia reveal that photosynthesis occurred at least 2.3 billion years ago and caused a rise in oxygen level in the atmosphere (Stal, 2002).

Differences in Body Forms and Structures

Unlike plants and algae, cyanobacteria have prokaryotic cells that have no nucleus or other membrane-bound organelles such as plastids and mitochondria. Cyanobacteria belong to the domain Bacteria whereas plants and algae belong to the domain Eukarya. Cyanobacteria, like all other prokaryotic bacteria, are unicellular and the cells aggregate to form filamentous or sheet-like colonies (Figure 20.3). Microscopic algae tend to be unicellular

Figure 20.1 Cyanobacteria in mineral water of Beauty Pool Hot Spring (Yellowstone National Park, Wyoming), giving it an intense coloration.

Figure 20.2a–b   (a) Stromatolites of fossilized cyanobacteria (Glacier National Park, Montana) and (b) stromatolites in Hamelin Pool (Shark Bay, Western Australia).

and their cells also aggregate to form colonies or filaments whereas larger algae tend to be multicellular. Plants are strictly multicellular and have complex tissues. Plants, algae, and cyanobacteria are distinct in cell type and body form (Table 20.1).

Photosynthetic Membranes

Recall from Chapter 3 that prokaryotic cells are distinctly different from eukaryotic cells. Photosynthesis in plants and algae takes place in specialized

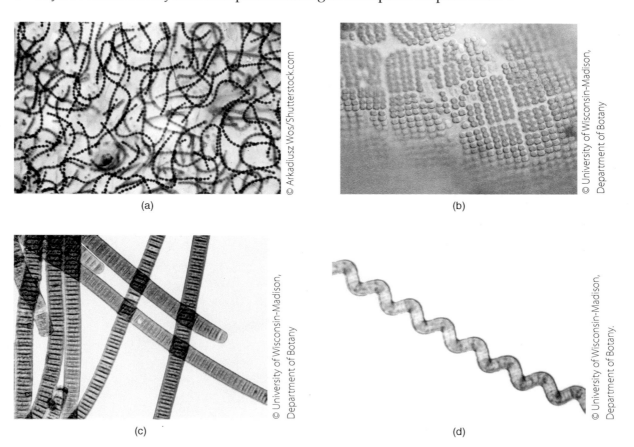

Figure 20.3a–d   Colonial forms of cyanobacteria: (a) *Nostoc*, string-like; (b) *Merismopedia*, sheet-like; (c) *Oscillartoria*, unbranched flat filament; and (d) *Spirulina*, unbranched coiled filament.

Table 20.1 Distinguishing Features among Cyanobacteria, Algae, and Plants

Features	Cyanobacteria	Algae	Plants
Cell type	Prokaryotic	Eukaryotic	Eukaryotic
Body form	Unicellular	Unicellular or multicellular	Multicellular
Cell wall main component	Peptidoglycan	Cellulose	Cellulose
Photosynthetic structure	Thylakoid membranes	Chloroplasts	Chloroplasts
DNA shape	Circular	Linear	Linear
DNA number	One	Many	Many
DNA associated with histones	No	Yes	Yes
Additional DNA	Plasmids	Mitochondrial and chloroplast DNA	Mitochondrial and chloroplast DNA
Cell division	Binary fission	Mitosis	Mitosis
Ribosome size	Smaller (70S)	Larger (80S)	Larger (80S)

S, sedimentation coefficient in Svedburg units.

thylakoid membrane system
Extensive folding and invagination of the plasma membrane found in photosynthetic bacteria.

peptidoglycan
Complex molecule of sugars and amino acids that make up the cell wall of some bacteria.

organelles or plastids called *chloroplasts*. Cyanobacteria do not have chloroplasts but photosynthesis occurs in the **thylakoid membrane system**, an extensive folding and invagination of the plasma membrane (Figure 20.4).

Prokaryotic Cell Wall

Like plants and most algae, cyanobacteria have cell walls but their cell walls are made up of **peptidoglycan**, a complex molecule of sugars and amino acids, not cellulose. Bacteria are divided into two major groups, gram-positive and gram-negative, based on the thickness of peptidoglycan in the cell wall

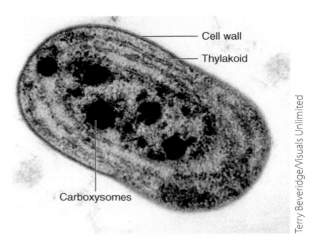

Figure 20.4 Thylakoid system of photosynthetic membranes in a marine cyanobacterium, *Synechococcus* viewed by a transmission electron microscope.

and the cell wall's reaction to the Gram's stain. Gram staining is a laboratory procedure used to identify bacteria. Cyanobacteria are **gram-negative bacteria**, which have less peptidoglycan in their cell wall and retain very little Gram's stain in the cell wall. Nitrogen-fixing bacteria such as *Rhizobium* are also gram-negative bacteria that form symbiotic relationships with plants. These bacteria convert atmospheric nitrogen into nitrites, nitrates, and ammonium, which are forms that plants can use. Other gram-negative bacteria include *Erwinia* and *Pseudomonas*, which cause plant diseases.

Prokaryotic Capsule and Biofilm

Many prokaryotic bacteria including cyanobacteria produce a **slimy polysaccharide coating** outside of the cell wall known as a capsule, mucilage coat, or slime sheath (Figure 20.5). The capsule of *Anabaena* allows these cyanobacteria to float, obtain nutrients from water, and even resist **phagocytosis** by organisms that may eat them. Nitrogen-fixing bacteria such as *Rhizobium* and *Agrobacterium* form symbiotic relationships with plants and use their capsule to attach themselves to plant surfaces. Many prokaryotes form **biofilms** by using their mucilage coat to glue themselves to protists, animal tissues, soil, and even solid surfaces. Dental plaque that forms on teeth and leads to tooth decay is a familiar example of a biofilm.

Prokaryotic DNA and Ribosomes

Although plants, algae, and cyanobacteria have DNA, there are major differences in the structure, shape, number, and location of their DNA. Like typical prokaryotes, cyanobacteria have one circular strand of DNA that is not associated with histone proteins (see Chapter 3). In contrast, plants and algae have many linear strands of DNA that are associated with histone proteins. Prokaryotic DNA is found in a region of cytoplasm called the *nucleoid* whereas eukaryotic DNA is found inside the nucleus. Prokaryotes often have additional circular strands of DNA called *plasmids*. Plasmids often carry genes that are used to produce toxins or confer resistance to antibiotics. Plants and algae also have mitochondrial DNA and chloroplast DNA in addition to nuclear DNA. Although plants, algae, and

gram-negative bacteria
Bacteria have less peptidoglycan in their cell wall and retain very little Gram's stain in the cell wall.

slimy polysaccharide coating
(capsule, mucilage, or slime sheath) Found outside of the cell wall of many bacteria; responsible for providing buoyancy, obtaining nutrients from water, and resisting phagocytosis by organisms that may eat them.

phagocytosis
A type of endocytosis in which large particulates or small organisms are taken up by a cell; often refers to as "cell eating".

biofilms
Many bacteria use their slimy polysaccharide coating to glue themselves to other organisms, soil, and solid surfaces.

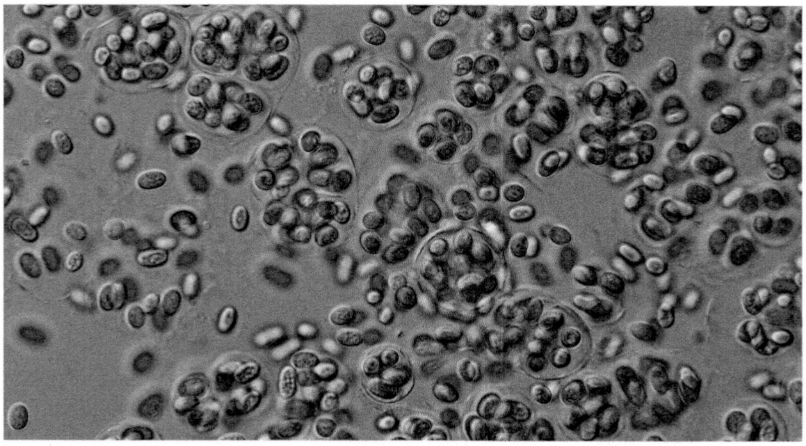

© University of Wisconsin-Madison, Department of Botany.

Figure 20.5   Colonies of freshwater cyanobacterial cells, Glaucocyctis spp., surrounded by a clear mucus sheath.

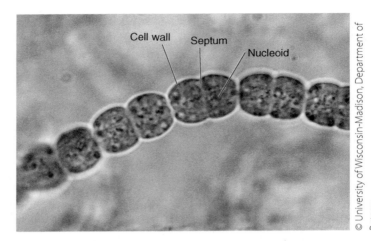

Figure 20.6   Binary fission of *Anabaena* viewed by an electron microscope.

cyanobacteria all have ribosomes, which are important to protein synthesis (see Chapter 3), ribosomes found in cyanobacteria are much smaller in size than those of plants and algae. Recall from Chapter 3 that mitochondria and chloroplasts are thought to have evolved from prokaryotic bacteria because their DNA and ribosomes are similar.

Differences in Cell Division

binary fission
Cell division process used by bacteria; the bacterial chromosome is duplicated as the cell grows in size.

Like typical bacteria, cyanobacteria reproduce by a cell division process called **binary fission**. In binary fission, the bacterial chromosome is duplicated as the cell grows in size. When the cell reaches a critical size, the cell wall and cell membrane begin to grow inward, forming a partition called a **septum** (Figure 20.6). Finally, the cell is pinched into two separate cells. Eukaryotes such as plants and algae reproduce by a more complex cell division process called *mitosis* (see Chapter 14). Mitochondria and chloroplasts found in eukaryotic cells not only have prokaryote-like DNA and ribosomes but also divide by binary fission.

septum
Near the end of binary fission, the cell wall and cell membrane of a bacterium begin to grow inward, forming a partition to separate the two new bacterial cells.

DIFFERENT TYPES OF CYANOBACTERIA

Cyanobacteria that Fix Nitrogen

heterocysts
Specialized cells found in nitrogen-fixing bacteria that are used to carry out nitrogen fixation.

Many cyanobacteria are able to not only carry out photosynthesis but also fix nitrogen. Like nitrogen-fixing bacteria, cyanobacteria such as *Nostoc* and *Anabaena* are able to convert atmospheric nitrogen into inorganic forms such as nitrites, nitrates, and ammonia. They are important contributors to nitrogen cycles by making nitrogen available to all other life forms. They use special cells called **heterocysts** to carry out nitrogen fixation (see Figure Box 20.1). Heterocysts only develop when available nitrogen is depleted in the environment. *Nostoc* and *Anabaena* often form symbiotic relationships with plants. *Nostoc* lives within the body of many nonvascular plants such as *Anthoceros* (hornwort) and *Blasia* (liverwort) (Adams, 2002) and also the roots of many cycads (gymnosperms) (Costa & Lindblad, 2002).

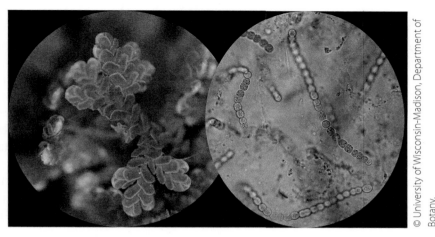

Figure 20.7   Water fern, *Azolla*, contains a symbiotic nitrogen-fixing cyanobacterium, *Anabaena azollae*.

Anabaena lives within the body of the water fern *Azolla* (Lechno-Yossef & Nierzwicki-Bauer, 2002) (Figure 20.7).

Cyanobacteria Acting as Chloroplasts

Glaucocystis nostochinearum is a colorless alga that forms endosymbiotic relationships with cyanobacteria (Figure 20.8). Recall from Chapter 2 that an endosymbiotic relationship is an intimate association between two different kinds of organisms, one living inside the other one. Numerous studies have shown that the cyanobacteria living within *Glaucocystis* act as chloroplasts providing food for the algal host (Echlin, 1967). These cyanobacteria have lost their cell wall and are surrounded by a single membrane (Hall & Claus, 1967). This symbiotic association adds weight to the theory that chloroplasts originated from cyanobacteria living within the cells of other organisms.

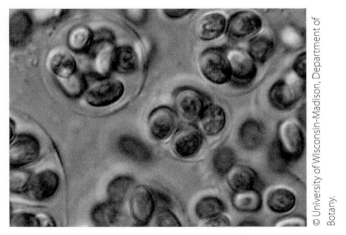

Figure 20.8   *Glaucocystis* (Cyanophyta or Glaucophyta), a colorless algae with cyanobacteria living within its body.

The "Other" Cyanobacteria: Prochlorophytes

Prochlorophytes are a group of cyanobacteria that carry out photosynthesis and produce oxygen. Unlike other cyanobacteria, prochlorophytes use chlorophyll a and b but not phycobilins as their photosynthetic pigments. The use of chlorophyll a and b for photosynthesis are characteristic of the chloroplasts within plants and green algae. In 1975, the discovery of *Prochloron didemni*, the first prochlorophyte, sparked research efforts to establish evolutionary ties between the prochlorophytes and green chloroplasts. However, molecule evidence from additional prochlorophyte species supported a closer relationship between prochlorophytes and cyanobacteria rather than chloroplasts. This has led some to consider the prochlorophytes as another type of cyanobacteria (Matthijs, van der Staay, & Mur, 1994).

TAKING A CLOSER LOOK AT ALGAE

Like cyanobacteria, algae are also important members of the Green World. They are aquatic and live in oceans, freshwater ponds, lakes, and streams. Algae (in Latin means "sea weeds") refer to a diverse group of photoautotrophic protists. Algae were once classified as plants because they are eukaryotic and have cell walls, one or more chloroplasts, and chlorophyll a as their main photosynthetic pigment. However, algae were later moved to the kingdom Protista and referred to as plant-like protists.

What Are Protists?

Protists are the approximately 200,000 species of eukaryotic organisms that are not plants, fungi, or animals (Table 20.2). These organisms are mostly microscopic and unicellular, colonial, filamentous, or simple multicellular eukaryotes. Simple multicellular protists do not form specialized tissues like animals or plants. In addition to algae, there are animal-like protists that are heterotrophic, called protozoa. Like animals, protozoa are motile and do not have cell walls or chloroplasts. Fungi-like protists such as water molds

chlorophyll a
A type of green pigment used by all photosynthetic organisms.

chlorophyll b
A type of green pigment used by all plants and some algae.

Table 20.2 Distinguishing Features among Protists, Fungi, Animals, and Plants

Features	Protists	Fungi	Animals	Plants
Body form	Unicellular, colonial, or filamentous or simple multicellular	Unicellular or multicellular filamentous	Multicellular with tissues	Multicellular with tissues
Cell wall component	Cellulose, silica; some do not have cell wall	Chitin	No cell wall	Cellulose
Motility	By means of cilia or flagella; some are nonmotile	Nonmotile; some produce motile spores	Motile	Nonmotile; few produce motile gametes
Energy acquisition	Autotrophic or heterotrophic	Heterotrophic (absorption)	Heterotrophic (ingestion)	Autotrophic
Habitat	Mostly aquatic; some terrestrial	Mostly terrestrial; some aquatic	Terrestrial and aquatic	Mostly terrestrial; some aquatic

and slime molds are heterotrophic and were previously classified as fungi. All these protists were once classified into a single kingdom, Protista. However, the kingdom Protista is being abandoned because recent phylogenetic analyses have revealed that protists do not form a monophyletic group. Instead, many protist phyla are now classified within several eukaryotic supergroups. The phylogenetic relationships among protists are expected to change as new information becomes available.

DIFFERENT TYPES OF ALGAE

In this chapter, we explore six major algal groups: green algae, red algae, brown algae, diatoms, dinoflagellates, and euglenoids. Their features are summarized in Table 20.3. Comparative analysis of ultrastructures and DNA sequences reveal phylogenetic relationships among these algal groups (Figure 20.9). Green algae and red algae are the closet relatives to plants and are currently placed in the supergroup Archaeplastida. Brown algae and diatoms are closely related and are placed in the same supergroup Stramenopila. Dinoflagellates are placed in the supergroup Alveolata with ciliates and disease-causing protozoa such as *Plasmodium* (causes malaria). Euglenoids are placed in the supergroup Excavata with disease-causing or pathogenic protozoa such as *Giardia* and *Trichomonas*.

Table 20.3   Distinguishing Features among Six Major Algal Groups

Group (Phylum)	Body Form	Motility	Cell Wall Component	Photosynthetic pigments	Storage
Green algae (Chlorophyta)	Unicellular, colonial; multicellular filamentous, or sheet-like	Some nonmotile; some have flagellates	Cellulose	Chlorophyll a and b, carotenoids	Starch
Red algae (Rhodophyta)	Most multicellular; some unicellular	Nonmotile	Carrageenan, agar	Chlorophyll a and d, carotenoids, phycobilins	Floridean starch
Brown Algae (Phaeophyta)	Multicellular	Nonmotile; flagellates on gametes	Cellulose, algin	Chlorophyll a and c, carotenoids, fucoxanthin	Laminarin
Diatoms (Bacillariophyta)	Unicellular; some colonial	Most nonmotile; some move by gliding	Silica	Chlorophyll a and c, carotenoids, fucoxanthin	Oil droplets or soluble carbohydrates
Dinoflagellates (Dinophyta)	Unicellular, some colonial	Two flagella (one trails along the body, one wraps around)	Cellulose; some do not have cell wall	Chlorophyll a and c, carotenoids, fucoxanthin	Starch
Euglenoids (Euglenophyta)	Unicellular	Two flagella (one long, one very short)	No cell wall	Chlorophyll a and b, carotenoids	Paramylon

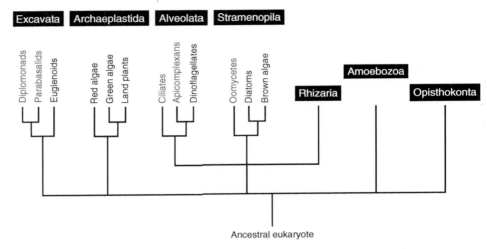

Figure 20.9 A simplified phylogenetic tree showing the major eukaryotic supergroups (see black boxes) and placement of the six major algal group discussed in this chapter. Some evolutionary relationships are still uncertain or underdisputed.

Green Algae Are the Closest Relatives of Plants

Green algae belong to the phylum Chlorophyta and are placed in the same supergroup with plants. It is generally accepted that plants evolved from green algal ancestors because they shared many characteristics. Like plants, green algae have cell walls made up of cellulose, store their energy or food as starch, and their chloroplasts use chlorophyll a, chlorophyll b, and **carotenoids** to capture light energy. Chloroplasts of green algae have two membranes and may have originated from a process called *primary endosymbiosis* (see Chapter 2, Box 2.1). **Primary endosymbiosis** refers to the capture of cyanobacteria by heterotrophic host cells and the formation of endosymbiotic relationships between the two. Green algae are the most diverse group of algae in terms of size and body form (Figure 20.10) and are found in a variety of habitats.

Unicellular Green Algae

The smallest eurkaryote is a green alga, *Micromonas*, which is only 1 µm in diameter. Most unicellular green algae are considerably larger than *Micromonas*. The largest is a marine alga, *Acetabularia*, commonly called mermaid's wineglass. This alga is made of a single giant cell that is about 5 cm long and shaped like a wine glass with a stem (Figure 20.10a). *Chlamydomonas* is the genus of green algae commonly found in freshwater pools and each cell is about 25 µm in diameter. These green algae have two flagella and are active swimmers (Figure 20.10b). The most prominent feature of *Chlamydomonas* is its single cup-shaped chloroplast that has one or two circular pyrenoids. **Pyrenoids** are structures responsible for starch production and storage. *Chlamydomonas* reproduces asexually by mitosis, but they may also reproduce sexually by forming gametes that fuse together to form zygotes.

Colonial Green Algae

Volvox is a genus of green algae that are unicellular but their cells aggregate to form colonies. A colony of *Volvox* resembles a hollow ball that

carotenoids
Yellow and orange pigments that are used for photosynthesis.

Primary endosymbiosis
Capture of cyanobacteria by heterotrophic host cells and the formation of endosymbiotic relationships between the two.

pyrenoids
Structures responsible for starch production and storage.

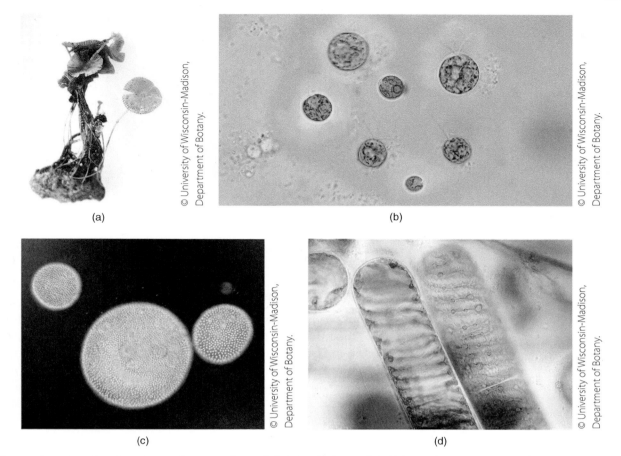

Figure 20.10a–d   A variety of green algae: (a) *Acetabularia*, (b) *Chlamydomonas*, (c) *Volvox*, and (d) *Spriogyra*.

can range in size from several hundred to several thousand cells (Figure 20.10c). Individual colonies can reproduce both asexually and sexually with daughter colonies formed inside them. Individual *Volvox* cells resemble *Chlamydomonas* cells and are flagellated. The flagella of individual cells beat separately to propel the whole colony.

Multicellular Green Algae

Spirogyra or watersilk is a multicellular green alga frequently found floating in masses on the surface of still waters. The cells form long unbranched filaments and each cell has one or two long ribbon-like chloroplasts containing pyrenoids (Figure 20.10d). *Spirogyra* reproduces asexually by **fragmentation** or breaking up of existing filaments and can also reproduce sexually by conjugation. **Conjugation** is when two individual filaments pair up and form conjugation tubes between their cells. The conjugation tube allows adjacent cells to fuse together to form zygotes. *Chara* or stonewort lives in shallow, freshwater lakes and ponds. This alga resembles small horsetail plants and has a more complex body form with short lateral branches in whorls along an axis (Figure 20.11). In addition, *Chara* has a more complex vegetative growth and reproduction. It is generally accepted that this group of green algae is more closely related to mosses than algae. A number of botanists and phycologists have even given *Chara* its own phylum, Charophyta.

fragmentation
Type of asexual reproduction by breaking up of existing filaments to produce new individuals.

conjugation
Type of sexual reproduction when two algal filaments pair up and form tubes or connections between their cells allowing adjacent cells to fuse together to form zygotes.

University of Wisconsin-Madison, Department of Botany.

Figure 20.11 A common stonewort, *Chara vulgaris.*

Lichen: Unique Partnership between a Green Alga and a Fungus

Lichens are commonly found growing on tree branches and rocks. They look like individual organisms but actually are symbiotic relationships between two very different organisms. The spongy **thallus** (flat undifferentiated body) of a lichen consists of a fungus, usually a sac fungus, and a green alga (or a cyanobacterium in the tropic). The algal partner supplies food for both organisms and the fungal partner protects the algal partner against desiccation and also absorbs water and minerals for both organisms. This symbiotic relationship seems to be mutualistic, which means both partners benefit from the relationship. However, physiological evidence suggests that the fungal partner parasitizes the algal partner and may even destroy some of the algal cells.

Lichens have been loosely grouped into three growth forms: crustose, foliose, and fruticose (Figure 20.12). **Crustose** lichens are often brightly colored crusty patches that are tightly attached to tree branches or rocks. **Foliose** lichens have flat leaf-like thalli and are weakly attached to the substrate. **Fruticose** lichens resemble miniature shrubs and their thalli grow upright and are usually branched.

Red Algae Are Close Relatives of Green Algae and Plants

Red algae belong to the phylum Rhodophyta. They are mostly large multicellular algae or macroalgae that live in warm and deep tropical oceans (Figure 20.13). They have no flagella and are nonmotile. Instead, they use a holdfast to attach themselves to rocks. Like green algae and plants, their chloroplasts have two membranes and may have originated from primary endosymbiosis. However, there are a number of significant differences between red algae and green algae. In addition to chlorophyll a and carotenoids, red algae use **phycoerythrin**, a red pigment, and **phycocyanin**, a blue pigment, as their photosynthetic pigments. Phycobilins such as phycoerythrin and phycocyanin are found in cyanobacteria but not in green algae or plants. Unlike green algae, red algae store their energy or food as **floridean starch**, a polysaccharide similar to glycogen. Their cell walls often contain thick, sticky polysaccharides such as agar or carrageenan, which have great

thallus
Flat and undifferentiated body found mostly in algae and some early land plants.

crustose
Type of lichen growth form that looks like brightly colored crusty patches; usually found tightly attached to tree branches or rocks.

foliose
Type of lichen growth form that looks like flat leaves; usually weakly attached to the substrate.

fruticose
Type of lichen growth form that resembles miniature shrubs.

phycoerythrin
Red pigment used by red algae and photosynthetic bacteria for photosynthesis.

phycocyanin
Blue pigment used by red algae and photosynthetic bacteria for photosynthesis.

floridean starch
Polysaccharide similar to glycogen that is used by some algae to store food.

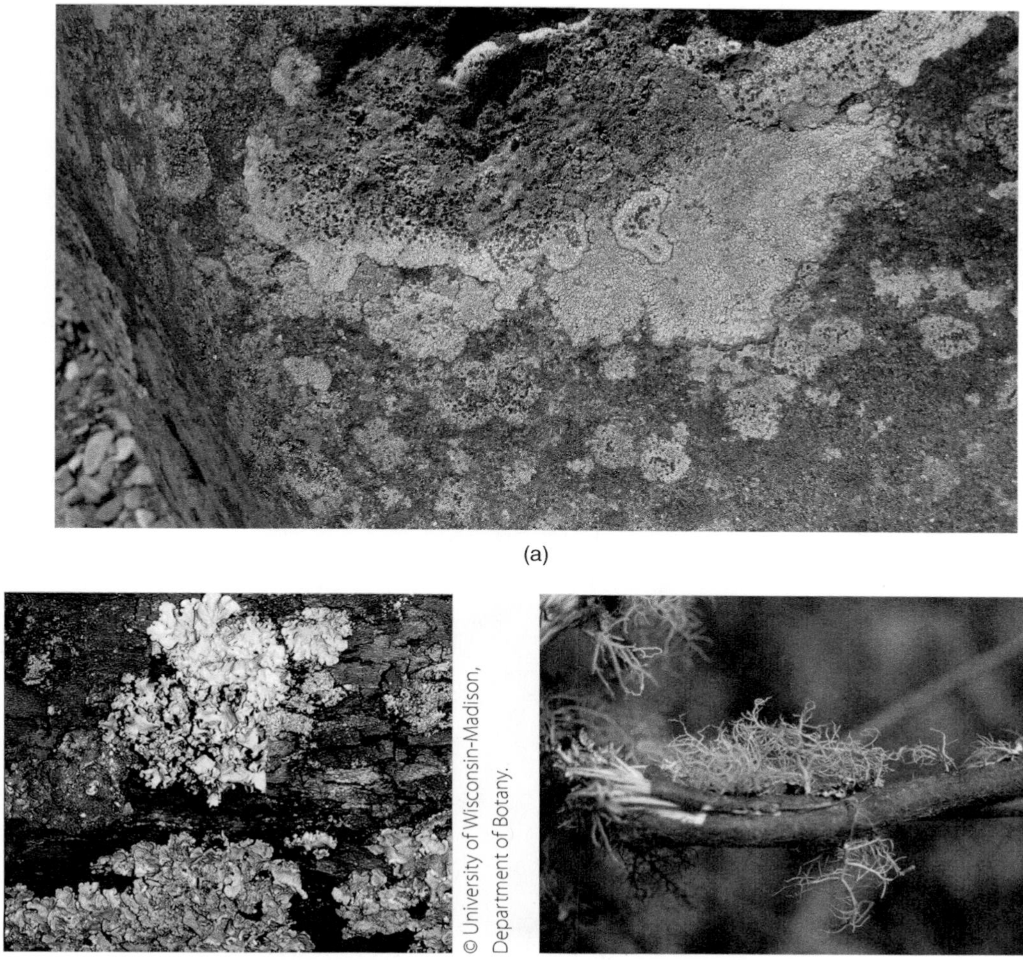

(a)

(b)

(c)

Figure 20.12a–c   Three growth forms of lichens: (a) crustose, (b) folicose, and (c) fructicose.

commercial value. **Agar** produced by *Gelidium* is used extensively as culture media for growing bacteria colonies or plant tissues. **Carrageenan** produced by *Chondrus crispus*, often called *Irish moss*, is mainly used to stabilize and thicken processed foods such as ice creams and chocolate milk (Figure 20.13a). It is also used to stabilize cosmetics, pharmaceuticals, and paints.

The body form and reproduction of red algae are much more complex than other algae. *Polysiphonia*, a common red alga of marine waters, produces three types of thallus structures: male gametophyte, female gametophyte, and tetrasporophyte (Figure 20.13b). Male and female gametophytes produce gametes and tetrasporophyte produces tetraspores. The alternation between the sexual generation and asexual generation closely resembles the life cycle of plants. Red algae are placed in the same supergroup, Archaeplastida, as green algae and plants.

Brown Algae Are the Giants of the Algal World

Brown algae belong to the phylum Phaeophyta. They are all large multicellular algae or macroalgae that live in colder and shallower ocean waters than red algae. Their bodies range in size from several centimeters to

agar
Thick, sticky polysaccharide found in the cell wall of some red algae.

carrageenan
Thick, sticky polysaccharide found in the cell wall of some red algae.

(a)

(b)

Claus Lunau/Science Source.

Gerd Guenther/Science Source.

Doug Sokell/Visuals Unlimited.

(c)

Figure 20.13a–c Red algae in the phylum Rhodophyta. (a) *Chondrus crispus*, (b) *Polysiphonia*, and (c) *Porphyra*.

chloropyll c
A type of green pigment used by some algae.

fucoxanthin
Brown photosynthetic pigment found in brown algae.

Secondary endosymbiosis
Refer to the capture of green or red algae by eukaryotic protist hosts and retention of the algal plastids in the host cells.

75 meters in length. The giant kelps, *Macrocystis pyrifera*, the largest brown algae, can reach 274 meters in length and may be found in water 30 meters or deeper. Many brown algae have a thallus body that is differentiated into flat, leaf-like blades, a stem-like stipe, and an anchoring holdfast (Figure 20.14). They often have gas-filled bladders or floats for buoyancy. Large masses of *Sagassum*, a floating brown alga, are often washed up on the shores along the Gulf of Mexico after storm surges. In addition to chlorophyll a and carotenoids, brown algae also use **chloropyll c** and **fucoxanthin**, a brown pigment, as photosynthetic pigments. Their chloroplasts have more than two membranes and may have originated from secondary endosymbiosis. **Secondary endosymbiosis** refers to the capture of green or red algae by eukaryotic protist hosts and the retention of the algal plastids in the host cells (Falkowski et al., 2004) (Figure 20.15). Chloroplasts of brown algae may

© University of Wisconsin-Madison, Department of Botany.

(a)   (b)   (c)

Figure 20.14a–c   (a) The giant kelps, *Macrocysti pyrifera*, in the phylum Phaeophyta. (b) Floating algae. and (c) A typical thallus of brown algae on the right showing leaf-like blades, a step-like stipe, and bladders along the stipe.

have originated from red rather than green algal ancestors. Brown algae store their energy or food as **laminarin**, a type of carbohydrates. Their cell walls are made up of **algin**, or alginic acid. Like agar and carrageenan, algin is used as a stabilizer and thickener for food, cosmetics, and pharmaceuticals. Reproduction in brown algae is complex and highly varied. Although brown algae are nonmotile, they do produce motile biflagellated gametes and zoospores.

laminarin
Type of carbohydrates used by brown algae to store food.

algin or alginic acid
Type of carbohydrate found in the cell wall of brown algae.

Diatoms Are Algae with Glass Shells

Diatoms are classified in the phylum Bacillariophyta. Most of them are unicellular and unflagellated. They are found in both freshwater and saltwater but are particularly abundant in colder marine waters. Diatoms are protected by a shell made up of two halves that fit together like a Petri dish. The main component of these shells is a glass-like material called **silica**. These glass shells have intricate patterns and come in a variety of shapes (Figure 20.16). Their shells also have fine grooves and pores along the edges. Most diatoms float in water, but some can glide on rocks and hard surfaces. Diatoms are closely related to brown algae and both are placed in the supergroup Stramenopila. Like brown algae, chloroplasts of diatoms may have originated from red algal ancestors via secondary endosymbiosis. Each diatom may have one, two, or many chloroplasts. In addition to chlorophyll a and carotenoids, these chloroplasts use chlorophyll c and fucoxanthin, the same brown pigment found in brown algae.

silica
Glasslike material that makes up the shells of diatoms.

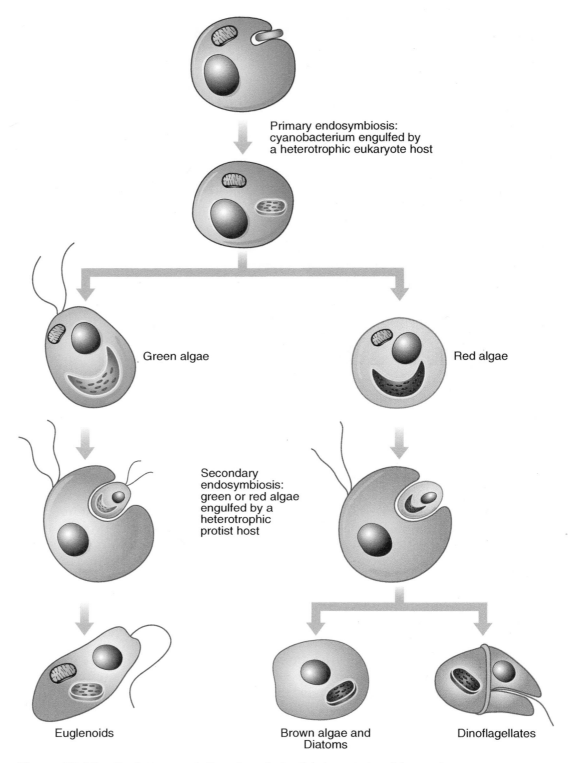

Figure 20.15   Evolution and diversity of plastids in autotrophic protists.

Diatoms often reproduce asexually by mitosis. When a diatom divides, the two halves of its shell separate and become the larger half of the shell for a new diatom. The diatom cells that receive the smaller halves of the shells become progressively smaller with each division. Eventually, the diatoms

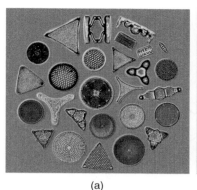 

© University of Wisconsin-Madison, Department of Botany.

(a)   (b)

Figure 20.16a–b   (a) Diatoms with intricate patterns on their glass shells and come in a vaierty of shapes. (b) Diatoms have fine grooves along the edges of the glass shells.

become so small that they switch to sexual reproduction. The small diatoms undergo meiosis to produce gametes. The gametes fuse together to form zygotes. These zygotes grow substantially before producing a new shell. Sexual reproduction restores diatoms to their original size. When diatoms die, their shells accumulate and form deposits called **diatomaceous earth**. Diatomaceous earth is used as filtering agents, metal polishes, insulation, and soundproofing materials.

diatomaceous earth
Deposits formed by accumulating dead diatoms.

Dinoflagellates Are Troublemakers of the Algal World

Dinoflagellates belong to the phylum Dinophyta. These algae are unicellular and are biflagellated (have two flagella). The two flagella have a distinct perpendicular arrangement. The flagellum found in the longitudinal groove trails behind the cell and acts like a rudder. The other flagellum encircles the transverse groove in the center of the cell and spins the cell. Dinoflagellates are covered with interlocking **cellulose plates** underneath the plasma membrane (Figure 20.17). Most of dinoflagellates store energy or food as **starch** and have two or more disc-shaped chloroplasts that contain chlorophylls a and c, carotenoids, and fucoxanthin. Their chloroplasts may have originated from red algal ancestors via secondary endosymbiosis. Some dinoflagellates may even have chloroplasts derived from tertiary endosymbiosis of diatoms. Many dinoflagellates live as endosymbionts inside the body of marine invertebrates such as jellyfish, coral, flatworms, and mollusks. These endosymbiotic dinoflagellates have lost their cellulose plates and flagella and are often called **zooxanthellae**. Some dinoflagellates have even lost their photosynthetic ability and have to obtain energy by ingesting food particles. A few dinoflagellates have become parasites that live off their invertebrate hosts. Dinoflagellates reproduce asexually by mitosis and sexual reproduction appears to be rare.

cellulose plates
Found underneath the plasma membrane of dinoflagellates.

starch
Very large polysaccharide made up of glucose molecules and is the primary storage molecules in plants and some algae.

zooxanthellae
Dinoflagellate that lives inside another organism and have lost their cellulose plates, and flagella.

Dinoflagellates are an important part of freshwater and marine environments because they generate oxygen and are food for aquatic organisms. Population explosions or blooms of these dinoflagellates frequently discolor water red, orange, or brown; hence these blooms are called *red tides*. Red tides have major ecological consequences because population of dinoflagellates can be so high that they create dead zones in the water.

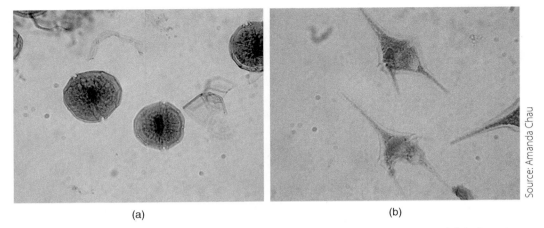

Source: Amanda Chau

(a)   (b)

Figure 20.17a–b The armor-plated dinoflagellates. (a) *Peridinium* and (b) *Ceratium*.

In addition, some dinoflagellates such as *Karenia* and *Pfisteria* produce powerful neurotoxins that may kill fish, pelicans, dolphins, manatees, and even humans.

Euglenoids Have Characteristics of Plants and Animals

pellicle
Fine parallel strips underneath the plasma membrane that spiral around the cell of euglenoids, providing a flexible cover.

eyespot
Found near the base of the flagellum of euglenoids; responsible for light detection.

Euglenoids in the phylum Euglenophyta are unicellular algae that live in freshwater environments. Like dinoflagellates, euglenoids have two flagella but their flagella are very different from those of dinoflagellates. One of the euglenoid flagella is very long and used for locomotion. The other one is so short that it cannot be seen outside of the cell. Like animals, euglenoids do not have a rigid cell wall. Euglenoids have fine parallel strips that spiral around the cell. These strips are underneath the plasma membrane and provide a flexible cover called a **pellicle** (Figure 20.18). The pellicle enables euglenoids to change shape as they move through water. Euglenoids have a distinct orange **eyespot** near the base of the flagellum that detects

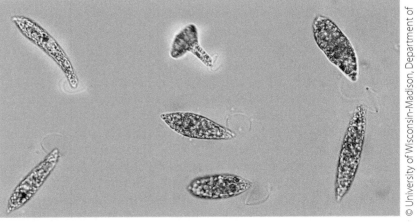

© University of Wisconsin-Madison, Department of Botany.

Figure 20.18 Euglena cells showing various unique structures.

light. Euglenoids reproduce asexually by mitosis. Sexual reproduction is suspected but has never been observed.

Most euglenoids have chloroplasts that contain carotenoids and chlorophylls a and b, the same pigments found in green algae and plants. Unlike green algae and plants, chloroplasts of euglenoids have more than two membranes and may have originated from green algae via secondary endosymbiosis. They store energy or food as **paramylon**, a type of polysaccharide. Euglenoids also have a **gullet** or groove through which food can be ingested. Some photosynthetic euglenoids lose their chlorophyll when kept in the dark. They turn into heterotrophs by ingesting organic matter. Some euglenoids are colorless and heterotrophic. Euglenoids are not closely related to green algae or plants and are placed in the supergroup Excavata because of their gullets.

Until the 1900s, only two kingdoms, plants and animals, were recognized. Euglenoids were classified in the plant kingdom because many of them carry out photosynthesis. At the same time, other biologists classified them in the animal kingdom because they move and feed. Euglenoids are one of the dilemmas of classification. Recall from Chapter 19 that Ernst Haeckel introduced a third kingdom, Protista, in 1894 to accommodate organisms such as euglenoids that are neither plants nor animals.

paramylon
Type of polysaccharides used by euglenoids to store food.

gullet or groove
Euglenoids use this groove to ingest food.

IMPORTANCE OF CYANOBACTERIA AND ALGAE TO HUMANS

Important Phytoplankton

Phytoplankton refers to a diverse group of mostly microscopic, unicellular, photosynthetic organisms that drift with the currents in freshwater or saltwater. Although they account for less than 1% of Earth's photosynthetic biomass, these microscopic organisms are responsible for more than 45% of our planet's net primary production (Falkowski & Raven, 1997). The vast majority of phytoplankton is made up of cyanobacteria. Phytoplankton is composed of cyanobacteria, diatoms, and dinoflagellates and produces food, oxygen, and nitrogen for aquatic organisms.

Toxic Blooms and Dead Zones

Cyanobacteria are abundant in both freshwater and saltwater. They can rapidly form thick mats or blooms when fertilizer-rich runoff from land is poured into rivers or oceans. The bloom can badly discolor water and affect water quality (Figure 20.19). Some cyanobacteria such as *Microcystis* and *Anabaena* produce cyanotoxins and their toxic blooms are a threat to people, animals, or the environment. As the cells in a cyanobacterial bloom die, they sink to the bottom and begin to decompose. Decomposition completely depletes oxygen in the water and creates dead zones, which have too little oxygen to sustain animal life. Every spring as conditions for algal and cyanobacterial blooms become favorable, the Gulf of Mexico Dead Zone extends from the mouth of the Mississippi River to beyond the Texas border.

Figure 20.19   Lake discolored by cyanobacterial blooms.

Like cyanobacteria, many algae such as dinoflagellates, diatoms, and euglenoids cause toxic blooms that are harmful to aquatic or human lives. The population of dinoflagellates can be so high that they create dead zones in the water.

Biofuel Producers

There is a growing interest in exploring aquatic phototrophs (cyanobacteria, algae, and diatoms) as alternatives to land-based crops for biofuel production. These aquatic phototrophs are more appealing than land-based crops because they are more efficient solar collectors, use less or no land, can be easily converted to liquid fuels, and offer secondary uses that fossil fuels do not provide (Dismukes, Carrieri, Bennette, Ananyev, & Posewitz, 2008). Research efforts are underway to overcome the technical challenges that hinder widespread use of these aquatic phototrophs for biofuel production.

Food and Industrial Products

With the exception of *Spirulina* and *Nostoc*, most cyanobacteria are rarely consumed by humans. For centuries, *Spirulina* has been used for food in the Lake Chad area of central Africa and areas around Mexico City. *Spirulina* is cultured commercially for its high vitamin content and sold in health food stores. *Nostoc* is consumed in Japan and China as a delicacy.

In Japan and other East Asian countries, green algae and red algae are widely eaten fresh, dried, or toasted. These algae are an important source of minerals and Vitamins A and C (Figure 20.20). *Nori* is the Japanese name for the edible red algae, *Porphyra*, which is added to soups and noodle dishes. During the Irish famine of 1845–1846, dulse, a red alga in the genus *Palmaria*, became an important substitute for the potato crop that was destroyed by blight. *Ulva*, sea lettuce, and *Chlorella* are the edible green algae. Green algae such as *Chlorella* are currently being explored not only as food sources but also portable oxygen generators and carbon dioxide scrubbers for deep-space exploration.

Figure 20.20   Edible seaweeds from left: green laver, also known as sea lettuce or aonori; dulse; devil's apron.

Agar, carrageenan, and algin (see previous sections on red algae and brown algae) are important stabilizers and thickeners for various industries. Algin is also used to coat frozen food cartons or grease-proof papers.

Key Terms

chlorophyll
photosystem II (PS II)
photosystem I (PS I)
water as the electron donor
phycobilins
thylakoid membrane system
peptidoglycan
gram-negative bacteria
slimy polysaccharide coating
biofilms
binary fission
septum
heterocysts
chlorophyll a
chlorophyll b

carotenoids
primary endosymbiosis
pyrenoids
fragmentation
conjugation
thallus
crustose
foliose
fruticose
phycoerythrin
phycocyanin
floridean starch
agar
carrageenan
chlorophyll c

fucoxanthin
secondary endosymbiosis
laminarin
algin
silica
diatomaceous earth
cellulose plates
starch
zooxanthellae
pellicle
eyespot
paramylon
gullet

Summary

- Plants, cyanobacteria, and algae are autotrophs capable of photosynthesis and oxygen production. They all use chlorophyll a as the main photosynthetic pigments, possess two photosystems, and use water as the electron donor.

- Unlike plants and algae, cyanobacteria have prokaryotic cells that have no nucleus or other organelles. Cyanobacteria do not have chloroplasts but photosynthesis occurs in the thylakoid membrane system. Like plants and most algae, cyanobacteria have cell walls but their cell walls are made up of peptidoglycan. Like many prokaryotic bacteria, cyanobacteria produce a slimy polysaccharide coating or capsule and have one circular strand of DNA. Cyanobacteria belong to the domain Bacteria whereas plants and algae belong to the domain Eukarya.

- Unlike plants and algae, cyanobacteria reproduce by a cell division process called binary fission. The bacterial chromosome is duplicated as the cell grows in size and then the cell is pinched into two separate cells.

- Many cyanobacteria are able to not only carry out photosynthesis but also fix nitrogen. They are important contributors to nitrogen cycles by making nitrogen available to all other life forms. Some cyanobacteria such as the ones living within Glaucocystis act as chloroplasts, providing food for the algal host. Unlike other cyanobacteria, prochlorophytes use chlorophyll a and b but not phycobilins as their photosynthetic pigments. The use of chlorophyll a and b for photosynthesis are characteristic of the chloroplasts within plants and green algae.

- Algae refer to a diverse group of photoautotrophic protists. Algae were once classified as plants because they are eukaryotic and have cell walls, one or more chloroplasts, and chlorophyll a as their main photosynthetic pigment. However, algae were later moved to kingdom Protista and referred to as plant-like protists.

- All protists were once classified into a single kingdom, Protista. However, the kingdom Protista is being abandoned because recent phylogenetic analyses have revealed that protists do not form a monophyletic group. Instead, many protist phyla are now classified within several eukaryotic supergroups.

- Green algae, red algae, brown algae, diatoms, dinoflagellates, and euglenoids are the six major algal groups covered in this chapter. These algal groups can be distinguished from one another by their body form, motility, cell wall component, photosynthetic pigments, and food storage.

- Green algae are placed in the same supergroup with plants. It is generally accepted that plants evolved from green algal ancestors because they shared many characteristics. Red algae are mostly large multicellular algae or macroalgae that live in warm and deep tropical oceans. Brown algae are all large multicellular algae or macroalgae that live in colder and shallower ocean waters than red algae.

- Most diatoms are unicellular and unflagellated. They are protected by a glass shell made up of two halves that fit together like a Petri dish. Dinoflagellates are unicellular and biflagellated. One flagellum found in the longitudinal groove trails behind the cell and acts like a rudder and the other encircles the transverse groove in the center of the cell and spins the cell. Euglenoids have two flagella but their flagella are very different from those of dinoflagellates. One of the euglenoid flagella is very long and used for locomotion. The other one is so short that it cannot be seen outside of the cell. Like animals, euglenoids do not have a rigid cell wall.

- Lichens are symbiotic relationships between a fungus, usually a sac fungus, and a green alga (or a cyanobacterium in the tropics). The algal partner provides food for both organisms and the fungal partner protects the algal partner against desiccation and also absorbs water and minerals for both organisms.

- Cyanobacteria and algae are important phytoplanktons that are responsible for more than 45% of our planet's net primary production. Some produce toxic blooms that are a threat to people, animals, or the environment. Some are being explored as alternatives to land-based crops for biofuel production. Some are cultured commercially as food or important stabilizers and thickeners.

Reflect

1. *Inhabitants of the Green World.* Name the three groups of organisms that make up the Green World. Describe the characteristics that they all have.

2. *What do you know about cyanobacteria?* Without using the book, write down as many characteristics that are unique to cyanobacteria. Write down as many characteristics that cyanobacteria share with algae.

3. *Prochlorophytes: Are they cyanobacteria or not?* Explain how prochlorophytes are different from other cyanobacteria.

4. *What are protists?* Explain how protists are similar to fungi, animals, and plants and how protists are different from the other three groups.

5. *What type of alga is this?* Describe the characteristics you use to distinguish the six major algal groups. Develop a dichotomous key for the six algae groups using these characteristics.

6. *Algae are in everything?!* Algae and their derivatives are in many of our consumer products. Take a trip to the grocery store and see how many items you can find that are made up of algae or their derivatives. Make a list and name the type of alga or derivative used in each item or product.

References

Adams, D. (2002). Cyanobacteria in symbiosis with hornworts and liverworts. In A. Rai, B. Bergman, & U. Rasmussen (Eds.), *Cyanobacteria in symbiosis* (pp. 117–135). New York: Kluwer Academic.

Costa, J., & Lindblad, P. (2002). Cyanobacteria in symbiosis with cycads. In A. Rai, B. Bergman, & U. Rasmussen (Eds.), *Cyanobacteria in symbiosis* (pp. 195–205). New York: Kluwer Academic.

Dismukes, G., Carrieri, D., Bennette, N., Ananyev, G., & Posewitz, M. (2008). Aquatic phototrophs: efficient alternatives to land-based crops for biofuels. *Current Opinion in Biotechnology, 19,* 235–240.

Echlin, P. (1967). The biology of Glaucocystis nostochinearum. *British Phycological Bulletin, 3*(2), 225–239.

Falkowski, P., Katz, M. E., Knoll, A. H., Quigg, A., Raven, J., Schofield, O., et al. (2004). The evolution of modern eukaryotic phytoplankton. *Science, 305,* 354–360.

Falkowski, P., & Raven, J. (1997). *Aquatic photosynthesis.* Oxford, UK: Blackwell Scientific.

Hall, W., & Claus, G. (1967, March). Ultrastructural studies on the cyanelles of *Glaucocystis nostochinearum Itzigsohn. Journal of Phycology, 3*(1), 37–51.

Lechno-Yossef, S., & Nierzwicki-Bauer, S. (2002). Azolla-Anabaena symbiosis. In A. Rai, B. Bergman, & U. Rasmussen (Eds.), *Cyanobacteria in symbiosis* (pp. 153–178). New York: Kluwer Academic.

Matthijs, H., van der Staay, G., & Mur, L. (1994). Prochlorophytes: the "other" cyanobacteria? In D. Bryant (Ed.), *The molecular biology of cyanobacteria* (pp. 49–64). New York: Kluwer Academic.

Schopf, J. W. (2002). The fossil record: tracing the roots of the cyanobacterial lineage. In B. Whitton & M. Potts (Eds.), *The ecology of cyanobacteria: Their diversity in time and space* (pp. 13–35). New York: Kluwer Academic.

Schulz, H. N., & Jørgensen, B. B. (2001) Big bacteria. *Annual Review of Microbiology, 55,* 105–137.

Stal, L. (2002). Cyanobacterial mats and stromatolites. In A. Whitton & M. Potts (Eds.), *The ecology of cyanobacteria* (pp. 61–120). New York: Kluwer Academic.

Whitton, B. A., & Potts, M. (Eds.). (2002). *The ecology of cyanobacteria: their diversity in time and space.* New York: Kluwer Academic.

Chapter 21

The Plant Kingdom

© Mark Carthy/Shutterstock.com

Learning Objectives

- Discuss the evidence supporting the idea plants and green algae share a common ancestor
- Explain the evolutionary relationships among the different groups of embryophytes
- Discuss the challenges of living on land and how plants overcome these challenges
- Name the key features distinguishing kingdom Plantae from other kingdoms
- Diagram a typical plant life cycle and explain the term *alternation of generations*
- Distinguish among the four major plant groups

embryophytes or land plants
Embryos of plants depend on maternal protection and resources for their development.

nonvascular plants or bryophytes
Group of plants that has no vascular tissues, true roots, true stems, or true leaves; has gametophytes that are dominant and sporophytes that are small and dependent on gametophytes for survival; includes liverworts, hornworts, and mosses.

vascular seedless plants
Group of plants that have vascular tissues, true roots and stems; use spores for dispersal; have sporophytes that are dominant and gametophytes that are small but independent from sporophytes; includes club mosses, ferns, and their relatives.

vascular seed plants
Group of plants that have vascular tissues, true roots, true stems, and true leaves; use seeds for dispersal; have sporophytes that are dominant and gametophytes that are small and dependent on sporophytes for survival; includes gymnosperms and angiosperms.

gymnosperms or cone-bearing plants
Group of vascular seed plants that produce seeds and their seeds are not enclosed within a fruit (*gymnos* in Greek means "naked" and *sperma* means "seed"); includes cycads, ginkgo, gnetophytes, and conifers.

Plants are autotrophic organisms carrying out photosynthesis to produce their own food and generate oxygen. They are the dominant primary producers on land and also important contributors to the global carbon and nitrogen cycles. What are plants? Where did they come from? How many different types or groups of plants are there? In this chapter, we explore the origin and diversity of plants, discuss the evidence suggesting plants and green algae share a common ancestor, and examine key events in plant evolution.

EVOLUTIONOFLANDPLANTS(EMBRYOPHYTES)

Plant taxonomists today only recognize liverworts, hornworts, mosses, club mosses, ferns and their relatives, gymnosperms, and angiosperms as embryophytes (land plants) in the kingdom Plantae. Land plants are known as **embryophytes** because of the dependency of their embryos on maternal protection and resources. These green organisms share a number of characteristics distinguishing them from other autotrophic organisms such as green algae. To be classified as embryophytes, the organisms must:

- Be multicellular eukaryotes
- Use chloroplasts to carry out photosynthesis
- Use chlorophyll a, chlorophyll b, and carotenoids (β-carotene and xanthophylls) as photosynthetic pigments
- Store energy or food as starch
- Have cell walls made up of cellulose
- Have waterproof coatings to prevent desiccation
- Be composed of tissues produced by apical meristem
- Have multicellular embryos that develop within female sex organs
- Have multicellular reproductive structures
- Have a life cycle alternating between haploid and diploid generations
- Be sedentary

The first five of these characteristics were inherited from algal ancestors and are also found in some green algae. However, the last six characteristics are plant innovations to live on land and are absent from even closely related charophyceans, a group of freshwater green algae. We discuss these innovations later in this chapter.

Major Plant Groups

More than 300,000 species of living plants are found in the kingdom Plantae. They are classified into 10 distinct phyla that can also be divided into informal groupings based on presence or absence of certain characteristics (Table 21.1). These informal groupings are **nonvascular plants** (bryophytes), **vascular seedless plants** (club mosses, ferns, and their relatives), and **vascular seed plants**. Vascular seed plants are subdivided into two major groups: **gymnosperms** (cone-bearing plants) and **angiosperms** (flowering plants) (Figure 21.1). We cover nonvascular plants, the bryophytes, in the next chapter and the other three groups in Chapters 23–25.

Table 21.1   Four Informal Groupings of Plants and Their Characteristics

Informal Groupings	Characteristics	Plant Phyla
Nonvascular plants	• No vascular tissues • No true leaves, stems, or roots • Dominant gametophyte generation and dependent small sporophyte generation • Rely on water for sexual reproduction • Use spores for dispersal	Phylum Marchantiophyta (liverworts) Phylum Anthocerophyta (hornworts) Phylum Bryophyta (mosses)
Vascular seedless plants	• Have vascular tissues • Have true stems and roots • Dominant sporophyte generation and independent small gametophyte generation • Rely on water for sexual reproduction • Use spores for dispersal	Phylum Lycopodiophyta (club mosses) Phylum Pteridophyta (ferns and their relatives)
Vascular seed plants: gymnosperms	• Have vascular tissues • Have true leaves, stems, and roots • Dominant sporophyte generation and dependent microscopic gametophyte generation • Do not need water for sexual reproduction • Use naked seeds for dispersal	Phylum Cycadophyta (cycads) Phylum Ginkgophyta (ginkgo) Phylum Gnetophyta (gnetophytes) Phylum Coniferophyta (conifers)
Vascular seed plants: angiosperms	• Have vascular tissues • Have true leaves, stems, and roots • Produce flowers and fruits • Dominant sporophyte generation and dependent microscopic gametophyte generation • Do not need water for sexual reproduction • Use enclosed seeds for dispersal	Phylum Anthophyta (angiosperms)

Ancestors of Plants

Green algae in the phylum Chlorophyta, discussed in Chapter 20, are the closest relatives of plants and are placed in the same supergroup (see also Figure 20.9). It is generally accepted that plants evolved from green algal ancestors because plants and green algae share many characteristics.

Cellular and Biochemical Traits Both plants and green algae have cell walls made up of cellulose and double-membrane chloroplasts that may have originated from primary endosymbiosis (see Chapter 20). Their chloroplasts use chlorophyll a, chlorophyll b, and carotenoids to capture light

angiosperms or flowering plants
Group of vascular seed plants that produce seeds and their seeds are enclosed within a fruit, which is a mature or ripened ovary. *Angeion* in Greek means "vessel" and *sperma* means "seed."

Figure 21.1a–d Representatives from the four major plant groups: (a) nonvascular plants (moss), (b) seedless vascular plants (fern), (c) gymnosperms (pine), and (d) angiosperms (magnolia).

energy for photosynthesis. They store their food as starch. Ultrastructural studies of reproductive cells provide evidence linking modern charophyceans and embryophytes (Pickett-Heaps, 1975). Like embryophytes, charophyceans use phragmoplasts during cell division (see Chapter 14). A phragmoplast is a persistent spindle apparatus formed during telophase with microtubules oriented perpendicular to the plane of cell division (Mattox & Stewart, 1984).

Molecular and Structural Data A large body of molecular sequence data from nuclear, plastid, mitochondrial DNA, and ribosomal RNA (rRNA) indicates charophyceans are closely related to embryophytes and they probably share a recent common ancestor (for reviews, see Bhattacharya & Medlin, 1998; Lewis & McCourt, 2004). Complex charophyceans in the orders of Charales and Coleochaetales have branched filaments with apical growth or thalli made up of parenchyma-like cells (Figure 21.2). The multicellular organization of complex charophyceans and land plants are similar. Like plants, complex charophyceans also have plasmodesmata in their cells for communication. The mounting evidence suggests charophyceans are more closely related to embryophytes than other green algae. As a result, the

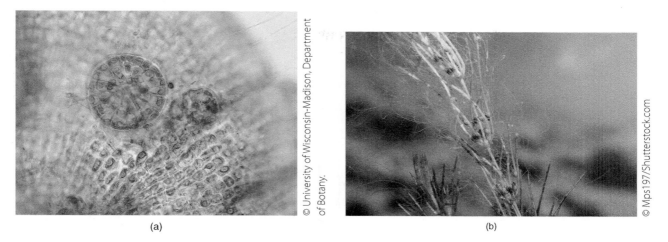

(a) © University of Wisconsin-Madison, Department of Botany.

(b) © Mps197/Shutterstock.com

Figure 21.2a–b   Complex multicellular charophyceans. (a) *Coleochaete* and (b) *Chara*.

phylum Streptophyta was created to include charophyceans with embryophytes and separate charophyceans from other green algae in the phylum Chlorophyta (for a review, see Becker & Marin, 2009).

Emergence and Diversification of Land Plants

The colonization of dry land by descendants of primeval charophyceans began approximately 490 to 476 million years ago during the mid-Ordovician period (for reviews, see Kenrick & Crane, 1997; Sanderson, Thorne, Wikstrom, & Bremer, 2004) and was one of the most important steps in the evolution of life on earth (Bateman et al., 1998; Graham, 1993; Kenrick & Crane, 1997). The emergence and diversification of land plants led to major changes in atmospheric oxygen and carbon dioxide concentrations (Berner, 1999; Mora, Driese, & Colarusso, 1996), caused the development of terrestrial ecosystems, and set the stage for evolution of other terrestrial organisms.

Divergence Times for Major Plant Groups   Plant biologists have used many methodologies to estimate the divergence times and evolutionary relationships among various plant groups. Paleobotanical data (fossil record) from the Silurian and Devonian periods and sequence data have been particularly useful in constructing plant phylogenies and estimating when these plant groups first appeared (for reviews, see Kenrick & Crane, 1997; Sanderson et al., 2004) (Figure 21.3). Liverworts diverged from charophycean algae around the mid-Ordovician period and were believed to be the earliest land plants. Hornworts, mosses, and early vascular plants appeared around 432 million years ago in the early Silurian period. True vascular plants (**eutracheophytes**) appeared next around the mid-Silurian and diversified substantially around the early Devonian period, 398 million years ago. Seed plants appeared around the late Devonian period. The Ordovician-Devonian radiation of embryophytes and vascular plants is the terrestrial equivalent of the Cambrian "explosion" in the evolution of Metazoa (multicellular animals). The "Cambrian explosion" refers to the appearance of multicellular animal body plants

eutracheophytes or true vascular plants
Land plants that have lignified tracheids to transport water and minerals (*eu-* in Greek means "true" or "well"); all vascular plants.

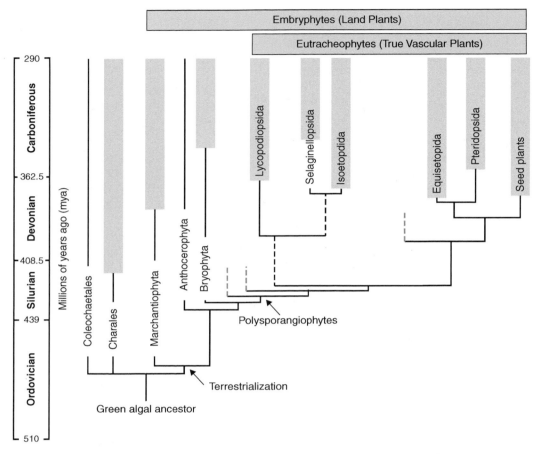

Figure 21.3   Simplified phylogenetic relationships among modern land plants showing minimum stratigraphic ranges based on megafossils (gray bars) and their minimum implied range extensions (black lines). Adapted from Kenrick and Crane (1997) and Vanderpoorten and Goffinet (2010).

during a comparatively brief period of time, from 530 to 520 million years ago. It was during this phase of evolution that the basic characteristics of land plants were established. By knowing the order in which the plant groups evolved, we can compare particular characteristics within and between plant groups and gain insights into how plants adapted to life on dry land.

Key Evolutionary Phases   Based on paleobotanical data and comparative studies of plant groups, four key phases were identified as pivotal to plant evolutionary history. We can also deduce the driving forces that may have shaped biochemical, anatomical, and morphological innovations of land plants (for a review, see Bateman et al., 1998; Friedman, Moore, & Purugganan, 2004; Kenrick & Crane, 1997).

- *Transition to dry land*. Dry land is drastically different from aquatic environments in a number of ways. Although carbon dioxide and sunlight are more readily available on land, pioneering land plants were subjected to desiccation, heat, harmful ultraviolet radiation, and microbe invasion. During this period of time, plants manufactured complex polymers (made up of many subunits called *monomers*) such as lignin, sporopollenin, and cutin to resist desiccation and ultraviolet damage and even provide structural support. Multicellularity and tissue formation, together with the alternation between independent gametophytic and dependent sporophytic generations, emerged. In the next section and also in Chapter 22, we explore these innovations that define the byrophytic life history.

- *Rise of vascular plants.* Modification of body form and tissue arrangements enabled land plants to more efficiently exploit land-based resources. Sporophytes began to branch and grow separately from gametophytes. The branched sporophytes enabled early land plants to harvest more light energy for photosynthesis. Having multiple growth tips allowed land plants to better survive attack by herbivores. The emergence of stems and roots with lignified tracheids enhanced transportation and absorption of water and minerals. Aerial stems with tough cuticles made up of cutin provided resistance to desiccation and the stomata on their epidermis provided gas exchange. In Chapter 23, we explore these innovations that define the life history of vascular plants.

- *Emergence of seed plants.* Modification of tissue and organ arrangements enabled vascular plants to greatly diversify their morphological form and size. The emergence of seeds and true leaves (**euphylls**) were the main innovations during this period of time. We explore these innovations that define the life history of seed plants in Chapter 24.

- *Diversification of flowering plants.* The formation of flowers for sexual reproduction and fruits for seed dispersal in some seed plants enabled coevolutionary associations with animals. We explore these inventions and their diversifications that define the life history of flowering plants in Chapter 25.

CHARACTERISTICS OF LAND PLANTS (KINGDOM PLANTAE)

When early land plants made the transition to land, they faced a number of major environmental challenges that required innovative solutions. Desiccation or drying-out is the first major challenge for any organisms living on dry land. They need to find ways not only to minimize water loss but also to obtain enough water to sustain life. Land organisms are also subjected to extreme temperatures and harmful ultraviolet radiation and have to find ways to minimize damage due to temperature fluctuation and ultraviolet exposure. Unlike water, air does not provide support for organisms. Land organisms have to find ways to raise their bodies up against the pull of gravity. Compared to aquatic environments, nutrients are not readily available in dry land. Land animals can move around to find water and food but land plants are sedentary and have to find ways to obtain sufficient water and nutrients. Innovations for land-based living such as manufacturing complex polymers, producing multicellular structures, alternating between sporophyte and gametophyte generations, and forming associations with fungi are found only and are obligatory for plants, members of the kingdom Plantae (Figure 21.4).

polymers
Complex and big molecules made up of many subunits called monomers.

euphylls or true leaves
Leaves with a branched vascular system and leaf gaps in the stem.

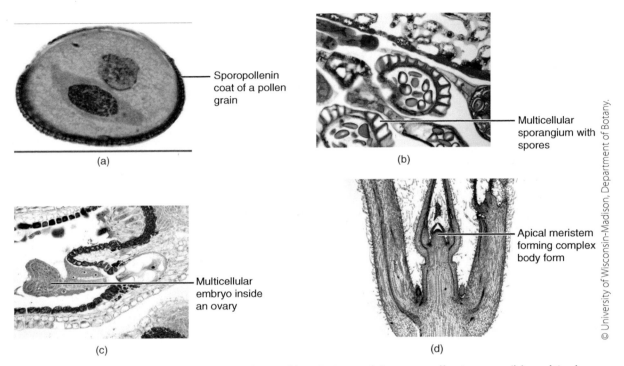

(a)

Sporopollenin coat of a pollen grain

(b)

Multicellular sporangium with spores

(c)

Multicellular embryo inside an ovary

(d)

Apical meristem forming complex body form

© University of Wisconsin-Madison, Department of Botany.

Figure 21.4a–d   Some major innovations of land plants: (a) sporopollenin coat, (b) multicellular sporangia, (c) multicellular embryo, and (d) tissues produced by apical meristem.

Complex Polymers

cuticle
Found on the surface of stems and leaves; made up of cutin and wax, which make plant surfaces waterproof and reflective.

Cutin and Wax Almost all land plants have a waxy coating called a **cuticle** on the surfaces of their stems and leaves. The cuticle is made up of two special types of lipids, cutin and wax. These molecules do not mix with water and are used to waterproof plant surfaces. The cuticle not only reduces excessive water loss but also impedes oxygen and carbon dioxide uptake. Recall from Chapter 3 that special openings called stomata are found on plant surfaces to allow gas exchange. Cutin and wax also make plant surfaces reflective to minimize ultraviolet light and heat damage. The cuticle and stomata are two important innovations by plants for land-based living.

Sporopollenin In addition to cutin and wax, land plants produced a complex polymer called **sporopollenin** on the outer walls of their spores and pollen grains (Figure 21.4a). Sporopollenin not only prevents spores and pollens from desiccation but also resists decomposition. Sporopollenin helps to preserve spores or pollens in soils and sediments for a long time and is the reason why plant spores and pollens are commonly found fossilized in rocks.

sporopollenin
Major component of the outer walls of plant spores and pollen grains; it helps spores and pollen grains to prevent excess water loss and resist decomposition.

Lignin Land plants also produce another complex polymer called *lignin*. Recall from Chapter 3 that lignin is found in the secondary cell walls of plant cells such as tracheids, vessels, and sclereids. Lignin not only waterproofs the plants but also strengthens them. Air and water have very different physical properties. Air provides little support for organisms because it is less dense than water. In order to live on land, animals developed bony

(a)   (b)

Figure 21.5a–b   Two forms of fossilized ferns (a) a sword fern and (b) a fan fern.

skeletons to support their bodies whereas land plants support their bodies by producing these tough polymers. These tough polymers are difficult to break down and are largely responsible for preserving entire plants and plant parts as fossils (Figure 21.5).

Multicellular Structures

Multicellular Sporangia and Gametangia The **sporangia** (spore-producing structures) and **gametangia** (gamete-producing structures) of land plants are multicellular. To resist desiccation, these structures have a layer of surrounding cells that form a protective cover (Figure 21.4b). By developing compact multicellular structures, land plants greatly reduce exposure of their surface areas to the air and thus further minimize water loss. Unlike plants, gametangia in algae are unicellular and do not have a protective cover.

sporangia
(singular *sporangium*)
Structure that produces spores.

gametangia
(singular *gametangium*)
Structures that produce gametes or sex cells.

Multicellular Embryos When organisms reproduce sexually, their gametes or sex cells come together to form zygotes. Plant zygotes develop into multicellular embryos (baby plantlets) within maternal tissues that originally surrounded the egg (Figure 21.4c). Similar to the sporangia and gametangia, the embryos are protected from the environment and from desiccation. Similar to some animals, internalization of vital functions and organs by plants may have been an adaptation for land-based living.

Multicellular Body and Apical Growth Recall from Chapters 4 and 5 that the apical meristem is indeterminate and produces primary growth for plants (Figure 21.4d). Indeterminate apical growth coupled with the ability to branch were the developmental hallmarks of land plants and played a key role in the evolution of complex three-dimensional body and tissue systems. Based on developmental patterns found in nonvascular plants and charophyceans, early land plants developed apical growth only in their **gametophytes** (adult forms producing gametes). Early tracheophytes

gametophytes
Adult plants produce gametes by mitosis in their gametangia.

sporophytes
Adult plants that produce spores by meiosis in their sporangia.

haploid
Having one set of chromosomes in the cell(s).

diploid
Having two sets of chromosomes in the cell(s).

antheridia
(singular *antheridium*)
Male gametangia for producing haploid sperm.

archegonia
(singular *archegonium*)
Female gametangia for producing haploid eggs.

alternation of generations
Unique life cycle of plants; plants spend part of their lives as multicellular haploid gametophytes and part as multicellular diploid sporophytes.

began to develop apical growth in their **sporophytes** (adult forms producing spores) and sporophyte of eutracheophytes became even more complex and much larger.

Unique Life Cycle Alternation of Generations

Unlike green algae, land plants spend part of their lives in a **haploid** (having one set of chromosomes) stage and another part in a **diploid** (having two sets of chromosomes) stage (Figure 21.6). The haploid gamete-producing plants are called *gametophytes* whereas the diploid spore-producing plants are called *sporophytes*. Gametophytes use gametangia to produce gametes by the process of mitosis (see Chapter 14). Male gametangia, called **antheridia**, produce sperms whereas female gametangia, called **archegonia**, produce eggs. Sporophytes use sporangia to produce spores by the process of meiosis. Land plants alternate between the gametophyte and sporophyte generations, hence this unique life cycle is known as **alternation of generations**.

Phylogenetic studies indicate land plants may have inherited a multicellular gametophyte from their green algal ancestors but their multicellular sporophyte evolved during the transition to life on land. Unlike green algae, which undergo meiosis to produce only four spores, the plant zygote develops into a multicellular sporophyte maximizing spore production when it undergoes meiosis. As plant groups diverged, the sporophyte became more complex and large whereas the gametophyte became smaller and less prominent.

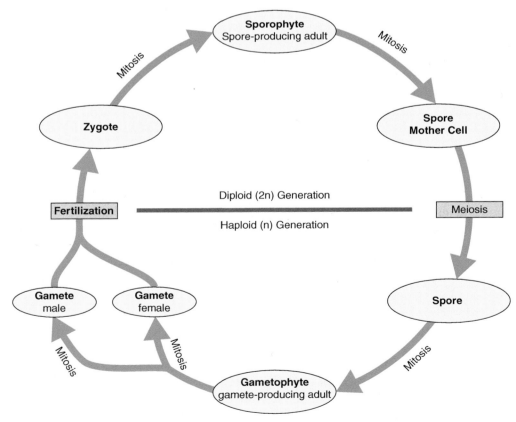

Figure 21.6 A typical plant life cycle: alternation of generations.

Associations with Mycorrhizal Fungi

Aside from desiccation, early land plants faced water and nutrient shortages when they made the transition to land. Plants developed intimate associations with fungi in their roots or rhizoids to form **mycorrhizas** (recall from Chapter 4). These modified roots or rhizoids enhance plant tolerance to biotic stresses or abiotic stresses and facilitate water and nutrient uptake. Mycorrhizas are found in all plant groups (Wang et al., 2010) and may have been a key innovation for the acquisition of water and nutrients from soil.

mycorrhizas
Symbiotic relationships between fungi and plant roots or rhizoids; enhance plant tolerance to stresses and facilitate water and nutrient uptake.

Key Terms

embryophytes	polymers	sporophytes
nonvascular plants	euphylls	haploid
vascular seedless plants	cuticle	diploid
vascular seed plants	sporopollenin	antheridia
gymnosperms	sporangia	archegonia
angiosperms	gametangia	alternation of generations
eutracheophytes	gametophytes	mycorrhizae

Summary

- Only liverworts, hornworts, mosses, club mosses, ferns and their relatives, gymnosperms, and angiosperms are recognized as embryophytes (land plants) in the kingdom Plantae. They are divided into four informal groups: nonvascular plants (liverworts, hornworts, and mosses), vascular seedless plants (club mosses, ferns, and their relatives), gymnosperms (cone-bearing plants), and angiosperms (flowering plants).
- Embryophytes have 11 distinct characteristics:
  - Characteristics inherited from green algal ancestors include:
    - Be multicellular eukaryotes
    - Use chloroplasts to carry out photosynthesis
    - Use chlorophyll a, chlorophyll b, and carotenoids (β-carotene and xanthophylls) as photosynthetic pigments
    - Store energy or food as starch
    - Have cell walls made up of cellulose
  - Characteristics that are plant innovations to live on land include:
    - Have waterproof coatings to prevent desiccation
    - Be composed of tissues produced by apical meristem
    - Have multicellular embryos that develop within female sex organs
    - Have multicellular reproductive structures
    - Have a life cycle alternating between haploid and diploid generations
    - Be sedentary
- Land plants and green algae, especially the charophyceans, have similar cellular, biochemical, and molecular traits. The multicellular organization of complex charophyceans and land plants are also very similar. It is generally accepted that plants evolved from green algal ancestors because plants and green algae share many characteristics.

- The colonization of dry land by early land plants began approximately 490 to 476 million years ago during the mid-Ordovician period and was one of the most important steps in the evolution of life on Earth.

- Liverworts diverged from charophycean algae around the mid-Ordovician period and were believed to be the earliest land plants. Hornworts, mosses, and early vascular plants appeared around 432 million years ago in the early Silurian period. True vascular plants (eutracheophytes) appeared next around the mid-Silurian and diversified substantially around the early Devonian period, 398 million years ago. Seed plants appeared around the late Devonian period.

- Four key phases were identified as pivotal to plant evolutionary history:
  - Transition to dry land
  - Rise of vascular plants
  - Emergence of seed plants
  - Diversification of flowering plants

- When early land plants made the transition to land, they faced a number of major environmental challenges that required innovative solutions:
  - Manufacturing complex polymers
  - Producing multicellular structures
  - Alternating between sporophyte and gametophyte generations
  - Forming associations with fungi

Reflect

1. *Land plants evolved from green algal ancestor.* What evidence is there to support this claim? Describe at least eight characteristics plants and green algae both have.

2. *What makes a plant, a plant?* Name all the distinct features of plants. Compare and contrast the four major plant groups.

3. *What does it take to live on dry land?* Describe the challenges early land plants faced when they made the transition to dry land from aqueous environments. Discuss the innovative solutions plants have to overcome these challenges.

4. *Alternation of generation.* Plants have a unique life cycle called alternation of generations. Use a diagram to show this unique life cycle and make sure you include the process that allows plant cells to change from haploid (n) to diploid (2n) and the one that allows plant cells to change from 2n to n. Make sure you give the first and last cells of each generation and its adult form.

References

Bateman, R. M., Crane, P. R., DiMichele, W. A., Kenrick, P. R., Rowe, N. P., Speck, T., et al. (1998). Early evolution of land plants: phylogeny, physiology, and ecology of the primary terrestrial radiation. *Annual Review of Ecology and Systematics, 29,* 263–292.

Becker, B., & Marin, B. (2009). Streptophyte algae and the origin of embryophytes. *Annals of Botany, 103,* 999–1004.

Berner R. A. (1999). Atmospheric oxygen over Phanerozoic time. *Proceedings of the National Academy of Sciences of the USA, 96,* 10955–10957.

Bhattacharya, D., & Medlin, L. (1998). Algal phylogeny and the origin of land plants. *Plant Physiology, 116,* 9–15.

Friedman, W. E., Moore, R. C., & Purugganan M. D. (2004). The evolution of plant development. *American Journal of Botany, 91,* 1726–1741.

Graham, L. E. (1993). *Origin of land plants.* New York: Wiley.

Kenrick, P., & Crane, P. R. (1997). The origin and early evolution of plants on land. *Nature, 389,* 33–39.

Lewis, L. A., & McCourt, R. M. (2004). Green algae and the origin of land plants. *American Journal of Botany, 91,* 1535–1556.

Mattox, K. R., & Stewart, K. D. (1984). Classification of the green algae: a concept based on comparative cytology. In D. E. G. Irvine, & D. John (Eds.), *Systematics of the green algae* (pp. 29–72). London: Academic Press

Mora, C. I., Driese, S. G., & Colarusso, L. A. (1996) Middle to late Paleozoic atmospheric CO_2 levels from soil carbonate and organic matter. *Science, 271,* 1105–1107.

Pickett-Heaps, J. D. (1975). *Green algae: Structure, reproduction and evolution in selected genera.* Sunderland, MA: Sinauer Associates.

Sanderson, M. J., Thorne, J. L.,Wikstrom, N., & Bremer, K. (2004). Molecular evidence on plant divergence times. *American Journal of Botany, 91,* 1656–1665.

Wang, B., Yeun, L. H., Xue, J. Y., Liu, Y., Ané, J. M., & Qiu, Y. L. (2010). Presence of three mycorrhizal genes in the common ancestor of land plants suggests a key role of mycorrhizas in the colonization of land by plants. *New Phytologist, 186,* 514–525.

Chapter 22

Bryophytes

© Collipicto/Shutterstock.com

Learning Objectives

- Discuss how bryophytes as a group differ from other plant groups
- Distinguish the strategies used by bryophytes to deal with drought and desiccation
- Explain the evolutionary relationships among bryophytes
- Compare and contrast liverworts, hornworts, and mosses
- Compare and contrast sexual and asexual reproduction by bryophytes
- Describe the importance of bryophytes to humans

The term *bryophytes* (*bryo* in Greek means "swell") refers to the three groups of nonvascular plants: liverworts (Marchantiophyta), hornworts (Anthocerophyta), and mosses (Bryophyta) (Figure 22.1). The terms *bryophytes* and *Bryophyta* are not synonymous. The term *bryophytes* is used informally to refer to all nonvascular plants whereas *Bryophyta* is the phylum name for mosses. There are about 17,000 species of living bryophytes in the world and they form the second largest group of land plants. Bryophytes are smaller and simpler than other land plants. They are common in moist habitats and can grow on rocks or tree bark. Some bryophytes develop an ability to withstand desiccation and can be found in much drier habitats.

Recall from Chapter 21 that early land plants made the transition from aquatic environments to dry land. The main challenges they faced were desiccation, acquisition of water and nutrients, temperature fluctuation, and ultraviolet exposure. In this chapter, we take a closer look at the bryophytes, whose ancestors were the earliest land plants, and examine their unique characteristics.

EVOLUTIONARY RELATIONSHIPS AMONG BRYOPHYTES

Bryophytes played a key role in land plant evolution because their ancestors made the transition to land. However, the phylogenetic relationships among the three groups of bryophytes are still under debate. Based on morphological characteristics, liverworts are the sister group to all the remaining embryophytes (see Figure 21.3), mosses share a common ancestor with the tracheophytes, and hornworts bridge the gap between liverworts and

(a)

(b)

(c)

Figure 22.1a–c Modern bryophytes: (a) liverworts (Marchantiophyta), (b) hornworts (Anthocerophyta), and (c) mosses (Bryophyta).

BOX 22.1   Dealing with Desiccation and Drought

Desiccation is a major challenge for any organisms living on dry land. Water is vital to living organisms because of its ability to dissolve many substances and allow all metabolic processes of life to take place. Water is also the main component of all cells (see Appendix). Extreme desiccation is lethal to almost all plants and animals. How do land plants cope with periods of drought? To deal with drought, land plants adapt one of three basic strategies: drought escape, dehydration avoidance, or desiccation tolerance (Vanderpoorten & Goffinet, 2010).

To escape periods of drought, annual mosses and thalloid liverworts complete their life cycle during the wet season and produce desiccation-tolerant spores that remain in the soil until conditions are favorable for germination. This strategy is referred to as **drought escape**. **Dehydration avoidance**, on the other hand, refers to the ability to maintain a favorable internal water balance during drought. Vascular plants rely mainly on this strategy to cope with drought conditions. They maintain favorable water content in their cells by using roots to absorb water from the soil and vascular tissues to efficiently transport water within their bodies. Unlike vascular plants, bryophytes do not have roots to absorb water or vascular tissues to efficiently transport water. Instead, they take up water directly through the surfaces of their body and their water content is tied directly to the ambient humidity. When the air is dry, bryophytes are dry. When it rains, bryophytes take up water and become rehydrated. Unlike vascular plants, bryophytes do not use dehydration avoidance. Instead, most bryophytes have the ability to deal with repeated drying and wetting. This strategy is referred to as **desiccation tolerance**, which allows bryophytes to exploit a wide range of habitats and not just humid or moist habitats.

Bryophytes vary greatly in tolerance to desiccation (Alpert, 2000). The pulvinate dry rock moss *Grimmia pulvinata*, found commonly in sun-exposed habitats, can tolerate severe desiccation for up to 60 days and quickly recover when rehydrated. The desert moss *Syntrichia caninervis* can remain desiccated for up to six years and still resume metabolic activity and growth when rehydrated. The most extreme case of desiccation tolerance comes from the crystalwort *Riccia*. *Riccia* specimens that were kept in a herbarium for 23 years became viable after rehydration.

Bryophytes adapt to desiccation at both cellular and biochemical levels. Their cell membranes and cytoplasmic components can withstand major water loss, remain intact, and reconstitute quickly when rehydrated. At the biochemical level, bryophytes cease all metabolic activities such as photosynthesis and cellular respiration when desiccated but resume these processes when rehydrated. Desiccation-tolerant bryophytes rely on both protective mechanisms and repair mechanisms to cope with repeated drying and wetting. They produce protective proteins to help stabilize and anchor cytoplasmic components when desiccated.

drought escape
Ability to escape drought periods by completing the life cycle within a short period of time during the wet season.

dehydration avoidance
Ability to maintain a favorable internal water balance during drought periods.

desiccation tolerance
Ability to deal with repeat drying and rewetting.

Water is essential in maintaining the integrity of cell membranes by preserving protein structures and spacing between phospholipids. However, when water is scarce, bryophytes accumulate sucrose to form a high-viscosity liquid substituting for water. Nevertheless, even desiccation-tolerant bryophytes sustain some cellular damage due to desiccation or rapid rehydration and rely heavily on specialized proteins such as hydrins and rehydrins to repair these damages. Attempts have been made to locate the genes responsible for desiccation repair in the hope of engineering desiccation tolerance in crop and forage species.

mosses. In contrast, gene sequence data suggest that tracheophytes share a common ancestor with hornworts instead of mosses. Other alternative phylogenetic relationships have been based on ultrastructure of male gametes and other molecular characteristics (for a full discussion, see Vanderpoorten & Goffinet, 2010). However, most of these phylogenetic relationships seem to agree that liverworts form the sister group to the rest of embryophytes and mark the transition to land.

CHARACTERISTICS OF BRYOPHYTES

Most bryophytes are small, ranging in size from 0.5 millimeters to a few centimeters, because they have no vascular tissues to efficiently transport water, minerals, and food within their bodies. Some mosses and liverworts have undifferentiated parenchyma cells or specialized conducting cells that can transport water and nutrients in a limited fashion. They have no true organs such as roots, stems, or leaves. Bryophytes use **rhizoids** for anchorage. Rhizoids are not roots but are simple multicellular filaments produced by epidermal cells. Bryophyte bodies are only a few cell layers thick, which allow them to take up water and minerals by simple diffusion.

Bryophytes are the only land plants with a dominant, free-living, branched gametophyte. The small and unbranched sporophyte depends on the gametophytes for resources and produces only a single sporangium (Figure 22.2). Bryophytes reproduce asexually by production of specialized **propagules** (structures that germinate into individual plants) and fragmentation. They can also reproduce sexually by formation of gametes. Their male gametophytes use antheridia to produce sperms and their female gametophytes use archegonia to produce eggs. Bryophytes produce flagellated sperms, which rely on water for dispersal.

Although liverworts, hornworts, and mosses share many similar characteristics, they have distinct morphological and anatomical characteristics and are easily distinguished from each other.

Taking a Closer Look at Liverworts

There are approximately 5,000 species of liverworts in the phylum Marchantiophyta. The word *wort* (derived from the Old English word *wyrt*) means "plant" or "herb." *Liver* refers to the resemblance of the liverwort's flattened lobed body to the lobes of a human liver. During medieval times, liverworts were thought to be useful in treating liver problems because of

rhizoids
Are not roots but are simple multicellular filaments produced by epidermal cells.

propagules
Structures that germinate into individual plants.

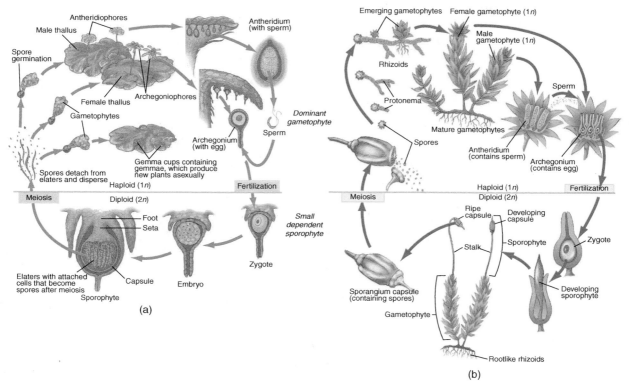

Figure 22.2a–b   Life cycle of bryophytes: (a) liverworts and (b) mosses.

their resemblance to human liver. Liverworts can be found on every continent and in a wide range of habitats.

The Gametophyte Generation

Like all bryophytes, the liverwort gametophyte generation is the dominant generation and it begins with a haploid spore. Growth takes place through the mitotic activity of a single apical cell and the body form of the gametophyte is determined by the arrangement of the daughter cells. The liverwort gametophyte body comes in two basic forms: thalloid (flattened and lobed) and leafy (branchy and leaf-like) (Figure 22.3). Although 80% of liverworts

Figure 22.3a–b   Two basic body forms of liverworts: (a) thalloid and (b) leafy.

are leafy, the best-known examples of liverworts in the genus *Marchantia* are all thalloid. The liverwort gametophyte body or thallus has little or no elaborate differentiation and lacks true roots, stems, and leaves. Unlike hornworts and mosses, liverworts do not have stomata. Instead, they use simple pores for gas exchange. These pores are visible on the thalloid body and each pore leads into an air chamber filled with photosynthetic parenchyma cells (Figure 22.4). Layers of parenchyma cells that are used for storage are found underneath these air chambers. Rhizoids found on the underside of the thallus provide anchorage. The lack of stomata in liverworts provide evidence liverworts are the earliest land plants and are the sister group to all other land plants.

Most liverworts are **dioecious**, which means male and female gametangia are found on different thalli. Liverworts in the genus *Marchantia* have umbrella- or disc-shaped gametangia on top of slender stalks. These miniature tree-like structures are called **gametophores** (Figure 22.5). The **antheridiophores**, gametophores with disc-shaped gametangia, contain sperm-producing antheridia. The **archegoniophores**, gametophores with umbrella-shaped gametangia, contain egg-producing archegonia. Liverworts produce flagellated sperm, which swim through water to reach the eggs within the archegonia. Fertilization and development of the embryo occur within the archegonia.

The Sporophyte Generation

After fertilization, the zygote develops into an embryo and matures into a sporophyte with a foot, an unbranched seta, and a single sporangium or capsule (Figure 22.6). The **foot** attaches the sporophyte to the gametophyte and

dioecious
Male and female sex organs found on distinct individuals.

gametophores
Structures with umbrella-shaped gametangia sitting on top of slender stalks; found only in liverworts.

antheridiophores
Gametophores with disc-shaped gametangia, contain sperm-producing antheridia.

archegoniophores
Gametophores with umbrella-shaped gametangia, contain egg-producing archegonia.

foot
Attaches the liverwort sporophyte to the gametophyte and facilitates nutrients and water uptake by the sporophyte.

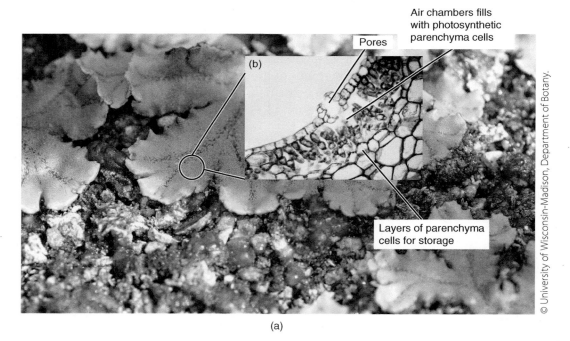

Air chambers fills with photosynthetic parenchyma cells

Pores

(b)

Layers of parenchyma cells for storage

(a)

© University of Wisconsin-Madison, Department of Botany.

Figure 22.4 Simple pores distributed all over this thalloid liverwort and each pore leads in to an air chamber fills with photosynthetic parenchyma cells.

(a)

(b)

Sperms

Antheridium

Antheridiophore with Antheridium

(c)

Archegonium

Neck   Egg

Archegniophore with Archegonia

Ventor

(d)

© University of Wisconsin-Madison, Department of Botany.

Figure 22.5a–d   Antheridiophores (a and c) on male gametophytes and archegoniophores (b and d) on female gametophytes of liverworts.

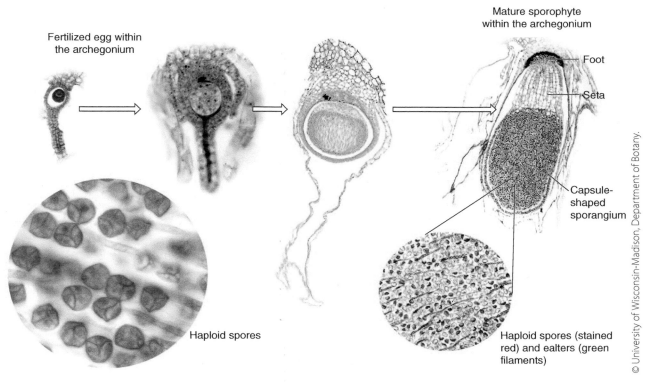

Fertilized egg within the archegonium

Mature sporophyte within the archegonium

Foot

Seta

Capsule-shaped sporangium

Haploid spores

Haploid spores (stained red) and ealters (green filaments)

© University of Wisconsin-Madison, Department of Botany.

Figure 22.6   Longitudinal sections showing progressive development of a sporophyte from a fertilized egg to its adult form within the archegonium.

seta
Found near the base of liverwort sporophyte connecting the foot to the capsule; responsible for spore dispersal.

capsule or sporangium
Found in liverworts and mosses; contains spore mother cells or sporocytes, which undergo meiosis to produce haploid spores.

spore mother cells or sporocytes
Undergo meiosis to produce haploid spores.

elater mother cells or elaterocytes
Develop into elaters to help in spore dispersal; elaters are elongated spindle-shaped cells that aids in spore dispersal.

gemma (singular *gemmae*)
Specialized structures for asexual reproduction; used by liverworts to produce individual gametophytes.

gemma cups
Cup-shaped structures found on the surface of liverwort gametophytes; contain gemma.

facilitates nutrient and water uptake by the sporophyte. After spore maturation, the **seta** elongates and aids in spore dispersal. The **capsule** contains spore mother cells or sporocytes and elater mother cells or elaterocytes. **Spore mother cells** undergo meiosis to produce haploid spores. **Elater mother cells** develop into elaters, which are long, spindle-shaped cells used for spore dispersal. Like typical bryophytes, the sporophyte generation is small and depends on the gametophyte generation for protection and resources.

Asexual Reproduction

Liverworts can also reproduce asexually by producing specialized propagules called **gemma**. In *Marchantia* liverworts, disc-shaped gemma are contained in **gemma cups** on the surface of the thallus (Figure 22.7). The gemma are dislodged by raindrops splashing into the gemma cups and each gemmae can germinate into a gametophyte.

TAKING A CLOSER LOOK AT HORNWORTS

Hornworts in the phylum Anthocerophyta are the smallest group of bryophytes. There are about 300 known species of hornworts worldwide. They are called hornworts because of their long conspicuous horn- or rod-like sporophytes (Figure 22.1b). Compared to liverworts and mosses, hornworts are simpler and less diverse in their body form. However, some phylogenetic studies suggest hornworts share a common ancestor with tracheophytes.

The Gametophyte Generation

The hornwort gametophyte is basically thalloid, which looks like either a rosette or a ribbon. Its growth form is determined by the geometric shape of the single apical cell. The hornwort thallus is not elaborately differentiated

Figure 22.7a–b (a) Gemma cups on the surface of liverwort thallus; (b) cross section shows individual gemma.

but it has pores and chambers filled with mucilage instead of air. **Mucilage** is a complex carbohydrate that is essential for water retention. Gametangia are found on the upper surface of the thallus. Most hornworts are **monoecious**, having both antheridia and archegonia on the same thallus. Rhizoids on the lower surface provide anchorage. Many hornworts maintain intimate symbiotic relationships with nitrogen-fixing cyanobacteria and mycorrhizial fungi.

The Sporophyte Generation

Typical of bryophytes, the hornwort sporophyte is small and depends on the gametophyte for resources. Unlike other bryophytes, the hornwort sporophyte has no seta between the foot and the sporangium and its sporangium grows continuously from the apical meristem at its base. The hornwort sporangium is distinct because it is horn-like rather than capsule-like. Each sporangium has a single elater mother cell to produce elaters and spore mother cells to produce spores. Like mosses, hornworts have stomata found only on their sporophytes. The stoma of hornworts and mosses is bounded by two guard cells and are morphologically similar to the stoma of vascular plants. However, unlike vascular plants, the stoma of hornworts and mosses are always open (Figure 22.8) and facilitate spore dispersal instead of gas exchange and transpiration. The presence of stomata on hornworts and mosses indicates they are more closely related to other land plants than liverworts.

Asexual Reproduction

Most hornworts have no specialized means of asexual reproduction and rely primarily on fragmentation. However, a few species have gemma on the upper surface of the gametophyte for asexual reproduction.

TAKING A CLOSER LOOK AT MOSSES

The largest group of bryophytes is the mosses (approximately 12,000 species) in the phylum Bryophyta. Mosses tend to grow in dense patches on

mucilage
Complex carbohydrate that is used to retain water.

monoecious
Male and female sex organs are found on the same individual.

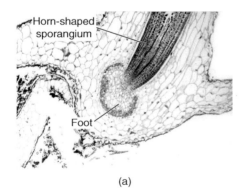
(a)

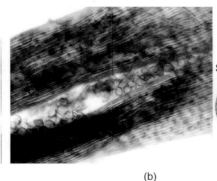
(b)

© University of Wisconsin-Madison, Department of Botany.

Figure 22.8a–b   Hornwort sporophyte has (a) a foot and horn-shaped sporangium but no seta, (b) stomata on its surface to facilitate spore dispersal.

moist soil, rock, tree bark, rotting wood, and even animal remains. They are found in every ecosystem except the ocean. Many true mosses are easily recognizable by their sporophyte, which has an elevated capsule on a long sturdy stalk or seta (Figure 22.1c). However, organisms such as reindeer moss, Irish moss, Spanish moss, and club mosses are not true mosses. Reindeer moss and Irish moss are not even plants. Reindeer moss is a lichen and Irish moss is a red alga (see Chapter 20). Spanish moss is a flowering plant in the phylum Anthophyta (see Chapter 25) and club mosses are vascular plants in the phylum Lycopodiophyta (see the next chapter). Some phylogenetic studies suggest mosses share a common ancestor with vascular plants and are more advanced than liverworts.

The Gametophyte Generation

Like other bryophytes, the moss gametophyte germinates from a haploid spore and growth results from the mitotic activity of a single apical cell. The immature gametophyte is a long, branching strand of green cells called a **protonema**. Although the protonema resembles a filamentous green alga, it forms buds that grow into individual gametophytes. The mature moss gametophyte is leafy and looks like a miniature bush with branches and leaf-like blades (Figure 22.9b). The gametophyte body consists of mostly parenchyma cells. Some of these parenchyma cells specialize in conducting water and are called **hydroids**. Others specialize in distributing food made by photosynthesis and are called **leptoids**. Unlike xylem and phloem in vascular plants, hydroids and leptoids are not complex tissues and do not efficiently transport water and nutrients. Peat mosses, in the genus *Sphagnum*, store water in their blades within large dead cells called **hyaline cells** or hyalocysts (Figure 22.10). Rhizoids found on the branches of some mosses provide anchorage. Most mosses are dioecious with antheridia at the tips of male plants and archegonia at the tips of female plants.

The Sporophyte Generation

Like liverworts, the moss sporophyte consists of a foot, an unbranched seta, and a terminal sporangium or capsule. However, the moss seta lengthens into a long sturdy stalk to raise the capsule above the branches and blades of the gametophyte. Unlike other bryophytes, the moss capsule is more complex and has additional structures like a calyptra, an operculum, and a peristome (Figure 22.11). The **calyptra** or hood initially covers and protects the immature capsule. A dome-shaped **operculum** or lid and a **peristome** shield the capsule opening. When the capsule matures, its calyptra and operculum fall off the capsule to enable spore dispersal. The peristome consists of one or two rows of teeth that control spore release. Most mosses rely on wind for spore dispersal and their spores are released passively and gradually. *Sphagnum* mosses, on the other hand, implode their capsules to physically expel the spores. Other mosses such as the ones in the family Splachnacaea, the dung mosses, rely on insects to disperse their spores. The stoma of mosses and hornworts is surrounded by two guard cells and is always open. The presence of stomata on mosses indicates mosses are closely related to vascular plants.

---

protonema (plural protonemata)
Immature moss gametophyte, which is a long, branching strand of green cells that superficially resembles a filamentous green alga.

hydroids
Parenchyma cells specialize in conducting water; found only in moss gametophytes.

leptoids
Parenchyma cells specialize in distributing food made by photosynthesis; found only in moss gametophytes.

hyaline cells or hyalocysts
Large dead cells found in moss blade for water storage.

calyptra
Hood-shaped structure that covers and protects the immature capsule; found only in mosses (Bryophyta).

operculum
Dome-shaped lid covers the capsule opening; found only in mosses (Bryophyta).

peristome
Structure found around the capsule opening; consists of one or two rows of teethlike cells; responsible for controlling spore release.

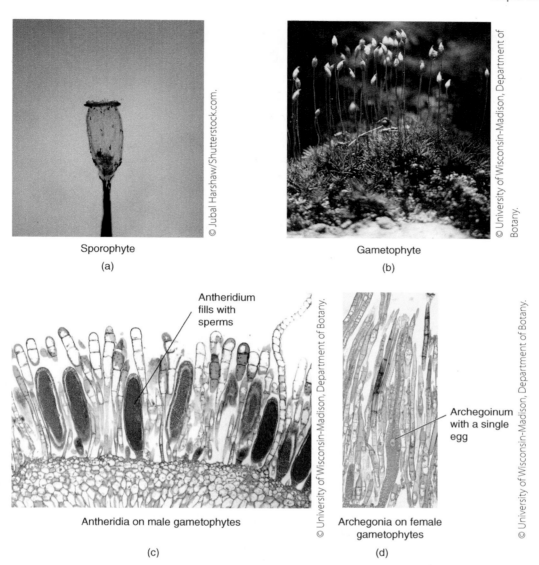

Sporophyte
(a)

Gametophyte
(b)

Antheridium fills with sperms

Antheridia on male gametophytes

(c)

Archegoinum with a single egg

Archegonia on female gametophytes

(d)

Figure 22.9a–d   (a) sporophytes, (b) gametophytes of a hair cap moss, (c) capsule antheridia, (d) archegonia.

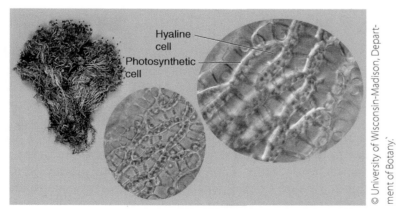

Hyaline cell

Photosynthetic cell

Figure 22.10   Surface view of a peat moss, *Sphagnum*, blade showing alternating strands of dead cells (hyaline) and living photosynthetic cells.

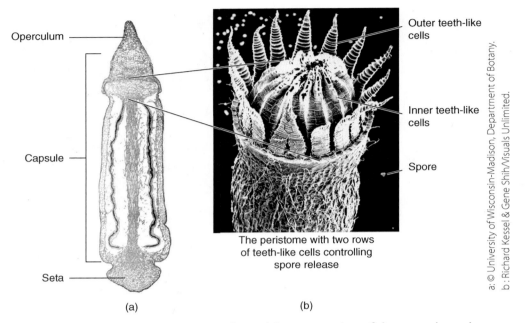

Operculum

Capsule

Seta

Outer teeth-like cells

Inner teeth-like cells

Spore

The peristome with two rows of teeth-like cells controlling spore release

(a)

(b)

a: © University of Wisconsin-Madison, Department of Botany.
b: Richard Kessel & Gene Shih/Visuals Unlimited.

Figure 22.11a–b The moss sporophyte: (a) cross section of the capsule and (b) scanning electron micrograph of the peristome teeth and spores.

Asexual Reproduction

Most mosses can reproduce asexually by fragmentation. Propagules are also used for asexual reproduction and can be found on the rhizoids, branches, or leaf-like blades of the gametophyte.

IMPORTANCE OF BRYOPHYTES TO HUMANS

Although bryophytes are not usually thought of as economically important plants, they have a variety of horticultural, household, industrial, medical, and ecological uses (for a review, see Glime, 2007). In this last section, we highlight some of these uses.

Horticultural Uses

In horticulture, bryophytes such as *Sphagnum* mosses are commonly used as soil additives, ground cover, seed beds, and for growing ferns, orchids, and mushrooms (Figure 22.12a). Bryophytes are often used to make decorative baskets and wreaths, and to cover floral containers. Pots made up of compressed peat are used for starting seeds (Figure 22.12b). Mosses have long been used to condition soil because they store and gradually release large amounts of water and nutrients. Nurseries often use wet *Sphagnum* mosses as packing materials to ship live plants. Many bryophytes are good ground cover because they decompose much slower than other plant materials. Bryophytes can also be used to create serene landscapes and moss gardens that are very popular in Japan and Great Britain.

Figure 22.12a–c   Bryophyte uses: (a) peat used to cultivate mushroom, (b) pot made with compressed peat for starting seeds, and (c) peat used for fuel.

Household and Industrial Uses

Mosses have been used in many household and industrial items. *Sphagnum* mosses are used to line hiking boots, diapers, and sanitary napkins because they are soft and absorb moisture and odor. *Sphagnum* mosses are also used to stuff pillows, mattresses, furniture, and dolls. In the United States, mosses are used as packing materials for vegetables and mushrooms. Biological supply companies use mosses as packing materials for live plants and animals. Mosses are also used to cushion and protect fragile items such as china. Dried mosses are widely used to decorate store windows and to make wreaths, floral arrangements, and even artworks.

Fuel Production

In waterlogged environments such as wetland bogs and peat swamp forests, peat deposits form when plant materials fail to decompose fully under acidic and anaerobic conditions. Peat has a long history of use as a fuel and is now the most important source of fuel for European countries such as Ireland, Finland, Germany, Sweden, and Poland (Figure 22.12c). Peat is often viewed as a promising alternative to fossil fuel because of its higher energy content (more than 8,000 BTU per dry pound) than wood and produces fewer pollutants when it burns. Moss harvesting has become a concern for ecologists and bryologists (scientists who study bryophytes). Overharvesting of peat, loss of peatlands to agriculture, and environmental pollution have resulted in a major decline in peatlands. In Germany, attempts at peatland restoration have been made but the recovery process is very slow and may not be able to offset depletion.

Medical Uses

Bryophytes have been used for more than 400 years as herbal medicines in many countries such as China, India, Europe, and North America. The Chinese used mosses such as *Polytrichum* and *Fissidens* as diuretics and hair growth stimulants. Native Americans used *Mnium* and *Bryum* to treat burns and wounds. In France, *Marchantia* liverworts were used as a diuretic. Many liverworts and mosses have antibacterial, antifungal, antiviral, and even anticancer properties. During World War I, *Sphagnum* mosses were used extensively as surgical dressings by the British Army, Canadian Red Cross, and United States Army. The wounds covered by *Sphagnum* dressings healed faster and had little or no secondary infections. *Sphagnum* dressings absorbed and held more liquid than cotton dressings and did not need to be changed as often. Bryophytes produce numerous compounds such as aliphatic compounds, prenylquinones, sugar alcohol, and aromatic and phenolic compounds and their medical properties are yet to be discovered.

Ecological Importance and Uses

Bryophytes are pioneer plants colonizing disturbed habitats and help control erosion before larger plants become established. They form symbiotic relationships with nitrogen-fixing cyanobacteria, which convert atmospheric nitrogen to ammonia and nitrates. The excess fixed nitrogen is often released into the soil for other land plants to use. Bryophytes play an important role in ecological succession and establishment of plant communities.

Many bryophyte species are used as environmental indicators for heavy metal deposits, acidity, soil quality, and air pollution. "Bryometers" or moss bags are used widely in Japan and Europe to monitor levels of air pollution, lead uptake, and heavy metals. In North America, bryophytes are used to monitor air pollutants such as sulfur dioxide, hydrogen fluoride, and ozone. The absorbent properties of *Sphagnum* mosses also make them excellent agents for bioremediation. *Sphagnum* mosses are used to decontaminate radioactive water, remove acid and toxic metals from factory runoffs, and clean up sewage or oil spills.

Key Terms

drought escape	foot	protonema
dehydration avoidance	seta	hydroids
desiccation tolerance	capsule	leptoids
rhizoids	spore mother cells	hyaline cells
propagules	elater mother cells	calyptra
dioecious	gemma	operculum
gametophores	gemma cups	peristome
antheridiophores	mucilage	
archegoniophores	monoecious	

Summary

- The bryophytes refer to three groups of nonvascular plants, liverworts (Marchantiophyta), hornworts (Anthocerophyta), and mosses (Bryophyta), and form the second largest group of land plants.
- Most of the phylogenetic studies seem to agree that liverworts form the sister group to the rest of embryophytes and mark the transition to land. It is still under debate whether mosses or hornworts are closely related to the tracheophytes.
- All bryophytes are:
  - Small and have no vascular tissues, roots, stems, or leaves. Rhizoids are used to provide anchorage.
  - The only land plants with a dominant, free-living, branched gametophyte.
  - The small and unbranched sporophyte depends on the gametophytes for resources and produces only a single sporangium.
- In bryophytes, growth takes place through the mitotic activity of a single apical cell.
- Bryophytes reproduce both asexually and sexually. They produce flagellated sperm and require water to swim to the eggs.
- The liverwort gametophyte can be thalloid or leafy. Most liverworts are dioecious. Liverworts in the genus *Marchantia* produce two types of gametophores:
  - The antheridiophores, gametophores with disc-shaped gametangia, contain the sperm-producing antheridia.
  - The archegoniophores, gametophores with umbrella-shaped gametangia, contain the egg-producing archegonia.
  - Their sporophyte consists of a foot, an unbranched seta, and a single sporangium or capsule. Liverworts reproduce asexually using gemma.
- The hornwort gametophyte is basically thalloid. They are mostly monoecious. The hornwort sporophyte has no seta between the foot and the sporangium and its sporangium grows continuously from the apical meristem at its base. Like mosses, hornworts have stomata found only on their sporophytes.
- Mosses in the phylum Bryophyta are the largest group of bryophytes. Like other bryophytes, protonema, immature moss gametophyte, germinates from a haploid spore. Unlike other bryophytes, mosses have hydroids and leptoids to transport water and food. Some mosses have hyaline cells in their blades to store water.
- The moss sporophyte consists of a foot, an unbranched but greatly elongated seta, and a terminal sporangium or capsule. Like hornworts, mosses have stomata only on their sporophytes. Unlike other bryophytes, the moss capsule has additional structures such as a calyptra, an operculum, and a peristome to protect the capsule and to control spore release.
- Bryophytes have a variety of horticultural, household, industrial, medical, and ecological uses even though they are not considered to be economically important plants.

Reflect

1. *Sticking close to water.* Based on the characteristics unique to bryophytes, explain why most bryophytes are found in humid or moist environments.
2. *Is it a liverwort, hornwort, or moss?* You discover small, soft green plants growing on rocks right next to a creek and want to know what type of bryophytes they are. Discuss the characteristics you used to distinguish among liverworts, hornworts, and mosses.

3. ***What are mosses?*** Are mosses nonvascular plants or early vascular plants? Explain why mosses blur the line between nonvascular plants and vascular plants.

4. In your opinion, which of the bryophytes has the most complex sporophyte? Which has the most complex gametophyte? Explain your answers.

5. ***Bryophytes in our daily life*** Do Google searches and find out how bryophytes are used in household, horticultural, industrial, and medical products. Give the brand name of the product and its bryophyte component.

References

Alpert, P. (2000). The discovery, scope, and puzzle of desiccation tolerance in plants. *Plant Ecology, 151,* 5–17.

Glime, J. M. (2007). Economic and ethnic uses of bryophytes. In R. H. Zander & P. M. Eckel (Eds.), *Flora of North America: North of Mexico: Vol. 27. Bryophytes: Mosses, Part 1* (pp. 14–41). New York: Oxford University Press

Vanderpoorten, A., & Goffinet, B. (2010). *Introduction to bryophytes.* New York: Cambridge University Press.

Chapter 23

Lycophytes and Ferns

© Aleksandr Petrunouskiyi/Shutterstock.com

Learning Objectives

- Discuss the evolutionary changes in vascular plants and how they differ from nonvascular bryophytes
- Compare and contrast microphylls and megaphylls
- Summarize the features distinguishing seedless vascular plants from nonvascular plants and seed vascular plants
- Explain the evolutionary relationships between lycophytes and ferns
- Compare and contrast club mosses, quillworts, spike mosses, whisk ferns, horsetails, and ferns
- Explain why whisk ferns and horsetails are classified as ferns
- Compare the alternation of generations between homosporous plants and heterosporous plants and give examples for each
- Describe the importance of seedless vascular plants to humans

Many people are very familiar with ferns such as Boston ferns and maidenhair ferns because they tolerate low light and make great houseplants. Ferns are also used as landscape and container plants and are found in interior environments such as atriums of hotels, shopping malls, and office buildings, as well as outdoor environments such as backyards and gardens. Ferns belong to a group of plants called **seedless vascular plants**. These plants do not produce seeds and have vascular tissues, hence the name.

After making the transition to dry land, early land plants began to develop characteristics allowing them to efficiently exploit resources on land. Plants began to develop vascular tissues, xylem and phloem, for efficient transport of water and food. As a result, their sporophyte became large and complex, which enabled them to outcompete other plants for light. In this chapter, we explore the origin and diversity of vascular plants, take a closer look at the lycophytes and ferns, the seedless vascular plants, and examine characteristics or innovations that are unique to them.

EVOLUTION OF VASCULAR PLANTS (EUTRACHEOPHYTES)

Recall from Chapter 21 that early vascular plants appeared in the early Silurian period. True vascular plants (**eutracheophytes**) appeared around the mid-Silurian and greatly diversified around the early Devonian period (see Figure 21.3). Vascular plants are subdivided into two main groups, the ones with spores for dispersal (seedless vascular plants) and those with seeds for dispersal (seed vascular plants). Based on the leaf structure, vascular plants can be divided into lycophytes having simple leaves and euphyllophytes (ferns and seed plants) having complex leaves (Figure 23.1). During the Silurian-Devonian period, characteristics unique to vascular plants began to emerge and these characteristics enabled early vascular plants to more efficiently exploit and compete for land-based resources.

seedless vascular plants
Plants such as lycophytes and ferns do not produce seeds and have vascular tissues; dispersed by spores.

eutracheophytes or true vascular plants
Land plants that have lignified tracheids to transport water and minerals (*eu-* in Greek means "true" or "well"); all vascular plants.

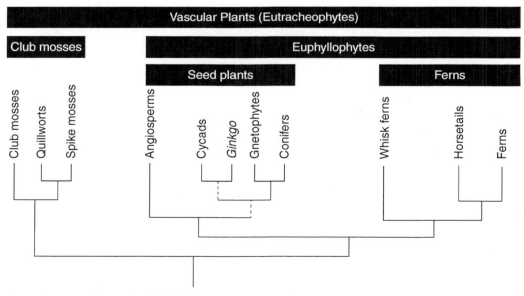

Figure 23.1 Simplified phylogenetic relationships among vascular plants. Adapted from Schuettpelz and Pryer (2008).

The Sporophyte Became Dominant

Unlike bryophytes, the sporophyte body of vascular plants became larger and branched. Branching of the sporophyte body enabled early vascular plants to harvest light more efficiently for photosynthesis. The sporophyte body grew independently from the gametophyte body and became dominant (for a review, see Graham, Cook, & Busse, 2000). Multiple growth tips on the branched sporophyte enabled vascular plants to develop complex body form and to better survive herbivorous attacks. The sporophyte body also began to produce multiple sporangia, which greatly increased spore production. Like bryophytes, early vascular plants used spores for dispersal and had a single apical cell in each growing tip.

Emergence of Vascular Tissues, Stems, and Roots

Early vascular plants produced vascular tissues to enhance transport of water and nutrients and also provide structural support. Recall from Chapter 3 that xylem and phloem are the main components of vascular tissues. Xylem, composed of lignified tracheids, transports water and minerals. Phloem transports sugars made by photosynthesis. True stems and roots with vascular tissues emerged during the Silurian-Devonian period. The stems of early vascular plants enabled them to grow toward the sun and to anchor photosynthetic organs, as well as spore-producing sporangia. Their stems were also covered with tough cuticles to minimize water loss and stomata to permit gas exchange and transpiration.

Evolution of Leaves

Recall from Chapter 6 that leaves are the main photosynthetic organs. The leaves of vascular plants have vascular tissues, a determinate growth, a bilateral symmetry, and a definite arrangement on the stem. There are two basic types of leaves, microphylls (simple leaves) and megaphylls (complex leaves). **Microphylls** found in lycophytes are often small and have only a single, unbranched vascular vein in their leaves. In contrast, **megaphylls** or **euphylls** found in ferns and seed plants are much larger and have a complex system of vascular veins. Phylogenetic and paleobotanical data suggest microphylls and megaphylls evolved independently and were very different from each other. Microphylls may have originated from sterilized sporangia or vascularized enations (flips of green tissues with no veins). Megaphylls, on the other hand, may have originated from flattened stems developing photosynthetic cells between their branches (Figure 23.2). As a result, megaphylls become flat and broad with many veins (for a review, see Piazza, Jasinski, & Tsiantis, 2005). Fossilized megaphylls appeared and became abundant during the late Devonian and early Carboniferous period. Some botanists believe megaphylls evolved in response to the dramatic drop (90%) in the concentration of carbon dioxide in the atmosphere around that time (Beerling, Osborne, & Chaloner, 2001). The large surface area and high stomata density of megaphylls could have been adaptations to enhance carbon dioxide uptake. Megaphylls found in ferns and seed plants are believed to have evolved independently of each other. Vascular plants having megaphylls are often referred to as euphyllophytes.

microphylls
Small leaves that have only a single unbranched vascular vein; found only in lycophytes.

megaphylls or euphylls
Large leaves that have a complex system of vascular veins; found in ferns and seed plants.

euphylls or true leaves
Leaves with a branched vascular system and leaf gaps in the stem.

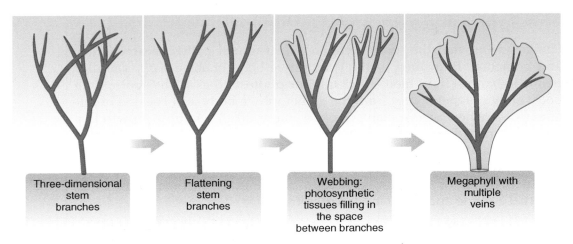

Figure 23.2 Evolution of megaphylls. Adapted from Moran (2004).

CHARACTERISTICS OF SEEDLESS VASCULAR PLANTS

Seedless vascular plants are the lycophytes (Lycopodiophyta) and the ferns (Pteridophyta). Like typical vascular plants, they have vascular tissues and true stems (aerial stems and rhizomes). The aerial stems are green and photosynthetic. Most seedless vascular plants also have true roots and either microphylls or megaphylls. Their sporophyte is dominant, large, and branched. They produce many sporangia and disperse by spores and not by seeds. Their gametophyte is smaller than the sporophyte but self-sustaining. The gametophyte uses antheridia to produce sperms and archegonia to produce eggs. Like nonvascular plants, all seedless vascular plants produce flagellated sperms requiring water to reach the egg. Lycophytes consist of club mosses, quillworts, and spike mosses (Figure 23.3). Ferns consist of whisk ferns, horsetails, and true ferns. Whisk ferns, horsetails, and ferns are morphologically distinct and were classified as three separate phyla. Recent DNA sequence data suggest whisk ferns and horsetails are more closely related to ferns than lycophytes. As a result, whisk ferns and horsetails are now placed in the same phylum (Pteridophyta) as ferns. Based on morphological characteristics and molecular sequence data, lycophytes are the sister group to all the remaining vascular plants (ferns and seed plants) and ferns share a common ancestor with the seed plants (Schuettpelz & Pryer, 2008) (Figure 23.1).

Taking a Closer Look at Lycophytes

There are about 1,000 living species of lycophytes, which represent less than 1% of vascular plants. Lycophytes such as club mosses, quillworts, and spike mosses are distinctly different from ferns. The leaves of lycophytes and ferns are very different. All lycophytes have microphylls whereas ferns have megaphylls.

Lycophytes have true roots, stems (both aerial stems and horizontal underground rhizomes), and microphylls. They have similar life cycles but produce different types of spores. Club mosses such as *Lycopodium* are **homosporous**, which mean they produce one type of spore

homosporous
Production of one type of haploid spore that germinates into a bisexual gametophyte.

Table 23.1   Seedless Vascular Plants and Their Key Characteristics

Seedless Vascular Plants	Sporophyte	Gametophyte	Representatives
Lycopodiophyta • Club mosses • Quillworts • Spike mosses	• Dominant • Large and branched • Photosynthetic • Aerial stems and rhizomes • True roots • Microphylls • Many sporangia • Terminal strobilus • Homosporous—club mosses • Heterosporous—quillworts and spike mosses	• Small • Photosynthetic • Can be bisexual or unisexual • Produce flagellated sperms • Require water for sexual reproduction	Club moss—*Lycopodium* Quillworts—*Isoetes* Spike mosses—*Selaginella*
Pteridophyta • Ferns	• Dominant • Large and branched • Photosynthetic • Rhizomes • True roots • Megaphylls or fronds • Many sporangia • Sori • Homosporous	• Small • Photosynthetic • Bisexual prothallus • Produce flagellated sperms • Require water for sexual reproduction	*Pteridium*
Pteridophyta • Horsetails	• Dominant • Large, some branched • Photosynthetic • Aerial stems and rhizomes • Silica crystals • True roots • Reduced megaphylls • Many sporangia • Terminal strobili • Homosporous	• Small • Photosynthetic • Bisexual • Produce flagellated sperms • Require water for sexual reproduction	*Equisetum*
Pteridophyta • Whisk ferns	• Dominant • Dichotomously branched • Photosynthetic • Aerial stems and rhizomes • No roots • No leaves but have enations • Many sporangia • Clusters of three • Homosporous	• Tiny • Underground • Partnership with mycorrhizal fungi • Produce flagellated sperms • Require water for sexual reproduction	*Psilotum*

Figure 23.3a–d The seedless vascular plants: (a) club mosses (Lycopodiophyta), (b) ferns (Pteridophyta), (c) horsetails (Pteridophyta), and (d) whisk ferns (Pteridophyta).

heterosporous
Production of two types of haploid spore: microspore that germinates into a male gametophyte and megaspore that germinates into a female gametophyte.

microspores
Germinate into male gametophytes, which only produce antheridia.

megaspores
Germinate into female gametophytes, which only produce archegonia.

strobili or cones
Clusters of sporophylls bearing sporangia.

(Figure 23.4). These spores germinate into bisexual gametophytes, which have both antheridia and archegonia. Quillworts (*Isoetes*) and spike mosses (*Selaginella*), on the other hand, are **heterosporous**, which mean they produce two types of spores: **microspore** and **megaspore**. (Figure 23.5). Microspores germinate into male gametophytes, which only produce antheridia. Megaspores germinate into female gametophytes, which only produce archegonia. Therefore, their gametophytes are unisexual. Lycophyte sporangia are found on the upper surface of leaves called *sporophylls*. Their sporangia are clustered at the tips of the stems to form **strobili** or cones. In heterosporous lycophytes such as quillworts and spike mosses, their strobili have two types of sporangia: **microsporangia** and **megasporangia**. Microsporangia are sporangia producing microspores. Megasporangia are sporangia producing megaspores.

Most lycophytes are found in moist and shady environments, but some can withstand desiccation and are found in arid environments. For example, the resurrection plant, *Selaginella lepidophylla*, is native to the deserts of Texas and Mexico. Like desiccation-tolerant bryophytes (Box 21.1), the

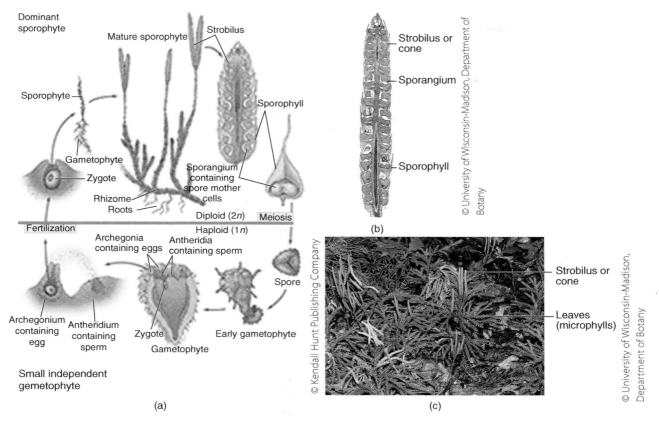

Figure 23.4a–c   Club moss (a) life cycle showing various stages, (b) longitudinal section of a strobilus, and (c) a photo showing what the structures like on the plant.

resurrection plant relies on both protective and repair mechanisms to cope with repeated drying and wetting (Figure 23.6).

Taking a Closer Look at Ferns

There are about 12,000 living species of ferns, horsetails, and whisk ferns in the phylum Pteridophyta, which account for little more than 3% of vascular plants. Like lycophytes, ferns have true roots and rhizomes, a dominant sporophyte, and a small but independent gametophyte (Figure 23.7). Unlike lycophytes, ferns have relatively large and conspicuous megaphylls called **fronds**. Fronds are produced by the underground rhizomes. When fronds emerge from the soil, they are tightly coiled and resemble the scroll or top of a fiddle or violin, hence the name *fiddleheads*. As fiddleheads grow, they unfold and expand into large fronds (Figure 23.8). Like the compound leaf of some angiosperms, a frond blade is subdivided into smaller sections. In ferns, these smaller sections are called **pinnae**.

Like lycophytes, ferns produce many sporangia and use spores for dispersal. However, fern sporangia are clustered on the underside of the fronds to form **sori** (singular *sorus*) (Figure 23.7). The **annulus** is a row of cells with thickened walls running down the medial plane of the fern sporangium. This unique structure is used to tear open the sporangium and catapult spores into the air (Figure 23.9). Unlike lycophytes, ferns are homosporous and these spores germinate into bisexual gametophytes.

microsporangia
Sporangia that produce microspores that germinate into male gametophytes.

megasporangia
Sporangia that produce megaspores, which germinate into female gametophytes.

fronds
Megaphylls or leaves of ferns.

pinnae
Individual sections or leaflets of a frond.

sori (singular *sorus*)
Clusters of sporangia found on the underside of fronds.

annulus
Row of cells with thickened cell walls running down the medial plane of the fern sporangium; used to tear open the sporangium and to catapult the spores into the air.

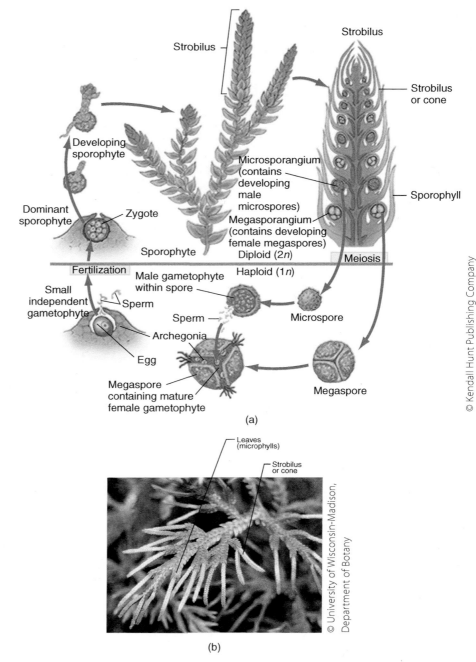

(a)

(b)

Figure 23.5 Life cycle of a heterosporous spike moss, *Selaginella*.

prothallus *(plural pro-thalli)*
Heart-shaped fern gametophyte; small, flat, and does not have vascular tissues, true roots, stems, or leaves; rhizoids are used for anchorage.

The fern gametophyte is often heart-shaped and small. It does not have vascular tissues, true roots, stems, or leaves and is often flat, hence the name **prothallus**. Simple rhizoids are used for anchorage. The prothallus carries out photosynthesis and is completely independent from the sporophyte (Figure 23.7).

Most ferns are found in terrestrial environments but some are found only in freshwater environments. Ferns are often thought of as decorative and used as houseplants, landscape plants, and aquarium plants. When the habitat or ecosystem is disturbed, some ferns grow out of control and cause major economic and ecological concerns (**Box 23.1**).

David Sieren/Visuals Unlimited

Figure 23.6   The resurrection plant Selaginella lepidophylla when rehydrated (*left*) and when desiccated (*right*).

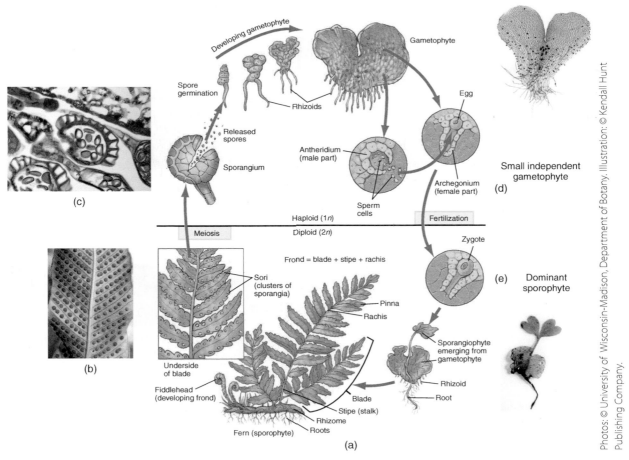

Photos: © University of Wisconsin-Madison, Department of Botany. Illustration: © Kendall Hunt Publishing Company.

Figure 23.7a–e   A fern showing (a) its life cycle, (b) sori on the underside of the leaf, (c) cross section of sporangia showing the spore within, (d) small independent gametophyte, and (e) the dominant sporophyte emerges from the gametophyte.

(a)   (b)

Figure 23.8a–b   (a) Closeup of a fiddle heads and (b) and unfolding fiddlehead.

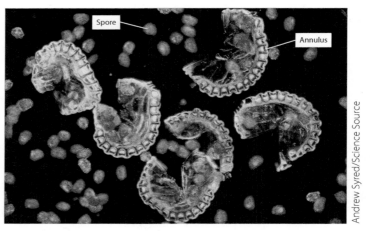

Figure 23.9   Closeup of a fern sporangia showing the annulus on each sporangium and spores.

BOX 23.1   Problem Ferns

At least 60 species of ferns can be problematic in terrestrial or aquatic environments and become weeds (Figure Box 23.1-1) (for a review, see Robinson, Sheffield, & Sharpe, 2010). Weeds are plants growing where they are unwanted. Some ferns are deliberately or accidentally introduced by humans into a non-native environment and these introduced or alien species spread rapidly and become major problems. Some native ferns become problematic after their environments are altered by human activities. Problem ferns tend to spread rapidly by spores, rhizome growth, or fragmentations and become widespread and abundant. In terrestrial

Figure Box 23.1-1a–d   Representatives of invasive ferns: (a) bracken (*Pteridium*), (b) giant salvinia (*Salvinia molesta*), (c) horsestail (*Equisetum*), and (d) climbing fern (*Lygodium*).

environments, problem ferns disrupt land productivity, food production, and local habitats. In aquatic environments, problem ferns interfere with water flow, flood management, navigation, and transport. They also negatively affect water quality and aquatic environments. Some problem ferns are harmful to humans and animals because of their toxicity or carcinogenicity. A few of these ferns even harbor pests carrying human or animal diseases.

Alien Ferns Gone Bad

Giant salvinia (*Salvinia molesta*), an aquatic fern native to Brazil, was introduced into many countries as a freshwater ornamental. Giant salvinia rapidly forms thick, dense, floating mats across water surfaces. These thick mats cut off light to other aquatic plants and greatly deplete nutrient availability. Oxygen level and water quality are also greatly reduced when older parts of the mats decompose. The oxygen level of the lake becomes so low that it can no longer support fish and other aquatic organisms. These floating mats greatly reduce water flow and block spillways, drainage systems, and irrigation canals. They also form physical barriers, making it difficult to travel by boat (Figure Box 23.1-3).

Figure Box 23.1-2 Thick bracken (*Pteridium aquicinum*) understory in an English woodland.

Figure Box 23.1-3 Infestation of *Salvinia* in backwater canals.

Native Ferns Gone Bad

Brackens, ferns in the genus *Pteridium*, are the most common vascular plants in the world and also the most troublesome. They tend to occupy pastures and cause major economic loss of livestock. Brackens are toxic and even carcinogenic and can cause acute poisoning, chronic illness, and cancer in animals when consumed. Brackens tend to form thick and dense vegetation, which are ideal habitats for ticks. Some ticks carry pathogenic organisms causing diseases such as tick-borne fever, babesiosis, and Lyme disease in humans, wild animals, and livestock (Figure Box 23.1-2). A thick bracken stand also hinders conservation efforts, recreation activities, and reforestation programs.

Management of Problem Ferns

There are five strategies to manage problem ferns: preemptive, eradication, containment, releasing or temporary control, and long-term, low-level equilibrium stasis (Robinson, 2007). Preemptive measures prevent problems. Such measures can include declarations of prohibited noxious weeds, outright import or export bans, and legislation to regulate ornamental suppliers. Eradication can be an effective option when the affected area is small and the infestation of alien species is in its early stage. However, eradication becomes more difficult and expensive when the affected area becomes too large. Containment is commonly used to manage native ferns such as *Pteridium* by applying periodic peripheral treatments to keep the ferns in check. Releasing or temporary control reduces the dominance of a problem fern and improves the competitiveness of vegetation targeted for conservation. Long-term, low-level equilibrium stasis is a strategy to maintain problem ferns at an acceptable level by using their natural enemies such as herbivores or pathogens. The use of natural enemies to control a pest population is referred to as *biological control*. The goal of biological control is to keep the problem fern in check by establishing a persistent and sustainable balance between the fern and its natural enemies.

Taking a Closer Look at Horsetails

Even though horsetails and ferns have similar molecular features and are classified in the same phylum, they look nothing alike. All living horsetails belong to the single genus *Equisetum* (*equus* in Latin means "horse" and *saeta* means "bristle"). There are about 15 living species of horsetails or scouring rushes and their sporophytes either have or do not have whorled branches (Figure 23.10). Species having whorled branches are often called *horsetails*. Scouring rushes, on the other hand, have no whorled branches. Horsetail stems have significant silica deposits and their rough texture make them ideal scrubbing materials used by early pioneers for cleaning pots and pans. Horsetails are found mostly in wet, marshy habitats but can also be found in arid environments.

Like lycophytes and ferns, horsetail sporophytes are dominant and have vascular tissues, true roots, stems (both aerial stems and rhizomes), and greatly reduced megaphylls. These reduced megaphylls fuse to form a whorl around each node on the stems (Figure 23.11). Horsetail sporophytes

(a)   (b)

© University of Wisconsin-Madison, Department of Botany

Figure 23.10a–b   Horsetails (Equisetum) with (a) whorled branches and (b) scouring rushes without whorled branches.

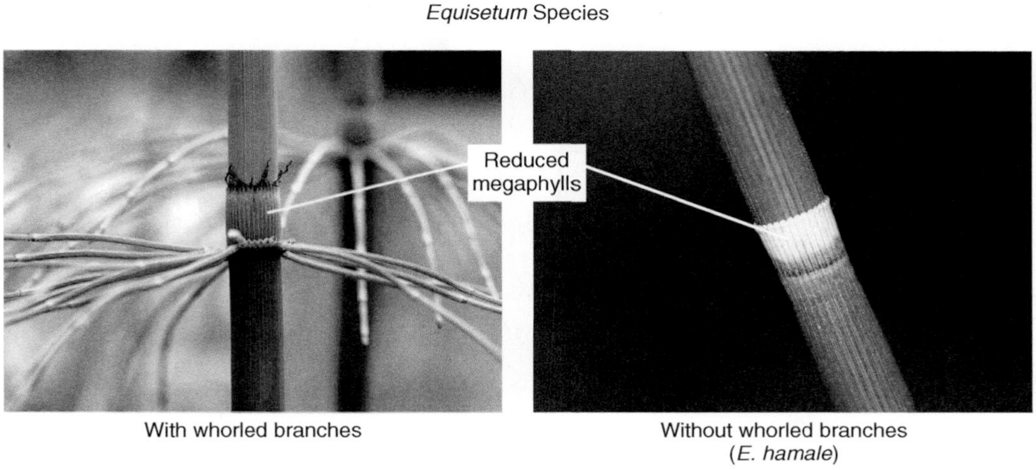

Equisetum Species

With whorled branches

Without whorled branches
(*E. hamale*)

Reduced megaphylls

© University of Wisconsin-Madison, Department of Botany

Figure 23.11   Horsetail megaphylls are greatly reduced and fused to form a whorl around the node.

sporangiophore
Umbrella-shaped structure that has 5–10 sporangia attached to it and a stalk that connects the cluster of sporangia to the common axis of the strobilus; found only in horsetails and scouring rushes.

produce many sporangia, which cluster to form terminal strobili. However, each strobilus is made up of many sporangiophores. The **sporangiophore** contains an umbrella-shaped structure with 5–10 sporangia attached to it and a stalk connecting the cluster of sporangia to the common axis of the strobilus (Figure 23.12a). Horsetail spores are distinct from other plant spores. Each spore has four ribbon-like elaters to facilitate dispersal by wind (Figure 23.12b). When the air is dry, the elaters extend like wings to help the spore to stay in the wind current. When the air becomes moist, which usually indicates favorable germination conditions, the elaters retract and cause the spore to drop to the ground. Like ferns, horsetails are homosporous. The horsetail gametophytes are small and self-sustaining.

Taking a Closer Look at Whisk Ferns

Whisk ferns do not resemble ferns, although they have similar molecular features and are both classified as pteridophytes. There are about 12 species of whisk ferns in the two genera *Psilotum* and *Timesipteris*. *Psilotum* whisk ferns are found mostly in tropical and subtropical regions. In the United States, they can be found in Hawaii, Arizona, and along the Gulf Coast states from Texas to Florida. *Timesipteris* whisk ferns are found only in Australia and other South Pacific islands.

enations
Simple flaps of green tissue that have no vascular veins; are not true leaves; found only in whisk ferns.

Like other seedless vascular plants, whisk fern sporophytes are dominant and branched. They have vascular tissues and true stems (both aerial stems and rhizomes) but do not have true roots and leaves. Simple rhizoids are found on the rhizomes for anchorage. Whisk ferns are the only seedless vascular plants without true leaves. Tiny scale-like projections called *enations* are found on their stems. **Enations** are simple flaps of green tissue having no vascular veins and are not considered to be true leaves. The aerial stems always branch or fork dichotomously (into two), which gives them a very distinctive appearance (Figure 23.13). Whisk ferns produce many sporangia, which tend to form clusters of three. The clusters turn yellow and are found all over the sporophyte.

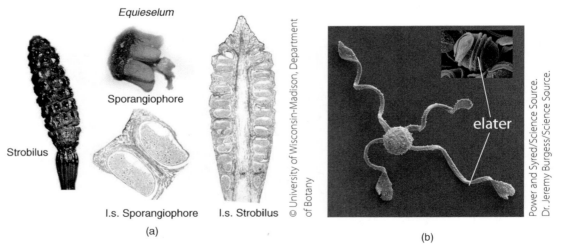

Figure 23.12a–b (a) A closeup and longitudinal-section views of the horsetail strobilus and sporangiophore; (b) colored scanning electron micrographs of a horsetail spore with the extended elaters or retracted elaters.

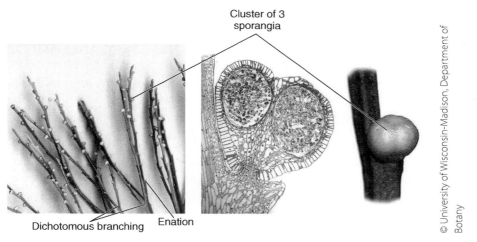

Cluster of 3 sporangia

Dichotomous branching   Enation

© University of Wisconsin-Madison, Department of Botany

Figure 23.13   Whisk ferns showing dichotomous branching, clusters of sporangia, and enations, and a close up look at a cluster of sporangia.

Whisk fern gametophytes are tiny (about 2 mm wide and no more than 6 mm long) and complete their development underground. They have no pigments and do not photosynthesize. They form symbiotic relationships with mycorrhizal fungi (see Chapter 21) and acquire food and minerals from their fungal partner.

IMPORTANCE OF SEEDLESS VASCULAR PLANTS TO HUMANS

Ecologically, ferns and lycophytes are important because their underground rhizomes and roots or rhizoids help to hold soil in place and prevent erosion. Their symbiotic relationships with nitrogen-fixing bacteria and mycorrhizal fungi enhance nutrient availability in soil. Like bryophytes, ferns and lycophytes are not considered to be economically important but they do have a variety of horticultural, agricultural, household, and medical uses. Some ferns are also used as food. Ancient and now extinct lycophytes and ferns were economically important plants because they contributed to soft or hard coal formation. Coal is one of our important fossil fuels (for reviews, see Croft, 1999; Moran, 2004; Srivastava, 2007).

Horticultural and Agricultural Uses

Ferns such as Boston fern, maidenhair fern, and staghorn fern are popular houseplants because of their aesthetic appeal and tolerance to low light condition. Other ferns are used as ornamentals for landscapes, gardens, and conservatories. Rhizome and root bark of tree fern and royal fern are favorable media for growing epiphytes such as orchids, bromeliads, and staghorn ferns. Although the mosquito fern in the genus *Azolla* is the world' smallest floating fern, it is also the world's most economically important fern (Moran, 2004). It is commonly used as an organic fertilizer for rice paddies in southeastern Asia, especially China and Vietnam. It is also used to grow other aquatic crops such as wild rice, taro, and arrowhead. *Azolla* fern is rich in nitrogen because of its symbiotic relationship with nitrogen-fixing cyanobacteria (Recall from Chapter 20). Aside from being used as an organic fertilizer, *Azolla* fern is used to supplement animal feeds for cattle, hogs, ducks, chickens, and carp. Like giant salvinia (see Box 23.1), *Azolla* fern can form a thick dense mat over water surfaces. This floating mat can prevent female mosquitoes from laying eggs in the water and mosquito larvae from coming up to the surface to breathe, hence the name *mosquito fern*.

Household and Industrial Uses

In many countries, ferns are used as construction materials for houses. Native Americans and other people groups have used climbing ferns such as *Lygodium* as binding and lashing twines for basketry and weaving. In countries such as New Guinea, the dense but soft and silky yellow hairs of the terrestrial fern, *Cystodium sorbifolium*, are used to stuff pillows, mattresses, and upholstery (Croft, 1999). Scouring rushes are frequently used to clean cooking and eating utensils because of the abrasive action of the silica crystal deposits on their stems. They are also used as sandpaper to smooth and shape tools.

Food

Certain ferns are used as flavorings and are eaten for their greens and starches. In North America, young fiddleheads or croziers of ostrich fern are often harvested, boiled, and eaten as food. In southeastern Asia, *Azolla* is often used as a raw or cooked vegetable. Native Americans often baked fern rhizomes in stone pits and ate the starchy stems. In areas of India and New Guinea where salt is not readily available, vegetable salts prepared from the ashes of *Asplenium* fern or from fronds are used to flavor food (Croft, 1999; Srivastava, 2007). Like most vegetable salts, salts prepared from ferns have higher potassium content than common salt.

Medical Uses

Ferns and lycophytes have been used in folk medicine as treatments for fractures, boils, ulcers, wounds, fevers, headaches, stomach pains, dysentery, labor pain, and even poisoning (Croft, 1999; Srivastava, 2007). The Chinese club moss *Huperzia serrate* contains an alkaloid for controlling epilepsy. In Europe, the male fern contains a drug effective in expelling intestinal worms such as tapeworms. Native Americans of the Pacific Northwest have used the licorice fern for treating sore throats and coughs. Ferns and lycophytes could be a potential source for new drugs and therapeutic compounds.

Coal Formation

Fossil fuels (natural gas, oil, and coal) supply over 85% of our energy today. The rate of coal formation was greatest during the Carboniferous period, which ranged from 362 to 290 million years ago. During this period, significant amounts of carbon were deposited in the earth and eventually became the coals (lignite, bituminous, and anthracite) we use today, hence the name *carboniferous*. The lands were covered with swamps dominated by five plant groups (Raven, Evert, & Eichhorn, 2005). Three of these plant groups, the now extinct lycophyte trees, giant horsetails (calamites), and tree ferns, were seedless vascular plants (Figure 23.14). Coal was used as

SPL/Science Source

Figure 23.14   Carboniferous swamp showing tree ferns, lycophyte trees, and giant horsetails. Artwork from the ninth edition of *Mosses and Geology* (1886; Samuel Kinns, London).

the main energy source for the Industrial Revolution from 1750 to 1850. Coal instead of wood was used to power steam locomotives and engines (National Energy Technology Laboratory, n.d.). Today, coals are mainly used by power plants to generate electricity.

Key Terms

seedless vascular plants	microspores	sori
eutracheophytes	megaspores	annulus
microphylls	strobili	prothallus
megaphylls	microsporangia	sporangiophore
euphylls	megasporangia	enations
homosporous	fronds	
heterosporous	pinnae	

Summary

- Vascular plants appeared after the ancestors of nonvascular plants made the transition to land around the mid-Silurian. Vascular plants became greatly diversified around the early Devonian period.
- Distinctive innovations of vascular plants include: (1) a dominant, branched sporophyte; (2) vascular tissues, true stems, and roots; (3) true leaves; (4) many sporangia; and (5) a reduced but self-sustaining gametophyte.
- Vascular plants are subdivided into two groups: seedless vascular plants (see this chapter) and seed vascular plants (see Chapters 24 and 25).
- Microphylls and megaphylls have very different origins. Microphylls are found in lycophytes and megaphylls are found in ferns and seed plants.
- The two main groups of seedless vascular plants are the lycophytes (Lycopodiophyta) and the ferns (Pteridophyta) such as true ferns, horsetails, and whisk ferns.

- Key characteristics of seedless vascular plants include: (1) dispersal by spores; (2) presence of enations, microphylls, or megaphylls; and (3) sporangia in clusters
- Lycophytes are the sister group to all the remaining vascular plants (ferns and seed plants) and ferns share a common ancestor with the seed plants.
- Horsetails and whisk ferns share similar molecular characteristics with ferns and are now classified in the same phylum, Pteridophyta.
- Lycophytes such as club mosses, quillworts, and spike mosses have true roots, stems (both aerial stems and rhizomes), and microphylls. Club mosses are homosporous whereas quillworts and spike mosses are heterosporous.
- Ferns have relatively large and conspicuous megaphylls or fronds, which are produced by the underground rhizomes. Fern sporangia are clustered on the underside of the fronds to form sori. Ferns are homosporous and produce a heart-shaped prothallus.
- Horsetails come in two forms: with or without whorled branches. They all have true roots, stems (both aerial stems and rhizomes), greatly reduced megaphylls, and terminal strobili.
- Whisk ferns have vascular tissues and true stems (both aerial stems and rhizomes) but do not have true roots and leaves. Unique characteristics include dichotomously branched stems, enations, and numerous yellow clusters of sporangia. Gametophytes are tiny and develop underground. They cannot photosynthesize and must acquire food and minerals from their mycorrhizal fungal partner.
- Seedless vascular plants have a variety of horticultural, agricultural, household, industrial, and medical uses. Some lycophytes and ferns are used for food. The fossil fuels we use today were formed by extinct lycophytes and ferns.

Reflect

1. *Vascular plants are more advanced and complex!* Do you agree with this statement? Why or why not? Describe at least eight characteristics that are unique to all vascular plants.
2. *Enations, microphylls, and megaphylls are all leaves.* Is this statement correct or incorrect? Compare and contrast enations, microphylls, and megaphylls.
3. *Alternation of generation.* Use a diagram to show the homosporous life cycle of a seedless vascular plant and another diagram to show the heterosporous life cycle of another seedless vascular plant. Make sure you describe the unique features of the sporophyte, the gametophyte(s), and the spore(s). Define terms like *homosporous* and *heterosporous*.
4. *What is this?* A friend asks you to identify a plant and you know it may be a seedless vascular plant. How do you determine whether it is a fern, a whisk fern, a horsetail, a club moss, or a spike moss?
5. *Lycophytes and ferns.* Compare and contrast lycophytes and ferns. Discuss their phylogenetic relationship to nonvascular plants and seed vascular plants.
6. *Horsetails, whisk ferns, and ferns—one big happy phylum.* If you have seen these plants, you know they look very different from one another. Why do plant scientists place them in the same phylum instead of separate phyla?
7. *Cranking up spore production.* Seedless vascular plants produce many sporangia to greatly increase spore production. Describe the unique sporangium arrangement in lycophytes, ferns, horsetails, and whisk ferns.
8. *Seedless vascular plants.* Describe how seedless vascular plants are used by humans.

References

Beerling, D. J., Osborne, C. P., & Chaloner, W. G. (2001). Evolution of leaf-form in land plants linked to atmospheric CO_2 decline in the late Palaeozoic era. *Nature, 410,* 352–354.

Croft, J. (1999). *Ferns and man in New Guinea.* Retrieved April 15, 2012, from www.anbg.gov.au/fern/ferns-man-ng.html.

Graham, L. E., Cook, M. E., & Busse, J. S. (2000). The origin of plants: body plan changes contributing to a major evolutionary radiation. *Proceedings of the National Academy of Sciences of the United States of America, 97,* 4535–4540.

Moran, R. C. (2004). *A natural history of ferns.* Portland, OR: Timber Press.

National Energy Technology Laboratory. (n.d.). *History of coal use.* U.S. Department of Energy. Retrieved from www.netl.doe.gov/keyissues/historyofcoaluse.html.

Piazza, P., Jasinski, S., & Tsiantis, M. (2005). Evolution of leaf developmental mechanisms. *New Phytologist, 167,* 639–710.

Raven, P. H., Evert, R. F., & Eichhorn, S. E. (2005). *Biology of plants* (7th ed.). New York: W.H. Freeman.

Robinson, R. C. (2007). Steps to more effective bracken management. *Aspects of Applied Biology, 82,* 143–155.

Robinson, R. C., Sheffield, E., & Sharpe, J. M. (2010). Problem ferns: their impact and management. In K. Mehltreter, L. R. Walker, & J. M. Sharpe (Eds.), *Fern ecology* (pp. 255–322). New York: Cambridge University Press

Schuettpelz, E., & Pryer. K. M. (2008) Fern phylogeny. In T. A. Ranker & C. H. Haufler (Eds.), *Biology and evolution of ferns and lycophytes* (pp. 395–416). New York: Cambridge University Press

Srivastava, K. (2007). Ethnobotanical studies of some important ferns. *Ethnobotanical Leaflets, 11,* 164–172.

Chapter 24

Gymnosperms

© Joseph Scott Photography/Shutterstock.com

Learning Objectives

- Discuss the evolutionary changes in seed vascular plants and how they differ from nonvascular bryophytes and seedless vascular plants
- Describe how seeds are different from spores and discuss the evolutionary significance of seeds
- Summarize the features that distinguish gymnosperms from angiosperms
- Explain the evolutionary relationships among the four groups of living gymnosperms
- Outline the alternation of generations of pines and discuss its key features
- Compare and contrast conifers, ginkgo, cycads, and gnetophytes
- Explain the difference between monoecious plants and dioecious plants and give examples for each group of plants
- Describe the importance of cycads, ginkgo, conifers, and gnetophytes to humans

Conifers, cycads, and ginkgo are popular ornamentals found in yards, parks, and urban areas. Many conifers are commonly used for bonsai, a Japanese art form producing miniature trees in small containers. Other conifers such as low-growing yews and junipers are popular groundcovers and hedges. Many conifers are famous for being the tallest, the largest, and the oldest trees in the world. Conifers also dominate the taiga or boreal forest in the Northern Hemisphere and are responsible for replenishing atmospheric oxygen and removing carbon dioxide from the atmosphere. Conifers, cycads, and ginkgo belong to a group of plants called gymnosperms. Unlike nonvascular plants and seedless vascular plants (see Chapters 22 and 23), these plants produce seeds. In this chapter, we explore the origin and diversity of seed plants and examine the advantages of using seeds instead of spores for dispersal. We take a closer look at the gymnosperms, their importance to humans, and their unique characteristics or innovations.

EVOLUTION OF SEED VASCULAR PLANTS (SPERMATOPHYTES)

Recall from Chapter 23 that vascular plants began to diversify around the early Devonian period. Based on paleobotanical data, the earliest seed plants from the late Devonian and early Carboniferous period showed major structural changes in their sporangia that significantly modified their sexual reproduction (Rowe, 1997). Characteristics unique to seed vascular plants began to emerge and these characteristics enabled early seed vascular plants to become abundant and widespread (for reviews, see Raven, Evert, & Eichhorn, 2005; Linkies, Graeber, Knight, & Leubner-Metzger, 2010).

Sporangia Became Indehiscent

pollen grains
Immature male gametophytes of seed vascular plants.

Unlike bryophytes and seedless vascular plants, the sporangia of seed vascular plants do not split open. Their spores are not released into the environment. Seed vascular plants are all heterosporous. Both microspores and megaspores are retained within their respective sporangia. Microspores develop within microsporangia into immature male gametophytes or **pollen grains**, which are then released into the environment. Megaspores develop within megasporangia into female gametophytes. In seed vascular plants, megaspore production is significantly reduced. Each megasporangium has a single megaspore mother cell, which eventually produces a single female gametophyte.

Gametophytes Became Dependent

The life cycle of seed vascular plants is completely opposite to the life cycle of nonvascular plants. In seed vascular plants, the gametophytes are microscopic and complete their development within the dominant sporophytes. In nonvascular plants or bryophytes, their gametophytes are dominant whereas their sporophytes are very small and depend on gametophytes for protection and nourishment. Recall from Chapter 23 gametophytes of seedless vascular plants are smaller than nonvascular plants but remain self-sufficient. Their sporophytes are dominant and large. The life cycle of seedless vascular plants exhibits a transitional pattern between nonvascular plants and seed vascular plants.

Pollen Development

In seed vascular plants, microspores remain in the microsporangia and eventually develop into immature male gametophytes or pollen grains. Pollens are a key innovation by seed vascular plants that allow them to reproduce without water. The pollen grains are released into the environment and transferred by wind or other means to female gametophytes. This transfer process is called **pollination**. After pollination, pollen grains produce a pollen tube to deliver the two sperm to the eggs. Unlike bryophytes and seedless vascular plants, seed vascular plants do not form antheridia for sperm production. Sperm are produced and delivered to the eggs by pollen grains. Water is no longer required as a medium for sperm delivery. As a result, seed vascular plants are able to greatly expand their habitat range.

Seed Development

In seed vascular plants, megasporangia possess an apical opening (**micropyle**) and an inner compartment (**pollen chamber**) to trap and direct windborne pollen grains toward the female gametophyte (Figure 24.1). A female gametophyte produces several archegonia and each archegonium yields a single egg. The megasporangium is surrounded by one or two layers of sporophyte tissue called the **integument**, which develops into a protective seed coat. The integument-covered megasporangium is called an **ovule**, which develops into a **seed** after fertilization.

Each seed contains an embryonic sporophyte and a food reserve. Seeds are a key innovation by seed vascular plants for land-based living and are reproductively superior to spores. Seed development occurs within sporophyte

pollination
Transfer of pollen grains male gametophytes to or near the female gametophytes.

micropyle
Apical opening of an ovule for pollen grains to enter.

pollen chamber
Inner compartment of an ovule for trapping and directing wind-borne pollen grains toward the female gametophyte.

integument
Thick layer of sporophyte tissue surrounding the megasporangium that develops into the seed coat to protect the seed.

ovule
Integument-covered megasporangium that develops into a seed after fertilization.

seed
Reproductive structure that contains an embryonic sporophyte, food reserve, and a protective seed coat.

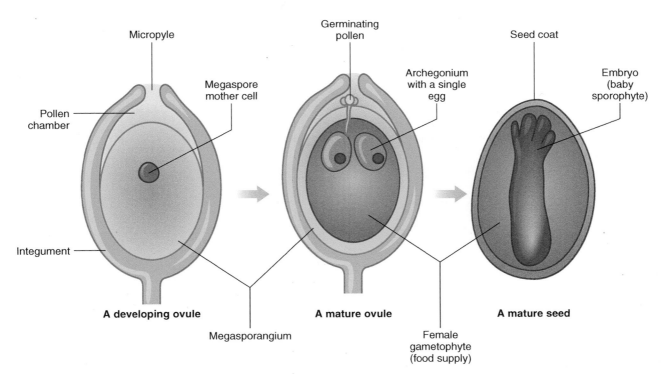

Figure 24.1   Seed development in gymnosperms. Adapted from Graham, Graham, and Wilcox (2006).

tissues and is supported and protected by the sporophyte. Spore development is relatively simple and does not receive the same level of sporophyte support and protection. A seed contains a well-developed embryonic sporophyte whereas a spore contains only a single cell. The well-developed embryonic sporophyte has a much better chance to complete its development after germination than a spore. Each seed contains ample food reserve to support the germinated sporophyte until it becomes self-sufficient. However, a spore does not have any food reserve to support the sporophyte after germination. The multicellular seed coat, often hard and tough, protects the seed from unfavorable conditions and allows it to survive for an extended period of time.

Sporophytes Became Woody

Seed vascular plants have a vascular cambium that produces secondary xylem or wood. Recall from Chapter 5 that vascular cambium and cork cambium are the lateral meristems increasing the girth of roots and stems. Vascular cambium produces secondary xylem toward the center and secondary phloem toward the surface. As a result, seed vascular plants tend to have a substantial stem and complex body (Donoghue, 2005; Judd, Campbell, Kellogg, Stevens, & Donoghue, 2008).

CHARACTERISTICS OF GYMNOSPERMS

Based on the ovule and seed structure, seed vascular plants can be divided into two major groups: gymnosperms (*gymnos* in Greek means "naked" and *sperma* means "seed") and angiosperms (*angeion* in Greek means "a vessel") (Figure 24.2). We examine gymnosperms in this chapter and angiosperms in the next chapter.

There are about 820 species of living gymnosperms, which divide into four phyla: Cycadophyta (cycads), Ginkgophyta (ginkgo), Coniferophyta (conifers), and Gnetophyta (gnetophytes) (Figure 24.3). Although gymnosperms are not a very large plant group, they make up most of the earth's biomass. The largest, tallest, and oldest known trees are all gymnosperms in the phylum Coniferophyta (Earle, 2011). The largest tree is a giant sequoia,

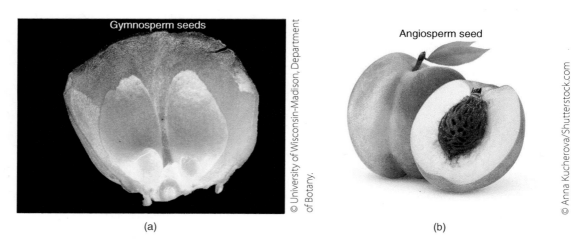

Figure 24.2a–b Naked versus enclosed seeds: (a) two "naked" seeds on the upper surface of a scale and (b) a single seed enclosed inside a fruit.

(a)   (b)   (c)   (d)

© University of Wisconsin-Madison, Department of Botany.

Nature's Images/Science Source.

Figure 24.3a–d   Four groups of living gymnosperms: (a) pine (Coniferophyta), (b) ginkgo (Ginkgophyta), (c) cycad (Cycadophyta), and (d) ephedra (Gnetophyta).

Sequoiadendron giganteum, named "General Sherman" (Figure 24.4a). It has the largest known stem volume, 1473.4 m³ (52,032.6 ft³) and is found in Sequoia National Park in California. In Northern California, a coastal red-wood, *Sequoia sempervirens*, named "Hyperion" is among the world's tallest trees, measuring 115.6 m (379.3 ft) in height and 4.84 m (15.9 ft) in stem diameter. As of 2012, the oldest tree is a Great Basin bristlecone pine, *Pinus longaeva*, named the "Methuselah" tree. The Methuselah tree is found in the White Mountains of California (Figure 24.4b) and was determined to be 4,789 years old in the summer of 1957.

Unlike angiosperms, gymnosperm ovules are surrounded by just one layer of integument (see **Table 24.1**). Gymnosperms also do not produce endosperm in their seeds as food supply for their embryos. Instead, the female gametophyte provides food for the developing embryo. Unlike angiosperms, gymnosperms do not have flowers and their seeds are not

(a)   (b)

Vanessa Vick/Science Source.

Gerald & Buff Corsi/Visuals Unlimited.

Figure 24.4a–b   Record-setting gymnosperms: (a) "General Sherman," the largest sequoia, *Sequoiadendron giganteum*, in Sequoia National Park in California, and (b) the "Methuselah" tree, the oldest bristlecone pine, *Pinus longaeva*, in the White Mountains of California.

Table 24.1 Seed Vascular Plants and Their Characteristics

Seed Vascular Plants	Naked or Exposed Seeds	Endosperm as Food Supply	Single Layer of Integument	Seed Cones	Flowers and Fruits	Pollen Tube for Sperm Delivery	Flagellated Sperm
Cycadophyta	✓	✗	✓	✓	✗	✗	✓
Ginkgophyta	✓	✗	✓	✗	✗	✗	✓
Coniferophyta	✓	✗	✓	✓	✗	✓	✗
Gnetophyta	✓	✗	✓	✓	✗	✓	✗
Anthophyta (Angiosperms)	✗	✓	✗	✗	✓	✓	✗

enclosed within fruits. Gymnosperm seeds are found on the surfaces of sporophylls, which cluster together to form seed cones. After fertilization, ovules become seeds and these seeds are exposed to the environment, hence the name "naked" seeds (Figure 24.2).

Evolutionary Relationships among Gymnosperms

The evolutionary relationships among the four living gymnosperms as well as their relationships to the angiosperms are still uncertain. Based on morphological characteristics, conifers and ginkgo are a sister group whereas cycads and gnetophytes form a separate group. However, other scientists have come to a different conclusion and suggest cycads are the most archaic seed plants whereas conifers and ginkgo are two separate groups. Based on molecular sequence data, conifers and cycads form a clade and are the earliest seed plants. Numerous hypotheses have been proposed to shed light on the evolutionary relationships among seed plants but none has been widely accepted. However, most of these phylogenetic relationships indicate conifers, ginkgo, cycads, gnetophytes, and angiosperms form a monophyletic group and all angiosperms also form a monophyletic group (Figure 23.1). Most phylogeny studies also indicate cycads are the most ancient group of gymnosperms (for reviews, see Chaw, Zharkikh, Sung, Lau, & Li, 1997; Crane, Herendeen, & Frils, 2004).

Taking a Closer Look at Cycads and Their Uses

Cycads are popular indoor and outdoor plants because of their interesting shape, colorful cones, and palm-like leaves (Figure 24.5). The word *cycad* comes from a Greek word meaning "palm." Most cycads have a squat trunk-like stem and a crown of large pinnately compound leaves. Their compound leaves resemble those of tree ferns or palms (Figure 24.5a). Many people have trouble distinguishing cycads from palms. For example, the Sago palm, *Cycas revoluta*, is actually a cycad and not a palm. True palms produce flowers and belong to the phylum Anthophyta. Cycads, on the other hand, produce very large and brightly colored pollen cones and seed cones. They are dioecious (see Chapter 22), meaning the two types of cones are

Cynthia McKenney.

Cynthia McKenney.

Dick Keen/Visuals Unlimited.

(a)

(b)

(c)

Figure 24.5a–c   Cycads: (a) Sago palm, *Cycas revoluta*; (b) Cone of chestnut dioon, Dioon edule; and (c) coontie cycad with cones, *Zamia floridana*.

found on separate plants. Although cycads are popular ornamental plants, they contain toxins and can cause serious neurological disorders in livestock, pets, and even humans when consumed (Schneider, Wink, Sporer, & Lounibos, 2002). Cycad toxins are believed to be essential in the development and maintenance of pollination by insects.

Unlike other gymnosperms, most cycads are pollinated not by wind but by insects such as curculionoid beetles (weevils) and thrip. Coevolutionary relationships between pollinators and angiosperms have been well studied and documented. The use of insect pollinators by gymnosperms, in particular cycads, has only been recently verified (for a review, see Schneider, Wink, Sporer, & Lounibos, 2002). Mutualistic relationships between insect pollinators and cycads may have developed during the Mesozoic era.

Cycads were most diverse and abundant in the Mesozoic era from the Triassic to the Cretaceous (265 to 250 million years ago), a period often referred to as the "Age of Cycads." Many cycads became extinct and to date there are about 300 species of living cycads left. Most of them are found in the tropics and subtropics. Cycads are believed to be an ancient group of seed plants because they retain a number of primitive characteristics. Unlike other seed plants, cycads produce very limited secondary xylem, thus they do not produce a substantial trunk. Cycads also produce large and motile sperm. These sperm are produced only after pollination. Cycad pollens do not produce long pollen tubes to deliver the sperm directly to the eggs in the ovules. Instead, cycad pollen tubes function as haustoria, which are specialized structures used by parasitic plants to extract water and nutrients from their hosts. Cycad pollen tubes extract water and nutrients from the female gametophytes for sperm development. Cycad pollens release their two large and motile sperm in the pollen chambers. A cycad sperm may have up to 40,000 flagella and swims to the egg (Fernando, Quinn, Brenner, & Owens, 2010). Only one sperm will fertilize the egg. Fertilization in cycads exhibits a transitional pattern between primitive plant groups (bryophytes and seedless vascular plants) and advanced plant groups (conifers, gnetophytes, and angiosperms). Many believe cycads are the most primitive living seed plants.

Taking a Closer Look at Ginkgo and Its Uses

Like cycads, ginkgoes were most abundant in the Mesozoic era and many of them became extinct. There is only one living species, *Ginkgo biloba*, left in the phylum Ginkgophyta. Ginkgo or the maidenhair tree is easily recognized by its distinctive leaves. Ginkgo leaves are notched and fan-shaped with very fine dichotomously branched veins (Figure 24.6) and will turn bright yellow in the fall before being shed. Ginkgo is a beautiful, regal tree that can reach 30 meters or more in height. It was native to eastern China but has become a popular landscape tree for parks and urban areas in many temperate regions of the world. Ginkgo is particular suitable for urban environments because of its tolerance to air pollution. Unfortunately, ginkgo is presumed to be extinct in nature and is only found in cultivation.

Ginkgo is dioecious and has separate male and female trees. Ginkgo seeds are found on the female trees and are exposed and not enclosed. Gingko seeds resemble fruits because of their thick, fleshy seed coats. As the seeds mature, the fleshy seed coats give off a nauseating odor and can cause skin irritation in some people. As a result, only male trees are used as ornamentals in parks and urban areas. In China and Japan, ginkgo seeds are widely used for food. Ginkgo has long been used for medicinal purposes. Extracts from the leaves are believed to increase blood flow to the brain and improve the memory and cognitive health of older adults. However, most clinical studies have been unable to verify such claims (Solomon, Adams, Silver, Zimmer, & DeVeaux, 2002; or visit *Journal of American Medical Association*, n.d., for more clinical studies).

Like cycads, ginkgo produces large and motile sperm after pollination. Ginkgo and cycad pollen tubes absorb water and nutrients for sperm development. Their pollen tubes are not used for sperm delivery. Ginkgo pollens release their two sperm in the pollen chambers. A ginkgo sperm may have up to 1,000 flagella and swims to the egg. Like cycads, only one of the sperm

(a)   (b)   (c)

Suzanne Carter/Science Source.

© University of Wisconsin-Madison, Department of Botany.

Figure 24.6a–c   The maidenhair tree, *Ginkgo biloba*: (a) popular landscape trees in urban areas; (b) closeup of a fan-shaped leaf with very fine, dichotomously branched veins; (c) a branch bearing leaves and fleshy seeds.

will fertilize the egg. Ginkgo and cycads are believed to be closely related because they share many similar features. They are also believed to be more ancient than conifers and gnetophytes.

Taking a Closer Look at Conifers and Their Uses

Conifers are the largest and most diverse group of gymnosperms. There are 627 living species of conifers, which include pines, cedars, redwoods, junipers, cypress, firs, and spruces. Like typical seed vascular plants, conifers have vascular cambium and cork cambium producing secondary tissues (wood and bark). Thus, conifers tend to have a substantial woody stem and woody bark. The largest and the tallest known trees of the world are all conifers. Conifer wood or xylem is made up of tracheids (see Chapter 3) and is referred to as *softwood*. Angiosperm wood, on the other hand, is often referred to as *hardwood* because its wood is made up of tracheids and vessels. Unlike cycads and ginkgo, most conifers are monoecious, which means separate male cones and female cones are found on the same tree. The word *conifer* (*coni* in Greek means "cone" and *fer* means "bear") means "cone-bearing."

Many conifers have a system of ducts or canals connecting their roots, stems, and leaves (Figure 24.7). These ducts store and transport oleoresin or resin, which is a mixture of monoterpenes, sesquiterpenes, and resin acids. **Resin** is produced by specialized secretory tissues in roots, stems, and leaves. It repels invading pests and pathogens. Crystalized resin plugs wounds and minimizes desiccation or loss of plant sap (Lewinsohn, Gijzen, & Croteau, 1991). Some conifers such as pines have preformed resin and a resin duct system. Others, such as firs in the genus *Abies*, produce resin and resin ducts when attacked.

Most conifers are called **evergreens** because they do not shed their leaves at the end of the growing season. A few conifers such as the bald cypress, the dawn redwood, and larches in the genus *Larix* are **deciduous**, which mean they shed all their leaves in the fall and regrow them in the spring. Conifer leaves or megaphylls are very distinctive (Figure 24.8). Some conifers such as pines have leaves that are long, narrow, and needle-like. Others such as cedars and giant sequoia have scale-like leaves. Junipers have awl-like leaves whereas yews have flat, feather-like leaves.

Life Cycle of Pines There are 100 species of pines in the genus *Pinus*, the largest genus of conifers. Conifers such as pines, spruces, and firs dominate the taiga of the Northern Hemisphere. Pines are also important in the

resin (oleoresin)
Mixture of monoterpenes, sesquiterpenes, and resin acids used to repel invading pests and pathogens and to seal the wound.

evergreens
Coniferous trees that bear their leaves throughout the year and do not shed all their leaves in the fall.

deciduous
Plants that shed all their leaves in the fall and regrow them in the spring.

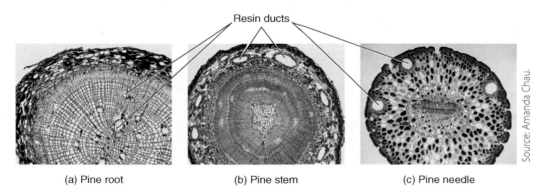

Resin ducts

(a) Pine root   (b) Pine stem   (c) Pine needle

Source: Amanda Chau.

Figure 24.7a–c   Resin ducts are found all over the (a) roots, (b) stems, and (c) leaves of pines.

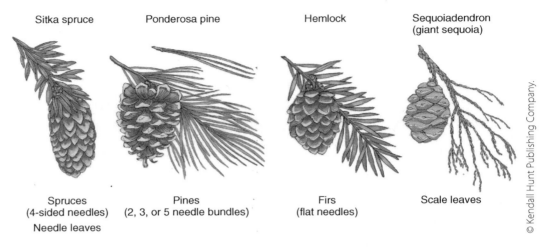

Sitka spruce   Ponderosa pine   Hemlock   Sequoiadendron (giant sequoia)

Spruces
(4-sided needles)   Pines
(2, 3, or 5 needle bundles)   Firs
(flat needles)   Scale leaves
Needle leaves

© Kendall Hunt Publishing Company.

Figure 24.8   Various types of conifer leaves or megaphylls.

Southern Hemisphere. We now take a closer look at the life cycle of pines, which is typical of conifers, and some unique adaptations that allow them to survive the frozen north.

Like typical seed vascular plants, pine sporophytes are dominant and heterosporous (Figure 24.9). Pines are monoecious, producing separate male (pollen) cones and female (seed) cones on the same tree (Figure 24.10). Inside the pollen cones, pine microspores develop into pollen grains, the immature male gametophytes (Figure 24.11). Pines, like most gymnosperms, rely on wind for pollination. Each pollen grain has two large air sacs acting like wings to aid in its dispersal (Figure 24.12). Unlike cycads and ginkgo, pines do not produce large and flagellated sperm and pine pollen tubes do not function as haustoria. Instead, pine pollen grains develop long pollen tubes to deliver the two non-motile sperm directly to the eggs. The egg is fertilized by only one of the sperm.

Inside the seed cones, pine megaspores develop within the ovules and become the microscopic female gametophytes (Figure 24.13). Each female gametophyte produces two archegonia, containing a single egg each. When the ovules are ready to receive pollen grains, they secrete sticky sugary drops (**pollination drops**) out of their micropyles. Pollination drops help to draw the pollen grains into the ovules. Pollination drops and **floral nectar** found in angiosperms have similar chemical composition and are used in plant reproduction (Nepi, von Aderkas, Wagner, Mugnaini, Coulter, & Pacini, 2009). However, these two types of sugary secretions differ in their function. The pollination drop provides a landing site for airborne pollen grains whereas the floral nectar provides nutrients as rewards for pollinators.

Not all conifers produce woody seed cones like pines. Yews in the genus *Taxus* produce ovules that are surrounded by fleshy, cup-shaped coverings called aril (Figure 24.14). Junipers in the genus *Juniperus* produce seed cones with fleshy scales resembling berries. Conifer cones also vary greatly in size (Flora of North America, 2008; Earle, 2011). The Eastern hemlock, *Tsuga canadensis*, produces the smallest cones, measuring no more than 2.5 cm (1 in.) in length. The sugar pine, *Pinus lambertiana*, produces the longest cone, which can grow up to 50 cm (1.6 ft) in length. The Coulter pine, *Pinus coulteri*, produces the largest and also the heaviest seed cones, which can grow up to 35 cm (1.1 ft) long and weigh up to 4.5 kg (10 lbs) (Figure 24.15).

pollination drops
Sticky, sugary drops secreted by gymnosperm female gametophyte to provide landing site for air-borne pollen grains and to help draw them into the ovules.

floral nectar
Sugary secretion produced by angiosperm flowers used as rewards for pollinators.

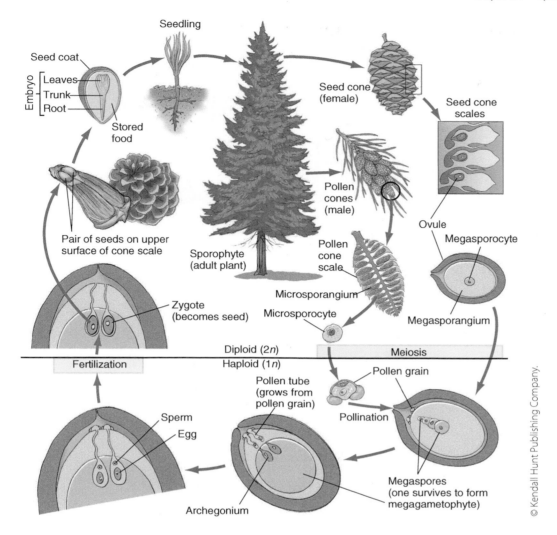

Figure 24.9   The alternation of generations of pines, *Pinus*.

Figure 24.10   Closeup of two male cones (*left*) and a female cone (*right*).

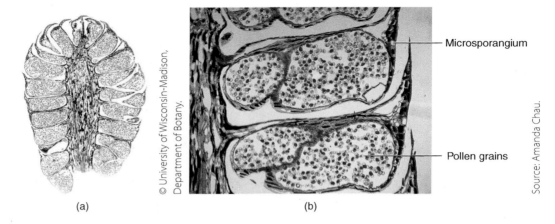

© University of Wisconsin-Madison, Department of Botany.

Source: Amanda Chau.

Microsporangium

Pollen grains

(a)

(b)

Figure 24.11a–b Longitudinal sections of (a) a pollen (male) cone and (b) individual microsporangia with pollen grains.

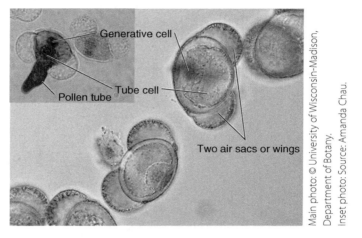

Generative cell

Tube cell

Pollen tube

Two air sacs or wings

Main photo: © University of Wisconsin-Madison, Department of Botany.
Inset photo: Source: Amanda Chau.

Figure 24.12 Pine pollen grains or immature male gametophytes with a generative cell, a tube cell, and two air sacs. (Insert shows a germinating pollen grain or mature male gametophyte with a pollen tube.)

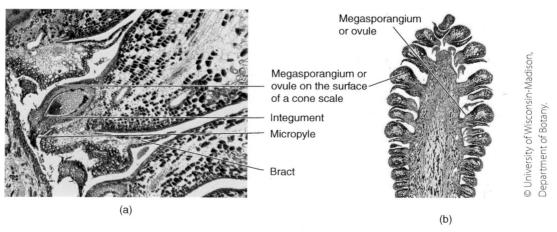

Megasporangium or ovule

Megasporangium or ovule on the surface of a cone scale

Integument

Micropyle

Bract

© University of Wisconsin-Madison, Department of Botany.

(a)

(b)

Figure 24.13a–b Longitudinal sections of (a) a seed (female) cone and (b) one of the megasporangia on the surface of a cone scale.

© Dmitri Melnick/Shutterstock.com
© Torsten Lorenz/Shutterstock.com
© Cindy Lee/Shutterstock.com

(a)   (b)   (c)

Figure 24.14a–c   Conifer seeds and their housing structures: (a) woody cones of pines, *Pinus*; (b) fleshy-coated seeds of yews, *Taxus*; and (c) fleshy cones of junipers, *Juniperus*.

Pines have unique adaptations allowing them to survive and thrive in cold climates such as the taiga. In cold climates, water usually exists as ice or snow, which is unavailable for plant use. To survive in cold climates, plants have to deal with not only freezing temperatures but also drought. Pine trees are generally cone-shaped and have flexible branches allowing snow to slide off easily. Pine megaphylls or needles are another unique adaptation to cold and dry conditions. Pine needles have a very small surface area and a thick cuticle, which greatly minimize water loss through evaporation (Figure 24.16). In addition, pine stomata are not on the leaf surface but are recessed or sunken in small pits. These sunken stomata further minimize water loss. To protect against freezing temperatures, pine needles have an additional layer called **hypodermis** just beneath their epidermis. Hypodermis consists of several layers of thick-walled cells protecting the photosynthetic tissues beneath.

Importance of Conifers Conifers are economically important because of their wood and resin. White pine woods are often used for crates, matchsticks, furniture, flooring, and paneling. Lodgepole pines have long straight

hypodermis
Pine needle that has several layers of thick-walled cells just beneath its epidermis; functions to protect photosynthetic tissues underneath from freezing temperatures.

20 cm

© University of Wisconsin-Madison, Department of Botany.

(a)   (b)

Figure 24.15a–b   (a) Record-setting pine cone: the Coulter pine, *Pinus coulteri*, produces (b) the largest and also the heaviest seed cone in the world.

Figure 24.16   Cross section of a pine needle.

trunks and are commonly used by Native Americans and First Nations peoples for building their tepees, lodges, and buildings. Lodgepole pines are also used for telephone poles, fence posts, mine timbers, railway ties, and furniture. White spruce is the main source of pulpwood for papers. Red spruce wood has a good resonance property and is used as soundboards for musical instruments. Wood of coastal redwoods is used for making wine barrels because of their antifungal and antibacterial properties. Turpentine and rosin are two important commercial products extracted and refined from conifer resin. Turpentine is considered a premier paint and varnish thinner. Rosin is used for coating medicines and chewing gums, waterproofing sailing ships, priming violin bows, and improving grip for bull riders and baseball pitchers.

Some conifers are prized for their medicinal properties and others as food or additives. The bark of the Pacific yew contains powerful compounds (marketed under the trade name *taxol*) for treating ovarian cancer, breast cancer, and lung cancer. Juniper berries are used to flavor gin. Many parts of conifers such as inner bark (phloem and cambium) and seeds are edible and have been used as food by Native Americans for centuries.

BOX 24.1   Wildfires and Fire-Response Traits in Plants

Like climate and soil, fire plays a key role in the distribution and composition of ecosystems. Fire exists when three essential elements, oxygen, fuel, and an ignition source, are present. The origin of fire may be tied to the origin of photosynthetic organisms because they produce oxygen and fuel, two of the three elements essential to support fire (for a review, see Pausas & Keeley, 2009). By the beginning of the Paleozoic era, photosynthetic organisms raised the atmospheric oxygen to a level that supported fire. However, there was no fuel for fire until the Silurian period when land plants began to appear. Charcoal and charred coprolites (fossilized feces) discovered in the Silurian period provide evidence for the earliest wildfires (Glasspool, Edwards, & Axe, 2004). Lightning and volcanic eruption are often the ignition sources in nature.

Figure Box 24.1–1   The 1988 "Summer of Fire" in Yellowstone National Park, Wyoming, and regeneration of lodgepole pines after the fires.

Most ecosystems on Earth have been scorched by wildfires in the past and will burn again in the future (Figure Box 24.1). Plant composition in an ecosystem is one of the factors determining the intensity and frequency of fires or fire regime. Fire regimes in turn determine the extent to which plant populations are affected. Frequent low-intensity surface fires, sometime called **understory fire regimes**, rarely damage the soil or kill overstory trees. Ecosystems subjected to understory fire regimes recover quickly as vegetation resprouts after the fire. Such fire regimes are typical of open woodland, savannah-type habitats such as longleaf pine–wiregrass ecosystems. Infrequent high-intensity fires, sometime called **stand-replacing fire regimes**, often kill all or most overstory trees and sometimes damage the soil. Ecosystems subjected to stand-replacing fire regimes recover very slowly as seeds from serotinous cones germinate on fire-prepared seedbeds. Stand-replacing fire regimes are typical of jack pine, Pinus banksiana, stands in the boreal forests, sand pine, P. clausa, stands in Florida, and Table Mountain, P. pungens, and pitch pine, P. rigida pitch pine stands in the southern Appalachians. In fire-prone environments, plants develop traits increasing their survival to specific fire regimes.

Adaptations to Understory Fire Regimes

North American conifers subjected to frequent lightning-ignited surface fires developed thick bark to protect the living tissues beneath from heat damage. They also self-prune lower dead branches to prevent surface fires from reaching the living canopy above. Some conifers, particularly pines, can produce new foliage from adventitious buds on tree branches or trunk to replace scorched foliage or branches. Longleaf pines produce dense sheaths of moisture-rich needles around their buds to protect them from heat damage during a fire.

Adaptations to Stand-Replacing Fire Regimes

A few conifers can resprout from root collar buds when the above ground portion of the tree is killed by fire. Most conifers produce **serotinous cones**, which do not release their seeds when mature. Instead, they require heat to open and release their seeds. Serotinous cones protect the seeds during the fire and disperse them after the fire moves through the area. Cone serotiny is an adaptive

understory fire regimes
Frequent low-intensity surface fires that rarely damage the soil or kill overstory trees.

stand-replacing fire regimes
Infrequent high-intensity fires often kill all or most overstory trees and sometimes damage the soil.

serotinous cones
Conifer cones requiring heat to open and release their seeds.

response to stand-replacing fire regimes. In addition, most conifer seeds germinate readily on fire-prepared seedbeds. Seed germination in some plant species is triggered by heat shock and smoke from fires. Smoke and smoke-water are used in horticulture to stimulate seed germination of wildflowers and vegetable crops such as lettuce and celery (Brown & van Staden, 1998).

Adaptations to Affect Fire Activity

Some conifers such as longleaf pines and jack pines develop traits to enhance flammability and predispose the plant communities to recurrent fire. Some conifers produce volatile compounds that are highly combustible. Other traits such as retention of dead leaves and branches greatly increase fuel load to support fire. Increased flammability may enhance reproductive success of plant species that rely on fire for seed release and remove competitors for light and space.

Sustainability and composition of some ecosystems are negatively affected when humans alter the frequency and intensity of wildfires. Increased urbanization in areas prone to stand-replacing fire regimes exposes more people and property to wildfires. Massive wildfires in California, Colorado, Texas, Greece, Chile, and Australia have made headlines recently because of loss of human lives and property. Human activities often increase fire frequency and have profound impacts on fire regimes, which can lead to permanent and irreversible changes to ecosystems.

Taking a Closer Look at Gnetophytes and Their Uses

There are 75 living species of gnetophytes (Gnetophyta) in the three genera: *Ephedra*, *Gnetum*, and *Welwitschia*. *Ephedra* stems are brewed into "Mormon tea" and have been used medicinally for a long time. The drug *ephedrine*, extracted from a Chinese species, *Ephedra sinica*, has been used for treating asthma and bronchitis for centuries. In North America, ephedrine was used as a dietary supplement to aid weight control and boost sports performance and energy. However, in 2004, the Food and Drug Administration (FDA) banned the use of ephedrine in all dietary supplements because of its serious side effects.

The three genera of gnetophytes look distinctly different from each other. *Ephedra* gnetophytes or joint firs are shrubby plants found in arid regions of the southwestern United States. They have jointed stems with tiny scale-like leaves (Figure 24.17a). *Gnetum* gnetophytes (melinjo tree) are tropical vines with simple broad, flat leaves oppositely arranged on the stem (Figure 24.17b). They produce fleshy-coated seeds that resemble angiosperms. *Welwitschia mirabilis* (tree tumbo) is the only living species in this genus and is a strange-looking plant found only in the deserts of southwestern Africa. It has a short, wide stem aboveground and a long taproot underground (Figure 24.17c). It produces only two strap-like leaves, which continue to grow throughout the plant's life. As the leaves flap about in the wind, they become tattered and split length-wise, giving the appearance of many leaves. Although *Ephedra*, *Gnetum*, and *Welwitschia* do not resemble one another, their placement in a monophyletic group is based on molecular and morphological data.

(a)

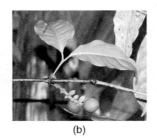
(b)

(c)

© Martynova Anna/Shut-terstock.com

© University of Wisconsin-Madison, Department of Botany.

© Pichugin Dmitry/Shut-terstock.com

Figure 24.17a–c   The three main groups of gnetophytes: (a) *Ephedra*, (b) *Gnetum*, and (c) *Welwitschia*.

Gnetophytes are a bizarre group of gymnosperms because they have angiosperm characteristics, making them distinct from other gymnosperms. Like angiosperms, gnetophyte xylem is composed of both tracheids and vessels. Typical gymnosperm xylem, on the other hand, is composed of only tracheids. Gnetophyte cones form clusters resembling inflorescences or clusters of flowers. Like conifers, gnetophyte pollen grains produce long pollen tubes to deliver the two nonmotile sperm to the egg. In some gnetophytes, both sperm fuse with different egg cells of the female gametophyte to produce multiple embryos. **Double fertilization** refers to the use of both sperm in reproduction, which is unique to angiosperms but also takes place in some gnetophytes. Although both eggs are fertilized within the ovule, only one zygote develops into an embryo. Like angiosperms, the female gametophytes no longer use archegonia to produce their eggs. Gnetophyte seeds have two layers of integument instead of one like typical gymnosperms. The presence of these angiosperm-like characteristics in some gnetophytes has led some botanists to suggest gnetophytes and angiosperms share a common ancestor. However, phylogenetic studies based on sequence data support the sister group relationship between gnetophytes and conifers. Phylogenetic relationships among gnetophytes, angiosperms, and other gymnosperms remain an unsettled issue (for a review, see Chaw et al., 1997).

double fertilization
Use of both sperms in reproduction; a unique feature of angiosperms and gnetophytes.

Key Terms

pollen grains
pollination
micropyle
pollen chamber
integument
ovule

seed
resin
evergreens
deciduous
pollination drops
floral nectar

hypodermis
understory fire regimes
stand-replacing fire regimes
serotinous cones
double fertilization

Summary

- Characteristics unique to seed vascular plants emerged from the late Devonian and early Carboniferous period. Major changes to the sporophyte and gametophyte generations enabled early seed vascular plants to become abundant and widespread.

- Seeds are a key innovation by seed vascular plants for land-based living and are reproductively superior to spores. Each seed contains an embryonic sporophyte, food reserve, and a protective seed coat.

- Based on ovule and seed structure, seed vascular plants can be divided into two major groups: gymnosperms and angiosperms.
- The four phyla of gymnosperms are Cycadophyta (cycads), Ginkgophyta (ginkgo), Coniferophyta (conifers), and Gnetophyta (gnetophytes).
- Most phylogeny studies agree cycads are the most ancient gymnosperms.
- Most cycads have a squat trunk-like stem and a crown of large pinnately compound leaves. Their compound leaves resemble those of tree ferns or palms.
- Insect pollinators formed obligate mutualistic relationships with some cycads and may have pollinated cycads since the Mesozoic era.
- There is only one living species, *Ginkgo biloba*, in the phylum Ginkgophyta. Ginkgo or the maidenhair tree is easily recognized by its distinctive leaves.
- Cycads and ginkgo are both dioecious and have separate male trees and female trees.
- Unlike other seed plants, cycad and ginkgo pollens do not produce long pollen tubes for sperm delivery. Instead, they produce large and motile sperm, which swim to reach the egg. Only one sperm will fertilize the egg.
- Fertilization in cycads and ginkgo exhibits a transitional pattern between primitive plant groups (bryophytes and seedless vascular plants) and advanced plant groups (conifers, gnetophytes, and angiosperms).
- Conifers are the largest and most diverse group of gymnosperms, which include pines (the largest group of conifers), cedars, redwoods, junipers, cypress, firs, and spruces.
- Like typical seed vascular plants, conifers produce secondary tissues (wood and bark) and tend to have a substantial woody stem and woody bark.
- Many conifers have a system of ducts or canals connecting roots, stems, and leaves for transportation and storage of oleoresin (resin).
- Like sporophytes of seed vascular plants, pine sporophytes are dominant and heterosporous. Their male and female gametophytes are microscopic. The male gametophytes are the pollen grains and the female gametophytes develop inside the ovules.
- Unlike cycads and ginkgo, most conifers are monoecious and have separate male cones and female cones on the same tree.
- Unlike cycads and ginkgo, pines use pollen tubes to deliver their two nonmotile sperm directly to the eggs.
- Unique adaptations of pines allow them to survive and thrive in cold and arid climates.
- Conifers are economically important because of their wood and resin.
- The gnetophytes (Gnetophyta) are divided into three genera: *Ephedra*, *Gnetum*, and *Welwitschia* which look distinctly different from each other.
- Gnetophytes are a bizarre group of gymnosperms because they have angiosperm characteristics making them distinct from other gymnosperms.

Reflect

1. *How unique are seed vascular plants?* Create a table to compare and contrast characteristics of nonvascular plants, seedless vascular plants, and seed vascular plants. Pay special attention to the characteristics of their sporophytes and gametophytes, and how spores and gametes are produced.

2. *Seeds are reproductively superior to spores.* Do you agree with this statement? Why or why not? Describe at least four characteristics that distinguish seeds from spores.

3. *Gymnosperms and angiosperms.* Compare and contrast gymnosperms to angiosperms. Discuss their phylogenetic relationship to nonvascular plants and seedless vascular plants.

4. *The four phyla of gymnosperms.* Compare and contrast cycads, ginkgo, conifers, and gnetophytes.

5. *Alternation of generation.* Use a diagram to show the life cycle of a pine. Make sure you describe the unique features of the sporophyte, the gametophyte(s), and the spore(s).

6. *Monoecious and dioecious.* Define the two terms and give examples.

7. *Gymnosperms and their uses.* Describe how gymnosperms are used by humans.

References

Brown, N. A. C., & van Staden, J. (1998). Plant-derived smoke: An effective seed pre-soaking treatment for wildflower species and with potential for horticultural and vegetable crops. *Seed Science and Technology, 26,* 669–673.

Chaw, S. M., Zharkikh, A., Sung, H. M., Lau, T. C., & Li, W. H. (1997). Molecular phylogeny of extant gymnosperms and seed plant evolution: Analysis of nuclear 18S rRNA sequences. *Molecular Biology and Evolution, 14,* 56–58.

Crane, P. R., Herendeen, P., & Frils, E. M. (2004). Fossils and plant phylogeny. *American Journal of Botany, 91,* 1683–1699.

Donoghue, M. J. (2005). Key innovations, convergence, and success: Macroevolutionary lessons from plant phylogeny. *Paleobiology, 31,* 77–93.

Earle, C. J. (2011). *Gymnosperms of Alta California.* Retrieved December 30, 2018, from https://www.conifers.org/topics/caltrees.php.

Fernando, D. D., Quinn, C. R., Brenner, E. D., & Owens, J. N. (2010). Male gametophyte development and evolution in extant gymnosperms. *International Journal of Plant Development Biology, 4,* 47–63.

Flora of North America. (2008). *Flora of North America.* Retrieved December 30, 2018, from http://floranorthamerica.org/.

Glasspool, I. J., Edwards, D., & Axe, L. (2004). Charcoal in the Silurian as evidence for the earliest wildfire. *Geology, 32,* 381–383.

Graham, L. E., Graham, J. M., & Wilcox, L. W. (2006). *Plant biology* (2nd ed.). Upper Saddle River, NJ: Pearson Prentice Hall.

Journal of the American Medical Association (n.d.). *Journal of the American Medical Association.* Retrieved December 30, 2018, from https://jamanetwork.com/.

Judd, W. S., Campbell, C. S., Kellogg, E. A., Stevens P. F., & Donoghue, M. J. (2008). *Plant Systematics* (3rd ed.). Sunderland, MA: Sinauer Associates.

Lewinsohn, E., Gijzen, M., & Croteau, R. (1991). Defense mechanisms of conifers: Differences in constitutive and wound-induced monoterpene biosynthesis among species. *Plant Physiology, 96,* 44–49.

Linkies, A., Graeber, K., Knight, C., & Leubner-Metzger, G. (2010). The evolution of seeds. *New Phytologist, 186,* 817–831.

Nepi, M., von Aderkas, P., Wagner, R., Mugnaini, S., Coulter, A., & Pacini, E. (2009). Nectar and pollination drops: how different are they? *Annals of Botany, 104,* 205–219.

Pausas, J. G., & Keeley, J. E. (2009). A burning story: The role of fire in the history of life. *BioScience, 59,* 593–601.

Raven, P. H., Evert, R. F., & Eichhorn, S. E. (2005). *Biology of plants* (7th ed.). New York: W.H. Freeman.

Rowe, N. P. (1997). Late Devonian winged preovules and their implications for the adaptive radiation of early seed plants. *Palaeontology, 40,* 575–595.

Schneider, D., Wink, M., Sporer, F., & Lounibos, P. (2002). Cycads: Their evolution, toxins, herbivores and insect pollinators. *Naturwissenschaften, 89,* 281–294.

Solomon, P. R., Adams, F., Silver, A., Zimmer, J., & DeVeaux, R. (2002). Ginkgo for memory enhancement, a randomized controlled trial. *Journal of the American Medical Association, 288,* 835–840.

Chapter 25

Angiosperms

© Valentyn Volkov/Shutterstock.com

Learning Objectives

- Discuss the evolutionary changes in angiosperms and how they differ from gymnosperms
- Describe the evolutionary significance of flowers and fruits
- Discuss the evolutionary trends among flowers and distinguish between primitive flowers and advanced flowers
- Discuss how floral traits are adapted to increase pollination efficacy
- Distinguish between basal angiosperms and core angiosperms
- Summarize the features that distinguish monocots and eudicots and give examples for each class
- Distinguish among the 10 families of flowering plants and name three examples for each family
- Describe the importance of angiosperms to humans

Flowering plants or angiosperms are the largest, the most diverse, and the most familiar plant group. Flowering plants are popular ornamentals for our homes, gardens, offices, and parks (Figure 25.1). Floral displays are created to convey our love, sympathy, and gratitude. Flowers are used to celebrate special occasions and holiday seasons such as roses for Valentine's Day, Easter lilies for Easter, and poinsettias for Christmas. Flowering plants are also the most economically important group because they provide food, fibers, medicine, and many other products.

Flowering plants are extremely diverse, representing one of the greatest terrestrial radiations. The earliest fossils of flowering plants appeared around the late Jurassic to early Cretaceous period, about 125 million years ago. By the end of the Cretaceous period, flowering plants became the dominant plant group in many habitats. In a letter to J.D. Hooker dated July 22, 1879, Charles Darwin referred to the origin and early diversification of angiosperms as "an abominable mystery" (Darwin & Seward, 1903). In this chapter, we explore the origin and diversity of flowering plants and examine their unique innovations. We take a closer look at some selected families of flowering plants and their importance to humans.

CHARACTERISTICS OF ANGIOSPERMS

Flowering plants or angiosperms are the largest and most diverse group of land plants, with at least 260,000 living species classified in more than 460 families (Angiosperm Phylogeny Group III [APG III], 2009). All angiosperms are classified in the phylum Anthophyta (*anthus* in Greek means "flower" and *phyto* means "plant"), hence the name "flowering plants." Most angiosperms are found in terrestrial habitats and others are found floating or rooted in aquatic habitats. Angiosperms vary greatly in the way they acquire energy and food. Most angiosperms carry out photosynthesis to make their own food whereas a few angiosperms such as dodders and mistletoes are **parasitic**, stealing water and nutrients from their host plants. Some angiosperms such as orchids and bromeliads are **epiphytic**. Epiphytic plants grow on the surfaces of other plants and acquire water and nutrients from the air but not from the other plant.

parasitic
Organisms that steal water and nutrients from other plants.

epiphytic
Organisms that grow on the surfaces of other plants and acquire water and nutrients from the air.

Source: Cynthia McKenney.

Source: Amanda Chau.

Figure 25.1 Plants beautify our homes, gardens, and parks.

A few angiosperms such as Indian pipe are **saprophytic**, obtaining their nutrients from dead organic matter. Angiosperms also vary greatly in body form and size. The smallest angiosperms are duckweeds in the family Lemnaceae, measuring no more than 4 millimeters (5/32 in.) in width (U.S. Forest Service, 2011) (Figure 25.2). The common watermeals in the genus *Wolffia* are the smallest duckweeds, about the size of a pinhead, which is less than 1 millimeter (1/32 in.) in width. The tallest angiosperm is a Mountain ash or swamp gum, *Eucalyptus regnans*, nicknamed Centurion, found in the Tasmania State Forest in Australia. The current measurements for Centurion are 99.6 meters (326.8 ft) in height and 405 centimeters (13.3 ft) in diameter (Taylor, 2013).

Although angiosperms are distinctly different from one another, they form a monophyletic group and share many key characteristics. Like gymnosperms, angiosperm sporophytes are dominant and heterosporous. Their male and female gametophytes are microscopic and dependent. Unlike gymnosperms, angiosperms develop both male and female gametophytes within flowers. Male gametophytes are the pollen grains that are used to deliver nonmotile sperm to the ovule. Female gametophytes do not have archegonia and produce eggs inside the ovule. Angiosperms have carpel(s) enclosing one or more ovules and each ovule is usually surrounded by two layers of integument. Angiosperms do not produce cones to house their seeds. Instead, flowers develop into fruits enclosing their seeds (Figure 25.3). All angiosperms produce flowers and fruits, undergo double fertilization, and develop endosperm in their seeds.

saprophytic
Organisms that obtain their nutrients from dead organic matter.

Flower Development

Recall from Chapter 7 that flowers have four basic parts: sepals, petals, stamens, and pistils. Flowers are reproductive organs responsible for producing gametes and promoting pollination. Fertilization and subsequent embryonic development also take place within the flowers. Angiosperm

(a)   (b)

Figure 25.2a–b   The smallest and the largest angiosperms. (a) The smallest angiosperms are the duckweed in the family *Lemnaceae*. (b) The tallest angiosperm is a mountain ash (*Eucalyptus regnans*), nicknamed Centurion.

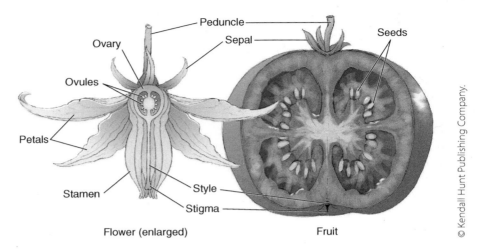

© Kendall Hunt Publishing Company.

Figure 25.3   Tomato flower and fruit.

flowers vary in the numbers and arrangement of their parts and are distinctly different among angiosperm families. However, the basic arrangement of the flower within a particular family tends to be fixed. Later in this chapter, we survey flowers from 10 angiosperm families.

Evolutionary Trends among Flowers

The oldest angiosperm fossils were discovered in Liaoning, China, and these extinct flowering plants are classified in the family Archaefructaceae. Members of Archaefructaceae were aquatic herbaceous plants that appeared around the late Jurassic to early Cretaceous period, about 125 million years ago. Their flowers looked very different from modern angiosperm flowers. *Archaefructus* flowers, unlike modern angiosperm flowers, had no sepals or petals (Sun et al., 2002) and had separate carpels and stamens on the same flowering shoot. As angiosperms diversified and coevolved with their pollinators, angiosperm flowers became more specialized (Evert & Eichhorn, 2013) and these specializations include:

- **Reduction in number of floral parts**: early or primitive flowers tend to have numerous parts whereas specialized or advanced flowers tend to have fewer parts.

- **Changes in arrangement of floral parts**: parts of early flowers are spirally arranged around the axis and are not fused together whereas parts of specialized flowers are not spirally arranged around the axis and are often fused together.

- **Changes in ovary position**: ovaries of early flowers are often in a superior position whereas ovaries of specialized flowers are often in an inferior position.

- **Flowers become bilaterally symmetrical**: early flowers are radially symmetrical (or actinomorphic) whereas specialized flowers are bilaterally symmetrical (or zygomorphic).

A magnolia flower is an example of an early flower (Figure 25.4a). It has numerous flower parts and the parts are arranged spirally around the

Figure 25.4a–b   Evolution of the flower: (a) a magnolia flower showing numerous stamens and pistils; (b) an orchid flower showing fused petals and fewer flower parts.

flower axis and are not fused together. It has a superior ovary and is radially symmetrical. An orchid flower, on the other hand, is an example of a specialized flower (Figure 25.4b). It has few flower parts and some parts are fused together. It has an inferior ovary and is bilaterally symmetrical.

Flowers and Their Pollinators   Unlike animals, plants cannot move to find their mates for sexual reproduction. Most gymnosperms and some angiosperms are pollinated by wind (Figure 25.5) while most angiosperms are pollinated by animals. Recall from Chapter 6 that flowers that are used to attract animal pollinators often produce nectar as a reward for their service. These flowers are often brightly colored and fragrant. Pollinator-specific floral traits began to emerge to enhance pollination efficacy (see Table 7.1). Animal pollinators play a key role in the evolution and diversity of flowers and their evolutionary success (Figure 25.6).

Figure 25.5a–e   Pollination by (a) wind or animals such as (b) honey bees, (c) carrion flies, (d) birds, and (e) bats.

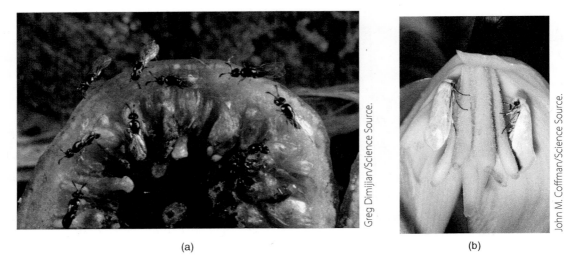

(a)   (b)

Greg Dimijian/Science Source.

John M. Coffman/Science Source.

Figure 25.6a–b Examples of obligate mutualistic relationships between flowering plants and their pollinators. (a) Fig wasps in the genus *Pegoscapus* on a strangler fig (*Ficus costaricana*). (b) Two yucca moths (*Tegeticula yuccasella*) on the stamens of a yucca flower.

Seed Development

Recall from Chapter 24 that seeds are a key innovation by seed vascular plants and are reproductively superior to spores. Although both gymnosperms and angiosperms produce seeds, angiosperms carry out double fertilization and produce an endosperm in their seeds.

Double Fertilization and Endosperm Production Double fertilization, a defining feature of angiosperms, refers to the use of both sperms of the pollen grain in reproduction. After pollination, each pollen grain develops a long pollen tube to deliver the two sperms to the ovule (Figure 25.7). One of the sperms fuses with the egg to produce a zygote and the other fuses with the two polar nuclei to produce an endosperm. Although a few gymnosperms such as *Ephedra* and *Gnetum* also carry out double fertilization, they produce two zygotes instead (see Chapter 24). In their seeds, angiosperms produce endosperm as food for the germinated sporophytes. In contrast, gymnosperms rely on the female gametophyte to provide food for the germinated sporophyte.

Fruit Development

As the ovules develop into seeds, the ovary and sometimes the receptacle or other floral parts develop into a fruit. Like flowers, fruits are a unique feature of angiosperms. Recall from Chapter 8 that fruits are used for seed dispersal. Seed dispersal is essential because it allows plants to colonize new areas, reduces resource competition between parents and their offspring, and lowers the risk of seed predation. Angiosperm fruits vary in their shape, color, and size (see Chapter 8), which allow them to be dispersed by a variety of agents (Figure 25.8).

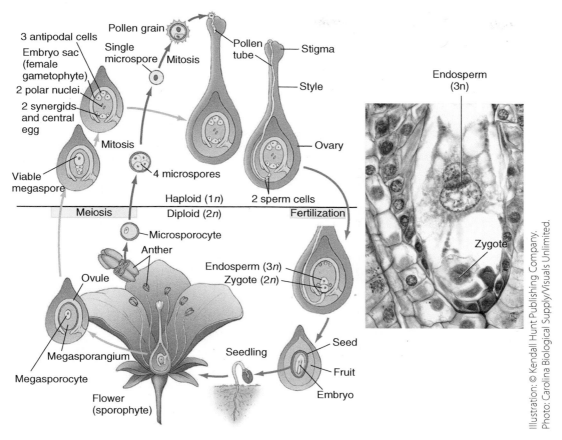

Illustration: © Kendall Hunt Publishing Company.
Photo: Carolina Biological Supply/Visuals Unlimited.

Figure 25.7   The alternation of generations in flowering plants.

EVOLUTIONARY RELATIONSHIPS AMONG ANGIOSPERMS

Although the closest relatives of angiosperms remain a mystery (Chapter 24), all angiosperms form a monophyletic group with *Amborella* (Amborellales), water lilies (Nymphaeales), and star anise (Austrobaileyales) being the basal groups. Phylogenetic studies strongly support the sister relationship of *Amborella* followed by water lilies and star anise to all other angiosperms which are known as the core angiosperms (Figure 25.9) (for reviews, see Judd, Campbell, Kellogg, Stevens, & Donoghue, 2008; Soltis, Bell, Kim, & Soltis, 2008; APG III, 2009).

Basal Angiosperms

The **basal angiosperms** (*Amborella*, water lilies, and star anise) are the most primitive groups of angiosperms. Most basal angiosperms do not have vessels in their xylem. Their flowers display primitive characteristics such as radial symmetry, a superior ovary, and flower parts that are not fused, spirally arranged, and numerous. Basal angiosperms are distinct from other angiosperms. Their carpels are tube-like whereas the carpels of other angiosperms resemble a leaf folded down the middle. In the following section, we take a closer look at three families of basal angiosperms (for more detailed descriptions, see Judd et al., 2008).

basal angiosperms
Three most primitive groups of angiosperms: *Amborella*, water lilies, and star anise.

Figure 25.8a–d   Seed dispersal mechanisms: (a) maple samara dispersed by wind; (b) coconuts dispersed by water; (c) fleshy fruits such as berries ingested and dispersed by animals; and (d) cocklebur fruits dispersed by grabbing onto animal fur.

The Amborella Family (Amborellaceae) *Amborella* is the only genus in the amborella family that has only one species, *A. trichopoda*. *Amborella* plants are shrubs native to New Caledonia, an island in the South Pacific. They have simple leaves, imperfect flowers, and their male flowers and female flowers are found on different trees. Their flowers form inflorescences and rely on wind and insects, in particular beetles, for pollination. They produce drupes that are dispersed by birds.

The Water Lily Family (Nymphaeaceae) There are eight genera and 70 species in the water lily family. Water lilies are aquatic herbaceous plants found from the tropic to temperate regions in calm freshwater environments such as ponds, lakes, and wetlands. Their leaves may be submerged, floating, or raised above the water surface. Water lilies produce perfect flowers and their flowers have a long pedicel, raising them above the water surface (Figure 25.10). Their flowers vary in size and shape from inconspicuous to large and showy. They produce a variety of fruit types from simple fleshy fruits such as berries to simple dry fruits such as nuts and seed pods. *Cabomba* is a popular aquatic plant

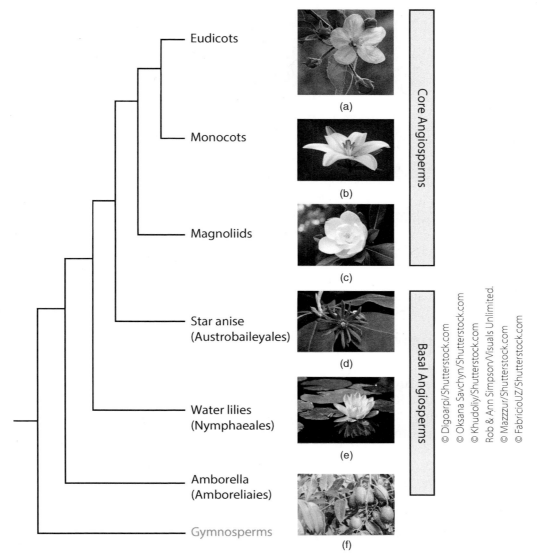

(a)

(b)

(c)

(d)

(e)

(f)

Core Angiosperms

Basal Angiosperms

Eudicots

Monocots

Magnoliids

Star anise
(Austrobaileyales)

Water lilies
(Nymphaeales)

Amborella
(Amboreliaies)

Gymnosperms

© Digoarpi/Shutterstock.com
© Oksana Savchyn/Shutterstock.com
© Khudoliy/Shutterstock.com
Rob & Ann Simpson/Visuals Unlimited.
© Mazzur/Shutterstock.com
© FabricioUZ/Shutterstock.com

Figure 25.9a–f Simplified phylogenetic relationships among major groups of angiosperms including (a) Eudicots represented by an apple blossom, (b) Monocots represented by an Asiatic lily, (c) Magnoliids represented by a southern magnolia, (d), Austrobaileyales represented by star anise, (e) Nymphaeales represented by water lilies and (f) ambroealiaies represented amborella. Adapted from Soltis et al. (2008) and APG III (2009).

for aquariums. Water lily (*Nymphaea*), yellow water lily (*Nuphar*), and Amazon water lily (*Victoria*) are commonly used as ornamental plants in ponds and pools. Roots, young leaves, flower buds, and seeds of some water lilies are used as food by Native Americans and also in Africa and Asia.

The Star Anise Family (Illiciaceae) The star anise family is one of four families in the order Austrobaileyales. There are 37 species of star anise and they all belong to the same genera, *Illicium.* Star anises are trees or shrubs with simple leaves found mainly in warmer regions such as southeastern Asia, southeastern United States, Cuba, Hispaniola, and Mexico. Their perfect flowers form inflorescences. Their follicles form star-shaped aggregated

© Trombax/Shutterstock.com

Figure 25.10 Star anise is an example of a basal angiosperm and has unique star-shaped aggregated fruits. The oil from the seeds is used as a flavoring in sweets.

core angiosperms
All angiosperms except basal angiosperms; divided into three main groups: magnoliids, monocots, and eudicots.

magnoliids
One of three classes of core angiosperms; more primitive than monocots and eudicots.

monocotyledons or monocots
One of three classes of core angiosperms; have one cotyledon in their seeds; flower parts come in threes or multiples of three; leaves have parallel veins.

eudicotyledons or eudicots
One of three classes of core angiosperms; have two cotyledons in their seeds; flower parts come in fours or fives, or in multiples of four or five; leaves have netted veins.

fruits (Figure 25.10). Anise oil extracted from *Illicium verum* are commonly used in sweets and cough drops. Some star anises are used medicinally and a few are used as ornamentals.

Core Angiosperms

The **core angiosperms** are divided into three main monophyletic groups: the **magnoliids** (Magnoliidae), the **monocotyledons** or **monocots** (Monocotyledonae), and the **eudicotyledons** or **eudicots** (Eudicotyledonae) (Figure 25.9). Phylogenetic studies suggest that the monocots and the eudicots form a sister group with the magnoliids basal to them (for a review, see Judd et al., 2008). The monocots and the eudicots form the largest group of angiosperms and represent 97% of flowering plants. They are also the most familiar and economically important plants, although some can cause major economic and ecological harm (**Box 25.1**). In the remaining chapter, we examine selected families from the three groups of core angiosperms, explore their flowers using floral formulae (see Chapter 7, Box 7.1), and discuss their importance to humans (for more detailed descriptions, see Hickey & King, 1997; Judd et al., 2008).

BOX 25.1 When Plants Go Bad . . . Real Bad

Angiosperms are an incredibly diverse and important plant group to humans, but some can cause serious ecological and economic harm. Many angiosperms are important food plants while others are used as ornamentals, medicines, fibers, industrial products, and construction materials. Many angiosperms are intentionally introduced to new biogeographic regions for food and forage production, erosion control, and as ornamentals. Some "introduced," "exotic," or "alien" species such as maize, wheat, rice, and others are beneficial and provide most of the world's food supply. A few introduced species, however, become serious problems. They displace native species, disrupt community and ecosystem processes, and cause major economic losses. These harmful introduced species are often referred to as "invasive species."

The term *weeds* is often used to refer to plants, which can be native or introduced, growing where they are not wanted. Ross and Lemi (2009) further define *weeds* as plants that interfere with the growth of desirable plants, are unusually persistent, damage managed and natural systems, and even disrupt human activities. In agricultural settings or managed ecosystems, weeds compete with crop plants for resources. Pimental and colleagues (2001) estimated that weeds may cause a 12% reduction of crop yields, which represents about a $33 billion loss in crop production per year. Pimental (1993) also estimated 73% of the 80 major crop weeds were introduced species and it is likely that about $27.9 billion out of the $33 billion in crop losses were due to these introduced species. Native species, of course, can also disrupt ecosystem processes, cause economic losses, and even threaten human health (Box 23.1).

Invasive species not only cause problems in managed ecosystems but also in natural ecosystems. They disrupt native species and significantly reduce biodiversity in the area. The European purple loosestrife (*Lythrum salicaria*) was introduced in the early 19th century as an ornamental plant (Malecki et al. 1993) (Figure Box 25.1a). It has been spreading at a rate of 115,000 ha (1,150 km^2) per year and is altering wetland vegetation and endangering wildlife such as bog turtles that depend on native plants. Purple loosestrife is found in 48 states and costs some $45 million per year in control costs and forage losses (Pimental, Zuniga, & Morrison, 2005). Kudzu (*Pueraria lobata*), a perennial vine native of Japan and China, was first introduced in the United States in 1876 as a forage plant and an ornamental. From 1935 to the mid-1950s, farmers in the south were encouraged to plant kudzu to reduce soil erosion. Kudzu is now common throughout most of the southeastern United States and is found as far north as Pennsylvania (Bergmann & Swearingen 2005) (Figure Box 25.1b). Kudzu spreads at a rate of 150,000 acres (607 Km^2) per year and costs some $6 million per year to control (USDA/Agricultural Research Service, 2009). Kudzu leaves form a solid blanket, smothering other plants and displacing native species in the area. Kudzu is often referred to as "the vine that ate the south." Renowned biologist E. O. Wilson states, "On a global basis . . . the two

(a)   (b)

Figure Box 25.1a–b   Invasive species destroy biodiversity: (a) purple loosestrife (*Lythrum salicaria*) displaces other native plants in wetland habitats; (b) kudzu (*Pueraria lobata*) plants blanket the area and smother other plants.

great destroyers of biodiversity are, first, habitat destruction and second, invasion by exotic species."

A maxim of medicine, "an ounce of prevention is worth a pound of cure," provides great insights into dealing with introduced species. Prevention is the first and most cost-effective defense against invasive species. Once an introduced species become established, it is extremely difficult, if not impossible, to eradicate. Recall from Chapter 23 (Box 23.1) the five strategies used to manage invasive ferns are also effective against other invasive species. Public education is an essential part of prevention and management programs. You and I can help prevent the establishment and spread of invasive species. Explore these online resources for more information:

1. **Learn more about invasive species.** Here are a few useful resources that provide information on invasive species, their distribution, and how to identify them: USDA National Invasive Species Information Center, University of Florida/IFAS Center for Aquatic and Invasive Plants, the University of Georgia Center for Invasive Species and Ecosystem Health, Cornell University Ecology and Management of Invasive Plants Program, Plant Conservation Alliance's Alien Plant Working Group.

2. **Prevent invasive species from hitchhiking on you.** Many invasive species are introduced and spread accidentally. Pieces of plants or seeds can hitch a ride on your clothing, socks, and hiking boots. They can also hitch a ride on your vehicle, boat, and bicycle. Check your clothing, belongings, and vehicles carefully.

3. **Use native plants as alternatives for your gardens.** Many invasive species were introduced as ornamentals. If you enjoy gardening, follow the guidelines provided by the Be PlantWise and Lady Bird Johnson Wildflower Center and help prevent introducing non-native species in gardens, parks, and natural areas.

4. **Be careful with aquatic plants.** Many invasive aquatic species were introduced as ornamentals for water gardens and aquariums. Do not dump aquatic plants down drains. Dispose of them properly (see the next bullet).

5. **Dispose of invasive species properly.** When disposing invasive plant materials, bag all the plant materials including seeds, fruits, or cuttings that could resprout. Use heavy black or white plastic bags. Allow the bags to sit in the sun for several weeks. If it is permissible and safe, burn all the plant materials.

6. **Contain your invasive plant.** If you cannot part with your invasive plant, take special precautions to keep it contained in your garden and prevent it from spreading. Insert root barriers, and regularly harvest and remove fruits or seeds before they spread.

7. **Be part of a monitoring program.** Check your state and see if they have any monitoring programs for invasive species. Help track the distribution of invasive species by becoming a volunteer in programs such as the Invaders of Texas.

Floral formula: * T -6-∞- **A** ∞ **G** ∞

(a)   (b)   (c)

Figure 25.11a–c   The magnolia family (Magnoliaceae): (a) flower of a purple magnolia, (b) floral formula, (c) cone of the southern magnolia.

TAKING A CLOSER LOOK AT MAGNOLIIDS AND THEIR IMPORTANCE

Magnoliids consist of 20 families which include magnolias, pawpaws, nutmegs, laurels, winter's barks, and peppers (APG III, 2009). They are trees, shrubs, or vines with alternate or opposite leaves. Their flowers are primitive, having numerous flower parts that are spirally arranged. Some magnoliids have monocot-like characteristics such as scattered vascular bundles in their stems, one aperture on their pollen grains, and seeds with a large endosperm.

The Magnolia Family (Magnoliaceae)

The two main genera in the magnolia family are *Magnolia* (218 species) and *Liriodendron* (2 species). Members of the magnolia family are commonly found in temperate regions of eastern North America and eastern Asia and also in tropical regions of South America. Their wood is used for timber. Some members such as the tulip tree, *Liriodendron tulipifera*, and a few Magnolia species are popular and important ornamentals. Magnolias are trees or shrubs with simple leaves alternately arranged on the stem. They have perfect flowers found at the terminal of their stems. Their sepals look identical to their petals (Figure 25.11). They have numerous flower parts spirally arranged and not fused together. Their flowers produce no nectar. Their fruits are an aggregate of follicles, fleshy and berry-like, or an aggregate of samaras.

TAKING A CLOSER LOOK AT MONOCOTS AND THEIR IMPORTANCE

There are approximately 65,000 species of monocots, making them the second largest group of angiosperms. Monocots include families such as orchids, grasses, palms, bromeliads, lilies, irises, and many other familiar and important plants. Recall from previous chapters that monocots are distinct from eudicots and their differences are summarized in Table 25.1. Monocots (*mono* in Greek means "one") have one cotyledon in their seeds, hence the name *monocotyledons* or *monocots*. When you pull up a monocot plant, it usually has a fibrous root system. Monocot flowers are distinct from eudicot flowers because their parts are always in threes or multiples of three (Figure 25.12a). Monocot leaves have parallel veins. All monocots are herbaceous, with the exception of palms, and do not produce secondary growth. Recall from Chapter 5 that palms produce pseudobark and do not have the same type of secondary growth as eudicots. For other monocot features, see **Table 25.1**. We now examine three of the monocot families and their importance to humans.

The Lily Family (Liliaceae)

The number of genera and species in the lily family (Liliaceae) ranges from 635 species in 16 genera (Judd et al., 2008) to 3,500 species in 220 genera (Hickey & King, 1997) depending on the classification of some genera. Hickey and King (1997) consider the Liliaceae in a broad sense whereas Judd and colleagues (2008) restrict the family to *Lilium* and 15 closely related genera. Judd and colleagues elevate the rest of the genera to families. Lilies are found worldwide. Some lilies are mainly found in temperate regions of the Northern Hemisphere and Europe, while others are found in temperate to tropical regions of Africa and Asia. Most lilies are important spring-blooming plants in open communities such as prairies and meadows. Tulips (*Tulipa*), lilies (*Lilum*), fritillary (*Fritillaria*), and mariposa lily (*Calochortus*) are popular and important ornamentals. When the lily family is considered in a broad sense (see Hickey & King, 1997), it also includes hyacinths, lilies-of-the-valley, daffodils, *Trilliums*, *Aloes*, onions, and asparagus. Hyacinths, lilies-of-the-valley, daffodils, and *Trilliums* are popular ornamentals. Plant

(a)

(b)

© Lurin/Shutterstock.com

Source: Andrew Chow.

Figure 25.12a–b Typical monocot flower and eudicot flower: (a) A monocot flower, such as red trillium (*Trillium erectum*), has three green sepals, three red petals, and six yellow stamens (flower parts in threes or multiple of three). (b) A eudicot flower, such as pink evening primrose (*Oenothera speciosa*), has four green sepals, four pink petals, and eight yellow stamens (flower parts in fours or multiples of four).

Table 25.1   Distinguishing Features of Monocots and Eudicots

Features	Monocots	Eudicots
Seeds	Embryo with one cotyledon	Embryo with two cotyledons
Food storage in seeds	Endosperm	Cotyledons
Root system	Fibrous root system	Tap root system
Vascular tissue arrangement in root	Alternating patches of xylem and phloem, pith in the center	Xylem core in the center, patches of phloem between arms, no pith
Vascular bundle arrangement in stem	Scattered, no pith	Arranged in a circle around the pith
Lateral meristems (vascular cambium and cork cambium)	Absent	Vascular cambium in both herbaceous and woody eudicots, cork cambium in woody eudicots
Leaf venation	Parallel	Netted
Stomatal distribution on leaves	Both upper and lower leaf surfaces	Lower leaf surface
Flower parts	In threes or multiples of three	In fours or fives or multiples of four or five
Pollen grains	One aperture or groove	Three apertures or grooves

saps produced by several *Aloe* species have medicinal properties and are used in shampoos, cosmetics, and for the treatment of burns. Onions and asparagus are economically important vegetables.

Lilies are mostly perennial herbaceous plants with rhizomes or bulbs and contractile roots. Their leaves have parallel veins and grow from the base of the stem and have alternate, spiral, or whorled arrangement on the stem. They form an inflorescence or a single flower at the terminal. Their flowers are perfect and have six tepals. Sepals and petals that look alike are referred to as tepals (Figure 25.13). Lilies attract insect pollinators by providing nectar at the base of tepals. Lily flowers have six stamens and three fused carpels. The ovary is superior and has numerous ovules. Their pollen grains have one aperture. Most lilies produce capsules but some produce berries.

The Orchid Family (Orchidaceae)

The orchid family (Orchidaceae) consists of some 19,500 species and is the second largest angiosperm family. Orchids are found worldwide and are most diverse in tropical regions. Orchids are perennial herbaceous plants with rhizomes, corms, or root-tubers. Orchids in temperate regions are terrestrial whereas those in the tropics are often epiphytic. Their leaves are simple and have parallel veins. Most orchids have leaves growing from the base of the stem and others have their leaves arranged alternately or spirally on the stem. Orchids are economically important because many of them are popular ornamentals such as corsage orchids (*Cattleya*), lady's slippers (*Paphiopedilum*),

Source: Cynthia McKenney.

Wally Eberhart/Visuals Unlimited.

(a)   (c)

Floral formula: * **T** -6- **A** 6 **G** $\overline{(3)}$

(b)

Figure 25.13a–c   The lily family (*Liliaceae*): its (a) flower, (b) floral formula, and (c) fruits.

dancing ladies (*Oncidiums*), *Phalaenopsis*, *Dendrobium*, and *Epidendrum*. One particular orchid, *Vanilla plantifolia*, is the source of vanilla extract, which is a popular flavoring for confectionery, baked goods, ice creams, and beverages.

Orchid flowers are conspicuous and considered to be highly specialized and highly advanced phylogenetically (Figure 25.14). They have perfect bilaterally symmetrical flowers. Some have inflorescences and others have a single axillary or terminal flower. Orchid flowers have six tepals and the center tepal forms a lip, which differentiates it from the other tepals. This lip-shaped tepal acts as a landing platform for insect pollinators. They have one or two stamens, which are fused with the pistil to form a column. They have three fused carpels and the ovary is inferior. Pollen grains are packed into masses or pollinia (singular *pollinium*), which stick to the pollinators. Orchids produce capsules containing many small seeds, which are dispersed by wind. Unlike other monocot seeds, orchid seeds have a tiny embryo and little or no endosperm.

Orchid flowers are extremely diverse and attract a wide range of pollinators such as bees, wasps, moths, butterflies, flies, and even birds. Some orchid flowers provide nectar as a reward for pollination but others trick their pollinators. Orchid flowers from the genera *Ophrys* and *Cryptostylis* mimic the shape, color, and scent of female bees, wasps, or flies to attract the males. When the males attempt to mate with the flowers, they pick up the pollinia and transfer them to another orchid.

The Grass Family (Poaceae)

The grass family (Poaceae) consists of about 9,700 species and is the most economically important family. Four of the top 10 world food crops (wheat, rice, maize or corn, and sorghum) belong to the grass family. Wheat, rice, and maize provide 60% of the world's food energy intake and are staples for over 4,000 million people (Food and Agriculture Organization, 1995). Cereal grains such as wheat, barley, oats, and sorghum have been in cultivation for at

(a)

(b)

(c)

(d)

Floral formula: ↑ **T** 5+1 **A** 1 or 2 **G** (3)

(e)

Figure 25.14a–e The diversity of flowers in the orchid family (*Orchidaceae*): (a) corsage orchids (*Cattleya*), (b) lady's slippers (*Paphiopedilum*), (c) pansy orchids (*Miltonia*), and (d) *Dendrobium* and (e) the typical floral formula for the family.

least 10,000 years. Grasses are also important as livestock feed. Some grasses such as Bermudagrass, ryegrasses, and bluegrasses are cultivated as turf for lawns and golf courses. Other grasses such as feather reedgrass (*Calamagrostis acutiflora*), fountaingrasses (*Pennisetum*), switchgrass (*Panicum virgatum*), and cordgrass (*Spartina pectinata*) are grown as ornamentals. Grasses are important ecologically for providing erosion control and food for wild animals. A major source of table sugar comes from the sugarcane (*Saccharum officinale*), which is also a grass. Grasses also provide a sugar source for the production of alcoholic beverages such as beer and whiskey. In many tropical areas, bamboos provide materials for construction, fiber for paper, and pulp for rayon. Young bamboo shoots are used as food for humans as well as a constituent of everything from clothes to flooring. Grasses like maize and switchgrass are also used for the production of renewable biofuels. Grasses are found

Nigel Cattlin/Visuals Unlimited.

Dr. Jeremy Burgess/Science Source.

Source: Cynthia McKenney.

(a) (b) (c)

Floral formula: * **T** -2- **A** (1-)3(-6) **G** (2-3)

(d)

Figure 25.15a–d The grass family (Poaceae): (a–b) the flowers, (c) the fruits, and (d) the floral formula.

culm
Stem or central axis of the mature grass shoot, comprised of nodes and internodes.

sheath
Lower part of the leaf that encloses the internode on the culm.

florets
Small flowers that make up an inflorescence.

lemma
Larger, outer bract that, along with the palea, serves to contain and protect the floret(s) within.

palea
Shorter, upper bract that, along with the lemma, serves to contain and protect the floret(s) within. See *lemma*.

worldwide and in all types of habitats ranging from desert to aquatic environments. Communities such as the North American prairie and plains, South American pampas, African veldt, and Eurasian steppes are dominated by grasses, which account for 24% of the vegetation on earth.

Grasses are herbaceous annual or perennial plants with simple leaves having parallel veins. Their long, narrow leaves are alternately arranged on the stem-like **culm**. A grass culm differs from most plant stems by being hollow except where the segments join together at a node. The base of the leaves forms a **sheath** around the stem before attachment. These sheaths and the collar on the inside of the sheath are one of the primary ways that you can identify different grass species. Turfgrasses often form tufts or mats. Given the crown of this plant are at or just below the soil line, these plants may be repeatedly mowed without damaging the plant itself. Their flowers form inflorescence (spikes, racemes, or panicles) at the terminal of the stem and are highly specialized. Individual flowers or **florets** are very small and are often found between two bracts, the **lemma** and the **palea** (Figure 25.15). Some grasses have perfect flowers and others have imperfect flowers. Some grasses with imperfect flowers are monoecious but others are dioecious. Grass flowers are pollinated by wind and often lack conspicuous petals. The floral parts are also greatly reduced. Grasses form a single-seeded grain or caryopsis and their seeds have a single cotyledon and a large endosperm.

TAKING A CLOSER LOOK AT EUDICOTS AND THEIR IMPORTANCE

Eudicots are the largest group of angiosperms, with at least 160,000 species. This huge group includes families such as legumes, sunflowers, buttercups, nightshades, mustards, pumpkins, roses, oaks, citrus, berries, cacti, and many other familiar and economically important plants. Eudicots (*eu* in Greek means "true" and *di* means "two") have two cotyledons in their seeds, hence the name *eudicotyledons* or *eudicots*. Unlike monocot flowers, eudicot flower parts are in fours or fives, or in multiples of four or five (Figure 25.12b). Eudicot leaves

(a)   (c)

Floral formula: * **K** (5) **C** 5 **A** (10-∞) **G** 1

(b)

Figure 25.16a–c   The bean or legume family (*Fabaceae*): its (a) flowers, (b) floral formula, and (c) fruits.

have netted veins and most eudicots have a tap root system. Some eudicots are herbaceous but others are woody. All eudicots produce secondary growth. See Table 25.1 for other distinguishing eudicot features. We now examine six eudicot families and their importance to humans.

The Bean or Legume Family (Fabaceae)

The legume family (Fabaceae) consists of some 18,000 species, is the third largest angiosperm family, and is second to Poaceae in economic importance. Many legumes are important food for humans and livestock. Peanuts (*Arachis*), soybeans (*Glycine*), chickpeas (*Cicer*), lentils (*Lens*), beans (*Phaseolus*), peas (*Pisum*), and tamarind (*Tamarindus*) are important food for humans. In addition, peanut seeds provide a source of oil and licorice roots (*Glycyrrhiza*) provide a source of the confectionary flavoring of licorice. Alfafa (*Medicago*), sweet clover (*Melilotus*), clover (*Trifolium*), and vetch (*Vicia*) are important forage for livestock. The nectar provided by these forage plants are collected by bees to make honey. Although many legumes are important food plants, a few legumes such as rose pea (*Abrus*) and locoweed or milkvetch (*Astragalus*) are highly poisonous to humans and livestock.

Many legumes are cultivated as ornamentals or used for soil enrichment and industrial products. Some popular ornamental legumes are red buds (*Cercis*), mimosa (*Albizia*), powder puff (*Calliandra*), sensitive plants (*Mimosa*), lupins (*Lupinus*), and *Wisteria*. Clovers develop root nodules containing nitrogen-fixing bacteria and are often plowed into the soil to replenish and enrich the nitrogen level in agricultural fields. Legumes such as *Indigofera* (blue dye) and *Acacia* and *Hymenaea* (commercial gums and resins) are used to produce industrial products.

Legumes are found worldwide and in a wide range of habitats. They are herbaceous plants, climbers, shrubs, or trees. Their leaves are often compound and alternately arranged on the stem. Some leaves and stems are modified into tendrils for climbing. Legume flowers are perfect but extremely diverse in size, form, and color. Some form inflorescences while others have single terminal flowers. Their flowers can be radially or bilaterally symmetrical. They often have five sepals and five petals, which can be free or fused together within the whorl or between the whorls. The uppermost petal is often large and forms a banner, while the two lower petals often fuse or stick together, forming a boat-shaped structure (Figure 25.16). The middle two petals form two wings covering the two lower petals. Their flowers often have 10 or more stamens. They have one carpel with two or more ovules and their ovary is superior. Their flowers usually produce nectar and are pollinated by a wide range of insects. Almost all members of this family produce dry dehiscent fruits, which are referred to as pods or "legumes," hence the name *legume* family. Pods are two-sided fruits with a suture on each side and several seeds on the interior. This dehiscent pod contains just one carpel and provides protection to the seeds held within. A few members of the legume family produce samara, follicle, berry, or drupe. Their seeds have two cotyledons but no endosperm.

The Rose Family (Rosaceae)

There are at least 3,000 species in the rose family (Rosaceae) found worldwide. This family is economically important because of their edible fruits and their popularity as ornamentals. Important edible fruits include:

- Pome fruits such as apples (*Malus*) and pears (*Pyrus*)
- Aggregated fruits such as strawberries (*Fragaria*), blackberries, loganberries, and raspberries from the genus *Rubus*
- Stone fruits in the genus *Prunus* such as peaches, cherries, apricots, plums, and almonds

Important ornamentals include:

- Herbaceous plants such as lady's mantle (*Alchemilla*), avens (*Geum*), meadowsweet (*Filipendula*), and cinquefoil (*Potentilla*)
- Woody plants such as roses in the genus (*Rosa*), crab apples (*Malus*), flowering quince (*Chaenomeles*), Japanese flowering cherries (*Prunus*), mountain ash (*Sorbus*), shadbush (*Amelanchier*), jetbead (*Rhodotypos*), and bridal wreath (*Spiraea*)

Others products include:

- Fragrant oil from which perfumes are made
- Furniture and cabinets made with wood from the black cherry, *Prunus serotina*

Members in the rose family form a monophyletic group even though they vary greatly in forms, flowers, and fruits. Most members are perennial woody plants such as shrubs or trees. A few members are perennial herbaceous plants and climbers. Roses often have palmately or pinnately compound leaves alternately arranged on the stem. Stipules are often found at the base of the petioles. Rose flowers are unspecialized, often perfect, and radially symmetrical (Figure 25.17). Most flowers are showy and produce nectar. Some form inflorescences while others have single terminal flowers. The receptacle is often hollowed, forming a cup. Rose flowers usually have five sepals and five petals. They have numerous stamens and one to many carpels that are free

Floral formula: * **K** 5 **C** 5 **A** 10-∞ **G** 1-∞

(g)

Figure 25.17a–g The rose family (*Rosaceae*) fruits and flowers: apple (a and d), raspberry (b and e), and cherry (c and f); and (g) their floral formula.

or fused. Most roses have superior ovaries while some have inferior ovaries. Species with smaller flowers are usually pollinated by flies while species with larger flowers are pollinated by bees, wasps, butterflies, moths, and beetles. Roses produce many different fruit types such as follicles, achenes, pomes, drupes, or aggregate fruits. Like other eudicots, their seeds have two cotyledons and no endosperm. Another identifying characteristic of the family is the presence of lenticels in the bark (see Chapter 5).

The Pumpkin Family (Cucurbitaceae)

The pumpkin family (Cucurbitaceae) consists of about 825 species and is also economically important because of its edible fruits and seeds. Pumpkins are found in the temperate regions but are most diverse in the tropics and subtropics. Genera producing edible fruits and seeds include *Cucurbita* (pumpkins, winter and summer squashes, and gourds), *Cucumis* (cantaloupes, muskmelons, honeydew melons, and cucumbers), *Citrullus* (watermelons), *Benincasa* (wax gourds), and *Sechium* (chayote). Dried fruits of *Lagenaria* (bottle gourd) are used as containers and dried fruits of *Luffa* (loofah) are used as vegetable sponges. Others such as *Momordica* (balsam apple) are used medicinally.

Members of the pumpkin family are herbaceous or soft-woody vines with tendrils. Their leaves are often simple and alternately or spirally arranged on the stem. They do not have stipules at the base of the petioles. Some species have inflorescences while others produce single flowers at

a: Inga Spence/Visuals Unlimited.
b: Robert Calentine/Visuals Unlimited.
c: Matt Johnston/Science Source.
d: Robert Calentine/Visuals Unlimited.
© BestPhotoStudio/Shutterstock.com

(a)   (b)   (c)   (d)

Male flowers
Anthers

Female flowers
Stigma
Ovary

Pepo

(e)

* K 5 C (5) A 5 G 0   * K 5 C (5) A 0 G (3)

(f)

Figure 25.18a–f The pumpkin family (Cucurbitaceae): (a–b) male flowers, (c–d) female flowers, (e) fruit, and (f) floral formula.

the terminal. Their flowers are radially symmetrical and usually imperfect. Some species are monoecious while others are dioecious. The flowers have five fused sepals that are often reduced and five fused petals form a tubular corolla (Figure 25.18). Their male flowers have three to five stamens and their female flower have three fused carpels. Most flowers have an inferior ovary. Most members of the pumpkin family produce modified berries or pepos while a few produce capsules or berries. Their seeds are flattened and have a fleshy outer seed coat. Like other eudicots, their seeds have two cotyledons but no endosperm.

The Mustard Family (Brassicaceae)

The mustard family (Brassicaceae) is found worldwide but is most diverse in the Mediterranean region, southwestern and central Asia, and western North America. There are about 4,130 species of mustards and many species are important food plants. Common vegetables such as cabbage, kale, broccoli, cauliflower, brussel sprouts, and kohlrabi all belong to a single species of mustard, *Brassica oleracea*. Other vegetables such as radish (*Raphanus sativus*), capers (*Capparis spinosa*), and Chinese cabbage (*B. rapa*) also belong to this family. The seeds of some mustard species are used as condiments such as Chinese mustard (*B. juncea*), black mustard (*B. nigra*), white mustard (*Sinapis alba*), and horseradish (*Armoracia rusticana*). The common table mustard is prepared by mixing the seeds of white mustard with either black or Chinese mustard. Vegetable oils such as canola oil and rapeseed oil are extracted from the seeds of *B. napus*. Many mustard species are also popular ornamentals such as spider flower (*Cleome*), wallflower (*Erysimum*), honesty, money plant (*Lunaria*), sweet alyssum (*Lobularia*), golden alyssum (*Aurinia*), rock cress (*Arabis*), and dame's violet (*Hesperis*).

Most mustard species are annual or perennial herbaceous plants but some are trees or shrubs. Their leaves are mostly simple and usually alternately or spirally arranged on the stem. Their flowers form inflorescences and are radially symmetrical. Each flower has four free sepals, four free petals, six or more stamens, and two fused carpels (Figure 25.19). The ovary is superior. The two outer stamens are shorter than the four inner stamens. The petals are opposite to each other, forming a cross or cruciform. Their fruits are mostly siliques while some are capsules or berries. Their seeds, similar to other eudicots, have two cotyledons and no endosperm.

The Nightshade Family (Solanaceae)

The nightshade or potato family (Solanaceae) consists of about 2,510 species and is found worldwide but is most diverse in Central and South America. Most members contain tropane or steroid alkaloids, making them highly toxic. Some species such as belladonna (*Atropa*), tobacco (*Nicotiana*), and jimsonweed (*Datura*) produce powerful pharmaceutical drugs or narcotics. Because of the reputation of deadly nightshade, one family member, the tomato, was considered poisonous until starving soldiers in the civil war ate them and did not die. Surprisingly, some members of the family are important food plants. Some species produce edible fruits such as tomatoes (*Solanum lycopersicum*), eggplants (*S. melongena*), tomatillos (*Physalis ixocarpa*), and peppers (*Capsicum*). Red, green, or cayenne peppers contain the alkaloid capsaicin, providing a "hot" flavor. Tubers of potato plants (*S. tuberosum*) are an important food for humans. Some species are used as ornamentals including *Petunia*, nightshade (*Solanum*), tobacco, Chinese

(a)

(c)

Gail Jankus/Science Source.

© Madlen/Shutterstock.com

Floral formula: * **K** 4 **C** 4 **A** (2-) 6-∞ **G** (2)

(b)

Figure 25.19a–c   The mustard family (Brassicaceae): its (a) flowers (white mustard), (b) floral formula, and (c) the seeds used to make various types of mustard.

Floral formula: * **K** (5) **C** (5) **A** 5 **G** (2)

(g)

Figure 25.20a–g The nightshade family (Solanaceae): (a) potato flower, (b) potato tubers, (c) tomato flower, (d) tomato fruit, (e) sweet pepper flower (f) sweet pepper fruit, and (g) the floral formula.

lantern (*Physalis alkekengi*), night-blooming jessamine (*Cestrum*), lady-of-the-night (*Brunfelsia*), and angel's trumpet (*Datura*).

Members of the family are herbaceous, vines, shrubs, or trees. Their leaves are simple, often in pairs, and alternately or spirally arranged on the stem. Their flowers form inflorescences or single terminal flowers and are radially symmetrical. Their flowers are usually showy and attract a wide range of pollinators. Each flower has five fused sepals, five fused petals forming a funnel-shaped or tubular corolla, five stamens, and two fused carpels (Figure 25.20). The five stamens are fused to the corolla. The ovary is superior. Their fruits are usually berries but some produce capsules. Seeds are often flattened.

The Sunflower Family (Asteraceae)

The sunflower family (Asteraceae) consists of some 23,000 species and is the largest angiosperm family. They are found worldwide, especially in open and dry habitats. Sunflowers (*Helianthus*) are important food plants because of their edible seeds and oils. Other food plants include artichoke (*Cynara*), Jerusalem artichoke (*H. tuberosus*), dandelion green (*Taraxacum*), lettuce (*Lactuca*), endive (*Cichorium endivia*), and chicory (*Cichorium*). Some members like tarragon and wormwood (*Artemisia*) are spice plants. A few members such as fleabane (*Pulicaria*) and pyrethrum (*Tanacetum*) contain active compounds with insecticidal properties. Many members of this family are popular ornamentals including pot marigolds (*Calendula*), chrysanthemum (*Dendranthema*), China aster (*Callistephus*), sunflower (*Helianthus*), garden dahlia, (*Dahlia*), French marigold (*Tagetes*), tickseed (*Coreopsis*), dusty miller

(a)

(b)

$$ * \, K \, \infty \, C \, (5) \, A \, 5 \, G \, \overline{(2)} \qquad \uparrow K \, \infty \, C \, (5) \, A \, 0 \, G \, \overline{(2)} $$

(c)

Figure 25.21a–c   The sunflower family (Asteraceae): its (a) flowers, (b) disc and ray florets, and (c) floral formula.

(*Senecio*), *Zinnia,* and many others. Ragweed (*Ambrosia*) pollen is a major allergen to many people and causes hay fever.

Members of the sunflower family are mainly herbaceous plants. Their simple leaves are frequently found near the base of the stem with alternate or opposite arrangement. Their flowers are perfect and radially or bilaterally symmetrical. They form inflorescences or heads surrounded by green bracts. Each head consists of two types of florets: **disc florets** in the center and petal-like **ray florets** around the outside (Figure 25.21). Each disc floret consists of many sepals and its five fused petals form a tubular corolla. In comparison, each ray floret consists of many sepals but its five fused petals form a tongue-like corolla. Their florets have two fused carpels and five stamens that are fused to the corolla. Their ovaries are inferior and contain a single ovule. They produce achenes containing a single seed. The development of the composite head with ray and disc flowers is considered to be a significant advancement as insect pollinators are able to pollinate many flowers at one landing when compared to flying to individual flowers.

disc florets
Small flowers found in the center of Asteraceae inflorescence or head; each floret has many sepals and five fused petals form a tubular corolla.

ray florets
Small flowers surrounding the disc florets in Asteraceae inflorescence or head; each floret resembles a petal and has many sepals but its five fused petals form a tongue-like corolla; are usually sterile or pistillate.

Key Terms

parasitic	monocotyledons	florets
epiphytic	monocots	lemma
saprophytic	eudicotyledons	palea
basal angiosperms	eudicots	disc florets
core angiosperms	culm	ray florets
magnoliids	sheath	

Summary

- Angiosperms vary greatly in their size, form, and habitats as well as the way they acquire energy and food. They form a monophyletic group and belong to the phylum Anthophyta.

- Angiosperms and gymnosperms share a similar alternation of generations.

- Angiosperms are distinctly different from gymnosperms. Angiosperms produce flowers and fruits, undergo double fertilization, and develop endosperm in their seeds.

- The four generalized trends in flower evolution are: (1) reduction in number of floral parts, (2) changes in arrangement of floral parts, (3) changes in ovary position, and (4) change from radial to bilateral symmetry.

- Animal pollinators play a key role in the evolution and diversity of flowers and are an important factor in the evolutionary success of flowering plants.

- Angiosperms carry out double fertilization and produce endosperm in their seeds.

- Fruits are a unique feature of angiosperms. Their fruits vary in their shape, color, and size, which allow them and the seeds within to be dispersed by a variety of agents.

- The basal angiosperms (*Amborella*, water lilies, and star anise) are the most primitive groups of angiosperms. The basal angiosperms are distinct from other angiosperms. They do not have vessels in their xylem and their flowers are primitive.

- The core angiosperms are divided into three main monophyletic groups: the magnoliids, the monocots, and the eudicots. The monocots and the eudicots form a sister group with the magnoliids basal to them.

- Magnoliids consist of 20 families, which include magnolias, pawpaws, nutmegs, laurels, winter's barks, and peppers. The magnolia family (Magnoliaceae) consists of two genera: *Magnolia* and *Liriodendron*. Some are popular and important ornamentals.

- Monocots include families such as orchids, grasses, palms, bromeliads, lilies, irises, and many other familiar and important plants. The three monocot families covered in this chapter are the lily family (Liliaceae), the orchid family (Orchidaceae), and the grass family (Poaceae).

- Monocots are distinct from eudicots. Their differences are found in their seeds, roots, stems, leaves, and flowers.

- Eudicots are the largest group of angiosperms, with at least 160,000 species, and include families such as legumes, sunflowers, buttercups, nightshades, mustards, pumpkins, roses, oaks, citrus, berries, cacti, and many other familiar and economically important plants.

- The six eudicot families covered in this chapter are the legume family (Fabaceae), the rose family (Rosaceae), the pumpkin family (Cucurbiaceae), the mustard family (Brassicaceae), the nightshade family (Solanaceae), and the sunflower family (Asteraceae).

Reflect

1. *Angiosperms are distinct from other land plants.* Describe at least five characteristics that distinguish angiosperms from all other land plants.

2. *Animal pollinators are an important factor in the evolutionary success of flowering plants.* Discuss the pros and cons of animal pollination. Explain how plants are adapted for animal pollination.

3. *Basal and core angiosperms.* Name at least three basal angiosperms and three core angiosperms and give at least two main differences between the two groups.

4. *Is it a monocot or eudicot?* You discover a patch of unique flowers along a hiking trail. In order to identify the plant, it will help if you can tell whether it is a monocot or eudicot. Discuss the characteristics you will use to distinguish between monocots and eudicots.

5. *What is a floral formula?* Explain the different parts of a floral formula and how to construct one (see Chapter 6). Discuss whether it is a useful tool or not for identification.

6. *Angiosperms and their uses.* Describe how angiosperms are used by humans. Name four families that are used as popular ornamentals. Name four families that are important food plants. Make sure you can give at least three specific examples from each family.

7. *Plant families.* Describe several plant families and their characteristics.

References

Angiosperm Phylogeny Group (APG III). (2009). An update of the Angiosperm Phylogeny Group classification for the orders and families of flowering plants: APG III. *Botanical Journal of the Linnean Society, 161,* 105–121.

Bergmann, C., & Swearingen, J. M. (2005). *Kudzu.* Plant Conservation Alliance's Alien Plant Working Group. Retrieved December 30, 2018, from http://www.doc-developpement-durable.org/file/Culture-plantes-alimentaires/FICHES_PLANTES/kudzu/Pueraria%20montana%20var%20lobata%20-%20National%20Park%20Service.pdf.

Darwin, F., & Seward, A. C. (1903). *More letters of Charles Darwin, a record of his work in a series of hitherto unpublished letters* (Vol. 2). New York: Appleton.

Evert, R. F., & Eichhorn, S. E. (2013). *Raven biology of plants* (8th ed.). New York: W.H. Freeman.

Food and Agriculture Organization. (1995). *Staple foods: What do people eat?* Retrieved December 30, 2018, from www.fao.org/docrep/u8480e/U8480E07.htm#Staple foods What do people eat.

Hickey, M., & King, C. (1997). *Common Families of Flowering Plants.* Cambridge, UK: Cambridge University Press.

Judd, W. S., Campbell, C. S., Kellogg, E. A., Stevens P. F., & Donoghue, M. J. (2008). *Plant systematics* (3rd ed.). Sunderland, MA: Sinauer Associates.

Pimentel, D. (1993). Habitat factors in new pest invasions. In K. C. Kim & B. A. McPheron (Eds.), *Evolution of insect pests* (pp. 165–181). New York: Wiley.

Pimental, D., McNair, S., Janecka, J., Wightman, J., Simmonds, C., O'Connell, C., et al. (2001). Economic and environment threats of alien plant, animal, and microbe invasions. *Agriculture, Ecosystem and Environment, 84,* 1–20.

Pimental, D., Zuniga, R., & Morrison, D. (2005). Update on the environmental and economic costs associated with alien-invasive species in the United States. *Ecological Economics, 54,* 273–288.

Ross, M. A., & Lemi, C. A. (2009). *Applied weed science including the ecology and management of invasive plants* (3rd ed.). Upper Saddle River, NJ: Pearson Education.

Soltis, D. E., Bell, C. D., Kim, S., & Soltis, P. (2008). Origin and early evolution of angiosperms. *Annals of the New York Academy of Sciences, 1133,* 3–25.

Stebbins, G. L. (1970). Adaptive radiation of reproductive characteristics in angiosperms, I: Pollination mechanisms. *Annual Review of Ecology and Systematics, 1,* 307–326.

Sun, G., Ji, Q., Dilcher, D. L., Zheng, S., Nixon, K. C., & Wang, X. (2002). Archaefructaceae: A new basal angiosperm family. *Science, 296,* 899–904.

Taylor, A. (2013) Australia's tallest trees. Retrieved December 30, 2018, from https://www.australiangeographic.com.au/topics/science-environment/2013/02/australias-tallest-trees/

USDA/Agricultural Research Service. (2009, July 19). Controlling kudzu with naturally occurring fungus. *ScienceDaily.* Retrieved December 30, 2018, from www.sciencedaily.com/releases/2009/07/090719185107.htm.

U.S. Forest Service. (2011). *Celebrating wildflowers—plant of the week—common duckweed* (Lemna minor). Retrieved December 30, 2018, from https://www.fs.fed.us/wildflowers/plant-of-the-week/lemna_minor.shtml.

Chapter 26

Fungi: Friends or Foes of the Green World

© Sergiy Telesh/Shutterstock.com

Learning Objectives

- Distinguish the kingdom Fungi from other living organisms
- Compare and contrast major groups of fungi and give examples
- Discuss the evolutionary relationships between fungi and other major groups of organisms and among fungal groups
- Discuss the importance of fungi to the green world
- Describe the importance of fungi to humans

Most people associate fungi with molds, mildews, grocery-store mushrooms, rusts, and even athlete's foot. Fungi are neither plants nor photosynthetic. In fact, fungi are closely related to animals and are placed in the same supergroup. What are fungi? Why do we study fungi in plant biology? Although fungi are not part of the Green World, they are important partners of many plants and some algae. Some fungi are decomposers helping release nutrients from dead plants and animals. Others, however, are detrimental to plant health and production. A few fungi even cause debilitating human diseases. In this chapter, we investigate unique characteristics of fungi and their importance to plants and humans.

CHARACTERISTICS OF FUNGI

Recall from Chapter 1 that fungi, like all living organisms, have cells, acquire energy and materials, grow and develop, reproduce, respond to stimuli, and adapt to their environment. However, fungi are very different from other living organisms and are placed in their own kingdom, the kingdom Fungi.

Evolutionary Relationships

Kingdom Fungi is one of the six major groups of living organisms (see Table 1.1). Fungi have eukaryotic cell(s) and are classified in the domain Eukarya with plants, protists, and animals. Together with kingdom Animalia and a few closely related protists, they are placed in a eukaryotic supergroup known as Opisthokonta (Figure 1.11). It is generally agreed that kingdom Fungi is monophyletic and closely related to kingdom Animalia. Like animals, fungi are heterotrophic (see Table 1.1) and acquire their nutrients by absorption. Fungi grow on a wide range of **substrates**, which include soil, rotting logs, living tissues, and breads. They secrete enzymes to break down complex organic materials into smaller simpler molecules, which are absorbed. Like animals, fungi store excess food as glycogen, a complex carbohydrate, in their cells.

Based on molecular phylogenetic analyses, members of the kingdom Fungi are now divided into seven major phyla: Microsporidia, Blastocladiomycota, Neocallimastigomycota, Chytridiomycota, Glomeromycota, Ascomycota, and Basidomycota (Figure 26.1) (Hibbett et al., 2007). Members of the phylum Zygomycota are now in other phyla. Slime molds (Myxomycota) and oomycetes (Oomycota) are now considered to be protists and were removed from the kingdom Fungi. It is generally accepted that fungi evolved from aquatic protists. Recent molecular studies suggest the amoeboid protists in the genus *Nuclearia* are the closest relatives to fungi (Steenkamp, Wright, & Baldauf, 2006) (Figure 26.1). Traditional views of fungal phylogeny indicate fungi with flagellated spores such as chytrids (Chytridiomycota) are the sister group to the remaining nonflagellated fungi (Zygomycota, Glomeromycota, Ascomycota, and Basidiomycota). However, recent molecular phylogenetic analyses show the basal phyla Chytridiomycota and Zygomycota are polyphyletic and their phylogenetic relationships remain poorly resolved (Tanabe, Saikawa, Watanabe, & Sugiyama, 2004; James, Porter, Leander, Vilgalys, & Longcore, 2000). Resolving the phylogenetic relationships among the basal groups and how they relate

substrate
1. Reactants that bind to the active site of an enzyme. 2. Organic materials that fungi use as food.

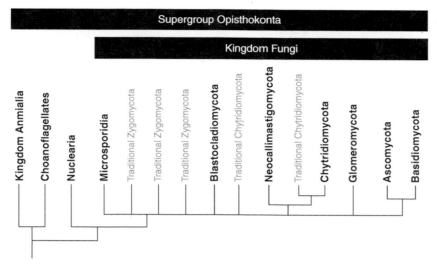

Figure 26.1   Simplified phylogenetic relationships among fungi. Adapted from Brooker, Widmaier, Graham, and Stiling (2011) and Hibbett et al. (2007).

to Ascomycota and Basidomycota are essential to our understanding of how fungi evolved from aquatic living to land-based living.

Unique Cell Structure and Body Form

Although fungi are a monophyletic group, they differ greatly in size, form, and habitats or substrates. Some fungi are aquatic and have flagellated spores while most are terrestrial with nonflagellated spores. Some are microscopic single-cell yeasts with no **fruiting bodies** while others are multicellular with large fruiting bodies such as mushrooms, bracket fungi, and puffballs (Figure 26.2). *Yeast* is a general term referring to all unicellular fungi that reproduce by budding. Except for yeast cells, most fungal

fruiting bodies
Large and conspicuous fungal reproductive structures that are composed of densely packed hyphae growing out of the substrate.

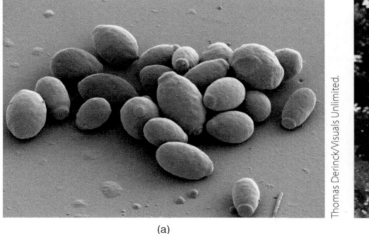

(a)   (b)

Figure 26.2a–b   Fungal cells vary in size and shape: (a) microscopic yeast cells (*Saccharomyces cerevisiae*) as seen in this scanning electron micrograph at 4500×, and (b) a giant puffball mushroom (*Calvatia gigantae*).

hyphae
(singular *hypha*) Thread-like cells of fungi.

mycelium
(pleural *mycelia*) Collection or mass of highly branched filaments know as hyphae.

mycology
Study of fungi.

septate
Fungal hyphae that have cross walls and are divided into many smaller sections.

septa
(singular *septum*) Cross walls that divide most fungal hyphae into many smaller sections or cells.

dikaryons
Fungal cells that have two genetically different nuclei.

aseptate
Fungal hyphae that have no cross walls.

chitin
Nitrogen-containing complex carbohydrate that forms the exoskeletons of many insects and the cell walls of fungi.

cells form thread-like **hyphae**. Hyphae often branch extensively and form a mass called a **mycelium** (Figure 26.3). The words *mycelium* and **mycology** (the study of fungi) are derived from the Greek word, *myke–s*, meaning "fungus." Most fungal hyphae are **septate**, divided by cross-walls or **septa** into many smaller sections or cells. Each cell may have one nucleus or two genetically different nuclei (Figure 26.4a). Fungal cells with two genetically different nuclei are called **dikaryons** (meaning "two nuclei" in Greek) (Figure 26.4b). Fungal hyphae of early-diverging or basal fungi are **aseptate**, without cross walls. Aseptate fungal hyphae have multiple nuclei sharing a common cytoplasm (Figure 26.4c).

Unlike animal cells, which have no cell wall, fungal cells have tough cell walls composed of **chitin**, a complex carbohydrate containing nitrogen. This tough cell wall prevents fungal cells from engulfing food particles and being mobile. In contrast, animal cells take in food particles by enclosing them with their cell membrane and move by changing their cell shape.

Unique Reproduction

Many fungi are able to reproduce sexually and asexually by using microscopic spores. Sexual reproduction allows two genetically different fungi to combine their genetic materials and produce recombinant fungi. These recombinant fungi will then be able to explore and colonize new habitats. In contrast, asexual reproduction produces genetically identical fungi, which allow them to rapidly colonize favorable habitats. Some fungi only reproduce asexually and are classified informally as imperfect fungi or deuteromycetes.

Sexual Reproduction and Fruiting Bodies Like other eukaryotes, sexual reproduction of fungi involves the process of meiosis, the union of

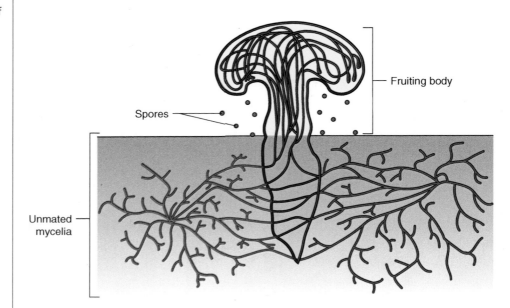

Figure 26.3 Sexual reproduction of two fungi (red and blue) producing a fruiting body that produces and disperses spores. Unmated mycelia are found within the substrate and the fruiting body made of mated hyphae is formed above the substrate.

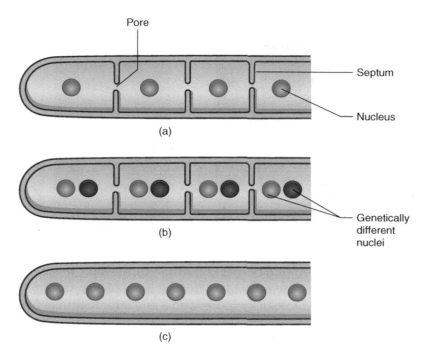

Figure 26.4 a–c   (a) Fungal hyphae can be septate with a single nucleus in each cell or (b) two genetically different nuclei in each cell, called dikaryotic. (c) Some hyphae are aseptate with multiple nuclei sharing a common cytoplasm.

gametes, and the formation of zygotes. The hyphal cells of most fungi function as gametes. There are multiple types of mating hyphae and they differ biochemically. The compatibility of mating types is controlled by particular genes. Two compatible mating hyphae will fuse and undergo plasmogamy and karyogamy. **Plasmogamy** refers to the process when the cytoplasms of the two mating hyphal cells are fused together whereas **karyogamy** refers to the process when the two nuclei are fused together (Brooker et al., 2011). In many fungi, karyogamy does not occur right after plasmogamy. The two haploid nuclei remain separate for a long period of time. Hyphae having two unfused haploid nuclei are called *dikaryons* (Figure 26.4b). Some fungi remain as dikaryons for hundreds of years.

Under favorable conditions, a dikaryotic mycelium may produce a fleshy fruiting body above the substrate (Figure 26.3). All the cells in the fruiting body are dikaryotic. When the fruiting body matures, the two haploid nuclei will fuse together or undergo karyogamy to form zygotes. These zygotes undergo meiosis to produce haploid spores, which are protected by tough chitin walls. These spores are then dispersed by wind, rain, or animals (Figure 26.5) and grow into haploid mycelia.

Asexual Reproduction   Many fungi reproduce asexually by using asexual spores known as **conidia** (*konis-* in Greek means "dust") (Figure 26.6a). Conidia develop at the tips of hyphae and germinate into mycelia when they reach suitable substrate. Some fungi produce asexual spores known as **sporangiospores** within sporangia. These sporangia develop on specialized stalks called **sporangiophores**. Neither compatible mating

plasmogamy
Process when the cytoplasms of two mating hyphal cells fuse together.

karyogamy
Process when the two nuclei of mating hyphal cells are fused together.

conidia
(singular *conidium*) Asexual spores produced at the tips of fungal hyphae.

sporangiospores
Asexual spores produced within fungal sporangia.

sporangiophore
Specialized stalks where fungal sporangia attached.

Figure 26.5a–b   (a) Fruiting bodies of puffball, *Lycoperdon*, release spores into the air to be dispersed by wind, and (b) fruiting body of common stinkhorn, *Phallus impudicus*, produce sticky spores and a foul odor to attract flies to disperse its spores.

budding
Asexual reproduction by yeasts; daughter cells or buds are split from mother cells.

types nor fruiting bodies are required for asexual reproduction. Unicellular yeasts reproduce asexually by **budding** (Figure 26.6b).

DIFFERENT TYPES OF FUNGI

As we mentioned earlier, fungi differ greatly in their range of habitats or substrates and type of reproductive spores. Although fungi are currently divided into seven major phyla, we will investigate the unique characteristics and reproduction of only four: Chytridiomycota, Glomeromycota, Ascomycota, and Basidiomycota.

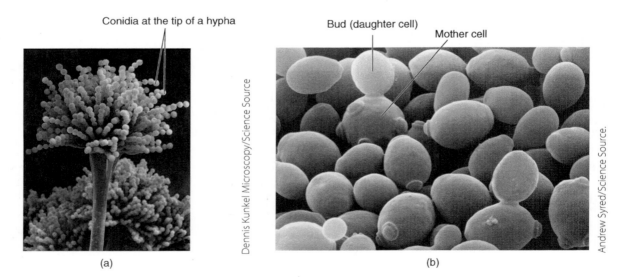

Figure 26.6a–b   Asexual reproduction of fungi: (a) asexual spores, conidia, form at the tip of hyphae (*Aspergillus versicolor*) and (b) budding yeasts (*Saccharomyces cerevisiae*).

Taking a Closer Look at Chytridiomycota

Molecular evidence indicates Chytridiomycota (chytrids) were among the earliest fungi to appear. The polyphyletic chytrids consist of about 790 species and are the simplest fungi (Evert & Eichhorn, 2013). Most are single cells producing hyphae while others produce branched, aseptate hyphae. Chytrids often produce flagellate cells (zoospores and gametes) for reproduction and are the only fungi that produce flagellate cells. The presence of a single, posterior flagellum on reproductive spores links fungi to ancestral forms of protists and animals.

Most chytrids live in aquatic habitats and some in moist soil. Most chytrids are **saprotrophs** or **decomposers**, which obtain their nutrients from dead organisms and help release minerals back into the habitat. Some are parasites of protists and others are pathogens of algae, plants, animals, and even other fungi. The chytrid *Batrachochytrium dendrobatidis*, which causes the skin of amphibians to thicken, has been associated with the decline of frog populations worldwide.

saprotrophs
Organisms that acquire their nutrients and energy by breaking down dead organic matter.

decomposers
See *saprotrophs*.

Taking a Closer Look at Glomeromycota

The phylum Glomeromycota consists of 200 species and is, by far, the most important partners of plants and cyanobacteria (Evert & Eichhorn, 2013). A few glomeromycetes grow only in association with cyanobacteria. Most

Table 26.1   Distinguishing Features among the Four Phyla of Fungi

Phylum Name (Common Name)	Habitat	Type of Hyphae	Reproductive Spores	Ecological Role
Chytridiomycota (Chytrid)	Primarily aquatic or moist soil	Aseptate	Flagellate spores	Mostly decomposers, some plant pathogens, some parasites
Glomeromycota (Arbuscular mycorrhizal fungi)	Terrestrial	Aseptate	Large, distinctive, nonflagellate, multinucleate asexual spores	Form mutually beneficial associations with plant roots (mycorrhizae)
Ascomycota (Ascomycetes)	Mostly terrestrial	Septate	Asexual conidia; nonflagellate sexual spores (ascospores) in sacs (asci) on fruiting bodies	Decomposers, pathogens; many form mutually beneficial associations with algae or cyanobacteria (lichens), or with plant roots (mycorrhizae)
Basidiomycota (Basidiomycetes)	Terrestrial	Septate	Several asexual spores; nonflagellate sexual spores (basidiospores) on club-shaped basidia on fruiting bodies	Decomposers; pathogens; many form mycorrhizae with plant roots; some form lichens with algae or cyanobacteria

glomeromycetes grow only in association with plant roots, forming mycorrhizae, which literally mean "fungus roots." Mycorrhizae are involved in nutrient cycling and even protect plants from stress such as heat, drought, salinity, heavy metal toxicity, and plant pathogens (see Chapters 4 and 13). When early land plants made the transition to land, they faced a number of major environmental challenges (see Chapter 22). Mycorrhizae are found in all plant groups and may have been a key innovation for the acquisition of water and essential nutrients from soil.

The type of mycorrhiza formed by glomeromycetes is called an **<u>a</u>rbuscular <u>m</u>ycorrhiza** (AM) and that is why glomeromycetes are sometimes referred to as AM fungi. Hyphae of AM fungi grow in the spaces between the cell wall and the plasma membrane of root cells. These hyphae form highly branched and bushy arbuscules (from the word "arbor" meaning "tree") (Figure 26.7). Because the hyphae grow inside root cells, this type of mycorrhizae is called **endomycorrhizae** (*endo* in Greek means "inside"). AM fungi have aseptate hyphae and reproduce only asexually by using large multinucleate spores. AM fungi often form associations with the roots of important crop plants such as apple trees, coffee shrubs, legumes, grasses, tomatoes, strawberries, and peaches.

Taking a Closer Look at Ascomycota

It is generally accepted ascomycetes and basidiomycetes are the more recently evolved groups of fungi. The ascomycetes (Ascomycota) consist of 32,300 species that include many familiar and economically important fungi (Evert & Eichhorn, 2013). Many molds that cause food spoilage are ascomycetes. Many major plant diseases such as powdery mildews, brown rot, chestnut blight, and Dutch elm disease are caused by ascomycetes. Yeasts important to the brewers and bakers are mostly ascomycetes. Delicacies such as edible morels and truffles are also ascomycetes. Some ascomycetes form mycorrhizal partnerships with plant roots while most form partnerships with algae or cyanobacteria to form lichens. Asexual fungi such as *Penicillium* and *Aspergillus* are commercially important ascomycetes.

arbuscular mycorrhizae
Hyphae of AM fungi grow in the spaces between root cell walls and plasma membranes, often forming highly branched and bushy arbuscules (from the word arbor meaning "tree").

endomycorrhizae
Hyphae of mycorrhizal fungi grow inside root cells of plants.

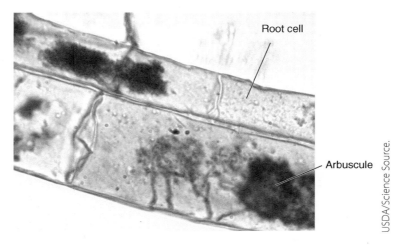

Figure 26.7   Hyphae of glomeromycetes or AM fungi grow in the spaces between cell walls and plasma membranes, forming highly branched arbuscules.

Ascomycetes are either multicellular or unicellular (yeasts). Multicellular ascomycetes produce septate hyphae and reproduce asexually by forming conidia. In comparison, unicellular ascomycetes do not produce hyphae and reproduce asexually by budding. Multicellular ascomycetes can also reproduce sexually when the two nuclei of mated hyphae undergo karyogamy to form zygotes. The zygotes undergo meiosis to form **ascospores** within saclike structures called **asci** (singular *ascus*) (Figure 26.8). Ascomycetes are often referred to as the sac fungi because of their asci. Both ascospores and asci are unique structures that distinguish ascomycetes from all other fungi.

Taking a Closer Look at Basidiomycota

Like ascomycetes, basidiomycetes have dikaryotic hyphae and are classified in the same subkingdom, Dikarya. The basidiomycetes (Basidiomycota) consist of 30,000 species that include many important decomposers and important mycorrhizal partners of plant roots (Evert & Eichhorn, 2013). Basidiomycetes produce a wide range of large fruiting bodies including mushrooms, puffballs, stinkhorns, and bracket fungi. However, two

ascospores
Sexual spores produced within an ascus; found only in ascomycetes.

asci
(singular *ascus*) Saclike structures used by ascomycetes to produce sexual spores (ascospores).

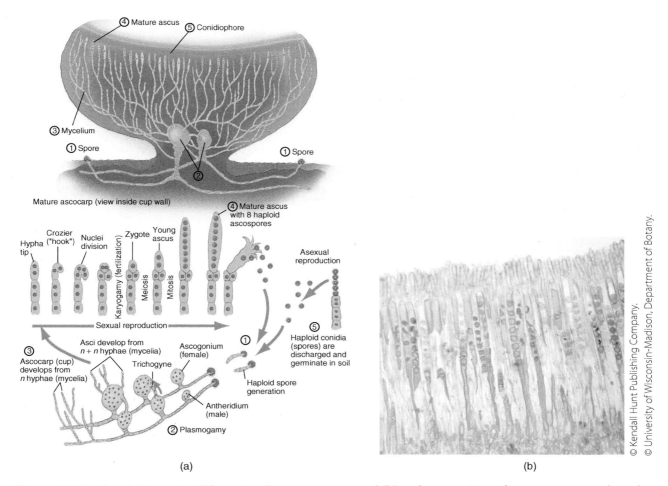

© Kendall Hunt Publishing Company.
© University of Wisconsin-Madison, Department of Botany.

(a)   (b)

Figure 26.8a–b   (a) A typical life cycle of ascomycetes and (b) a closeup view of ascospores produced inside asci of the cup fungus (*Peziza*).

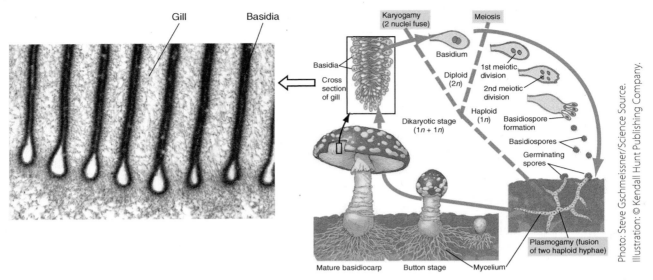

Figure 26.9 A typical life cycle of basidiomycete, showing basidia with basidiospores line the gills of a mushroom.

important groups of plant pathogens, the rusts and smuts, do not produce fruiting bodies. One of the largest and oldest living organisms is a basidiomycete, *Armillaria gallica* (= *bulbosa*). A single *A. bulbosa* fungus found near the Michigan–Wisconsin border was estimated to occupy a minimum of 15 hectares, weigh in excess of 10,000 kg, and be at least 1,500 years old (Smith, Bruhn, & Anderson, 1992). This fungus was recorded in the *Guinness Book of World Records* in 1998 as the biggest and oldest single organism. An individual *Armillaria ostoyae* found in Washington State was estimated to be even bigger, covering 600 hectares, but much younger (between 400 and 1,000 years old) (Anonymous, 1992).

Basidiomycetes reproduce asexually using asexual spores and sexually using basidiospores. **Basidiospores** form externally on **basidia** (singular *basidium*). Each basidium bears four basidiospores. Basidia are found lining the gills or pores of fruiting bodies, which are generally known as **basidiocarps** (Figure 26.9). When the basidiospores mature, they are forcefully discharged from the basidia (Figure 26.5a). Basidiospores produced by stinkhorns are not discharged into the air. Instead, they produce sticky basidiospores and foul odor to attract flies to disperse their spores (Figure 26.5b). Basidiospores, basidia, and basidiocarps are unique structures distinguishing basidiomycetes from other fungi.

basidiospores
Sexual spores produced on a basidium; found only in basidiomycetes.

basidia
(singular *basidium*) Short stalk-like structures found on the gills or spores of fruiting bodies of basidomycetes; used to produce sexual spores, basidiospores.

basidiocarps
Also known as fruiting bodies; found only in basidiomycetes.

IMPORTANCE OF FUNGI TO THE GREEN WORLD

Many fungi are important symbiotic partners of plants, algae, and cyanobacteria. The fungi and their "green" partner live together and form mutualistic associations. Many fungi are important decomposers and are responsible for nutrient cycling in the habitats. Other fungi, however, cause devastating diseases of crop plants and some even human diseases. In the following section, we explore the various relationships fungi have with the Green World.

Mycorrhizae: Partnerships between Fungi and Plant Roots

The word *mycorrhizae* was coined for the symbiotic partnership between plant roots and fungi. The first example was discovered in the mid-1800s when scientists tried to determine how Indian pipes in the genus *Monotropa* survived without chlorophylls (Rayner, 1927). It was determined the roots of *Monotropa* were infected with a fungus that helped to transport nutrients and water to the plant. About 80% of plants form mycorrhizae; however, most plants have chlorophylls and are photosynthetic. Fungi from phyla Glomeromycota, Ascomycota, and Basidiomycota are all known to form partnerships with plant roots. Mycorrhizal fungi greatly increase the surface area of plant roots, enabling greater absorption of water and nutrients. In return, mycorrhizal fungi receive photosynthetic products as their food.

Mycorrhizae are divided into two main types: ectomycorrhizae (*ecto* in Greek means "outside") and endomycorrhizae (Figure 26.10) (see Chapter 4). Ectomycorrhizae are common on many conifers (Figure 26.11) as well as some deciduous plants. Most mycorrhizae are endomycorrhizae or arbuscular mycorrhizae. Experiments on *Prunus cerasifera*, an important rootstock for peaches, have shown the beneficial effects of AM fungi such as *Glomus mosseae* or *G. intraradices* on plant growth (Berta et al., 1995). Compared to peach plants without AM fungi, plants with AM fungi have increased root, stem, and leaf weights. They also have increased leaf area and root length. Uptake of phosphorus and its concentration in leaves are higher in plants with AM fungi than those without AM fungi. The enhancement of plant growth by AM fungi can be explained by their production of growth-regulating substances (Figure 26.12). Arbuscular mycorrhizae fungi have been shown to increase the level of cytokinins, a class of growth hormone, in most plants (Allen, Moore, & Christensen, 1980). In culture, AM fungus such as *G. mosseae* produces gibberellin and cytokinin-like substances, which regulate plant growth and development (Barea & Azcón-Aguilar, 1982). Arbuscular mycorrhizae fungi also interact synergistically with other rhizobacteria to enhance plant growth by making phosphorus and nitrogen readily available (Artursson, Finlay, & Jansson, 2006).

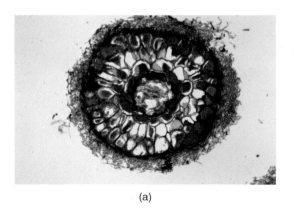

(a)

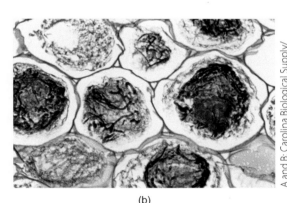

(b)

A and B: Carolina Biological Supply/ Visuals Unlimited.

Figure 26.10a-b   Two main types of mycorrhizae: (a) ectomycorrhizae and (b) endomycorrhizae.

Figure 26.11 Ectomycorrhizae: partnerships between fungi and Douglas fir roots.

Endophytes: Partnerships between Fungi and Plants

endophytes
Fungi that live within the leaf and stem tissues of plants without causing overt signs of tissue damage.

Some fungi live within the leaf and stem tissues of plants either intercellularly (between cells) or intracellularly (within the cell) and are known as **endophytes**. Unlike mycorrhizae, most endophytes live in aboveground plant tissues and do not form external hyphae. Endophytes are ascomycetes that have been found to form mutualistic partnerships with most plant species (Saikkonen, Faeth, Helander, & Sullivan, 1998). Endophytes obtain their nutrients from plants and, in return, produce toxins or antibiotics protecting the plants from herbivores, pathogens, and various abiotic stresses. Endophytes in woody plants such as *Theobroma cacao* (chocolate) significantly decrease leaf necrosis and mortality caused by a *Phytophthora* pathogen (Arnold et al., 2003). Endophytes in grasses have been associated with toxicosis and hoof gangrene of grazing livestock.

Figures 26.12 Arbuscular mycorrhizae improve plant growth and uptake of nutrients and water compared to the control (without AM fungi) on the left.

Endophytes also alter plant physiology and stimulate plant growth and reproduction. Grasses infected with endophytes produce more inflorescences and seeds than uninfected plants. Seedlings from infected seeds germinate and grow faster than seedlings from uninfected seeds (for more on fungal endophytes of grasses, see Clay, 1990). Although endophytic fungi live inside plant tissues, they do not cause any overt signs of tissue damage.

Pathogens: Parasites of Plants

Some fungi living in plant tissues are parasites because they steal nutrients and water from their hosts without providing a benefit. These fungi tend to cause tissue damage and diseases and are considered to be parasites of plants. More than 70% of major crop diseases are caused by parasitic fungi. Many of these fungal diseases have devastating impacts on human society and a few have even changed human history.

Destroyers of Crops *Phytophthora infestans*, the fungal pathogen causing late blight of potato and also tomato, was responsible for the infamous Irish famine during the 1840s. Late blight devastated the potato crops all over Ireland and caused severe famine (Figure 26.13a). More than 1 million people died from starvation or famine-related diseases and 1.5 million emigrated from Ireland (for more on history of late blight, see Levetin & McMahon, 2006; Hudler, 1998). Late blight also altered the course of world history. *Phytophthora infestans* completely devastated potato crops and caused major famine in Germany during the 1910s. Some historians argued the victory of the Allies in World War I was aided by weakened German resolve resulting from this famine (Hudler, 1998).

The rust fungi attack a wide range of crop plants such as cereal crops, coffee, apple, and pine trees and cause devastating plant diseases. Rust fungi are basidiomycetes that do not produce fruiting bodies. *Puccinia graminis* subspecies *tritici* is responsible for stem rust of wheat and is the most important rust pathogen in North America (Figure 26.13b). It is estimated that over 1 million metric tons of wheat are lost annually to stem rust worldwide (Levetin & McMahon, 2006).

Like rust fungi, the smut fungi are major pathogens causing significant losses of grain crops. They are also basidiomycetes lacking fruiting bodies. *Ustilago maydis* (corn smut) is the most important smut pathogen of corn. It forms galls replacing kernels on the ears (Figure 26.13c) and completely devastates corn crops. The corn smut is a widespread disease and can be found wherever corn is grown. It is more common on sweet corn than on other varieties (Levetin & McMahon, 2006).

Destroyers of Trees Fungal pathogens cause devastating diseases in not only crop plants but also native trees. The native American chestnut, *Castanea dentata*, was once the dominant tree in the eastern hardwood forest. Some of the chestnut trees could grow to 120 feet (37 meters) tall and trunk 1.5 meters wide. They had been called the "redwoods of the East." Chestnut wood was ideal for construction because of its straight-grained wood and resistance to decay. It was used to frame and side many of the houses, bridges, and barns built in the 19th century. Its nuts were also an

(a)

(b)

(c)

(d)

Nigel Cattlin/Visuals Unlimited.

Nigel Cattlin/Visuals Unlimited.
Dr. Keith Wheeler/Science Source.

Dayton Wild/Visuals Unlimited.

Mary Thacher/Science Source.
Dr. Jeremy Burgess/Science Source.

Figure 26.13a–d   Devastating fungal plant diseases: (a) late blight of potato caused by *Phytophthora infestans* showing damages to leaves and tubers; (b) wheat rust caused by *Puccinia graminis* subspecies *tritici*, showing fungi broke open on leaf epidermis (see insert); (c) corn smut caused by *Ustilago maydis*; and (d) Dutch elm disease caused by *Ophiostoma ulmi* and the breeding galleries of elm bark beetles (see insert).

important food source for wildlife and people. American chestnut trees were devastated by the fungus *Cryphonectria parasitica*, which causes chestnut blight. In 1904, this fungal pathogen began killing chestnut trees in the New York City Zoological Park in the Bronx. It was later discovered this fungal pathogen was introduced to the "Bronx zoo" on infected Oriental chestnut trees a few years earlier. By 1908, chestnut blight had caused millions of dollars of damage to trees in New York City (Money, 2007). *Cryphonectria parasitica* attacks the living bark of chestnut trees by killing the cambium and preventing transport of water and nutrients (Hudler, 1998). It continues to grow on dead trees until all resources are exhausted. Most of the American chestnut trees from Maine to Georgia have been killed by chestnut blight. This pathogen also attacks other trees such as post oaks, live oaks, shagbark hickory, red maple, and staghorn sumac. It even causes other diseases such as cankers of soybeans and peach trees, stem-end rot of citrus fruits, and bitter rot of grape (for more on history of chestnut blight, see Hudler, 1998; Money, 2007).

Destroyers of Urban Landscapes *Ophiostoma ulmi*, the fungal pathogen causing Dutch elm disease, attacks all elm species but it is most destructive to the native America elm, *Ulmus americana* (Figure 26.13d). America elm was a popular shade tree in the United States and valued for its overarching growth and broad canopy. *Ophiostoma ulmi* attacks the outermost wood and eventually prevents the transport of water and nutrients. Dutch elm disease can spread quickly because the fungal spores are vectored by a complex of elm bark beetles. These beetles form extensive breeding galleries or tunnels between bark and wood of diseased or dying trees. When the new generation of bark beetles emerges, they spread the fungal spores to other healthy elm trees. The fungal pathogen was first discovered in Ohio in 1930 and was probably introduced to North America on contaminated logs. In the decades following its introduction, Dutch elm disease quickly spread across the continent and killed over 77 million elms. Other costs associated with the disease were the millions of dollars spent each year to remove diseased or dead elms (Levetin & McMahon, 2006).

Lichens: Partnerships between Fungi and Algae or Cyanobacteria

Fungi also are important partners to some green algae or cyanobacteria. Lichens are unique associations between a fungal partner, usually a sac fungus, and a photosynthetic partner, usually a green alga (or a cyanobacterium in the tropics). The fungal partner provides its algal partner with water, minerals, carbon dioxide, and also protection. The algal partner in return provides its fungal partner with photosynthetic products such as food and oxygen. Some cyanobacterial partners can even fix or convert atmospheric nitrogen into bioavailable forms (for more details, see Chapter 20).

Lichens have three growth forms: crustose, foliose, and fruticose. There are at least 25,000 species of lichens and they live in diverse types of habitats. They can even be found in habitats where most plants cannot survive such as deserts, the Arctic, the Antarctic, and mountaintops. In these desolate habitats, lichens serve as an important food source for many organisms such as reindeer. Lichens help to improve soil quality for other plants by making nitrogen available. Lichens are also important monitors for air quality because they are very sensitive to pollutants such as sulfur oxide. High levels of air pollutants can severely injure the photosynthetic partners and eventually kill the lichens. Lichens can also concentrate radioactive fallout. Within eight months of the nuclear accident in Chernobyl in 1986, reindeer meat in Norway and Sweden had over 10 times the legal limit of measurable radiation. In some areas, the radiation level of the meat was 20 times or more over the limit (Hudler, 1998).

BOX 26.1 Fungi Cleaning Up the Environment

Bioremediation is the use of organisms for the treatment of pollution. Since the mid-1980s, fungi have been investigated extensively for bioremediation because of their high tolerance and ability to degrade toxic chemicals. Fungi can use a wide variety of substrate and their degradative enzymes can break down a much wider range of pollutants in contaminated soil than other microorganisms. The white-rot fungi *Phanerochaete chrysosporium* and its relatives have remarkable degradative properties and can degrade complex molecules such as lignin, aromatic hydrocarbons, chlorinated organics, pesticides, or dyes. Introducing white-rot fungi to contaminated sites has been shown to effectively clean up contaminants such as pentachlorophenol (PCP) and polynuclear aromatic hydrocarbons (PAHs) (Pointing, 2001). Augmenting autochthonous (native) fungi in historically contaminated sites has also been shown to be a successful bioremediation approach (Vogel, 1996; D'Annibale, Rosetto, Leonardi, Federici, & Petroccioli, 2006). The use of fungi in bioremediation is often referred to as **mycoremediation** (for more on the role fungi play in bioremediation, see Gadd, 2001).

Phytoremediation is a remediation approach using plants and microorganisms inhabiting the rhizosphere (soil environment surrounding the roots) to degrade and contain pollutants such as heavy metals, hydrocarbons, pesticides, and chlorinated solvents in soils, sediments, ground water, and even the atmosphere. In areas with low levels of contaminants, phytoremediation alone may be the most cost-effective remediation approach. In highly contaminated areas, phytoremediation is often used as the final mop-up step in a remediation project. Increased attention has been given to management and manipulation of rhizosphere communities to enhance the efficiency of phytoremediation (see review by Susarla, Medina, & McCutcheon, 2002).

Rhizosphere microorganisms such as bacteria and fungi affect nutrient and water uptake and also toxicity tolerance of plants. Ectomycorrhizal fungi produce densely packed mycelia, sheathing plant roots and protecting them from direct contact with toxic pollutants. The large surface area and cation-exchange capacity of these mycelia help to absorb and degrade toxic compounds and further reduce their toxicity to the host plant (for a review, see Susarla et al., 2002). Arbuscular mycorrhizal fungi have been shown to enhance phytoremediation. Their hyphae help to absorb and contain pollutants such as heavy metals. The heavy metals are then translocated from the roots to the shoots for storage. Since these fungi also enhance plant growth, they also further increase plant capacity to remove and accumulate heavy metals (for a review, see Göhre & Paszkowski, 2006).

mycoremediation
Use of fungi for the treatment of soil pollution.

phytoremediation
Using plants for the reduction or removal of contaminants from soil, water, or air; also known as bioremediation.

Important Decomposers and Biogeochemical Transformers: Recycling Nutrients for Plants

Many fungi are able to grow in a wide range of substrates and break down a wide range of complex organic matters such as cellulose, lignin, lignocellulose, and keratin. As a result, many fungi are important decomposers preventing litter or organic debris buildup. They release CO_2 back into the atmosphere and recycle minerals back to the soil and water. Plants and algae can then take up these nutrients for growth and development. Without decomposers, valuable nutrients would remain bound up in dead organisms and unavailable for living organisms to use. Fungi also play a key role in soil maintenance and stabilization because of their extensive hyphae growth. Their ability to grow on many substrates and form mutualistic partnerships with plants, algae, and cyanobacteria make them key players in biogeochemical transformation and redistribution of elements such as carbon, nitrogen, phosphorus, and other metals in both aquatic and terrestrial environments (Gadd, 2007).

Key Terms

substrates
fruiting bodies
hyphae
mycelium
mycology
septate
septa
dikaryons
aseptate
chitin

plasmogamy
karyogamy
conidia
sporangiospores
sporangiophores
budding
saprotrophs
decomposers
arbuscular mycorrhiza (AM)
endomycorrhizae

ascospores
asci
basidiospores
basidia
basidiocarps
endophytes
myocoremediation
phytoremediation

Summary

- Members of kingdom Fungi have eukaryotic cell(s) and are classified in the domain Eukarya with plants, protists, and animals. Together with kingdom Animalia and a few closely related protists, they are placed in a eukaryotic supergroup known as Opisthokonta.

- Fungi are now divided into seven major phyla: Microsporidia, Blastocladiomycota, Neocallimastigomycota, Chytridiomycota, Glomeromycota, Ascomycota, and Basidomycota. In this chapter, we reviewed the unique characteristics and reproduction of only four: Chytridiomycota, Glomeromycota, Ascomycota, and Basidiomycota.

- Fungi differ greatly in size, form, and habitats or substrates. Most fungal cells form thread-like hyphae whereas yeasts are unicellular and do not form hyphae. Most fungal hyphae are septate while some are aseptate. All fungal cells have cell walls composed of chitin.

- Sexual reproduction of fungi involves two compatible mating hyphae undergoing plasmogamy and karyogamy. Karyogamy does not immediately follow plasmogamy and the two haploid nuclei remain separate for a long period of time. Hyphae with two unfused haploid nuclei are called dikaryons.

- Under favorable conditions, a dikaryotic mycelium may produce a fleshy fruiting body above the substrate. When the fruiting body matures, the two haploid nuclei will undergo karyogamy to form zygotes. These zygotes produce haploid spores, which are protected by tough chitin walls and are then dispersed by wind, rain, or animals.

- Many fungi reproduce asexually by using asexual spores known as conidia. Conidia develop at the tips of hyphae and germinate into mycelia when they reach suitable substrate. Unicellular yeasts reproduce asexually by budding.

- Most chytrids (Chytridiomycota) are saprotrophs or decomposers while some are parasites of protists and others are pathogens of algae, plants, animals, and even other fungi.

- Most glomeromycetes (Glomeromycota) grow only in association with plant roots forming endomycorrhizae while a few grow only in association with cyanobacteria. Glomeromycetes are sometimes referred to as arbuscular mycorrhiza (AM) fungi because their hyphae form highly branched and tree-like arbuscules in the spaces between the cell wall and the plasma membrane of root cells.

- The ascomycetes (Ascomycota) include many familiar and economically important fungi. Some ascomycetes form mycorrhizal partnerships with plant roots while most form partnerships with algae or cyanobacteria to form lichens.

- Ascomycetes are either multicellular or unicellular (yeasts). Unicellular ascomycetes do not produce hyphae and reproduce asexually by budding. In comparison, multicellular ascomycetes produce septate hyphae and reproduce asexually by forming conidia. They can also reproduce sexually and form ascospores within asci. Ascomycetes are often referred to as the sac fungi because of their asci.

- The basidiomycetes (Basidiomycota) include many important decomposers and important mycorrhizal partners of plant roots. Basidiomycetes produce a wide range of large fruiting bodies. However, the rusts and smuts, two important groups of plant pathogens, do not produce fruiting bodies.

- Basidiomycetes reproduce asexually using asexual spores and sexually using basidiospores. Basidiospores form externally on basidia. Basidia line the gills or pores of basidiocarps. Basidiospores, basidia, and basidiocarps are unique structures distinguishing basidiomycetes from other fungi.

- Many fungi are important symbiotic partners of plants, algae, and cyanobacteria. The fungi and their "green" partner live together and form mutualistic associations. Many fungi are important decomposers and responsible for nutrient cycling. Other fungi, however, cause devastating diseases of plants. A few fungi even changed human history.

Reflect

1. *Characteristics of fungi.* Describe characteristics fungi share with all living organisms and those that distinguish fungi from other organisms.

2. *The four phyla of fungi.* Compare and contrast chytrids, glomeromycetes, ascomycetes, and basiodiomycetes. Give examples for each group of fungi.

3. *Fungi reproduce in unusual ways.* Do you agree with this statement? Describe asexual and sexual reproduction by fungi. Compare and contrast these reproductive methods among the four phyla of fungi.

4. *Fungi changed the world.* Describe how some fungi changed the course of human history.

5. *Are fungi friends or foes of the Green World?* Discuss whether fungi are beneficial or detrimental to plants, green algae, and cyanobacteria.

References

Allen, M. F., Moore, Jr., T. S., & Christensen, M. (1980). Phytohormone changes in *Bouteloua gracilis* infected by vesicular-arbuscular mycorrhizae: I. Cytokinin increases in the host plant. *Canadian Journal of Botany, 58*, 371–374.

Anonymous. (1992). The freat fungus. *Nature, 357*, 179.

Arnold, A. E., Mejia, L. C., Kyllo, D., Rojas, E. I., Maynard, Z., Robbins, N., et al. (2003). Fungal endophytes limit pathogen damage in a tropical tree. *Proceedings of the National Academy of Sciences of the USA, 100*, 15649–15654.

Artursson, V., Finlay, R. D., & Jansson, J. K. (2006). Interactions between arbuscular mycorrhizal fungi and bacteria and their potential for stimulating plant growth. *Environmental Microbiology 8*, 1–10.

Barea, J. M., & Azcón-Aguilar, C. (1982). Production of plant growth-regulating sustances by the vesicular-arbuscular mycorrhizal fungus *Glomus mosseae*. *Applied and Environmental Microbiology 43*, 810–813.

Berta, G., Trotta, A., Fusconi, A., Hooker, J. E., Munro, M., Atkinson, D., et al. (1995). Arbuscular mycorrhizal induced changes to plant growth and root system morphology in *Prunus cerasifera*. *Tree Physiology 15*, 281–293.

Brooker, R. J., Widmaier, E. P., Graham, L. E., & Stiling, P. D. (2011). *Biology* (2nd ed.). New York: McGraw-Hill.

Clay, K. (1990). Fungal endophytes of grasses. *Annual Review of Ecology and Systematics, 21*, 275–297.

D'Annibale, A., Rosetto, F., Leonardi, V., Federici, F., & Petruccioli, M. (2006). Role of autochthonous filamentous fungi in bioremediation of a soil historically contaminated with aromatic hydrocarbons. *Applied and Environmental Microbiology, 72*, 28–36.

Evert, R. F., & Eichhorn, S. E. (2013). *Raven biology of plants* (8th ed.). New York: W.H. Freeman.

Gadd, G. M. (2001). *Fungi in bioremediation*. New York: Cambridge University Press.

Gadd, G. M. (2007). Geomycology: Biogeochemical transformations of rocks, minerals, metals, and radionuclides by fungi, bioweathering and bioremediation. *Mycological Research 111*, 3–49.

Göhre, V., & Paszkowski, U. (2006). Contribution of the arbuscular mycorrhizal symbiosis to heavy metal phytoremediation. *Planta, 223*, 1115–1122.

Hibbett, D. S., Binder, M., Bischoff, J. F., Blackwell, M., Cannon, P. F., Eriksson, O. E., et al. (2007). A higher-level phylogenetic classification of the fungi. *Mycological Research 111*, 509–547.

Hudler, G. W. (1998). *Magical mushrooms, mischievous molds*. Princeton, NJ: Princeton University Press.

James, T. Y., Porter, D., Leander, C. A., Vilgalys, R., & Longcore, J. E. (2000). Molecular phylogenetics of the Chytridiomycota supports the utility of ultrastructural data in chytrid systematics. *Canadian Journal of Botany 78*, 336–350.

Levetin, E., & McMahon, K. (2006). *Plants and society*. New York: McGraw-Hill.

Money, N. P. (2007). *The triumph of the fungi: A rotten history*. New York: Oxford University Press.

Pointing, S. B. (2001). Feasiblity of bioremediation by white-rot fungi. *Applied Microbiology and Biotechnology, 57*, 20–33.

Rayner, M. C. (1927). *Mycorrhiza: An account of non-pathogenic infection by fungi in vascular plants and bryophytes*. London: Wheldon and Wesley.

Saikkonen, K., Faeth, S. H., Helander, M., & Sullivan, T. J. (1998). Fungal endophytes: a continuum of interactions with host plants. *Annual Review of Ecology and Systematics, 29*, 319–343.

Smith, M. L., Bruhn, J. N., & Anderson, J. B. (1992). The fungus *Armillaria bulbosa* is among the largest and oldest living organisms. *Nature, 356*, 428–431.

Steenkamp, E.T., Wright, J., & Baldauf, S.L. (2006). The protistan origins of animals and fungi. *Molecular Biology and Evolution 23*, 93–106.

Susarla, S., Medina, V. F., & McCutcheon, S. C. (2002). Phytoremediation: An ecological solution to organic chemical contamination. *Ecological Engineering, 18,* 647–658.

Tanabe, Y., Saikawa, M., Watanabe, M. M., & Sugiyama, J. (2004). Molecular phylogeny of Zygomycota based on EF-1α and RPB1 sequences: Limitations and utility of alternative markers to rDNA. *Molecular Phylogeneics and Evolution, 30,* 438–449.

Vogel, T. M. (1996). Bioaugmentation as a soil bioremediation approach. *Current Opinion in Biotechnology, 7,* 311–316.

SECTION VI

THE GREEN WORLD IN THE WEB OF LIFE

Chapter 27   Ecosystems and Biomes

Chapter 28   Dynamics of Plant Communities and Populations

Chapter 29   Plants as Food, Commercial Products, and Pharmaceuticals

Ecosystems and Biomes

Source: Ursula Schuch.

Learning Objectives

- Define ecology
- Outline global climate patterns and their main characteristics
- Describe the effect of mountains and oceans on climate
- Explain the main characteristics of a biome and how they are distinguished
- Identify three major biomes of the world and discuss their major characteristics
- Compare the arctic tundra, alpine tundra, and boreal forest biomes
- Discuss why some biomes are desirable for human settlements
- Discuss the general principles of species richness in different biomes

Have you ever wondered why the rainforest is found in the tropics close to the equator and other moist locations (Figure 27.1)? Why most of the conifer trees are located in the Northern Hemisphere in cool climates? Why certain animals are living in habitats with certain types of plants? The field of ecology explores these and other related questions. Ecology investigates the relationship between organisms and their environment; how plants, animals, and other organisms interact in nature and in environments modified or built by humans. Ecosystems include all living organisms, and the physical environment such as rocks, soil, water, and the atmosphere. The biosphere is the largest ecosystem and includes all areas on Earth populated by organisms. Ecologists study the distribution and quantity of organisms in ecosystems, their interactions, and the factors limiting and supporting their inhabitants. These studies can encompass different levels ranging from a limited geographic area of a riparian woodland to an area as vast as the Sahara desert. Chapter 28 takes a closer look at the organizational levels of ecosystems and the increasingly complex interactions between individual organisms, populations, and communities.

Climate is the most important factor determining where species can live and how many of a species can support themselves in a particular environment. In this chapter, we examine first how global climate patterns shape the physical environment in different locations. Then, we examine the 14 major plant communities' characteristic for their distinct vegetation such as desert, tropical rain forest, and coniferous forest.

GLOBAL CLIMATE PATTERNS

climate
Long-term weather conditions based on temperature, rainfall, humidity, and wind.

The **climate** in a location is affected by global climate patterns. Factors shaping the climate are sun exposure and seasonality, global wind patterns, and the height, expanse, and orientation of mountain ranges (Raven, Evert, & Eichhorn, 1999). Climate in different regions on Earth depends primarily on the amount of sunlight an area receives. Areas near the equator receive

(a)   (b)

Figure 27.1a–b   (a) Vegetation in the temperate rainforest of the Olympic Peninsula in Washington is adapted to high rainfall and temperatures above freezing and up to 81°F (27°C) and features conifer trees covered with epiphytic club mosses, ferns, and moss on the ground. (b) The warmer, moist rainforest of Bolivia has similar lush vegetation with ferns and epiphytes. Plants in both rainforests bear resemblance, but species are different.

the most sunlight in a year while areas near the poles receive no direct overhead sunlight at all (Figure 27.2). Seasons are caused by the earth's tilt and the resulting change in day length. The axis of the earth is tilted by 23.5°. As earth moves around the sun over the course of a year, it gradually increases exposure on the northern hemisphere from March to June, reversing from June to September. On March 20 the sun is overhead at the equator, and on June 21 it is overhead at 23.5° N latitude. The sun then moves back toward the equator and is again overhead on September 21. By December 21 the sun is overhead at 23.5° S latitude, bringing summer to the Southern Hemisphere.

Water has high **specific heat** and changes temperature more slowly than land. With more land in the Northern Hemisphere and more oceans in the Southern Hemisphere, heat distribution is irregular around the globe. This leads to atmospheric circulation patterns affecting ocean currents and temperatures (Figures 27.3).

Global patterns of air circulation are called *cells*. The **Hadley cell** is located between the equator and 30° latitude (Figure 27.3). There are two more cells located toward the poles in the Northern and Southern Hemisphere. The Hadley air circulation pattern is responsible for the moist equatorial tropics and the desert areas 30° to the north and south of the equator. Energy from the sun heats up the tropics, resulting in warm air rising. As it moves higher, moist air cools and loses its ability to hold as much moisture, leading to ample rainfall. The cooler air moves toward the poles and as it sinks it warms and absorbs moisture, resulting in the belt of deserts north and south of the equator. The warm air is then pushed again toward

specific heat
Amount of heat required to raise the temperature of 1 gram of a substance by 1.8°F (1°C).

Hadley cell
Atmospheric air circulation pattern where warm, moist air rises near the equator, resulting in rain and then flows toward the poles, sinking at about 30 degrees latitude north or south.

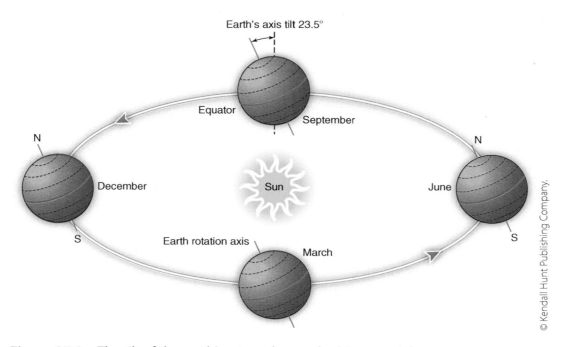

© Kendall Hunt Publishing Company.

Figure 27.2   The tilt of the earth's axis and annual orbit around the sun create seasons. The sun is directly overhead the equator on March 20 and September 21. The Northern Hemisphere is most exposed to the sun in June and the Southern Hemisphere receives the most direct sun in December.

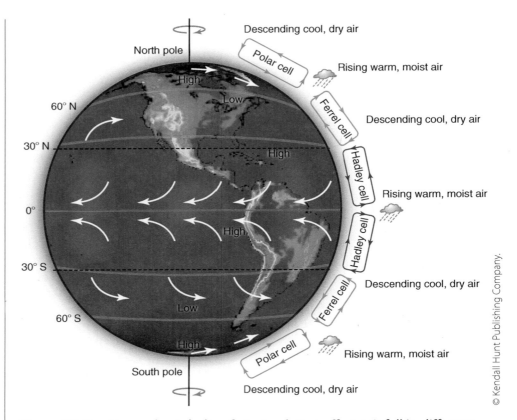

Descending cool, dry air

North pole

Polar cell

Rising warm, moist air

High

Low

60° N

Ferrel cell

Descending cool, dry air

30° N

Hadley cell

High

Rising warm, moist air

0°

Hadley cell

High

30° S

Ferrel cell

Descending cool, dry air

Low

60° S

High

Polar cell

Rising warm, moist air

South pole

Descending cool, dry air

© Kendall Hunt Publishing Company.

Figure 27.3 Atmospheric belts of air circulation affect rainfall in different areas. The tropics are wet because warm air rises and releases moisture. High pressure systems form around 30° latitude where most of the world's deserts are found.

the equator and the cycle repeats. Adjacent cells follow similar patterns, resulting in higher rainfall around 60° north and south of the equator. In the Hadley cell trade winds move air in an easterly direction because of the earth's rotation direction. At the equator where the cells meet, winds are often calm. Cells adjacent to the Hadley cells have wind movement in a westerly direction.

Oceans have a moderating effect on regional climate because water can store large amounts of heat. In summer, ocean waters can absorb heat from the warmer atmosphere, providing relief from high temperatures in coastal areas. In winter, the ocean can release heat to warm up the colder air, moderating freezing temperatures. Proximity to these bodies of water greatly impacts weather patterns.

Mountain ranges affect wind and rainfall patterns because they alter the directions of air movement. In western North America, wind blows prevalently from the Pacific Ocean toward the east and rises first over the Coast Range and then over the Cascades. As moist air coming from the ocean rises, it cools and drops rain on the western slopes of the mountains (Figure 27.4). The eastern slopes are in the **rain shadow** and receive much less rainfall. This pattern is repeated over the Cascades. Wind moving further east will pick up some moisture and rise over the Rocky Mountains. Due to the drier inland atmosphere less rain will fall on the western slopes of the Rocky Mountains than the western slopes of the Cascades.

rain shadow
Dry back side of a mountain, located behind the ridge on the side of the mountain where warm, moist air rises and rain is deposited before air crosses the ridge.

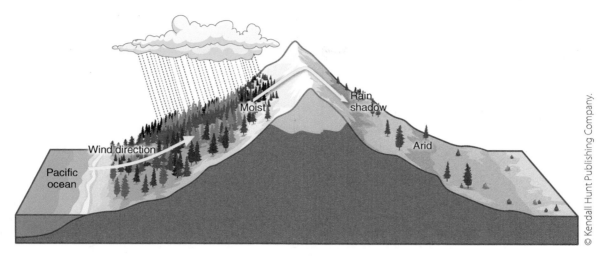

Figure 27.4 Mountain ranges affect rainfall patterns and create a dry environment in the rain shadow.

BIOMES OF THE WORLD

Biomes are communities of plants, animals, and other organisms inhabiting an area characterized by distinct vegetation. Biomes are **biogeographical** regions where climate and especially temperatures and rainfall distinguish the different biomes. Current classification identifies 14 unique terrestrial biomes worldwide (Pidwirny, Draggan, McGinley, & Frankis, 2007; Olson et al., 2001) (Figure 27.5). General temperature and rainfall conditions for major biomes are shown in Figure 27.6 (Whittaker, 1975). Aquatic

biogeography
Geographical distribution of species, populations, and ecosystems on Earth.

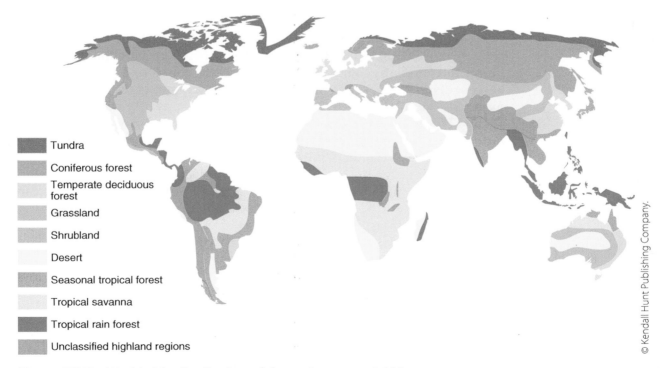

Tundra
Coniferous forest
Temperate deciduous forest
Grassland
Shrubland
Desert
Seasonal tropical forest
Tropical savanna
Tropical rain forest
Unclassified highland regions

Figure 27.5 Worldwide distribution of the major terrestrial biomes.

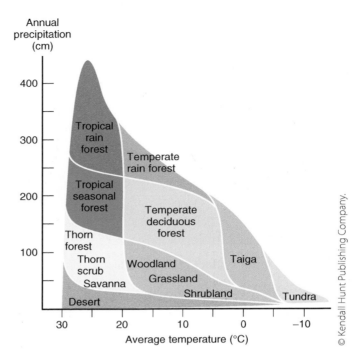

Figure 27.6 General characteristics of rainfall and temperature of major terrestrial biomes.

environments are distinguished based on freshwater or saltwater. Their biomes are further classified by depth, proximity to land, water flow, and organisms inhabiting them. Some aquatic and terrestrial environments interface at coastlines and lakeshores. Next we concentrate on the terrestrial biomes and their characteristic plants.

Plant and animal life on earth has evolved to adapt to diverse climatic conditions. Comparable ecosystems have evolved in similar climates found in different geographic locations. Different communities of plants have evolved based on weather factors such as temperature and rainfall, moisture, wind, and sunlight. Average temperatures between 68 and 86°F (20 and 30°C) favor tropical vegetation, more lush under high rainfall conditions, turning **xeric** in water-limiting environments (Figure 27.6). Biomes supporting tree growth require a minimum growing season and rainfall below which only shrubs, grasses, and other smaller plants can be supported. For example, plants from the desert and xeric shrubland biome can withstand long periods without rain due to their morphological and physiological adaptations. This biome is found on five continents and plants have similar adaptive features, although they are often unrelated.

Biomes are not uniform and can vary significantly, especially when they cover vast areas of land. The map of the world biomes has a coarse scale and generalizes over large areas where regional and local differences in altitude, climate, and soil can change the characteristics of the vegetation. Geographically limited biomes such as mangroves or flooded grasslands may be interspersed with other biomes. Even within the broadly classified biomes there is diversity. Adjacent biomes are often not separated distinctly but may be connected by areas of transition where vegetation types are mixed.

xeric
Dry habitat, adapted to little moisture.

Figure 27.7a–d   (a) An aerial view of the treeless tundra with ponds thawed in summer; (b) bearberry (*Arctostaphylos alpine*), (c) lichen and grass, and (d) cushion plant (*Laretia compacta*) in alpine tundra in Chile.

The following discussion describes eight major biomes of the world: arctic tundra, boreal forest, temperate coniferous forest, temperate broadleaf and mixed forest, temperate and tropical grasslands, savanna and shrubland, deserts and xeric shrubland, and the tropical and subtropical moist broadleaf forest.

ARCTIC TUNDRA

The arctic tundra is the terrestrial biome closest to the poles not covered permanently with ice. This biome covers large areas of land north of 65° latitude in North America and Eurasia. Very little land is occupied by arctic tundra in the Southern Hemisphere, which supports only a few plant species in Antarctica. During the growing season, which lasts only up to three months, frost can occur at any time. The soil in most of the tundra is frozen year-round, a condition called **permafrost**. In summer the snow melts and soil near the surface thaws, resulting in marshy, waterlogged conditions. Those areas are covered with bogs and ponds in the summertime because of the lack of drainage (Figure 27.7a). Rainfall in the tundra is low with less than 9.8 inches (25 cm) per year, but there is little evaporation because of the cool temperatures. Average annual temperatures in the brief summer are between 37 and 54°F (3 and 12°C), and in winter around –29°F (–34°C) (Pidwirny et al., 2012).

The number of plant species in the tundra biome is low, with about 2,000. No trees grow in the tundra. Low-growing shrubs such as willows (*Salix* sp.),

permafrost
Soil frozen year-round where only a small portion of the profile near the surface thaws briefly in summer.

birch (*Betula* sp.), blueberry (*Vaccinium* sp.), Labrador tea (*Rhododendron* sp.), and bearberry (*Arctostaphylos alpina*) grow no taller than grasses that are found in drier parts (Figure 27.7b–c). Other plants include sedges, mosses, lichen, and many small perennial flowering plants (Figure 27.7c). The topsoil is very shallow and extremely fragile. The root zone of plants is limited by permafrost and wet conditions. Soils are nutrient poor because of low microbial activity in the cool environment. Many plants carry the majority of their biomass underground as storage bulbs or succulent roots (Raven et al., 1999).

Alpine tundra is found above the tree line in mountain areas. In the tropics, alpine tundra occurs only at high elevations above 14,764 feet (4,500 m). In North America, most alpine tundra is located in the western mountains 3,281 feet (1,000 m) lower than in the tropics. In the eastern United States this biome is found on Mount Washington in New Hampshire at elevations around 5,906 feet (1,800 m). The vegetation is similar to the arctic tundra but permafrost is absent and drainage is better. The alpine meadows are covered with low-growing grasses, sedges, and herbs as well as perennial plants flowering spectacularly in the brief summer. Cushion plants (Figure 27.7d) form low-growing mats of dense woody plants that grow extremely slow, often less than one millimeter per year, and can be hundreds of years old. Summer days can be warm, but nights are always cold. Extreme weather with high winds, snow, and hail can occur at any time.

alpine tundra
Treeless biome similar at high elevations above the tree line. See *tundra*.

BOX 27.1 Shift in Biomes Due to Warmer Climate

Global warming has increased the earth's average temperature by 1.4°F (0.8°C) in the last 130 years, with accelerated warming since the 1970s. Temperature changes vary regionally, with northern latitudes and higher elevations experiencing more warming than other areas. Receding or disappearing glaciers and arctic ice attest to the changes. Phenological observations revealed that plants started growing and flowering earlier in spring and continued later into the fall (Walther et al., 2002). When researchers measured the effects of warming climate on plants, they found in the 18 years before the year 2000 the growing season in North America had increased by 12 days and in Eurasia by 18 days (Zhou et al., 2001). Studies in the tundra in northern Europe showed that plants respond to climate warming, but not all plants respond the same (Aerts, Cornelissen, & Dorrepaal, 2006). Shrubs and herbs can move further north and can start to grow earlier in spring and longer in fall. Along with the phenological changes, substantial increases in nitrogen mineralization were measured when some locations were artificially warmed to simulate further temperature increases. The higher temperatures lead to greater availability of nitrogen which will further change the dynamics of the tundra ecosystem.

Early spring flowering plants are most affected and respond to a greater degree to earlier, warmer temperatures, while plants flowering in mid and late summer show little response (Walther et al., 2002). A shift toward northern latitudes and higher elevations in response to the warmer climate has been documented for many plant species. In Antarctica, the two species of higher plants have increased significantly in number and mosses have spread extensively on previously bare ground. Global warming influences both the composition of communities and the interaction among members. The diversity in different biomes and the subtle differences in climates make exact predictions of how individual plant species will adapt to the warmer climate challenging. The question of how this will change the interaction between different plant species and other organisms is even more uncertain.

BOREAL FOREST

The boreal forest or **taiga** is located between 50°N and 70°N latitude, south of the arctic tundra and north of the temperate coniferous forest or grasslands. The boreal forest extends across Alaska, most of Canada, Northern Europe, and Russia, comprising the largest terrestrial biome and one third of the world's forest area. Summers are short with a growing season up to 4 months. Temperature fluctuations are extreme with summers warmer than in the arctic tundra. Annual rainfall varies from 9.8" (25 cm) to over 39.4" (100 cm) in boreal forests of Western North America, creating a moist, cool to cold climate where little water is lost to evaporation. The boreal forest is dominated by a few species of conifer trees, which can tolerate the environment (Figure 27.8a). Soils are very high in organic matter and the forest floor is covered in a thick layer of litter. Decomposition rates are slow and soils are often acidic and have mineral deficiencies. Some areas have waterlogged soils and ponds are often found in summer when soils thaw in colder regions (Figure 27.8b). Fires caused by lightning or humans are common and vital to the regeneration of the boreal forest. Fires release nutrients from vegetative litter and increase growth of subsequent vegetation. Fires change the vegetation composition and contribute to biodiversity of the flora and fauna in the ecosystem (Tyrell, 2018).

The main genera of trees growing in the boreal forest are spruce (*Picea*), pine (*Pinus*), fir (*Abies*), larch (*Larix*), birch (*Betula*) and aspen (*Populus*). The

taiga
Wide belt of conifer forest located south of the polar ice. See *tundra*.

(a)

(b)

A and B: © Pi-Lens/Shutterstock.com

Figure 27.8a–b   (a) Boreal forest in Yukon Territory, Canada. (b) Pond from thawing permafrost in the boreal forest.

evergreens are well adapted and have the ability to resume photosynthesis whenever temperatures are favorable. Understory vegetation is sparse in the densely shaded forest with low growing shrubs of wild berries and some herbaceous species. Productivity is low as trees grow slow and live for a long time. Common conifers in the North American boreal forest include black spruce (*Picea mariana*), white spruce (*Picea glauca*), jack pine (*Pinus banksiana*), lodgepole pine (*Pinus contorta*), tamarack (*Larix laricina*), white cedar (*Thuja occidentalis*), and balsam fir (*Abies balsamea*). Some species emerge especially after a disturbance such as timber harvesting, insect outbreaks, storm throw, or fire. These early colonizers include deciduous birch and aspen trees. Cones of jack pine and lodgepole pine are covered with a waxy coating and rely on fire to discharge their seeds.

Boreal forests differ in species composition and fire ecology based on their geographic location. Boreal forests influenced by the moderating climate of the Pacific or Atlantic oceans are generally more productive due to greater availability of moisture and less extreme temperatures. Interior boreal forests withstand extreme cold and dry conditions and are often afflicted by widespread intense fires. Trees in boreal forests in Eurasia have a longer lifespan, 400–600 years on average, than boreal forest species in North America. They live on average about 150–200 years, which is primarily due to the more frequent and widespread wildfires (Tyrell, 2018).

TEMPERATE CONIFEROUS FOREST

The temperate climate with annual temperatures from 41°F to 68°F (5°C to 20°C and higher rainfall supports the temperate coniferous forest (Figure 27.9a). In North America, two distinct areas are in the Pacific Northwest and in the Southeast (Pidwirny, Draggan, McGinley, & Frankis, 2007, Revised on January 2012). This biome is also found in Europe, Asia, and South America with trees of similar phenotypes to those found in North America, but different species. Temperate coniferous forests produce the highest amount of biomass of any terrestrial biome. They can consist entirely of conifers, or they can be a mix of broadleaf evergreen and conifers. In general, these forests are composed of trees in the overstory and of small shrubs and herbaceous plants in the understory.

Coniferous forest in the Northwest close to the ocean enjoy ample rainfall with more than 250 cm and boast highly productive evergreens such as Douglas fir (*Pseudotsuga menziesii*) (Figure 27.9a), red cedar (*Thuja plicata*), western hemlock (*Tsuga heterophylla*), sitka spruce (*Picea sitchensis*), and redwood (*Sequoia sempervirens*). This area is sometimes referred to as temperate rainforest.

Further inland on the east side of the mountains, the climate becomes more continental and drier supporting forests dominated by ponderosa pine (*Pinus ponderosa*), Engelmann spruce (*Picea engelmannii*), and lodgepole pine (*Pinus contorta*). Summer drought and fires occur regularly. Giant sequoias (*Sequoiadendron giganteum*), the largest organisms in the world, grow in a small area on the western Sierra Nevada in California. In the Southeastern United States, low productivity conifer forests grow on nutrient poor, sandy soils. Typical plants adapted to this environment are pitch pine (*Pinus rigida*), longleaf pine (*Pinus palustris*), and slash pine (*Pinus elliottii*), all adapted to the frequent fires in this habitat.

Figure 27.9a–c   (a) temperate conifer forest of Douglas fir (*Pseudotsuga menziesii*); (b) temperate mixed forest in fall; (c) bluebells (*Hyacinthoides*) in the broadleaf forest in England flower in early spring before the canopy of emerging leaves casts too much shade on the forest floor.

In the southern hemisphere, the temperate coniferous forest covers the lower elevations of the south-central Andes in Chile into Argentina, and the monkey puzzle tree (*Araucaria araucana*) is the dominant conifer. The Valdivian temperate rain forest extends from the Pacific coast to the base of the Southern Andes and the Patagonian cypress (*Fitzroya cupressoides*) is one example of the characteristic large conifers growing in this cool, moist climate.

TEMPERATE BROADLEAF AND MIXED FORESTS

The temperate broadleaf and mixed forest is also known as the temperate deciduous forest. This biome is located north and south of 30° latitude. The climate is characterized by warm summers, cold winters, and relatively high amounts of rainfall throughout the year (Figure 27.6). In many areas, rainfall is distributed throughout the year, although in some regions a dry season can occur in summer or winter. This temperate climate has less annual temperature fluctuation than the biomes located further north and supports many species of deciduous broadleaf trees. The agreeable climate has led to much deforestation for urban development and agriculture.

Temperate broadleaf and mixed forests dominate in the northern hemisphere and cover large areas of the Northeastern United States and Southeastern Canada, Western Europe into Western Asia, and Eastern Asia including Japan. In the southern hemisphere, this biome is found in the southern coastal region of South America, along the eastern coast of Australia and in Tasmania and New Zealand. In Europe and Eastern North

America, deciduous trees dominate this biome, although conifers are mixed among the broadleaf trees (Figure 27.9a). Broadleaf tree species include maple (*Acer*), beech (*Fagus*), oak (*Quercus*), elm (*Ulmus*), poplar (*Populus*), and willow (*Salix*), and conifer species include pine (*Pinus*), fir (*Abies*), and spruce (*Picea*). In the southern hemisphere, broadleaf deciduous trees dominate the overstory of this biome almost exclusively. Dominant large trees include many Eucalyptus species in the warmer areas and beech (*Nothofagus*) in the cooler areas. Commonly up to 25 different tree species are found, generally more in warmer locations and fewer in cooler regions of this biome. Distinct seasons are marked by the drop of leaves in fall and flowering of the rich herbaceous understory plants in spring before the leaves of deciduous trees and shrubs cast a dense shade (Figure 27.9c). Smaller trees form a subcanopy under the taller trees. The large quantities of litter in fall are quickly decomposed into humus. The soil in this forest is nutrient rich and supports diverse plant communities.

TEMPERATE GRASSLANDS, SAVANNAS, AND SHRUBLANDS

Temperate grasslands are located in latitudes similar to the temperate deciduous forest. They are often found between deserts and temperate forest. Transition zones from grassland have features of either forest where rainfall is higher or desert in arid climates. They dominate the interior of North America, where they are known as prairie (Figure 27.10a and b), and Eurasia where they are named steppe. In South America, these grasslands are known as pampas and in Southern Africa they are called veld. Climate of grasslands has pronounced seasons with warm to hot summers and cool to cold winters similar in temperature to the temperate deciduous forest. Temperatures can range from 104°F (40°C) in summer to −40°F (−40°C) in winter. Annual rainfall of 50–90 cm largely falls in spring and early summer and is followed by an extended dry period. In grasslands receiving more moisture, trees are generally eliminated by recurring fires (Figure 27.10c).

The grass blanketing the prairies of the Midwestern United States before the arrival of settlers was bluestem (*Andropogon* spp.) (Figure 27.10d) growing in dense stands up to 6.6' (2.0 m) (Pidwirny et al., 2007, Revised on January 2012). Further west in drier conditions, low growing buffalo grass (*Buchloe dactyloides*) and other drought adapted grasses are common. Some herbaceous plants and a few trees are found along streams. Soils in this biome are extremely deep and fertile due to the abundant biomass of grass roots, weathered rock, and rainfall which does not promote leaching of nutrients from the soil. Few natural grasslands remain because their ideal soil, flat topography, and climate conditions have led to agricultural or grazing uses (Figure 27.10e). Dominant grasses of the grasslands surrounding the Gobi Desert in Mongolia include feather grass (*Stipa*) species. The majority of the prairie in central North America and the steppe in Eurasia have been converted to crop land and are utilized to grow the majority of wheat, corn, and other grains important for our food supply. In other areas, native grasses have been replaced with introduced species and grasslands serve as pasture or for urban development.

Savannas are grasslands with trees spaced further apart, either solitary or in groups and grass growing between the trees (Figure 27.11a). **Shrublands**

savanna
Grasslands interspersed with some trees.

shrubland
Grasslands with widely spaced shrubs.

Figure 27.10a–e   (a) A bison herd in the grasslands of Oklahoma; (b) wildflowers on a prairie; (c) fires are common in grasslands; (d) Big bluestem grass once dominated the natural prairie in central North America; (e) many accessible grasslands have been converted to agricultural production.

are similar with shrubs widely spaced and grass growing between the shrubs. Woodlands are forests where the trees are spaced apart so they do not form a closed canopy (Figure 27.11b). Savannas (Figure 27.11c) and their related biomes (Figure 27.11d) are found in geographic areas adjacent to deserts, between grasslands and temperate deciduous forest, or in colder areas between grasslands and the boreal forest. Rainfall is lower than the adjacent forest biomes and higher than the grasslands. These areas are often exposed to a seasonal drought when plants go into dormancy and drop their leaves. In Australia, several eucalyptus (*Eucalyptus*) species and some acacias (*Acacia*) are trees growing in the savanna.

In California, the **chaparral** (Figure 27.11a) is a shrubland with shrubs up to 9.8′ (3.0 m) tall. The Mediterranean climate is characterized by mild winter weather with rains and hot, dry summers. Growth occurs primarily

chapparal
Vegetation type found in a Mediterranean climate consisting of broadleaf often thorny shrubs able to withstand long summer droughts.

Figures 27.11a–d (a) Microclimate separates the north-facing grassland savanna and the south-facing chaparral in California; (b) oak woodland savanna in spring; (c) marshy grassland in Florida; (d) Mediterranean scrub is called chaparral in California and *macquis*, as shown here in Spain.

during the cool, moist winter and many plants enter dormancy in the dry summer. Dead plant material accumulates and feeds the frequent wildfires. Many plants are adapted to fire and seeds and underground storage organs will readily resprout. Shrublands are also abundant near the Mediterranean Sea where they are called maquis (Figure 27.11d), in the interior of Chile where they are known as matorral and in South Africa where they are named fynbos (Raven, Evert, & Eichhorn, 1999).

Savannas are found at higher elevation in the drier climate of the Southwestern United States. The semi-arid climate with 9.8" to 19.6" (25 to 50 cm) annual rainfall supports the slow-growing pinyon pine (*Pinus edulis*) and small juniper trees (*Juniperus* spp.).

The tropical and subtropical grasslands, savannas, and shrublands are often just called tropical savanna. This biome is probably best known by the large mammals like giraffe, buffalo, zebra, gazelle, and elephants that graze these areas in Eastern Africa. Sizable areas of tropical savanna are located also in Australia and South America. Perennial grasses grow up to 3.3'–6.6' (1–2 m) and the sparse trees do not grow taller than 32.8' (10 m) because of the year-round hot temperatures and the dry season which limits growth. Annual rainfall during the wet summer and dry winter season amounts to a total of 39.4" to 59.1" (100 to 150 cm). Monthly temperatures vary less than in the desert biome, but more than in the rainforest.

DESERTS AND XERIC SHRUBLANDS

The desert and xeric shrubland biome is found primarily around 30°N and 30°S latitude. Within the desert biome, different communities have developed based on temperature and rainfall. The desert biome is very diverse as is evident from the broad range of temperatures it occupies, ranging from a few degrees below freezing to 86°F (30°C) (Figure 27.12). The Gobi desert in Asia, featuring primarily rock, is a cold desert with an annual mean temperature just around freezing in some locations and annual extreme temperatures of −40°F (−40°C) and 120°F (49°C). Daily temperature fluctuations in the desert are extreme because the dry land with little vegetation does

Figure 27.12a–f   (a) Mohave desert with Joshua tree; (*Yucca brevifolia*) (b) the Sonoran desert is one of the lushest desert biomes and is characterized by the saguaro cactus (*Carnegia gigantea*) and palo verde tree (*Parkinsonia*), (c) ocotillo (*Fouquieria splendens*), cacti and other drought-tolerant plants. (d) Quiver trees (*Aloe dichotoma*) in the rocky Namib desert are well adapted. (e) Bunch grasses in the Australian desert are spaced apart because of plant competition for water. (f) Many desert wildflowers are ephemerals.

not store much heat. Hot and dry deserts are found in North America's west and include the Mohave, Sonoran (Figure 27.12a–c), and Chihuahuan Desert, and a cold desert, the Great Basin. Desert biomes are also found in Australia, Africa, South and Central America, and Southern Asia. The Sahara desert is the largest desert. Australia is the driest continent inhabited by humans and covered with 70% of desert.

Climate is typically dry with very little rain except for the North American deserts, which receive about 10.2" (26 cm) annually. Some deserts receive as little as a few centimeters of rain a year or none in some years (Figure 27.12d). Humidity is low most of the year and with sparse or no vegetation, heat is lost readily at night, resulting in large diurnal temperature ranges. Annual rainfall in deserts is low, but the distribution varies by region. In the continental climate, rain is evenly distributed throughout the year. In the Mediterranean climate, fall and winter are the moist seasons, while closer to the equator summer is the rainy season. Rainfalls can be intense and cause flash floods and erosion.

Plants adapted to the deserts are generally low growing, with small leaves, and spines for protection. Some plants have thick, fleshy leaves with a thick cuticle to prevent water loss and to store water to endure long periods without rain. Some desert plants use the crassulacean acid metabolism (CAM) photosynthesis opening stomata at night for gas exchange to conserve moisture loss to transpiration. Few herbaceous species grow in the desert. Some of them are ephemeral annuals completing their life cycle within a couple of weeks when moisture is available (Figure 27.12e and f). Typical plants in North American deserts include the mesquite tree (*Prosopis* spp.), Joshua tree (*Yucca brevifolia*), creosote bush (*Larrea divaricata*), sagebrush (*Artemisia tridentata*), succulent perennials like cacti, and agave (*Agave* spp.) (Raven et al., 1999). Soils in the desert are not much weathered and can contain much gravel or sand. Plant growth is very slow in most desert environments because of the limiting moisture available to plants. There is little or no accumulation of organic matter because decomposition rates are low due to low humidity and low moisture conditions.

TROPICAL AND SUBTROPICAL MOIST BROADLEAF FORESTS

This biome is also known as the rainforest and is rich in plant and animal species. Tropical and subtropical moist broadleaf forests occur near the equator within 23.5° northern and southern latitude. Due to this location, there is almost no deviation from the 12-hour daylength throughout the year. Temperatures in the tropical rainforest are almost constant year-round between 68°F and 86°F (20°C and 30°C), and the difference between day and night temperature can be larger than the average change in annual temperature. Annual rainfall is very high, ranging from 79" to 236" (200 to 600 cm) precipitation. Distribution throughout the year can vary with no dry months in the tropical rainforest where trees maintain their evergreen canopy throughout the year. The largest areas covered by tropical moist broadleaf forests are found in the Congo basin in central Africa, the Amazon basin in South America, and Malaysia and Indonesia. In seasonal rainforests, dry periods in winter cause leaf drop of some species although temperatures remain warm. These forests occur in coastal West Africa, parts of the Indian subcontinent and Southeast Asia, and South and Central America.

(a)

(b)

(c)

Figure 27.13a–c   (a) Aerial view of tropical rainforest in the Amazon. (b) Kapok tree (*Ceiba pentandra*) grows to 197 to 230 feet (60 to 70 m) above the general canopy in the Amazon rainforest and is estimated to be more than 800 years old. (c) Australian temperate rainforest.

Tropical rainforests house the largest number plant and animal species of all the biomes (Figure 27.13a and b). Up to 300 tree species may be found in one hectare (10,000 m²) while in a temperate rainforest (Figure 27.13c) between 20 and 30 tree species are growing in a comparable area. Some very tall trees grow in the tropical rain forest with just a few towering up to 197' (60 m) above the canopy of trees below (Figure 27.13b). Most trees are evergreen and have broad leaves. Multiple layers of larger and smaller trees, shrubs, vines including lianas, and plants such as orchids, bromeliads, ferns, and mosses contribute to the extraordinary diversity of this environment. With so many layers of canopy, the forest floor is often dark with few plants. Low growing plants on the forest floor are often shade tolerant and a greater number of plants is found where the canopy has been broken up, admitting more light. Litter on the ground does not accumulate because it is quickly decomposed by microorganisms. Soils in the tropics are exposed to a fast weathering process with abundant rains leaching nutrients and soils turning acidic.

Rainforests house more than half of the plant and animal species living on earth. These forest ecosystems are fragile and are often destroyed for agriculture, logging, and mining. Plants from rainforests show great promise for medicinal purposes and more than 2000 species show anticancer properties. This number is less than 1% of all the plant species found in the tropical forest.

Key Terms

climate	biogeography	taiga
specific heat	xeric	savanna
Hadley cell	permafrost	shrubland
rain shadow	alpine tundra	chapparal

Summary

- Global climate patterns are primarily a result of the amount of sunlight a region is exposed to over a year. The uneven heating of land and oceans or large bodies of water results in air circulation patterns over the globe causing moist and dry areas.

- The Hadley cell circulation explains why the equatorial areas are receiving ample amounts of rain and areas around 30° latitude are deserts. Oceans have moderating effects on the climate of nearby land.

- Mountain ranges in Western North America receive more rainfall on the western slopes as moist warm air rises and little rain on the eastern slopes as air cools and sinks. The eastern slopes are in the rain shadow and characterized by a dry environment.

- Biomes are characterized by communities of plants, animals, and other organisms located in an area with distinct vegetation. Temperature and rainfall contribute to the differences between biomes.

- Plants in a biome on different continents are adapted to the climate and often look similar because of morphological characteristics, but are often unrelated.

- Currently 14 different terrestrial biomes are classified worldwide. Biomes are not uniform, but vary and can be diverse.

- Environments that support tree growth require a minimum in annual average temperature and a minimum in rainfall.

- The most northern biome is the arctic tundra with few plant species, brief, cool summers, and permafrost. Alpine tundra is found at higher elevations in mountains at lower latitudes with similar climate and vegetation.

- The boreal forest or taiga borders the tundra to the south and covers vast areas of Russia, Northern Europe, Canada, and Alaska. Few species of conifer trees dominate in the boreal forest and growing seasons are short and temperatures low.

- Temperate coniferous forests receive more rain and are located in a milder climate. Forests in the Pacific Northwest are an example with highly productive trees.

- Temperature broadleaf and mixed forests are located north and south of 30° latitude and consist mainly of several species of deciduous hardwood trees and a few conifers. This biome has warmer, longer summers, and cold winters.

- Grasslands, savannas, and shrublands are located between deserts and temperate forest, they occur both in temperate and tropical regions. Grass is the main vegetation with savannas and shrubland characterized by a few trees or shrubs.

- Deserts and xeric shrublands are located around the 30° latitude north and south. This biome is characterized by vegetation able to withstand long periods of drought and temperature extremes from cold to hot.

- The tropical and subtropical moist broadleaf forest or rainforest biome is richest in the number of plant and animal species among all biomes. Annual temperatures vary little and the climate is warm with ample rainfall.

Reflect

1. *Global climate patterns shape vegetation and biomes.* Describe the major factors influencing global climate patterns and why deserts are located in a two bands around the globe.

2. *There are 14 different biomes recognized worldwide.* Explain why some cover vast areas while others are confined to very small areas.

3. *Biomes can be very diverse.* Use the savanna to discuss diversity within this biome.

4. *The rainforest occurs in different latitudes.* Explain the difference between the tropical rainforest and the rainforest located on the Olympic Peninsula in Washington.

5. *Plants have developed different strategies to cope with extreme climate conditions.* Give some examples of how plants in the tundra cope.

6. *Oceans play an important role for global climate.* Explain and give some examples of ocean effects.

7. *Humans have changed some biomes almost entirely from their original state.* Which ones are they and why have they been changed?

References

Aerts, R., Cornelissen, J. H. C., & Dorrepaal, E. (2006). Plant performance in a warmer world: General responses of plants from cold, northern, biomes and the importance of winter and spring events. *Plant Ecology, 182,* 65–77.

Olson, D. M., Dinerstein, E., Wikramanayake, E. D., Burgess, N. D., Powell, G. V. N., Underwood, E. C., et al. (2001). Terrestrial ecoregions of the world: A new map of life on earth. *BioScience, 51*(11), 933–938.

Pidwirny, M., Draggan, S., McGinley, M., & Frankis, M. (2007, Revised on January 2012). Terrestrial biome. In C. J. Cleveland (Eds.). *Encyclopedia of Earth.* Washington, D.C.: Environmental Information Coalition, National Council for Science and the Environment.

Raven, P., Evert, R. F., & Eichhorn, S. E. (1999). *Biology of plants.* New York, NY: W.H. Freeman and Co.

Tyrell, M. (2018). *Global forest atlas.* Yale School of Forestry and Environmental Studies. https://globalforestatlas.yale.edu/boreal-zone Accessed July 4, 2018.

Walther, G. R., Post, E., Convey, P., Menzel, A., Parmesan, C., Beebee, T. J. C., et al. (2002). Ecological responses to recent climate change. *Nature, 416,* 389–395.

Whittaker, R. H. (1975). *Communities and ecosystems.* New York: MacMillan Publishing. Company.

Zhou, L., Tucker, C. J., Kaufmann, R. K., Slayback, D., Shabanov, N. V., & Myneni, R. B. (2001). Variations in northern vegetation activity inferred from satellite data of vegetation index during 1981 to 1999. *Journal of Geophysical Research. Atmospheres: JGR, 106,* 20069–20083.

Dynamics of Plant Communities and Populations

© Iiseykina/Shutterstock.com

Learning Objectives

- Describe the levels of ecological study: organismal, population, community, and ecosystem ecology
- Outline the trophic levels found in ecosystems
- Compare a food chain versus a food web
- Discuss how energy moves through ecosystems
- Explain how human activities have changed natural processes in ecosystems and the consequences
- Identify the main processes in the global carbon cycle
- Analyze which steps in the nitrogen cycle result in products that contribute to plant growth and which ones make nitrogen unavailable to plants.
- Compare and describe primary and secondary succession in a temperate climate

Ecology studies how organisms interact with each other and their environments. These relationships can be investigated at many different levels. In the previous chapter we examined the characteristics of global ecosystems or biomes. Now we take a closer look at how organisms, populations, and communities interact and what affects these dynamics. The next time you go for a walk in nature, look around and think about all the organisms populating the environment and how they depend on certain conditions. Starting with the soil, what is the geological material underlying the soil and how did climate and geological age and events shape the current soil so the organisms can live there? The soil supports certain plants, which depend on the level of mineral nutrients, chemical factors such as pH and salinity, and physical factors such as porosity and water-holding capacity. Plants are dependent on the amount of sunlight, temperatures, and rainfall they receive in their habitats. Plants support herbivores, which in turn support organisms feeding on herbivores.

Changing one factor in a system affects other organisms in this environment. For example, pines in a conifer forest stressed by many years of drought eventually become susceptible to the bark beetle, which can kill them. Healthy pine trees can fend off the bark beetles and offer food and shelter to many insects and animals, while dying or dead pine trees provide these services for other organisms. Dead pine trees become a fire hazard and extensive fires have consumed large areas covered with dead trees. Fire will clear the area of existing vegetation, recycle some nutrients to the soil, result in losses of some nutrients due to erosion, and begin a new cycle of organisms establishing and developing over time into an ecosystem. Will this be the same as the previous one dominated by pine trees? It all depends on how the environmental conditions, the soil and water resources, and the species populating the environment will interact. This chapter covers the different organizational levels and dynamics within an ecosystem, populations, communities, and organisms; how energy flows through ecosystems; global nutrient cycles; and developmental changes in these systems over time.

LEVELS OF ECOLOGICAL STUDIES

Ecology studies organisms and their environments at different levels or hierarchy, starting with the organism or a species, moving to a population, a community, an ecosystem, and finally a biome (Figure 28.1). The biosphere includes all terrestrial and aquatic biomes, the atmosphere, soil, and water of the earth.

organismal ecology
The study of behavior, physiology, and morphology of individual organisms of a species.

population
All individuals of the same species that live and interact in the same area.

- **Organismal ecology** investigates individual organisms of a species; their behavior, physiology, and morphology; and how they respond to their environment. This can be a honey bee or Russian thistle.
- **Population ecology** studies a population of individuals able to interbreed that live at the same time in the same location. A population is a group of individuals of the same species. Areas of interest are factors influencing the abundance and genetic composition of a population, and how to predict a population's response to changes in their environment. Redosier dogwood (*Cornus sericea*) has a wide distribution throughout North America except for the southeast and

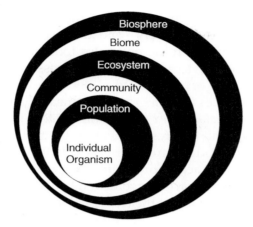

Figure 28.1   Levels of ecology start with individual organisms and progress through increasingly complex levels up to the biosphere.

south-central states, but geographic populations are distinct based on their response to environmental cues such as day length and temperature. Adaptations to different climatic conditions have allowed the plant to survive in locations from Alaska to Newfoundland and California to Virginia.

- **Community ecology** focuses on interactions between populations of all living organisms in one place. Community structure and organization are affected by how individuals or groups of organisms interact with each other and the environment.

- **Ecosystem ecology** looks at the interaction of all the organisms in a large region and their physical environment, including all biotic and abiotic interactions. This area of study is very broad and involves entire systems such as nutrient cycling and energy flows. Streams or ponds are aquatic ecosystems, while the ponderosa pine forest is an example of a terrestrial ecosystem.

INTERACTIONS BETWEEN ORGANISMS

Organisms in a community or ecosystem interact in different ways. As we have previously discussed in Chapter 4, these interactions can be classified by whether they are beneficial to both species, called **mutualism**. If the relationship between two species is beneficial for one and neither harmful nor beneficial to the other, then it is **commensalism**. When both organisms are harmed they are in **competition**. In interactions where one member is negatively affected and the other organism gains from the relationship, it is **predation**.

Mutualism

Mutualism, relationships between organisms where both benefit, has been described before and includes the *Rhizobium* bacteria fixing nitrogen for host plants in the legume family and receiving organic compounds in return (Chapter 4). Mycorrhizal associations benefit both the fungus that also

community
Populations of all organisms that live and interact in the same area.

ecosystem
Community and its physical environment in a particular area.

mutualism
Symbiotic relationship in which both organisms benefit and neither are harmed.

commensalism
Relationship of two species where one benefits and the other is not harmed and does not benefit.

competition
When two organisms or species attempt to use the same resources such as land, light, water, or nutrients.

predation
Individuals of one species eating living individuals of another species.

receives carbohydrates and the host plant that has increased capacity for water and mineral uptake (Chapters 4 and 27). Insects or animals pollinating flowers generally benefit from the pollen for food and flowers benefit from increasing their reproductive capacity through cross-pollination (Chapter 7).

A complex example of mutualism is between ant species and plants known as **myrmecophytes** (*myrmeko* in Greek means "ant" and *phyton* means "food"). The ants (*Pseudomyrnex ferruginea*) protect the bull's-horn acacia (*Acacia cornigera*) from herbivores by stinging them. The ants remove fungi from the trees to protect them from disease. Plants germinating under the canopy of the bull's-horn acacia or neighboring plants' extending branches that touch the bull's-horn acacia are removed by the ants, most likely to prevent shading. Ants further assist the trees by dispersing their seeds (Janzen, 1966). The ants benefit from the plant by receiving shelter in enlarged woody stipules called **domatia**, internal plant structures suitable for housing ants (Figure 28.2). Ants also receive food from nectar and from protein and lipids in food bodies called **Beltian bodies** attached to the terminal end of leaflets. Another feature of acacias essential for coevolution with ants is year-round leaf production, which is possible in the tropical climate. So codependent are the two species that the tree will die if the ants are removed. The ants, however, can live on any acacia with swollen thorns (Janzen, 1966).

Commensalism

Commensalism is the relationship between organisms where one species is the beneficiary and the other one is not affected positively or negatively. Epiphytes, described in Chapter 4, are examples where the tree provides the location for species such as bromeliads, orchids, and lichens, but experiences no harm or benefit from the epiphyte. The advantage for epiphytes is

myrmecophytes
Mutualistic relationship between ants and plants.

domatia
Enlarged woody stipules.

beltian body
Food bodies consisting of lipids and proteins that are attached to the terminal end of leaflets on acacias.

© University of Wisconsin-Madison, Department of Botany.

Figure 28.2 Bull's-horn acacia provide shelter for ants in the enlarged woody stipules. Closeups show where ants can get proteins and lipids from the Beltian bodies (*top*) and nectar (*bottom*).

the improved location on a well-drained surface, access to light, and protection from ground-dwelling herbivores.

Some plants have developed relationships facilitating a plant's success. Several tree and shrub species in the Sonoran desert serve as nurse trees for germinating and growing cacti such as the saguaro (*Carnegia gigantea*) and barrel cactus (*Ferrocactus* sp.), which thrive under their canopies (Figure 28.3a). This relationship is based on the improved microclimate and nutrient availability as well as protection from herbivores and cold temperatures for the young cacti.

The distribution of plant seeds or vegetative parts by animals or humans is another example of commensalism. Seeds relying on this mechanism for distribution have appendages that facilitate sticking to anything mobile that passes close by. The organism or vehicles transporting the seeds or vegetative parts that will readily regrow in a new location experience in general little or no negative effects while dispersal of seeds is a bonus for the plant's reproduction and survival.

Competition

Plants compete for light, water, and nutrients in the soil. When resources are limited the competing plants will not grow as well compared to only one plant occupying the space claimed by two or more. In competition situations all organisms are usually negatively affected. Invasive plants taking the place of native plants in an ecosystem compete successfully when displacing the native species. Buffelgrass (*Pennisetum ciliare*) is a drought-tolerant grass native to Africa. It was brought to the southwestern United States to control soil erosion and for cattle grazing. In the last 20 years the grass started to spread rapidly. It is outcompeting the native vegetation because it forms dense mats of continuous vegetation. The native plants grow some distance apart, never providing much fuel for wildfires. This change in vegetation cover favors hot wildfires from the large amount of grass fuel, killing seedlings of the native vegetation. Invasion of buffelgrass is changing a diverse desert habitat into savanna grassland dominated by fire.

Parasitism or Predation

Relationships where one organism takes all the energy it requires from the other organism can be parasitic. Mistletoe growing on trees or shrubs enters the plant to tap water and minerals from its host. Dodder (*Cuscuta* sp.) is a parasitic vine that attaches to a wide range of host plants and enters their vascular system with haustoria, which are modified roots. Dodder can overwhelm entire plants and can cause great damage to crops (Figure 28.3b).

Animals or insects consuming plants are characterized as predators or herbivores and benefit from the plants' energy. **Herbivory** is common as most organisms rely on plants as the source of energy. Some plants have developed specialized protection against herbivory, including spines, thorns, and waxy cuticles. Other plants produce chemicals, rendering the plant poisonous or unpalatable. A special relationship exists between some plants and their predators such that the predator can tolerate the poison of a plant. This type of evolution is called **coevolution** because the two species have developed together. Monarch caterpillars can consume the leaves of

herbivory
When an organism is eating a plant. Herbivores subsist exclusively on plants.

coevolution
Evolution of two species that interact with each other and continuously shape each other's adaptations.

(a) (b)

Figure 28.3a–b (a) The palo verde tree (*Parkinsonia microphylla*) and creosote bush (*Larrea tridentata*) are nurse plants for the barrel cactus (*Ferrocactus*) in a commensal relationship, with the cactus receiving shelter and the tree and shrub not being affected by the cactus. (b) Dodder, a parasitic vine, overwhelms other plants.

the common milkweed (*Asclepias syriaca*), which contain alkaloids and are poisonous to most vertebrate herbivores. The caterpillars benefit by having few competitors that can feed on the leaves. The caterpillars ingest the toxins in the plant material and accumulate it in their bodies, making them unpalatable or toxic to other predators.

POPULATION ECOLOGY

Populations are often described by their demographics, their numbers, and age distribution (Figure 28.4). Ecologists also develop models to describe how a population changes if one or more factors in their environment changes. These models are used to predict how populations will develop in a habitat over time, and how to protect endangered or threatened species and to manage or restore habitats.

Plant populations are often characterized by their distribution, which can be random, by age, location, or number. A forest planted after a fire often has an even age distribution with all the trees of the same species and the same age (Figure 28.5a). Evenly spaced trees occupy a similar amount of space. Such a forest is extremely uniform in many aspects. In contrast, old-growth

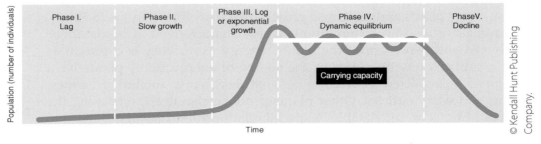

Figure 28.4 Typical growth of a population over time. The carrying capacity of a population is reached in phase IV.

forest that developed naturally without human intervention may be at a stage where a tree hundreds of years old and very large in size can be close to decade-old trees, decaying stumps, and seedlings. This forest can be characterized as heterogenous or random in age distribution (Figure 28.5b) but likely also in species diversity and space distribution. However, on a smaller scale there may be small pockets of homogenous seedlings that germinated under favorable conditions from seeds of a remaining mature tree.

Environmental resources such as light, water, mineral nutrition, and soil characteristics shape the size of populations. Species growing without competition can grow to larger numbers compared to when they have to share resources (Figure 28.6). The presence of herbivores, disease organisms, and competing species further influence how large a population can grow. When the number of organisms in an environment has reached its maximum size and no more organisms can be supported, this is called the **carrying capacity** (Figure 28.4). Populations of different species change size based on the amount of resources available.

In arid environments consistent moisture at certain times of the year results in abundant plant cover, which supports a growing number of herbivores. If these conditions persist over a period of time, populations of predators such as hawks and coyotes will increase to feed on the abundant herbivore populations. This can continue until the carrying capacity of the habitat is reached and the amount of vegetation cannot increase and

carrying capacity
Maximum number of organisms of a species a habitat can support.

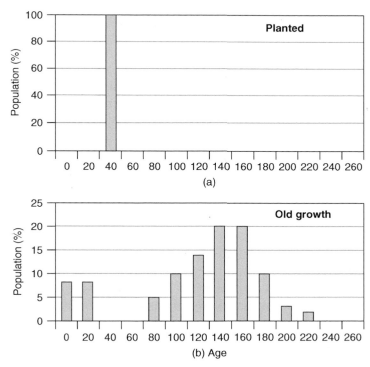

Figure 28.5a–b   Age dynamics in two forest populations. (a) Age distribution in the forest plantation is around 40 years for all trees planted at the same time. (b) Distribution of trees in the old-growth forest includes some very young and some very old ones, with the majority between 100 and 180 years old.

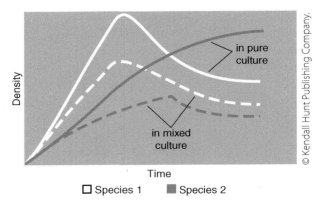

© Kendall Hunt Publishing Company.

☐ Species 1 ■ Species 2

Figure 28.6 Two species of clover grow denser with more individuals per unit area without competition (solid lines) compared to when they are competing in mixed culture (broken lines). *Trifolium repens* (species 1) grows faster to greater density than *Trifolium fragiferum* (species 2) in pure culture. Lower numbers of each species are growing when the plants compete for resources such as light, nutrients, and water.

support more herbivores. At that time equilibrium is reached and populations will stabilize. If a drought period begins the amount of vegetation will shrink, quickly reducing the number of herbivores it can support. As a consequence numbers of predators will quickly decline because their food source is diminished.

One way ecologists characterize populations is by the **r/K selection theory** of reproduction developed in the 1960s (MacArthur & Wilson, 1967). The theory is based on the idea that the combination of traits of an organism will favor different patterns of reproduction and thus the species richness in an area. In a sparsely occupied area that has a greater carrying capacity conditions are favorable for populations that can rapidly reproduce. Organisms ready to colonize this often unstable environment have the traits characteristic of **r-selection**. These include fast maturation and reproduction at an early age, and large numbers of offspring with few survivors. Offspring are often dispersed over a large area. These species often have short lifespans and encounter relatively little competition for which they are poorly adapted. Examples are bacteria, algae, weeds, and lizards.

Characteristics of **K-selected populations** have a longer time until they mature and reproduce. They are subject to competition in a stable environment that has reached its carrying capacity. They have fewer offspring with greater survival rates. K-selected organisms generally have a long lifespan. Examples of plants include long-lived oak trees and agave, which reproduce only once and then die. Humans and other large mammals also fall in this category.

This theory has been used to predict how populations evolve in a location over time. Reproduction, however, is only one factor determining population size. Rate of death is the other determining factor and populations will increase in size only if the rate of reproduction is greater than the rate of death. Alternatively, if members of the same species from other

r/K selection theory
Theory in ecology that relates to the traits in reproduction based on whether large or small numbers of offspring are produced with different survival rates and lifespans. See also *K-selection traits* and *r-selection traits.*

r-selection traits
Typical of organisms with a short life, reproducing rapidly and living in an unstable environment.

K-selection traits
Organisms in these populations mature slowly in a stable environment, have relatively few offspring, and a long lifespan.

areas migrate into an area, the population numbers may increase as well. This demonstrates the dynamic nature of populations and how the physical environment sets limits to the increase of population sizes.

ECOSYSTEM DYNAMICS AND HUMAN ACTIVITY

The majority of ecosystems on the planet have been altered by humans in their attempt to inhabit and use the resources of an ecosystem to provide for their livelihood. They have changed the physical environment such as soil and water, the abundance and distribution of native plants and animals, and have introduced new organisms into the system. Thus many interactions of the original inhabitants of an ecosystem have been changed or disrupted permanently.

There are many examples of how human use of natural resources has not only depleted the natural resource but also damaged the remaining land and wildlife. Logging of old growth, the original forest, is the harvest of this timber for use as building materials. In many cases areas are clear cut and are not replanted in a timely manner. This leaves barren mountains prone to soil erosion (Figure 28.7a). Forest soils and nutrients are lost and can negatively impact streams in which they are deposited. This in turn can affect fish and other wildlife. In the Pacific Northwest, vast areas of old growth have been harvested and only part of the land has been replanted. Harvest has proceeded much faster than regrowth of the second-growth forest. Old-growth forests represent a rich ecosystem with great species diversity whereas the second-growth forests are less diverse.

Loss of mature old-growth forest started in the 1850s to clear land for agriculture and housing but accelerated since the 1950s due to logging. The Northern Spotted Owl (*Strix occidentalis caurina*) living in this habitat is inextricably linked to the health of old-growth forest (Figure 28.7b–c) and has been identified as endangered, threatened, and a species of special concern by different agencies managing public lands (Thomas et al., 1990). Action to study and protect the species stems from a concern over the loss of biodiversity. The loss of biodiversity is caused by changes in an organism's environment and many organisms require specific conditions and are

(a)
© Christopher Kolaczan/Shutterstock.com

(b)
Art Wolfe/Science Source.

(c)
© Timothy Epp/Shutterstock.com

Figure 28.7a–c   (a) Old-growth forest harvesting started in the 19th century and accelerated in the last century, leaving only remnants of the original old growth. (b) Old growth in the Pacific Northwest of the United States is the preferred habitat for the Northern Spotted Owl (*Strix occidentalis caurina*), a threatened species because of habitat destruction. (c) A large old-growth tree which is also threatened by harvesting.

conservation biology
Study of populations, communities, and ecosystems with the goal to preserve and restore them.

not able to adapt to subtle or obvious changes in their habitats. The spotted owl is only one of many species threatened by the loss of old-growth forest.

The field of **conservation biology** uses the different levels of ecological studies to understand why some species are declining in numbers or are reduced in the areas they occupy. This field of study also looks at why other species may flourish, outcompete other organisms, or overpopulate an area. Conservation biology studies the environment and the species, their populations and communities. Preservation of threatened species and restoration of their environments by altering land and resource management is the goal of many efforts, including the study of the spotted owl and its habitat.

The preferred habitat of the spotted owl is located primarily in old-growth forest with multiple layers of canopy and different tree species, including some densely vegetated areas as well as some open patches (Thomas et al., 1990). Owls thrive in areas with dense litter on the forest floor, large decaying logs and stumps, and some large standing trees with various levels of decay. The combination of these characteristics is typical for old-growth forest 150–200 years or older. Younger forest stands supporting breeding owl population generally have remnants of old-growth stands. In addition to habitat loss, **fragmentation** of habitat, which isolates some owl populations from others, is putting them, their habitat, and their dispersal or movement at greater risk (Thomas et al., 1990; Lamberson, McKelvey, Noon, & Voss, 1992). Fragmentation of habitats threatens plants and wildlife populations, communities, and ecosystems. Scientists concluded that larger areas of habitat with several owl pairs are superior to smaller ones with only a single pair and less fragmented habitat is better for owls to disperse safely. Studies have evaluated models to predict how populations of the spotted owls will survive under different land management models. Depending on the model and its assumptions, the future of the spotted owl can be secured for the next 100 years. This case of the spotted owl demonstrates how many different organisms and their actions and interactions affect an ecosystem. Humans have started to realize the importance of ecosystems and many efforts are under way to protect fragile systems such as coastal wetlands, the rainforest, or mountain ecosystems.

fragmentation
A large habitat of a population is divided into smaller ones by disturbances resulting in isolated smaller patches of the original habitat, now separated by other habitats.

BOX 28.1 Wetlands—A Change in Perspective from Destruction to Conservation

wetland
Area covered with shallow water on a regular or permanent basis.

Wetlands can be freshwater or saltwater habitats with water either stagnant or moving (Figure Box 28.1a–b). Water covers the area to different depths either permanently or for long periods during a year, resulting in very low concentrations of oxygen in the soil, especially below the stagnant water of bogs. Wetlands differ from lakes or ponds because they are colonized by plants that grow above the water. Estuaries, marshes, swamps, and bogs are wetlands and home to a large number of species of plants, invertebrates, fish, birds, and insects. Wetlands have among the richest species diversity among the known ecosystems of the world.

For a long time wetlands were considered wasteland because they could not be used for agricultural production or housing. As a result, the majority of natural wetlands in North America have been drained and converted to

(a)

(b)

Figure Box 28.1a–b   Two types of natural wetlands: (a) salt marsh in Maine and (b) red mangroves in the Florida Everglades.

other land uses between 1780 and 1980 (Dahl, 1990). In colonial times, the area that is now the continental United States had an estimated 221 million acres of wetlands. By the 1980s slightly over half the wetlands have been lost, with California sustaining the greatest loss at 91% of all states. Florida lost the greatest acreage, over 9.3 million acres during those 200 years. Many states lost more than half of their original wetlands.

Wetlands have important ecological and biological functions, many of which were not recognized until they were drained and not carrying out their function. Wetlands prevent shoreline erosion and rivers from submerging the adjacent bottomlands because they slow down the flow rate of water and can temporarily buffer the influx of large amounts of water from flooding. They are a natural water filter for sediments, nutrients, and other pollutants. Wetlands are part of the global water cycle and affect local conditions by recharging groundwater, releasing water to adjacent streams, and influencing the local climate through evaporation. Wetlands are important carbon sinks because dead plant material will decompose very slowly under low oxygen conditions.

Wetland plants are adapted to the low oxygen conditions and have developed special root and shoot systems to deal with water inundation. Some plants have **pneumatophores**, roots that grow above the water level to allow oxygen uptake. Buttress roots hold larger trees in place in a wetland. Grasses, sedges, and rushes have stems with **aerenchyma** for air to move from the leaves to the roots. Oxygen discharged from the roots is used for some biochemical processes such as the breakdown of organic compounds. Some wetland plants are adapted to the saline water of estuaries.

The understanding of the multiple functions of wetlands has led to their protection and preservation. Some wetlands previously altered have been restored, however, many species and their habitats cannot be recreated to their natural state through such efforts. Even 100 years after restoration biological and chemical functions are still lower than in comparable wetlands in their original state (Moreno-Mateos, Power, Comín, & Yockteng, 2012). Small wetlands with stagnant water in cold climates recover slowest after disturbance whereas wetlands larger than 100 hectares with faster water flow in warm regions recover faster from disturbances. Some wetlands are artificially created by flooding an area and populating it with wetland plants. These habitats with low species diversity are not comparable in species richness to the natural wetlands.

pneumatophores
Roots growing above the water level to allow for oxygen uptake.

aerenchyma
Specialized cells allowing oxygen movement from leaves to roots.

ENERGY FLOW IN ECOSYSTEMS
The Food Chain and Food Web

Ecosystems gain their energy from the sun and harness it through photosynthesis. The autotroph organisms are the basis of all food in the system. The energy they produce passes through different levels of the food chain and food web before nutrients are recycled again (Figure 28.8). The **food chain** and **food web** describe who eats whom, the hierarchy of feeding, and the connections between the different organisms (Pimm, Lawton, & Cohen, 1991). The **trophic** pyramid depicts the relationship between the different levels of consumers and the food web shows the interrelationships between the different organisms in an ecosystem

The simplest food chain has three trophic levels: plants or autotrophs (Figure 28.9), herbivores or primary consumers, and carnivores or secondary consumers. Producers are plants and other organisms performing photosynthesis. Primary consumers feed on the plants and are herbivores such as grazing animals or leaf-eating insects. Secondary consumers obtain their energy by eating or feeding on primary consumers. These include predators such as bear, lion, or bobcat, or mosquitoes, snakes, and birds. Some

food chain
Single-line hierarchy of which organisms eat each other at different trophic levels in an ecosystem.

food web
Complex interactions between all organisms at different trophic levels in an ecosystem describing the movement of energy.

trophic level
Level within the hierarchy of feeding in an ecosystem.

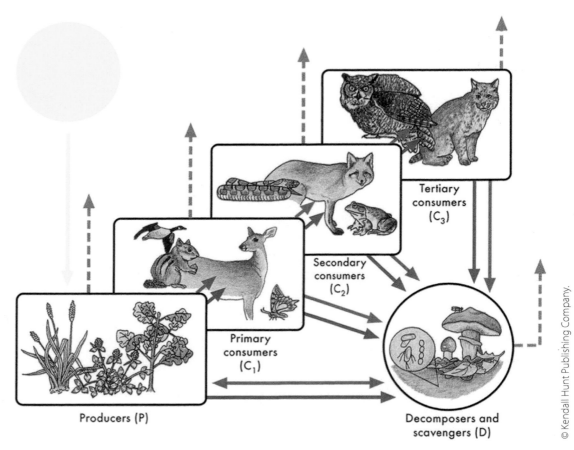

© Kendall Hunt Publishing Company.

Figure 28.8 Energy production and consumption in an ecosystem. Energy is transferred from the sun to autotroph organisms and then through different trophic levels (blue arrows) to the top of the food chain. At each step energy is lost through respiration (brown dashed arrows). Organisms and biomass are reused in the ecosystem (red arrows).

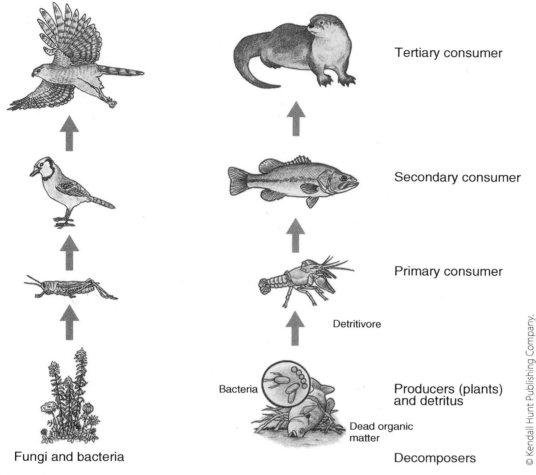

Tertiary consumer

Secondary consumer

Primary consumer

Detritivore

Bacteria

Producers (plants) and detritus

Dead organic matter

Fungi and bacteria   Decomposers

© Kendall Hunt Publishing Company.

Figure 28.9   Fungi and bacteria break down organic matter in the soil. The terrestrial food chain (*left*) starts with Queen Anne's lace (*Daucus carota*), yarrow (*Achillea millefolium*), and clover (*Trifolium*) producing green biomass. Herbivorous grasshoppers are primary consumers and serve as food sources for the carnivorous blue jay, a secondary consumer. Cooper's hawk is a tertiary consumer and feeds on the blue jays. The aquatic food chain (*right*) starts with dead matter that is broken down by bacteria. Crayfish are primary consumers that feed on the detritus. Larger fish such as bass are secondary consumers eating the crayfish. Bass are food for otters, tertiary consumers at the top of the food chain. All energy consumed at the different trophic levels can be traced back to the autotroph producers.

of the secondary consumers are eaten by tertiary consumers and several of the secondary consumers feed on other secondary consumers. Humans and lions are examples of the top of the food chain because they consume the energy accumulated by several levels of consumers below them but are not hunted by other organisms. Decomposers such as fungi and bacteria use the waste from living organisms and remainders of dead organisms, breaking down the material for detritus feeders or uptake by producers. Detritus feeders such as earthworms, pill bugs, or insect larvae decompose or consume organic matter and in turn are eaten by secondary consumers such as birds or moles. Food chains are a simple way to describe a limited number of organisms involved in the ecosystem. Most food chains have no more than three to four links.

Food webs describe the complex relationships of many species in a community or an ecosystem (Pimm et al., 1991). Food webs describe how different food chains are linked and which species feed at which trophic level (Figure 28.10). Food webs also start with producers but take into account that many different food chains may start with the same plant species and that food chains intersect at different levels. Food webs generally have three to four trophic levels. Omnivores feed at several trophic levels. They may feed on their prey, this species' prey, and on the lower level of this food chain as well as different food chains. For example, a bear may feed on a fox, a rabbit, a salmon, a ground squirrel, or berries.

Ecosystem ecology examines how different organisms regulate each other. For example, overgrazing is a problem when too many herbivores feed on a limited supply of grass, herbs, and other foliage. If herbivores increase in population beyond what the vegetation can support, carnivores feeding on them may decrease the number of herbivores, reducing grazing pressure, and allowing the plant population to thrive again. Conversely, if carnivores do not reduce the number of herbivores the plant population may degrade, land may become prone to erosion, and herbivores have to move on or will decline in numbers.

A classic example of populations regulating each other through a predator–prey relationship is that of the Canadian snowshoe hare (*Lepus americanus*) and the lynx (*Lynx canadensis*) (Figure 28.11). Cycling of hare

Phytoplankton   Zooplankton

Invertebrates

Figure 28.10   This example of a basic food web shows the interconnected food chains in an ecosystem and how energy is produced, used, and recycled in a wetland and surrounding terrestrial communities in the eastern United States.

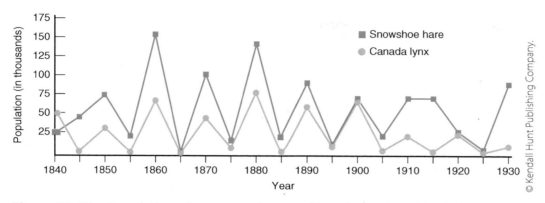

Figure 28.11   Population of snowshoe hare and lynx in northern North America. Populations fluctuate at approximately 10-year cycles.

populations was thought to be a result of the availability of vegetation they preferred or could tolerate. Lynx populations then followed the increase and decline of the hare population. New analyses of the food web data of this boreal forest community revealed a more complex relationship between hare, lynx, and other species (Stenseth, Falck, Bjørnstad, & Krebs, 1997). Lynx are specialized predators on hares and are indeed regulated by the amount of hares, their primary prey. Hares, however, are regulated from below, their multiple sources of vegetation, and from above, their multiple predators. Hares feed on grasses, forbs, and trees, which they compete for with other herbivores. They are also targeted by hawks, owls, lynx, coyote, red fox, and wolves. The analysis of the food web involving hare and lynx showed that the simple predator prey food chain is not the only factor responsible for fluctuations in their populations.

Energy Transfer in Ecosystems

The amount of energy conserved at each successive level of the food chain or web diminishes and energy use is less efficient the longer the food chain. At each step of the consumption, only about 10% of the energy is passed on to the next trophic level. Most of the energy is lost through respiration as it passes through the food chain. This loss of energy is a concern as more people need to be fed with the resources available on the planet. When humans eat meat, the animals they consume are often fed grain that is raised specifically for animal feed. This process is energy-inefficient and many people cannot afford to eat at the secondary consumer level but obtain their energy from vegetables. A vegetarian diet uses more of the energy stored in plants than if the plant is fed first to an animal that is subsequently consumed by humans.

Ecosystem structures can be illustrated on the basis of biomass, energy, and numbers. The biomass pyramid shows the large amount of biomass in a Panamanian rainforest necessary to sustain primary and secondary consumers (Figure 28.12a). In an ecosystem where turnover of biomass is rapid, decomposers are essential to recycle nutrients. The pyramid of energy shows how energy captured from the sun is passed to primary, secondary, and tertiary consumers in a freshwater aquatic system in Florida (Figure 28.12b). Of the 20,800 kilocalories per square meter per year captured by autotrophs in the community through photosynthesis, only 21 kilocalories per square meter or 1 percent are passed to the tertiary consumer. The pyramid of numbers shows that millions of plants per area are necessary to support primary consumers in the grassland ecosystem (Figure 28.12c). Smaller numbers of secondary consumers feed on these herbivores and very few carnivores are supported at the tertiary trophy level. These relationships demonstrate how food chains and food webs are dependent on healthy producers in ecosystems. When producers cannot thrive, the whole community and ecosystem will be compromised.

BIOGEOCHEMICAL CYCLES

Chemical elements available for life on earth are present in limiting amounts. Elements like carbon, nitrogen, oxygen, and phosphorus can neither be created nor destroyed and are constantly recycled in the biosphere. The element cycles involve the physical environment and living organisms; therefore, they are called biogeochemical cycles. The chemical elements move between these organic and inorganic reservoirs. Elements contained in organic material of living organisms are lost through respiration, decomposition, or excretion and move to the atmosphere, soil or water. Chemical elements in air, soil, or water become parts of the building blocks of living organisms when they are assimilated or used in photosynthesis. The atmosphere, soil, and water gain chemical elements in inorganic form when rocks weather and erode, when fossil fuels are burned, or during volcanic eruptions. Conversely, inorganic chemical elements in air, soil and water will become unavailable during sedimentary rock formation.

Ideally, all nutrients remain in an ecosystem to be used repeatedly. Some nutrients may be lost to leaching below the root zone, erosion into streams or the ocean, or volatilization into the atmosphere. Once nutrients are lost

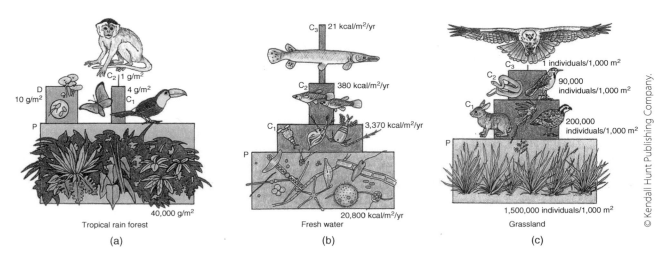

Figure 28.12a–c Pyramids showing biomass, energy, and number of organisms at each trophic level. (a) A large amount of biomass (grams per square meter) is required in the Panamanian rainforest to support a small amount of consumer mass while decomposer biomass is twice as large in this ecosystem where organic matter is recycled rapidly. (b) The Florida freshwater community shows that only about 10% of the energy is passed on at each trophic level with much energy lost as heat. (c) Fewer numbers of organisms can be supported at each trophic level, from primary consumers to the top of the food chain.

from an ecosystem, productivity will decline until new nutrient sources become available. Major nutrient loss occurs when a wildfire burns all existing vegetation and subsequent rains cause erosion of the top layer of soil. Human activities of harvesting crops and timber also remove substantial amounts of nutrients from a system. Whether caused by nature or humans, these lost nutrients remain in the biosphere and will migrate to different locations for reuse or they might be locked temporarily into sediments at the bottom of a lake or the ocean.

The rate of nutrient cycling is primarily dependent on the rate of decomposition. Environmental factors such as temperature and rainfall affect how fast nutrients can be recycled and extreme conditions of either factor generally slow down decomposition rates. The quality of organic matter also influences how fast microorganisms recycle the nutrients. Next, we examine the carbon and nitrogen cycles, which are of great importance for the growth of plants, the primary producers. Both carbon and nitrogen cycle through the atmosphere, land, and water via physical and chemical processes.

The Carbon Cycle

The main carbon reservoir is found in the ocean, followed by fossil fuels stored in the earth's crust, and third in plants and soil. Carbon moves from nonliving into living reservoirs in the cycle (Figure 28.13). The fossil fuels we burn for energy today started with the accumulation of organic matter that was deposited at the bottom of lakes and the ocean and was not decomposed. Geologic forces converted the material into natural gas, oil, and coal under large amounts of sediments. Major contributions to carbon dioxide release into the atmosphere are the burning of fossil fuels, forest fires, and methane production by anaerobic bacteria.

Carbon dioxide from the atmosphere is taken up or sequestered though photosynthesis and is stored in plant tissue. When passing through primary consumers and into the different trophic levels, carbon is stored in the tissue of herbivores and carnivores. When an organism dies, carbon is released through decomposition. Carbon dioxide is also released through the process of cellular respiration by all organisms. Human activities have significantly altered the global carbon cycle and have resulted in major changes affecting ecosystems. Recent increases in carbon dioxide concentration in the atmosphere and the consequences are discussed in Chapter 10 and in Box 27.1.

The Nitrogen Cycle

Nitrogen occurs in various forms in nature. The most abundant amount of nitrogen is found in the atmosphere which contains 78% by volume. Nitrogen in the atmosphere is present as nitrogen gas (N_2) and cannot be used by most living organisms. Nitrogen is a key component in life's basic processes and is found in amino acids, proteins, and nucleic acids. Nitrogen is the element necessary in greatest amounts for plant growth. It is the most limiting nutrient in most ecosystems and crop production systems and fertilization with nitrogen increases crop productivity significantly. Plants can take up nitrogen only in two forms, as nitrate (NO_3^-) or as ammonium (NH_4^+), with nitrate being the preferred source.

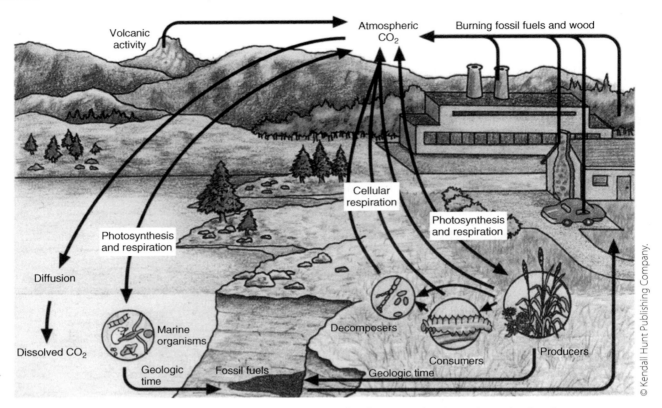

Figure 28.13 The carbon cycle reuses carbon through photosynthesis and respiration in ecosystems. Volcanic activity and carbonate rocks are natural sources of carbon dioxide while burning of wood and fossil fuels are anthropogenic sources of CO_2.

nitrogen fixation
Atmospheric nitrogen gas fixed into ammonia by bacteria in root nodules or by lightning.

mineralization
Conversion of organic material into inorganic forms by microorganisms.

nitrification
Two-step conversion of ammonia to nitrite and then nitrate by bacteria.

denitrification
Nitrates in the soil are converted into atmospheric nitrogen gas by anaerobic bacteria.

Nitrogen becomes accessible to plants through the processes of **nitrogen fixation**, **mineralization**, and **nitrification** (Figure 28.14). The processes of **denitrification**, immobilization, and volatilization renders nitrogen inaccessible to plants, but these pools of nitrogen remain in the biochemical nitrogen cycle and can become available again for plant growth at a later time or in a different location.

Biological nitrogen fixation is the process where nitrogen fixing bacteria (*Rhizobium* sp.) living symbiotically on the roots of plants or cyanobacteria (*Anabaena* sp.) convert the atmospheric nitrogen (N_2) to ammonia (NH_3). The ammonia forms ammonium (NH_4^+) when it dissolves in water. Industrial fixation of nitrogen from the atmosphere uses energy to break the triple covalent bond between the two nitrogen atoms and produces synthetic nitrogen fertilizer. Small amounts of ammonium are generated during forest fires and volcanic emissions and nitrates are generated during lightning. These natural sources of nitrogen enter the soil through rain and are available for uptake by plant roots. Nitrogen fixation in the rhizosphere and in the atmosphere converts the inorganic nitrogen from the atmosphere into a form that can be used by plants. Mineralization or ammonification is the process when nitrogen in organic material is converted by bacteria or fungi into ammonium. This catabolic process is accelerated under warm, moist soil conditions.

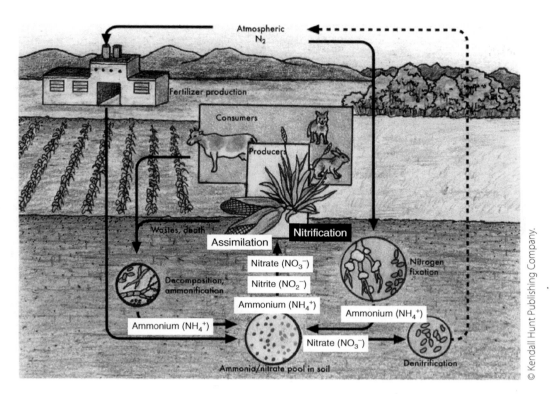

Figure 28.14   The nitrogen cycle. Atmospheric N$_2$ is fixed through bacteria in roots and converted to ammonium. Ammonium is released when decomposers break down organic matter (mineralization). Soil bacteria change ammonium to nitrite and nitrate (nitrification). Soil bacteria convert nitrate to atmospheric N$_2$ (denitrification). Plant roots take up nitrate and ammonium (assimilation).

Nitrification converts ammonium to nitrate by bacteria living in the soil. The two-step process starts with bacteria (*Nitrosomonas* sp.) oxidizing ammonium (NH$_4^+$) to nitrites (NO$_2^-$). High concentrations of nitrite in the soil can be toxic to plants therefore nitrite is immediately oxidized into nitrates (NO$_3^-$). Most plants prefer nitrogen in the form of nitrates and many commercial fertilizers are a mix of nitrate and ammonium. Because of their negative charge, nitrates do not adsorb to the clay and organic particles and are very mobile in the soil. They are easily leached below the root zone where they become inaccessible for plant roots and can accumulate in the environment. High levels of nitrogen pollute groundwater and aquatic communities and pose a health risk for infants and young animals.

Denitrification is the process where bacteria reduce nitrates to nitrogen gas (N$_2$). Losses to denitrification are relatively small and occur primarily under anaerobic conditions. In the gas form, nitrogen is again inaccessible for plant uptake. Immobilization of nitrogen occurs when microbes take up and assimilate ammonium or nitrates. Volatilization is the loss of ammonia from the terrestrial ecosystem to the atmosphere. This can be significant when ammonium containing fertilizers are not incorporated into the soil and become prone to volatilization.

Assimilation of nitrogen is the process when plants take up ammonium or nitrate and incorporate the nitrogen into amino acids, chlorophyll or

assimilation
Uptake of inorganic nitrogen as nitrate or ammonium by plant roots and conversion into organic nitrogen.

other molecules. Once part of the plant, nitrogen enters the food chain and is incorporated into different organisms. The reverse process of assimilation is mineralization.

Human activities have increased the amount of nitrogen added to the global nitrogen cycle such that it exceeds nitrogen fixation through natural sources. The sources contributing to increased nitrogen in the global nitrogen cycle are nitrogen fertilizer, nitrogen-fixing crops such as legumes, and fossil fuel. Burning fossil fuel releases nitrogen into the atmosphere and with oxygen produces nitric oxide (NO). This compound is a component of acid rain and photochemical smog. Nitrous oxide (N_2O) and nitric oxide have increased due to intensified agricultural use of fertilizer, which stimulates microbial nitrification and denitrification. Nitrous oxide is a greenhouse gas much more effective in trapping heat than carbon dioxide and contributes to global climate change.

The additional nitrogen in ecosystems leads to increased growth of some plant species, loss of biodiversity in terrestrial and aquatic ecosystems, acidification of soils, streams, and lakes, and pollution of coastal waters. Eutrophication is caused by excessive nutrients, mainly nitrogen and phosphorus, in water bodies and results in bloom of phytoplankton and excessive algae growth. These harmful algal blooms can cause direct mortality of fish and other organisms due to toxins or change species composition and food web dynamics of communities (Anderson, Glibert, & Burkholder, 2002). Red tides, another name for harmful algal blooms, have occurred with increasing frequencies all over the world in estuaries, coastal waters, and inland lakes and seas and have harmed aquatic life, ecosystems, and businesses. Improving the efficiency of fertilizer in agricultural production and decreasing the overall nitrogen fixation resulting from the burning of fossil fuels can help mitigate some of the negative consequences of human activities on the nitrogen cycle.

ECOLOGICAL SUCCESSION

succession
Gradual process of colonizing an area with organisms.

pioneer species
First species to colonize a bare area.

climax community
Stable ecosystem after several stages of succession

primary succession
Gradual process where an area without soil is first colonized by plants and other organisms.

Ecosystems change over time in a process called **succession** describing the gradual establishment of different species in a habitat. This starts with few organisms, called **pioneer species**, colonizing a bare area and over time evolving into a more complex ecosystem. The system progresses through several stages with changes in the dominant species. The final stage is a stable ecosystem known as **climax community**, which will persist until a disturbance rebalances the equilibrium between the species and their population size.

The two types of ecological succession that are distinguished are primary and secondary succession. **Primary succession** is the process when an area without existing soil previously not occupied by vegetation is colonized for the first time (Figure 28.15). This can be bare bedrock, a new lava field, or pond. If left undisturbed, bare rock will be colonized first by species such as mosses and lichen, which are known as primary successors. Soil begins to form as rocks under the lichen and mosses start to dissolve and organic matter is accumulating. Following, grasses, herbs, and shrubs can find a foothold in the shallow soil. As more and larger plants colonize an area, more soil develops; roots hold the soil in place and allow for greater water-holding capacity, which in turn improves the conditions for more plants to grow. Some of the plants during early succession will be replaced

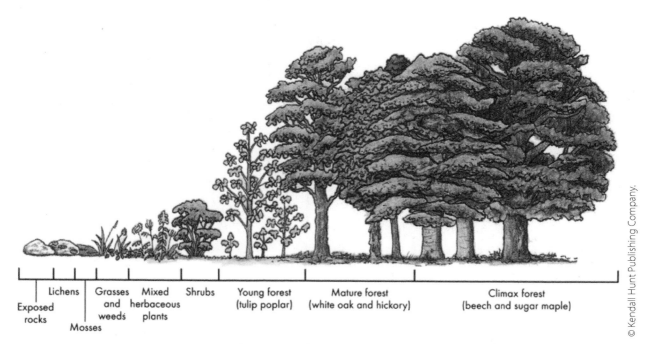

Exposed rocks | Lichens | Mosses | Grasses and weeds | Mixed herbaceous plants | Shrubs | Young forest (tulip poplar) | Mature forest (white oak and hickory) | Climax forest (beech and sugar maple)

© Kendall Hunt Publishing Company.

Figure 28.15   Primary succession in a temperate deciduous biome.

by different organisms from subsequent communities. The next succession step is the appearance of tree seedlings that cannot tolerate shading, such as tulip poplar trees. Other hardwood species follow in subsequent stages such as white oak and hickory. The stable climax community contains beech and sugar maple and once established can occupy the area for a long time. Each stage is characterized by the dominant species conferring the name to the system and accounting for the majority of organisms; however, other species of plants and other organisms are an integral part of the community.

Secondary succession is the process when vegetation is getting reestablished after existing vegetation is cleared by humans or natural disasters and is claimed again by natural vegetation. Soil is already present when secondary succession starts on abandoned farmland, surface mines, landslides, and flooded or burned areas where existing communities and ecosystems are disrupted. Plants will move into the disturbed habitat and develop through several succession steps until a climax community becomes reestablished (Figure 28.16). In a temperate climate, the first stage is typically dominated by annual weeds, which progress after a few years to perennial weeds and grasses. Then shrubs and larger woody plants become established. After hundreds of years a mature forest with different dominant species than the early tree successors will establish, often in a multilayered community.

Succession is a slow process and occurs over long periods of time. Primary succession starting with the buildup of soil to a climax community takes thousands of years. Secondary succession of abandoned farmland returning to a prairie can be complete within a few decades or for mature forests can take hundreds of years. The speed of succession differs based on the biomass production and is dependent on climate and availability of other resources. Aquatic ecosystems go through the process of succession just like terrestrial ecosystems (Figure 28.17). Animals and other organisms are

secondary succession
Gradual process where a community was disturbed and is being recolonized on existing soil.

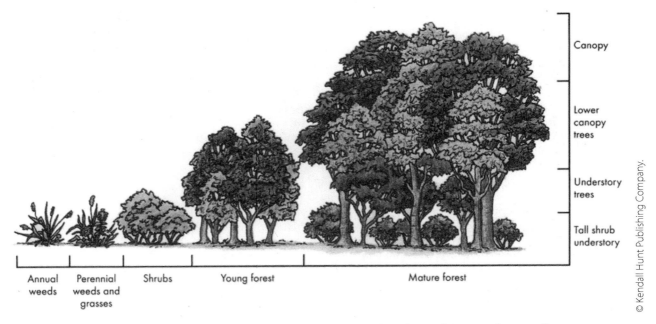

Annual weeds

Perennial weeds and grasses

Shrubs

Young forest

Mature forest

Canopy

Lower canopy trees

Understory trees

Tall shrub understory

Figure 28.16 Secondary succession occurs in a temperate deciduous biome where soil was present and a new community became established following disturbance.

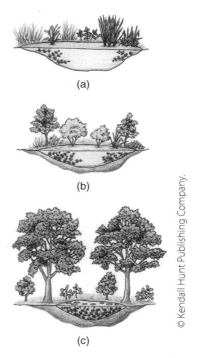

(a)

(b)

(c)

Figure 28.17a–c Succession in beach ponds on Lake Erie in Pennsylvania. (a) Pioneer species are bullrush, cattail, and chara. (b) After 50 years, cottonwood, bayberry, and willow are established. (c) After 100 years, ponds are smaller and cottonwood, *Myriophyllum*, yellow water lily, and pondweed are dominating.

intricately linked to the process of ecological succession as they participate in maintaining or changing the size of plant populations. Climax communities are more of a theoretical concept describing changes in a community. Biotic and abiotic factors will affect the equilibrium of different organisms. If the climate warms, different species will be favored in reproduction and competition. If water becomes scarce different species will become more competitive. Herbivores, diseases, and severe weather events can further change the composition of a community and populations.

Succession on Mount St. Helens Following a Volcanic Eruption

On May 18, 1980, Mount St. Helens in Washington State erupted and denuded 212 square miles (550 square kilometers) of forest and other ecosystems within a few hours. The debris from the eruption mixed with rock and melted snow and ice caused massive flows down the mountainside, choking rivers and streams and killing all vegetation in its way. Some areas were covered up to 148 feet (45 meters) thick with ash, pumice, and debris (Perkins, 2010). Other areas are still covered with dead trees killed from the intense heat of the eruption (Figure 28.18a).

The area affected by the volcanic eruption has been a living laboratory to study succession over the last four decades. Succession is not a straight development from one point to another; rather, it starts at different stages and its development can be changed by new disturbances (Mazza, 2010). Scientists found several different paths of succession (Figure 28.18b–e). Some started from organisms surviving the eruption, some migrated from the perimeter of less disturbed areas, and some were blown in by wind. In some areas no succession has been observed because of severe primary destruction and continuous follow-up disturbances. Proximity to riparian areas generally resulted in earlier and faster revegetation. Grasses, fireweed (*Chamaerion augustifolium*), and lupines (*Lupinus* sp.) were among the first plants to revegetate some areas within a few months after the eruption. Lupines enriched the soil with their nitrogen-fixing roots and resulted in more and diverse plants than areas where no lupines established. Red alder (*Alnus* sp.), a primary successor, is the dominant tree growing on debris of the avalanche deposits.

In the blowdown zone where trees were killed by the intense heat, the vegetation is more diverse than on the bare debris deposits. Some Pacific silver fir (*Abies amabilis*) and mountain hemlock (*Tsuga mertensiana*) have already grown 16.4–30 feet (5–9 meters) tall. These shade-tolerant trees dispersed in different patches are biological legacies from the former old-growth forest and survived the eruption as seedlings under a cover of snow. Other survivors include huckleberry (*Vaccinium parvifolium*) and vine maple (*Acer circinatum*), which are usually in the understory. Succession created a patchwork of plant and animal communities that include primary successors through species from previous old growth.

A new finding was that volcanic ash on the fallen trees rendered them less flammable and protected them from insects (Mazza, 2010). Quickly implemented logging to remove the dead trees in order to prevent an outbreak of insects was probably not as urgent as first believed. Human intervention affected revegetation when quick aerial seeding with non-native

Figure 28.18a–e (a) Trees in the blast zone at Mount St. Helens. (b) Early succession on a new lava field. (c) Herbaceous pioneer plants on Mount St. Helens. (d) Small patches of trees survived the blast under snow cover and have grown tall, but some patches are still with little or no vegetation. (e) Soil condition affects where plants can grow on Mount St. Helens.

plants to prevent erosion resulted in the establishment of some exotic plants in the area. Management practices have since changed to use native seeds in revegetation. Success of the natural revegetation calls into question whether artificial seeding should be done at all.

Key Terms

organismal ecology	coevolution	trophic
population ecology	carrying capacity	nitrogen fixation
community ecology	r/K selection theory	mineralization
ecosystem ecology	r-selection	nitrification
mutualism	K-selected populations	denitrification
commensalism	conservation biology	assimilation
competition	fragmentation	succession
predation	wetlands	pioneer species
myrmecophytes	pneumatophores	climax community
domatia	aerenchyma	primary succession
Beltian bodies	food chain	secondary succession
herbivory	food web	

Summary

- Ecology studies the relationships between organisms and their environment. Ecology is divided into four areas based on the level: organismal, population, community, and ecosystem ecology. Interactions between organisms include mutualism, commensalism, competition, and predation.

- Populations can be described based on their reproductive characteristics in the r/K model. Populations with r characteristics generally move into new, unstable habitats and can support a large number of organisms. Organisms from K populations are living in mature, stable communities, reproduce in small numbers, and have a long lifespan.

- Ecosystems are dynamic and in a state of flux. Humans have altered many ecosystems and have changed habitats of many species with the consequent loss of biodiversity. Conservation biology uses information from organismal to ecosystem studies to understand ecosystem dynamics with the goal of restoring habitat and preserving biodiversity.

- Energy in ecosystem flows through food chains and food webs describing who eats whom. Food chains usually have three to four links describing the chain from one organism that consumes another organism. Food webs are more complex and describe many interrelationships of energy transfer between organisms in an ecosystem.

- The trophic pyramid starts with producers that are autotroph plants, followed by primary consumers, secondary consumers, and so on until consumers at the top of the food chain. Decomposers convert organic matter back into usable energy for producers.

- Energy is lost at each step in the food chain and only 10 percent are passed from producers to primary consumers and subsequent consumers. Highest energy efficiency is achieved by herbivores. Ecosystems can be described in pyramids of biomass, energy, and number of organisms.

- Biogeochemical cycles describe recycling of nutrients or other resources in the ecosystem. The rate of nutrient recycling is regulated by the rate of decomposition.

- The carbon cycle describes how carbon moves from the major reservoirs to the ocean, the earth's crust through plants, soil, and other organisms. Photosynthesis is the major process to store carbon in plant tissue and introduce it into the food chain.

- The nitrogen cycle describes how the different forms of nitrogen are moved between the atmosphere and living organisms with the aid of bacteria. The major processes in the nitrogen cycle are nitrogen fixation, mineralization, nitrification, and denitrification.

- Ecological succession is the development of communities over time, gradually changing the composition and number of organisms in a community.

- Primary succession is the process when organisms start to colonize an area previously not inhabited by other organisms and devoid of soil. Secondary succession is the development of an ecosystem after a previous one has been disturbed. Succession stages start with pioneer plants and move through a number of stages to a stable, self-perpetuating climax community until another disturbance changes the equilibrium of the system.

Reflect

1. Ecosystems can be studied at different levels. Explain the role of each level.
2. Why are the global nitrogen and carbon cycle important for human existence? Describe the critical components and how humans have changed the natural cycles.
3. An infestation of leaf-eating larvae is developing without control in a temperate broadleaf forest. Provide a brief description of how a conservation biologist might study what is happening in this ecosystem.

4. Energy moves through ecosystems in food chains and food webs. Describe similarities and differences between a food web and a food chain.

5. Succession is an important process that has been altered by humans. Give examples of how humans have changed ecosystems in your area and how succession has claimed some of them back from human use.

6. What are some of the lessons we have learned studying succession after a volcanic eruption or after restoring a wetland?

7. What are some changes humans can consider to increase the food supply for the rapidly growing world population? Name three practices that should change and describe the long-term effect of the changes on sustainable food production.

8. Compare the timeline and organisms in the primary ecological succession in a short grass prairie versus a tropical rainforest.

References

Anderson, D. M., Glibert, P. M., &and Burkholder, J. J. (2002). "Harmful algal blooms and eutrophication: Nutrient sources, composition, and consequences." *Estuaries, 25* (2002):, 704–726.

Dahl, T. E. (1990). *Wetlands losses in the United States, 1780's to 1980's*. Report to Congress.

Janzen, D. H. (1966). Coevolution of mutualism between ants and acacias in Central America. *Evolution, 20*(3), 249–275.

Lamberson, R. H., McKelvey, R., Noon, B.R., & Voss, C. (1992). A dynamic analysis of northern spotted owl viability in a fragmented forest landscape. *Conservation Biology, 6*(4), 505–512.

MacArthur, R. H., & Wilson, E. O. (1967). *The theory of island biogeography*. Princeton, NJ: Princeton University Press.

Mazza, R. (2010). Mount St. Helens 30 years later: A landscape reconfigured. *Science Update*, pp. 1-9.

Moreno-Mateos D., Power, M. E., Comín, F. A., & Yockteng, R. (2012). Structural and functional loss in restored wetland ecosystems. *PLoS Biol, 10*(1), e1001247.

Pimm, S. L., Lawton, J. H., & Cohen, J. E. (1991). Food web patterns and their consequences. *Nature, 350*, 669–674.

Stenseth, N. C., Falck, W., Bjørnstad, O. N., & Krebs, C. J. (1997). Population regulation in snowshoe hare and Canadian lynx: asymmetric food web configurations between hare and lynx. *Proceedings of the National Academy of Sciences, 94*, 5147–5152.

Thomas, J. W., Forsman, E. D., Lint, J. B., Meslow, E. C., Noon, B. B. & Verner, J. (1990). A conservation strategy for the Northern Spotted Owl: Report of the Interagency Scientific Committee to Address the Conservation of the Northern Spotted Owl. Portland, OR.

Chapter 29

Plants as Food, Commercial Products, and Pharmaceuticals

© ifong/Shutterstock.com

Learning Objectives

- Describe the factors that led to the domestication of plants by humans
- Explain the significance of plants in the grass family for the world population
- Discuss the characteristics common to many of the major food staples in the world
- Explain the role fruits, nuts, and vegetables play for humans
- Discuss the consequences when people rely almost exclusively on one crop for food
- Summarize the importance of herbs and spices in food preparation and for other uses
- Describe the features of plants that make them useful for building materials, fuel, and for textiles
- Explain why modern medicine still relies heavily on plants

Plants are essential for humans to survive and thrive. Humans have used plants for many different purposes throughout history. Plants produce oxygen; store carbon; and provide food and beverages, fibers for many uses, building materials and fuel, chemicals for medicines, and spices and herbs. Different cultures have deep attachments to some plants for cultural or religious ceremonies. Plants have bestowed power and fortunes or have inflicted misery or death. Plants have changed history. Plants beautify our landscapes and homes and are used to express emotions. They inspire us to collect and breed them. Some plants are used for a single purpose such as food; others such as bamboo are used for a multitude of purposes from art supplies to furniture to building materials for houses to bamboo shoots for food. Agaves are another versatile plant used for fiber for clothing, ropes, rugs and baskets as well as making the liquor tequila.

HISTORY OF PLANT DOMESTICATION FOR FOOD AND OTHER USES

Plants and other organisms have supported humans to develop and thrive for thousands of years. Our primitive ancestors lived on plant parts such as roots, leaves, seeds, berries, and fruits and supplemented their vegetarian diet with game, fish, and eggs. Humans relied on foraging and hunting until about 10,500 years ago when they began to cultivate plants as crops. This allowed them to store food and have resources during times of crop failure. They could also feed more people that were living in larger settlements or towns. Today there is only a very small population of nomadic people in some deserts and rainforests who forage and hunt for food.

Crop cultivation started with the collection and preservation of desirable seeds, seedbed preparation and planting, plant protection from herbivores and competing vegetation, and gathering, processing, and storing crop products. Rice and soybean remnants in Thailand from the Neolithic period about 10,000 years ago are interpreted as evidence for the first domestication of plants. The grains emmer and barley were found at a site in Iran dating to 6750 B.C. Early Chinese agriculture ventures started in the valleys of the Yellow River and Yangtze River with millet production around 7500 B.C., soon followed by the cultivation of rice. In the New World crops such as corn, squash, chili peppers, and potatoes were found among the earliest crops in the highlands of Central and South America.

Egyptian agriculture is documented in harvest scenes of hieroglyphs, and specimens of seeds and plants in ancient tombs from 5000 to 3400 B.C. The Fertile Crescent, the area from the Mediterranean Sea to the Persian Gulf, is the origin of agriculture (Figure 1.4). Wheat, barley, and legumes were the first foods cultivated upon which the Egyptian and Sumerian civilizations were built. With some people free to pursue activities other than hunting and gathering food, cultures started to develop. The arid climate, fertile soils, and access to river water brought many advances in agriculture. Written documentation of plants and their properties on papyrus were found dating to 1550 B.C. Greek and Roman scholars wrote about plants and agricultural production.

Important plants during this early agricultural period were wheat, corn, rice, and barley. These grains formed the basis of staple foods and were supplemented with legumes and meat. Different geographical areas specialized in crops appropriate to their environment. In the first century A.D., Romans became the first commercial bakers. Yams were planted as a staple food starting 10,000 years ago in Africa. Lentils and fava beans were cultivated about 6,500 B.C. in the Mediterranean and dates were cultivated in Pakistan. Linen, a cloth produced from flax, was made as early as 5000 B.C. while cotton started to be cultivated in Mexico around 1300 A.D. More than 5,000 years ago silkworms were domesticated in China. They were selected to consume mulberry (*Morus alba*) leaves and to produce precious silk (Levetin & McMahon, 2012).

From 1450 to 1650 trade and discovery of plants and plant products played an important role in nutrition, medicine, and the spice trade (Howell, 2009). During this period gardens became established in different parts of the world. Sugar was introduced to different countries. Many spices were brought from Asia to Europe with different countries vying for dominance of the trade. Spices such as black pepper, nutmeg, and cinnamon became important goods. Vanilla and chocolate came from the Americas and were originally only available to the wealthy. Although paper was invented in 105 A.D. in China, it was not widely available in Europe for another 1,500 years until the printing press was invented. Soon after, more books on plants, including botanical descriptions, medicinal and herbal properties, and agricultural methods were published. From the middle of the 17th century until the middle of the 18th century botanical explorers roamed the different continents and published natural histories including descriptions of the flora about different geographical areas. Plants of economic value such as spices and sugar and plants having medicinal and ornamental interest were moved from their native countries to different countries. Coffee and tea became popular beverages outside their native countries. Subsequent periods are marked by more trade of plants between countries and continents. The rise of significant commercial products such as sugar, pineapple, coffee, and plant nurseries raising plants for sale supported the practice of growing more plants outside their indigenous locations. Advances in science in the last 150 years have led to greater understanding of different plant properties and ways to cultivate and propagate plants. New uses for plants are continuously discovered.

The edible plants our primitive ancestors consumed are different from today's modern plants, which are produced in small- or large-scale agricultural operations. Domesticated plants have been selected over time for specific traits such as larger cob size and larger kernel size of corn, nonshattering heads of grains, and larger tubers, fruits, and seeds. Corncobs from 8,000 years ago were very short and had just a fraction of the 600–1,200 kernels found in a contemporary ear of corn (see Box 1.1). In their wild form, heads of wheat and rice shattered easily when ripe and scattered their seeds widely to reproduce. Early gatherers selected the seeds of those plants that remained on the head and were nonshattering.

Food security is the basis of a healthy population and today still varies greatly in different areas of the world. In their quest to provide enough food, people have changed natural ecosystems and have altered natural biogeological cycles to where some ecosystems cannot fulfill their natural functions anymore. Hunters and gatherers and early agricultural societies learned about the perils of overharvesting food in one area and degrading the land. Today, slash and burn, clearing vegetation and using the fertile soil for a short time before it is eroded or degraded, is still practiced in some parts of the world. While this chapter is focused on plants and their use as food, pharmaceuticals, and commercial products, we have to remember plants filter our air, regulate the hydrologic cycle, and protect soils from degradation, to name a few ecosystem services.

FOOD PLANTS ESSENTIAL TO HUMANS

Of the more than 300,000 plant species known, scarcely over 100 cultivated species supply more than 90 percent of the energy eaten as food by people (Howell, 2009). Maize, rice, and wheat are the most important food staples, providing more than 60 percent of the calories for humans.

Grasses: Maize, Rice, Wheat, and Other Grasses

Grains or cereal crops that were the basis of ancient agriculture are still the most important food staples for the world today. Maize, rice, and wheat account for more than half of the food calories consumed by the world population (Food and Agriculture Organization, 2010). These grains provide carbohydrates, protein, and essential minerals and vitamins. Other crops from the grass family ranking among the top 35 food crops are barley, sorghum, and millet. Oats and rye are also of importance regionally. Cereals are cultivated worldwide in cool, temperate, subtropical, and tropical climates. Some cereals are warm-season C_4 plants and include maize, millet, and sorghum. Cool-season C_3 cereals are rice, wheat, barley, rye, oat, and wild rice.

Cereal crops are in the Poacea or grass family (see Chapter 25) and produce a seed packed with energy. The center of the seed, or botanically the dry fruit, consists of the embryo or germ, which is a good source of oils, enzymes, and vitamins, and the starchy endosperm. This is surrounded by the protein-, vitamin-, and oil-rich aleurone layer. The surrounding seed coat is fused to the outer fruit wall and is known as the bran, providing fiber from the grain. The refining process produces foods such as white flour or white rice and removes the chaff or bracts around the grain and the bran and aleurone layer. What is remaining is the endosperm, high in starch and with fewer nutrients compared to whole-grain foods where only the chaff is removed.

Table 29.1 Production of Major Food Crops and Areas of Harvested Land Worldwide, 2010

Crop	Production (1,000 tons)	Area Harvested (1,000 ha)
Cereals	2,476,416	693,701
Vegetables	1,044,380	55,598
Starchy roots and Tubers	747,740	53,578
Oil Crops	170,274	269,680
Pulses	68,829	78,311

Source: FAO Production Yearbook 2013. January 2, 2018, from http://www.fao.org/docrep/018/i3107e/i3107e.PDF

The dry grains of cereal crops can be easily transported, stored, and milled. This is an advantage over tubers and starchy fruit such as sweet potato, cassava, yam, and plantain, which have shorter shelf-lives. Cereal grains provide carbohydrates, oils, and many essential vitamins for human and animal nutrition.

The seeds of some broadleaf plants are used like cereal grains, although they are not in the *Poaceae* family. Those plants are called pseudocereals and include buckwheat (*Fagopyrum esculentum*), amaranth (*Amaranthus caudatus*), and quinoa (*Chenepodium quinoa*).

Maize   Maize (*Zea mays*) is native to America and was domesticated from the grass teosinte (*Zea diploperennis*) around 5500 B.C. in Mesoamerica-(Levetin & McMahon, 2012). According to archaeological evidence, cultivation of the grain spread throughout the American continent from Mexico starting about 5,000 years ago. Maize is known as corn in North America and Australia. It is the most widely grown cereal in the world. Corn is consumed as whole grain, coarsely ground or as ground cornmeal in many cooked or baked foods. Corn is converted to high-fructose corn syrup, Bourbon whiskey, cooking oil, breakfast cereals, and cornstarch (Figure 29.1d). Corn not only serves as food for humans, it is a major source of animal feed for livestock, the biofuel ethanol, and other industrial products such as plastics, fabrics, and pharmaceutical products. Sweet corn cultivars are grown for human food; field corn is cultivated for animal feed. Overall, about 40 percent of the United States' corn crop is grown for animal feed and 33 percent for biofuel. In contrast, corn for human consumption accounts for only 1.5 percent and for seed corn only 0.2 percent. Corn is also used for industrial, pharmaceutical, export, and other purposes. Ancient people prepared maize with a lime treatment to make niacin or Vitamin B_3 available. Without this process a diet with corn as the main staple can cause pellagra, a disease due to vitamin deficiency.

Teosinte is believed to be the ancestor of the modern corn plant, but evidence is not definite. Teosinte is a short grass plant with multiple terminal male inflorescences and several small ears of corn carrying about 6–10 kernels each. Kernels on the ears were surrounded by a hard fruit case and arranged on a spike that shattered upon maturity. Modern corn plants consist of a single stem 2–3 m tall with one broad, long leaf emerging from several nodes along the stem. Corn plants are monoecious with the tassel, the staminate flowers, located at the terminal end of the stem (Figure 29.1a). The lateral female inflorescence is a spike and entirely covered by leaves called *husks*. Silks look like soft hair and are elongated styles, with the stigma emerging from the leaves that cover the flowers. Each silk needs to be pollinated to develop into one grain on the ear of corn. The center of the ear is the stem of the female flower and unique to corn among the grasses. Pollen released from the tassel will spread by wind and pollinate nearby silks. Ripened kernels remain attached to the center of the spike, the corncob, and will not shatter (Figure 29.1b–c). When the ears are immature they are usually consumed fresh. As they mature they dry out and are unsuitable as fresh food, but are then used for milling and other purposes.

Corn has been used worldwide as a model plant for genetic studies for many decades. Each kernel is pollinated separately and is therefore a different genotype. A whole population is present on one ear of corn. Great

Figure 29.1a–d (a) Corn plants with terminal staminate inflorescence and female spike developing into an ear of corn. (b) Dried corn ready for harvest. (c) Multicolored corn gets its color from the pigments in the aleurone layer and the pericarp. Yellow corn has the pigment in the endosperm. (d) Sampling of products from corn.

aflatoxins
Compounds in foods such as nuts, corn, or cotton seed caused by the infection with *Aspergillus flavus* or related fungi that produce toxic metabolites.

variability is found in corn and the crop is easily grown in about 5 months. The large corn chromosomes are easily visible even under a light microscope. One of the major genetic discoveries using multicolored Indian corn (Figure 29.1c) were transposable elements or jumping genes for which Barbara McClintock was awarded the Nobel Prize in 1983 for her work in the 1950s. McClintock, an American scientist, studied genes on chromosomes of corn and found the genes responsible for the different colors could move from one chromosome to another. The genome of corn has more than 32,000 genes and has been completely mapped in 2009.

Extensive breeding experimentation has been conducted on maize to produce thousands of modern cultivars with desirable traits to enhance the genetic base of corn. Traits of interest are overall biomass production; kernel size; grain composition such as protein, oil, starch, and water content; sugar content of stems; tolerance to drought, high temperature, insects, and disease; and **aflatoxin** resistance, to name a few. Adaptations to different climatic conditions and length of growing seasons are also important to accommodate the wide geographic and climatic range where corn is produced. Genetically modified (GM) corn is widely grown in most countries other than in Europe. GM corn has been engineered for herbicide resistance, insect resistance, and increased content of the naturally low amino acids tryptophan and lysine.

Rice Rice (*Oryza sativa*) is the main staple for people in Asia and feeds more than half the earth's population (Figure 29.2a). Rice has been cultivated for more than 10,000 years, starting in China. Upland rice is grown in regular fields but makes up only a small percentage of cultivated rice. About 90 percent of rice is grown in flooded fields or paddies where the water is 5–10 cm deep (Figure 29.2b). The rice plant is an annual monocot that grows with several stalks to a height of 1.0–1.5 m. Grains are surrounded by bracts and are arranged in a terminal panicle. Once the rice is harvested it has to be dried before being milled where the husk is removed. Today China and India are the top rice-producing countries, although rice is cultivated worldwide in areas where temperatures are favorable.

There are thousands of rice cultivars differing in grain size, shape, color, flavor, and consistency of the cooked product. The two most important subspecies are *indica* and *japonica*. Long-grain rice (*O. sativa* subsp. *indica*) is high in amylose starch and separates into individual grains after cooking. Long-grain rice is grown in tropical and subtropical lowlands, primarily in flooded fields. Short-grain rice (*O. sativa* subsp. *sativa*) is also known as japonica or sticky rice because the grains will adhere to each other when cooked. This rice as well as medium-grain rice is high in amylopectin but low in amylose starch. It is cultivated in upland tropical and temperate areas. Wild rice (*Zizania sp.*), with long grains dark brown to black in color, is a different genus from *Oryza sativa* rice. It is native in the upper Midwest and Eastern Canada and China.

Conventional breeding continues to develop new cultivars with desirable traits related to environmental adaptation, yield, and nutritional improvements. The International Rice Research Institute, devoted to rice research worldwide, has a collection of more than 100,000 rice accessions. The rice genome has been sequenced in 2002, the first of the cereals. Rice has the smallest genome of the cereals, with 430 Mb (millions of base pairs). This makes rice a good model system for genetic research and genetic modification. GM rice has been developed for enhanced nutrition, flood tolerance, insect resistance, and herbicide tolerance. Golden rice synthesizes beta carotene, a precursor to Vitamin A, which is lacking in regular rice (see Chapter 17). People who rely on rice as their main staple suffer from Vitamin A deficiency, which leads to blindness. Although rice is grown in shallow flooded paddies, the plant is intolerant of being covered entirely by water for more than a few days. GM rice with flood tolerance is now available and will reduce losses to flooding worldwide. Resistance to insects and herbicides in GM crops boosts yields and reduces the input for the rice crop.

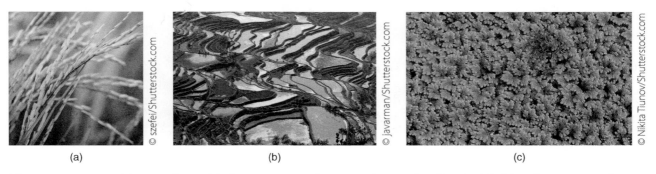

(a)   (b)   (c)

Figure 29.2a–c   (a) Rice plant. (b) The majority of rice is cultivated in flooded fields, often terraced on mountain slopes. (c) The nitrogen-fixing water fern azolla is often cultivated with rice.

Rice is often grown with azolla (*Azolla spp.*) (Figure 29.2c), which are several species of aquatic ferns fixing nitrogen. The blue-green algae *Anabaena azollae* has a symbiotic relationship with the weed azolla (see Chapter 20). The small ferns float on the surface of rice paddies and other aquatic crops and add nitrogen, shade out other weeds, and add organic matter to the production areas.

Rice is often milled, which removes the bran and germ, resulting in white, polished grains. Brown rice, only hulled and not milled, contains more minerals and vitamins, similar to unmilled wheat. The lack of thiamine, Vitamin B_1, in white rice causes a nutrient deficiency known as *beriberi*, when people rely on white rice as their main nutrition source. Rice is **gluten** free, an important attribute for food sought by people with gluten allergies. Although rice is prepared in many ways, it is also made into gluten-free noodles.

gluten
Protein found in the endosperm of wheat and other cereals, making dough elastic and helping it to rise.

Wheat Bread wheat (*Triticum aestivum*) (Figure 29.3) is also known as the staff of life because when ground, mixed with water, and baked, bread has served as a staple for thousands of years. Leavened bread was discovered by Egyptians. They added yeast to the dough, fermenting it and trapping carbon dioxide, which resulted in a light bread. Leavening requires a certain amount of gluten, proteins that make dough elastic. Only wheat and rye flour contain enough gluten for leavening; barley and oat grains lack sufficient gluten to allow the dough to rise.

Grain was originally milled by stones to break down the grain and the whole grain was used for preparing bread. Later on, the milling process improved and steel rollers refined the grain, removing all of the bran and germ and leaving mostly the starchy endosperm. Longer shelf-life of bread resulted from this change because the oils from the germ were taken out. Along with it essential minerals and vitamins were removed, leaving refined wheat flour a nutritionally inferior product compared to the whole-grain type. Today, many grain products are enriched or fortified with the minerals and vitamins lost in the refining process.

There are different types and species of wheat. The first cultivated species of wheat was einkorn (*T. monococcum*), a diploid species cultivated first in Turkey (Bacon, 2008). Emmer wheat (*T. dicoccon*) and durum wheat

(a)

(b)

Figure 29.3 Wheat is the second most important cereal grain and is grown from extreme northern locations to tropical climates due to many cultivars developed by breeders. Modern cultivars are short to prevent lodging.

(*T. durum*), both tetraploid species from a cross between wild wheat and wild goatgrass (*Aegilops sp.*), were the important grains for Mediterranean civilizations 3,000 years ago. Today's bread wheat (*T. aestivum*) and spelt (*T. spelta*) are hexaploids and originated about 6,000 B.C. as a cross between goatgrass (*Aegilops tauschii*) and tetraploid wheat species. Early wheat species are hulled with tough glumes surrounding the grain. After threshing, the spikelets require further processing to remove the hulls. Bread and durum wheat are free threshing; they bear naked fruit yielding the grains after threshing without further need for processing. Other important food grains with hulled fruit are oats, barley, and rye.

Bread wheat accounts for the majority of wheat grown today, which is used for bread, cereals, and pastries. Durum wheat is the second most important wheat species after bread wheat. It is the basis for Italian pasta because of the high protein content. Semolina is the endosperm of the ground durum wheat or other grains after being crushed and sieved. Bulgur, a parboiled whole-wheat product, and couscous, made of wheat semolina, are important staples from durum wheat in North Africa. Breeders have developed thousands of modern wheat cultivars to maximize yields under different environmental conditions and to provide wheat suitable for different purposes. These cultivars vary in protein content, gluten content, adaptability to various environmental conditions, and disease resistance. In the United States red and white wheat are distinguished based on their grain color. Soft and hard wheat differ in protein content. Hard wheat has high protein content and is commonly used for bread whereas soft wheat with lower protein content is favored for cakes and pastries.

Other Grains Important to Humans Sorghum (*Sorghum bicolor*; Figure 29.4a) is a C_4 grass with grains in a terminal inflorescence. Sorghum is important as food in Africa, India, and Asia. It is used primarily for forage, biofuel, and syrup production in North America. *Millet* is a term for a number of species of small-grain cereals. Pearl millet (*Pennisetum glaucum*; Figure 29.4b) is the most important millet grown and has been an important staple in India and some areas of Africa for thousands of years. Among the cereals, millet is most tolerant to arid conditions, poor soil fertility, and harsh environmental conditions.

Barley (*Hordeum vulgare*), similar in appearance to wheat (Figure 29.4c), is among the earliest cereals domesticated in the Fertile Crescent. It is used for animal feed and is a main ingredient in beer and other fermented beverages. A very small amount of barley is milled and used for human consumption. Barley production has decreased in the last decades. Leading production areas are Eastern Europe and Canada.

Rye (*Secale cereal*) has been domesticated more recently, only for a few thousand years. Rye is similar in appearance to wheat and thrives in colder, drier regions than wheat, making it popular in Northern Europe. The grain has a lower gluten content, rendering heavier bread such as pumpernickel. Rye is also used as animal feed and for alcoholic beverages. Breeders have crossed rye with wheat, which yielded triticale (× *Triticosecale*) about 100 years ago. Triticale combines the best features of both grains with high protein and lysine content, higher yields, disease resistance, and cold hardiness.

Oat (*Avena sativa*) grains grow in an open panicle with several florets (Figure 29.4d). Oats were cultivated more recently and thrive in temperate

Figure 29.4a–d Other important grains for food, animal feed, and energy are (a) sorghum, (b) pearl millet (Purple Majestic variety), (c), barley, and (d) oats.

regions such as northwest Europe and Canada. They are used for food, in beverages, and as horse fodder. Oatmeal is considered a health food due to its cholesterol-lowering properties.

Legumes

Legumes are plants in the Fabaceae family (see Chapter 25) used for food, forage, and cover crops. Legume refers to the fruit of a plant in this family, which is commonly called a *pod* and botanically a simple dry fruit that usually dehisces. Peanuts (*Arachis hypogaea*) are an exception in that they are an indehiscent legume fruit. **Pulse** is another term used for legumes and refers to plants harvested for their dry seed. Legumes produce their own nitrogen through *Rhizobium* bacteria in their roots and are rich in protein and oils. These plants are well adapted to arid climates and thrive in hot, dry conditions.

Beans, peas, lentils, and peanuts are legumes and are an important source of protein, especially for people who choose not to eat or cannot afford animal protein. Beans, peas, and lentils have been cultivated for thousands

pulse
Dried seeds of legumes.

of years. Several species of beans are grown with the common bean (*Phaseolus vulgaris*), which include haricot, black, red, white, brown or mottled seeds, and are the most cultivated species in the world (Bacon, 2008). Other bean species are the large-seeded broad or fava bean (*Vicia faba*) from the Mediterranean or southwestern Asia; the tepary bean (*Phaseolus acutifolius*) with rich, nutty flavor originating from Central America; and the adzuki bean (*Vigna angularis*) from tropical Asian origins and popular in Japan and China. Chickpeas or garbanzo beans (*Cicer arietinum*) were cultivated in the Fertile Crescent more than 6,000 years ago and are an important staple in India, North Africa, and the Middle East. Chickpeas have the third highest protein content after soybeans and peanuts and are prepared into well-known Middle Eastern dishes such as hummus and falafel. Common beans are cooked and eaten in their pods when young, mature beans are larger and hard and require soaking before cooking. Mung beans (*Vigna radiata*), originating in India, are very small and do not require soaking before cooking. They are used as sprouts; their seeds are cooked or ground for starch to produce vermicelli noodles. Beans are an ingredient in many signature dishes of different cultures and include refried beans, chile, or sweet bean paste in desserts.

Soybeans (*Glycine max*; Figure 29.5a) are native to East Asia and have been grown there for 3,000 years. They were introduced to other continents in the 18th century (Bacon, 2008). The United States adopted soybeans as a major crop in 1915 and is now leading world production followed by Brazil, Argentina, and China. Soybeans mature in two to four months under optimum growing temperatures of 68–86°F (20–30°C). Genetically modified soybeans were introduced in 1995 to tolerate the herbicide Roundup. Today, Roundup Ready® soybeans are almost exclusively cultivated in North and South America, but not in European countries. Mapping of the soybean genome was completed in 2010.

Soybeans contain 40 percent protein, the highest protein content among plants, and 20 percent oil, thus making it a very valuable commodity for animal feed, industrial use, and food. Soybeans are traditional staples in the

(a)   (b)

Figure 29.5a–b   (a) Soy beans and (b) peanuts are important legumes for food because of their high protein content and as oil crops.

human diet in Asian cultures. They require cooking or fermenting because they are indigestible when consumed raw. Typical food products include tofu, soy sauce, soy milk, and fermented soybean products such as tempeh and miso (Levetin & McMahon, 2012). Soybeans are manufactured into many dairy-type products and are important to people who cannot tolerate lactose in cow's milk. Soybeans are also processed into textured vegetable protein, spun soy fibers flavored and shaped to resemble meat products. Soy products are also valued for the **isoflavones** they contain, phytoestrogens attributed to have several health benefits such as lowering cholesterol and preventing or reducing recurrence of certain cancers. Soy oil is used in many food and industrial applications, including cooking oil, ink, and biodiesel.

Peanuts originated in South America and likely were already cultivated in pre-Inca times. Peanuts develop in the ground on an annual herbaceous plant (Figure 29.5b) and contrary to their name are not nuts but seeds of a legume in a pod. Flowers grow aboveground and die after pollination. The flower stalk or pedicel turns toward the ground, pushing the developing fruit into the soil. Fruits develop one to four ovules, mostly two per pod within four to five months (Bacon, 2008). Peanuts contain about 25 percent protein, the second highest level after soybeans and more than any nut, and have an oil content of about 45 percent. The nutritional value of peanuts is high, especially in essential nutrients such as several vitamins and minerals. Peanuts are used worldwide for their oil, in sweet and savory dishes, and the majority of peanuts in the United States are made into the popular peanut butter. In the United States, peanuts were grown more widely, especially in the South, based on the work of George Washington Carver (1864–1943), who devised many industrial and culinary uses for peanuts and promoted their cultivation. The leading peanut-producing country is China. Allergies to peanuts and contamination with aflatoxins can pose health risks through food and animal feed. Peanuts can cause allergies with mild to severe symptoms, affecting 1–2 percent of the population in the United States. Aflatoxins develop when peanuts, but also corn or cottonseed, are infected by *Aspergillus flavus* or related species, fungi that produce toxic metabolites.

Potatoes, Cassava, Sweet Potatoes, and Other Starches

Underground storage organs rich in carbohydrates are an important food source for humans and animals. Potatoes (*Solanum tuberosum*; Figure 29.6a) are stem tubers high in starch, originating in the Andean highlands of southern Peru where people domesticated wild plants about 10,000 years ago. This plant became the staple of the Inca civilization in South America. Potatoes were introduced to Spain in the 16th century and became a major food staple throughout Europe about 200 years later. Potatoes are the third largest staple food in the world and about one-third of world potato production is in China and India. There are thousands of potato cultivars known and tubers come in different sizes, shapes, and colors of white, yellow, or purple. Potatoes are a cool-season crop that can be cultivated from sea level to elevations above 4,500 meters. Potatoes are very efficient in using water and produce higher yields per area than cereals.

isoflavones
Phytoestrogens that have possible health benefits for lowering cholesterol or that are used in cancer treatment.

All aboveground parts of the potato plant are poisonous due to the **glycoalkaloid** solanine. Solanine is found in plants of the nightshade or Solanaceae family (see Chapter 25), such as tomato and eggplant. Tubers of potatoes are not poisonous, except when they are exposed to excessive light after harvest and develop large green patches, which contain solanine. The nutritional value of potatoes lies in its carbohydrates, fiber, protein, vitamins, and minerals. From its introduction to Europe until the middle of the 19th century, potatoes became the main food for peasants in Ireland. The pathogen *Phytophthora infestans* destroyed stems and leaves of potatoes, killing plants in a short time and causing widespread famine and death among the Irish population from 1845 to 1849 (see Chapter 26). Potato production and use have surged in developing countries in the last 50 years. Potatoes are used fresh and are fried, boiled, or baked. They are also used for their starch as animal feed and seed potatoes and are fermented into alcoholic beverages such as vodka.

Cassava (*Manihot esculenta*) is a woody shrub producing edible tuberous roots (Figure 29.6b). It is a crop of great importance to people in tropical and subtropical areas, especially Africa, Asia, and South America. Cassava originated in South America and is widely grown in tropical lowlands. The plants can tolerate a wide range of soil and moisture conditions, making them a popular crop in drier regions. Cassava takes 8–16 months until harvest after stem cuttings are planted. Cassava belongs to the Euphorbiaceae or spurge family, the only food crop in this family of plants containing a white milky sap that can be an irritant. The roots and leaves of cassava are eaten, but proper cooking is required to remove the toxic cyanides. Cassava is classified into sweet or bitter depending on the level of cyanides. The roots contain high levels of carbohydrates and minerals but are low in protein. Cassava is often ground into flour and then cooked with other ingredients or used as tapioca. It is also fried or fermented into alcoholic beverages.

The sweet potato (*Ipomoea batatas*) is a tuberous root vegetable (Figure 29.6c). This herbaceous perennial vine in the Convulvulacea or morning glory family is native to Peru. The plants grow well in poor soil with little water and have become an important food staple in Africa and Asia (Bacon, 2008). This crop needs a warm growing season and cannot tolerate frost. The tubers vary in color from beige, yellow, orange, red, and purple. They are prized for high carbohydrate levels, sugar, Vitamin A and beta-carotene, and minerals. Sweet potatoes are used boiled, fried, baked, or ground into flour. George Washington Carver developed more than 100 products from sweet potato, including glue, starch, and dehydrated sweet potato.

True yams (*Dioscorea sp.*; Figure 29.6d) are sometimes confused with sweet potato. The plant, an herbaceous, perennial vine with underground tubers, is an important food source in many tropical countries. The white- to orange-colored tubers are rich in starch. Tubers can vary from the size of a potato to over 2 m long and weighing over 40 kg.

Bananas (*Musa acuminata, M. balbisiana*; Figure 29.6e) grow from a corm and are herbaceous perennial plants in spite of their shape and size resembling a tree. What appears to be the trunk of the plant are the tightly packed petioles broadening into a sheath below the leaf blade. This **pseudostem** produces one inflorescence with up to several hundred bananas developing from one flower. The stem then dies and new pseudostems emerge at the plant base. Bananas contain high amounts of starch and sugar and are

glycoalkaloid
Alkaloids in plants from the nightshade (Solanaceae) family.

pseudostem
Structure appearing like a stem, but composed of folded or rolled petioles and leaf blades; found on the herbaceous banana plant.

Figure 29.6a–e (a) Potato plants in the Solanaceae family are toxic except for the tubers. (b) Cassava tubers contain toxic cyanides that are removed by special preparation. (c) Sweet potato and (d) yams are important starch crops. (e) Banana bunches carry many fruits from one flower.

the fourth most important fruit worldwide. Native to Southeast Asia, many banana cultivars are grown in tropical and subtropical regions. Some are consumed raw such as the Cavendish banana popular in North America and Europe; many other cultivars are cooked, fried, baked, or dried. Bananas are a good source of some vitamins and minerals. Fibers from the plant have been used to make textiles and paper.

Fruits, Nuts, and Vegetables

Humans require macronutrients such as carbohydrates, proteins, and fats, which are partially supplied by many of the grains and starchy tubers discussed earlier. Vitamins and minerals are essential micronutrients in the human diet, which are also partially provided by these staples. Fruits, nuts, and vegetables are a major source of energy and micronutrients (Figure 29.7).

(a)   (b)

Figure 29.7   Fruits and vegetables are important for a healthy diet because they provide vitamins, minerals, and phytochemicals, in addition to calories for energy.

However, they also contribute **phytochemicals**, a group of compounds produced by plants that have biological activity (Higdon & Drake, 2013). Phytochemicals are considered beneficial for human health, but not essential. These compounds include carotenoids from yellow, orange, and red fruits and vegetables; curcumin from the spice turmeric; indole-3-carbinol and isothiocyanates from cruciferous vegetables such as broccoli, kale, and cabbage; resveratrol from grapes, red wine, peanuts, and certain berries; and isoflavones from soybeans. Diets rich in plant-based foods have many health benefits. However, the specific effect of individual phytochemicals to prevent a specific disease or condition or to treat one has not been conclusively proven.

Antioxidants are chemical compounds preventing oxidation of molecules and the production of free radicals. These free radicals can cause reactions damaging cellular DNA, lipids, and proteins or killing cells and initiating disease. The consumption of fresh fruits and vegetables that are rich in antioxidants prevents cardiovascular disease and is thought to prevent several other diseases. Antioxidants found in many fruits, vegetables, and nuts are Vitamins E and C and carotenoids. The antioxidants resveratrol and **flavonoids** are contained in coffee, tea, chocolate, and some spices and foods discussed earlier.

Members of the Rosaceae or rose family (see Chapter 25) are important sources of stone fruit and pome fruit. Remember from Chapter 8 that stone fruits are plums, peaches, apricots, and cherries; apples and pears are pome fruit. Aggregate fruit such as strawberries, raspberries, and blackberries also belong to this family. Apples (*Malus × domestica*) are native in Europe, Asia, and North America and are the most consumed fruit worldwide (Bacon, 2008). Thousands of cultivars have been bred and grow well in temperate climates. Apples are eaten fresh, cooked, or used in pastries and for alcoholic beverages such as apple cider, cider vinegar, and apple juice. The European pear (*Pyrus communis*) and the Asian pear (*Pyrus pyrifolia*) also grow well in temperate climates. They are mostly eaten as fresh fruit. Pears can be picked before fully ripening to increase longevity in cold storage. Apples and pears are good sources of fiber and carbohydrates.

phytochemicals
Compounds produced from plants that have biological activity and are thought to be beneficial but are not essential to human health.

antioxidant
Chemical compound preventing oxidation of molecules and production of free radicals, which can damage cells.

flavonoids
Secondary plant metabolites with potential protective properties for human health.

Of the stone fruits, apricots (*Prunus armeniaca*) were cultivated 4,000 years ago in their native China. Apricots grow on small deciduous trees in climates with temperate, cool winters. The perishable fruit is eaten fresh but is often dried, poached, or baked in desserts. Peaches are also native to China and were cultivated there for at least 3,000 years. Nectarines have been around for over 2,000 years. Both fruits are excellent for eating fresh when they are ripe from the tree, but also popular baked, poached, or grilled. Cherries originated near the Caspian Sea and sweet cherries (*Prunus avium*) and sour cherries (*P. cerasus*) were cultivated by 300 B.C. Sweet cherries are mostly eaten fresh; sour cherries are cooked and used in many baked pies or other desserts and jams.

Avocados (*Persea americana*) originated in Central America and are cultivated in Mediterranean climates and tropical areas. They are nutritious with high levels of monounsaturated fats, antioxidants, and minerals. This tree fruit is sometimes considered a fruit and sometimes a vegetable. It is eaten fresh and prepared as guacamole, and is added in salads and desserts such as avocado ice cream. Another highly nutritious fruit is the date (*Phoenix dactilyfera*). Dates have been an important food staple for Middle Eastern cultures since prehistoric times (Bacon, 2008). Dates grow on the date palm, which is also widely grown as an ornamental tree. Dates are used in sweet and salty dishes; when dried they contain more than 70 percent sugar. Dried dates are the food of ancient travelers as the fruit contains many essential macro- and micronutrients. Other parts of the plant are used for fiber and wood.

Pineapple (*Ananas comosus*) belongs to the bromeliad family and is valued for their fruit high in carbohydrates, manganese, and Vitamin C. Each fruit is composed of many fruitlets and can weigh from 1–3 kg. Origins of this plant are the South American tropics. Pineapples are used as fresh fruit, juice, and in savory Thai and other Asian dishes. The pineapple represents hospitality and welcome in some cultures.

Citrus fruits are in the rue family or Rutaceae and originated in Southeast Asia. The many citrus species—including oranges, lemons, limes, grapefruits, kumquats, mandarins, and tangerines—grow on broadleaf evergreen trees or shrubs, some of them with spiny branches. They need a warm subtropical or tropical climate and most cannot tolerate frost. Citrus fruits are treasured for their high juice content and for the essential oils in the exocarp, which is used as zest to flavor baked goods or other dishes. The sour flavor is typical for citrus fruits, which are often eaten fresh or the pressed juice is used for many culinary purposes. Citrus is high in Vitamin C and flavonoids. Citrus trees are also grown as ornamental plants. In the 17th and 18th century citrus trees were favored by royalty in northern Europe who grew the trees in large containers, moving them into glass houses called *orangeries* to protect them from cold weather.

Important fruit growing on vines are watermelons, kiwis, melons, and grapes. Watermelons (*Citrullus lanatus*) are an annual trailing or climbing plant. Their fruit can weigh up to 20 kg, are composed of 92 percent water and 6 percent sugar, and contain Vitamin C, beta-carotene, and lycopene. Watermelons are native to Africa and were cultivated in ancient Egypt about 1,000 B.C. Watermelons are one of the most popular fruits, botanically they are a pepo, a berry with a mostly red fleshy mesocarp and endocarp and a thick exocarp (rind). Watermelons are the quintessential summer fruit and are consumed fresh or used in salads and chilled desserts. The rind is cooked or pickled in some areas. Different cultivars of *Cucumis melo* are the cantaloupe, honeydew, and muskmelon, belonging to the pumpkin or Cucurbitaceae family (see Chapter 25) like the watermelon. Their wild ancestors are from subtropical and tropical areas in Africa and Asia and they grow well in arid climates. Melons are an important commercial crop worldwide.

Wild ancestors of grapes (*Vitis vinifera*) were growing in the Northern Hemisphere 23 million years ago as evidenced by fossil leaves (Howell, 2009). In ancient Egypt wine was produced in 2400 B.C. and later was brought to Greece where wine production and consumption flourished. Grapes were brought to China by 100 B.C. and to Northern Europe by the Romans. Grapes in North America were cultivated by European settlers who used the native Fox or Concord grape to cross with European varieties. Phylloxera (*Dactylosphaera vitifoliae*), a tiny, sap-sucking insect feeding on roots and leaves of grapevines, was introduced from America and destroyed major parts of the European wine-growing area, especially France, in the 1860s. The partially resistant Fox grape and other grapes native to North America were used as root stock for grafting desirable cultivars onto or as a parent in hybridization with the European grapes. Grapes are important for today's worldwide wine industry, the fresh market, juice, or dried as raisins (Figure 29.8a–b). Grape leaves are used in Middle Eastern cuisine and grapevines are also cultivated as ornamentals for their colorful fall foliage.

Nuts are an important addition to the human food palette, providing protein, unsaturated fats, fiber, vitamins, essential minerals, and antioxidants (Table 29.2). What is called an edible nut is botanically a nut, drupe, or seed (Figure 29.9a–c). Coconuts account for more than half of the world nut production, peanuts for one-third, and all other nuts for the remainder (Food and Agriculture Organization, 2010). Coconut flesh developed from the endosperm is eaten fresh or is used dried in many sweet and savory dishes (Figure 29.9a). Liquid coconut water is used as a drink, often sold in the coconut with a straw inserted. Coconut milk is the liquid pressed or extracted from the grated coconut meat and is high in saturated fat. Coconut oil is extracted after processing the meat and used in cooking and cosmetics. The remainder of the coconut husk, fronds, and stems are used for growing media, construction materials, fuel, brooms, and dye.

(a)

(b)

Figure 29.8a–b   (a) Grapes are used for eating fresh, (b) dried as raisins, pressed as juice, and fermented as wine.

Table 29.2 Nuts Used for Culinary Purposes and Their Major Nutritional Properties

Latin Name	Common Name	Fruit Type	Nutritional Properties
Corylus avellana	Hazelnut	Nut	Protein, fiber, unsaturated fat, Vitamin B
Juglans regia, J. nigra	Walnut	Drupe	Protein, unsaturated fat, antioxidants, manganese
Prunus amygdalus	Almond	Drupe	Protein, fiber, unsaturated fat, Vitamins E and B, minerals
Anacardium occidentale	Cashew	Drupe	Protein, carbohydrates, unsaturated fat, essential minerals
Cocos nucifera	Coconut	Drupe	Saturated fat, fiber, minerals
Macadamia integrifolia	Macadamia nut	Seed	Unsaturated fat, vitamins
Pinus spp. (P. edulis, P. pinea, P. koraiensis)	Pine nut	Seed	Protein, unsaturated fat, Vitamins E and B, manganese
Carya illinoinensis	Pecan	Drupe	Protein, unsaturated fat, Vitamin B

Vegetable is a term used to describe plants that are entirely or in part edible for human consumption. Vegetables are usually differentiated from fruit as having a savory flavor, whereas fruit has a sweet flavor. These descriptions are arbitrary and not related to botanical definitions. Vegetables are categorized based on whether they are taxed by law as a vegetable. According to this definition the United States Supreme Court ruled in 1893 that the tomato is considered a vegetable and is taxed as one according to the 1883 U.S. Tariff Act, although botanically it is a fruit. Mushrooms, for example, are considered a vegetable, although they are not even classified in the plant kingdom. Leafy vegetables are easily recognized and include lettuce,

(a)

(b)

(c)

Figure 29.9a–c (a) Coconuts, (b) cashews, and (c) almonds are a rich source of fat, fiber, and minerals.

kale, cabbage, spinach, beets, and mustard greens. Common root vegetables are beets, onions, carrots, radishes, potatoes, and sweet potatoes, the last two of which require cooking. Leafy stems of celery and rhubarb are eaten; however, rhubarb stems are the only edible part of the plant, the leaves are poisonous. Tomatoes, eggplant, squash, and pumpkins are fruits. Corn is a grain commonly served as a vegetable. Legumes such as French beans or sugar peas are cooked as a vegetable in the whole while black beans, lentils, and split peas are prepared without the pods.

Vegetables are parts of herbaceous plants consumed raw or cooked, with some requiring cooking to make them edible. The different parts of vegetables eaten include leaves, stems, roots, bulbs, fruits, flowers, and flower buds. Vegetables are important for the human diet because they contain fiber, vitamins, essential minerals, antioxidants, and phytochemicals. Diets rich in vegetables and fruits are recommended to prevent or treat some common diseases related to a diet lacking a wide variety of plants. **Vegan** diets rely exclusively on vegetables and other plant-based foods.

vegan
Diets using plant-based food only, excluding also egg and dairy products.

BOX 29.1 Moringa—Another Superfood

Moringa oleifera Lam. is a fast growing tree native to northern India along the Himalayas. Common names for *M. oleifera* include moringa, horseradish tree due to the roots tasting like horseradish, drumstick tree due to its long, narrow shaped seed pods, and benoil tree due to the oil that is extracted from the seeds. Moringa is classified in the family Moringaceae that contains only one genus, Moringa. It is widely cultivated in tropical and subtropical areas of southeast Asia, Central America, and Africa. The plant requires mean annual temperatures of 59°F to 86°F (15°C to 30°C) and annual rainfall between 76 to 225 cm. Long dry periods or cold temperatures stunt leaf growth, lead to leaf abscission, and prevent flowering and fruit set. The plants are well adapted to a wide range of soil conditions and thrive with sufficient irrigation and fertilization.

Different parts of the moringa plant make it versatile as food, medicine, and for horticultural and industrial uses. The leaves contain between 20% and 35% protein on a dry matter basis, and are high in essential amino acids, vitamins A and C, calcium, and potassium. This makes them an ideal source of food to alleviate malnutrition in tropical countries. Moringa powder produced from the leaves is often used as a nutritional supplement in Africa.

Moringa's reputation as a superfood is not only based on the valuable nutritious characteristics but also on the many traditional medicinal uses. Bark and roots of the tree contain isothiocyanate compounds that account for the horseradish scent lending one of the common names of the plant. These compounds have been shown to have antimicrobial and antifungal properties and extracts of the plant have been used to treat tumors and cancer.

The large leaves are tripinnate and are 30–60 cm long. Flowers are creamy white and have yellow stamens. They develop into triangular pods, tapered at both ends, and a length of 30–120 cm. Immature, green pods are consumed as vegetables; once pods mature and turn brown, dry seeds can be processed for oil extraction. Moringa oil is prized for its stability and can be stored for a long time without losing quality. The oil is used for industrial, culinary, and cosmetic purposes. In addition to serving as human food, leaves are also used as supplemental animal fodder.

Moringa trees can be propagated from seed or cuttings. Under optimum conditions, some plants selected for early flowering can produce marketable pods within 6 months, while others may take one to two years to yield a large number of pods. As trees increase in size, they are often cut at a height of 50 to 100 cm and new branches with harvestable pods can develop within 6 months.

Moringa has great potential to expand into a more prominent specialty tree crop in climates where it grows well. The undemanding cultivation requirements and the many benefits from this plant offer great promise as an industrial crop, cultivating on small acreage, or as backyard tree. The evaluation of different Moringa species and continuous selection and breeding of *M. olifeira* for desirable traits further improves opportunities to develop greater production capacity.

(a)   (b)

Source: Ursula Schuch

Figure Box 29.1 Moringa leaves, flowers (left), and immature, green pods (right) are rich in nutrients and popular in vegetable dishes.

Plant Oils and Sugar

Plants are the source of oils used for many different purposes and have been used by humans for thousands of years. Vegetable oils are used for food preparation, as preservatives, as fuel, in cosmetics, and as lubricants. Vegetable oils have **triglycerides** as their main component. Triglycerides from plants are primarily unsaturated, are liquid at room temperature, and are considered healthy in the human diet (see Appendix). Fat from animals is primarily composed of saturated triglycerides, which are solid at room temperature and potentially less healthy for humans. Oils liquid at room temperature spoil and turn rancid sooner than those solid at room temperature. The process of hydrogenating vegetable oils renders them solid at room temperature, but has negative health effects on human blood chemistry.

Oil from palms and soybeans are extensively used worldwide. Important oils for use in cooking are extracted from rapeseed (canola oil), corn, sunflower seeds, peanuts, olives, coconuts, and sesame seeds. Biodiesel is

triglyceride
Lipid composed of three fatty acids connected to a glycerol molecule; main component of vegetable oils.

produced from soybeans, rapeseed, castor, other oil crop plants, algae, and animal fats. Biodiesel is an alternative to petroleum-based fuel and is considered environmentally cleaner. Increased demand for these oil crops cause concern they may compete for land use and food production.

Essential oils are extracted from plants for their volatile aromatic compounds and they contain a specific scent related to the plant. The oils are extracted by distillation or pressing. Mint, eucalyptus, citrus, roses, lavender, almond, and cloves are used for harvesting essential oils. Some plants such as mint or lavender are used entirely for distillation. Peels of citrus fruit are pressed for their oils while rose petals and orange blossoms are distilled. Essential oils are used in perfumes, cosmetics, as flavoring in food or beverages, as scent in many different products, and in alternative medicine such as aromatherapy.

essential oils
Oils harvested from plants for their volatile aromatic compounds.

Sugar is a luxury in the human diet because it is not essential for survival (Bacon, 2008). Major sugar-producing plants are sugar cane (*Saccharum officinarum*) and sugar beets (*Beta vulgaris*) (Figure 29.10a–b). The cultivation of sugar cane in the 18th century brought with it slavery, war over the sugar trade, domination of colonial powers, and forced migration of people. The increase in refined sugar consumption has been linked to increased obesity and several other diseases. Sugar cane cultivation requires a tropical climate, ample water, and labor. Sugar cane provides most of the world with raw, brown, or white sugar, molasses, or rum. Sugar beets are a root crop cultivated in cooler climates. They provide the majority of European table sugar.

(a)

(b)

(c)

Figure 29.10a–c   (a) Sugar cane, (b) sugar beets, and (c) sweetleaf or stevia satisfy the world's sugar demand.

Sweetleaf or stevia (*Stevia rebaudiana*; Figure 29.10c) is a shrub grown for the high sugar concentration in its leaves. The plant is native to South America and the leaves have 30–45 times the sweetness of sucrose, regular table sugar. Stevia contains no carbohydrates and no calories, making it popular for people on diets and those avoiding sugar. Stevia is used widely in many parts of the world but to a lesser extent in Europe and North America. Other sources of sugar are maple syrup, which is the boiled xylem sap from *Acer saccharum* harvested in late winter primarily in southeast Canada and the northeastern United States. The carob tree (*Ceratonia siliqua*) in the legume family produces seeds high in sugar and a chocolate-like flavor. Sugar is also extracted from agave, sorghum, and many other plants used regionally.

COMMERCIAL PRODUCTS

Flavored and Fermented Beverages

Humans have long discovered improvements to beverages to satisfy the need for water. Many flavored beverages such as tea, coffee, and cocoa are based on water flavored with the extract of leaves or beans. Fermented drinks such as beer and wine rely on the fermentation of grains and grapes. These beverages have been used by humans for thousands of years and have become important in some cultural and religious ceremonies. Sharing a cup of tea or coffee is a social ritual many people participate in. Caffeine contained in these beverages is a stimulant, making them a daily habit for many people. For ancient Greeks and Romans wine was an important part of daily life; in colder regions beer was a common beverage in the Middle Ages. Juice extracted from fruit or used for flavoring beverages has been popular for thousands of years. Fruit-flavored waters have come on the market recently.

Tea from the perennial shrub *Camellia sinensis* (Figure 29.11a) is the most consumed beverage after water worldwide. Tea originated in East Asia where new growth of leaves and buds are harvested up to 30 times a year. The different types of tea—white, green, oolong, and black—derive their distinct color and flavor from processing, the amount of oxidation the leaves are exposed to before drying stops oxidation. Many different cultivars are grown today in tropical, subtropical, and even temperate climates. Herbal teas do not contain tea, but fruit or leaves from other plants; they are also free of caffeine. Tea has beneficial health effects based on phytochemicals and vitamins that protect from certain diseases.

Coffee is brewed from roasted seeds of the evergreen shrub *Coffea arabica* (Figure 29.11b). This species dominates today's coffee trade. *Coffea canephora*, producing the more bitter "Robusta" coffee but being more resistant to disease, plays a secondary role. *Coffea arabica* grows best at middle and higher altitudes in Africa, Southeast Asia, and Central and South America. Robusta coffee thrives at lower elevations where coffee leaf rust (*Hemileia vastatrix*) decimates *C. arabica*. This plant, native to Ethiopia, was first commercially cultivated in the 15th century in Arabia (Bacon, 2008). The Arabian monopoly of the increasing coffee production and trade was broken when viable seeds were smuggled out of the area and the Dutch established coffee plantations in Java. With increasing demand for coffee

Figure 29.11a–d   (a) Tea, (b) coffee, and (c–d) cocoa are used worldwide for beverages and cocoa for chocolate. The plants contain alkaloids, which are stimulating.

worldwide, plantations were established in many subtropical and tropical countries and today Brazil and Vietnam are the major coffee producers.

Coffee berries are red, yellow, or purple when they are ripe about 8 months after flowering. Hand labor is necessary to pick only the ripe berries as they mature for best flavor. The flesh of berries is removed and they need to be dried before roasting. For decaffeination, the oils containing the caffeine are extracted from the green seeds with hot water and solvents. The final flavor, body, and color of coffee are affected by the intensity of roasting. The roasted beans are ground and brewed by many different methods into various coffee beverages or are used as flavoring for other drinks or food.

Cocoa is a product of the seeds of the cocoa tree (*Theobroma cacao*; Figure 29.11c–d) and is used to make chocolate. Cocoa trees originate in tropical America and were cultivated first around 1000 B.C. The small evergreen trees grow flowers in clusters on their trunks, which are pollinated by flies. Cocoa pods develop to contain about 40 seeds. Seeds and surrounding pulp are taken out of the pod, fermented, dried, and roasted. About half of the roasted seeds are composed of cocoa butter, which is used with milk and sugar to prepare chocolate. Theobromine in cocoa is an alkaloid similar to but milder than caffeine.

Herbs and Spices

Humans have used herbs and spices for thousands of years in food preparation to change, add to, or mask the flavor and aroma of dishes. Herbs and spices are generally used in dried form and in very small quantities because their flavor and aroma are so strong (Figure. 29.12a–b); they often are expensive and considered luxury items because they are not essential to human nutrition. Herbs are most often the leafy part of a plant, used fresh or dried for flavoring and garnish of food, as medicine, and for fragrance. Spices are the dried fruit, seed, root, or bark of plants. Spices often have antimicrobial properties and are used for flavoring, in religious ceremonies, for medicinal purposes, and in cosmetics (Table 29.3).

International trade of spices, parts of plants native primarily in tropical and subtropical Asia, became widespread and profitable in the 15th century when maritime European powers brought the desired goods from the Asian continent to Europe. Prominent trading routes brought wealth to the cities lying on the crossroads such as Alexandria in Egypt. During wars over the dominance of the spice trade many indigenous people were killed in Southeast Asia and colonial powers were established over spice monopolies.

Important spices include black pepper, cinnamon, clove, nutmeg, allspice, hot or chile peppers, ginger, vanilla, and saffron (Table 29.3). Many herbs belong to the mint (*Lamiacae*) and parsley (*Apiaceae*) family. They are prized for their **secondary metabolites**, which give the plants their specific flavor. Essential oils are extracted to capture the fragrance, flavor, or aroma of the plants. Herbs also have been used traditionally for medicinal purposes in addition to culinary use.

secondary metabolites
Chemical compounds produced by plants that are not essential for plant survival but assist in primary metabolism and defense mechanisms against pests, pathogens, and herbivores. See also alkaloids.

Paper, Cloth, and Wood

Plants classified as fiber crops provide people with material to make paper, cloth, baskets, and ropes. Fibers are the elongated sclerenchyma cells providing support in plants (see Chapter 3). They have thick lignified walls and

(a)

© Elena Schweitzer/Shutterstock.com

(b)

© Sandra Caldwell/Shutterstock.com

Figure 29.12a–b (a) Spices are almost always used dried in small quantities, many of which originated in tropical Southeast Asia. (b) Herbs are generally the leaves or shoots of plants, often treasured for their aromatic compounds, and are used fresh and dried.

Table 29.3   Properties of Common Spices and Herbs

Latin Name	Common Name	Plant Part	Properties and Uses
Capsicum annuum	Hot pepper	Fruit	Hot- and mild-flavored peppers (bell pepper) are the same species. Capsaicin gives hot peppers the heat, used fresh or dried in many savory dishes; potential medicinal use.
Cinnamomum verum	Cinnamon	Bark	Medicinal and ceremonial use in ancient times, contemporary use for flavoring for sweet and savory dishes and beverages.
Crocus sativa	Saffron	Stigma of flowers	Medicinal use, yellow dye for fibers, culinary use for flavoring and color.
Laurus nobilis	Bay laurel	Leaves	Bay wreaths were symbols of victory in Roman times, for scholars and poets in Middle Ages and later, culinary use in many cooked dishes.
Mentha spp.	Mint	Leaves	Medicinal use, essential oils for fragrance, fresh or dried in beverages, Asian dishes, desserts. Many different flavors such as peppermint, spearmint, apple mint, and pineapple mint.
Myristica fragrans	Nutmeg and mace	Nutmeg seed for nutmeg, pericarp for mace	Used since ancient times for the nutty flavor and warm, sweet aroma in Middle Eastern and Indian dishes, both sweet and savory.
Piper nigrum	Black pepper	Fruit	Produces black, pink, green, and white pepper. Used before refrigeration to improve flavor of salted meat. Universal spice for savory dishes and food flavoring.
Rosmarinus officinalis	Rosemary	Leaves	Medicinal and ceremonial use, signifies remembrance; preservation of meat, condiment, used fresh and dried in savory dishes.
Vanilla planifolia	Vanilla	Pods	Most labor-intensive crop due to hand pollination. Whole processed pods or "beans" used as flavoring in baked goods and desserts.
Zingiber officinale	Ginger	Rhizome	Consumed in ancient China until now for medicinal purposes, used fresh in many Asian cooked dishes and beverages, as pickles, and in desserts, used dried in baked goods.

bast fibers
Sclerenchyma cells sur-rounding the vascular tissue or contained in the bark or cortex of fiber plants.

are dead once they mature. Flexible fibers have thinner cell walls with less lignin such as yucca leaves compared to inflexible fibers, which are found in hardwood such as oak. One type of fiber cells, the **bast fibers** or soft fibers, occur as bundles or sheaths around the vascular system in stems, bark, or cortex. Bast fibers are in stems of jute (*Corchorus* spp.), which is used to produce coarse cloth such as burlap, jute rope, mats, carpets, and woven chair coverings. Jute is the least expensive fiber and used second in quantities produced after cotton. It is also useful for its biodegradable properties as a ground cloth to temporarily stabilize soil. Other bast fibers are found in flax (*Linum usistatissimum*), used to produce precious linen fabric, banknotes, and rope; hemp (*Cannabis sativa*) to make rope, cloth, and pulp; and ramie (*Boehmeria nivea*) for fishing nets and fire hoses because of its high strength when the fiber is wet.

Fibers from cotton (*Gossypium hirsutum*) are trichomes growing from the seed coat and covering the cotton seeds in the cotton boll or seed capsule (Figure 29.13a). They are harvested to produce many different textiles and cloth. Cotton is the most important fiber crop in the world and has been used for 7,000 years. Cotton fibers are well suited for textiles because they are soft and smooth and can be woven into thin or thick fabric, they are naturally white and readily accept dye, and they hold up to repeated washing without breaking.

Leaf fibers are harvested from several plants in the agave (*Agavaceae*) family. Sisal fibers are from *Agave sisalana* and are used for crafts, twine, and rough carpeting. Abaca is a leaf fiber from *Musa textilis* in the banana (Musaceae) family and is manufactured into specialty paper for tea bags, bank notes, and specialty textiles.

Wood fibers are important for pulp production, which is the basis of paper (Figure 29.13b). Wood fibers consist of cellulose, hemicellulose, and lignin, with cellulose being the desired component for paper. Fibers or tracheids from conifers are longer and are preferred for paper production over the fibers from hardwood. Wood is used for many other purposes including fuel, timber for construction, furniture, posts, railroad ties, veneers, and plywood, to name a few.

(a)

(b)

Figure 29.13a–b Cotton field ready to harvest. (a) The white fibers in the boll are almost pure cellulose. (b) The spruce logs will be made into pulp, which separates the cellulose fibers from the lignin. Cellulose fibers are used to manufacture paper.

BOX 29.2   Bamboo—Versatile for over 3,000 Years

Bamboo is a fast-growing, woody evergreen grass in the Poaceae family and Bambusoidae subfamily (Farrelly, 1984) (Figure Box 29.2a). The more than 1,500 species of bamboo are adapted to diverse environmental conditions from lowlands to high elevation, and humid, hot-to-cold climates. Bamboo is native to Asia, Australia, and North and South America. Bamboo is a flowering monocot and grows either **sympodial** or **monopodial** based on their rhizome structure. Sympodial bamboo species are native to the tropics, sensitive to frost, and grow in defined clumps. Monopodial bamboo, also known as free-standing or running bamboo, grows in colder environments and some species tolerate temperatures as low as –20.2°F (–29°C). Underground rhizomes of monopodial bamboo grow significant distances and form extensive root systems whereas rhizomes from sympodial species are confined and will not spread widely. The root system of bamboo can potentially become invasive, which in some areas is desirable to prevent soil erosion. Buds on the rhizomes emerge aboveground (Figure Box 29.2b), giving rise to a culm or stem that reaches its final height within one to four months. The nodes on culms give bamboo its characteristic appearance. Adventitious roots can grow on nodes at lower culm height and thin branches can emerge on nodes further up. The culm diameter remains the same throughout its lifespan. The internodes of culms are hollow and give great strength when culms are used as building materials (Figure Box 29.2h). They also allow this shallow rooted plant great resilience in high winds, making the plant a desirable windbreak. Individual culms live from 5 to 20 years. Plant propagation occurs primarily vegetative as flowering of most bamboo species is infrequent, several decades to over 100 years apart. Many bamboo species flower **gregariously**, meaning their vegetative growth stops and all plants start synchronized flowering, which spreads over wide areas and continents, taking one to several years. Many species die after flowering, requiring years to regenerate new culms and mature groves.

Bamboo has one of the fastest growth rates in the world, making it an ideal plant for sustainable production of many materials. In optimum growing environments bamboo groves can increase biomass by 10–30 percent annually while biomass in a forest increases only 2–5 percent during that time (Farrelly, 1984). Daily height growth of larger species can be 0.1–0.4 m, with the fastest growth recorded on *Phyllostachys edulis* in Japan with 1.2 m in 24 hours. Bamboo varies in height from 0.1 m up to 40 m, with some notable exceptions reaching twice this height. Culm diameter varies as well and larger species can grow to 0.2 m in diameter. Taller, thicker culms grow in warmer, tropical regions.

For more than 2,000 years bamboo played an important role in China and Southeast Asia for its many uses. Bamboo is likely the most versatile plant utilized for commercial, practical, medicinal, and edible purposes, with more than 1,000 uses listed (Farrelly, 1984) (Figure Box 29.2c–g). Bamboo is used to construct buildings, scaffolding, furniture, skin on airplanes, ancient water systems, many different musical instruments, bicycle frames, sandals, matting, and chopsticks. Other uses include needles, arrows, baskets, boats, shelters, laquerware, waterwheels, and as medicine against stomach ailments and respiratory problems. Around the 10th century a Buddhist

sympodial
Growth resulting in lateral branching from secondary axis.

monopodial
Growth of a plant from a single growing point in one direction.

▶▶▶

monk authored a cookbook devoted to recipes with bamboo shoots (Laws, 2010). Contemporary cooks in Asia prepare young bamboo shoots in a variety of different dishes. Strips of bamboo have been used in ancient China to maintain records before paper was invented. The short fiber of bamboo has been used to manufacture paper and rayon fabric. Musical instruments from bamboo include flutes produced since ancient times and more recently the ukulele. Implements used in the Japanese tea ceremony like the whisk and the tea ladle are traditionally made from bamboo. Calligraphy and painting have relied on bamboo brushes. Bamboo is also used as livestock forage for cattle, sheep, and horses, is the primary food source for the giant panda in China, and is consumed by other wildlife. Bamboo plants grow in commercial bamboo forests and as ornamental plants in many areas outside their native range. Bamboo has great significance in Asian cultures, symbolizing longevity, resilience, friendship, strength, and honorability.

(a)

(b)

(c)

(d)

(e)

(f)

(g)   © El Greco/Shutterstock.com

(h)   © kaman985shu/Shutterstock.com

Figure Box 29.2a–g   (a) Bamboo forest in Taiwan; (b) bamboo shoot among larger culms; (c) bamboo scaffolding; (d) bamboo bridge; (e) dish with bamboo shoots; (f) bamboo steamer; (g) weaving of a bamboo basket; and (h) cut bamboo culms showing hollow interior.

MEDICINAL PLANTS

Plants have been used for medicinal purposes since ancient times. Ethnobotany studies the relationship people of different cultures have with plants related to many areas of life, one of them the medicinal use of plants. Many different plants are used for healing a great number of diseases and symptoms by using plant parts as a whole, as extracts, as poultice, brewed, or as mixtures of different plant parts. Studies of four groups of Indians indigenous to southern Mexico showed their use of more than 200 species of plants for medicinal purposes (Heinrich, 2000). People from geographically different areas use some of the same species to treat common ailments such as gastrointestinal problems. However, not all species were effective to treat a problem when tested. The compounds that are effective as medicine are often produced by the plant for defensive purposes. They include secondary metabolites such as alkaloids and other compounds that are the basis for modern pharmaceuticals, now often synthesized in the lab instead of being extracted from the plant. Researchers analyze plants from all over the world to discover new medicinal efficacy.

Aloe vera heals burns when the succulent leaves are cut and applied to an injury. Quinine, isolated from the bark of the cinchona tree (*Cinchona officinalis*), is a potent remedy to treat malaria. The bark of willow trees (*Salix* spp.) used to be chewed for pain relief and contains salicin, the ingredient later used to develop the analgesic aspirin. The common ornamental purple foxglove (*Digitalis purpurea*) contains compounds effective in treating irregular heartbeat. Taxol is found in the bark of yew trees (*Taxus brevifolia*) and in the needles of yews and is a powerful drug against breast cancer. Other drugs against cancer are based on compounds isolated from vinca (*Catharanthus roseus*) and star anise (*Illicium verum*). Plants such as opium poppies (*Papaver somniferum*), containing the active alkaloid morphine, and marijuana (*Cannabis sativa*), containing tetrahydrocannabinol (THC), have medicinal value to treat pain and some other ailments. They also affect the central nervous system and can cause hallucinations. Many other plants are used for healing purposes; however, the effectiveness of most has not been validated.

Key Terms

gluten
pulse
isoflavones
aflatoxins
glycoalkaloid
pseudostem

phytochemicals
antioxidants
flavonoids
vegan
triglycerides
essential oils

secondary metabolites
bast fibers
sympodial
monopodial
gregariously

Summary

- Plants are essential for human survival for food, fuel, shelter, fiber, and medicine.
- Cultivation of crops started about 10,500 years ago when hunters and gatherers started to settle and grow plants for food crops and other purposes.
- Criteria for domestication were selection of larger grain size, nonshattering heads of grains and larger tubers, fruits, and seeds.
- Staple foods such as wheat, rice, yams, and legumes were among the first crops that were cultivated for food. Plants for food, medicine, and spices were introduced to different locations over time. Civilizations started to develop with growing food security.
- The cereal grains maize, rice, and wheat supply more than 60 percent of the calories consumed by humans worldwide at this time. Cereal grains are easy to store, mill, and transport and provide carbohydrates, oils, and many essential vitamins for human nutrition.
- Legumes in the *Fabaceae* family include beans, peanuts, and lentils. Their seeds are important protein sources for food and other uses.
- Starch crops cultivated for their underground storage organs include potatoes, cassava, sweet potatoes, yams, and the herbaceous perennial bananas producing fruit aboveground.
- Fruits and vegetables are a major source of energy and minerals, vitamins, and antioxidants. They contain phytochemicals believed to be beneficial for human health.
- Nuts are rich in protein, unsaturated fats, minerals, and vitamins.
- Oil crops include oil palm, soybeans, coconut palms, rapeseed, and peanuts. Oils are important for food preparation, biodiesel, preservatives, cosmetics, and industry. Essential oils are extracted from plants such as citrus, eucalyptus, mint, and lavender for their volatile aromatic compounds.
- Sugar is nonessential for human nutrition and is produced from sugar cane, sugar beets, and sweetleaf or stevia, and maple trees.
- Grapes, fruit, and cereals are used to produce fermented beverages; tea, coffee, and cocoa plants are processed for flavored beverages.
- Herbs and spices are generally the dried part of a plant and are used in very small quantities for their strong flavor and aroma. Many contain secondary metabolites and are sometimes used for medicinal purposes, fragrance, or ceremonial purposes.
- Fiber crops include cotton, wood, agave, jute, flax, and bamboo. They are used for many purposes including building materials, textiles, furniture, rope, and paper, to name a few.
- Plants have been used for medicinal purposes since ancient times and provide the basis for many current pharmaceuticals. Important medicinal plants include the leaves of aloe vera; the bark of the cinchona, yew, and willow tree; and the seeds of opium poppies. Active ingredients from these plants are now often synthesized in a lab instead of being extracted from plant parts.

Reflect

1. *Plants providing the majority of nutrition to humans belong to one family.* What are the characteristics of this family and why are these plants so important to humans worldwide?

2. *Tools for improving crop production have changed since the beginning of agriculture.* Describe the criteria used in the beginning of food cultivation and contemporary goals. What are some important tools available to modern plant breeders?

3. *Legumes are important as human food and several other purposes.* Explain why legumes are used all over the world and give some examples of their uses.

4. *Underground storage organs are the second most important food staples in the world.* Explain why they are important for human nutrition and give some examples of plants.

5. *Diets rich in plants are recommended for good health.* Discuss the important constituents in plants and which ones are needed and which ones are desired by humans.

6. *Flavored beverages are consumed worldwide.* Describe what makes them attractive in so many different cultures.

7. *You have discovered a new plant that you think will become important as a fiber crop.* Explain which properties of this plant are important as a fiber crop and which products it might be used for.

References

Bacon, J. (2008). *Edible: An illustrated guide to the world's food plants.* Washington, DC: National Geographic.

Farrelly, D. (1984). *The book of bamboo.* San Francisco: Sierra Club.

Food and Agriculture Organization (FAO). (2013). Food production yearbook. Rome: FAO. Retrieved January 2, 2018, from http://www.fao.org/docrep/018/i3107e/i3107e.PDF

Heinrich, M. (2000). Enthnobotany and its role in drug development. *Phytotherapy Research, 14,* 479–488.

Higdon, J., Drake, V. J. (2013). An evidence-based approach to phytochemicals and other dietary factors (2nd ed.). New York: Thieme.

Howell, C. H. (2009). *Flora mirabilis.* Washington, DC: National Geographic.

Laws, B. (2010). *Fifty plants that changed the course of history.* Hove, UK: Quid Publishing.

Levetin, E., & McMahon, K. (2012). *Plants and society* (6th ed.). New York: McGraw-Hill.

Glossary

3-phosphoglycerate (PGA) Molecule containing three carbon atoms, first stable compound in the Calvin cycle.

abscission Physiological process that causes leaves to shed from a plant.

absorption Ability of plant roots to take up water and absorb nutrients from the soil.

acid growth hypothesis Based on auxin causing an increase in proton pumps in the cell wall, acidification of the cell wall leading to loosening, and subsequent expansion of the cell through incoming water.

actin filaments Thin protein fibers that make up of the cytoskeleton.

actinomorphic flowers See *radial symmetry*.

action spectrum Range of the light spectrum that causes particular responses in plants.

adhesion Ability of different molecules to stick to each other.

aerenchyma Specialized cells allowing oxygen movement from leaves to roots.

aerial roots Adventitious roots arising from branches and other aboveground structures that grow vertically down to the soil where they root and provide additional support for the plant.

aerobic respiration Release of stored energy in organic compounds in the presence of oxygen.

aflatoxins Compounds in foods such as nuts, corn, or cotton seed caused by the infection with *Aspergillus flavus* or related fungi that produce toxic metabolites.

after ripening When seed is held in appropriate storage conditions to provide time for the embryo to develop.

alcohol fermentation Process of fermentation where pyruvate is converted to ethanol.

alpine tundra Treeless biome similar at high elevations above the tree line. See *tundra*.

alternation of generations Unique life cycle of plants; plants spend part of their lives as multicellular haploid gametophytes and part as multicellular diploid sporophytes.

amino acids Small units that make up a polypeptide or protein.

amylopectin Form of starch made up of long and branched chains of glucose molecules.

amyloplasts Leucoplasts that synthesize and store starch.

anabolic Process resulting in the biosynthesis or construction of larger molecules; opposite of catabolic.

anaphase Where duplicated chromosomes begin to separate and move to opposite ends of the cell.

anchorage Support provided by the roots of a plant so it does not fall over.

angiosperms or flowering plants Group of vascular seed plants that produce seeds and their seeds are enclosed within a fruit, which is a mature or ripened ovary. *Angeion* in Greek means "vessel" and *sperma* means "seed."

anion Negatively charged ion. See also *cation*.

annulus Row of cells with thickened cell walls running down the medial plane of the fern sporangium; used to tear open the sporangium and to catapult the spores into the air.

antennae complex Pigment molecules in photosystem I and II that absorb light and direct it to the reaction center.

anther Portion of the stamen that bears pollen.

antheridia (singular *antheridium*) Male gametangia for producing haploid sperm.

antheridiophores Gametophores with disc-shaped gametangia, contain sperm-producing antheridia.

antioxidant Chemical compound preventing oxidation of molecules and production of free radicals, which can damage cells.

antiporter Membrane protein transporting two molecules or ions across a membrane in opposite directions. One ion or molecule moves along its electrochemical gradient as a different ion or molecule moves *in the opposite direction* against its own electrochemical gradient. See also *symporter*.

apomixis Asexual reproduction where seeds develop without fertilization.

apoplast Pathway through cell walls and intercellular spaces.

aquaporins Channel proteins in the plasma membrane to facilitate water transport.

arbuscules Highly branched fungal mats of endomycorrhizal fungi formed within the tissues of a plant' roots.

arbuscular mycorrhizae Hyphae of AM fungi grow in the spaces between root cell walls and plasma membranes, often forming highly branched and bushy arbuscules (from the word *arbor* meaning "tree").

arbuscular mycorrhizae Endomycorrhizae that grow into the root cells of a plant and create mats of highly branched fungal hyphae. Hyphae of AM fungi grow in the spaces between root cell walls and plasma membranes, often forming highly branched and bushy arbuscules (from the word *arbor* meaning "tree").

archegonia (singular *archegonium*) Female gametangia for producing haploid eggs.

archegoniophores Gametophores with umbrella-shaped gametangia, contain egg-producing archegonia.

asci (singular *ascus*) Saclike structures used by ascomycetes to produce sexual spores (ascospores).

ascospores Sexual spores produced within an ascus; found only in ascomycetes.

aseptate Fungal hyphae that have no cross walls.

asexual reproduction Does not involve gametes from two individuals and produces offspring that are genetically identical to their parent.

assimilation Uptake of inorganic nitrogen as nitrate or ammonium by plant roots and conversion into organic nitrogen.

atoms The smallest functional units of elements and the simplest level of biological organization; cannot be broken down further into other substances by ordinary chemical or physical means.

autotrophs Organisms that are able to make their own food using photosynthesis.

auxin Growth-stimulating hormone indoleacetic acid involved in phototropism, apical dominance, root formation in cuttings, and other processes.

avirulence gene Gene in a pathogen that produces elicitors that can be recognized by products from the corresponding R-gene in plants.

axillary buds Buds originating in the region between the stem and a petiole.

basal angiosperms Three most primitive groups of angiosperms: *Amborella*, water lilies, and star anise.

basal plate Compressed stem of a corm where the roots attach.

basidia (singular *basidium*) Short stalk-like structures found on the gills or spores of fruiting bodies of basidomycetes; used to produce sexual spores, basidiospores.

basidiocarps Also known as fruiting bodies; found only in basidiomycetes.

bast fibers Sclerenchyma cells surrounding the vascular tissue or contained in the bark or cortex of fiber plants.

beltian body Food bodies consisting of lipids and proteins that are attached to the terminal end of leaflets on acacias.

bilateral symmetry Expressed by flowers when a plane passed through a flower creates mirror images in one direction but not in another.

binary fission Asexual reproduction of a prokaryotic cell by "splitting in half".

biogeography Geographical distribution of species, populations, and ecosystems on Earth.

biome Community of animals, plants, and other organisms living in an environment classified by the type of vegetation and adaptations of organisms to the environment.

biosphere All the ecosystems on Earth; the highest and most complex level of organization in the biological world.

bracts Highly modified leaf frequently mistaken for a flower. It generally persists longer and attracts insect pollinators.

breeder's seed Intial seed resulting from a desired cross used to establish the seed line.

bud scales Leafy structure protecting young buds or shoots.

budding Asexual reproduction by yeasts; daughter cells or buds are split from mother cells.

budding The process of placing a bud from a desirable cultivar on the stock plant of another cultivar.

bulblets/bulbils Small daughter bulbs forming at the base of the mother bulb.

bundle scars Small marks left on the leaf scar by the separation of the vascular bundles of a leaf.

bundle sheath cells Specialized cells in C_4 plants surrounding the vascular bundles where the Calvin cycle occurs.

buttress roots Massive roots at the trunk base that produce large flares to supply increased support of a plant. They are frequently found in tropical locations.

C_3 plants Plants that use the Calvin cycle to fix CO_2 with the enzyme rubisco and produce 3-phosphoglycerate, a three-carbon compound, as the first stable molecule.

callus Mass of undifferentiated cells capable of developing into root and shoot tissue.

calyptra Hood-shaped structure that covers and protects the immature capsule; found only in mosses (Bryophyta).

calyx Refers to all sepals collectively.

capitulum Inflorescence term synonymous with composite head.

capsule or sporangium Found in liverworts and mosses; contains spore mother cells or sporocytes, which undergo meiosis to produce haploid spores.

carpel Flower structure from a modified leaf bearing ovules; multiples may fuse together.

carrier proteins or transporters Membrane proteins binding to molecules and undergoing a reversible change when moving the molecule across a membrane.

carrying capacity Maximum number of organisms of a species a habitat can support.

Casparian strip Layer of waterproof tissue containing suberin, making it impermeable to the movement of water to and from the vascular cylinder.

catabolic Reactions that break down complex molecules and release energy; opposite of anabolic.

catalysts Substances that speed up a chemical reaction by lowering its activation energy but do not take part in the reaction.

cation Positively charged ion. See also *anion*.

cation exchange Cations such as potassium or magnesium become available for uptake by plants after they are released from negatively charged soil particles and are replaced by protons or other cations.

centromere A structure or region that connects the two sister chromatids.

cell Basic unit of life; makes up all living organisms; where all life-sustaining processes take place and where the genetic blueprint, DNA, resides.

cell biology Study of cell structure and function.

cell cycle Composed of interphase and cell division.

cell division Divided into two stages: mitosis and cytokinesis; also referred to as the M phase.

cell plate Separates the cell into two compartments during cytokinesis.

cell wall Rigid coating made up of cellulose fibers that is found outside the plasma membrane of plant cells; responsible for protecting and supporting the plant cell.

cellular respiration Complex metabolic process that all organisms use to break down organic molecules and release the stored energy to sustain life.

cellulose A very large polysaccharide made up of glucose molecules which provides structure support for plants.

cellulose plates Found underneath the plasma membrane of dinoflagellates.

central vacuole Single large vacuole that takes up as much as 90 percent of the volume of mature plant cells; a unique cell structure found only in plant cells.

certified seed Seed line with less regulation and is established from registered seed, which maintains uniform cropping.

channels or channel proteins Proteins embedded in membranes allowing specific ions to pass through.

chapparal Vegetation type found in a Mediterranean climate consisting of broadleaf often thorny shrubs able to withstand long summer droughts.

chemiosmosis Synthesis of ATP that is coupled with an electrochemical gradient across a membrane producing the required energy; occurs in photosynthesis and aerobic respiration.

chemiosmosis The process where the electrochemical gradient across a membrane results in ATP production.

chemoautotrophs Autotrophic organisms that use chemical energy for food production; compare with photoautotrophs.

chimera When a plant is composed of two genetically different tissues growing directly next to each other.

chitin Nitrogen-containing complex carbohydrate that forms the exoskeletons of many insects and the cell walls of fungi.

chlorophyll Green photosynthetic pigments that are essential for photosynthesis.

chloroplasts Most common plastids found mostly in leaf and stem cells; responsible for carrying out photosynthesis to produce food and oxygen.

chromatin Complex of DNA molecules and histones.

chromoplasts Plastids that synthesize and accumulate color pigments such as carotenoids (*chroma* in Greek means "color").

chromosome Condensed and visible form of DNA/histone complex.

cladophyll Modified flattened stem functioning as a leaf.

climacteric Fruits that have high respiration rates during ripening and concurrent high ethylene production.

climate Long-term weather conditions based on temperature, rainfall, humidity, and wind.

climax community Stable ecosystem after several stages of succession.

clones Genetically identical plants arising asexually from the same mother plant.

codominant When there are multiple trunks or branches on a tree which prevents the development of a central leader.

coevolution Evolution of two species that interact with each other and continuously shape each other's adaptations.

cohesion Ability of similar molecules to adhere or stick to each other.

collenchyma tissue Made up of collenchyma cells that have thicker primary cell walls than parenchyma cells; mainly responsible for providing flexible support for herbaceous plants.

cohesion-tension theory Tension or negative pressure from transpiration pulls water from the roots through the xylem into the leaves and stomata.

coleoptiles First structure of a monocot seedling that emerges through the soil.

coleorhizae Portion of the germinating monocot seed that precedes the emerging radicle.

commensalism Relationship of two species where one benefits and the other is not harmed and does not benefit.

community Populations of all organisms that live and interact in the same area.

companion cell Narrower and more tapered than a sieve-tube element; responsible for providing life support system for sieve-tube elements.

competition When two organisms or species attempt to use the same resources such as land, light, water, or nutrients.

complete flower When a flower has all four primary parts including the sepals, petals, stamens, and pistils.

compound fruit Fruit that develops from an ovary with multiple carpels.

compound pistil Plant that has ovules enclosed in multiple chambers or carpels.

compression wood Reaction wood present in gymnosperms.

conidia (singular *conidium*) Asexual spores produced at the tips of fungal hyphae.

conservation biology Study of populations, communities, and ecosystems with the goal to preserve and restore them.

contractile roots Roots found on bulbs and corms that shrink or contract during adverse conditions, pulling the storage structure lower in the soil and preventing it from heaving out during freezing weather.

core angiosperms All angiosperms except basal angiosperms; divided into three main groups: magnoliids, monocots, and eudicots.

cork cambium Found in bark; responsible for producing cork cells and phelloderm.

cork cells Made up of the bark or periderm; cells that are dead when mature and their cell walls are heavily fortified with a waterproof substance called suberin.

corm Specialized underground stem with nodes on the external surface and no fleshy leaves internally.

cormels/cormlets Small daughter corms created near the basal plate of the mother corm.

corolla Refers to all petals collectively.

cortex Root tissue between the epidermis and vascular tissue; responsible for storage; allow oxygen to diffuse through.

cotransporters Membrane protein transporting two ions, one in the direction of its electrochemical gradient, while simultaneously transporting another ion *in the same or opposite direction* but against its electrochemical gradient. See also *symporter* and *antiporter*.

cristae Fingerlike projections formed by folding of the inner mitochondrial membrane; responsible for creating a larger surface area for aerobic respiration to take place.

cross-pollination When a plant is pollinated by the pollen of a different plant.

cryopreservation Seed held under liquid nitrogen for long-term storage.

crypts Depressions on the underside of some leaves where clustered of stomata are located.

culm Stem or central axis of the mature grass shoot, comprised of nodes and internodes.

cuticle Found on the surface of stems and leaves; made up of cutin and wax, which make plant surfaces waterproof and reflective.

cutin Lipid used by plants to prevent desiccation found in plant cell walls and cuticle.

cyclic photophosphorylation Electrons from PSI are diverted to cytochrome complex by ferredoxin then back to PSI producing ATP instead of NADPH.

cytology See *cell biology*.

cytokinesis Division of the cytoplasm follows mitosis

cytoplasmic division Refers to the physical separation of the original cell into two new cells during cytokinesis.

cytoplasmic streaming The movement of cytoplasm within the cell. Also called cyclosis.

cytoskeleton Intricate network of protein fibers that extends throughout the cytoplasm and provides structural support and internal transport for the cell.

cytosol Aqueous portion of the cytoplasm where many chemical reactions take place.

day-neutral plant Flowering of these plants is not affected by the relative length of day and night. See also *short-day plant* and *long-day plant*.

deciduous Plants that shed all their leaves in the fall and regrow them in the spring.

decomposers See *saprotrophs*.

decurrent Growth habit where codominant branches form the crown with no central leader.

deductive reasoning Makes a specific prediction from a general principle and then tests it; begins with a general principle and ends with specific studies.

dehiscent Fruiting structure that splits along a suture at maturity, releasing the seeds within.

dehydration avoidance Ability to maintain a favorable internal water balance during drought periods.

dehydration synthesis Chemical reaction combining two or more monomers together to form a polymer; water is also being produced.

denaturation Disruption of chemical bonds within a protein, causing it to unfold and lose its shape.

denitrification Nitrates in the soil are converted into atmospheric nitrogen gas by anaerobic bacteria.

deoxyribonucleic acid (DNA) "Blueprint" of life that is responsible for the storage, expression, and transmission of genetic information.

deoxyribonucleotides Four nucleotides that make up DNA and have deoxyribose as their pentose.

dependent variable Response to the independent variable that is being measured.

desiccation tolerance Ability to deal with repeat drying and rewetting.

determinate growth After an organism reaches a certain size, growth stops.

development Includes both growth and differentiation.

dictyosomes Flattened disc-shaped sacs found stacked together in the cytoplasm; responsible for collecting, processing, and delivering proteins for use outside of the cell.

differentiation Process when cells take on different shape and form, allowing them to perform different functions.

diffusion Movement of particles from an area of high concentration to low concentration. See also *facilitated diffusion*.

dikaryons Fungal cells that have two genetically different nuclei.

dioecious Male and female sex organs found on distinct individuals.

diploid Having two sets of chromosomes in the cell(s).

dipolar Molecule having an uneven distribution of electric charges in two different regions of the molecule.

disaccharides Carbohydrates that make up of two monosaccharides.

disc florets Small flowers found in the center of Asteraceae inflorescence or head; each floret has many sepals and five fused petals form a tubular corolla.

discovery-based science Collection and analysis of data without any preconceived hypothesis or expectation; generally leads to hypothesis testing.

distal Portion of the plant furthest from the central axis.

disulfide bridge Type of covalent bond formed between sulfur-containing amino acids.

division Practice of using a knife to cut an underground storage structure into multiple propagules.

division of labor Cells in different parts of the plants become adapted for specific tasks and take on different shape and form, which allows the entire organism to become more efficient.

domatia Enlarged woody stipules.

domestication Selection of particular plant characteristics for cultivation that eventually leads to morphological and physiological changes in a crop plant.

dormancy Form of rest for the seed where germination does not occur due to physiological or anatomical issues.

double dormancy When a seed experiences both physical and physiological dormancy issues.

double fertilization Use of both sperms in reproduction; a unique feature of angiosperms and gnetophytes.

drought escape Ability to escape drought periods by completing the life cycle within a short period of time during the wet season.

dry fruit Fruit in which the pericarp is not fleshy or succulent when developed.

ecosystem Community and its physical environment in a particular area.

ectomycorrhizae Fungi that grow on the external surface of a plant's root in a mutualistic relationship.

elaioplasts Leucoplasts that synthesize and store oils.

elater mother cells or elaterocytes Develop into elaters to help in spore dispersal.

electromagnetic spectrum Entire range of radiation of different wavelengths from shortest wavelength with highest energy (gamma rays) to longest wavelength with lowest energy (radio waves).

electron transport chain This final step in aerobic respiration occurs in the mitochondrial membrane and results in energy as electrons are moving through protein complexes.

electronegative The attraction of a given atom for the electrons of a covalent bond.

elements Substances that cannot be broken down further by ordinary means.

elicitors Compound produced in response to a stimulus and only recognized by plants with a specific gene.

embolism Air or water vapor filling a void in the xylem tissue.

embryophytes or land plants Embryos of plants depend on maternal protection and resources for their development.

enations Simple flaps of green tissue that have no vascular veins; are not true leaves; found only in whisk ferns.

endergonic Chemical process that requires energy input; the opposite of exergonic.

endocarp Innermost layer of the pericarp of a fleshy fruit.

endodermis Innermost layer of cells in the cortex that regulate mineral absorption.

endogenous When a material is found within a plant.

endomycorrhizae Hyphae of mycorrhizal fungi grow inside root cells of plants.

endophytes Fungi that live within the leaf and stem tissues of plants without causing overt signs of tissue damage.

endoplasmic reticulum (ER) Network of membranes that connects to the outer membrane of the nuclear envelope; responsible for manufacturing life-sustaining molecules and membrane components of the cell.

enzymes Special class of proteins that speeds up chemical reactions.

enzyme-linked receptors Transmembrane protein activated in signal transduction and responsible for starting the phosphorylation cascade, which amplifies signals received by the cell.

epicotyls Shoot that develops above the cotyledons.

epidermis Made up of mostly unspecialized cells forming a single layer covering the surface of herbaceous plants for protection.

epigeous Type of germination in which eudicot seeds develop a hypocotyl arch that emerges first through the soil before the cotyledons.

epiphyte Plant that attaches by its roots to another plant for support and does not need to have its roots in the soil.

epiphytic Organisms that grow on the surfaces of other plants and acquire water and nutrients from the air.

essential oils Oils harvested from plants for their volatile aromatic compounds.

eudicotyledons or eudicots One of three classes of core angiosperms; have two cotyledons in their seeds; flower parts come in fours or fives, or in multiples of four or five; leaves have netted veins.

eukaryote Organism that is made up of one or more eukaryotic cells (cells with a nucleus).

eukaryotic cells Cells with a true nucleus (*eu-* in Greek means "well," *kary* means "nut" or "nucleus").

euphylls or true leaves Leaves with a branched vascular system and leaf gaps in the stem.

eutracheophytes or true vascular plants Land plants that have lignified tracheids to transport water and minerals (*eu-* in Greek means "true" or "well"); all vascular plants.

evaporation See *vaporization*.

evaporative cooling Heat is removed as water evaporates, which helps to cool down surfaces or tissues.

evergreens Coniferous trees that bear their leaves throughout the year and do not shed all their leaves in the fall.

excurrent Growth habit where a central leader forms the crown of a tree.

exocarp Outer layer of the pericarp.

exogenous When a material is found or applied on the outside of a plant.

expansin Protein in the cell wall involved in loosening cellulose microfibrils.

eyespot Found near the base of the flagellum of euglenoids; responsible for light detection.

fibers Long and tapered sclerenchyma cells often found in clumps.

fibrous proteins Provide structure support and shape for organisms; have mostly α-helical and β-pleated sheet folding.

fibrous roots Several initial embryonic roots that develop into multiple primary roots, creating a dense root system.

field capacity Percentage of water held in a soil saturated with water and drained.

flavonoids Secondary plant metabolites with potential protective properties for human health.

fleshy fruit Fruit in which the pericarp is thick and succulent when developed.

floral formula Shorthand summary of floral features for a specific flower.

floral nectar Sugary secretion produced by angiosperm flowers used as rewards for pollinators.

florets Small flowers that make up an inflorescence.

fluid-mosaic model Model proposed by Singer and Nicolson in 1972 to explain how phospholipids and proteins come together to form membranes.

food chain Single-line hierarchy of which organisms eat each other at different trophic levels in an ecosystem.

food storage roots Roots that serve to store excess carbohydrates until they are needed at a later time.

food web Complex interactions between all organisms at different trophic levels in an ecosystem describing the movement of energy.

foot Attaches the liverwort sporophyte to the gametophyte and facilitates nutrients and water uptake by the sporophyte.

foliar application Applying water-soluble fertilizer to leaves in a spray solution.

foundation seed Seed from breeder's seed that is used to maintain a very pure seed line.

fragmentation A large habitat of a population is divided into smaller ones by disturbances resulting in isolated smaller patches of the original habitat, now separated by other habitats.

fronds Megaphylls or leaves of ferns.

fructans A polymer made up of fructose molecules; primary storage molecule in leaves and stems of wheat, rye, and barley.

fruit Ripened ovary.

fruiting bodies Large and conspicuous fungal reproductive structures that are composed of densely packed hyphae growing out of the substrate.

G_1 phase Part of the cell cycle where cells increase in size; begins right after a nucleus has divided and ends before DNA replication.

G_2 phase Part of the cell cycle where cells produce more proteins, replicate organelles, and accumulate energy in form of ATP; begins right after DNA replication is completed.

G-proteins Amplify signal in signal transduction and produce large amounts of second messengers.

gametophytes Adult haploid plants produce gametes by mitosis in their gametangia.

gametangia (singular *gametangium*) Structures that produce gametes or sex cells.

gametes Sex cells (1n) that fuse together to create a zygote.

gametes Haploid reproductive cells such as sperms or eggs. Gametes unite during sexual reproduction to produce a diploid zygote.

gametophores Structures with umbrella-shaped gametangia sitting on top of slender stalks; found only in liverworts.

gemma (singular *gemmae*) Specialized structures for asexual reproduction; used by liverworts to produce individual gametophytes.

gemma cups Cup-shaped structures found on the surface of liverwort gametophytes; contain gemma.

gene-for-gene hypothesis Explains how plants sense pathogen infection and how they activate the hypersensitive response. Resistant genes in plants produce receptors that will recognize and neutralize corresponding avirulence gene products from the pathogen.

geophytes Storage stems located below the soil surface.

germination When a seed goes through multiple physical and chemical changes that results in growth of the enclosed embryo.

germplasm Collection of seed that provides specific genetic resources.

globular proteins Proteins that have either a tertiary or quaternary structure.

gluten Protein found in the endosperm of wheat and other cereals, making dough elastic and helping it to rise.

glyceraldehyde 3-phosphate (G3P) Product of the reduction phase of the Calvin cycle; one molecule of G3P is produced after three completed cycles.

glycoalkaloid Alkaloids in plants from the nightshade (Solanaceae) family.

glycolysis The phosphorylation of glucose in the first step of aerobic respiration resulting in energy and the final product glyceraldehyde-3-phosphate; occurs in the cytosol.

Golgi apparatus See *dictyosomes*.

graft compatibility Degree of success attained when two plant portions are placed in direct contact and allowed to heal together.

graft junction Point of contact between the scion and the rootstock.

grafting Process of creating an improved plant by placing the rootstock of one plant in direct contact with the scion of another plant and allowing them to heal together.

grana (singular *granum*) Disc-shaped structures found within chloroplasts; have numerous interconnected discs that look like stacks of coins.

gravitropism Movement of a plant in response to gravity.

ground meristem Primary tissue that produces parenchyma cells, creating the cortex.

guard cells Found beside the stoma; responsible for opening and closing the stomatal pore.

gullet or groove Euglenoids use this groove to ingest food.

guttation Droplets of water are forced out of leaves as a result of root pressure.

gymnosperms or cone-bearing plants Group of vascular seed plants that produce seeds and their seeds are not enclosed within a fruit (*gymnos* in Greek means "naked" and *sperma* means "seed"); includes cycads, ginkgo, gnetophytes, and conifers.

Hadley cell Atmospheric air circulation pattern where warm, moist air rises near the equator, resulting in rain and then flows toward the poles, sinking at about 30 degrees latitude north or south.

haploid Having one set of chromosomes in the cell(s).

haploid Having one set (n) of chromosomes in the cell(s).

hardwood cutting When a cutting is made from stiff, inflexible wood from the previous season's growth.

heartwood Wood at the center of a tree stem that is darkened and has become inactive.

heat of vaporization Heat required to change 1 gram of a substance from its liquid form to gaseous form.

herbaceous plant A plant which is not woody and freezes down during the winter; perennial will return the following spring; annuals will not return.

herbivory When an organism is eating a plant. Herbivores subsist exclusively on plants.

heterosporous Production of two types of haploid spore: microspore that germinates into a male gametophyte and megaspore that germinates into a female gametophyte.

heterotrophs Organisms that cannot produce their own food but have to depend on other organisms for food.

homosporous Production of one type of haploid spore that germinates into a bisexual gametophyte.

hyaline cells or hyalocysts Large dead cells found in moss blade for water storage.

hydroids Parenchyma cells specialize in conducting water; found only in moss gametophytes.

hydrolysis Chemical reaction that splits a polymer into individual monomers by adding water.

hypersensitive response Plant defense mechanism where cells intruded by a pathogen will kill themselves and surrounding cells to starve the pathogen.

hypertonic Solution with a higher solute concentration or potential and a lower water concentration than a comparison solution. See also *isotonic* and *hypotonic*.

hyphae (singular *hypha*) Thread-like cells of fungi.

hypocotyls Shoot that develops below the cotyledons.

hypodermis Pine needle that has several layers of thick-walled cells just beneath its epidermis; functions to protect photosynthetic tissues underneath from freezing temperatures.

hypogeous Type of germination in which monocot seeds have the plumule emerge through the soil first and the cotyledon remains below the soil surface.

hypothesis testing See *scientific methods*.

hypotonic Solution with a lower solute concentration or potential, and a higher water concentration than a comparison solution. See also *isotonic* and *hypertonic*.

imbibition When a seed absorbs moisture resulting in initiation of growth activities.

imperfect flower When a flower has only male or female components but not both.

incomplete flower When a flower is missing one or more of the four primary parts including the sepals, petals, stamens, and pistils.

indehiscent fruits When a dry fruit has sutures that do not split and the seed is distributed by other mechanisms.

independent variable One variable, condition, or factor being manipulated and tested in an experiment.

indeterminate growth Growth continues throughout an organism's life.

indoleacetic acid (IAA) See *auxin*.

inductive reasoning Generates a unifying explanation or general principle after carefully evaluating specific studies; begins with specific studies and ends with a general principle.

infiltration rate Measure of how much water per unit of time a soil can absorb.

inflorescence Floral stalk bearing multiple flowers or a flower cluster.

integument Thick layer of sporophyte tissue surrounding the megasporangium that develops into the seed coat to protect the seed.

internode Section of stem between two nodes.

interphase A preparatory stage or period in the cell cycle (*inter* in Latin means "between").

ion channels Channel protein allowing passage of specific ions through the plasma membrane along an electrochemical gradient.

isoflavones Phytoestrogens that have possible health benefits for lowering cholesterol or used in cancer treatment.

isotonic Solution with the same solute concentration and the same water concentration than a comparison solution. See also *hypotonic* and *hypertonic*.

K-selection traits Typical for organisms living long in a stable environment that reached its carrying capacity.

karyogamy Process when the two nuclei of mating hyphal cells are fused together.

kinetochore A structure of proteins attached to the centromere that connects each sister chromatid to the mitotic spindle

lactic acid fermentation Process of fermentation where pyruvate is converted to lactic acid because of the lack of oxygen.

lamina Blade of a leaf that is the main surface for absorbing energy from the sun.

laminate bulbs Vertical, compressed underground stem composed of fleshy storage leaves.

lateral buds see axillary buds.

leaching Loss of ions in the soil when they move with water below the root zone.

leaf cuttings When a leaf is used as the primary plant propagule for production of new plants.

leaf margin Outer edge of a leaf that varies in shape and provides a method of identification.

leaf scars Markings left by the attachment of a petiole to a branch.

lemma Larger, outer bract that, along with the palea, serves to contain and protect the floret(s) within.

lenticels Found on the surface of bark; areas where the cells are loosely arranged to allow gas exchange to take place.

leptoids Parenchyma cells specialize in distributing food made by photosynthesis; found only in moss gametophytes.

leucoplasts Plastids found commonly in seeds, roots, and stems and are colorless; mainly responsible for producing and storing starch, oils, or proteins.

liana Wood stem that is not able to support itself and resembles a vine.

light compensation point When photosynthesis and respiration are at an equilibrium and net photosynthesis is zero.

light saturation point Net photosynthesis remains steady even if light intensity is further increased.

lignin Hard substance found in the secondary cell wall in which the cellulose fibers are embedded; found in tracheids, vessels, and sclereids; provides waterproofing and structure support for plants.

loam Soil containing approximately one-third sand, one-third silt, and one-third clay particles.

locule Ovary chamber.

long-day plant Plant that flowers in response to short night periods. See also *day-neutral plant* and *long-day plant*.

lower epidermis Similar to the upper epidermis in providing protection to the inner tissues but differs in being the site for most of the stomata.

macromolecules Large organic molecules that are made by living organisms; the four major classes are proteins, lipids, carbohydrates, and nucleic acids.

magnoliids One of three classes of core angiosperms; more primitive than monocots and eudicots.

matrix Fluid portion inside the mitochondrion that contains enzymes, proteins, mitochondrial DNA, and ribosomes.

medullary rays Rays of xylem which cut across the growth rings of a tree at a perpendicular angel to allow radial transportation of water.

megaphylls or euphylls Large leaves that have a complex system of vascular veins; found in ferns and seed plants.

megasporangia Sporangia that produce megaspores, which germinate into female gametophytes.

megaspores Germinate into female gametophytes, which only produce archegonia.

meiosis A type of cell division that undergoes two rounds of divisions, meiosis I and meiosis II. Each division contains five stages; responsible for sexual reproduction

meristematic tissue Region of rapidly dividing tissue that allows for the formation of roots.

meristems Regions of rapidly dividing tissues that produce growth.

mesocarp Middle layer of the pericarp.

messenger RNA (mRNA) Delivers the genetic information or protein recipe to the ribosome, the protein factory of the cell.

metabolic process Series of chemical reactions carried out by living organisms to build or break down organic molecules and to store or release energy to power life.

metabolism All the chemical reactions that take place inside a cell.

metallothionein Protein binding to a metal ion and for storage or transport, preventing the metal from becoming toxic in cells.

metaphase Where chromosomes begin to line up in the middle of the cell.

microphylls Small leaves that have only a single unbranched vascular vein; found only in lycophytes.

micropyle Apical opening of an ovule for pollen grains to enter.

microsporangia Sporangia that produce microspores that germinate into male gametophytes.

microspores Germinate into male gametophytes, which only produce antheridia.

microtubules Thick protein fibers that make up the cytoskeleton.

middle lamella Layer of pectin that acts as a glue to hold adjacent plant cells together.

midrib Primary vein that runs the length of a leaf.

mitochondria (singular *mitochondrion*) Tiny rod-shaped membrane-bounded structures that carry out aerobic respiration to produce energy for the cell.

mitotic spindles Thick protein fibers called microtubules that are involved in the movement of chromosomes during nuclear division.

mitosis A type of cell division; divided into five stages: prophase, prometaphase, metaphase, anaphase, and telophase; responsible for asexual reproduction, growth, and repair.

monocotyledons or monocots One of three classes of core angiosperms; have one cotyledon in their seeds; flower parts come in threes or multiples of three; leaves have parallel veins.

monoecious Male and female sex organs are found on the same individual.

monomers Building blocks that make up a polymer.

monopodial Growth of a plant from a single growing point in one direction.

monosaccharides Simple carbohydrates or single sugars are building blocks or monomers of other carbohydrates.

mucigel Mixture of polysaccharides that provides a lubricant that eases the root tip through the soil.

mucilage Complex carbohydrate that is used to retain water.

multicellular Organisms that are made up of many cells (*multi* in Latin means "many").

mutualism Symbiotic relationship in which both organisms benefit and neither are harmed.

mycelium (plural *mycelia*) Collection or mass of highly branched filaments know as hyphae.

mycology Study of fungi.

mycoremediation Use of fungi for the treatment of soil pollution.

mycorrhizae Common soil fungi that form mutualistic relationships with the roots of plants.

mycorrhizas Symbiotic relationships between fungi and plant roots or rhizoids; enhance plant tolerance to stresses and facilitate water and nutrient uptake.

myrmecophytes Mutualistic relationship between ants and plants.

nitrogen fixation Atmospheric nitrogen gas fixed into ammonia by bacteria in root nodules or by lightning

nitrogen-fixing bacteria Microbes that grow within the roots of plants and are capable of fixing nitrogen from the atmosphere.

nitrification Two-step conversion of ammonia to nitrite and then nitrate by bacteria.

node Region on a stem where leaves attach.

noncertified seed Seed that has the least regulation and is used for general cropping.

noncyclic electron flow Flow of electrons from water to PSII, ETC, PSI, and ETC producing ATP and NADPH; also known as Z-scheme.

nonvascular plants Group of plants that has no vascular tissues, true roots, true stems, or true leaves; has gametophytes that are dominant and sporophytes that are small and dependent on gametophytes for survival; includes liverworts, hornworts, and mosses.

nuclear division Refers to the sorting and separation of duplicated chromosomes during mitosis and meiosis.

nuclear envelope Separates the content of a nucleus from the rest of the cell with two layers of membranes.

nuclear pores Structurally complex openings found all over the nuclear envelope and are responsible for selectively allowing molecules to enter or leave the nucleus.

nucleoplasm Network of protein fibers found within the nucleus.

nucleotides Basic units of building blocks that make up nucleic acids.

operculum Dome-shaped lid covers the capsule opening; found only in mosses (Bryophyta).

organic matter In soils, organic matter refers to once living material now decaying and recycling nutrients back into the soil.

organic molecules Complex molecules found in living organisms and containing carbon and hydrogen.

organism Different organs come together to make up a distinct living entity.

organismal ecology The study of behavior, physiology, and morphology of individual organisms of a species.

organs Group(s) of tissues that come together to perform specific functions.

osmosis Diffusion of water across a semipermeable membrane; water will move from an area of higher water concentration to an area of lower water concentration.

ovary Female flower component that is at the base of the style and contains the eggs.

ovule Integument-covered megasporangium that develops into a seed after fertilization.

oxidative phosphorylation The process when ATP is generated through chemiosmosis in the mitochondria in aerobic respiration.

P680 Reaction center molecules of photosystem II with peak absorption at 680 nm.

P700 Reaction center molecules of photosystem I with peak absorption at 700 nm.

palea Shorter, upper bract that, along with the lemma, serves to contain and protect the floret(s) within. See *lemma*.

palisade mesophyll Primary photosynthetic tissue of the leaf; tightly packed mesophyll cells found just below the upper epidermis.

palmately Compound leaflet arrangement where leaflets radiate from a central point like the fingers of a hand joining together at the palm.

paramylon Type of polysaccharides used by euglenoids to store food.

parasitic Organisms that steal water and nutrients from other plants.

parasitism Symbiotic relationship in which one organism is harmed and the other organism benefits.

parenchyma tissue Made up of parenchyma cells, which are the most abundant cells in the plant body; mainly responsible for storage.

parthenocarpic When fruit is developed without the process of pollination.

passive transport Movement of a molecule across a membrane from higher to lower concentration through channels, channel proteins, or by simple diffusion; requires no energy.

pectin Gelatin-like substance found in the cell wall; responsible for holding cellulose fibers together.

pedicels Small stalks which attach the individual florets to the main stalk of an inflorescence.

peduncle Flower stalk of a single flower or cluster of flowers.

pellicle Fine parallel strips underneath the plasma membrane that spiral around the cell of euglenoids, providing a flexible cover.

PEP carboxylase Enzyme in C_4 photosynthesis that fixes CO_2 to phosphoenolpyruvate to form oxaloacetate, a four-carbon organic acid.

perfect flower When a flower has both male and female components.

perianth Refers to the calyx and corolla collectively.

pericarp Ovary wall of a developing fruit.

pericycle Layer of cells that surrounds the vascular tissues in the root; gives rise to secondary roots.

periderm Forms a thick protective layer around the roots and stems of woody plants.

peristome Structure found around the capsule opening; consists of one or two rows of teethlike cells; responsible for controlling spore release.

permafrost Soil frozen year-round where only a small portion of the profile near the surface thaws briefly in summer.

permanent wilting point Percentage of soil moisture below which plants cannot absorb more water and begin to wilt.

petals Showy component of a flower that attracts insects and are collectively called the corolla.

petiole Structure attaching the leaf blade to the stem of the plant.

phelloderm Living parenchyma tissue found in the bark; mainly responsible for storage.

phosphoenolpyruvate (PEP) Molecule used in the CO_2 fixation step of the C_4 pathway.

phospholipid Type of lipid that has a modified phosphate and two fatty acids attached to a glycerol backbone.

phospholipids Structural basis of all membranes; contains Two fatty acid chains two fatty acid chains and a modified phosphate group attached to the glycerol.

phosphorylation cascade Number of phosphorylation (a process that adds phosphate to another molecule) events triggered by enzymes as a result of signal transduction to amplify a signal through the plasma membrane.

photoautotrophs Autotrophic organisms that use solar energy for food production; compare with chemoautotrophs.

photolysis Splitting of water molecules in photosystem II during the light reaction of photosynthesis

photomorphogenesis Growth and developmental responses of plants to light.

photon Elementary light particle.

photoperiod Number of hours of daylight in a 24-hour period.

photoperiodism Growth and development of a plant in response to the hours of day and night in a 24-hour period.

photophosphorylation Addition of an inorganic phosphate (P_i) to a molecule such as addition of P_i to ADP synthesizing ATP; using energy from the light.

photosynthesis Complex metabolic process that green organisms use to capture solar energy and convert it into chemical energy of organic molecules such as glucose or food.

Photosystems Pigments organized in membrane proteins having a reaction center with chlorophyll at the center surrounded by antennae pigments; they absorb light energy for photosynthesis.

photosystem I (PS I) Protein and pigment complex that is used to harness energy from light to make NADPH and occasionally ATP; an important component of light reactions of photosynthesis.

photosystem II (PS II) Protein and pigment complex that is used to harness energy from light to make ATP; an important component of light reactions of photosynthesis.

phototropins Blue light receptors in plants that cause growth responses.

phototropism Movement or growth of a plant in response to light.

photoreversibility One pigment exists in two forms; phytochrome absorbs red light and changes to the far-red conformation, in the far-red conformation it absorbs far-red light and reverts to the red conformation.

phyllotaxy Classification of leaf arrangement on a stem.

physical dormancy Type of dormancy where the hard seed coat prevents imbibitions of water to initiate germination.

physiological dormancy Internal dormancy in which a specific treatment must occur in order for the embryo to germinate.

phytochemicals Compounds produced from plants that have biological activity and are thought to be beneficial but are not essential to human health.

phytochrome Photoreceptor that absorbs red or far-red light; exists in two shapes and reverses shape in response to the illuminating wavelengths; affects flowering and seed germination.

phytoremediation Using plants for the reduction or removal of contaminants from soil, water, or air; also known as bioremediation.

pigment Molecules that absorb light of certain wavelengths.

pinnae Individual sections or leaflets of a frond.

pinnately Compound leaflet arrangement where leaflets are directly across from one another.

pioneer species First species to colonize a bare area.

pistillate Flower having only female components.

pistils Female flower component composed of the stigma, style, and ovary.

pith Ground tissue comprised of parenchyma and located in the center of roots and stems.

placentation Placement of the ovules in the locules.

plasmodesmata (singular *plasmodesma*) Tiny channels through the cell walls that connect the plasma membrane and cytoplasm of adjacent plant cells; responsible for cell communication

plasmogamy Process when the cytoplasms of two mating hyphal cells fuse together.

plastids Large organelles that are found in plant cells; the three main types are chloroplasts, chromoplasts, and leucoplasts.

plumule Rudimentary portion of a plant embryo giving rise to the shoot.

pneumatophores Roots growing above the water level to allow for oxygen uptake.

polar Add "Movement of auxin in one direction, with gravity, from shoot tips to the base and towards root tips.

polar covalent bond Chemical bond in which atoms share their electrons instead of exchanging them. However, the electrons are not shared equally among atoms due to different electronegativity, resulting in uneven distribution of electric charge.

pollen Male sex cells that originate in the anther.

pollen chamber Inner compartment of an ovule for trapping and directing wind-borne pollen grains toward the female gametophyte.

pollen grains Immature male gametophytes of seed vascular plants.

pollination Transfer of pollen grains male gametophytes to or near the female gametophytes.

pollination drops Sticky, sugary drops secreted by gymnosperm female gametophyte to provide landing site for air-borne pollen grains and to help draw them into the ovules.

polymers Complex and big molecules made up of many subunits called monomers.

polypeptide A single long chain of amino acids.

polysaccharides Complex carbohydrates made up of many monosaccharides.

population Populations of all organisms that live and interact in the same area.

population All individuals of the same species that live and interact in the same area.

predation Individuals of one species eating living individuals of another species.

pressure potential See *turgor pressure*.

prickle Extension of the epidermis making a sharp spine-like structure.

primary cell wall First layer of cell wall laid down by a growing cell.

primary dormancy When a seed is not capable of germination immediately after harvest.

primary growth Growth in height or length due to apical meristems.

primary structure Linear sequence of amino acids made up of a polypeptide.

primary succession Gradual process where an area without soil is first colonized by plants and other organisms.

procambium Primary root tissue that gives rise to the xylem and phloem tissue.

prokaryote Organism that is made up of a prokaryotic cell (cell without a nucleus).

prokaryotic cells Cells without (before) a nucleus (*pro-* in Greek means "before," *kary* means "nut" or "nucleus").

prokaryotic cells Cells without (before) a nucleus (*pro-* in Greek means "before," *kary* means a "nut" or "nucleus").

prometaphase Mitotic spindles begin to attach to each sister chromatid. The nuclear envelope is now completely fragmented. Chromosomes become more condensed.

prophase First stage of mitosis where chromosomes become visible, nuclear envelope and nucleolus start to disappear.

prop roots Adventitious roots that arise from the stem and grow vertically down to the soil where they root and supply support for the plant.

propagules Structures that develop into individual plants.

proplastids Small, colorless or pale green, simple organelles found in the meristemic; responsible for producing various plastids.

protein One of four organic molecules that are essential to life.

proteome Sum of all the proteins found in a cell or an organism.

proteomics Systematic study of proteomes encoded by genomes.

prothallus (plural *prothalli*) Heart-shaped fern gametophyte; small, flat, and does not have vascular tissues, true roots, stems, or leaves; rhizoids are used for anchorage.

protoderm Primary tissue that gives rise to the epidermis of the plant.

proton pumps Membrane proteins moving protons across a cell membrane against an electrochemical gradient, also known as H$^+$ ATPase.

protonema (plural *protonemata*) Immature moss gametophyte, which is a long, branching strand of green cells that superficially resembles a filamentous green alga.

pseudostem Structure appearing like a stem, but composed of folded or rolled petioles and leaf blades; found on the herbaceous banana plant.

pubescence Fine hairs covering the surface of a leaf.

pulse Dried seeds of legumes.

quaternary structure Consists of two or more polypeptides that come together to form a functional unit.

quiescence When a seed does not develop even when the appropriate environmental conditions are provided.

R-genes Disease-resistant genes in plants with the ability to recognize corresponding avirulence genes in pathogens.

r/K selection theory Theory in ecology that relates to the traits in reproduction based on whether large or small numbers of offspring are produced with different survival rates and lifespans. See also *K-selection traits* and *r-selection traits*.

r-selection traits Typical of organisms with a short life, reproducing rapidly and living in an unstable environment.

rachis Main axis where pedicels attached to in an inflorescence.

radially symmetrical flowers The petals are evenly distributed around a central point; mirror images created in multiple directions or planes.

radicle Initial embryonic root that protrudes through the seed coat and initiates the root system.

rain shadow Dry back side of a mountain, located behind the ridge on the side of the mountain where warm, moist air rises and rain is deposited before air crosses the ridge.

ray florets Small flowers surrounding the disc florets in Asteraceae inflorescence or head; each floret resembles a petal and has many sepals but its five fused petals form a tongue-like corolla; are usually sterile or pistillate.

reaction center Complex of chlorophyll molecule and proteins in the center of the photosystem that converts light energy into chemical energy.

reaction wood Secondary xylem created by a plant in response to mechanical stress.

receptacle Structure at the top of the stem to which the flower parts are attached.

region of cell division See root apical meristem.

region of cell maturity Portion of the root system where cells differentiate and begin to specialize in their tasks.

region of elongation Region of root system where newly divided cells expand in size; thus increasing the length of the root system.

registered seed Seed established from foundation seed and is kept at a level of purity that is high, but not at the level of foundation seed.

respiration Breakdown of organic molecules to yield energy (ATP) through oxidation of glucose by enzyme-mediated steps.

resin (oleoresin) Mixture of monoterpenes, sesquiterpenes, and resin acids used to repel invading pests and pathogens and to seal the wound.

rhizoids Are not roots but are simple multicellular filaments produced by epidermal cells.

rhizome Horizontal below-ground stem with nodes and internodes; exhitbit monopodial growth.

ribonucleotides Four nucleotides that make up RNA and have ribose as their pentose.

ribosomal RNA (rRNA) Ribonucleic acids that make up ribosomes.

Robert Hooke discovered tiny pores in cork and coined the word "cells"

root apical meristem (RAM) Region synonymous with the region of cell division; area of rapidly dividing cells, resulting in root growth and production of the root cap.

root cap The tip of the root that have direct contact with the soil and ease the small root into the soil; also perceive gravity and water gradients.

root cuttings A method of plant propagation where a portion of the root system is excised and used as propagule.

root hairs Simple extensions from individual epidermal cells in roots; responsible for increasing surface area of the root epidermis for absorption.

rough ER Endoplasmic reticulum that has ribosomes attached to its surface; responsible for producing and sorting proteins for different destinations within or outside of the cell.

rubisco Enzyme RuBP carboxylase/oxygenase mediates the first step in the Calvin cycle and is the most common enzyme on Earth.

RuBP (ribulose 1,5-bisphosphate) Five-carbon sugar with which the Calvin cycle starts and ends.

RuBP carboxylase/oxygenase or rubisco see rubisco

S phase Part of the cell cycle where DNA replication takes place and a duplicate copy of the blueprint is made.

saprophytic Organisms that obtain their nutrients from dead organic matter.

saprotrophs Organisms that acquire their nutrients and energy by breaking down dead organic matter.

sapwood Light-colored wood in a tree trunk that is the youngest tissue and conducts water and minerals.

saturated When a fatty acid has no double bonds between carbon atoms and contains the maximum number of hydrogen atoms.

savanna Grasslands interspersed with some trees.

scaling The process of gently removing each fleshy leaves from a scaly bulb for propagation.

scaly bulb Compressed, vertical underground stem in which the fleshy leaves are attached in a spiral pattern and are not enclosed in a tunic.

scarification Method of damaging an impervious seed coat to allow for imbibitions of water.

scion Top portion of a grafted plant that is the desirable cultivar selected for the best flowers or fruit.

sclereids or stone cells Type of sclerenchyma cells that are short and vary in shape.

sclerenchyma tissue Made up of sclerenchyma cells that are specialized for rigid structural support.

scooping/scoring Process of using cuttage to damage the basal plate of a bulb causing many smaller bulblets to form.

second messengers Small molecules produced rapidly to amplify signal transduction and response.

secondary cell wall Second layer of cell wall that has ligin and more cellulose than the primary cell wall.

secondary dormancy Induced by extreme stress or inappropriate storage.

secondary growth Growth in the girth of a plant as a result of the cambium.

secondary metabolites Chemical compounds produced by plants that are not essential for plant survival but assist in primary metabolism and defense mechanisms against pests, pathogens, and herbivores. See also *alkaloids*.

secondary structure When the amino acids in a polypeptide form hydrogen bonds between each other, the chain begins to fold into either a helical shape (α-helix) or a pleated sheet (β-pleated sheet).

secondary succession Gradual process where a community was disturbed and is being recolonized on existing soil.

secondary tissues Produced by the Vascular and cork cambium of a plant and result in girth growth.

seed Reproductive structure that contains an embryonic sporophyte, food reserve, and a protective seed coat.

seedless vascular plants Plants such as lycophytes and ferns do not produce seeds and have vascular tissues; dispersed by spores.

self-pollination When a flower is able to be pollinated by the pollen from the same plant; synonymous with the term *selfing*.

semihardwood cutting Portions of a stem that is the current season's growth but more mature and generally more stiff.

seminal roots Adventitious seed roots arising from the base of the plant that grow laterally and help uptake soil moisture for the new seedling.

senescence Period of aging that eventually leads to the death of the leaves.

sepals Leaf-like structures that surround the petal and are collectively called the calyx.

separation Process of physically pulling and removing daughter bulbs from the mother bulb.

septa (singular *septum*) Cross walls that divide most fungal hyphae into many smaller sections or cells.

septate Fungal hyphae that have cross walls and are divided into many smaller sections.

serotinous cones Conifer cones requiring heat to open and release their seeds.

sessile When florets, flowers, fruits, leaflets, or leaves attach directly to a stem or axis.

seta Found near the base of liverwort sporophyte connecting the foot to the capsule; responsible for spore dispersal.

sexual reproduction Union of the gametes produced by two individuals produces offspring that are genetically different from their parents.

sheath Lower part of the leaf that encloses the internode on the culm.

shoot Above ground portion of a plant including the leaves and stems.

short-day plant Flowering of this plant is initiated by exposure to long nights. See also *day-neutral plant* and *long-day plant*.

shrubland Grasslands with widely spaced shrubs.

sieve-tube elements Links of long, thin cells that are heavily perforated, allowing cytoplasm to extend from one sieve-tube element into the next; responsible for transporting food and essential materials throughout the plant.

signal deactivation When no more signal is received by the signal receptor and signal transduction stops.

signal receptor Specialized protein in plasma membrane receiving signals from outside the cell and binding to a signal molecule.

signal transduction Conversion of a signal from outside the cell to within the cell.

simple fruit Fruit that is developed from an ovary with a single carpel.

sink Location where sugars exit the phloem and are used in metabolic processes.

smooth ER Endoplasmic reticulum that is not associated with any ribosomes; responsible for adding carbohydrates to proteins and lipids.

softwood cuttings Portion of the stem that is composed of early spring growth, which is less mature and more flexible.

soil texture Proportion of different-sized soil particles used to classify physical soil characteristics.

solutes A substance that is dissolved in a solution.

solute or osmotic potential Also known as solute potential, osmotic potential is part of the water potential and refers to the difference in energy of water based on the solute concentration; values are negative.

solution A Liquid with one or more substances dissolved in it.

solvent Liquid in which a substance is dissolved.

source Location where sugars are produced by photosynthesis and enter the phloem.

specialized storage structures Modified compressed stems which are located under the ground and allow a plant to survive periods of adverse conditions by storing carbohydrates.

specific heat Amount of heat required to raise the temperature of 1 gram of a substance by 1.8°F (1°C).

spine Modified leaves forming a sharp projection frequently confused with a prickle.

spongy mesophyll Loosely organized tissue below the palisade mesophyll with air spaces and vascular bundles passing through it.

sporangia (singular *sporangium*) Structure that produces spores.

sporangiophore 1. Specialized stalks where fungal sporangia attached. 2. Umbrella-shaped structure that has 5 10 sporangia attached to it and a stalk that connects the cluster of sporangia to the common axis of the strobilus; found only in horsetails and scouring rushes.

sporangiophore Specialized stalks where fungal sporangia attached.

sporangiospores Asexual spores produced within fungal sporangia.

spore mother cells or sporocytes Undergo meiosis to produce haploid spores.

sporophytes Adult plants that produce spores by meiosis in their sporangia.

sporopollenin Major component of the outer walls of plant spores and pollen grains; it helps spores and pollen grains to prevent excess water loss and resist decomposition.

stamens Male floral structure composed of the anther and filament.

staminate Male flowers having no pistils.

stand-replacing fire regimes Infrequent high-intensity fires often kill all or most overstory trees and sometimes damage the soil.

starch Very large polysaccharide made up of glucose molecules and is the primary storage molecules in plants and some algae.

statolith hypothesis Role of statoliths in sensing gravity by sinking to the bottom of cells, triggering a gravity receptor, and causing reorientation of the cell.

stele Vascular cylinder of the root that contains the xylem, phloem, and, in monocots, the pith.

stem cuttings A method of plant propagation where portion of the stem is excised and used as propagule.

sterols Made up of four interconnected hydrocarbon rings that may have a hydrocarbon chain attached to one of the rings.

stigma Female floral structure at the tip of the style that receives pollen.

stock Roots and stem portion of a grafted plant that is selected for the tolerance to soils or disease or may be selected for its ability to dwarf the newly created plant.

stolon Horizontal stem found on or above the soil surface with nodes and internodes; not fleshy.

stomata (singular *stoma*) Tiny openings surrounded by two guard cells; responsible for gas exchange and regulation of transpiration.

storage Capacity of a plant to be able to put back and store surplus carbohydrates and other materials for use by the plant later.

stratification Seed treatment in which seed is chilled in a moist media.

strobili or cones Clusters of sporophylls bearing sporangia.

stroma Liquid portion of the chloroplast that contains enzymes essential for photosynthesis, chloroplast DNA, and ribosomes.

style Female floral structure below the stigma where the pollen tube grows through to the ovary.

suberin Major component of the walls of cork cells in woody plants and root endodermal cells; helps to restrict water and mineral movement in the roots.

substrate 1. Reactants that bind to the active site of an enzyme. 2. Organic materials that fungi use as food.

substrate Reactants that bind to the active site of an enzyme.

succession Gradual process of colonizing an area with organisms.

symbiotic relationship Close biological relationship between two different species.

symplast Continuous pathway in the roots leading through cells and plasmodesmata.

sympodial Growth resulting in lateral branching from secondary axis.

systematics Study of biological diversity and evolutionary relationships among organisms.

system-acquired resistance Development of resistance to pathogen invasion of an entire plant in response to a localized pathogen attack.

systemin Hormone that is produced after wounding of a plant; results in the production of proteinase inhibitors, making plants unpalatable to herbivores.

taiga Wide belt of conifer forest located south of the polar ice. See *tundra*.

taproot Type of root structure that consists of a single primary root and much smaller secondary roots.

taxonomy Study of describing, naming, and grouping of organisms.

TCA or Krebs cycle This part of the aerobic respiration includes several reactions that releases energy starting with acetyl Co-A as substrate and occurs in the matrix of mitochondria.

telophase Last stage of mitosis where chromosomes begin to decondense and become invisible; nuclear envelope and nucleolus start to reform.

tendrils Modified leaf that aids in plant support by wrapping around a structure or another plant.

tension wood Reaction wood created by eudicots.

tepals Petals and sepals when they are indistinguishable from one another.

tertiary structure When the secondary structure folds to form a complex shape, which is maintained mainly by disulfide bridges.

testa Hard outer seed coat frequently called the integument.

theory of endosymbiosis Proposed by Margulis in 1971 to explain the origins of mitochondria and chloroplasts.

thigmomorphogenesis Changes in growth and development of a plant in response to touch or mechanical stimuli.

thigmotropism Movement or growth of a plant in response to touch.

thorn Modified stems or branches creating a sharp projection.

thylakoids Hollow coin-like structures inside the chloroplasts.

tip cuttings A method of plant propagation in which the tip or apical growing portion is excised and used as propagule.

tissues Group(s) of cells coming together to perform specific functions.

totipotency Capability of cells to divide and grow into a complete mature organism.

tracheids Long, tapered cells with thick secondary cell walls that make up the xylem; often found in patches or clumps.

transfer RNA (tRNA) Transfers specific amino acids to the ribosome during protein synthesis.

translocation Transport of organic molecules and sugars in the phloem.

transpiration Loss of water vapor through the stomata in the leaves.

trichomes Specialized epidermal cells that are used by some plants for protection.

triglyceride Lipid composed of three fatty acids connected to a glycerol molecule; main component of vegetable oils.

trophic level Level within the hierarchy of feeding in an ecosystem.

tuber Fleshy underground storage structure located at the end of a rhizome with nodes on the surface such as an Irish potato.

tuberous roots Root modification in which a swollen fibrous root system stores food for later distribution throughout the plant.

tunic Dry, papery covering on a laminate bulb that provides protection for the bulb.

turgor pressure When water enters the central vacuole and causes it to swell and exert pressure against the cell wall; responsible for providing strength to keep nonwoody plants upright.

understory fire regimes Frequent low-intensity surface fires that rarely damage the soil or kill overstory trees.

unicellular (*uni-* in Latin means "one") Organisms that are made up of just one cell.

unsaturated When a fatty acid has one or more double bonds among its carbon atoms and contains less hydrogen atoms.

vaporization When a substance is transformed from its liquid form to gaseous form.

vascular bundles Xylem and phloem tissues that transport water and nutrients in the plant.

vascular cylinder Composed of primary xylem and phloem; located in the center of the root.

vascular seed plants Group of plants that have vascular tissues, true roots, true stems, and true leaves; use seeds for dispersal; have sporophytes that are dominant and gametophytes that are small and dependent on sporophytes for survival; includes gymnosperms and angiosperms.

vascular seedless plants Group of plants that have vascular tissues, true roots and stems; use spores for dispersal; have sporophytes that are dominant and gametophytes that are small but independent from sporophytes; includes club mosses, ferns, and their relatives.

vegan Diets using plant-based food only, excluding also egg and dairy products.

vessel elements Much wider in diameter than tracheids and tend to be open or heavily perforated at both ends of the cell; make up the xylem.

vestigial seeds Rudimentary seeds that develop in a fruit that is considered seedless.

vine Herbaceous stem that is not woody and cannot support itself.

water-holding capacity Amount of water a soil can hold based on the soil texture.

water potential Energy of water in an environment; in plants is the sum of the solute or osmotic potential and pressure potential.

water storage roots Type of root system found primarily on arid and semiarid plants that retains water, allowing the plant to withstand periods of drought.

waxes Lipids used by plants to prevent desiccation found in plant cell walls and cuticles.

waxy bloom Layer of material that reduces water loss from the leaf.

wetland Area covered with shallow water on a regular or permanent basis.

woody plant A plant in which supportive tissues and lignin combine to provide a hard stem capable of supporting itself.

wounding When a cutting is intentionally damaged, resulting in a concentration of hormones in the affected region.

xeric Dry habitat, adapted to little moisture.

zeaxanthins Carotenoid pigment responding to blue light and regulating stomatal opening.

zooxanthellae Dinoflagellate that lives inside another organism and have lost their cellulose plates, and flagella.

zygomorphic See *bilateral symmetry*.

zygote Single cell formed by the union of the male and female gametes.

Index

Note: Page numbers followed by a *f* denote figures and those followed by a *t* denote tables

A

Abies balsamea, 578
Abscisic acid (ABA), 255
Abscission, 136
Absorption spectrum, 209*f*
Accessory pigments, 208–209
Acclimation, 388
Acid growth hypothesis, 248, 250*f*
Actin filaments, 32, 34
Actinomorphic flowers, 149, 149*f*
Action spectrum, 208
Active ion exclusion, 283
Active transport, 283, 282*f*
Adaptation, 291
Adaptive radiation, 394–395, 395*f*
Adenosine triphosphate (ATP), 32–33, 212, 212*f*, 214*f*
Adhesion, 27, 289
Aequorea victoria, 355
Aerenchyma, 599
Aerobic respiration, 230
Aflatoxins, 620
Agar, 436
Agrobacterium tumefaciens, 357, 357*f*
Agronomy, 16
Alcohol fermentation, 239
Algae
  biofuel producers, 444
  brown algae, 437–438
  colonial green algae, 434–435
  diatoms, 439–441, 441*f*
  dinoflagellate, 441–442, 442*f*
  euglenoids, 442–443, 442*f*
  food and industrial products, 444–445
  lichen, 436
  multicellular green algae, 435

  phytoplankton, 443
  protists, 432–433
  red algae, 436–437, 437*f*
  toxic blooms and dead zones, 443–444
  unicellular green algae, 434
Algin/alginic acid, 439
Allopatry, 398
Allopolyploid organisms, 399
Aloe vera, 643
α-amylase, 38
Alpine tundra, 576
Alternation of generations, 310, 458*f*, 458
Amborella family, 526
Amino acids, 35
Amylopectin, 31
Amyloplasts, 65, 262
Amylose, 31, 32
Anabolic reactions, 55, 202
Anaerobic cellular respiration, 237
Ananas comosus, 224
Anaphase, 306
Anaphase I, 309
Anatomy, 16
Anchorage, 78
Aneuploid, 343
Angiosperms, 15, 450, 452*f*
  basal, 525–526, 527*f*, 528
  characteristics, 520–525
  eudicots, 536–543
  evolutionary relationships among, 525–530
  magnoliids, 531–536
Anions, 280
Annulus, 485, 488*f*
Antennae complex, 211, 211*f*
Anther, 146, 146*f*
Antheridia, 458

Antheridiophores, 468
Anticodon, 340
Antioxidants, 629
Antiporters, 283
Apomixis, 194, 399
Apoplast pathway, 286, 287f
Aquaporins, 265
Arabidopsis thaliana, 359
Araucaria araucana, 579
Arbuscular mycorrhiza (AM), 92, 554, 554f, 558f
Arbuscules, 92
Archaeplastida, 14, 14f
Archegonia, 458
Archegoniophores, 468
Arctic tundra, 575–576
Asci, 555, 555f
Ascospores, 555
Aseptate, 550
Asexual reproduction, 11, 11f, 144
  advantages, 183, 184f
  apomixis, 194
  budding, 191, 192f
  clones, 182, 183f
  disadvantages, 184
  grafting, 109–191, 190f, 191f
  layering, 192, 193f
  micropropagation, 194, 195f
  underground plant parts, 193, 193f
  vegetative cuttings (*see* Vegetative cuttings)
Assimilation, 607
Atoms, 7, 8f
Autotrophs, 2, 61, 203, 204f
Auxin
  acid growth hypothesis, 248, 250f
  expansin, 249
  gravitropism, 252, 253f
  indoleacetic acid (IAA), 248
  low concentrations of, 252
  phototropism, 248, 249f, 252, 253f
  polar movement, 249, 249f
  vascular tissue development, 252
Avirulence gene (Avr), 266
Axillary buds, 102
Azolla, 621–622

B

Bacillus subtilis, 369
Bamboo, 641–643
Bananas, 628f, 627

Barley, 623, 624f
Basal angiosperms, 525–526, 527f, 528
  amborella family, 526
  star anise family, 527
  water lily family, 526, 527f, 527
Basal plate, 111
Basidia, 556, 556f
Basidiocarps, 556, 556f
Basidiospores, 556, 556f
Bast fibers, 640
Beltian bodies, 592
Beverages, 636–637
Bilateral symmetry, 149
Binary fission, 302, 430
Biochemistry, 16
Biogeochemical cycles
  carbon cycle, 605, 606f
  nitrogen cycle, 605–608, 607f
Biogeography, 573
Biolistic transformation, 354, 356
  plant cells, 359
  plant plasmids, 359
Biological species concept, 396
Biomes, 9, 573–574
Biopharming, 328, 369
Bioremediation, 562
Biosphere, 9, 10
Boreal forest, 577–578, 577f
Bottleneck effect, 391
Bracts, 127, 127f
Brassinosteroids, 258
Breeder's seed, 177
Bryology, 16
Bryophytes
  characteristics, 466–470, 467f
  definition, 464
  ecological importance and uses, 476
  evolutionary relationships, 464, 467f
  fuel production, 475–476, 475f
  hornworts, 470–471, 471f
  horticultural uses, 474
  household and industrial items, 475
  liverworts, 466–470, 469f, 469f, 470f
  medical uses, 476
  mosses, 471–472, 473f, 474f
Bud scales, 102
Budding, 191, 192f, 552
Bulbils, 110
Bulblets, 110
Bundle scars, 103
Bundle sheath cells, 219

C

Callus, 253, 357
Calvin–Benson–Bassham cycle, 216
Calvin-Benson cycle, 212, 213f
Calvin cycle, 216–218
Calyptra, 472, 474f
Calyx, 145, 145f
Capillary action, 289
Capitulum, 152, 153
Capsule, 470
Carbohydrates, 29–32
Carbon cycle, 205–206, 205f, 605, 606f
Carnivorous plants, 128–129, 128f
Carotenoids, 434
Carrier proteins/transporters, 282
Carrying capacity, 594f, 595
Casparian strip, 86
Cassava, 627
Catabolic reactions, 55
Catalysts, 37, 38
Cation exchange capacity, 281
Cations, 281
Cavitation, 291
Cell biology, 16, 48
Cell cycle, 302
    stages of, 305f
Cell division, 302
Cell plate, 306
Cell theory, 48, 48f
Cell wall, 65–66, 66f
Cells, 7
    anaphase, 306
    animal cell, 66, 67f
    cell theory, 48, 48f
    cell wall, 65, 66f
    chloroplasts, 63–64, 46f
    chromoplasts, 65, 64f
    cytokinesis, 306
    cytoplasm-metabolic center, 55
    deoxyribonucleic acid (DNA) molecules, 55, 56f
    dictyosomes/Golgi apparatus, 58
    Elodea cells, 46, 46f
    endoplasmic reticulum, 58, 57f
    eukaryotic, 50, 52f, 53t
    internal transport, 55
    interphase, 303, 305f
    leucoplasts, 65, 64f, 65
    light microscopes, 49–50, 50f, 51f
    metaphase, 305
    mitochondria, 59–60, 60f, 62f
    nucleus, 56–58, 57f
    parenchyma cells, 46f
    plant cell, 60, 63f, 66, 67f
    plasma membrane, 53, 55, 63f
    prokaryotic, 50, 52f, 53t
    prophase, 304, 304f
    red pepper cells, 46f
    ribosomes, 60
    Robert Hooke, 47–48, 47f
    sizes, 49, 49f
    stomata, 46f
    stone cells, 46f
    telophase, 306
    tracheids, 46f
    vacuoles, 58
Cellular respiration, 3, 4, 231f
    types of, 238f
Cellulose, 31, 32
    plates, 441
Centers for Disease Control and Prevention (CDC), 372
Central vacuole, 58
Centromere, 304
Cercis mexicana, 204f
Cereal crops, 618–619
Certified seed, 177
Channel proteins, 282
Chapparal, 581, 582f
Chemiosmosis, 213, 236
Chemoautotrophs, 9, 204
Chimeric restriction enzymes, 351
Chitin, 65, 550
Chlorophyll, 208, 424
    chlorophyll c, 438
Chloroplasts, 63, 64f, 208, 210, 210f
Chromatin, 57, 58
Chromoplasts, 63–64, 65f
Chromosome, 57, 58, 302
Cladogram, 411
Cladophylls, 113, 114f
Climacteric fruit, 256
Climate, 570–571
    global patterns, 571, 571f, 572f
Climax community, 608
Clustered regulatory interspaced short palindromic repeat (CRISPR)/Cas-based RNA-guided DNA endonucleases, 351
Cocoa, 637, 637f
Coevolution, 396, 397f, 593
Coffee, 636–637, 637f

Cohesion, 27, 290
Cohesion–tension theory, 290–291, 290*f*
Coleoptiles, 175
Coleorhizae, 175
Coleus tip cuttings, 185, 185*f*
Collenchyma tissue, 69, 69*f*
Commensalism, 90–91, 91*f*, 591–593, 594*f*
Community, 9, 10*f*
    ecology, 591
Companion cells, 70
Competition, 591, 593
Complete dominance, 325
Compound fruit, 165
Compound pistil, 147
Compression wood, 98
Cone-bearing plants. *See* Gymnosperms
Cones. *See* Strobili
Conidia, 551
Coniferous forest, 578–579, 579*f*
Conifers, 507–514, 507*f*–513*f*
Conjugation, 435
Conservation biology, 598
Contractile roots, 111
Convergent evolution, 395–396
Core angiosperms, 528
Cork cambium, 73
Cork cells, 72
Cormlets/cormels, 112
Corms, 111–112, 111*f*
Corn. *See* Maize
Corolla, 144, 145*f*
Cortex, 98
Cotransporters, 265, 283
C₄ Photosynthesis, 218, 219*f*
C₃ plants, 216
    vs C₄ pathways, 220, 221*t*
Crassulacean acid metabolism (CAM),
    221–222, 222*f*, 223*f*
Cristae, 60
Crop cultivation, 616
Crossing over, 308, 309
Cross-pollination, 146
Crustose lichens, 436
Cryopreservation, 178
Crypts, 122, 123
Culm, 536
Cuticle, 40, 41, 71, 121, 456
Cutin, 40
Cyanobacteria
    biofuel producers, 444
    cell division, 430

as chloroplasts, 431
colonial forms of, 427, 427*f*
colonies of *Nostoc* with heterocysts, 426, 425*f*
food and industrial products, 444–445
in mineral water, 426, 426*f*
nitrogen fixing, 430–431
photosynthetic membranes, 428, 428*f*
phycobilins, 424
phytoplankton, 443
prochlorophytes, 432
prokaryotic capsule and biofilm, 429, 429*f*
prokaryotic cell wall, 428–429
prokaryotic DNA and ribosomes, 429–430
stromatolites, fossilized cyanobacteria,
    427, 427*f*
toxic blooms and dead zones, 443–444
Cycads, 504–505, 505*f*
Cyclic electron flow, 216
Cyclic photophosphorylation, 215
Cytokinins, 252–253, 254*f*, 303, 306
Cytology, 48
Cytoplasmic division, 303
Cytoplasmic streaming, 55
Cytoskeleton, 55
Cytosol, 55

D

Day-neutral plants, 260
Deciduous, 507
Decomposers, 553
Deductive reasoning, 18
Dehydration
    avoidance, 465
    synthesis, 30
Denaturation, 37, 38
Denitrification, 606
Deoxyribonucleic acid (DNA), 7
Deoxyribonucleotides, 32
Dependent variable, 17
Dermal tissue system, 71–73, 71*f*, 72*f*, 73*t*
Desert, 583–584, 583*f*
Desiccation tolerance, 465
Determinate growth, 9
Development, 10
Diatomaceous earth, 441
Diatoms, 439–441
Dichotomous Key, 414–420
Dictyosomes, 58
Differentiation, 10
Diffusion, 282

Dihybrid cross, 322
Dikaryons, 550
Dioecious, 468
  flower, 147
Diploid, 306
Dipolar, 27
Direct gene transfer, 353
Direct microinjection, 354
Disaccharides, 29
Disc florets, 543, 543f
Discovery-based science, 17
Disease-resistant genes (R-genes), 266
Disulfide bridge, 37
DNA ligases, 348
DNA-mediated gene transfer, 353
Domatia, 592
Domestication, 4–7, 4f, 6f
Dominant trait, 318
Dormancy, 176
Double dormancy, 177
Double fertilization, 515, 524, 525f
Drought escape, 465
Dry dehiscent fruit, 167, 168t
Dry indehiscent fruit, 167, 169t
Duplicated chromosome, 304f

E

Ecology, 16, 570
Economic botany, 16
Ecosystem, 9
  biogeochemical cycles, 604–608
  dynamics and human activity, 597–598
  ecology, 590–591
  energy flow, 600–604
  energy transfer, 603–604, 604f
  succession, 608–612
Ectomycorrhizae, 92, 557, 557f, 558f
Elaioplasts, 65
Elater mother cells, 470
Electromagnetic spectrum, 207
Electronegative, 26
Electron transport chain, 234, 235f
Electroporation, 353, 354f
Elements, 29
Elicitors, 266
Embolism, 291
Embroyophytes/land plants, 15, 450–455, 454f
Enations, 492
Endergonic process, 202
Endocarp, 164

Endodermis, 86
Endomycorrhizae, 554, 554f
Endophytes, 558
Endoplasmic reticulum (ER), 58, 59f
Endosperm production, 524, 525f
Endosymbiosis theory, 61
Energy transfer, 603–604, 604f
Enzyme-linked receptors, 246
Enzymes, 37, 38
Epicotyls, 174
Epidermis, 71, 72f, 98, 121
Epigeous germination, 174
Epiphytes, 78, 520
Escherichia coli, 302, 348
Essential oils, 635
Ethnobotany, 16, 643
Ethylene
  abscission, 136
  in epicotyl growth, 255, 256f
  fruit ripening, 256
Eudicotyledons/Eudicots, 528
  legume family, 537–538, 537f
  mustard family, 540–541, 541f
  nightshade/potato family, 541–542, 542f
  pumpkin family, 539–540, 540f
  roots, 90, 90f
  rose family, 538–539, 539f
  sunflower family, 542–543, 543f
Eukaryotes, 13, 50, 52f, 53t
Eukaryotic cell cycle, 304f
Euphylls, 455. See also Megaphylls
Eutracheophytes, 453, 480–482, 480f–482f
Evaporation, 28
Evaporative cooling, 28
Evergreens, 507
Evolution
  adaptive radiation, 394–395, 395f
  Book of Genesis, 382
  coevolution, 396–397, 397f
  convergent evolution, 395
  EPSPS gene, 394
  evidence in, 385–387
  gene flow, 390
  genetic composition and, 387–388
  genetic drift, 390–391
  glyphosate-resistant weeds, 394
  Hardy–Weinberg law, 388–389
  Lamarck's theory, 383
  microevolution and macroevolution,
    392–393
  mutation, 390, 390f

Evolution *(continued)*
  natural selection, 383, 391, 392*f*
  punctuated equilibrium, 393
  random mating, 391
  speciation, 396–400
  transmutation, 383
  variation beaks, 383, 384*f*
Exocarp, 164
Expansin, 249
Extinction, 385
Eyespot, 442

F

Feedback inhibition, 237
Fermentation, 238
Ferns, 485, 489–495, 489*f*–495*f*
Fibers, 69
Fibrous proteins, 37
Fibrous root systems, 79, 80*f*
Field capacity, 274
Filament, 146, 146*f*
First filial (F1) generation, 318
Fitness, 387
Fitzroya cupressoides, 579
Flavonoids, 629
Fleshy fruits, 166, 167*t*
Floral nectar, 508
Florets, 536, 536*f*
Floridean starch, 436
Flower(s), 47, 521–522, 522*f*
  corolla types, 150–152, 151*t*
  floral formula calculation, 153–154, 154*f*
  food and dye source, 156–157, 156*f*–157*f*
  fragrances and perfumes, 155, 155*f*
  inflorescence morphology, 152, 152*t*–153*t*
  insecticides, 159, 159*f*
  medicinals, 157–158, 158*f*, 159
  petals, 144, 145*f*
  pistils, 147
  pollinators, 147, 148*f*, 149*t*
  sepals, 144, 145*f*
  stamens, 146–147, 146*f*
  symmetry, 149–150, 150*f*
  tepals, 145, 145*f*
Flowering plants. *See* Angiosperms
Fluid-mosaic model, 53, 54*f*
Foliar application, 284
Foliose lichens, 436
Food chain, 600–603, 600*f*, 602*f*
Food security, 617–618

Food web, 600–603, 602*f*
Foot, 468
Forcing flowering, 262
Forensic botany, 16
Forestry, 16
Fossil records, 385, 385*f*
Foundation seed, 177
Founder effect, 391
Fragmentation, 435, 598
Fronds, 485, 487*f*
Fructans, 31
Fruiting bodies, 549
Fruits, 628–633
  compound fruit, 165
  development, 164
  dry dehiscent fruit, 167, 168*t*
  dry indehiscent fruit, 167, 169*t*
  fleshy fruits, 166, 167*t*
  seeds (*see* Seeds)
  simple fruit, 165
  structural components, 164–165, 165*f*
  unique fleshy fruits, 166, 168*t*
Fruticose lichens, 436
Fucoxanthin, 438
Fungi
  in bioremediation, 562
  characteristics, 548–552
  importance of, 556–563
  lichens, 561
  as parasites, 559–561
  reproduction, 550–552
  structure, 549–550
  types, 552–556, 553*t*

G

Gametangia, 457
Gametes, 144, 306
Gametophores, 468
Gametophytes, 310, 458
Gemma, 470, 470*f*
  gemma cups, 470, 470*f*
Gene flow, 390
Gene-for-gene hypothesis, 266
Gene stacking, 363
Genetically modified (GM), 352, 357
Genetically modified organisms (GMO), 348
Genetic drift, 390
Genetics, 15
  complex pattern of inheritance, 325–327
  DNA replication, 338–339

DNA structure and organization, 336–338
gene expression, 339–342
Mendelian inheritance, 316–318
Mendel's principles of inheritance, 322–325
mutations, 343–344
one-character inheritance/single-factor crosses, 318–321
plant breeding, 328–330
two-character inheritance or two-factor crosses, 321–322
Genotype, 319
Geophytes, 108
Germination, 168–175, 174f
Germplasm, 178
Gibberellins (GA)
hormone activity of, 253
internode elongation, 254, 255f
seed germination, 254
shoot elongation, 254
stem length reduction, 255
Ginkgo, 506–507, 506f
Global agriculture, 5–7
Global warming, 576–577
Globular proteins, 37
Gluten, 37, 622
Glyceraldehyde 3-phosphate (G3P), 216
Glycoalkaloid, 627
Glycolysis, 231, 232f
GM foods, 348
GM plants, risks of, 370–374
GM Traits
abiotic stress, 368–369
examples, 362t–363t
product quality, 365–368
stacking, 363, 364f
Gnetophytes, 514–515, 515f
Golgi apparatus, 58
G_1 phase, 303
G_2 phase, 303
G-proteins, 245
Grafting, 190–191, 190f, 191f
Gram-negative bacteria, 429
Grana, 210
Grasses, 618–624
Grass family, 534–536, 536f
Grasslands, 580, 581f
Gravitropism, 83, 252, 253f
Green fluorescent protein (GFP), 355, 355f
Greenhouse effect, 3, 4
Gregarious flowering, 641
Ground meristem, 84, 85f, 98

Ground tissue system, 68–69, 69f
Guanosine triphosphate (GTP), 245
Guard cells, 71, 72f
Guttation, 288, 289f
Gymnosperms, 15, 450
characteristics, 502–504, 502f–503f, 504t
conifers, 507–512, 507f–512f
cycads, 504–505, 505f
evolutionary relationships, 504
ginkgo, 506–507, 506f
gnetophytes, 514–515, 515f

H

Hadley cell, 571
Haploid, 306, 458
Hardwood cuttings, 186
Hardy–Weinberg law, 388–389
Heartwood, 99
Heat of vaporization, 28
Herbaceous plants, 100, 101f
Herbicide tolerance (HT), 364f, 371
Herbivory, 593
Herbs, 638, 638f, 639t
Heterocysts, 430–431
Heterosporous, 484, 485f
Heterotrophs, 61, 204, 484
Hinged leaf traps, 128–129, 128f
Homologous chromosomes, 308
Homology, 386
Homosporous, 482–483, 485f
Hormonal and environmental stimuli
defenses against herbivores, 268
environmental signals, 265
gravity responses, 262–263, 262f
hypersensitive response, 266–267
internal and external factors influencing growth, 244, 244f
light responses, 259–262
mechanical stimuli, 263–265
plant hormones (see Plant hormones)
signal reception and transduction, 244–247, 245f, 246f
system-acquired resistance, 267–268, 267f
Hornworts
asexual reproduction, 471
gametophyte generation, 470–471
sporophyte generation, 471, 471f
Horsetails, 491–492, 491f–492f
Horticulture, 16
Humus, 272

Hyaline cells, 472
Hybridization, 316–317
Hybrid speciation, 399
Hybrid zones, 399
Hydroids, 472
Hydrolysis, 31
Hypersensitive response (HR), 266–267, 266*f*
Hypertonic solution, 286
Hyphae, 550
Hypocotyls, 174, 175
Hypodermis, 511
Hypogeous germination, 174
Hypothesis testing, 17
Hypotonic solution, 286

I

Imperfect flowers, 147
Incomplete dominance, 326–327
Independent variable, 17
Indeterminate growth, 9
Indoleacetic acid (IAA), 248
Inducible defensive response, 266
Inductive reasoning, 18
Infiltration rate, 274
Inflorescence morphology, 152, 152*t*–153*t*
Integument, 501, 501*f*
Internodes, 102
Interphase, 302
Invasive species, 528–530
Ion channels, 245–246
Ipomoea batata, 204*f*
Isoflavones, 626
Isolated populations, 399–400
Isotonic solution, 286

K

Karyogamy, 551
Kinetochore, 304
Krebs cycle, 233, 234*f*
K-selected populations, 596

L

Lactic acid fermentation, 239
Lamina, 120
Laminarin, 439
Laminate bulbs, 109–110, 109*f*, 110*f*
Larix laricina, 578
Lateral buds, 102

Leaching, 280
Leaves
    air pollution, 137
    carnivorous plants, 128–129, 128*f*
    complexity, 129–130, 130*f*
    cuttings, 186, 186*f*
    dyes and medicinal uses, 133–134, 134*f*, 135*f*
    external morphology, 120–121, 121*f*
    food source, 134–135, 136*f*
    heat and drought, 136, 137*f*
    internal morphology, 121–123, 122*f*, 123*f*
    margin, 121
    photosynthesis, 123
    plant protection, 126, 126*f*
    plant reproduction, 135, 136*f*
    pollination, 127, 127*f*
    scars, 103
    shape and color, 131–132, 131*f*, 132*f*
    structural uses, 132, 133*f*
    support, 126, 127*f*
    transpiration, 123
    venation, 130, 131*f*
    water retention, 124–125, 124*f*–125*f*
    weather changes, 136–137, 137*f*
Legumes, 537–538, 537*f*, 624–626
Lemma, 536, 536*f*
Lenticels, 73, 73*f*, 103
Leptoids, 472
Leucoplasts, 65, 65*f*
Liana, 100, 101*f*
Lichens, 436, 561
Light compensation point, 225
Light reactions, 212
Light responses
    blue light, 259
    red and far-red light, 259–262
Light saturation point, 225
Lignin, 66, 456–457, 457*f*
Lily family, 532–533, 534*f*
Linkage, 327
Lipids, 38–41
Liverworts
    asexual reproduction, 470, 470*f*
    body forms, 467*f*
    definition, 466–467
    gametophyte generation, 467–468
    life cycle, 467*f*
    sporophyte generation, 468, 469*f*, 470
Loam, 273
Locules/carpels, 147
Long-day plants, 260

Lower epidermis, 122, 123
Lycophytes, 482, 484, 485, 486f

M

Macromolecules, 7
Magnoliids, 528
    magnolia family, 531, 531f
Maize, 619–620, 620f
Maple (Acer), 580
Mass selection, 329
Matrix, 60
Medicinal plants
    Aloe vera, 643
    ethnobotany, 643
Medullary rays, 99
Megaphylls, 481, 482f
Megasporangia, 484
Megaspores, 484, 486f
Meiosis, 303, 306
Meiosis I, 307f
Meiosis II, 307f
Mendelian inheritance
    characters, 317–318
    hybridization experiments, 316–317
    traits, 318, 317f
    true-breeding line, 318
Mendel's law of independent assortment, 325
Mendel's law of segregation, 325
Mesembryanthemum crystallinum, 224
Mesocarp, 164
messenger RNA (mRNA), 32, 340
Metabolic process, 2
Metabolism, 55
Metallothioneins, 283
Metaphase, 305
Metaphase I, 309
Microphylls, 481
Micropyle, 501, 501f
Microsporangia, 484
Microspores, 484, 486f
Microtubules, 55
Middle lamella, 66, 66f
Midrib, 121
Mineralization, 606
Mitochondria, 59–62, 60f, 62f
Mitosis, 303–306, 305f, 304f
    vs Meiosis, 307t, 308f
Mitotic spindles, 304
Monoclonal antibody (mAb), 370
Monotropa uniflora, 204

Molecular biology, 16
Molecular dating, 386
Monocotyledons/Monocots, 528
    vs. eudicots, 533t
    grass family, 534–536, 536f
    lily family, 532–533, 534f
    orchid family, 533–534, 535f
    root, 90, 90f
Monoecious, 471
    flower, 147, 148
Monohybrid cross, 319
Monomers, 30
Monophyletic taxa, 410
Monopodial bamboo species, 641
Monosaccharides, 29
Moringa oleifera Lam, 633–634
Morphology, 16
Morphospecies concept, 398
Mosses, bryophytes
    asexual reproduction, 470
    definition, 472
    gametophyte generation, 472, 473f
    sporophyte generation, 472, 474f
Mucigel, 83
Mucilage, 471
Multicellular organism, 7, 10f, 50
Mustard family, 540–541, 541f
Mutation, 390, 390f
Mutualism, 90, 91f, 591, 592
Mycelium, 550
Mycology, 550
Mycoremediation, 562
Mycorrhizae, 459, 554, 557–558
    fungi, 92
Myrmecophytes, 592

N

National Agricultural Statistics Service
    (NASS), 364f
Natural selection, 383, 391, 392f
Neottia nidus-avis, 204
Nicotinamide adenine dinucleotide phosphate
    (NADPH), 212, 214f
Nightshade/potato family, 541–542, 542f
Nitrification, 606
Nitrogen cycle, 605–608, 607f
Nitrogen fixation, 606
Nodes, 102
Noncertified seed, 177
Noncyclic electron flow., 215

Nontarget organisms, 372–373
Nonvascular plants, 15, 450
Nuclear division, 303
Nuclear envelope, 56
Nuclear pores, 57
Nucleic acids, 32–33
Nucleolus, 57, 57f
Nucleoplasm, 57
Nucleotides, 32, 55
Nutrient deficiency
    examples of, 278, 278f
    hydroponic growing systems, 278–280
    mobility of elements, 278
    plant symptoms for, 280, 281t
    plant tissue analysis, 278
Nutrients
    symbiotic microorganisms, 284
    water uptake, 281
Nuts, 631, 632t

O

Oats, 623, 624
Okazaki fragments, 352
Oleoresin, 507
Operculum, 472, 474f
Opisthokonta, 14, 14f
Orchid family, 533–534, 535f
Organic molecules, 28, 42t
Organism, 7, 8, 8f
Organismal ecology, 590
Organs, 9
Origin of replication, 302
Osmosis, 286, 287f
Ovary, 147
Ovule, 501, 501f
Oxidative phosphorylation, 236

P

P$_{680}$, 211–213
P$_{700}$, 211
Palea, 536, 536f
Paleobotany, 16
Palindromic, 349
Palisade mesophyll, 122
Palmately compound leaf, 129, 130f
Palynology, 16
Papaya ringspot virus (PRSV), 365
Paper, 638, 640
Paraphyletic group, 410

Parasitic plants, 520
Parasitism, 91, 91f. *See* Predation
Parenchyma tissue, 68, 69, 69f
Parental (P) generation, 318
Parkinsonia microphylla, 204f
Parthenocarpic fruit, 164
Particulate theory of inheritance, 323
Passive ion exclusion, 283
Passive transport, 282
Peanuts, 625, 626
Pedicels, 152, 153
Peduncle, 152
PEG-mediated transformation, 354
Pellicle, 442
PEP carboxylase, 219
Peptidoglycan, 65, 428
Perfect flowers, 147
Perianth, 145
Pericarp, 164
Pericycle, 87
Periderm, 71
Peristome, 472, 472f
Permafrost, 575
Permanent wilting point, 275
Petals, 144, 145f
Petiole, 121
Phagocytosis, 429
Pharmaceutical products, 5
Phelloderm, 72
Phloem loading and unloading, 293–295, 294f
Phosphodiester bond, 351
phosphoenolpyruvate (PEP), 219
3-Phosphoglycerate (PGA), 216
Phospholipids, 40, 53
Phosphorylation cascade, 40
Photinus pyralis, 357
Photoautotrophs, 9
Photolysis, 213
Photomorphogenesis, 259
Photons, 207
Photoperiod, 260
Photoperiodism, 260
Photophosphorylation, 214
Photorespiration, 218
Photoreversibility, 260
Photosynthesis, 2, 123, 202, 202f, 203f
    autotrophs, 203
    C$_4$ pathway, 218–220
    C$_3$ *vs.* C$_4$ pathways, 220–221
    Calvin Benson cycle, 216–218
    carbon cycle and global warming, 205–206

carbon dioxide, 226
chemoautotrophs, 204
chloroplasts and photosystems, 208
crassulacean acid metabolism (CAM), 221–224
environmental factors, 225–226
function of light, 207–208
heterotrophs, 204
light, 207–208
light intensity, 225
light reactions, 212–216
oxidation and reduction, 202, 202f
photorespiration, 218
pigments role, 208–209
role on earth, 203–205
temperature, 226
Photosystems, 210–211, 210f
photosystem I (PS I), 211, 214f
photosystem II (PS II), 211, 212, 214f, 424
Phototropins, 259
Phototropism, 249, 249f, 252, 253f, 259
Phragmoplast, 306
Phycobilins, 424
Phycocyanin, 436
Phycoerythrin, 436
Phylogenetic species concept, 397
Phylogeny and taxonomy, 410
binomial nomenclature, 406–410
cladistics, 411
classification systems, 406
domains, 411–420
hierarchical classification and taxonomy, 406, 407t–408t, 409f
systematics, 410–411
Physical dormancy, 176–177
Physiological dormancy, 176
Physiology, 15
Phytochemicals, 629
Phytochrome, 259, 259f
Phytoremediation, 283, 562
Picea glauca, 578
Picea mariana, 578
Pigments, 208
Pinnae, 485
Pinnately compound leaf, 129, 130f
Pinus banksiana, 578
Pinus contorta, 578
Pioneer species, 608
Pistillate, 147
Pistils, 146, 147
Pitfall traps, 128f, 129

Pith, 98
Placentation, 147
Plant(s)
adaptations, 11–12, 12f
breeding, 328
characteristics, 15
domestication, 616–618
fiber crops, 638, 640
life cycle, 309f
medicinal purposes, 643
oils, 634–635
sugar, 635–636
Plant growth elements
macronutrients and micronutrient, 275, 276t–277t
nitrogen, 278
nutrient availability and cation exchange, 280–281
nutrient deficiency, 278–279
Plant hormones
abscisic acid, 255
auxin (see Auxin)
brassinosteroids, 258
chemical structures of, 248, 248f
cytokinins, 252–253, 254f
ethylene, 255–257, 256f, 257f
FLAVR SAVR tomato, 257–258
gibberellins (GA), 253–255, 255f
Plant systematics, 16
Plant transport
active ion exclusion, 283
active transport, 283, 282f
capillary action, 289–290
cohesion–tension theory, 290–291, 290f
in dry/saline soils, 291
evaporative demand, 288, 288f
hypertonic solution, 286
hypotonic solution, 286
isotonic solution, 286
nutrient uptake, leaves, 281
osmosis, 286, 287f
passive ion exclusion, 283
passive transport, 282
root pressure, 288
solute/osmotic potential, 285
symplast and apoplast path, 286, 287f
translocation, 291–295
transpiration, 285
water and solute transport, 284–285
water potential, 285
Plantae, kingdom, 455–459

Plasmids, 352, 353*f*

Plasmodesmata, 66, 66*f*, 86, 86*f*

Plasmogamy, 551

Plastids, 65, 65*f*, 66

Pleiotropy, 326

Plumule, 174

Pneumatophores, 12, 13, 599

Poinsettia cuttings, 185, 185*f*

Polar covalent bond, 26

Pollen, 146, 146*f*

Pollen chamber, 501, 501*f*

Pollen grains, 500

Pollination, 501

Pollination drops, 508

Pollinators, 523, 524*f*

Polyethylene glycol(PEG), 354

Polygenic Inheritance, 326

Polymers, 30, 455

Polypeptides, 36

Polyploidy, 343, 399

Polysaccharides, 29

Population(s), 9, 10, 10*f*, 570
    ecology, 591, 594–597

Portulaca oleracea, 224

Post-harvest processing, 371–372

Postzygotic isolation, 397

Potatoes, 626, 627

Potato leaf roll virus (PLRV), 365

Predation, 591, 594*f*, 593

Pressure potential, 285

Pressure-flow hypothesis, 292, 292*f*

Prezygotic isolation, 397

Prickles, 104, 105*f*

Primary cell wall, 65

Primary dormancy, 176

Primary endosymbiosis, 434

Primary structure, 36

Primary succession, 608, 609*f*

Principle of maximum parsimony, 411

Procambium, 84, 85*f*, 98

Prokaryotes, 13, 50, 52*f*, 53*t*, *302*

Prokaryotic cell cycle, 302*f*

Prometaphase, 304

Prometaphase I, 308, 309

Propagules, 182, 466

Prophase, 304

Prophase I, 308

Proplastids, 63

Proteins, 33–38, 53

Proteome, 34

Proteomics, 34

Prothallus, 486

Protoderm, 84, 85*f*, 98

Proton pumps, 283

Protonema, 472

Pseudostem, 627

Pteridology, 16

Pteridophyta. *See* Ferns

Pubescence, 124

Pulse, 624

Pumpkin family, 539–540, 540*f*

Punctuated equilibrium, 393

Punnett square, 320–321

Pure-line selection, 329–330

Pyrenoids, 434

Pyruvate to Acetyl coenzyme A conversion,
    232–233

Q

Quaternary structure, 37

Quiescence, 176

R

Rachis, 130, 152, 153

Radially symmetrical flower, 149, 150*f*

Radicle, 79

Radiocarbon dating, 386

Rainforest, 484–485, 585*f*

Rain shadow, 572, 573*f*

Random mating, 391

Ray florets, 543, 543*f*

Reaction center, 210

Reaction wood, 115

Receptacle, 144

Recessive trait, 319

Recombinant DNA (rDNA) molecules, 348

Registered seed, 177

Reporter genes, 355

Resin, 507

Respiration, 230
    aerobic, 230–231
    definition of, 230
    electron transport chain, 234–236
    fermentation, 238–239
    fresh produce, 239–240
    glycolysis, 231–232
    Krebs cycle, 233
    pyruvate conversion to acetyl coenzyme A,
        232

Restriction endonucleases, 349

Restriction enzyme digest, 349
Restriction enzymes, 348, 349f, 350t, 351f
Rhizoids, 466
Rhizomes, 112, 112f
Ribonucleotides, 32
ribosomal RNA (rRNA), 32
Ribulose 1,5-bisphosphate (RuBP), 217
Rice, 622–623
r/K selection theory, 596
Robert Hooke, cell discovery, 47–48, 47f, 48f
Root(s)
    absorption, 79
    aerial roots, 81, 81f
    apical meristem, 84–85, 85f
    buttress roots, 81, 82f
    cap, 83
    cassava, 88
    cell elongation, 85
    contractile roots, 82, 83f
    cuttings, 186, 188f
    eudicot roots, 90, 90f
    evening primrose, 88
    fibrous root systems, 79, 80f, 81, 82f
    food storage roots, 81
    functions, 78–79
    ginseng root, 88
    gravitropism, 83
    hairs, 71–72
    monocot, 90, 90f
    mucigel, 83–84
    mutualism, 90
    mycorrhizae fungi, 92
    nitrogen-fixing bacteria, 92
    prop roots, 80, 81f
    purple coneflower, 88
    region of maturation, 85–87, 86f
    root apical meristem, 84–85, 85f
    seminal roots, 79, 80f
    storage, 79
    sugar beets, 87
    symbiotic relationship, 90–92, 91f
    taproot system, 79, 79f, 82, 82f
    water storage roots, 82, 83f
    yucca plant, 89
Rose family, 538, 539f
Rough endoplasmic reticulum (ER), 58, 59f
r-selection traits, 596
Rubisco, 218
RuBP carboxylase/oxygenase, 218
Runners, 113
Rye, 623

S

Saprotrophs, 521, 553
Sapwood, 99
Saturated fatty acids, 39
Savanna, 580, 582f
Scaling, 110
Scaly bulbs, 110, 110f
Scarification, 177
Scientific methods, 17–18, 18f
Scion, 190
Sclereids, 69
Sclerenchyma tissue, 69f, 69
Scooping/scoring, 110
Secondary cell wall, 66, 68
Secondary dormancy, 176
Secondary endosymbiosis, 438
Secondary metabolites, 638
Secondary structure, 37
Secondary succession, 609, 610f
Secondary tissues, 87
Second filial (F2) generation, 318
Second messengers, 245
Seedless vascular plants
    characteristics, 482, 483t
    coal formation, 494–495, 495f
    ferns, 485–486–489f
    as food, 494
    horsetails, 491–492, 491f
    horticultural and agricultural uses, 493–494
    household and industrial uses, 494
    lycophytes, 482, 484, 485, 486f
    medical uses, 494
    whisk ferns, 492–493, 493f
Seed(s), 501
    certification, 177
    coat dormancy, 176
    dispersal, 526f
    dispersal mechanisms, 165, 166t
    double dormancy, 177
    energy source, 171–172, 173f
    germination, 168, 174, 175, 174f, 175f
    oil source, 170, 172f
    physical dormancy, 176–177
    physiological dormancy, 177
    preservation, 177–178
    propagation, 167
    protein source, 172, 173f
    quiescence, 176
    spices and flavorings, 169, 170f, 171f
    vascular plants (see Gymnosperms)

Self-pollination, 146
Semihardwood cuttings, 186
Seminal roots, 79, 80f
Senescence, 137
Sensitive plants, 11, 12, 12f
Sepals, 144, 145f
Septa, 550
Septate, 550
Septum, 430
Serotinous cones, 513
Sessile, 152
Seta, 470
Sexual hybrids, 330
Sexual reproduction, 10–11, 11f, 144
Sheath, 536
Shifty photosynthesis, 224
Short-day plants, 261, 261f
Shrubland, 580–581
Sieve-tube elements, 70
Signal deactivation, 246
Signal receptors, 244
Signal transduction, 244
Silica, 439
Simple fruit, 165
Slimy polysaccharide coating, 429
Smooth endoplasmic reticulum (ER), 58, 59f, 60
Softwood cuttings, 168
Soil(s)
    in horizons, 272, 273f
    organic matter, 274
    structure, 273–275
    texture, 273, 274f
Solanine, 627
Solutes, 27
Solutions, 27
Solvent, 26
Somatic cells, 306
Somatic hybrids, 330
Sorghum, 623, 624
Sori, 485, 487f
Soybeans, 625–626
Speciation
    allopatry, 398
    definition, 396
    isolated populations, 399–400
    reproductive isolation, 397
    sympatry, 399
Specific heat, 571
S phase, 303
Spices, 638, 638f, 639t

Spines, 104, 105f
Spongy mesophyll, 122
Sporangia, 457
Sporangiophores, 492, 492f, 551
Sporangiospores, 551
Spore mother cells, 470
Sporophytes, 310, 458
Sporopollenin, 456, 456f
Stamens, 146–147, 146f
Staminate flowers, 147
Stand-replacing fire regimes, 513
Star anise family, 527–528
Starch, 31, 32
Start codon (AUG), 342
Statolith hypothesis, 262
Stele, 90
Stems
    axillary buds, 102
    bark, 104–108, 105f–108f
    bud scales, 102
    bundle scars, 103
    cladophylls, 113, 114f
    classification, 100–101, 101f, 102f
    corms, 111–112, 111f
    definition, 96
    eudicots and gymnosperms secondary
        growth, 99, 100f
    functions, 96–98, 96f, 98f
    geophytes, 108, 109f
    internode, 102
    laminate bulbs, 109–110, 109f, 110f
    lateral buds, 102
    leaf attachment, 103, 103f, 104f
    leaf scars, 103
    lenticels, 103
    monocots lack secondary growth, 100f
    nodes, 102
    primary growth, 98
    response to environment, 114–116, 114f, 115f
    rhizomes, 112, 112f
    runners, 113
    specialized storage structures, 108
    stolons, 112–113, 113f
    tubers, 112, 113f
    vascular tissues, 98–99, 99f
    woody twigs, 102, 102f
Sterols, 41
Sticky traps, 129f, 129
Stigma, 147
Stock, 190, 190f
Stolons, 112–113, 113f

Stomata, 71, 122, 122*f*
Stratification, 176
Strobili, 484
Stroma, 64
Stroma lamella, 210
Style, 147
Suberin, 40, 73
Substrates, 38, 548
Succession, 608–609, 611
Sugar, 635, 635*f*
Sunflower family, 542–543, 543*f*
Sweet potato, 627
Sympatry, 399
Symplast pathway, 286, 287*f*
Sympodial bamboo species, 641
Syringa sp., 220*f*
System-acquired resistance (SAR), 258, 267–268, 267*f*
Systematics, 13
Systemin, 268

T

Taiga, 577
Taproot system, 79, 79*f*
Taxonomy, 13, 16
Tea, 636, 637*f*
Telophase, 306
Telophase I, 309
Tendrils, 126, 127*f*
Tension wood, 115, 115*f*
Teosinte, 619
Tepals, 144–145, 145*f*
Tertiary structure, 36–37
Test cross, 321
Testa, 174
Tetraploid, 343
Thigmomorphogenesis, 263
Thigmotropism, 263, 263*f*
Thorns, 104, 105*f*
Thuja occidentalis, 578
Thylakoid membranes, 210, 210*f*
Thylakoids, 64, 428, 428*f*
Ti plasmid, 358*f*
Tip/stem cuttings, 185
Tissues
  complex, 68
  dermal tissue system, 71–73, 72*f*, 72*f*, 73*t*
  division of labor, 68
  epidermis, 46–47
  ground tissue system, 68–69, 68*f*

simple, 68
  vascular plants, 68
  vascular tissue system, 69–71, 70*f*, 71*f*
  xylem, 47
Totipotency, 253, 352
Tracheids, 70, 70*f*
Traditional/Linnaean classification, 13
Transcription, 339
Transcription activatorlike effector nucleases (TALENs), 351
transfer RNA (tRNA), 32
Transformation/transfection, 352
Transgenes, 352
Transgenic, 352
Transgenic tobacco plant, 358*f*
Translocation
  phloem loading and unloading, 293–295
  pressure-flow hypothesis, 292, 292*f*
  sink, 291
  source, 291
Transpiration, 123, 285
Tricarboxylic acid (TCA), 230, 233, 234*f*
Trichomes, 71, 124
Triglycerides, 634
Triploid, 343
Trophic level, 600
Tubers, 112, 113*f*
Tumor-inducing (Ti), 357
Tunic, 109
Turgor pressure, 58, 285

U

Understory fire regimes, 513
Unicellular organism, 7, 50
Unique fleshy fruits, 166, 167*t*
United States, GM crops, 360
Unsaturated fatty acids, 39

V

Vacuoles, 58
Vaporization, 28
Vascular bundles, 122
Vascular cylinder, 87
Vascular seedless plants, 15, 450
Vascular seed plants, 15, 450
Vascular tissue system, 69–71, 70*f*, 71*f*
Vector, 352
Vegan, 633
Vegetable, 633

Vegetative cuttings
  coleus tip cuttings, 185, 185f
  crop covers, 185
  environmental and physiological conditions,
    188
  flowers removal, 185
  good tip/stem cutting, 189–190, 189f
  hardwood cuttings, 186
  leaf cuttings, 186, 186f
  poinsettia cuttings, 185, 185f
  root cuttings, 187, 187f
  semihardwood cuttings, 186
  softwood cuttings, 186
  tip/stem cuttings, 185
Vegetative propagation. *See* Asexual reproduction
Venus flytrap *(Dionaea muscipula)*, 263, 264f
Vessel elements, 70
Vestigial seeds, 165
Vine, 101
Virulence, 357
Vitamin A deficiency (VAD), 371

W

Water, 26–28
  imbibition, 168
  potential, 285
  retention, 124–125, 124f–125f

Water lily family, 526, 527f, 527
Water-holding capacity, 274, 275f
Waxes, 40
Waxy bloom, 121
Wetlands, 598–599, 599f
Wheat, 622–623
Whisk ferns, 493, 493f
Wild species, 373–374
Woody plant, 99

X

Xeric, 575

Y

Yams, 628f, 627

Z

Zea mays, 220f
Zeaxanthins, 259
Zinc-finger nucleases (ZFNs), 351
Zooxanthellae, 441
Zygomorphic flowers, 149
Zygote, 144